VOLUME FIVE HUNDRED AND SIXTY FOUR

Methods in
ENZYMOLOGY

Electron Paramagnetic Resonance
Investigations of Biological Systems by
Using Spin Labels, Spin Probes, and
Intrinsic Metal Ions, Part B

METHODS IN ENZYMOLOGY

Editors-in-Chief

JOHN N. ABELSON and MELVIN I. SIMON
Division of Biology
California Institute of Technology
Pasadena, California

ANNA MARIE PYLE
Departments of Molecular, Cellular and Developmental
Biology and Department of Chemistry Investigator
Howard Hughes Medical Institute
Yale University

DAVID W. CHRISTIANSON
Roy and Diana Vagelos Laboratories
Department of Chemistry
University of Pennsylvania
Philadelphia, PA

Founding Editors

SIDNEY P. COLOWICK and NATHAN O. KAPLAN

VOLUME FIVE HUNDRED AND SIXTY FOUR

METHODS IN
ENZYMOLOGY

Electron Paramagnetic Resonance Investigations of Biological Systems by Using Spin Labels, Spin Probes, and Intrinsic Metal Ions, Part B

Edited by

PETER Z. QIN
Department of Chemistry and
Department of Biological Sciences
University of Southern California
Los Angeles, California, USA

KURT WARNCKE
Department of Physics
Emory University
Atlanta, Georgia, USA

AMSTERDAM • BOSTON • HEIDELBERG • LONDON
NEW YORK • OXFORD • PARIS • SAN DIEGO
SAN FRANCISCO • SINGAPORE • SYDNEY • TOKYO
Academic Press is an imprint of Elsevier

Academic Press is an imprint of Elsevier
225 Wyman Street, Waltham, MA 02451, USA
525 B Street, Suite 1800, San Diego, CA 92101-4495, USA
The Boulevard, Langford Lane, Kidlington, Oxford OX5 1GB, UK
125 London Wall, London, EC2Y 5AS, UK

First edition 2015

Copyright © 2015 Elsevier Inc. All rights reserved.

No part of this publication may be reproduced or transmitted in any form or by any means, electronic or mechanical, including photocopying, recording, or any information storage and retrieval system, without permission in writing from the publisher. Details on how to seek permission, further information about the Publisher's permissions policies and our arrangements with organizations such as the Copyright Clearance Center and the Copyright Licensing Agency, can be found at our website: www.elsevier.com/permissions.

This book and the individual contributions contained in it are protected under copyright by the Publisher (other than as may be noted herein).

Notices

Knowledge and best practice in this field are constantly changing. As new research and experience broaden our understanding, changes in research methods, professional practices, or medical treatment may become necessary.

Practitioners and researchers must always rely on their own experience and knowledge in evaluating and using any information, methods, compounds, or experiments described herein. In using such information or methods they should be mindful of their own safety and the safety of others, including parties for whom they have a professional responsibility.

To the fullest extent of the law, neither the Publisher nor the authors, contributors, or editors, assume any liability for any injury and/or damage to persons or property as a matter of products liability, negligence or otherwise, or from any use or operation of any methods, products, instructions, or ideas contained in the material herein.

ISBN: 978-0-12-802835-3
ISSN: 0076-6879

For information on all Academic Press publications
visit our website at http://store.elsevier.com/

CONTENTS

Contributors   xiii
Preface   xix

Section I
Spin Labeling Studies of Proteins

1. **Saturation Recovery EPR and Nitroxide Spin Labeling for Exploring Structure and Dynamics in Proteins**   3
   Zhongyu Yang, Michael Bridges, Michael T. Lerch, Christian Altenbach, and Wayne L. Hubbell

   1. Introduction   4
   2. Theoretical Background and the Measurement of T_{1e} with SR   6
   3. Instrumentation and Practical Considerations   9
   4. Applications of Long-Pulse SR   11
   5. Summary and Future Directions   22
   References   23

2. **High-Pressure EPR and Site-Directed Spin Labeling for Mapping Molecular Flexibility in Proteins**   29
   Michael T. Lerch, Zhongyu Yang, Christian Altenbach, and Wayne L. Hubbell

   1. Introduction   30
   2. Variable-Pressure CW EPR   37
   3. Pressure-Resolved DEER   42
   4. Example Applications and Perspectives   44
   5. Summary and Future Directions   52
   References   53

3. **Exploring Structure, Dynamics, and Topology of Nitroxide Spin-Labeled Proteins Using Continuous-Wave Electron Paramagnetic Resonance Spectroscopy**   59
   Christian Altenbach, Carlos J. López, Kálmán Hideg, and Wayne L. Hubbell

   1. Introduction   60
   2. Site-Directed Spin Labeling   62
   3. Analysis of EPR Spectra in Terms of R1 Nitroxide Motion   63
   4. Measurement of R1 Solvent Accessibility With Power Saturation   68
   5. CW-Based Interspin Distance Measurements   72

6.	Determination of Secondary Structure and Features of the Tertiary Fold from R1 Mobility and Accessibility	80
7.	Backbone Dynamics	85
8.	Identifying Slow Conformational Exchange and Detecting Conformational Changes	90
9.	Other Applications of CW EPR	93
10.	Conclusion and Future Directions	94
References		95

4. Bifunctional Spin Labeling of Muscle Proteins: Accurate Rotational Dynamics, Orientation, and Distance by EPR   101

Andrew R. Thompson, Benjamin P. Binder, Jesse E. McCaffrey, Bengt Svensson, and David D. Thomas

1.	Introduction	102
2.	Methods of BSL Labeling	104
3.	Rotational Dynamics	105
4.	Orientation	108
5.	Distance	113
6.	Discussion	116
References		120

5. EPR Distance Measurements in Deuterated Proteins   125

Hassane El Mkami and David G. Norman

1.	Introduction and Overview	126
2.	Deuteration	143
3.	The Pulse EPR Experiment	145
4.	Data Analysis	147
5.	Concluding Remarks	149
Acknowledgments		150
References		150

6. Spin labeling and Double Electron-Electron Resonance (DEER) to Deconstruct Conformational Ensembles of HIV Protease   153

Thomas M. Casey and Gail E. Fanucci

1.	Perspective—Conformational Sampling of HIV-1 Protease	154
2.	Site-Directed Spin-Labeling Electron Paramagnetic Resonance Spectroscopy	156
3.	DEER Distance Profiles Reflect the Fractional Occupancies of HIV-1PR Conformational States	165
4.	"DEERconstruct," a Tool for Statistical Analysis of DEER Distance Profiles	170
References		183

Section II
Spin Labeling Studies of Membrane and Membrane-Associated Proteins

7. Ionizable Nitroxides for Studying Local Electrostatic Properties of Lipid Bilayers and Protein Systems by EPR   191

Maxim A. Voinov and Alex I. Smirnov

1. Introduction   192
2. EPR Characterization of pH-Sensitive Thiol-Specific Nitroxide Labels for Mapping Local Protein Electrostatics   196
3. pH-Sensitive Spin-Labeled Lipids for Measuring Surface Electrostatics of Lipid Bilayers by EPR   202
4. Conclusions and Outlook   211
Acknowledgments   212
References   213

8. Peptide–Membrane Interactions by Spin-Labeling EPR   219

Tatyana I. Smirnova and Alex I. Smirnov

1. Introduction   220
2. Peptide Labeling with EPR Active Probes and Preparation of Membrane Mimetic Systems   221
3. EPR Measurements of Membrane Peptide Binding   227
4. Analysis of Binding Isotherms   233
5. Topology of Membrane-Associated Peptides: Paramagnetic Relaxers Accessibility EPR Experiments   236
6. Detecting Membrane-Induced Aggregation and Agglomeration of Peptides   242
7. Conclusions and Outlook   250
Acknowledgments   251
References   251

9. Structural Characterization of Membrane-Curving Proteins: Site-Directed Spin Labeling, EPR, and Computational Refinement   259

Mark R. Ambroso, Ian S. Haworth, and Ralf Langen

1. Introduction   260
2. SDSL, EPR, and Computational Refinement Are a Powerful Combination for Studying Membrane-Curving Proteins Bound to Membranes of Defined Curvature   261
3. Sample Preparation Methodology for SDSL   263
4. EPR Measurements   270
5. Computational Refinement   277

6. Outlook	284
Acknowledgment	285
References	285

10. Determining the Secondary Structure of Membrane Proteins and Peptides Via Electron Spin Echo Envelope Modulation (ESEEM) Spectroscopy   289

Lishan Liu, Daniel J. Mayo, Indra D. Sahu, Andy Zhou, Rongfu Zhang, Robert M. McCarrick, and Gary A. Lorigan

1. Introduction	290
2. Integration of Membrane Peptides into Lipid Bilayer	294
3. ESEEM Spectroscopy on Model Peptides in a Lipid Bilayer	298
4. Development of ESEEM Secondary Structure Determination Approach	302
5. Summary and Future Direction	308
Acknowledgments	308
References	309

11. Spin Labeling Studies of Transmembrane Signaling and Transport: Applications to Phototaxis, ABC Transporters and Symporters   315

Johann P. Klare and Heinz-Jürgen Steinhoff

1. Introduction	316
2. Applications to Membrane Proteins	322
3. Outlook	340
Acknowledgments	341
References	341

12. Navigating Membrane Protein Structure, Dynamics, and Energy Landscapes Using Spin Labeling and EPR Spectroscopy   349

Derek P. Claxton, Kelli Kazmier, Smriti Mishra, and Hassane S. Mchaourab

1. Introduction	350
2. An EPR Primer	354
3. Principles of DEER Spectroscopy to Uncover Conformational Dynamics	361
4. Practical Considerations in Sample Preparation	372
5. Perspective	379
Acknowledgments	381
References	381

13. Spin Labeling of Potassium Channels   389

Dylan Burdette and Adrian Gross

1. Introduction	390
2. General Considerations for Spin Labeling Studies of Potassium Channels	391

3. Examples of Labeling Methods	395
References	399

Section III
Spin Labeling Studies of Nucleic Acids

14. Advanced EPR Methods for Studying Conformational Dynamics of Nucleic Acids — 403
B. Endeward, A. Marko, V.P. Denysenkov, S.Th. Sigurdsson, and T.F. Prisner

1. Introduction	404
2. Spin Labels	407
3. Theory for Orientation-Selective PELDOR	408
4. Experimental Procedure for Multifrequency/Multifield PELDOR	410
5. Analysis of Multifrequency/Multifield PELDOR Data	415
6. Summary and Outlook	418
Acknowledgments	421
References	421

15. An Integrated Spin-Labeling/Computational-Modeling Approach for Mapping Global Structures of Nucleic Acids — 427
Narin S. Tangprasertchai, Xiaojun Zhang, Yuan Ding, Kenneth Tham, Remo Rohs, Ian S. Haworth, and Peter Z. Qin

1. Introduction	428
2. SDSL of Nucleic Acids Using a Nucleotide-Independent Nitroxide Probe	431
3. Measuring Inter-R5 Distances Using Double Electron–Electron Resonance Spectroscopy	439
4. Integration of DEER-Measured Distances with Computational Modeling	443
5. Conclusions	449
Acknowledgment	450
References	450

Section IV
EPR-NMR Methods and Applications

16. Overhauser Dynamic Nuclear Polarization Studies on Local Water Dynamics — 457
Ilia Kaminker, Ryan Barnes, and Songi Han

1. Introduction	458
2. Theory	461
3. Hardware	469

4.	Data Acquisition	472
5.	Data Analysis	475
6.	Examples of ODNP	477
7.	Summary	479
	Acknowledgments	479
	References	479

17. Practical Aspects of Paramagnetic Relaxation Enhancement in Biological Macromolecules 485
G. Marius Clore

1.	Introduction	485
2.	Paramagnetic Labels for PRE Measurements	487
3.	Measurement of the PRE	487
4.	Using the PRE in Structure Determination	489
5.	Using the PRE to Detect Transient Sparsely Populated States	490
	Acknowledgments	495
	References	495

Section V
In Vivo EPR Oxymetry and Imaging

18. *In Vivo* pO_2 Imaging of Tumors: Oxymetry with Very Low-Frequency Electron Paramagnetic Resonance 501
Boris Epel and Howard J. Halpern

1.	Introduction: The Importance of Imaging Molecular Oxygen in Cancer Therapy	502
2.	Summary	521
	Acknowledgments	522
	References	522

19. Direct and Repeated Measurement of Heart and Brain Oxygenation Using *In Vivo* EPR Oximetry 529
Nadeem Khan, Huagang Hou, Harold M. Swartz, and Periannan Kuppusamy

1.	Introduction	530
2.	Principles of EPR Oximetry	532
3.	Paramagnetic Oxygen-Sensitive Probes for pO_2 Measurements by EPR	535
4.	Application of EPR Oximetry in Ischemic Pathologies	540
5.	Summary	546
	Acknowledgments	546
	References	547

20. Free Radical Imaging Using *In Vivo* Dynamic Nuclear Polarization-MRI 553
Hideo Utsumi and Fuminori Hyodo

1. Introduction 554
2. Principle of DNP 556
3. Apparatus 558
4. Spin-Probe Technique for DNP-MRI 562
5. Application of DNP-MRI to the Intrinsic Molecular Imaging 566
6. Conclusion 568
Acknowledgments 568
References 568

Author Index 573
Subject Index 603

CONTRIBUTORS

Christian Altenbach
Department of Chemistry and Biochemistry, Jules Stein Eye Institute, University of California, Los Angeles, California, USA

Mark R. Ambroso
Department of Biochemistry and Molecular Biology, Zilkha Neurogenetic Institute, University of Southern California, Los Angeles, California, USA

Ryan Barnes
Department of Chemistry and Biochemistry, University of California Santa Barbara, Santa Barbara, California, USA

Benjamin P. Binder
Department of Biochemistry, Molecular Biology and Biophysics, University of Minnesota, Minneapolis, Minnesota, USA

Michael Bridges
Department of Chemistry and Biochemistry, Jules Stein Eye Institute, University of California, Los Angeles, California, USA

Dylan Burdette
Department of Biochemistry and Molecular Biology, Rosalind Franklin University of Medicine and Science, North Chicago, Illinois, USA

Thomas M. Casey
Department of Chemistry, University of Florida, Gainesville, Florida, USA

Derek P. Claxton
Department of Molecular Physiology and Biophysics, Vanderbilt University School of Medicine, Nashville, Tennessee, USA

G. Marius Clore
Laboratory of Chemical Physics, National Institute of Diabetes and Digestive and Kidney Diseases, National Institutes of Health, Bethesda, Maryland, USA

V.P. Denysenkov
Institute of Physical and Theoretical Chemistry and Center of Biomolecular Magnetic Resonance, Goethe University Frankfurt am Main, Frankfurt am Main, Germany

Yuan Ding
Department of Chemistry, University of Southern California, Los Angeles, California, USA

Hassane El Mkami
School of Physics and Astronomy, University of St. Andrews, St. Andrews, United Kingdom

B. Endeward
Institute of Physical and Theoretical Chemistry and Center of Biomolecular Magnetic Resonance, Goethe University Frankfurt am Main, Frankfurt am Main, Germany

Boris Epel
Center for Electron Paramagnetic Resonance Imaging *In Vivo* Physiology, Department of Radiation and Cellular Oncology, University of Chicago, Chicago, Illinois, USA

Gail E. Fanucci
Department of Chemistry, University of Florida, Gainesville, Florida, USA

Adrian Gross
Department of Biochemistry and Molecular Biology, Rosalind Franklin University of Medicine and Science, North Chicago, Illinois, USA

Howard J. Halpern
Center for Electron Paramagnetic Resonance Imaging *In Vivo* Physiology, Department of Radiation and Cellular Oncology, University of Chicago, Chicago, Illinois, USA

Songi Han
Department of Chemistry and Biochemistry, and Department of Chemical Engineering, University of California Santa Barbara, Santa Barbara, California, USA

Ian S. Haworth
Department of Pharmacology and Pharmaceutical Sciences, School of Pharmacy, University of Southern California, Los Angeles, California, USA

Kálmán Hideg
Institute of Organic and Medicinal Chemistry, University of Pécs, Pécs, Hungary

Huagang Hou
Department of Radiology, EPR Center for the Study of Viable Systems, Geisel School of Medicine at Dartmouth, Norris Cotton Cancer Center, Dartmouth-Hitchcock Medical Center, Lebanon, New Hampshire, USA

Wayne L. Hubbell
Department of Chemistry and Biochemistry, Jules Stein Eye Institute, University of California, Los Angeles, California, USA

Fuminori Hyodo
Innovation Center for Medical Redox Navigation, Kyushu University, Fukuoka, Japan

Ilia Kaminker
Department of Chemistry and Biochemistry, University of California Santa Barbara, Santa Barbara, California, USA

Kelli Kazmier
Department of Molecular Physiology and Biophysics, Vanderbilt University School of Medicine, Nashville, Tennessee, USA

Nadeem Khan
Department of Radiology, EPR Center for the Study of Viable Systems, Geisel School of Medicine at Dartmouth, Norris Cotton Cancer Center, Dartmouth-Hitchcock Medical Center, Lebanon, New Hampshire, USA

Johann P. Klare
Physics Department, University of Osnabrück, Barbarastr. 7, Osnabrück, Germany

Periannan Kuppusamy
Department of Radiology, EPR Center for the Study of Viable Systems, Geisel School of Medicine at Dartmouth, Norris Cotton Cancer Center, Dartmouth-Hitchcock Medical Center, Lebanon, New Hampshire, USA

Ralf Langen
Department of Biochemistry and Molecular Biology, Zilkha Neurogenetic Institute, University of Southern California, Los Angeles, California, USA

Michael T. Lerch
Department of Chemistry and Biochemistry, Jules Stein Eye Institute, University of California, Los Angeles, California, USA

Lishan Liu
Department of Chemistry and Biochemistry, Miami University, Oxford, Ohio, USA

Gary A. Lorigan
Department of Chemistry and Biochemistry, Miami University, Oxford, Ohio, USA

Carlos J. López
Department of Chemistry and Biochemistry, Jules Stein Eye Institute, University of California, Los Angeles, California, USA

A. Marko
Institute of Physical and Theoretical Chemistry and Center of Biomolecular Magnetic Resonance, Goethe University Frankfurt am Main, Frankfurt am Main, Germany

Daniel J. Mayo
Department of Chemistry and Biochemistry, Miami University, Oxford, Ohio, USA

Jesse E. McCaffrey
Department of Biochemistry, Molecular Biology and Biophysics, University of Minnesota, Minneapolis, Minnesota, USA

Robert M. McCarrick
Department of Chemistry and Biochemistry, Miami University, Oxford, Ohio, USA

Hassane S. Mchaourab
Department of Molecular Physiology and Biophysics, Vanderbilt University School of Medicine, Nashville, Tennessee, USA

Smriti Mishra
Department of Molecular Physiology and Biophysics, Vanderbilt University School of Medicine, Nashville, Tennessee, USA

David G. Norman
Nucleic Acids Research Group, University of Dundee, Dundee, United Kingdom

T.F. Prisner
Institute of Physical and Theoretical Chemistry and Center of Biomolecular Magnetic Resonance, Goethe University Frankfurt am Main, Frankfurt am Main, Germany

Peter Z. Qin
Department of Chemistry, and Molecular and Computational Biology Program, Department of Biological Sciences, University of Southern California, Los Angeles, California, USA

Remo Rohs
Department of Chemistry, and Molecular and Computational Biology Program, Department of Biological Sciences, University of Southern California, Los Angeles, California, USA

Indra D. Sahu
Department of Chemistry and Biochemistry, Miami University, Oxford, Ohio, USA

S.Th. Sigurdsson
Department of Chemistry, Science Institute, University of Iceland, Reykjavık, Iceland

Alex I. Smirnov
Department of Chemistry, North Carolina State University, Raleigh, North Carolina, USA

Tatyana I. Smirnova
Department of Chemistry, North Carolina State University, Raleigh, North Carolina, USA

Heinz-Jürgen Steinhoff
Physics Department, University of Osnabrück, Barbarastr. 7, Osnabrück, Germany

Bengt Svensson
Department of Biochemistry, Molecular Biology and Biophysics, University of Minnesota, Minneapolis, Minnesota, USA

Harold M. Swartz
Department of Radiology, EPR Center for the Study of Viable Systems, Geisel School of Medicine at Dartmouth, Norris Cotton Cancer Center, Dartmouth-Hitchcock Medical Center, Lebanon, New Hampshire, USA

Narin S. Tangprasertchai
Department of Chemistry, University of Southern California, Los Angeles, California, USA

Kenneth Tham[*]
Department of Chemistry, University of Southern California, Los Angeles, California, USA

David D. Thomas
Department of Biochemistry, Molecular Biology and Biophysics, University of Minnesota, Minneapolis, Minnesota, USA

Andrew R. Thompson
Department of Biochemistry, Molecular Biology and Biophysics, University of Minnesota, Minneapolis, Minnesota, USA

Hideo Utsumi
Innovation Center for Medical Redox Navigation, Kyushu University, Fukuoka, Japan

Maxim A. Voinov
Department of Chemistry, North Carolina State University, Raleigh, North Carolina, USA

---

[*]Current address: School of Pharmacy, University of California San Francisco, San Francisco, California, USA.

Zhongyu Yang[†]
Department of Chemistry and Biochemistry, Jules Stein Eye Institute, University of California, Los Angeles, California, USA

Rongfu Zhang
Department of Chemistry and Biochemistry, Miami University, Oxford, Ohio, USA

Xiaojun Zhang
Department of Chemistry, University of Southern California, Los Angeles, California, USA

Andy Zhou
Department of Chemistry and Biochemistry, Miami University, Oxford, Ohio, USA

[†]Current address: Department of Chemistry and Biochemistry, North Dakota, State University, Fargo, North Dakota, USA.

PREFACE

Electron paramagnetic resonance (EPR, or electron spin resonance, ESR) spectroscopy is one of the few methods that selectively and directly detects species containing unpaired electrons (e.g., organic radicals, metal ions) and characterizes their interactions with the surrounding environment. EPR has long been used to investigate contributions of molecular structure and dynamics to function in biological systems, via characterizing paramagnetic species intrinsically present (e.g., metal centers, reaction intermediates) or extrinsically introduced (e.g., covalently attached spin labels or freely diffusing spin probes). The field continues to advance in response to the needs of the biomedical, biomaterials, and biotechnology communities for molecular-level information that ranges on spatial scales from macromolecules through whole cells to organisms, and on temporal scales from solvent fluctuations to physiological processes.

In organizing these two volumes, we aim to present to the EPR practitioners, as well as the broader scientific community, state-of-the-art EPR methodologies for studying relationships among structure, dynamics, and function in biological systems. It is challenging to distinguish categories, such as advances in technique, hardware, and software, from the applications and systems that drive development, which is a sign of the synergistic interplay of EPR spectroscopy and the science it enables. Thematic threads that run through the chapters, that reflect recent progress in EPR studies, include the following: (1) developments in instrumentation, experimental, and analytical approaches, particularly the use of multiple frequencies/magnetic fields outside of traditional X-band (e.g., ≥ 95 GHz), which expand the information content obtainable; (2) advances in incorporating stable paramagnets into biological targets, which expand the scope of systems and questions tractable by EPR approaches, (3) progress in characterizing structure and dynamics of biological molecules, and in particular, methods utilizing distances measured via dipolar interactions and efforts to improve the related data analysis and interpretation; (4) methodologies combining EPR and nuclear magnetic resonance (NMR), in the general area of sensitivity enhancement that enables access to previously veiled structural and dynamic information; and (5) EPR in the area of cellular, or *in vivo*, measurements, including the march toward EPR oximetry and imaging of radical reactions and tumors in humans.

The chapters present the principles and practices that underlie the various EPR approaches in the "hands-on" format, a hallmark of *Methods in Enzymology*. We sincerely hope that they promote understanding and straightforward application, for the continued impact of EPR methods on the understanding of biological structure, dynamics, and function.

Edited by PETER Z. QIN and KURT WARNCKE

SECTION I

Spin Labeling Studies of Proteins

CHAPTER ONE

Saturation Recovery EPR and Nitroxide Spin Labeling for Exploring Structure and Dynamics in Proteins

Zhongyu Yang, Michael Bridges, Michael T. Lerch, Christian Altenbach, Wayne L. Hubbell[1]

Jules Stein Eye Institute and Department of Chemistry and Biochemistry, University of California, Los Angeles, California, USA
[1]Corresponding author: e-mail address: hubbellw@jsei.ucla.edu

Contents

1. Introduction — 4
2. Theoretical Background and the Measurement of T_{1e} with SR — 6
3. Instrumentation and Practical Considerations — 9
4. Applications of Long-Pulse SR — 11
   4.1 Resolving Protein Secondary Structure via Solvent Accessibility — 11
   4.2 Measuring Interspin Distances with Relaxation Enhancement — 14
   4.3 Measuring Protein Conformational Exchange with T_1 Exchange Spectroscopy — 19
5. Summary and Future Directions — 22
References — 23

Abstract

Experimental techniques capable of determining the structure and dynamics of proteins are continuously being developed in order to understand protein function. Among existing methods, site-directed spin labeling in combination with saturation recovery (SR) electron paramagnetic resonance spectroscopy contributes uniquely to the determination of secondary and tertiary protein structure under physiological conditions, independent of molecular weight and complexity. In addition, SR of spin labeled proteins was recently demonstrated to be sensitive to conformational exchange events with characteristic lifetimes on the order of μs, a time domain that presents a significant challenge to other spectroscopic techniques. In this chapter, we present the theoretical background necessary to understand the capabilities of SR as applied to spin labeled proteins, the instrumental requirements, and practical experimental considerations necessary to obtain interpretable data, and the use of SR to obtain information on protein: (1) secondary structure via solvent accessibility measurements, (2) tertiary structure using interspin distance measurements, and (3) conformational exchange.

1. INTRODUCTION

Saturation recovery (SR) electron paramagnetic resonance (EPR) methodology was developed by Hyde in the 1970s to measure the electron spin–lattice relaxation time (T_{1e}) of paramagnetic species (Huisjen & Hyde, 1974; Percival & Hyde, 1975). It has been extensively employed with nitroxide spin labels to investigate the dynamics of lipids in membranes (Kawasaki, Yin, Subczynski, Hyde, & Kusumi, 2001; Popp & Hyde, 1982; Yin, Pasenkiewicz-Gierula, & Hyde, 1987) and to determine the diffusivity of oxygen in membranes (Kusumi, Subczynski, & Hyde, 1982; Subczynski, Hyde, & Kusumi, 1989, 1991; Yin & Hyde, 1987). This chapter focuses on recent developments that combine SR and site-directed spin labeling (SDSL) to explore structure and internal dynamics of proteins. The most commonly used nitroxide in SDSL is the side chain designated R1, introduced by cysteine substitution mutagenesis followed by reaction with a sulfhydryl-specific methanethiosulfonate reagent (Fig. 1A). Interpretation of the EPR spectra of R1 in proteins is greatly aided by the extensive information that is available on structure and dynamics for this side chain (Columbus & Hubbell, 2004; Columbus, Kálai, Jekö, Hideg, & Hubbell, 2001; Fleissner, Cascio & Hubbell, 2009). Applications of SR presented below will be restricted to proteins containing this side chain and derivatives thereof, although the principles are the same for other nitroxide side chains.

For an ensemble of paramagnets at equilibrium in an external magnetic field, there is a macroscopic magnetization vector with a longitudinal component M_z along the field direction, and transverse components M_x and M_y orthogonal to M_z. If the Boltzmann distribution of the spin population is perturbed in some way, thermodynamic equilibrium of M_z is restored through interactions with the surroundings (the "lattice") by the process of spin–lattice relaxation. In the case of an ensemble of noninteracting spins in dilute solution, its recovery to an equilibrium state is exponential according to,

$$M_z(t) = M_z(\text{eqm}) + [M_z(0) - M_z(\text{eqm})]\exp(-t/T_{1e}) \quad (1)$$

where $M_z(t)$, $M_z(\text{eqm})$, and $M_z(0)$ are the magnetization values at any time t, at equilibrium, and immediately after the perturbation, respectively, and T_{1e} is the spin–lattice relaxation time. The amplitude of a continuous wave (CW) EPR spectrum is proportional to M_z, and the recovery to equilibrium can be followed by recording the amplitude as a function of time. For R1 in

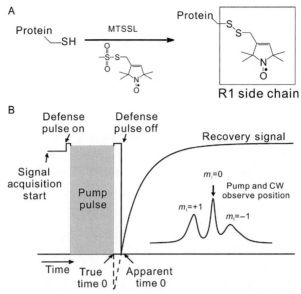

Figure 1 Site-directed spin labeling and saturation recovery EPR. (A) The introduction of the R1 side chain via the reaction of a nitroxide methanethiosulfonate reagent with cysteine (Berliner, Grunwald, Hankovszky, & Hideg, 1982). (B) The time course of a long-pulse saturation recovery experiment. An excitation (pump) pulse located at the field position of maximum intensity in the EPR absorption spectrum (see inset) causes saturation of the electron spin system. Low-power CW detection (CW observe) at the same field position records the signal recovery. The "defense pulse" is required to protect the detector electronics, and the earliest time points of the recovery signal beginning at the true time 0 are not recorded.

proteins, T_{1e} is on the order of microseconds at room temperature at X-band (~9.5 GHz) EPR frequencies. The microwave frequency dependence and mechanisms that give rise to nitroxide spin–lattice relaxation in fluid solution have been investigated (Froncisz et al., 2008; Mailer, Nielsen, & Robinson, 2005; Owenius, Terry, Williams, Eaton, & Eaton, 2004; Robinson, Haas, & Mailer, 1994; Sato et al., 2008), and interested readers should consult relevant references for details.

For topics discussed in this chapter, the most useful features that influence the apparent T_{1e} for a nitroxide attached to a protein are (1) the rotational correlation time (τ_R) of the nitroxide, in the range of 1–10 ns (Bridges, Hideg, & Hubbell, 2010; Sato et al., 2008); (2) the Heisenberg exchange frequency with another freely diffusing paramagnetic species "R_{ex}" (Altenbach, Froncisz, Hemker, McHaourab, & Hubbell, 2005; Pyka, Ilnicki, Altenbach, Hubbell, & Froncisz, 2005); (3) the distance-dependent enhancement of nitroxide T_{1e} due to the presence of a paramagnetic metal ion with relatively long T_{1e} (Eaton & Eaton, 2002); and (4) exchange between different environments that takes place on the time scale of T_{1e} (Bridges et al., 2010; Kawasaki et al., 2001). Both (1) and (2) depend on the local structure of the protein around R1, while (3) can provide information on global structure in a protein via distance measurements between a bound metal ion and nitroxide. Exchange effects (4) can arise in cases where two conformations of a spin labeled protein coexist at equilibrium. In such cases, the presence of exchange events can be identified from measurement of effective T_{1e} values. Thus, measurement of T_{1e} under the appropriate conditions can provide information on both static and dynamic features of a protein and applications that make use of the above dependencies will be outlined below.

2. THEORETICAL BACKGROUND AND THE MEASUREMENT OF T_{1e} WITH SR

The methodology and theory of SR has been considered in detail in the literature (Eaton & Eaton, 2005; Freed, 1974; Percival & Hyde, 1975; Robinson et al., 1994), and here, we outline the main results relevant to R1 in proteins and the applications mentioned above. Figure 1B illustrates a typical SR experiment for the commonly employed ^{14}N nitroxide, where an intense pulse of microwave radiation (the "pump") is applied to the absorption maximum of the central ($m_I=0$) resonance line. If the pump has sufficient microwave power, the spin system will be saturated in the sense that $M_z(0)$, and thus the CW-detected EPR signal amplitude is zero. The receiver of the spectrometer is protected from effects of the intense pump pulse with a defense pulse that transiently blocks the signal path; the length of the defense pulse sets the dead time of the experiment. Low-intensity CW microwave radiation (the "observe") monitors the return of the EPR signal to equilibrium, which is proportional to $M_z(t)$.

In general, the recovery of $M_z(t)$ for a single population of nitroxides, including the effects of Heisenberg exchange, is a sum of exponentials (Haas, Mailer, Sugano, & Robinson, 1992; Pyka et al., 2005) with the following time dependence:

$$f(t) = A_1 e^{-(2W_e + 2W_{ex})t} + A_2 e^{-(2W_e + 2W_{ex} + 3W_n)t} + A_3 e^{-(2W_e + 2W_{ex} + 2W_R)t} + A_4 e^{-(2W_e + 2W_{ex} + 3W_n + 2W_R)t} + \text{H.O.T} \qquad (2)$$

where the W values are relaxation rates of various processes that contribute to recovery from saturation: $W_e = (2T_{1e})^{-1}$ is the direct electron spin–lattice relaxation rate, W_{ex} is the Heisenberg exchange rate in the presence of R_{ex}, W_n is the rate of nitrogen nuclear relaxation, a process that couples the nitroxide hyperfine components $m_I = 1, 0, -1$ (Fig. 1B), and $W_R = \tau_R^{-1}$ is the rotational rate of the nitroxide; H.O.T are higher order terms considered to be insignificant (Pyka et al., 2005; Sugano, Mailer, & Robinson, 1987). Rotational motion of the nitroxide and nuclear relaxation lead to transfer of saturation to different regions of the spectrum, a process generally referred to as spectral diffusion (Fleissner et al., 2011; Haas, Sugano, Mailer, & Robinson, 1993).

For the majority of studies where T_{1e} is of interest, the correlation time of the nitroxide for R1 in a protein is in the range of 1–10 ns (the intermediate motional regime). For example, the shortest correlation time for R1 in a folded protein is dictated by the internal motion of R1 at surface sites, for which $\tau_R \approx 1.5$–2.5 ns (Columbus et al., 2001). For R1 at a buried site, or for a rigidly attached nitroxide side chain (Fleissner et al., 2011; Guo, Cascio, Hideg, Kálái, & Hubbell, 2007), τ_R is determined by the rotational diffusion of the entire protein, which for a small globular proteins of M.W. ≈ 20 kDa is approximately 6 ns in aqueous solution at room temperature. In this intermediate motional regime, W_R and W_n are on the order of 100 and 10 MHz, respectively (Robinson et al., 1994), while W_e is on the order of 0.1 MHz. Thus, the spectral diffusion processes are 2–3 orders of magnitude faster than W_e; if the duration of the saturating pump pulse is $\geq \sim 500$ ns, these processes are complete within the pump time and do not contribute significantly to the recovery curve. For this "long-pulse" SR experiment, with pump durations typically in the 1–4 μs range (Hyde, 1979), only the first term in Eq. (2) is important and a true spin–lattice relaxation time can be observed in the absence of Heisenberg exchange. For very fast ($\tau_R \lesssim 1$ ns) or very slow ($\tau_R > 10$ ns) nitroxide motions, spectral diffusion due to W_n will contribute to the recovery and must be considered; W_R contributes to the recovery if $\tau_R \gg 10$ ns (Fleissner et al., 2011; Haas et al., 1992). The spectral diffusion terms are themselves of interest to measure slow protein dynamics and can be extracted with SR measurements in tandem with the electron–electron double resonance (ELDOR) technique, a method closely related to SR but where pump and observe frequencies are different; a discussion of this method is beyond the scope of this chapter, and the interested reader is referred to the literature (Fleissner et al., 2011; Haas et al., 1993; Hyde, Chien, & Freed, 1968).

For a single-spin population, and in the absence of spectral diffusion and Heisenberg exchange, Eq. (2) predicts a single-exponential recovery of $M_z(t)$ following a saturating pulse. However, conventional EPR spectrometers do not measure $M_z(t)$ directly, but rather the CW observe beam monitors the transverse magnetization $M_y(t)$ that is proportional to $M_z(t)$, and the experimentally detected recovery signal, $S(t)$, includes additional time-dependent terms according to (Percival & Hyde, 1975):

$$S(t) \propto M_{y0} \exp\left(-\frac{t}{T_{2e}}\right) + \gamma_e B_1 T_{2e} \left[\frac{M_{z0} - M_{z\infty}}{1 + \gamma_e^2 B_1^2 T_{1e} T_{2e}}\right]$$

$$\exp\left(-\frac{t}{T_{1e}} - \gamma_e^2 B_1^2 T_{2e} t\right) + \frac{M_{z\infty} \gamma_e B_1 T_{2e}}{1 + \gamma_e^2 B_1^2 T_{1e} T_{2e}} - M_{z0} \gamma_e B_1 T_{2e} \exp\left(-\frac{t}{T_{2e}}\right) \quad (3)$$

where M_{y0} and M_{z0} are the magnetizations along on the y- and z-axes, respectively, at time zero after a pump pulse, $M_{z\infty}$ is the equilibrium magnetization along z, T_{2e} is the electron spin transverse relaxation time, γ_e the electron gyromagnetic ratio, and B_1 the magnetic field strength of the observe beam. The term highlighted in gray is the desired SR, and the third term is simply the steady-state CW EPR spectral amplitude, a term that does not affect the time course of the recovery. The first term, the "free induction decay" (FID), and the fourth term are undesirable and, in principle, may complicate analyses of the recovery curves. In practice, these terms that involve "$\exp(-t/T_{2e})$" cause little complication for experiments with nitroxides in the correlation time range of 1–10 ns where $T_{2e} \ll T_{1e}$, and these terms may vanish within the effective dead time of the instrumentation defined by the defense pulse. Under any circumstance, however, the FID term is removed by a phase cycling procedure implemented in the commercially available instruments from Bruker Biospin. Alternatively, a microwave source that is not coherent with the observe source can be used as a pump to eliminate the FID; the ELDOR arm in a Bruker Elexsys 580 (E580) so equipped is such a source. The fourth term can be eliminated by a sufficiently high pump power to achieve complete saturation, i.e., where $M_{z0} = 0$. It is prudent to confirm the absence of these terms in the recovery as described by Hyde (1979).

Of particular importance is the presence of the quantity $-\gamma_e^2 B_1^2 T_{2e}^t$ that adds to $-t/T_{1e}$ in the exponential of the SR signal and shortens the apparent recovery time constant. The magnetic field of the observe beam is related to the microwave power according to $B_1 = \sqrt{QP_1}$ where Q is the quality factor of the resonator and P_1 is the incident observe microwave power. B_1 multiplies all terms in Eq. (3) except the FID, and the observe microwave power should be as large as possible to insure acceptable signal to noise in the detected signal, but must sufficiently low not to substantially shorten the recovery time.

In summary, the pure SR signal is observed with phase cycling to remove the FID, with low observe power, and sufficiently high pump power so that $M_{z0} \approx 0$. Practical considerations for achieving these conditions are discussed in the next section.

3. INSTRUMENTATION AND PRACTICAL CONSIDERATIONS

Commercial instrumentation for SR is available as part of the E580 spectrometer from Bruker Biospin. In the SR spectrometer, unlike the conventional CW EPR spectrometer, magnetic field modulation is not employed in the detection of the CW observe signal, and a large number of transient SR curves are averaged, typically 1–2 million. The repetition rate is determined by the T_{1e} of the sample, and the time between pump pulses is conservatively selected to be $\approx 5 \times T_{1e}$. Field-independent instrumental artifacts due to switching transients and resonator heating from the pump pulse are canceled by subtracting signals recorded on- and off-resonance; the off-resonance signal is obtained by stepping the magnetic field at a rate suitable for field stabilization (≈ 1 Hz for the Bruker E580). The entire process of data acquisition, including phase cycling to remove the FID and field-stepping for baseline correction, is fully automated under software control.

The resonator used in this laboratory with the Bruker E580 is a 2-loop-1-gap resonator (LGR) with a sample loop of 1 mm diameter × 5 mm length (Hubbell, Froncisz, & Hyde, 1987). LGRs are well suited for SR measurement due to a short ring-down time for the pump pulse resulting from their relatively low Q (<1000), a high sample filling factor that gives good detection sensitivity, and a high efficiency for conversion of incident power to microwave magnetic field in the resonator (Hyde, Yin, Froncisz, & Feix, 1985). The short ring-down time $t_{RD} = Q/2\pi\nu \approx 20$ ns, where ν is the spectrometer frequency, means that data collection can occur ≈ 100 ns after the pump pulse termination, allowing for measurement of relatively short $T_{1e}s$ (≈ 0.5 μs). The high conversion efficiency means that relatively low-incident microwave powers are needed for complete saturation of the spin system. The Bruker split-ring resonator ER 4118X-MS2 has similar properties and is also well suited for SR.

The presence of paramagnetic oxygen in liquid samples shortens the apparent T_{1e} of a nitroxide by Heisenberg exchange. Dissolved oxygen in the sample is conveniently removed by flowing nitrogen gas over the sample contained in a gas-permeable sample tube made of TPX plastic or thin-walled Teflon. TPX capillaries of 0.6 mm (I.D.) × 0.8 mm (O.D.) are commercially available (Molecular Specialties, Inc. and Bruker Biospin) and are compatible with both the LGR and Bruker split-ring resonator; the active volume of the sample is ≈ 2.5 µL, and spin concentrations are typically in the range of 100–500 µM. Nitrogen flow is conveniently provided by the Bruker temperature control unit (ER 4131VT) which employs nitrogen gas as the heat transfer medium.

As is evident from the previous section and Eq. (3), proper selection of the pump pulse duration and power and the CW observe power is essential for interpretation of SR data. Detailed considerations for selecting these parameters have been published (Eaton & Eaton, 2005; Hyde, 1979), and the key points are summarized here. For all applications considered in this chapter, suppression of spectral diffusion is desired and long-pulse SR is employed. Typically, a 1–4 µs pump is sufficient for R1 in a protein where τ_R is in the 1–10 ns, but this should be confirmed by investigating the time course of recovery as a function of pump pulse length; the pump length is increased until the observed time constant(s) become independent of pulse length. The pump power should be sufficiently high to achieve saturation, thereby suppressing the fourth term in Eq. (3) by making $M_{z0} \approx 0$. Experimentally, the pump power is increased until the amplitude and time constant of the recovery signal become essentially independent of pump power. The amplitude of the recovery will increase with increasing pump power because the recovery amplitude is proportional to $|M_{z0} - M_{z\infty}|$ and is largest when $M_{z0} = 0$ at high power and complete saturation (2nd term, Eq. 3). Using the highest power available and very long pulses would appear to be best, but resonator heating becomes a problem at high powers and long durations, so optimization is necessary. Incident pump powers on the order of 200 mW are typical for the LGR.

As the power of the CW observe beam is increased, the amplitude of the recovery increases; note that the third term in Eq. (3) is the amplitude of the CW signal at $t = \infty$, and this is proportional to B_1, which is in turn proportional to the square root of observe power. On the other hand, B_1 shortens the apparent T_{1e} by the quantity $\gamma_e^2 B_1^2 T_{2e}$, so the observe power must also be optimized for maximum signal with an acceptable perturbation of T_{1e}. Again, the optimum observe power will depend on the resonator and is on the order of 100 μW for the LGR. Despite the effect on T_{1e}, high observing powers can be used in the SR determination of Heisenberg exchange rates with negligible error (Yin & Hyde, 1989); the enhancement of signal strength is dramatic.

With the above considerations, a good approximation to the true T_{1e} can be obtained. For a single-spin population, single-exponential recoveries should always be obtained when the pump and observe parameters are optimized.

4. APPLICATIONS OF LONG-PULSE SR

4.1 Resolving Protein Secondary Structure via Solvent Accessibility

Sequence-correlated R1 solvent accessibility encodes a remarkable amount of information on a protein fold, including the type of regular secondary structure and its orientation within the tertiary fold (Altenbach et al., 2005; Isas, Langen, Haigler, & Hubbell, 2002). For membrane-bound proteins, the topology of the structure with respect to the membrane surface can be determined (Hubbell & Altenbach, 1994; Hubbell, Gross, Langen, & Lietzow, 1998; Oh et al., 1996).

In SDSL, the solvent accessibility of R1 in a protein is measured via the collision rate of the nitroxide with a paramagnetic species in solution (R_{ex}) that has a T_{1e} much shorter than that for the nitroxide. The collision results in Heisenberg (electron) spin exchange, which appears as a spin–lattice relaxation event for the nitroxide, and the T_{1e} of the nitroxide is shortened in proportion to the Heisenberg exchange rate (W_{ex}),

$$W_{ex} = j_{ex} C_R = \left[\frac{1}{T_{1e}(R)} - \frac{1}{T_{1e}(0)} \right] \quad (4)$$

where $T_{1e}(R)$ and $T_{1e}(0)$ are the spin–lattice relaxation rates of the nitroxide in the presence and absence of a relaxation reagent, respectively, C_R is the concentration of R_{ex} and j_{ex} is the Heisenberg exchange rate constant. The spin–lattice relaxation rates are conveniently measured by SR with the considerations given above, and j_{ex} is determined from Eq. (4) as a direct measure of solvent accessibility. Although j_{ex} may be estimated from a single value of C_R, typically j_{ex} is determined from the slope of a plot of

W_{ex} versus C_R. A more detailed analysis of Heisenberg exchange as a measure of accessibility is provided by Altenbach et al. (2005), wherein the Heisenberg exchange rate constant was designated k_{ex}; here k_{ex} is reserved for conformational exchange and j_{ex} is used for Heisenberg exchange.

The choice of R_{ex} is determined by the information sought. The most commonly used species for mapping of solvent accessibility are the paramagnetic metal complex Nickel (II) EDDA (NiEDDA) and molecular oxygen, both of which have T_{1e} values that are significantly shorter than those of nitroxides (Bertini, Luchinat, & Parigi, 2001; Teng, Hong, Kiihne, & Bryant, 2001). In addition, both reagents are electrically neutral, thus ensuring that W_{ex} is not influenced by the local electrostatic potential of the protein. The two reagents differ in size and polarity, and the larger NiEDDA gives a higher contrast between exposed and partially buried sites, whereas O_2 provides improved contrast among partially buried sites (Isas et al., 2002). Typical concentrations of R_{ex} used to induce detectable changes in apparent T_{1e} relaxation are low, on the order of 0–1.5 mM for NiEDDA and 0–0.26 mM for oxygen. Under these conditions, these R_{ex} generally do not perturb the protein structure.

Heisenberg exchange with charged R_{ex} has been used to measure local electrostatic potential using CW methods (Lin et al., 1998; Shin & Hubbell, 1992), and extensions using SR are straightforward. Indeed, the many applications that have employed CW saturation methods for estimating Heisenberg exchange can in future studies enjoy the benefits of direct measure by SR. One of the most important advantages of SR compared to CW saturation is the ability to detect multiple populations of nitroxides with different $T_{1e}s$. For example, if a protein exists in two conformations where R1 has a different T_{1e} in each, in general the SR relaxation will be biexponential and the individual accessibilities can be determined (Pyka et al., 2005).

An example of using SR to measure solvent accessibility is shown in Fig. 2, where R1 was introduced sequentially (a "nitroxide scan") at sites 128 through 135 in an α-helix of T4 lysozyme (Fig. 2A). The oxygen- and NiEDDA-dependent j_{ex} values and the intrinsic spin–lattice relaxation rates for each site were determined using methods described above. A clear periodicity can be observed for both reagents, reflecting the helical structure of the protein (Fig. 2B). In this case, W_e values determined in the absence of R_{ex} show a similar periodicity (Fig. 2B) due to the approximately linear dependence of W_e on the nitroxide correlation time (Fig. 2C) (Bridges et al., 2010; Sato et al., 2008), which is modulated periodically in the helical structure. The data establish both the identity of the secondary structure and

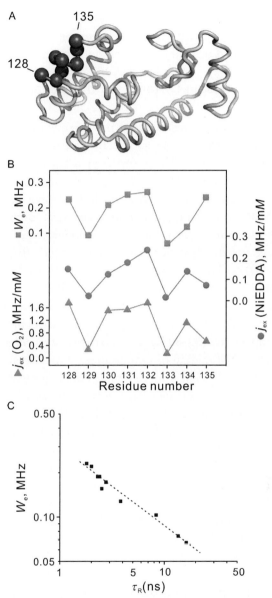

Figure 2 Determination of secondary structure with saturation recovery EPR. (A) Sites of sequential introduction of R1 along a helical segment in T4L; spheres mark the position of the C_α carbon atom of the labeled side chains. (B) The intrinsic spin–lattice relaxation rates measured in the absence of O_2 (W_e, blue (gray in the print version) squares) and rate constants obtained in the presence of O_2 (red (gray in the print version) triangles) or NiEDDA (green (gray in the print version) circles) are plotted as a function of residue number.

(Continued)

its orientation within the protein fold. Measurement of distances between residues in different secondary structural elements can provide information on the tertiary fold and that subject is considered next.

4.2 Measuring Interspin Distances with Relaxation Enhancement

Long-range (tens of Angstroms) distance measurement is a key tool in the elucidation of tertiary structure, structural changes, and structural heterogeneity in proteins, and SDSL-EPR spectroscopy is one of the most powerful techniques for this purpose. The most popular approach in this regard is pulsed dipolar spectroscopy (PDS), which includes double electron–electron resonance (DEER) and double-quantum coherence. Both spectroscopic techniques report the strength of the magnetic dipolar interaction between two spins using specifically tailored pulsed sequences (Borbat & Freed, 2014; Jeschke, 2012). The approximate distances measurable via PDS using R1 range from 12 to 70 Å. A unique feature of PDS is that the distance distributions may be determined in addition to most probable distances, but a disadvantage is that the method is typically carried out in frozen solution at cryogenic temperatures (50–80 K). We also note that recent studies utilizing the triarylmethyl (TAM) radical or spirocyclohexyl spin labels have reported progress in performing PDS near room temperature in proteins, although the practical upper limit of measurable distances using these radicals is relatively short (~35 Å) (Meyer et al., 2015; Yang et al., 2012).

An alternative to PDS for interspin distance measurements is the relaxation enhancement (RE) method. In RE, the enhancement of spin–lattice relaxation rate (W_e) for a spin label due to the presence of a second nearby spin is measured with SR. An endogenous or introduced paramagnetic metal is used as the second spin and the measured RE is proportional to r^{-6}, where r is the nitroxide-metal interspin distance. In contrast to PDS, RE provides only an average distance without resolving the distance distribution. However, RE can be measured with equal ease at both room and cryogenic temperatures, and PDS can be carried out on the same sample at cryogenic temperature. From the PDS distance distribution obtained,

Figure 2—Cont'd A clear periodicity is observed in all three cases, reflecting the helical nature of the T4L segment between residues 128–135. (C) The dependence of nitroxide W_e on the rotational correlation time (τ_R). W_e for R1 at various sites in T4L (differed from those shown in A) were obtained with a long-pulse SR experiment, and τ_R values were estimated from simulations of the CW spectra. *Replotted using data from Bridges et al. (2010).*

the expected RE can be computed and compared with experimental data. This capability can be used to test the effect of freezing on the protein structure and to aid in the interpretation of the RE data (Yang et al., 2014) (see example below). At room temperature, nitroxide/Cu^{2+} RE has measured interspin distances of up to 40 Å in proteins (Yang et al., 2014), although the use of TAM/Cu^{2+} pairs should permit measurement of distances up to ~50 Å (see below).

In the following, only RE distance measurements using a nitroxide spin label and a paramagnetic metal ion introduced site specifically are considered. The paramagnetic metal ion for RE-based distance measurements should have a T_{1e} within one or two orders of magnitude of ω^{-1}, where ω is the resonant frequency of the nitroxide. At X-band, $\omega \approx 10^{11}$ s^{-1} and Cu^{2+} with a T_{1e} of 1–5 ns is a suitable choice (Bertini et al., 2001; Jun, Becker, Yonkunas, Coalson, & Saxena, 2006). For a protein which does not contain an endogenous Cu^{2+}-binding site, a Cu^{2+}-binding motif compatible with the protein must be introduced site selectively. The affinity of the site must be much higher than that for nonspecific sites, which typically have K_d in the μM range. Recently, reported high-affinity Cu^{2+}-binding motifs (Fig. 3A) include a tripeptide sequence, GGH, which mimics the metal-binding site of Albumin (Yang et al., 2014); a double Histidine mutations in combination with an iminodiacetate ligand, dHis-IDA (Cunningham, Putterman, Desai, Horne & Saxena, 2015); and an EDTA-derivative Cu^{2+} chelate modified to react with a protein cysteine (Cunningham, Shannon, et al., 2015). The tripeptide GGH sequence provides a means for introducing Cu^{2+} in loop regions, while the dHis-IDA motif is suitable for helical or β-strand sequences (Cunningham, Putterman, et al., 2015); each has $K_d \ll 1$ μM. In addition to high-binding affinity, internal flexibility of the engineered Cu^{2+} motif should be minimal because high flexibility would reduce the spatial resolution of the method and possibly introduce another potential T_{1e} relaxation pathway (via dynamic modulation of the interspin distance) (Yang et al., 2014). The tripeptide and the EDTA Cu^{2+} complexes were reported to have an internal flexibility comparable to that of the commonly used R1 side chain (Cunningham, Shannon, et al., 2015; Yang et al., 2014), while the dHis-IDA-Cu^{2+} complex is apparently very well-localized spatially (Cunningham, Putterman, et al., 2015).

The nitroxide can be introduced using cysteine substitution mutagenesis followed by reaction with a sulfhydryl-specific nitroxide reagent. The choice of nitroxide side chain is important: the longer the T_{1e} of the nitroxide, the longer the maximum distance that can be determined by

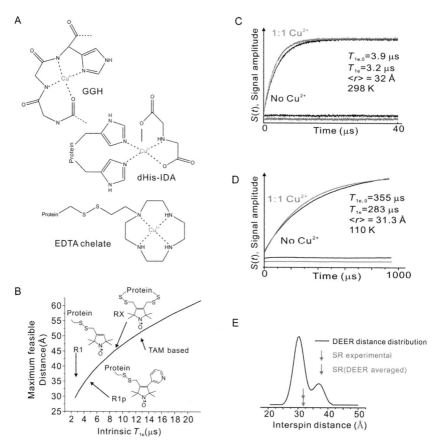

Figure 3 Determination of interspin distances in proteins with saturation recovery EPR. (A) Structures of ligands used to introduce Cu^{2+}-binding sites in proteins. (B) The maximum feasible Cu^{2+}/nitroxide distance that can be measured using the RE approach as a function of the intrinsic T_{1e} of the nitroxide. Structures of spin labels R1, R1p, RX are shown as insets, with arrows indicating their relaxation times and maximum distances feasible to measure. (C) and (D) Examples of RE of R1p by Cu^{2+} in the GGH loop-binding motif at 298 and 110 K, respectively; R1p is attached at site 131 in T4L, and GGH is inserted at site 23 (Yang et al., 2014). The relaxations in the presence (red (gray in the print version)) and absence (black) of Cu^{2+} were fit with single exponentials, and the residuals are shown. The mean distance measured at 298 K was computed according to the fast motion model while that at 110 K was obtained using the rigid limit model. Values of T_{1e} in the presence and absence of Cu^{2+} and the computed distances are shown in the inset. (E) The distance distribution of the Cu^{2+}-R1p distance in the same sample obtained from DEER at cryogenic temperature. The interspin distance calculated from room temperature RE experiments (red (light gray in the print version) arrow) is consistent with the "DEER average" distance corresponding to a sum of simulated exponentials weighted based on the experimental DEER distance distribution (gray arrow) (see Yang et al., 2014).

RE with Cu^{2+} (Fig. 3B; Yang et al., 2014). At room temperature, T_{1e} of an R1 side chain at surface sites is typically 1.5–2.5 μs depending on local structure and dynamics at the site, thus giving a maximal measurable interspin distance of ~25 Å (Jun et al., 2006). According to Fig. 2C, T_{1e} is inversely related to the correlation time of the nitroxide. Thus, nitroxide side chains with hindered internal motions are desired. These include R1p (Fawzi et al., 2011), V1 (Toledo Warshaviak, Khramtsov, Cascio, Altenbach, & Hubbell, 2013), and RX (Fleissner et al., 2011), which should allow for distance measurements up to 35 and 40 Å via RE at room temperature (Fig. 3B; Yang et al., 2014). In addition, with the long T_{1e} of the TAM radical (≈ 20 μs) (Owenius, Eaton & Eaton, 2005), it should be possible to measure distances up to 50 Å at room temperature. Spin-labeling reagents to introduce R1, R1p, V1, and RX at cysteine residues are now commercially available (Toronto Research Chemicals; Enzo life sciences), although the TAM-labeling reagent still requires custom synthesis (Yang et al., 2012). We also note that if a cysteine residue is used to introduce a Cu^{2+} chelate, an orthogonal chemistry is necessary to introduce the nitroxide side chain, for example, using an unnatural amino acid (Fleissner, Brustad, et al., 2009; Razzaghi et al., 2013). At the present time, nitroxides introduced in this fashion have high internal mobility and therefore short relaxation times, limiting the maximum distance measurable.

Once a nitroxide spin label and Cu^{2+} are introduced into a protein, interspin distance determinations require measurement of the nitroxide RE due to the metal ion, i.e., $RE = (T_{1e})^{-1} - (T_{1e}^0)^{-1}$, where T_{1e} and T_{1e}^0 are relaxation times in the presence and absence of Cu^{2+}, respectively. Interpretation of RE in terms of interspin distance depends on the rotational correlation time of the nitroxide–Cu^{2+} interspin vector, which is essentially the correlation time of the protein (τ_c) for the rigid nitroxide side chains and Cu^{2+} ligands considered above. Analytical expressions relating RE to interspin distance have been obtained for two models: in the fast motional limit the static magnetic dipolar interaction is completely averaged by rotational diffusion, while in the rigid limit, no such averaging occurs (Hirsh & Brudvig, 2007; Jun et al., 2006). In the case of Cu^{2+}/nitroxide interaction, the choice of model depends on both τ_c and the strength of the dipolar interaction (proportional to r^{-3}) according to (Yang et al., 2014):

$$5.3 \times 10^{11} \left[\frac{\tau_c}{r^3}\right] \begin{matrix} \ll 1 \text{ (fast motional limit)} \\ > 1 \text{ (rigid limit)} \end{matrix} \tag{5}$$

where r is the interspin distance in Angstroms. The fast motional limit is appropriate for small proteins (M.W. < 20 kDa) and peptides in solution for the distance range of 25–40 Å (Jun et al., 2006; Yang et al., 2014). For shorter distances or somewhat larger proteins, the system will be in between the fast motional and rigid limits, but the correlation time can be "tuned" by increasing the viscosity of the medium or by attaching the protein to a solid support (López, Fleissner, Brooks, & Hubbell, 2014) in order to move the system to the rigid limit.

Analytical expressions that relate RE to interspin distance are (Yang et al., 2014).

At the fast motional limit,

$$\frac{1}{T_{1s}} - \frac{1}{T_{1s}^0} = \frac{2\pi^2 g_s^2 g_f^2 \beta_e^4}{5\hbar^2 r^6} \left[\frac{T_{2f}}{1 + (\omega_f - \omega_s)^2 T_{2f}^2} + \frac{3T_{1f}}{1 + \omega_s^2 T_{1f}^2} + \frac{6T_{2f}}{1 + (\omega_f + \omega_s)^2 T_{2f}^2} \right] \quad (6)$$

and at the rigid limit,

$$\frac{1}{T_{1s}} - \frac{1}{T_{1s}^0} = \frac{4\pi^2 g_s^2 g_f^2 \beta_e^4}{\hbar^2 r^6} \left[\frac{T_{2f}}{1 + (\omega_f - \omega_s)^2 T_{2f}^2} (1 - 3\cos^2\theta)^2 \right.$$

$$\left. + \frac{3T_{1f}}{1 + \omega_s^2 T_{1f}^2} \sin^2\theta \cos^2\theta + \frac{6T_{2f}}{1 + (\omega_f + \omega_s)^2 T_{2f}^2} \sin^4\theta \right] \quad (7)$$

The parameters in Eqs. [6] and [7] are: μ_0, h, and β_e, the vacuum permeability, Planck constant, and the Bohr magneton, respectively; g_f and g_s, the g values for Cu^{2+} and the nitroxide (the fast and slowly relaxing species), respectively; ω_f and ω_s, the resonant frequencies for Cu^{2+} and the nitroxide, respectively; T_{1s} and T_{1s}^0, the spin–lattice relaxation times for the nitroxide in the presence and absence of Cu^{2+}, respectively; lastly, T_{1f} and T_{2f}, the spin–lattice and spin–spin relaxation times for Cu^{2+}. In the rigid limit, the calculation of the interspin distance requires integration over the angle θ, the angle between the interspin vector and the external magnetic field.

Examples of RE of R1p by Cu^{2+} chelated in the GGH loop in T4L are provided in Fig. 3C and D for 298 and 110 K, respectively, along with residuals for single-exponential fits to the data that give the indicated relaxation times. The distances computed from Eqs. [6] (at 298 K) and [7] (at 110 K) are in good agreement, suggesting little effect due to freezing on this sample. This is further supported by the Cu^{2+}/nitroxide DEER data obtained on the same sample at 80 K, which showed that the weighted average distance computed from the DEER distribution (Sarver, Silva, & Saxena, 2013) is identical to that determined by RE (Fig. 3E).

4.3 Measuring Protein Conformational Exchange with T_1 Exchange Spectroscopy

It is now recognized that molecular flexibility is an inherent property of proteins in solution, and it can account for many aspects of function, including promiscuity of protein–protein interaction (James, Roversi, & Tawfik, 2003), enzyme catalysis (Eisenmesser et al., 2005), protein evolution (James et al., 2003), and allostery (Hilser, Wrabl, & Motlagh, 2012). The time scale of flexibility extends from ps–ns for backbone and side chain fluctuations to μs–ms for conformational exchanges (Henzler-Wildman & Kern, 2007). Remarkably, it appears that the *lifetimes* of individual substates involved in the fluctuating structures may be as important as the structural differences (Manglik et al., 2015).

Recent developments in SDSL have shown that the CW EPR spectra of R1 at select sites can measure the dynamics of fast (ps–ns) backbone fluctuations (Columbus & Hubbell, 2004; López, Oga, & Hubbell, 2012) and that the existence of two conformational substates in slow exchange (μs–ms) may be revealed by two-component EPR spectra as indicated in Fig. 4A (López, Fleissner, Guo, Kusnetzow, & Hubbell, 2009; López et al., 2012). However, exchange between two rotamers of the nitroxide can also give rise to resolved spectral components (Fig. 4B). To enable SDSL-EPR to detect true conformational exchange, it is thus essential to distinguish protein conformational variation from R1 rotameric exchange as the origin of observed two-component EPR spectra. Evidence to date indicates that long-pulse SR is effective for achieving this goal as well as for estimating conformational exchange rates.

Consider a spin label in equilibrium exchange between two environments in a protein, α and β, with the nitroxide being free of protein contacts in α but immobilized due to local interactions in β (Fig. 4A):

$$\alpha \underset{k_r}{\overset{k_f}{\rightleftarrows}} \beta \qquad (8)$$

The exchange rate constant is defined as $k_{ex} = \frac{1}{2}(k_f + k_r)$, with k_f and k_r being the forward and reverse rate constants, respectively. If the exchange

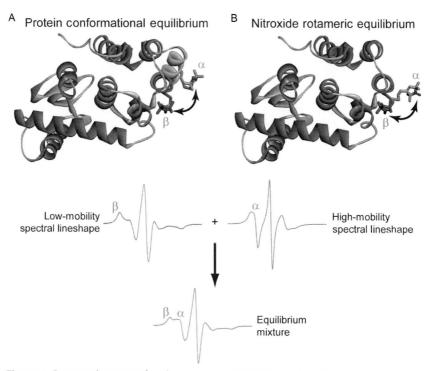

Figure 4 Structural origins of multicomponent CW EPR spectra. (A) Example of conformational exchange between two states related by the rotation of a helical segment. In one conformation (red (dark gray in the print version)), the nitroxide is immobilized due to interactions with the protein resulting in a broad spectral lineshape (β). In the other (rotated helix, cyan (light gray in the print version)), the interacting constraints are removed, resulting in a narrow resonance lineshape (α). In a slowly exchanging equilibrium mixture of these two states, the resulting spectrum is a sum of the two components in proportion to their population (green (gray in the print version) trace). (B) Two rotamers of a nitroxide side chain in a protein. In one rotamer, interaction with the protein results in an immobilized state (β), while in another, the constraints are removed, giving a mobile state (α). If the rotamer exchange is slow, the equilibrium mixture will consist of the sum of the two spectra, which is indistinguishable from the case of conformational exchange in (A).

lifetime, defined as $\tau_{ex} = k_{ex}^{-1}$, is ≥ 100 ns, the EPR spectrum of the mixture will be the algebraic sum of the spectra corresponding to α and β in proportion to the respective populations, and the lineshape will be unperturbed by the dynamic exchange event. On the other hand, the effective spin–lattice relaxation times will be strongly influenced by exchange events with lifetimes as long as ≈ 50 μs, providing an opportunity for SR to explore a time scale of motion nearly three orders of magnitude slower than that corresponding to lineshape effects. The analysis of SR data in terms of exchange kinetics is considered in detail by Bridges et al. (2010); the main points that allow the distinction between conformational and rotameric exchange is summarized below.

Because the T_{1e} of the nitroxide depends on its total rotational correlation time (Fig. 2C), the α and β states will, in general, have different intrinsic spin–lattice relaxation rates: $W_{e\alpha} = (2T_{1e\alpha})^{-1}$ and $W_{e\beta} = (2T_{1e\beta})^{-1}$. If the first-order exchange rate constant between the states is such that,

$$k_{ex} \ll |W_{e\alpha} - W_{e\beta}| \quad (9)$$

the process is in the slow exchange limit, and the SR recovery is biexponential with each relaxation rate equal to the intrinsic relaxation rate of the respective individual state.

In the fast exchange limit,

$$k_{ex} \gg |W_{e\alpha} - W_{e\beta}| \quad (10)$$

the SR relaxation will be monoexponential with an effective relaxation rate W_{eff} that is simply the population-weighted average of the individual rates:

$$W_{eff} = \frac{1}{2T_{1eff}} = f_\alpha W_{e\alpha} + f_\beta W_{e\beta} \quad (11)$$

where f_α and f_β are the fractional populations of the states.

In intermediate exchange,

$$k_{ex} \approx |W_{e\alpha} - W_{e\beta}| \quad (12)$$

Two-component EPR spectra and biexponential SR relaxations are observed, but the relaxation times are complex functions of k_{ex}, f_α, f_β, $W_{e\alpha}$, and $W_{e\beta}$ (Bridges et al., 2010).

Intrinsic T_{1e} values for mobile (α) and immobile (β) states of R1 are typically ≈ 2 μs and ≈ 6 μs, respectively, and $|W_{e\alpha} - W_{e\beta}| \approx 170$ kHz, setting the time scale for T_{1e} exchange spectroscopy roughly in the range of 20 kHz–2 MHz, or exchange lifetimes in the range of 0.5–50 μs.

Remarkably, all documented cases to date of two-component EPR spectra that arise from distinct rotamers of R1 in a single protein conformation fall in the fast exchange limit at temperatures near ambient, i.e., $\tau_{ex} < 0.5$ μs (Bridges et al., 2010). If this proves to be general, the observation of a single-exponential relaxation for a two-component EPR spectrum identifies rotamer exchange as the origin of the two-component spectrum.

Thus, SR complements osmotic perturbation SDSL-EPR (López et al., 2009) and high pressure SDSL-EPR (Lerch, Horwitz, McCoy, & Hubbell, 2013; McCoy & Hubbell, 2011) as strategies to distinguish rotamer exchange from conformational exchange in SDSL-EPR studies of protein dynamics.

If a two-component EPR spectrum and a biexponential SR relaxation are observed, it is possible in some cases to distinguish slow from intermediate exchange. Simulation of the two-component EPR spectrum using the microscopic order–macroscopic disorder model (Budil, Lee, Saxena, & Freed, 1996; Freed, 1976) provides values for the nitroxide rotational correlation times from which expected values of $T_{1e\alpha}$ and $T_{1e\beta}$ for slow exchange can be obtained (see example in Fig. 2C). If the experimentally determined time constants from the biexponential relaxation are comparable to the T_{1e} values obtained from simulations, the system is likely in slow exchange. Intermediate exchange will result in relaxation times significantly shorter than those estimated from the nitroxide rotational correlation times (Bridges et al., 2010).

A further test to distinguish the slow and intermediate exchange regimes is provided by establishing a dependence of W_{eff} for each component on the concentration of an exchange reagent that differentially modulates the two rates based on differences in solvent accessibility. In slow exchange, the two states are effectively isolated and a plot of W_{eff} versus C_R is linear for each with slopes equal to j_{ex} (Fig. 5A; Eq. 4); in intermediate exchange the plot is curved (Fig. 5B). However, the curvature is slight and its resolution requires SR data with a high signal-to-noise ratio. In ideal cases, it is possible to estimate the numerical value of k_{ex} in the 20 kHz–2 MHz regime from the W_{eff} versus C_R data by globally fitting the data set to analytical expressions for W_{eff} in intermediate exchange. The procedure is not robust due to the large number of parameters involved (f_α, f_β, k_{ex}, $W_{e\alpha}$, $W_{e\beta}$, $j_{ex\alpha}$, and $j_{ex\beta}$). The limitations and strategies involved for application of the method are discussed in detail by Bridges et al. (2010).

5. SUMMARY AND FUTURE DIRECTIONS

At the present stage of development, SR EPR in combination with SDSL is a unique tool for resolving sequence-specific secondary structure through measurement of solvent accessibility, for mapping global protein structure via long-range distance measurements (10–40 Å) at physiological temperatures, and for identifying slow (μs–ms) conformational exchange in

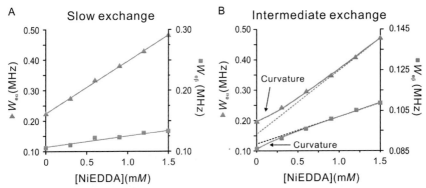

Figure 5 Distinguishing slow from intermediate exchange. (A) T4L 46R1 has a two-component EPR spectrum and a biexponential SR recovery due to slow exchange between two conformations of the protein. The relaxation rate for each component increases linearly with the concentration of an exchange reagent. (B) T4L 130R1 also has a two-component EPR spectrum and biexponential recovery. However, the relaxation rates are nonlinear functions of the exchange reagent concentration, a hallmark of intermediate exchange. The magenta (light gray in the print version) and dark blue (black in the print version) dashed lines are added to emphasize the nonlinearity of the fits. *Replotted with data from Bridges et al. (2010).*

proteins at equilibrium. In ideal cases, conformational exchange rates can also be estimated, but future developments in analytical methods will be required to improve the robustness of the measurements. For example, using SR relaxation amplitudes as well as rate data in a global analysis is one of the strategies under investigation. In addition, high pressure SR is an area for future development. Exchange events in general involve nonzero activation volumes, and pressure can be used to "titrate" exchange rates, providing an additional tool for exploring the time scale of exchange events. Finally, new spin labels with longer T_{1e}s are desired to extend the range of distance measurements by RE. In this perspective, the TAM radical is a promising platform for future development.

REFERENCES

Altenbach, C., Froncisz, W., Hemker, R., McHaourab, H., & Hubbell, W. L. (2005). Accessibility of nitroxide side chains: Absolute Heisenberg exchange rates from power saturation EPR. *Biophysical Journal, 89*(3), 2103–2112.

Berliner, L. J., Grunwald, J., Hankovszky, H. O., & Hideg, K. (1982). A novel reversible thiol-specific spin label: Papain active site labeling and inhibition. *Analytical Biochemistry, 119*(2), 450–455.

Bertini, I., Luchinat, C., & Parigi, G. (Eds.), (2001). In *Current Methods in Inorganic Chemistry: Vol. 2. Solution NMR of paramagnetic molecules: Applications to metallobiomolecules and models* (pp. 1–372). New York: Elsevier Science.

Borbat, P., & Freed, J. (2014). Pulse dipolar ESR: Distance measurements. In J. Harmer & C. Timmel (Eds.), *Structure and bonding: Vol. 152. Structural information from spin-labels and intrinsic paramagnetic centers in the biosciences* (pp. 1–82). New York, USA: Springer. Heidelberg, Germany.

Bridges, M., Hideg, K., & Hubbell, W. (2010). Resolving conformational and rotameric exchange in spin-labeled proteins using saturation recovery EPR. *Applied Magnetic Resonance, 37*(1–4), 363–390.

Budil, D. E., Lee, S., Saxena, S., & Freed, J. H. (1996). Nonlinear-least-squares analysis of slow-motion EPR spectra in one and two dimensions using a modified Levenberg-Marquardt algorithm. *Journal of Magnetic Resonance, Series A, 120*(2), 155–189.

Columbus, L., & Hubbell, W. L. (2004). Mapping backbone dynamics in solution with site-directed spin labeling: GCN4-58 bZip free and bound to DNA. *Biochemistry, 43*(23), 7273–7287.

Columbus, L., Kálai, T., Jekö, J., Hideg, K., & Hubbell, W. L. (2001). Molecular motion of spin labeled side chains in α-helices: Analysis by variation of side chain structure. *Biochemistry, 40*(13), 3828–3846.

Cunningham, T. F., Putterman, M. R., Desai, A., Horne, W. S., & Saxena, S. (2015a). The double histidine Cu^{2+}-binding motif: A highly rigid, site-specific spin probe for electron spin resonance distance measurements. *Angewandte Chemie, 54*(21), 6330–6334.

Cunningham, T. F., Shannon, M. D., Putterman, M. R., Arachchige, R. J., Sengupta, I., Gao, M., et al. (2015b). Cysteine-specific Cu2+ chelating tags used as paramagnetic probes in double electron electron resonance. *The Journal of Physical Chemistry. B, 119*(7), 2839–2843.

Eaton, G. R., & Eaton, S. S. (2002). Determination of distances based on T_1 and T_m effects. In L. Berliner, G. Eaton, & S. Eaton (Eds.), *Distance measurements in biological systems by EPR: Vol. 19* (pp. 348–381). New York: Kluwer Academic/Plenum Publisher.

Eaton, S. S., & Eaton, G. R. (2005). Saturation recovery EPR. In S. S. Eaton, L. J. Berliner, & G. R. Eaton (Eds.), *Biological magnetic resonance Vol. 24. Biomedical EPR part B: Methodology, instrumentation, and dynamics* (pp. 3–18). New York City: Springer.

Eisenmesser, E. Z., Millet, O., Labeikovsky, W., Korzhnev, D. M., Wolf-Watz, M., Bosco, D. A., et al. (2005). Intrinsic dynamics of an enzyme underlies catalysis. *Nature, 438*(7064), 117–121.

Fawzi, N., Fleissner, M., Anthis, N., Kálai, T., Hideg, K., Hubbell, W., et al. (2011). A rigid disulfide-linked nitroxide side chain simplifies the quantitative analysis of PRE data. *Journal of Biomolecular NMR, 51*(1–2), 105–114.

Fleissner, M. R., Bridges, M. D., Brooks, E. K., Cascio, D., Kálai, T., Hideg, K., et al. (2011). Structure and dynamics of a conformationally constrained nitroxide side chain and applications in EPR spectroscopy. *Proceedings of the National Academy of Sciences of the United States of America, 108*(39), 16241–16246.

Fleissner, M. R., Brustad, E. M., Kálai, T., Altenbach, C., Cascio, D., Peters, F. B., et al. (2009a). Site-directed spin labeling of a genetically encoded unnatural amino acid. *Proceedings of the National Academy of Sciences of the United States of America, 106*(51), 21637–21642.

Fleissner, M. R., Cascio, D., & Hubbell, W. L. (2009b). Structural origin of weakly ordered nitroxide motion in spin-labeled proteins. *Protein Science, 18*(5), 893–908.

Freed, J. H. (1974). Theory of saturation and double resonance in electron spin resonance spectra. VI. Saturation recovery. *The Journal of Physical Chemistry, 78*(12), 1155–1167.

Freed, J. H. (1976). Theory of slow tumbling ESR spectra for nitroxides. In L. J. Berliner (Ed.), *Spin labeling theory and applications* (pp. 53–132). New York, NY: Academic Press.

Froncisz, W., Camenisch, T. G., Ratke, J. J., Anderson, J. R., Subczynski, W. K., Strangeway, R. A., et al. (2008). Saturation recovery EPR and ELDOR at W-band for spin labels. *Journal of Magnetic Resonance, 193*(2), 297–304.

Guo, Z., Cascio, D., Hideg, K., Kálái, T., & Hubbell, W. L. (2007). Structural determinants of nitroxide motion in spin-labeled proteins: Tertiary contact and solvent-inaccessible sites in helix G of T4 lysozyme. *Protein Science, 16*(6), 1069–1086.

Haas, D. A., Mailer, C., Sugano, T., & Robinson, B. A. (1992). New developments in pulsed electron paramagnetic resonance: Direct measurement of rotational correlation times from decay curves. *Bulletin of Magnetic Resonance, 14*, 35–41.

Haas, D. A., Sugano, T., Mailer, C., & Robinson, B. H. (1993). Motion in nitroxide spin labels: Direct measurement of rotational correlation times by pulsed electron double resonance. *The Journal of Physical Chemistry, 97*(12), 2914–2921.

Henzler-Wildman, K., & Kern, D. (2007). Dynamic personalities of proteins. *Nature, 450*(7172), 964–972.

Hilser, V. J., Wrabl, J. O., & Motlagh, H. N. (2012). Structural and energetic basis of allostery. *Annual Review of Biophysics, 41*(1), 585–609.

Hirsh, D. J., & Brudvig, G. W. (2007). Measuring distances in proteins by saturation-recovery EPR. *Nature Protocols, 2*(7), 1770–1781.

Hubbell, W. L., & Altenbach, C. (1994). Investigation of structure and dynamics in membrane proteins using site-directed spin labeling. *Current Opinion in Structural Biology, 4*(4), 566–573.

Hubbell, W. L., Froncisz, W., & Hyde, J. S. (1987). Continuous and stopped flow EPR spectrometer based on a loop gap resonator. *Review of Scientific Instruments, 58*(10), 1879–1886.

Hubbell, W. L., Gross, A., Langen, R., & Lietzow, M. A. (1998). Recent advances in site-directed spin labeling of proteins. *Current Opinion in Structural Biology, 8*(5), 649–656.

Huisjen, M., & Hyde, J. S. (1974). A pulsed EPR spectrometer. *Review of Scientific Instruments, 45*(5), 669–675.

Hyde, J. S. (1979). Saturation recovery methodology. In L. Kevan & R. N. Schwartz (Eds.), *Time domain electron spin resonance* (pp. 1–30). New York City, NY: John Iley & Sons, INC.

Hyde, J. S., Chien, J. C. W., & Freed, J. H. (1968). Electron–electron double resonance of free radicals in solution. *The Journal of Chemical Physics, 48*(9), 4211–4226.

Hyde, J. S., Yin, J.-J., Froncisz, W., & Feix, J. B. (1985). Electron–electron double resonance (ELDOR) with a loop-gap resonator. *Journal of Magnetic Resonance, 63*(1), 142–150.

Isas, J. M., Langen, R., Haigler, H. T., & Hubbell, W. L. (2002). Structure and dynamics of a helical hairpin and loop region in annexin 12: A site-directed spin labeling study. *Biochemistry, 41*(5), 1464–1473.

James, L. C., Roversi, P., & Tawfik, D. S. (2003). Antibody multispecificity mediated by conformational diversity. *Science, 299*(5611), 1362–1367.

Jeschke, G. (2012). DEER distance measurements on proteins. *Annual Review of Physical Chemistry, 63*(1), 419–446.

Jun, S., Becker, J. S., Yonkunas, M., Coalson, R., & Saxena, S. (2006). Unfolding of alanine-based peptides using electron spin resonance distance measurements. *Biochemistry, 45*(38), 11666–11673.

Kawasaki, K., Yin, J.-J., Subczynski, W. K., Hyde, J. S., & Kusumi, A. (2001). Pulse EPR detection of lipid exchange between protein-rich raft and bulk domains in the membrane: Methodology development and its application to studies of influenza viral membrane. *Biophysical Journal, 80*(2), 738–748.

Kusumi, A., Subczynski, W. K., & Hyde, J. S. (1982). Oxygen transport parameter in membranes as deduced by saturation recovery measurements of spin–lattice relaxation times of spin labels. *Proceedings of the National Academy of Sciences of the United States of America, 79*(6), 1854–1858.

Lerch, M. T., Horwitz, J., McCoy, J., & Hubbell, W. L. (2013). Circular dichroism and site-directed spin labeling reveal structural and dynamical features of high-pressure states of myoglobin. *Proceedings of the National Academy of Sciences of the United States of America, 110*(49), E4714–E4722.

Lin, Y., Nielsen, R., Murray, D., Hubbell, W. L., Mailer, C., Robinson, B. H., et al. (1998). Docking phospholipase A2 on membranes using electrostatic potential-modulated spin relaxation magnetic resonance. *Science*, *279*(5358), 1925–1929.

López, C. J., Fleissner, M. R., Brooks, E. K., & Hubbell, W. L. (2014). Stationary-phase EPR for exploring protein structure, conformation, and dynamics in spin-labeled proteins. *Biochemistry*, *53*(45), 7067–7075.

López, C. J., Fleissner, M. R., Guo, Z., Kusnetzow, A. K., & Hubbell, W. L. (2009). Osmolyte perturbation reveals conformational equilibria in spin-labeled proteins. *Protein Science*, *18*(8), 1637–1652.

López, C. J., Oga, S., & Hubbell, W. L. (2012). Mapping molecular flexibility of proteins with site-directed spin labeling: A case study of myoglobin. *Biochemistry*, *51*(33), 6568–6583.

Mailer, C., Nielsen, R. D., & Robinson, B. H. (2005). Explanation of spin–lattice relaxation rates of spin labels obtained with multifrequency saturation recovery EPR. *The Journal of Physical Chemistry. A*, *109*(18), 4049–4061.

Manglik, A., Kim, T. H., Masureel, M., Altenbach, C., Yang, Z., Hilger, D., et al. (2015). Structural insights into the dynamic process of β2-adrenergic receptor signaling. *Cell*, *161*(5), 1101–1111.

McCoy, J., & Hubbell, W. L. (2011). High-pressure EPR reveals conformational equilibria and volumetric properties of spin-labeled proteins. *Proceedings of the National Academy of Sciences of the United States of America*, *108*(4), 1331–1336.

Meyer, V., Swanson, M. A., Clouston, L. J., Boratyński, P. J., Stein, R. A., McHaourab, H. S., et al. (2015). Room-temperature distance measurements of immobilized spin-labeled protein by DEER/PELDOR. *Biophysical Journal*, *108*(5), 1213–1219.

Oh, K. J., Zhan, H., Cui, C., Hideg, K., Collier, R. J., & Hubbell, W. L. (1996). Organization of diphtheria toxin T domain in bilayers: A site-directed spin labeling study. *Science*, *273*(5276), 810–812.

Owenius, R., Eaton, G. R., & Eaton, S. S. (2005). Frequency (250 MHz to 9.2 GHz) and viscosity dependence of electron spin relaxation of triarylmethyl radicals at room temperature. *Journal of Magnetic Resonance*, *172*(1), 168–175.

Owenius, R., Terry, G. E., Williams, M. J., Eaton, S. S., & Eaton, G. R. (2004). Frequency dependence of electron spin relaxation of nitroxyl radicals in fluid solution. *The Journal of Physical Chemistry. B*, *108*(27), 9475–9481.

Percival, P. W., & Hyde, J. S. (1975). Pulsed EPR spectrometer, II. *Review of Scientific Instruments*, *46*(11), 1522–1529.

Popp, C. A., & Hyde, J. S. (1982). Electron–electron double resonance and saturation-recovery studies of nitroxide electron and nuclear spin–lattice relaxation times and Heisenberg exchange rates: Lateral diffusion in dimyristoyl phosphatidylcholine. *Proceedings of the National Academy of Sciences of the United States of America*, *79*(8), 2559–2563.

Pyka, J., Ilnicki, J., Altenbach, C., Hubbell, W. L., & Froncisz, W. (2005). Accessibility and dynamics of nitroxide side chains in T4 lysozyme measured by saturation recovery EPR. *Biophysical Journal*, *89*(3), 2059–2068.

Razzaghi, S., Brooks, E. K., Bordignon, E., Hubbell, W. L., Yulikov, M., & Jeschke, G. (2013). EPR relaxation-enhancement-based distance measurements on orthogonally spin-labeled T4-lysozyme. *ChemBioChem*, *14*(14), 1883–1890.

Robinson, B. H., Haas, D. A., & Mailer, C. (1994). Molecular dynamics in liquids: Spin–lattice relaxation of nitroxide spin labels. *Science*, *263*(5146), 490–493.

Sarver, J., Silva, K. I., & Saxena, S. (2013). Measuring Cu2+-nitroxide distances using double electron–electron resonance and saturation recovery. *Applied Magnetic Resonance*, *44*(5), 583–594.

Sato, H., Bottle, S. E., Blinco, J. P., Micallef, A. S., Eaton, G. R., & Eaton, S. S. (2008). Electron spin–lattice relaxation of nitroxyl radicals in temperature ranges that span glassy solutions to low-viscosity liquids. *Journal of Magnetic Resonance, 191*(1), 66–77.

Shin, Y. K., & Hubbell, W. L. (1992). Determination of electrostatic potentials at biological interfaces using electron–electron double resonance. *Biophysical Journal, 61*(6), 1443–1453.

Subczynski, W. K., Hyde, J. S., & Kusumi, A. (1989). Oxygen permeability of phosphatidylcholine—Cholesterol membranes. *Proceedings of the National Academy of Sciences of the United States of America, 86*(12), 4474–4478.

Subczynski, W. K., Hyde, J. S., & Kusumi, A. (1991). Effect of alkyl chain unsaturation and cholesterol intercalation on oxygen transport in membranes: A pulse ESR spin labeling study. *Biochemistry, 30*(35), 8578–8590.

Sugano, T., Mailer, C., & Robinson, B. H. (1987). Direct detection of very slow two-jump processes by saturation recovery electron paramagnetic resonance spectroscopy. *The Journal of Chemical Physics, 87*(5), 2478–2488.

Teng, C.-L., Hong, H., Kiihne, S., & Bryant, R. G. (2001). Molecular oxygen spin–lattice relaxation in solutions measured by proton magnetic relaxation dispersion. *Journal of Magnetic Resonance, 148*(1), 31–34.

Toledo Warshaviak, D., Khramtsov, V. V., Cascio, D., Altenbach, C., & Hubbell, W. L. (2013). Structure and dynamics of an imidazoline nitroxide side chain with strongly hindered internal motion in proteins. *Journal of Magnetic Resonance, 232*, 53–61.

Yang, Z., Jiménez-Osés, G., López, C. J., Bridges, M. D., Houk, K. N., & Hubbell, W. L. (2014). Long-range distance measurements in proteins at physiological temperatures using saturation recovery EPR spectroscopy. *Journal of the American Chemical Society, 136*(43), 15356–15365.

Yang, Z., Liu, Y., Borbat, P., Zweier, J. L., Freed, J. H., & Hubbell, W. L. (2012). Pulsed ESR dipolar spectroscopy for distance measurements in immobilized spin labeled proteins in liquid solution. *Journal of the American Chemical Society, 134*(24), 9950–9952.

Yin, J.-J., & Hyde, J. S. (1987). Spin-label saturation-recovery electron spin resonance measurements of oxygen transport in membranes. *Zeitschrift für Physikalische Chemie, 153*, 57–65.

Yin, J.-J., & Hyde, J. S. (1989). Use of high observing power in electron spin resonance saturation-recovery experiments in spin-labeled membranes. *The Journal of Chemical Physics, 91*(10), 6029–6035.

Yin, J. J., Pasenkiewicz-Gierula, M., & Hyde, J. S. (1987). Lateral diffusion of lipids in membranes by pulse saturation recovery electron spin resonance. *Proceedings of the National Academy of Sciences of the United States of America, 84*(4), 964–968.

CHAPTER TWO

High-Pressure EPR and Site-Directed Spin Labeling for Mapping Molecular Flexibility in Proteins

Michael T. Lerch*, Zhongyu Yang*,[1], Christian Altenbach*, Wayne L. Hubbell*,[2]

*Department of Chemistry and Biochemistry, Jules Stein Eye Institute, University of California, Los Angeles, California, USA
[2]Corresponding author: e-mail address: hubbellw@jsei.ucla.edu

Contents

1. Introduction	30
1.1 The Thermodynamics of Proteins Under Pressure	31
1.2 SDSL-EPR Detection of Pressure Effects on Proteins	32
2. Variable-Pressure CW EPR	37
2.1 High-Pressure Cells for SDSL-EPR	37
2.2 Pressure Generation and Regulation	39
2.3 Practical Considerations	41
3. Pressure-Resolved DEER	42
3.1 Equipment	43
3.2 Practical Considerations	44
4. Example Applications and Perspectives	44
4.1 Identifying Local Compressibility and Structural Fluctuations in Helical Proteins	45
4.2 Characterizing Conformational Exchange Using Variable-Pressure CW EPR	47
4.3 Structural Characterization of the High-Pressure Conformational Ensemble with PR DEER	50
5. Summary and Future Directions	52
References	53

Abstract

High hydrostatic pressure is a powerful probe of protein conformational flexibility. Pressurization reveals regions of elevated compressibility, and thus flexibility, within individual conformational states, but also shifts conformational equilibria such that "invisible" excited states become accessible for spectroscopic characterization. The central aim of

---

[1] Current address: Department of Chemistry and Biochemistry, North Dakota State University, Fargo, North Dakota, USA

this chapter is to describe recently developed instrumentation and methodologies that enable high-pressure site-directed spin labeling electron paramagnetic resonance (SDSL-EPR) experiments on proteins and to demonstrate the information content of these experiments by highlighting specific recent applications. A brief introduction to the thermodynamics of proteins under pressure is presented first, followed by a discussion of the principles underlying SDSL-EPR detection of pressure effects in proteins, and the suitability of SDSL-EPR for this purpose in terms of timescale and ability to characterize conformational heterogeneity. Instrumentation and practical considerations for variable-pressure continuous wave EPR and pressure-resolved double electron–electron resonance (PR DEER) experiments are reviewed, and finally illustrations of data analysis using recent applications are presented. Although high-pressure SDSL-EPR is in its infancy, the recent applications presented highlight the considerable potential of the method to (1) identify compressible (flexible) regions in a folded protein; (2) determine thermodynamic parameters that relate conformational states in equilibrium; (3) populate and characterize excited states of proteins undetected at atmospheric pressure; (4) reveal the structural heterogeneity of conformational ensembles and provide distance constraints on the global structure of pressure-populated states with PR DEER.

1. INTRODUCTION

Proteins exist in an equilibrium mixture of conformational substates and, while a well-ordered native state typically predominates under physiological conditions, excursions to higher energy (excited) states may be required for function (Baldwin & Kay, 2009; Henzler-Wildman & Kern, 2007). The transition from native to excited state may involve local unfolding (Kitahara et al., 2012; Neudecker et al., 2012) or rigid-body motions (Bouvignies et al., 2011; Korzhnev, Religa, Banachewicz, Fersht, & Kay, 2010), and both the structural changes and the timescale of exchange may contribute to protein function (Manglik et al., 2015). Functional excited states are often only a few kcal/mol higher in energy than the native state, yet this difference results in excited state populations that are inaccessible to traditional spectroscopic detection.

One solution to this problem is the application of high hydrostatic pressure, which induces compression of individual states, but also shifts conformational equilibria toward "invisible" excited states, allowing for their spectroscopic characterization (Akasaka, 2006; Fourme, Girard, & Akasaka, 2012). For example, with this strategy folding intermediates (Kuwata et al., 2001) and low-lying excited states with functional roles (Kitahara, Yokoyama, & Akasaka, 2005) have been investigated. Even when protein conformations are of similar energy and the populations are readily observable, pressure-induced shifts in the populations can be used to confirm an equilibrium, extract thermodynamic properties, and define mechanistic roles of known states (Kitahara et al., 2000). In addition, the kinetics of conformational exchange may be determined by pressure-jump relaxation experiments (Torrent et al., 2006).

In this chapter, a brief introduction to the thermodynamics of proteins under pressure is considered, followed by a discussion of the principles underlying the detection of pressure effects in proteins using site-directed spin labeling (SDSL) and EPR. Recent technological advances and instrumentation will be considered along with practical considerations for high-pressure measurement. Finally, illustrations of data analysis will be discussed in the context of specific recent applications. The full information content from high-pressure SDSL-EPR experiments on protein remains to be explored, but important conclusions regarding conformational equilibria and dynamics can be extracted at the present level of understanding.

1.1 The Thermodynamics of Proteins Under Pressure

For a protein in equilibrium between two conformations, application of pressure at a constant temperature will shift the equilibrium toward the conformation with the smallest partial molar volume, while at the same time compressing the individual conformations. For an equilibrium between two conformations $A \leftrightarrow B$, the pressure dependence of the equilibrium constant K is given by (Akasaka, 2006):

$$\ln\left(\frac{K}{K(0)}\right) = -\frac{\Delta \bar{V}^{\circ}}{RT}(P) + \frac{\Delta \bar{\beta}_T}{2RT}(P)^2 \tag{1}$$

where K and $K(0)$ are the equilibrium constants at applied (gauge) pressure P and at atmospheric pressure ($P=0$), respectively, and

$$\Delta \bar{V}^{\circ} = \bar{V}_B^{\circ} - \bar{V}_A^{\circ} \tag{2}$$

$$\Delta \bar{\beta}_T = \bar{\beta}_{T,B} - \bar{\beta}_{T,A} \tag{3}$$

where the partial molar volume, $\bar{V}_i^{\circ}$, and isothermal compressibility, $\bar{\beta}_{T,i}$, are defined as follows:

$$\bar{V}_i^{\circ} = \left(\frac{\partial V}{\partial n_i}\right)_{T,P,n_{j \neq i}} \tag{4}$$

$$\bar{\beta}_{T,i} = -\left(\frac{\partial \bar{V}_i^{\circ}}{\partial P}\right)_T \tag{5}$$

Equation (1) can be obtained from a Taylor's expansion of ΔG in P for the $A \rightarrow B$ transition at constant temperature and ignoring third-order terms and higher; higher order terms account for the pressure dependence of the $\overline{\beta}_{T,i}$, which are assumed to be constant when using (1).

The first term in (1) typically dominates at low pressures due to the relative magnitudes of $\Delta \overline{V}^{\circ}$ and $\Delta \overline{\beta}_T$, but compressibility differences between states becomes important at higher pressures due to the quadratic pressure dependence of the second term. In many studies, $\Delta \overline{\beta}_T$ is assumed to be negligible, but unless confirmed this assumption may lead to errors in other thermodynamic parameters ($\Delta \overline{V}^{\circ}$ and ΔG°) determined from experimentally measured $K(P)$ and (1) (Prehoda, Mooberry, & Markley, 1998).

The $\Delta \overline{V}^{\circ}$ term contains contributions from the difference in molecular volumes of the conformers (i.e., the sum of atomic and solvent-inaccessible cavity volumes) and from differences in the volume of solvation water. Current evidence suggests that the latter contribution is small, and that $\Delta \overline{V}^{\circ}$ is dominated by a difference in effective cavity volumes between the conformers, due either to solvent entry in one conformer but not the other (Ando et al., 2008; Collins, Hummer, Quillin, Matthews, & Gruner, 2005; Nucci, Fuglestad, Athanasoula, & Wand, 2014; Roche et al., 2012), or to repacking of the core in one conformer that fills cavities with protein native side chains (Lerch et al., 2015). Similarly, $\Delta \overline{\beta}_T$ contains contributions from differences in intrinsic molecular compressibility of the protein and from differences in compressibility of solvent associated with differences in hydration. As considered in Section 1.2, effects of protein compression and conformational shift due to pressure perturbation can be detected with SDSL-EPR.

1.2 SDSL-EPR Detection of Pressure Effects on Proteins

SDSL-EPR is a powerful spectroscopic tool for exploring both structure and dynamics in proteins. Most applications have employed the extensively characterized nitroxide side chain designated R1 (Fig. 1A, inset) (Fleissner, Cascio, & Hubbell, 2009), and the present considerations will be restricted to use of R1 at helical sites in proteins where most information is available. Continuous wave (CW) spectral analysis and pulse-saturation recovery (SR) of single R1 residues introduced site-specifically reveal local structural and dynamical features of the protein fold (see Chapters 17 and 1 of current volume), while pairs of R1 spin labels are employed to provide information on global structure *via* interspin distance measurements using

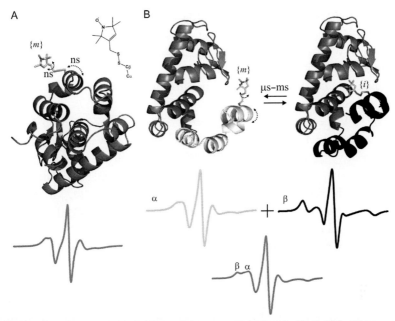

Figure 1 Protein dynamical modes and their manifestation in SDSL-EPR. (A) In a well-ordered protein with a single dominant conformational state (green ribbon), the CW spectral lineshape for R1 at noninteracting surface sites reflects weakly anisotropic motion of the nitroxide (green spectrum), which varies from site-to-site due to variations in local backbone motion (López et al., 2012). The structure of R1 is shown in an inset, and the side chain is shown in stick representation attached to the protein model. The nitroxide motion is directly influenced by backbone fluctuations, here illustrated by a helix rocking mode (dashed curved arrow) and internal R1 motions (solid curved arrow) on the nanosecond timescale. (B) Equilibrium exchange between conformations α and β on the microsecond and longer timescale. Provided that R1 is placed such that it has distinct motions and spectra in α and β, the spectrum of the equilibrium mixture is the weighted sum (purple spectrum) of the spectra reflecting each state. In this example, the nitroxide exhibits a weakly anisotropic motion in α (red), and is immobilized due to interaction with the local protein environment in β (blue). The contributions from α and β in the low-field resonance line of the composite spectrum are indicated. (See the color plate.)

CW EPR (Altenbach, Oh, Trabanino, Hideg, & Hubbell, 2001; Jeschke, 2012; Mchaourab, Steed, & Kazmier, 2011; Rabenstein & Shin, 1995) or the time domain methods of double electron–electron resonance (DEER) (Jeschke, 2012; Pannier, Veit, Godt, Jeschke, & Spiess, 2000) or double quantum coherence (DQC) (Borbat & Freed, 2013). There is no inherent size restriction on the systems that may be studied by SDSL-EPR, making it particularly attractive for studying integral membrane proteins and large protein complexes. At the present time, CW spectral analysis and DEER have been used to explore the effects of pressure on spin-labeled proteins, although in principle, the full complement of SDSL-EPR analytical tools, including SR, can be employed to monitor protein structure and dynamics at high pressure.

The CW spectrum of R1 in a protein encodes information on dynamic modes of the structure and the way in which they are revealed is determined by the relative timescales of the EPR experiment and protein motions. The spectral lineshape of a nitroxide is modulated by rotational correlation times (τ_R) and/or lifetimes for exchange between different states (τ_{ex}) in the range of $\sim$0.1–100 ns for X-band EPR spectroscopy ($\sim$9.5 GHz), thus defining the time window through which CW EPR views protein motions. This time window conveniently overlaps with that for fast backbone fluctuations occurring on the ns timescale, and sequence-specific variations in CW lineshape of R1 can provide a map of the relative amplitude and rate of fast backbone motions throughout a protein (Columbus & Hubbell, 2004; López, Oga, & Hubbell, 2012) (Fig. 1A). On the other hand, fluctuations between different conformations of a protein (conformational exchange) generally occur with τ_{ex} in the μs–ms range and are too slow to produce lineshape effects. However, if an R1 residue is placed at a site in the protein where the motion of the nitroxide is differentially modulated due to interactions with the protein, the EPR spectrum will have two resolved components, thus revealing the presence of conformational exchange but without information on τ_{ex} (López et al., 2012). Essentially, the CW spectrum represents a snapshot of the conformational equilibrium frozen on the EPR timescale, as illustrated in Fig. 1B.

The dynamical modes illustrated in Fig. 1 are pressure sensitive and serve to define the information content derived from high-pressure SDSL-EPR at the present stage of development. Consider first the influence of pressure on the ns timescale backbone motions illustrated in Fig. 1A. For an R1 residue on the solvent accessible surface of a helix or structured loop, where the nitroxide does not make contact with the surrounding macromolecular environment, the EPR spectrum consists of a single component that reflects an anisotropic motion of the nitroxide with effective rotational correlation time on the order of 1–2 ns (Columbus, Kálai, Jekö, Hideg, & Hubbell, 2001). Such sites are generally abundant in relatively rigid helical proteins (López et al., 2012), and the overall nitroxide motion has contributions from internal fluctuations in the R1 side chain itself and fast backbone fluctuations

(Columbus & Hubbell, 2004; Columbus et al., 2001). To analyze the pressure dependence of the overall motion, McCoy and Hubbell adopted a simple model wherein the R1 side chain and protein segment to which it is attached were considered to move as a single kinetic unit (McCoy & Hubbell, 2011). Here, we choose a model that considers R1 and protein motions as independent.

The EPR spectrum of R1 in a stable helix is primarily sensitive to rocking motions of the segment to which it is attached (Columbus & Hubbell, 2002). Such motion adds to the internal motion of R1 so that the observed overall correlation time of R1 on a helical segment (τ_o) is approximately related to the correlations times for R1 internal (τ_i) and protein (τ_P) motions by

$$\frac{1}{\tau_o} = \frac{1}{\tau_i} + \frac{1}{\tau_P} \quad (6)$$

Each reciprocal correlation time (i.e., rate) in (6) can be related to pressure through an activation volume, $\Delta V^{\ddagger}$, according to McCoy and Hubbell (2011), by

$$\frac{1}{\tau(P)} = \frac{1}{\tau(0)} e^{\frac{-P\Delta V^{\ddagger}}{RT}} \approx \frac{1}{\tau(0)} \left(1 - \frac{P\Delta V^{\ddagger}}{RT}\right) \quad (7)$$

where $\tau(0)$ is a correlation time at $P=0$ and the linearized approximate form is obtained by expansion of the exponential about $P=0$ and truncating higher order terms based on the typically small $\Delta V^{\ddagger}$ encountered (≤ 7 mL/mol) and limiting analysis to moderate pressures (≤ 2 kbar). Combining (6) and the linearized form of (7) for each correlation time yields

$$\frac{1}{\tau_o(P)} = \frac{1}{\tau_o(0)} - \frac{P}{RT}\left(\frac{\Delta V_i^{\ddagger}}{\tau_i(0)} + \frac{\Delta V_P^{\ddagger}}{\tau_P(0)}\right) \quad (8)$$

For the noninteracting solvent-exposed surface sites being considered here, the internal motion of R1 has a rate $1/\tau_i(0) \approx 5 \times 10^8 \text{ s}^{-1}$ that is expected to be site-independent (Columbus & Hubbell, 2002, 2004). With this value and $\tau_o(0)$ from simulations, $\tau_P(0)$ can be computed from (6). For backbone motions that strongly influence the EPR lineshape, $\tau_P(0)$ must be on the order of $\tau_i(0)$ and experimental data suggest that in many cases they are similar. Under the condition that $\tau_P(0) = \tau_i(0)$, and incorporating the approximate value of $1/\tau_i(0)$, (8) becomes

$$\frac{1}{\tau_o(P)} \approx \frac{1}{\tau_o(0)} - \frac{P}{RT}(5 \times 10^8 \text{s}^{-1})\left(\Delta V_i^\ddagger + \Delta V_P^\ddagger\right) \qquad (9)$$

Equation (9) is appropriate for the examples considered below, but in particular cases where R1 and backbone motions have substantially different correlation times, (8) will be required. According to (9), a plot of $1/\tau_o(P)$ versus P will be linear with a slope proportional to $\left(\Delta V_i^\ddagger + \Delta V_P^\ddagger\right)$. The activation volume for nitroxide motion, $\Delta V_i^\ddagger$, likely has its origin in fluctuations of the solvent cage around the nitroxide (McCoy & Hubbell, 2011) and, like $\tau_i(0)$, is expected to be approximately site-independent. Thus, variations of the slope from site-to-site are taken to reflect differences in $\Delta V_P^\ddagger$, which in this simplified model corresponds to local volume fluctuations at frequency near $1/\tau_P(0) \approx 5 \times 10^8$ s^{-1}. Mean-squared volume fluctuations of a solvent-impenetrable particle, including a folded protein, are related to the compressibility according to Cooper (1984):

$$\langle(\delta V)^2\rangle = \beta_T kT \qquad (10)$$

where k is the Boltzmann constant. Thus, activation volumes from variable-pressure CW spectra of R1 at noninteracting helix surface sites may be viewed in the context of the relative compressibility of different regions, and can be used to define a site-specific map of relative molecular compressibility correlated with motions on the ns timescale.

Next consider the conformational equilibrium of Fig. 1B, where R1 has a different mobility, and hence EPR spectrum, in each conformation due to differential interactions of the nitroxide with the protein. For conformational exchange that is slow on the CW EPR timescale, the spectrum of an equilibrium mixture of states is the sum of spectra corresponding to the individual states weighted by their population (Fig. 1B). The populations of the individual components can be extracted from such composite spectra by methods described in Section 4.2 and the apparent equilibrium constant thus determined. In general, equilibrium constants are pressure-dependent and fitting data for the equilibrium constants as a function of pressure to (1) can provide the thermodynamic parameters that relate the conformational states. In situations where the equilibrium involves an excited state, one expects new spectral components to appear with applied pressure.

The distinction between "backbone fluctuations" and "conformational exchange" made in Fig. 1 is based on the EPR timescale. Fluctuations of sufficiently high frequency to directly influence spectral lineshapes and to produce spectral averaging of distinct states ($\geq 10^7$ s^{-1}) are classified as fast backbone motions within a given conformational state. On the other hand, fluctuations slow on the EPR timescale (frequency $<10^7$ s^{-1}) are detected by multiple components in the spectrum of R1 and are assigned as fluctuations between resolved protein states. These states could be two different conformations, as shown in Fig. 1B, but could also be two substates of similar energy within a particular conformation. For example, the simple helical rocking motion illustrated in Fig. 1A could in principle occur on a μs rather than an ns timescale and a properly placed nitroxide would sense different rotational states of the helix through local interactions, resulting in a multicomponent spectrum (López et al., 2012). In this case, due to the EPR timescale, even low-amplitude helix rocking would be classified as an exchange event between substates rather than fast backbone fluctuation.

A priori, distinguishing exchange between substates in a single global conformation from exchange between distinct conformations cannot be made on the basis of a single multicomponent CW EPR spectrum. However, the dependence of the CW spectra on pressure aids in this distinction, as will be discussed in Sections 4.1 and 4.2. Structural fluctuations between substates are expected to be of small amplitude compared to those that transition one conformation into another, and intramolecular distance measurements between pairs of R1 residues with double electron–electron resonance (DEER) spectroscopy (Jeschke, 2012) may distinguish the two as discussed in Section 4.3. The unique strength of DEER is that it provides a distance distribution, revealing each structure present at equilibrium along with its probability. With pressure-resolved DEER (PR DEER) (Lerch, Yang, Brooks, & Hubbell, 2014), it becomes possible to track the population shifts and the magnitude of structural transitions as a function of pressure.

The above discussion provides the motivation for high-pressure SDSL-EPR. The instrumentation for high-pressure SDSL-EPR is considered next, followed by examples that illustrate applications of the principles discussed.

2. VARIABLE-PRESSURE CW EPR

2.1 High-Pressure Cells for SDSL-EPR

Two types of high-pressure sample cells have been developed and implemented in X-band SDSL-EPR experiments. The first is fabricated from PTFE-coated fused silica capillary tubing (0.1 mm ID × 0.36 mm OD, Polymicro Technologies) folded into a continuous bundle with 15–25 segments of length ~50 mm (Fig. 2A). Fabrication of the cell requires

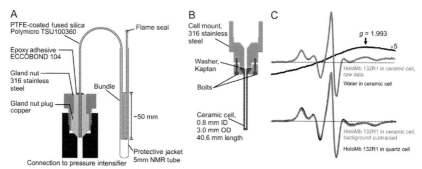

Figure 2 Sample cells used for variable-pressure CW EPR. Details of (A) the fused silica cell and (B) the ceramic cell assembly. (C) The EPR spectrum of holomyoglobin (holoMb) 132R1 at atmospheric pressure in the ceramic cell before subtraction of the background signal (red (gray in the print version), upper panel). The ceramic EPR signal (black, upper panel) is amplified five times with respect to that in the holoMb 132R1 spectrum; the g factor corresponding to the signal maximum is ∼1.993. The holoMb 132R1 spectrum after subtraction of the ceramic background signal (red (gray in the print version), lower panel) is superimposed on the spectrum collected in quartz (black, lower panel), where there is no background signal, to illustrate the accuracy of the background subtraction. The horizontal dashed line is added to indicate the magnitude of the baseline artifact before and after subtraction of the ceramic signal. The holoMb 132R1 spectra were previously reported in Lerch, Horwitz, McCoy, and Hubbell (2013) and López et al. (2012). Panel (A): Adapted from McCoy and Hubbell (2011).

strict attention to detail to prevent contamination or scratching of the capillary surface as described by McCoy and Hubbell (2011). The capillary bundle is inserted into a standard 5 mm OD NMR sample cell (Fig. 2A), which can be used with a variety of available resonators without modification, including the Bruker high-sensitivity cavity (McCoy & Hubbell, 2011) and a 5-loop 4-gap resonator (Eaton, Eaton, & Berliner, 2005; Lerch et al., 2013). The total internal volume of the cell is 12–16 μL (including ∼75 cm connection to the high-pressure system), and the sample volume in the ∼1 cm active region of a resonator is <2 μL, necessitating a relatively high sample concentration of ∼500 μM. In practice, the folded capillary bundle has a maximum pressure capability of ∼4 kbar. Different sizes of fused silica capillary tubing are readily available, so this cell is adaptable to resonators for different microwave frequencies, particularly higher frequencies where a reduction in the required sample volume may negate the need for folding the capillary into a bundle.

An alternative high-pressure cell for X-band SDSL-EPR is the commercially available yttria-stabilized zirconia ceramic cell (HUB440-Cer; Pressure BioSciences, Inc.) (Lerch et al., 2013) shown in Fig. 2B. A commercially available dielectric resonator with custom-designed support hardware for the HUB440-Cer cell is available from Bruker on request (ER4123D, Bruker Biospin). Alternatively, the 5-loop 4-gap resonator mentioned above for the capillary bundle cell may be used. The total sample volume of the cell is 19 μL, with ~5 μL in the ~1 cm active region of a resonator; sample concentrations of ~200 μM give satisfactory signal-to-noise. The typical maximum operating pressure of these cells is 2.4 kbar, although pressures up to 3 kbar are achievable with select cells.

The ceramic cell exhibits a broad EPR signal due to paramagnetic inclusions (Slipenyuk et al., 2004). While this signal overlaps that of nitroxides in CW spectra, it is pressure independent and is readily removed by subtraction after data collection (Fig. 2C). The larger inner diameter of the ceramic cells compared to the capillary bundle cell makes it more amenable to studies of proteins immobilized on solid support (López, Fleissner, Brooks, & Hubbell, 2014). Ceramic cells designed for use in high-pressure NMR, but which may be compatible with EPR, are available from other manufacturers including HiPer Ceramics GmbH and Daedalus Innovations, LLC.

2.2 Pressure Generation and Regulation

Syringe-style pressure generators and air-operated pressure intensifiers for application of hydrostatic pressure are available from various sources, including Pressure BioSciences, Inc. (PBI), High Pressure Equipment Company (HiP), and Daedalus. In the author's laboratory, air-operated pressure intensifiers from PBI are employed. These intensifiers are particularly attractive for on-line operation because they are self-regulating and maintain a set pressure even in the presence of small leaks. The PBI HUB440 pressure intensifier has a maximum operating pressure of 4 kbar and is suitable for use with the ceramic cells from the same company. The PBI HUB880 has a maximum operating pressure of 6.2 kbar and is employed in PR DEER experiments described in Sections 3 and 4. Water can be used as a pressurization fluid in these systems, and sample diffusion is sufficiently slow to eliminate the need for a separator between the sample and pressurization fluid. Both the HUB440 and HUB880 can operate under external computer control, enabling time-varying pressure profiles of arbitrary shape to be implemented.

A schematic of the high-pressure system used in this laboratory for both static and time-varying pressure experiments, including pressure-jump experiments, is given in Fig. 3A along with suppliers and part numbers

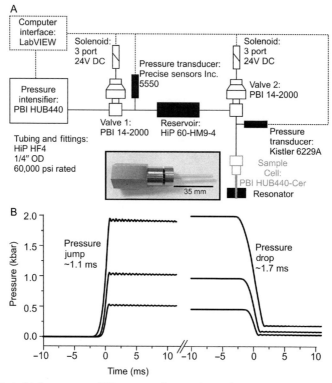

Figure 3 A high-pressure EPR system for static and pressure-jump experiments. (A) Schematic of the high-pressure system. A photograph of the HUB440-Cer ceramic cell is shown in an inset. (B) The time dependence of the system pressure monitored by the Kistler transducer during pressure jumps. Jumps of different magnitude were induced by opening of valve 2 with various initial pressures of the reservoir and sample (see Section 2.2); the pressure jump is ∼1.5 ms and is relatively independent of the jump magnitude and direction.

for key items. The valves are air-operated, where the airflow to each valve is controlled by a solenoid. For static and slowly varying pressure operation both valves are open and the pressure is controlled by the intensifier. For pressure-jump experiments the sample cell is pressurized to pressure P_1 with both valves open, valve 2 is closed, the reservoir is pressurized to P_2, and then valve 1 is closed. A pressure jump is initiated by opening valve 2, which connects the sample and reservoir thus triggering a rapid equilibration to an intermediate pressure; the final pressure is determined by P_1, P_2, and the relative volumes of sample and reservoir. The system is capable of kbar-magnitude pressure jumps with a 10–90% rise time of ∼1.5 ms (Fig. 3B). With computer control of the valves and pressure intensifier, the system can be reset to initial or other selected conditions and the cycle repeated with full automation. As indicated with part numbers, the pressure intensifier, solenoid-operated valves, and ceramic cell (pictured) are from PBI, and high-pressure tubing and fittings from HiP. Components including tubing, fittings, and pressure sensors of various designs are readily available from a number of manufacturers, including HiP, MAXPRO Technologies, Swagelok, and Parker Autoclave. The high-pressure cells introduced in Section 2.1 may be used in pressure-jump EPR experiments without further modification. The primary future use of the pressure-jump mode of operation will be to determine the relaxation time for conformational exchange on the timescale of ms.

2.3 Practical Considerations
2.3.1 Buffers
Proper buffer selection is important in any high-pressure experiment because buffer ionization is accompanied by volume changes due to electrostriction of water, giving rise to pressure-dependent changes in pH. Moreover, for conformational changes accompanied by protonation or deprotonation of titratable residues, the buffer ionization volume will directly contribute to the overall volume change (Lee, Heerklotz, & Chalikian, 2010; Rasper & Kauzmann, 1962; Zipp & Kauzmann, 1973). Thus, buffers with strongly pressure-dependent properties will invalidate the underlying assumption of constant solvent conditions in the thermodynamic treatment presented in Section 1.1, and buffers with small ionization volumes and minimal pH shifts with pressure are desired. Both the volume change of ionization and the associated pressure-dependent change in pH have been reported for many buffers, based on different theoretical and experimental approaches (El'yanov & Hamann, 1975; Kitamura & Itoh, 1987; Neuman, Kauzmann, & Zipp, 1973), including recent measurements using a pH meter capable of operation at high pressures (Samaranayake & Sastry, 2010). Some buffers useful with high pressure along with their corresponding properties are provided in Table 1.

2.3.2 Water Compressibility
Water compresses by ~12% at 4 kbar (Chen, Fine, & Millero, 1977), and the increased water content in the active volume of an EPR resonator under pressure causes a shift in the resonant frequency due to the high dielectric constant of water. In addition, water compression increases the effective concentration of spin in the sample. This is not an issue for EPR spectral

Table 1 Examples of Suitable Buffers for Use Under High Pressure (Kitamura & Itoh, 1987; Neuman et al., 1973; Samaranayake & Sastry, 2010)

Buffer	pH Range	$\Delta V°$(Ionization) (mL/mol)	$\Delta pH/\Delta P$ (kbar^{-1})
2-Amino-2-(hydroxymethyl)-1, 3-propanediol (TRIS)	7.0–9.0	+4.3	−0.03
2-(N-morpholino) ethanesulfonic acid (MES)	5.5–6.7	+3.9	−0.06
Sodium acetate	3.6–5.6	−11	−0.03

line shape analysis, but is important where accurate normalization of spectra is required. Even minor inaccuracies in subtraction of the ceramic cell background signal make normalization using double-integration of CW spectra unreliable. An effective alternative normalization strategy is to scale spectra by a correction factor based on the compression of water (Chen et al., 1977). Fused silica cells have little background signal, and with good signal-to-noise the standard normalization procedure using double-integration may be employed.

3. PRESSURE-RESOLVED DEER

The CW EPR spectrum of R1 is exquisitely sensitive to even small-amplitude structural fluctuations (Lerch et al., 2013; López, Yang, Altenbach, & Hubbell, 2013), but to determine the true magnitude of structural changes requires distance mapping with DEER or DQC using pairs of spin labels. By providing long range (~20–80 Å) distance distributions with angstrom-level resolution, DEER and DQC directly reveal the structure and heterogeneity of conformational states. DEER is the commonly used method for distance determination, and with one exception (Meyer et al., 2015), DEER data on nitroxide pairs has been collected at cryogenic temperatures (50–80 K), typically using samples flash-frozen in liquid nitrogen. As such, resolving pressure-populated changes in protein structure with DEER requires a different approach from that outlined above for CW EPR measurements. In PR DEER, samples are flash-frozen under pressure, kinetically trapping the high-pressure equilibrium for subsequent data acquisition at atmospheric pressure and cryogenic temperatures. This principle has been employed previously in high-pressure crystallography (Collins, Kim, & Gruner, 2011; Urayama, Phillips, & Gruner, 2002).

3.1 Equipment

The strategy used in PR DEER is shown schematically in Fig. 4 and discussed in detail in Lerch et al. (2014). Rather than loading samples into specialized high-pressure cells, standard sample cells are enclosed within a pressure bomb, pressurized, rapidly frozen in a dry ice/ethanol bath, depressurized to atmospheric pressure, and transferred to liquid nitrogen prior to DEER data acquisition. As shown in Fig. 4A, a standard borosilicate capillary cell (1.4 mm ID × 1.7 mm OD; VitroCom, Inc.) is modified by the addition of a silicone piston separating the sample from the pressurization fluid (ethanol) and a magnetic collar (416 stainless steel) glued to the upper portion of the cell. The key feature of the magnetic collar is that it allows the position of the cell within the pressure bomb to be controlled externally using a neodymium-iron-boron ring magnet. Standard high-pressure tubing may serve as the pressure bomb. The pressure bomb is sufficiently long to

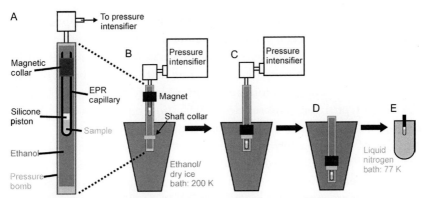

Figure 4 PR DEER methodology. (A) Detail of sample in the pressure bomb. A borosilicate capillary is modified by the addition of a magnetic collar near the top. A silicone piston (red) separates the sample (green) from the ethanol pressurization fluid (light blue) that fills the pressure bomb. (B) The bomb is connected to the pressure intensifier with the lower portion immersed in dry ice/ethanol (dark blue) at 200 K. The sample is held at the top of the bomb using the magnet and the temperature is controlled during pressurization. (C) Rapid cooling to 200 K is triggered while under pressure, when the sample is moved quickly to the bottom of the bomb. Addition of a shaft collar on the pressure bomb helps ensure proper positioning of the cell during cooling. (D) The bomb is depressurized, disconnected from the pressure intensifier, and submerged in the dry ice/ethanol bath. (E) The sample capillary is transferred to liquid nitrogen (purple) for cooling to 77 K in preparation for DEER data acquisition at 80 K. The magnetic collar is removed prior to transfer to the resonator. *Figure is adapted from Lerch et al. (2014).* (See the color plate.)

allow the bottom portion to be cooled to 200 K in the dry ice/ethanol bath while the sample is held in the upper portion at room temperature during equilibration at high pressure. Pre-fabricated tubing from HiP (60-HM6-12) and MAXPRO (101N6U-12-160) were used as pressure bombs in Lerch et al. (2014, 2015). Other high-pressure equipment required for PR DEER, including the pressure generator, pressure sensors, connectors, and tubing are as discussed in Section 2.2.

3.2 Practical Considerations

Control of the sample temperature in the upper portion of the bomb during equilibration at high pressure is of importance in PR DEER (Fig. 4B) due to heat transfer between the stainless steel pressure bomb and the dry ice/ethanol bath. Temperature control was previously accomplished using a heat gun, where heat gun placement and temperature setting were calibrated using a pressure bomb and cooling bath setup with a thermocouple in place of the sample (Lerch et al., 2014, 2015). This approach is sufficient for maintaining samples near room temperature prior to freezing under pressure, but alternative methods of temperature control allowing a wider range of holding temperatures is an area for future development.

As outlined above, the typical sample preparation process for PR DEER experiments requires cooling the sample from 200 to 77 K in liquid nitrogen after depressurization. Conformational movements cannot occur at 77 K, but the potential for conformational relaxation during the time held at 200 K, close to the glass transition temperature of proteins (Doster, 2010), is of concern. Potential relaxation of pressure-populated structural changes at 200 K is discussed in detail in Lerch et al. (2014), and experimental results therein shown that conformational relaxation does not occur on the experimental timescale for the cases investigated.

While transferring the sample from the pressure bomb to liquid nitrogen (Fig. 4E), it is important to remove ethanol from the outer surface and inside of the capillary cell. Residual ethanol on the outer surface of the cell will become solid upon cooling to 77 K and may prevent proper insertion of the sample into the resonator.

4. EXAMPLE APPLICATIONS AND PERSPECTIVES

The technological and methodological advances discussed above present an opportunity for high-pressure SDSL-EPR to provide unique insight into protein structure and flexibility. In the sections below, examples are provided that illustrate the analysis of variable-pressure CW EPR spectra and interpretation of the results in terms of protein compressibility and conformational equilibria. PR DEER examples highlight the ability of the method to resolve the individual structures comprising a high-pressure conformational ensemble and to determine the magnitude of the structural differences between states. A prerequisite for meaningful interpretation of changes observed with pressure application in either CW EPR or PR DEER is a demonstration of reversibility, i.e., the system must return to its original state following depressurization. Only with reversibility can one guarantee thermodynamic equilibrium between states observed under pressure, and that states populated by pressure are part of the protein conformational ensemble even at ambient conditions, although perhaps at a level below the detection limit. Reversibility was verified for all examples presented in this chapter.

4.1 Identifying Local Compressibility and Structural Fluctuations in Helical Proteins

Flexible regions in an otherwise rigid protein are often of functional significance (Berlow, Dyson, & Wright, 2015). As outlined in Section 1.2, variable-pressure CW EPR promises to be an important tool to identify such regions by monitoring the pressure dependence of the nitroxide correlation time (τ_o) for R1 at noninteracting helix surface sites where the spectra correspond to a simple anisotropic motion (Fig. 1A). Values of the correlation time are readily extracted at each pressure by spectral simulations using the microscopic-order-macroscopic-disorder (MOMD) model of Freed and coworkers (Budil, Lee, Saxena, & Freed, 1996; Freed, 1976) with only the correlation time as an adjustable parameter. Details of the spectral simulation are given in McCoy and Hubbell (2011, supplement).

According to the model presented in Section 1.2, the pressure dependence of τ_o can be analyzed to give a total volume of activation, $\Delta V_i^{\ddagger} + \Delta V_P^{\ddagger}$ from which $\Delta V_P^{\ddagger}$ can be determined if a value of $\Delta V_i^{\ddagger}$ is known. $\Delta V_P^{\ddagger}$ is of particular interest because it reflects the volume fluctuations and thus compressibility of the protein segment to which R1 is attached (see Eq. 10). To estimate $\Delta V_i^{\ddagger}$, the pressure dependence of τ_o is measured for an R1 residue at a noninteracting site on the solvent-exposed surface of a rigid helix, where the pressure dependence may be assigned exclusively to R1 internal motion. Holo-myoglobin (holoMb) provides one suitable model with rigid core helices. As an example, Fig. 5 shows

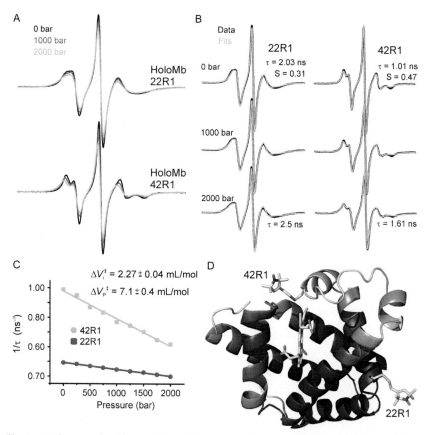

Figure 5 Protein backbone fluctuations revealed in variable-pressure CW EPR. (A) Variable-pressure CW spectra of holoMb 22R1 (top) and 42R1 (bottom) at pH 6, normalized to the same number of spins. (B) Pressure-dependence of the nitroxide τ was determined from fits to the spectra, and (C) a plot of $1/\tau$ versus pressure was fit to extract the indicated $\Delta V_i^{\ddagger}$ and $\Delta V_P^{\ddagger}$. (D) Models of the R1 side chain are shown at residue 22 and 42 of holoMb (PDB: 2MBW) in stick representation. The backbone is color coded according to crystallographic thermal factor from lowest (blue) to highest (red). In (A) and (B), spectra are color coded as indicated. HoloMb 22R1 and 42R1 were prepared and variable-pressure CW data were collected according to methods in Lerch et al. (2013). HoloMb 42R1 0 and 2000 bar spectra were previously reported in Lerch et al. (2013). (See the color plate.)

the behavior of holoMb 22R1 in the pressure range of 0–2 kbar; residue 22 is located in the core B helix, one of the most rigid regions of the sequence based on crystallographic thermal factors (Fig. 5D) and NMR data (Eliezer, Yao, Dyson, & Wright, 1998). It is evident from the superimposed spectra in Fig. 5A that the pressure dependence of R1 internal motion is small. Nevertheless, small changes in correlation time as a function of pressure can be extracted by MOMD simulations varying correlation time at constant order (Fig. 5B), and the data plotted according to (9) with $\Delta V_P^\ddagger = 0$ to give $\Delta V_i^\ddagger = 2.27\,\text{mL/mol}$ (Fig. 5C). With this value for $\Delta V_i^\ddagger$, the pressure dependence of R1 at other sites can be used to estimate $\Delta V_P^\ddagger$ and gauge the relative flexibility of the sequences to which R1 is attached.

As an example, Fig. 5 shows the pressure dependence of holoMb 42R1, located in the short C helix which has been identified to having significant flexibility on the basis of crystallographic thermal factors (Fig. 5D) and other studies (Frauenfelder et al., 1987; López et al., 2012; Phillips, 1990). The pressure dependence of the spectra is clearly greater than that for holoMb 22R1. Analysis of the data with (9), using the $\Delta V_i^\ddagger$ determined from analysis of 22R1, leads to an observed $\Delta V_P^\ddagger = 7.1\,\text{mL/mol}$. Thus, activation volumes from variable-pressure CW spectra report the relative flexibility of different regions of a protein and can be used to define a map of local ns timescale compressibility.

4.2 Characterizing Conformational Exchange Using Variable-Pressure CW EPR

As illustrated in Fig. 1B, conformational exchange on the μs–ms timescale generates multicomponent CW spectra. In the case of a two-state conformational equilibrium, where each state is characterized by a unique spectral component, spectral simulations using a two-component MOMD model can provide the relative population of each state as a function of pressure, thereby providing the pressure-dependent equilibrium constant (K) for determination of $\Delta \bar{V}^\circ$ and $\Delta \bar{\beta}_T$ using (1). An illustrative example is provided in Fig. 6A for T4 lysozyme (T4L) partially unfolded in 2 M urea. The relative amounts of folded and unfolded form can be monitored by T4L 118R1, which is buried and strongly immobilized (i) in the folded structure (Guo, Cascio, Hideg, Kálái, & Hubbell, 2007) but solvent exposed and highly mobile (m) in the unfolded state. The spectrum of the equilibrium mixture at any pressure can be simulated with a two-component MOMD model with effective correlation times and populations of each component as variables with fixed order parameters; variations in the correlation times of each component account for local compressibility effects (McCoy & Hubbell, 2011). The relative populations of folded and unfolded states from simulation provide the equilibrium constant as a function of

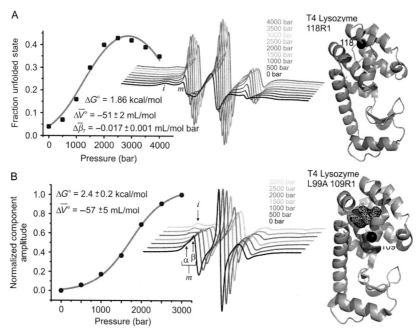

Figure 6 Protein conformational exchange revealed in variable-pressure CW EPR. (A) Variable-pressure CW spectra of T4L 118R1 in 2 M urea at pH 6.8. Contributions from *i* and *m* spectral components in the low-field resonance line are indicated. The fraction unfolded state (*m*) determined from spectral simulation is plotted versus pressure. A fit (red trace) to this plot using the two-state model of (1) yielded the indicated values of $\Delta G°$, $\Delta \bar{V}°$, and $\Delta \bar{\beta}_T$. (B) Variable-pressure CW spectra of T4L L99A 109R1 at pH 6.8. The normalized first component amplitude from SVD of the spectra is plotted versus pressure, and a fit (red trace) to this plot yielded the indicated values of $\Delta G°$ and $\Delta \bar{V}°$. EPR spectra are normalized to the same number of spins for each pressure series. Spectra are color coded as indicated. Structures are shown of wild-type T4L (PDB: 3LZM) and T4L L99A (PDB: 3DMV) with the spin-labeled sites indicated by orange spheres. EPR data in (B) were reported in McCoy (2011). Panel (A) is adapted from McCoy and Hubbell (2011). (See the color plate.)

pressure, from which the $\Delta \bar{V}°$ and $\Delta \bar{\beta}_T$ indicated in Fig. 6A were obtained from (1) (McCoy & Hubbell, 2011). The data showed that up to ~3 kbar pressure increases the population of the unfolded state due to the lower volume of this state, but above 3 kbar the greater compressibility of the native state dominates and pressure drives refolding. This study provided one of the few cases where a compressibility difference between states was determined spectroscopically; generally, $\Delta \bar{\beta}_T$ is assumed to be zero for simplicity.

The above example illustrates the behavior of a system at equilibrium where two conformations of different compressibility are resolved at atmospheric pressure and each has a single-component EPR spectrum. Such cases can be analyzed by spectral simulations. However, in general the EPR spectrum of each conformational state may have multiple and overlapping components due to μs exchange between substates within a given conformational well on the energy landscape. Moreover, the population of each conformation is in general unknown at any pressure, and the spectra of the "pure" states are unknown. In such cases, analysis by spectral simulation is impractical and singular value decomposition (SVD) may be a viable alternative strategy. SVD is commonly used in analysis of time-dependent changes in spectroscopic data (Henry & Hofrichter, 1992), and has been implemented in EPR (Keszler, Kalyanaraman, & Hogg, 2003; Lauricella, Allouch, Roubaud, Bouteiller, & Tuccio, 2004). As an example consider the L99A cavity-enlarging mutant of T4L, which has a well-ordered "ground state" (G) of population ~97% with a structure similar to the wild-type (WT) protein (Bouvignies et al., 2011; Eriksson et al., 1992; Lerch et al., 2015; Mulder, Mittermaier, Hon, Dahlquist, & Kay, 2001). Despite the preponderance of a single state of the protein, the CW spectrum of 109R1 in the flexible F helix of T4L L99A has two components of unknown structural origin at atmospheric pressure with populations of 80% and 20% (α and β Fig. 6B); the spectrum is indistinguishable from the WT protein at atmospheric pressure (McCoy, 2011). Upon pressurization, a third spectral component corresponding to an immobilized state of the nitroxide is resolved (i, Fig. 6B), indicating that pressure populates an "excited state" (E) whose population is below the detection limit at atmospheric pressure; the 109R1 spectrum of the WT protein is nearly pressure independent (McCoy, 2011). The first component amplitude (A) from SVD analysis reflects the shift in population from the G to E state and the data are well-fit using the two-state equilibrium model of (1) with $\Delta \bar{V}^\circ$ of -57 mL/mol and $\Delta \bar{\beta}_T = 0$ (Fig. 6B). This is close to the $\Delta \bar{V}^\circ$ of -56 mL/mol for a two-state transition detected in T4L L99A using high-pressure fluorescence measurements (Ando et al., 2008), which was attributed to hydration of the L99A cavity. Assuming the two techniques monitor the same transition, the SDSL-EPR data indicates a clear structural change accompanying hydration, although the magnitude of the structural transition must be determined by PR DEER.

4.3 Structural Characterization of the High-Pressure Conformational Ensemble with PR DEER

Although the CW EPR spectrum of an R1 side chain is an exquisitely sensitive reporter of a conformational change, the spectral change provides very little information on the magnitude of the change or on the full heterogeneity of an ensemble of states or substates. Providing a length scale to the energy landscape with SDSL-EPR is the domain of distance measurement by DEER. This includes reporting the amplitude of motion that accompanies a transition from one state to another and a high resolution view of the number of discrete conformations present; PR DEER extends SDSL-EPR technology to mapping of structure and heterogeneity of states stabilized by high pressure. Analysis of PR DEER data is unmodified from that at atmospheric pressure DEER, conveniently provided by available software (LongDistances, available at http://www.chemistry.ucla.edu/directory/hubbell-wayne-l and DEERAnalysis (Jeschke et al., 2006)). Applications of PR DEER to apo-myoglobin (apoMb) and to the L99A cavity-enlarging mutant of T4L will serve to illustrate the remarkable capabilities of the technology.

The molten globule (MG) state of apoMb populated at pH 6 and pressures of ~2 kbar was shown by NMR to be conformationally flexible on the ms timescale (Kitahara, Yamada, Akasaka, & Wright, 2002), but correlations of the fluctuations with their structural magnitude were not determined. However, PR DEER from seven doubly labeled mutants covering all eight native helices of apoMb revealed extremely broad distance distributions (>25 Å) beginning at ~2 kbar in five mutants; Fig. 7A shows examples of the PR DEER-derived distance distributions and the corresponding fits to the dipolar evolution functions for the spin pair 12R1/132R1 in apoMb. Remarkably, high-pressure circular dichroism revealed native levels of secondary structure content (Lerch et al., 2013), which together with the PR DEER data indicates widespread, nanometer-scale rigid-body motions of helices or helical segments of the protein in the MG state.

In striking contrast to the fluctuating global topology observed in apoMb under pressure, a second example illustrates the ability of PR DEER to monitor pressure-dependent shifts in population among discrete conformations and to estimate differences in free energy and partial molar volume between them. T4L L99A containing the additional mutations G113A and R119P was shown by NMR to exist in an equilibrium between two conformations, one with a structure similar to the WT protein (G state)

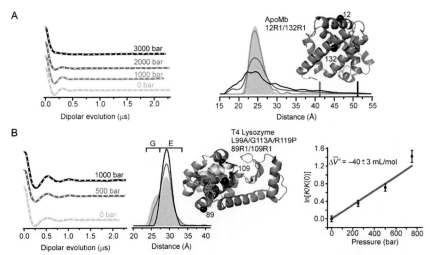

Figure 7 Pressure-populated structural changes reported by PR DEER. (A) (Left) Background-corrected dipolar evolution functions with fits to the data overlaid (dotted black); (right) area-normalized PR DEER distance distributions for apoMb 12R1/132R1 at pH 6. A crystal structure of holoMb (PDB: 2MBW) is shown in an inset. The spin-labeled sites are indicated with orange spheres. (B) (Left) Background-corrected dipolar evolution functions with fits to the data overlaid (dotted black); (center) area-normalized PR DEER distance distributions for T4 lysozyme L99A/G113A/R119P 89R1/109R1 at pH 6.8. The inset shows structures of wild-type T4L (gray, PDB: 3DMV) and the excited state of T4L L99A (green, PDB: 2LCB) overlaid to illustrate the difference in helix F position in the ground and excited states. The spin-labeled sites on both protein models are indicated with orange spheres at residue 109 and 89; (right) a plot of $\ln(K/K(0))$ versus pressure was fit (red trace) using a two-state model of (1) to determine $\Delta \bar{V}^\circ$ for the $G \leftrightarrow E$ exchange. In (A) and (B), the distance distributions and dipolar evolutions are color coded as indicated. Dipolar evolutions are vertically offset for clarity. The green and black bars in (A) indicate the upper limit of reliable shape and distance of the distribution (Jeschke, 2012). Absence of these bars in (B) indicates that the limit of reliability is beyond the maximum distance shown. *Panels (A) and (B) are adapted from Lerch et al. (2014, 2015).* (See the color plate.)

and another where the short helix F rotates to partially fill the cavity formed by the L99A mutations with a side chain (F114) (*E* state) (Bouvignies et al., 2011). The distance between a spin label in helix F, 109R1, and a reference site, 89R1, serves to monitor the position of helix F (Fig. 7B). The DEER-derived interspin distance distribution contains two discrete populations, with the most probable distances corresponding closely to the NMR models for the G and E conformations. Upon application of pressure the equilibrium is dramatically shifted toward the E state in which the cavity is filled with F114. The distance distributions for the spin-labeled mutant of T4L L99A/G113A/R119P were fit with multiple Gaussians to determine the relative populations of G and E, and hence the equilibrium constant, as a function of pressure (Lerch et al., 2015). From these data, $\Delta \bar{V}^\circ$ of this transition was obtained with (1), taking $\Delta \bar{\beta}_T = 0$ for the low-pressure regime used in the experiment. It is interesting to note that the $\Delta \bar{V}^\circ$ estimated based on PR DEER is similar to the molar volume of the Phe114 side chain occupying the L99A cavity in the excited state (42 mL/mol). Significantly, this was the first direct experimental observation of a structure-relaxation mechanism for volume reduction in response to pressure as opposed to cavity hydration (Lerch et al., 2015).

5. SUMMARY AND FUTURE DIRECTIONS

Current evidence indicates that pressure cleanly shifts the relative populations of states solely according to differences in partial molar volume without altering the shape of the energy landscape. Thus, variable pressure is a powerful tool for dissecting details of the landscape, and SDSL-EPR is an ideal strategy in terms of sensitivity and timescale to detect the effects of pressure and interpret them in terms of structure and dynamics. In this chapter, high-pressure and pressure-jump instrumentation suitable for SDSL-EPR of proteins in aqueous solution is described, and practical considerations were outlined for both variable-pressure CW EPR and PR DEER. Although high-pressure SDSL-EPR is in its infancy, the recent applications presented illustrate the considerable potential of the method to (1) identify compressible (flexible) regions in a folded protein; (2) determine thermodynamic parameters that relate conformational states in equilibrium; (3) populate and characterize excited states of proteins undetected at atmospheric pressure; (4) reveal the structural heterogeneity of conformational ensembles and provide distance constraints on the global structure of pressure populated states with PR DEER. Future developments of high-pressure SDSL-EPR include pressure-jump relaxation spectroscopy to determine the lifetime of conformational states in the ms range, and high-pressure saturation recovery exchange spectroscopy to enable measurement of lifetimes of states in the μs range. SDSL-EPR has unique advantages for the study of membrane proteins in their native environment under physiological conditions, and applications of high-pressure SDSL-EPR to explore the conformational equilibria and dynamics of integral membrane proteins is a high priority.

REFERENCES

Akasaka, K. (2006). Probing conformational fluctuation of proteins by pressure perturbation. *Chemical Reviews*, *106*(5), 1814–1835. http://dx.doi.org/10.1021/cr040440z.

Altenbach, C., Oh, K.-J., Trabanino, R. J., Hideg, K., & Hubbell, W. L. (2001). Estimation of inter-residue distances in spin labeled proteins at physiological temperatures: Experimental strategies and practical limitations. *Biochemistry*, *40*(51), 15471–15482. http://dx.doi.org/10.1021/bi011544w.

Ando, N., Barstow, B., Baase, W. A., Fields, A., Matthews, B. W., & Gruner, S. M. (2008). Structural and thermodynamic characterization of T4 lysozyme mutants and the contribution of internal cavities to pressure denaturation. *Biochemistry*, *47*(42), 11097–11109. http://dx.doi.org/10.1021/bi801287m.

Baldwin, A. J., & Kay, L. E. (2009). NMR spectroscopy brings invisible protein states into focus. *Nature Chemical Biology*, *5*(11), 808–814.

Berlow, R. B., Dyson, H. J., & Wright, P. E. (2015). Functional advantages of dynamic protein disorder. *FEBS Letters*. http://dx.doi.org/10.1016/j.febslet.2015.06.003.

Borbat, P., & Freed, J. (2013). Pulse dipolar electron spin resonance: Distance measurements. In C. R. Timmel & J. R. Harmer (Eds.), *Structural information from spin-labels and intrinsic paramagnetic centres in the biosciences: Vol. 152*. (pp. 1–82). Berlin Heidelberg: Springer.

Bouvignies, G., Vallurupalli, P., Hansen, D. F., Correia, B. E., Lange, O., Bah, A., et al. (2011). Solution structure of a minor and transiently formed state of a T4 lysozyme mutant. *Nature*, *477*(7362), 111–114. http://dx.doi.org/10.1038/nature10349.

Budil, D. E., Lee, S., Saxena, S., & Freed, J. H. (1996). Nonlinear-least-squares analysis of slow-motion EPR spectra in one and two dimensions using a modified Levenberg–Marquardt algorithm. *Journal of Magnetic Resonance Series A*, *120*(2), 155–189. http://dx.doi.org/10.1006/jmra.1996.0113.

Chen, C. T., Fine, R. A., & Millero, F. J. (1977). The equation of state of pure water determined from sound speeds. *The Journal of Chemical Physics*, *66*(5), 2142–2144. http://dx.doi.org/10.1063/1.434179.

Collins, M. D., Hummer, G., Quillin, M. L., Matthews, B. W., & Gruner, S. M. (2005). Cooperative water filling of a nonpolar protein cavity observed by high-pressure crystallography and simulation. *Proceedings of the National Academy of Sciences of the United States of America*, *102*(46), 16668–16671. http://dx.doi.org/10.1073/pnas.0508224102.

Collins, M. D., Kim, C. U., & Gruner, S. M. (2011). High-pressure protein crystallography and NMR to explore protein conformations. *Annual Review of Biophysics*, *40*(1), 81–98. http://dx.doi.org/10.1146/annurev-biophys-042910-155304.

Columbus, L., & Hubbell, W. L. (2002). A new spin on protein dynamics. *Trends in Biochemical Sciences*, *27*(6), 288–295. http://dx.doi.org/10.1016/S0968-0004(02)02095-9.

Columbus, L., & Hubbell, W. L. (2004). Mapping backbone dynamics in solution with site-directed spin labeling: GCN4-58 bZip free and bound to DNA. *Biochemistry*, *43*(23), 7273–7287. http://dx.doi.org/10.1021/bi0497906.

Columbus, L., Kálai, T., Jekö, J., Hideg, K., & Hubbell, W. L. (2001). Molecular motion of spin labeled side chains in α-helices: Analysis by variation of side chain structure. *Biochemistry*, *40*(13), 3828–3846. http://dx.doi.org/10.1021/bi002645h.

Cooper, A. (1984). Protein fluctuations and the thermodynamic uncertainty principle. *Progress in Biophysics and Molecular Biology*, *44*(3), 181–214. http://dx.doi.org/10.1016/0079-6107(84)90008-7.

Doster, W. (2010). The protein-solvent glass transition. *Biochimica et Biophysica Acta, 1804*(1), 3–14. http://dx.doi.org/10.1016/j.bbapap.2009.06.019.

Eaton, S. R., Eaton, G. R., & Berliner, L. J. (2005). *Biomedical EPR, part B: Methodology, instrumentation, and dynamics,* (Vol. 24) New York: Springer.

El'yanov, B. S., & Hamann, S. D. (1975). Some quantitative relationships for ionization reactions at high pressures. *Australian Journal of Chemistry, 28*(5), 945–954.

Eliezer, D., Yao, J., Dyson, H. J., & Wright, P. E. (1998). Structural and dynamic characterization of partially folded states of apomyoglobin and implications for protein folding. *Nature Structural Biology, 5*(2), 148–155.

Eriksson, A. E., Baase, W. A., Zhang, X. J., Heinz, D. W., Blaber, M., Baldwin, E. P., et al. (1992). Response of a protein structure to cavity-creating mutations and its relation to the hydrophobic effect. *Science, 255*(5041), 178–183.

Fleissner, M. R., Cascio, D., & Hubbell, W. L. (2009). Structural origin of weakly ordered nitroxide motion in spin-labeled proteins. *Protein Science, 18*(5), 893–908. http://dx.doi.org/10.1002/pro.96.

Fourme, R., Girard, E., & Akasaka, K. (2012). High-pressure macromolecular crystallography and NMR: Status, achievements and prospects. *Current Opinion in Structural Biology, 22*(5), 636–642. http://dx.doi.org/10.1016/j.sbi.2012.07.007.

Frauenfelder, H., Hartmann, H., Karplus, M., Kuntz, I. D., Kuriyan, J., Parak, F., et al. (1987). Thermal expansion of a protein. *Biochemistry, 26*(1), 254–261. http://dx.doi.org/10.1021/bi00375a035.

Freed, J. H. (1976). Theory of slow tumbling ESR spectra for nitroxides. In L. J. Berliner (Ed.), *Spin labeling theory and applications* (pp. 53–132). New York: Academic Press.

Guo, Z., Cascio, D., Hideg, K., Kálái, T., & Hubbell, W. L. (2007). Structural determinants of nitroxide motion in spin-labeled proteins: Tertiary contact and solvent-inaccessible sites in helix G of T4 lysozyme. *Protein Science, 16*(6), 1069–1086. http://dx.doi.org/10.1110/ps.062739107.

Henry, E. R., & Hofrichter, J. (1992). [8] Singular value decomposition: Application to analysis of experimental data. In M. L. J. Ludwig Brand (Ed.), *Methods in enzymology: Vol. 210.* (pp. 129–192). New York: Academic Press.

Henzler-Wildman, K., & Kern, D. (2007). Dynamic personalities of proteins. *Nature, 450*(7172), 964–972.

Jeschke, G. (2012). DEER distance measurements on proteins. *Annual Review of Physical Chemistry, 63,* 419–446. http://dx.doi.org/10.1146/annurev-physchem-032511-143716.

Jeschke, G., Chechik, V., Ionita, P., Godt, A., Zimmermann, H., Banham, J., et al. (2006). DeerAnalysis2006—A comprehensive software package for analyzing pulsed ELDOR data. *Applied Magnetic Resonance, 30*(3–4), 473–498. http://dx.doi.org/10.1007/BF03166213.

Keszler, A., Kalyanaraman, B., & Hogg, N. (2003). Comparative investigation of superoxide trapping by cyclic nitrone spin traps: The use of singular value decomposition and multiple linear regression analysis. *Free Radical Biology and Medicine, 35*(9), 1149–1157. http://dx.doi.org/10.1016/S0891-5849(03)00497-0.

Kitahara, R., Sareth, S., Yamada, H., Ohmae, E., Gekko, K., & Akasaka, K. (2000). High pressure NMR reveals active-site hinge motion of folate-bound Escherichia coli dihydrofolate reductase. *Biochemistry, 39*(42), 12789–12795. http://dx.doi.org/10.1021/bi0009993.

Kitahara, R., Simorellis, A. K., Hata, K., Maeno, A., Yokoyama, S., Koide, S., et al. (2012). A delicate interplay of structure, dynamics, and thermodynamics for function: A high pressure NMR study of outer surface protein A. *Biophysical Journal, 102*(4), 916–926. http://dx.doi.org/10.1016/j.bpj.2011.12.010.

Kitahara, R., Yamada, H., Akasaka, K., & Wright, P. E. (2002). High pressure NMR reveals that apomyoglobin is an equilibrium mixture from the native to the unfolded.

Journal of Molecular Biology, 320(2), 311–319. http://dx.doi.org/10.1016/S0022-2836(02)00449-7.

Kitahara, R., Yokoyama, S., & Akasaka, K. (2005). NMR snapshots of a fluctuating protein structure: Ubiquitin at 30 bar–3 kbar. *Journal of Molecular Biology, 347*(2), 277–285. http://dx.doi.org/10.1016/j.jmb.2005.01.052.

Kitamura, Y., & Itoh, T. (1987). Reaction volume of protonic ionization for buffering agents. Prediction of pressure dependence of pH and pOH. *Journal of Solution Chemistry, 16*(9), 715–725. http://dx.doi.org/10.1007/BF00652574.

Korzhnev, D. M., Religa, T. L., Banachewicz, W., Fersht, A. R., & Kay, L. E. (2010). A transient and low-populated protein-folding intermediate at atomic resolution. *Science, 329*(5997), 1312–1316. http://dx.doi.org/10.1126/science.1191723.

Kuwata, K., Li, H., Yamada, H., Batt, C. A., Goto, Y., & Akasaka, K. (2001). High pressure NMR reveals a variety of fluctuating conformers in β-lactoglobulin. *Journal of Molecular Biology, 305*(5), 1073–1083. http://dx.doi.org/10.1006/jmbi.2000.4350.

Lauricella, R., Allouch, A., Roubaud, V., Bouteiller, J.-C., & Tuccio, B. (2004). A new kinetic approach to the evaluation of rate constants for the spin trapping of superoxide/hydroperoxyl radical by nitrones in aqueous media. *Organic & Biomolecular Chemistry, 2*(9), 1304–1309. http://dx.doi.org/10.1039/B401333F.

Lee, S., Heerklotz, H., & Chalikian, T. V. (2010). Effects of buffer ionization in protein transition volumes. *Biophysical Chemistry, 148*(1–3), 144–147. http://dx.doi.org/10.1016/j.bpc.2010.03.002.

Lerch, M. T., Horwitz, J., McCoy, J., & Hubbell, W. L. (2013). Circular dichroism and site-directed spin labeling reveal structural and dynamical features of high-pressure states of myoglobin. *Proceedings of the National Academy of Sciences of the United States of America, 110*(49), E4714–E4722. http://dx.doi.org/10.1073/pnas.1320124110.

Lerch, M. T., López, C. J., Yang, Z., Kreitman, M. J., Horwitz, J., & Hubbell, W. L. (2015). Structure-relaxation mechanism for the response of T4 lysozyme cavity mutants to hydrostatic pressure. *Proceedings of the National Academy of Sciences of the United States of America, 112*(19), E2437–2446. http://dx.doi.org/10.1073/pnas.1506505112.

Lerch, M. T., Yang, Z., Brooks, E. K., & Hubbell, W. L. (2014). Mapping protein conformational heterogeneity under pressure with site-directed spin labeling and double electron–electron resonance. *Proceedings of the National Academy of Sciences of the United States of America, 111*(13), E1201–E1210. http://dx.doi.org/10.1073/pnas.1403179111.

López, C. J., Fleissner, M. R., Brooks, E. K., & Hubbell, W. L. (2014). Stationary-phase EPR for exploring protein structure, conformation, and dynamics in spin-labeled proteins. *Biochemistry, 53*(45), 7067–7075. http://dx.doi.org/10.1021/bi5011128.

López, C. J., Oga, S., & Hubbell, W. L. (2012). Mapping molecular flexibility of proteins with site directed spin labeling: A case study of myoglobin. *Biochemistry, 51*(33), 6568–6583. http://dx.doi.org/10.1021/bi3005686.

López, C. J., Yang, Z., Altenbach, C., & Hubbell, W. L. (2013). Conformational selection and adaptation to ligand binding in T4 lysozyme cavity mutants. *Proceedings of the National Academy of Sciences of the United States of America, 110*(46), E4306–E4315. http://dx.doi.org/10.1073/pnas.1318754110.

Manglik, A., Kim, T. H., Masureel, M., Altenbach, C., Yang, Z., Hilger, D., et al. (2015). Structural insights into the dynamic process of β2-adrenergic receptor signaling. *Cell, 161*(5), 1101–1111. http://dx.doi.org/10.1016/j.cell.2015.04.043.

McCoy, J., & Hubbell, W. L. (2011). High-pressure EPR reveals conformational equilibria and volumetric properties of spin-labeled proteins. *Proceedings of the National Academy of Sciences of the United States of America, 108*(4), 1331–1336. http://dx.doi.org/10.1073/pnas.1017877108.

McCoy, J. J. (2011). *High pressure EPR of spin labeled proteins.* 3451964 Ph.D, Los Angeles: University of California. Retrieved from, http://search.proquest.com/docview/863625911?accountid=14512.

Mchaourab, H. S., Steed, P. R., & Kazmier, K. (2011). Toward the fourth dimension of membrane protein structure: Insight into dynamics from spin-labeling EPR spectroscopy. *Structure, 19*(11), 1549–1561. http://dx.doi.org/10.1016/j.str.2011.10.009.

Meyer, V., Swanson, M. A., Clouston, L. J., Boratyński, P. J., Stein, R. A., McHaourab, H. S., et al. (2015). Room-temperature distance measurements of immobilized spin-labeled protein by DEER/PELDOR. *Biophysical Journal, 108*(5), 1213–1219. http://dx.doi.org/10.1016/j.bpj.2015.01.015.

Mulder, F. A., Mittermaier, A., Hon, B., Dahlquist, F. W., & Kay, L. E. (2001). Studying excited states of proteins by NMR spectroscopy. *Nature Structural Biology, 8*(11), 932–935. http://dx.doi.org/10.1038/nsb1101-932.

Neudecker, P., Robustelli, P., Cavalli, A., Walsh, P., Lundström, P., Zarrine-Afsar, A., et al. (2012). Structure of an intermediate state in protein folding and aggregation. *Science, 336*(6079), 362–366. http://dx.doi.org/10.1126/science.1214203.

Neuman, R. C., Kauzmann, W., & Zipp, A. (1973). Pressure dependence of weak acid ionization in aqueous buffers. *The Journal of Physical Chemistry, 77*(22), 2687–2691. http://dx.doi.org/10.1021/j100640a025.

Nucci, N. V., Fuglestad, B., Athanasoula, E. A., & Wand, A. J. (2014). Role of cavities and hydration in the pressure unfolding of T4 lysozyme. *Proceedings of the National Academy of Sciences of the United States of America, 111*(38), 13846–13851.

Pannier, M., Veit, S., Godt, A., Jeschke, G., & Spiess, H. W. (2000). Dead-time free measurement of dipole-dipole interactions between electron spins. *Journal of Magnetic Resonance, 142*(2), 331–340. http://dx.doi.org/10.1006/jmre.1999.1944.

Phillips, G. N., Jr. (1990). Comparison of the dynamics of myoglobin in different crystal forms. *Biophysical Journal, 57*(2), 381–383. http://dx.doi.org/10.1016/S0006-3495(90)82540-6.

Prehoda, K. E., Mooberry, E. S., & Markley, J. L. (1998). Pressure denaturation of proteins: Evaluation of compressibility effects. *Biochemistry, 37*(17), 5785–5790. http://dx.doi.org/10.1021/bi980384u.

Rabenstein, M. D., & Shin, Y. K. (1995). Determination of the distance between two spin labels attached to a macromolecule. *Proceedings of the National Academy of Sciences of the United States of America, 92*(18), 8239–8243.

Rasper, J., & Kauzmann, W. (1962). Volume changes in protein reactions. I. Ionization reactions of proteins. *Journal of the American Chemical Society, 84*(10), 1771–1777. http://dx.doi.org/10.1021/ja00869a001.

Roche, J., Caro, J. A., Norberto, D. R., Barthe, P., Roumestand, C., Schlessman, J. L., et al. (2012). Cavities determine the pressure unfolding of proteins. *Proceedings of the National Academy of Sciences of the United States of America, 109*(18), 6945–6950. http://dx.doi.org/10.1073/pnas.1200915109.

Samaranayake, C. P., & Sastry, S. K. (2010). In situ measurement of pH under high pressure. *The Journal of Physical Chemistry B, 114*(42), 13326–13332. http://dx.doi.org/10.1021/jp1037602.

Slipenyuk, A. M., Glinchuk, M. D., Bykov, I. P., Ragulya, A. V., Klimenko, V. P., Konstantinova, T. E., et al. (2004). ESR investigation of Yttria stabilized Zirconia powders with nanosize particles. *Ferroelectrics, 298*(1), 289–296. http://dx.doi.org/10.1080/00150190490423723.

Torrent, J., Font, J., Herberhold, H., Marchal, S., Ribo, M., Ruan, K. C., et al. (2006). The use of pressure-jump relaxation kinetics to study protein folding landscapes. *Biochimica et Biophysica Acta, 1764*(3), 489–496. http://dx.doi.org/10.1016/j.bbapap.2006.01.002.

Urayama, P., Phillips, G. N., Jr., & Gruner, S. M. (2002). Probing substates in sperm whale myoglobin using high-pressure crystallography. *Structure, 10*(1), 51–60. http://dx.doi.org/10.1016/S0969-2126(01)00699-2.

Zipp, A., & Kauzmann, W. (1973). Pressure denaturation of metmyoglobin. *Biochemistry, 12*(21), 4217–4228.

CHAPTER THREE

Exploring Structure, Dynamics, and Topology of Nitroxide Spin-Labeled Proteins Using Continuous-Wave Electron Paramagnetic Resonance Spectroscopy

Christian Altenbach[*], Carlos J. López[*], Kálmán Hideg[†], Wayne L. Hubbell[*,1]

[*]Department of Chemistry and Biochemistry, Jules Stein Eye Institute, University of California, Los Angeles, California, USA
[†]Institute of Organic and Medicinal Chemistry, University of Pécs, Pécs, Hungary
[1]Corresponding author: e-mail address: hubbellw@jsei.ucla.edu

Contents

1. Introduction — 60
2. Site-Directed Spin Labeling — 62
3. Analysis of EPR Spectra in Terms of R1 Nitroxide Motion — 63
   3.1 The EPR Timescale and Protein Rotational Diffusion — 63
   3.2 The Internal Motion of R1 in the Absence of Tertiary Interactions — 64
   3.3 Modulation of R1 Internal Motion by Local Protein Dynamics — 66
4. Measurement of R1 Solvent Accessibility With Power Saturation — 68
5. CW-Based Interspin Distance Measurements — 72
   5.1 Overview — 72
   5.2 Static Dipolar Interaction — 73
   5.3 Relaxation Dipolar Broadening — 78
   5.4 More Complicated Scenarios — 79
6. Determination of Secondary Structure and Features of the Tertiary Fold from R1 Mobility and Accessibility — 80
7. Backbone Dynamics — 85
8. Identifying Slow Conformational Exchange and Detecting Conformational Changes — 90
9. Other Applications of CW EPR — 93
10. Conclusion and Future Directions — 94
References — 95

Abstract

Structural and dynamical characterization of proteins is of central importance in understanding the mechanisms underlying their biological functions. Site-directed spin labeling (SDSL) combined with continuous-wave electron paramagnetic resonance (CW EPR) spectroscopy has shown the capability of providing this information with site-specific resolution under physiological conditions for proteins of any degree of complexity, including those associated with membranes. This chapter introduces methods commonly employed for SDSL and describes selected CW EPR-based methods that can be applied to (1) map secondary and tertiary protein structure, (2) determine membrane protein topology, (3) measure protein backbone flexibility, and (4) reveal the existence of conformational exchange at equilibrium.

1. INTRODUCTION

Proteins play a central role in a wide range of biological processes such as intracellular transport, catalysis, cell signaling, and gene regulation. The advent of X-ray crystallographic methods has enabled the determination of the atomic resolution structures of ~90,000 proteins to date (http://www.rcsb.org/pdb/statistics/holdings.do). The information provided by crystallography alone is not sufficient to describe molecular mechanisms underlying function because proteins often rely on molecular flexibility to accomplish their physiological role, a feature not revealed in the confines of a crystal lattice. Solution methods such as nuclear magnetic resonance (NMR) spectroscopy have shown that proteins in solution exhibit molecular flexibility over a wide range of time and length scales and a link between protein flexibility and function has been clearly established in many cases (Boehr, Nussinov, & Wright, 2009; Henzler-Wildman & Kern, 2007; Henzler-Wildman et al., 2007).

Thus, to understand molecular mechanisms underlying protein function, it is necessary to have experimental strategies that are capable of providing structural and dynamical information on a relevant timescale and under physiological conditions (i.e., out of the confines of the crystal lattice). NMR spectroscopy is particularly effective in providing atomic level information on dynamics for proteins in solution (Boehr, Dyson, & Wright, 2006; Mittermaier & Kay, 2006). However, applying solution NMR to complex systems such as high-molecular-weight proteins, membrane-bound proteins in a native lipid environment under physiological conditions, and nonequilibrium states that evolve in time is particularly challenging. The method of site-directed spin labeling (SDSL) combined with electron paramagnetic resonance (EPR) spectroscopy has been shown to be powerful for investigating such complex systems, and numerous reviews are available on aspects of SDSL-EPR technology and applications (Bordignon & Steinhoff, 2007; Hubbell & Altenbach, 1994; Hubbell, Altenbach, Hubbell, & Khorana, 2003; Hubbell, Cafiso, & Altenbach, 2000; Hubbell, Mchaourab, Altenbach, & Lietzow, 1996; Klare & Steinhoff, 2009).

This chapter will highlight the molecular information that can be obtained from protein SDSL and continuous-wave (CW) EPR spectroscopy. In principle, any paramagnetic species introduced into a protein can serve as a spin label, but this chapter will focus only on stable free radical spin labels based on heterocyclic nitroxides (Fig. 1). Even with this restriction, the field of CW nitroxide SDSL is extensive. Coverage here will be selective and emphasize well-documented and general strategies to explore

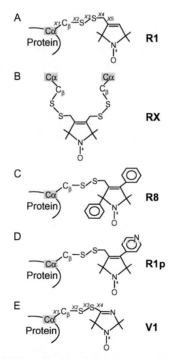

Figure 1 The structure of nitroxide side chains discussed in this chapter. (A) R1 is the most common and most well-studied nitroxide side chain. (B–E) Nitroxide side chains with restricted motion: (B) the cross-linking side chain RX requires two suitably placed cysteines. (C) R8, (D) R1p, and (E) the imidazolidine nitroxide side chain V1.

protein structure and dynamics: Section 2 introduces nitroxide spin labeling methods; Sections 3–5 present strategies for extracting information from CW EPR spectral lineshapes, while Sections 6–8 illustrate applications of these strategies to determine protein secondary and tertiary structure, investigate membrane protein topology, map sequence-specific backbone dynamics, and detect conformational fluctuations. We note that the powerful method of saturation transfer EPR has been extensively reviewed (Beth & Hustedt, 2005; Hyde & Thomas, 1980) and will not be included here. The emerging field of CW high-pressure SDSL-EPR is covered in a separate chapter (Lerch, Yang, Altenbach, & Hubbell, 2015). It is assumed that the reader has a basic working knowledge of nitroxide EPR spectroscopy. Where relevant, references to informative protocols will be provided.

2. SITE-DIRECTED SPIN LABELING

In the most common implementation of SDSL, a cysteine residue is engineered at a particular site using site-directed mutagenesis for subsequent modification with a sulfhydryl-specific nitroxide reagent. The most widely used reagent is the highly reactive 1-oxy-2,2,5,5-tetramethylpyrroline-3-methylmethanethiosulfonate (Berliner, Grunwald, Hankovszky, & Hideg, 1982) that generates a disulfide-linked side chain designated R1 (Fig. 1A). The methanethiosulfonate moiety has no detectable reaction with other native side chains (Mchaourab, Lietzow, Hideg, & Hubbell, 1996) and the reaction with a solvent-exposed cysteine is complete within minutes using stoichiometric quantities of reagent and protein in the micromolar range at neutral pH in the temperature range of 4–25 °C.

The R1 side chain produces remarkably little perturbation of protein structure and function when introduced at solvent-exposed sites, and in some cases at buried sites (Altenbach, Marti, Khorana, & Hubbell, 1990; Mchaourab et al., 1996). Derivatives of R1 with substituents on the nitroxide ring have been developed for specific purposes (Columbus, Kalai, Jeko, Hideg, & Hubbell, 2001), some of which are shown in Fig. 1B–D. A related disulfide-linked side chain based on an imidazolidine nitroxide has recently been characterized (Toledo Warshaviak, Khramtsov, Cascio, Altenbach, & Hubbell, 2013) and is shown in Fig. 1E.

Quantitative aspects of SDSL aimed at determining protein structure and dynamics require knowledge of the rotameric structure and internal motion of the spin-labeled side chain. Because R1 and related side chains are the most extensively characterized in these regards, the methods discussed in this chapter will be confined to the use of the structures shown in Fig. 1.

The CW EPR spectrum of a nitroxide side chain in a protein encodes information on the motion of the side chain, solvent accessibility, and strength of the distance-dependent magnetic dipolar interaction with another spin label. In Sections 3–5 to follow, methods to extract this information from spectra are considered.

3. ANALYSIS OF EPR SPECTRA IN TERMS OF R1 NITROXIDE MOTION

3.1 The EPR Timescale and Protein Rotational Diffusion

At X-band (i.e., 9–10 GHz), the CW EPR spectrum of a nitroxide spin-labeled protein directly reflects the rotational correlation time (τ) of the nitroxide in the picosecond to nanosecond timescale (0.1–100 ns). The overall τ includes contributions corresponding to the correlation times of side chain internal motions (τ_i), fast backbone fluctuations (τ_{bb}), and, for small proteins, the overall protein rotational diffusion (τ_r). To interpret EPR spectra of R1 exclusively in terms of the internal modes of motion that reflect protein structure and dynamics (i.e., τ_{bb}, τ_i), τ_r must be sufficiently long to have minimal effects on the lineshape.

For the relatively rapid anisotropic motion of R1 at solvent accessible sites (see below), it has been found that this condition is met for $\tau_r > 20$ ns. Assuming Stokes–Einstein behavior for soluble globular proteins at room temperature,

$$\tau_r \approx (3 \times 10^{-13}) \eta \times \text{MW}, \qquad (1)$$

where η is the solution viscosity in centipoise (Voet & Voet, 2004). Thus, for a protein with a molecular weight of 20 kDa at room temperature in water, $\tau_r \approx 6$ ns. For such small proteins, τ_r can be increased to ~ 20 ns by increasing the viscosity of the medium to 3 cP using 30% w/w sucrose or 25% w/w Ficoll (López, Fleissner, Guo, Kusnetzow, & Hubbell, 2009; Mchaourab et al., 1996). When using these common viscogens, one must be aware of the potential osmotic (for sucrose) and crowding (for Ficoll) effects that can influence the conformation of flexible structures (Bolen & Rose, 2008; López et al., 2009; Stagg, Zhang, Cheung, & Wittung-Stafshede, 2007).

For a rigid side chain such as the cross-linking RX (Fig. 1B; Fleissner et al., 2011) or the conformationally constrained R8 (López, Fleissner,

Brooks, & Hubbell, 2014; Sale, Sar, Sharp, Hideg, & Fajer, 2002), R1p (Fawzi et al., 2011), or V1 (Toledo Warshaviak et al., 2013; Fig. 1C–E), where there is little internal motion, rotational diffusion, and backbone fluctuations are the dominant motions that are sensed by the nitroxide. Indeed, the CW spectra of these side chains in proteins can be employed to measure protein rotational diffusion as slow as $\tau_r \approx 100$ ns. With these completely or partially immobilized nitroxides, attachment to a solid support (López et al, 2014) is an effective way to observe exclusively internal motions of the protein. Very slow motions beyond the range of sensitivity of conventional CW lineshape analysis can be analyzed by the method of saturation transfer EPR (Beth & Hustedt, 2005; Hyde & Thomas, 1980).

3.2 The Internal Motion of R1 in the Absence of Tertiary Interactions

Assuming that effects of overall rotational diffusion of the protein have been eliminated, the interpretation of EPR spectra in terms of structure and dynamics of the protein requires knowledge of the contribution from internal motions of the nitroxide side chain itself. For the R1 side chain, the structural basis of its internal motion on helices and ordered loops has been elucidated in detail through high-resolution crystal structures of spin-labeled proteins (Cunningham et al., 2012; Fleissner, Cascio, & Hubbell, 2009; Guo, Cascio, Hideg, & Hubbell, 2008; Guo, Cascio, Hideg, Kálái, & Hubbell, 2007; Kroncke, Horanyi, & Columbus, 2010), variation of side chain structure (Columbus et al., 2001; Mchaourab, Kalai, Hideg, & Hubbell, 1999), mutational analysis (Mchaourab et al., 1996), and quantum mechanical calculations (Warshaviak, Serbulea, Houk, & Hubbell, 2011).

The outcome of these studies is summarized with reference to Fig. 2. The main feature for R1 at solvent-exposed sites in helices and ordered loops is a ubiquitous interaction of the S_δ sulfur of the side chain with the backbone C_α hydrogen (i.e., HC_α), illustrated in Fig. 2A (Fleissner et al., 2009). This *intra-residue* S_δ–HC_α interaction has two important consequences: (1) it restricts the rotameric space of R1 to 3 preferred rotamers about the X_1 and X_2 angles (Fleissner et al., 2009) and (2) it restricts the dominant motion in the side chain to torsional oscillation about the terminal X_4 and X_5 dihedrals. This is known as the "X_4/X_5 model," as shown in Fig. 2B (Columbus et al., 2001). Thus, despite the presence of five bonds that could be sites of rotameric flexibility, R1 exhibits restricted internal motions. As a consequence, the purely internal motion of R1 is anisotropic.

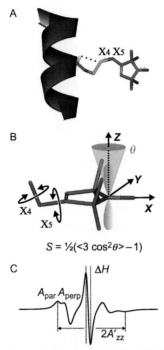

Figure 2 The nitroxide R1 side chain on a helix surface. (A) The structure of the spin label R1 on a helix surface site. The dotted line shows the S_δ–HC_α interaction stabilizing the structure. Only rotations of the X_4 and X_5 bonds contribute to the magnetic averaging. (B) Rotations around X_4 and X_5 leads to anisotropic averaging in a cone. The opening angle of the cone defines the order parameter S. $S=0$ for complete averaging and $S=1$ for no averaging. (C) A typical anisotropic single component spectrum of a surface site of a rigid α-helix as shown in (A); the definition of the central linewidth, apparent A'_{zz}, and spectral line splitting due to parallel and perpendicular components of the hyperfine interaction are shown.

A theory by Freed provides an important tool for computing EPR spectral lineshapes from motional models of the nitroxide (Budil, Lee, Saxena, & Freed, 1996), and thus for interpreting the lineshapes in terms of the details of nitroxide motion. The theory includes a macroscopic-order microscopic-disorder (MOMD) model that specifies spatial ordering potentials necessary to account for the anisotropic motion of R1. Fitting of an experimental spectrum to the MOMD model provides an order parameter, S, that describes the ordering (angular amplitude) of an anisotropic motion that lies between 0 (isotropic motion) and 1 (perfect order) (Fig. 2B), and a corresponding effective correlation time, τ_i. For R1 on the solvent-exposed surface of rigid helices (i.e., in the absence of backbone fluctuations) and in the absence of effects from overall protein rotational diffusion, the spectra can be simulated with a simple model wherein the nitroxide 2p orbital undergoes motion in a cone with $S \approx 0.5$ and $\tau_i \approx 2$ ns, consistent with the X_4/X_5 model (Columbus et al., 2001; Fig. 2C). These values correspond to the purely anisotropic internal motion of R1 and serve as references for interpreting spectra in terms of protein structure and dynamics. The resolved features in the spectrum (Fig. 2c), A_{par} and A_{perp}, do not arise from different dynamic modes of R1, but rather are a consequence of the anisotropic motion that leads to incomplete averaging of the hyperfine tensor (A) of the nitroxide. A_{par} arises from nitroxides where the z-axis of the nitroxide is parallel to the external magnetic field, and A_{perp} from those oriented with the external field perpendicular to z. A detailed description of anisotropic motion can be found in Griffith and Jost (1976).

3.3 Modulation of R1 Internal Motion by Local Protein Dynamics

The internal motion of R1 can be perturbed from the above reference values of ($S=0.5$, $\tau_i=2$ ns) in two ways, namely, by backbone fluctuations on the nanosecond timescale that *increase* the apparent mobility of R1 (decrease S and/or τ), or by interactions of the side chain with local protein structure that *decrease* the mobility (in general, increase S and τ); the term "mobility" will be used as a descriptor of motion that includes effects of both S and τ. As will be discussed below, these effects on S and τ relative to the reference can provide metrics for identifying backbone flexibility, for mapping sequence-specific secondary and tertiary structure, and for detecting conformational changes and equilibria.

Spectral simulations play an important role in quantitative analysis of lineshapes, but simple measures derived from the spectra can serve as monitors of R1 mobility. For example, for the simple anisotropic motion of R1 at noninteracting surface sites, the inverse of the central resonance linewidth (ΔH^{-1}, with ΔH defined as the peak-to-peak separation of the first derivative lineshape, see Fig. 2C) is a selective measure of the *rate* of nitroxide motion (i.e., proportional to τ^{-1}) and serves as a measure of contribution due to fluctuations of the backbone to which R1 is attached (Columbus & Hubbell, 2004). Large amplitude backbone fluctuations also reduce S; an example is shown in Fig. 3A for a dynamically disordered loop in rhodopsin (compare with Fig. 2C). Measurement of backbone dynamics is discussed further in Section 7. On the other hand, interactions of the nitroxide with the protein can dramatically reduce its mobility. Figure 3B

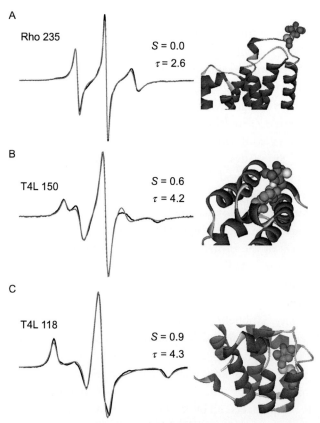

Figure 3 Spin label states as a function of protein structure. (A) A dynamically disordered loop site in rhodopsin exhibits a highly mobile spectrum, indicating large amplitude backbone fluctuations. (B) The presence of an H-bond to a nearby tyrosine causes reduced mobility, as reflected in the spectrum. (C) The EPR spectrum of a site buried in the protein interior shows an immobilized lineshape.

and C provide examples for a solvent-exposed sites, where the nitroxide makes a putative H-bond with a nearby tyrosine (T4L 150R1) (Fleissner, 2007); and for R1 buried in a protein interior (T4L 118R1) (Guo et al., 2007; PDB accession number: 2NTH) where many interactions are made that strongly dampen motion. In such cases, ΔH^{-1} is still employed to measure nitroxide mobility, but it is not simply related to influences of τ alone. In the low mobility regime, other spectral measures of mobility include the overall spectral splitting ($2A_{zz}'$) (Fig. 2C) and widths of resolved hyperfine extrema (Freed, 1976) or spectral moments (Marsh, 2014). The case of T4L 118R1 shows that R1 can be tolerated even at buried sites in stable proteins.

Remarkably, although the S_δ–HC_α interaction is strong enough to order the nitroxide motion in the absence of steric constraints (surface sites), it is also sufficiently weak to enable the side chain to adapt to confined spaces in the protein core without disrupting protein structure.

Examples shown in Figs. 2C and 3 illustrate one-component EPR spectra that arise from a single dynamic mode. Multicomponent spectra can arise when R1 experiences multiple environments in a protein (e.g., when more than one protein conformation exists, see Section 8) or adopts discrete rotamer states. Simulations are important for determining whether a given spectrum can be accounted for by a single dynamic mode, or whether multiple modes are required. In the case of multicomponent spectra, simulations provide orders and rates for each component as well as relative populations of each. Where it can be shown that the individual components correspond to distinct states of the protein, simulations can provide equilibrium constants. The program "MultiComponent" provides a user-friendly interface to Freed theory and the MOMD model for the simulation and fitting of EPR spectra, and is freely available (http://www.biochemistry.ucla.edu/biochem/Faculty/Hubbell/). Practical considerations for simulations of R1 in proteins can be found in Columbus et al. (2001) and Lietzow & Hubbell (2004).

Although the database of crystal structures of R1 in β-sheets is more limited than that for helices, the available X-ray structures suggest that the motion of R1 in β-sheets can be strongly influenced by the nearest neighbor side chains and the polarity of the local environment (Cunningham et al., 2012; Freed, Khan, Horanyi, & Cafiso, 2011; Lietzow et al., 2004).

4. MEASUREMENT OF R1 SOLVENT ACCESSIBILITY WITH POWER SATURATION

Solvent accessibility is a cross-disciplinary structural parameter that is not unique to EPR spectroscopy, but is well understood in the field of protein crystallography and fluorescence spectroscopy. In the context of this chapter, solvent accessibility is measured by the collision frequency (W_{ex}) of a paramagnetic reagent (R_{ex}) with spin-lattice relaxation time (T_1) much shorter than that for the nitroxide. In this case, both the nitroxide T_1 and spin–spin relaxation time (T_2) are shortened in proportion to W_{ex} by the process of Heisenberg exchange:

$$\Delta T_1^{-1} = \Delta T_2^{-1} = W_{ex}, \qquad (2)$$

where W_{ex} is proportional to the concentration of R_{ex} through a constant determined by the local protein structure around R1 that determines it accessibility to direct collision:

$$W_{ex} = j_{ex} \times [R_{ex}]. \tag{3}$$

As such, W_{ex}, and hence solvent accessibility, can be measured either by changes in nitroxide T_1 or T_2 caused by R_{ex} collisions. Changes in T_1 are measured either with methods of power saturation (Altenbach, Froncisz, Hemker, Mchaourab, & Hubbell, 2005) or saturation recovery (Pyka, Ilnicki, Altenbach, Hubbell, & Froncisz, 2005), while changes in T_2 can be measured by changes in spectral linewidth (Altenbach et al., 2005). For a typical spectrum of a nitroxide spin-labeled protein, T_1 and T_2 are in the μs and ns time domains, respectively, and collision rates that significantly reduce T_1 do not measurably change T_2. A significantly higher concentration of relaxing agent is needed to perceptibly change T_2 and thus linewidth. The principles of both methods are described in detail (Altenbach et al., 2005) and only an outline of the practical considerations for the power saturation method is presented here. A detailed experimental protocol of the measurement is available (Oh, Altenbach, Collier, & Hubbell, 2000). The use of pulsed saturation recovery to directly measure changes in T_1 is described as a separate chapter (Yang, Bridges, Lerch, Altenbach, & Hubbell, 2016).

For a simple solvent accessibility measurement, R_{ex} should be small and electrostatically neutral to avoid bias by the presence of nearby charges. One choice is dissolved molecular oxygen in equilibrium with air. In fact, it is typically present, but ignored in lineshape studies because it causes only negligible broadening. Oxygen concentration in the sample can be controlled by using a gas permeable TPX capillary surrounded by a gas mixture containing the desired partial pressure of O_2. In this way, measurements in the absence and presence of dissolved oxygen can be obtained sequentially on the same sample and with identical instrument settings. The oxygen collision rate depends on the product of concentration and diffusion rate, both dependent on temperature and buffer composition when in equilibrium with a given partial oxygen pressure. Oxygen is several times more soluble in a nonpolar medium, such as in the fluid membrane interior, compared to water.

A collision reagent complimentary to O_2 is the nickel complex Nickel(II) ethylenediamine-N,N'-diacetic acid (NiEDDA) (Isas, Langen,

Haigler, & Hubbell, 2002). Unlike oxygen, it has a low solubility in nonpolar solvents compared to water, and the O_2/NiEDDA pair finds use in the study of membrane protein topology (see Section 6). Other complexes such as the neutral NiAA (nickel acetyl acetonate) or the negatively charged CROX (chromium(III) oxalate) (Altenbach et al., 1990) have also been used. The use of charged R_{ex} will bias a measured W_{ex} when there is an electrostatic potential around R1 in the protein. Indeed, this effect has been used to measure the distance of R1 from a charged membrane surface, and to use the information to orient a peripheral membrane protein relative to the bilayer (Lin et al., 1998).

A convenient method to measure W_{ex} is power saturation, wherein EPR spectra are recorded as a function of microwave power (P), both in the presence and absence of the chosen R_{ex}. At low P, the amplitude of the signal increases linearly with the square root of the power ($\propto P^{\frac{1}{2}}$) (Fig. 4). As

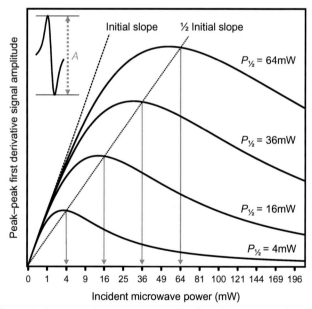

Figure 4 Theoretical power saturation curves. The first derivative peak-to-peak amplitude (insert) of the central line of and EPR spectrum is graphed as a function of incident microwave power. These curves are typical for the EPR spectra of spin-labeled proteins, where T_2 is too fast to be measurably affected by collisions. A line that has half the initial slope will intercept the curve at $P = P_{\frac{1}{2}}$. A change in T_2 due to collision would manifest itself as a drop in the initial slope and this is for example observed for a rapidly tumbling free spin label in solution where $T_2 \approx T_1$.

power is increased, the spin system starts to saturate and the amplitude of the EPR signal deviates from the linear dependence on $P^{1/2}$. The power at which this process becomes important depends on the effective T_1 of the nitroxide. For shorter T_1 values, more power is required to reach saturation. The saturation curve of first derivative peak–peak amplitude (A) versus power (P) is fit to the function

$$A = I \times P^{\frac{1}{2}} \left[1 + \left(2^{\left(\frac{1}{\varepsilon}\right)} - 1 \right) \left(\frac{P}{P_{1/2}} \right) \right]^{-\varepsilon}. \qquad (4)$$

The variables I, ε, and $P_{1/2}$ are adjustable parameters in the fitting routine: the scaling constant I is unimportant for analysis, ε is a factor in the range of 0.5–1.5 and depends on the homogeneity of the spectrum, and $P_{1/2}$ is the power where the amplitude of the first derivative EPR signal is reduced to 50% of its unsaturated value. Figure 4 shows some typical power saturation curves. Only $P_{1/2}$ is used for further analysis and is proportional to the inverse product of T_1 and T_2:

$$P_{1/2} \propto \frac{1}{(T_1 T_2)}. \qquad (5)$$

The change in $P_{1/2}$ due to nitroxide collisions with R_{ex} is defined as $\Delta P_{1/2} = P_{1/2} - P_{1/2}°$, where $P_{1/2}°$ is the value in the absence of R_{ex}. The concentration of R_{ex} is chosen to cause collisions on a timescale comparable to T_1, and under these conditions T_2 is approximately constant but determines the spectral linewidth of the sample. The $\Delta P_{1/2}$ value can be divided by the linewidth ΔH_{pp} in order to obtain a value that is proportional to simply T_1.

The degree of saturation directly depends on the magnitude of the microwave field, H_1, at the sample and the value of $P_{1/2}$ is strongly dependent on the resonator efficiency Λ (where $\Lambda = H_1/P^{1/2}$). A resonator with high efficiency, such as a loop gap resonator (Hubbell, Froncisz, & Hyde, 1987; Hyde & Froncisz, 1989) or a dielectric resonator (e.g., the Bruker ER 4123D), is required to achieve significant saturation, which facilitates an accurate fit of the data. A typical microwave cavity resonator has a much lower Λ and the saturation curve shows only a slight curvature even at the highest possible microwave power, rendering it difficult to accurately determine the important parameter $P_{1/2}$. To correct for hardware dependence, the measured value is further normalized using a DPPH sample measured on the same instrument (Altenbach et al., 2005).

$$\Pi = \left(\frac{\Delta P_{\frac{1}{2}}}{\Delta H_{pp}}\right)\left(\frac{\Delta H_{pp}(\text{DPPH})}{P_{\frac{1}{2}}(\text{DPPH})}\right). \quad (6)$$

The resulting dimensionless quantity Π (the "accessibility parameter") is directly proportional to W_{ex}, the measure of solvent accessibility:

$$W_{ex} = \alpha \times \Pi. \quad (7)$$

The constant α is known from calibration to be ≈ 1.8 MHz (Altenbach et al., 2005), although Π is generally used to report patterns of solvent accessibilities if all data is from the same instrument.

5. CW-BASED INTERSPIN DISTANCE MEASUREMENTS

5.1 Overview

CW EPR spectroscopy is capable of measuring distances and even detailed distance probability distributions between two spin labels (Altenbach, Oh, Trabanino, Hideg, & Hubbell, 2001; Mchaourab, Oh, Fang, & Hubbell, 1997; Rabenstein & Shin, 1995). Typically, two spin labels are simultaneously introduced into one molecule, but other scenarios, such as the study of dimers of singly labeled proteins, can be investigated in the same way. Two spin labels in close proximity experience a magnetic dipolar interaction depending inversely on the cubed interspin distance (i.e., $\propto r^{-3}$). For relatively weak interactions, this is manifested as a simple broadening of the resonance lines. In particular, for distances in the range of $\sim$8–25 Å, static dipolar broadening can be analyzed directly from the CW spectrum. For distances above 25 Å, dipolar broadening becomes very small and difficult to determine reliably from CW spectrum. Pulse methods (DEER, DQC, and SR) can measure distance probability distribution up to 80 Å and beyond under ideal conditions. (DQC: Borbat, Mchaourab, & Freed, 2002; Saxena & Freed, 1997; DEER: Jeschke & Polyhach, 2007; SR: Yang et al., 2014).

If the interspin vector is static or tumbles slowly on the timescale of the magnetic interaction (see Section 5.2), line broadening depends on the orientation of the interspin vector with respect to the magnetic field as well as the interspin distance. With specific assumptions regarding the orientation distribution, the interspin distance distribution can be determined. From a combination of such measurements, changes in the distance probability profile that accompany a conformational change can give precise information on the nature, magnitude, and direction of structural changes. For example, in rhodopsin such distance measurements in frozen solutions before and after photo bleaching indicated a light-activated ~8 Å rigid body movement of transmembrane helix 6 (TM6) that is required for the recruitment of the protein transducin to initiate the visual signal transduction pathway (Farrens, Altenbach, Yang, Hubbell, & Khorana, 1996). This structural change was later confirmed and refined using pulsed EPR methods. (Altenbach, Kusnetzow, Ernst, Hofmann, & Hubbell, 2008).

If the interspin vector tumbles rapidly on the timescale of the magnetic interaction, the dipolar interaction is averaged out, but broadening due to distance-dependent relaxation effects arise. See Fig. 5A for a cartoon of the two scenarios, which are discussed separately in the following sections.

5.2 Static Dipolar Interaction

Historically, the study of the static dipolar interaction in terms of interspin distance was carried out in frozen samples to insure the static condition (Rabenstein & Shin, 1995). However, it was shown by Altenbach (Altenbach et al., 2001) that sufficiently slow rotation of the interspin vector is the only prerequisite, while the localized fast motion of the nitroxide side chain guarantees a random relative orientation of nitroxides but does not significantly average the dipolar interaction. This insight expanded the usefulness of the method, allowing studies at room temperature and under native conditions. The condition for "sufficiently slow" depends on the strength of the dipolar interaction and hence the interspin distance; the correlation time for the interspin vector, which is that for overall protein tumbling, must be such that

$$\tau \gg \frac{r^3 h}{3\pi g^2 \beta^2 10^{24}}, \qquad (8)$$

where g is the electronic g factor of the two (identical) spins, β is the Bohr magneton, and h is the Planck constant; here the interspin distance, r, is in units of Å. For example, for an interspin distance of 10 Å, the protein must have a correlation time longer than about 2 ns. For a protein of 20 kDa, the correlation time is already ≈ 6 ns, so the static condition is met. On the other hand, for an interspin distance of 20 Å, near the limit of detection for CW dipolar broadening, the criterion is that the protein correlation time must be longer than ≈ 15 ns. Thus, the static condition is not met for such a long distance, and rotational diffusion averages the weak dipolar interaction. To detect broadening in this situation, the correlation time for the protein

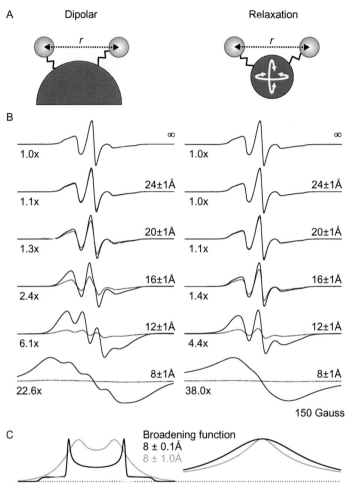

Figure 5 Distance-dependent spectral broadening due to spin interaction. (A) A cartoon of two typical scenarios. Static dipolar interaction is shown on the left and dynamic relaxation broadening is shown on the right. (B) A simulation of the distance-dependent broadening of a spin-labeled protein. The distance probability distribution is assumed to be Gaussian centered at the indicated distance and with a standard deviation of 1 Å (equivalent to a full width at half height of 2.36 Å). Left: Dipolar broadening in the case of a frozen or slowly rotating interspin vector and absence of orientation selection. Right: Relaxation broadening for a protein tumbling with a rotational correlation time of 6 ns. In both cases, the amplitude of the broadened first derivative spectra drops significantly with increasing interactions. All spectra are shown on the same vertical scale (dotted) or scaled to the same height (solid) and the applied scaling factor is shown in the bottom left of each spectrum. (C) The broadening function for an interspin distance of 8 Å and a distribution width of 1 and 0.1 Å as indicated. Only for a very narrow distance distribution are the Pake pattern (left) or Lorentzian (right) well resolved.

must be increased, either by addition of a viscogen to the solution or by immobilization of the protein through covalent attachment to a solid support (López et al, 2009). In the case of a viscogen, a ~30% w/w sucrose solution (with a viscosity of ~3 cP) is adequate (Mchaourab et al., 1996).

In the static limit for a protein in solution, the interspin vector is effectively frozen and randomly oriented with respect to the magnetic field. At the same time the individual spin labels are oriented randomly with respect to the interspin vector, as typically found at protein surface sites due to the rapid motion of the nitroxide about dihedrals X_4 and X_5 (Fig. 2).

CW spectral lineshapes depend on many structural and dynamical parameters; in particular, nitroxide motion has a large effect on spectral widths. Unless the dipolar broadening is substantial and distinct, it is often impossible to isolate the dipolar broadening from the underlying lineshape directly.

In order to accurately identify and measure dipolar broadening, a reference spectrum must be obtained under otherwise identical conditions, but in the absence of the dipolar interaction. When measuring distances in dimers of singly labeled proteins, this can be achieved by adding an excess of unlabeled protein, resulting in a solution of dimers that statistically contain, at most, a single spin. More typically, a doubly labeled protein is explored by the dipolar broadening method, however. Here, the noninteracting reference spectrum is most conveniently obtained by labeling the same protein with a mixture of the paramagnetic nitroxide label and an excess of a diamagnetic labeling compound of similar structure (Altenbach et al., 2001). The spectrum obtained will have an identical underlying lineshape to that of the doubly labeled protein, but where broadening due to the dipolar interaction is primarily absent (Fig. 6). Acquiring the lineshapes in the presence and absence of dipolar interaction allows one to determine the dipolar broadening function, and thus one can resolve the probability distribution of interspin distances using numerical methods.

In both of the above methods, the noninteracting reference spectrum obtained often exhibits a small fraction of the interacting spectrum due to the statistical probability of having two paramagnetic labels in the same structure. Similarly, the interacting spectrum can contain a small amount of noninteracting spectrum due to incomplete labeling. These artifacts can be eliminated by linearly combining the pairs of experimental spectra to extract the pure interacting and noninteracting spectra for further analysis. The noninteracting spectrum needs to be processed as described here since trace amounts of dipolar broadened spectrum can interfere with the distance

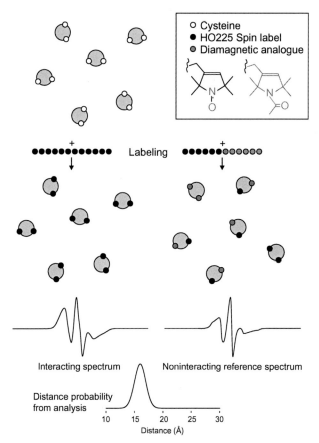

Figure 6 The recommended procedure to produce the fully labeled protein and the dilute labeled reference protein to determine distances based on broadening. A protein containing two labeling sites is either labeled simply with a nitroxide spin label (red (black in the print version)) or with a mixture of the nitroxide spin label and a diamagnetic analogue (yellow (light gray in the print version)). Because there are two paramagnetic centers per protein in the fully labeled proteins (left), a dipolar broadened spectrum is obtained. In contrast, use of the paramagnetically diluted labeling mixture generates labeled proteins that have only one paramagnetic center on average, thus revealing the underlying spectral lineshape in the absence of dipolar interaction (right). Comparison of the spectra in the presence and absence of dipolar interaction allows for canceling out all motional contributions to the lineshape, enabling the determination of the dipolar broadening function and thus the interspin distance probability distribution. The small statistical fraction of doubly labeled component in the dilute reference (indicated with *) can be removed by a suitable linear combination of the two spectra or minimized by reducing the fraction of spin label in the mixture.

analysis. However, the dipolar interacting spectrum can be used directly, as the minute spectral contribution from the noninteracting component will simply indicate an infinite distance in the analysis output. Simple statistical methods allow one to estimate the amount of doubly labeled impurity in the noninteracting spectrum.

A practical upper distance limit for the CW EPR method is about 25 Å. Above this distance, only trivial linewidth changes due to the dipolar interaction are observed which could be masked by minute variations in experimental conditions that can affect lineshape (e.g., noise, dissolved oxygen, temperature, viscosity, modulation broadening, and saturation broadening).

For interspin distances near the lower limit of the method (~8 Å), the resultant spectra become very broad and need to be recorded over several hundred Gauss to accurately determine the baseline and normalize the integrated area. At the same time, the absolute spectral amplitude becomes very small and trace contributions from noninteracting spins can dominate the lineshape, requiring careful analysis.

In the limits of random interspin vector orientation and random nitroxide orientation the dipolar broadening function is a Pake pattern (Pake, 1948). Typically, several distances are present and the broadening function is a weighted sum of Pake patterns, one for each distance. The real distance distribution is, of course, continuous, but it is sufficient to approximate it by a suitably spaced set of discrete distances, allowing efficient use of numerical methods for analysis. The corresponding broadened lineshape is the mathematical convolution of the noninteracting spectrum with the broadening function. This can be achieved by a simple matrix multiplication of the distance distribution with a matrix containing a library of convoluted spectra generated from the noninteracting spectrum and the broadening function for each distance. Note that the convolution can be performed directly on the first derivative CW lineshape to produce the first derivative of the broadened spectrum, thus avoiding integration/differentiation processing that can introduce errors. Figure 5B (left column) shows a typical spectrum of a spin-labeled protein and how it changes in the presence of dipolar interactions as a function of distance.

In practice, one needs to solve the inverse problem, i.e., computing the distance distribution given the above matrix and the interacting spectrum. This is an ill-posed problem but can be solved by, e.g., model-free Tikhonov regularization or approximated by model-based fitting. In model-based fitting, the distance probability distribution is assumed to be a sum of simple lineshapes. An analysis program named "ShortDistances" to obtain distance probability distributions from a set of spectra with and without dipolar interaction is available for download (http://www.biochemistry.ucla.edu/biochem/Faculty/Hubbell/). Note that this approach assumes that all parts of the spectrum experience the same dipolar broadening and thus will not work for composite spectra where each component is involved in a different distance distribution. Care must also be taken to remove any spectral impurities prior to analysis, such as those due to small amounts of unreacted free label in solution.

The deconvolution approach for CW-based distance measurements has been applied to a variety of proteins, including sensory rhodopsin (Bordignon et al., 2007), the inhibitory component of troponin (Brown et al., 2002), the SNARE complex (Kim, Kweon, & Shin, 2002), and the KcsA potassium channel (Liu, Sompornpisut, & Perozo, 2001).

For interested readers, a detailed step-by-step protocol for sample preparation and data collection for CW-based distance measurements using fast tumbling and the rigid limit theories has been published elsewhere by Cooke and Brown (2011).

5.3 Relaxation Dipolar Broadening

If the two spin labels are in a small protein with a short rotational correlation time such that

$$\tau \ll \frac{r^3 h}{3\pi g^2 \beta^2 10^{24}} \tag{9}$$

the dipolar interaction will average out, but relaxation effects can lead to a distance-dependent line broadening (Mchaourab et al., 1997). The broadening will be Lorentzian with a width that depends on r^{-6} as well as on the correlation time of the interspin vector, the latter of which can be estimated from the protein molecular weight according to the Stokes–Einstein formulism (Eq. 1). Figure 7 shows examples demonstrating a good agreement with the inverse sixth power dependence (Mchaourab et al., 1997). A more detailed analysis of the line broadening in terms of interspin distance can be carried out analogous to the method described for dipolar broadening. The primary difference is that a sum of Lorentzian broadening functions rather than a sum of Pake functions is applied. Figure 5B shows a simulated example of this situation (right column). This hypothetical system could be converted to one that exhibits the static dipolar interaction (discussed in

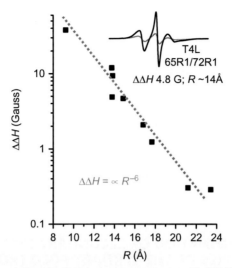

Figure 7 Short-range interspin distance measurements with CW EPR using relaxation. The incremental spectral broadening ($\Delta\Delta H$) as a function of interspin distance is shown as black squares. The dotted line shows the expected R^{-6} dependence and is in good agreement with the measured data (Mchaourab et al., 1997). Inset: example of one of the data points shown in the graph. The EPR spectrum of 65R1/72R1 is shown in red (light gray in the print version). The sum of the corresponding single spectra is shown in black. The observed spectral broadening increment and calculated interspin distance using Redfield relaxation theory are indicated.

Section 5.1) by simply slowing down the protein tumbling rate, e.g., by increasing the solution viscosity or by coupling the protein to a solid support (López et al., 2014).

5.4 More Complicated Scenarios

While the procedures described above are often sufficient to analyze the data in terms of a simple distance probability distribution, more complicated scenarios are briefly discussed here.

For interspin distances below 8 Å, through-bond or through-space spin exchange interactions arising from orbital overlap are present. The complicated underlying theory for the study of interspin distances based on exchange interaction will not be discussed here, but interested readers are referred to Closs, Forbes, and Piotrowiak (1992) and Fiori, Miick, and Millhauser (1993) and references therein for details on the theory and application of such strategy. This situation can be avoided by selecting more suitable labeling sites, but often a qualitative indication of "these sites are very close to one another" is sufficient to answer the structural questions with the desired accuracy.

If the two nitroxides have a fixed orientation with respect to the interspin vector, there exists a relationship between the orientation-dependent dipolar broadening and the orientation-dependent spectral line positions. In specially designed systems with rigidly attached nitroxides, detailed analysis allows extraction of not only the interspin distance, but also the six angles defining the two nitroxide system (Hustedt & Beth, 1999). The analysis can be further aided by the introduction of macroscopic orientation in the sample and recording the spectrum at various angles (Ghimire, Abu-Baker, et al., 2012; Ghimire, Hustedt, et al., 2012).

6. DETERMINATION OF SECONDARY STRUCTURE AND FEATURES OF THE TERTIARY FOLD FROM R1 MOBILITY AND ACCESSIBILITY

CW EPR spectral lineshapes and solvent accessibilities strongly correlate with the local protein structure, but even a very detailed analysis of a single site provides relatively limited structural information without the context of other sites in the same protein. The method of nitroxide scanning involves the study of a set of spin-labeled sites to obtain patterns that can be interpreted in terms of structure and dynamics. The set may consist of consecutive residues to determine local secondary and tertiary structure (Lietzow & Hubbell, 2004; Mchaourab et al., 1996), or may contain only surface residues to determine the depth of penetration of a sequence in a membrane interior (Altenbach, Greenhalgh, Khorana, & Hubbell, 1994). In most cases, a detailed lineshape analysis is not required and simplified parameters such as the inverse central linewidth (ΔH^{-1}) or solvent accessibility (Π) measured as a function of sequence position are sufficient. The sections below illustrate the use of nitroxide scanning together with these parameters to resolve structure in both soluble and membrane proteins.

For a protein in solution, the inverse linewidth and accessibility to molecular oxygen or NiEDDA are all highly correlated. Surface sites show narrow linewidths, while buried sites are immobile and yield broad lines. The solvent accessibility is high for surface sites and low for buried sites and scale directly with a structure-based parameter (Altenbach et al., 2005). Regular solvent-exposed secondary structure is directly revealed in the periodic variation of accessibility parameters in a continuous R1 scan along a sequence. Beta sheets have an alternating pattern that repeats every two residues while alpha helices repeat every 3.6 residues (100° angular increment). Typically, the pattern is immediately obvious and reveals the secondary structure. More detailed analysis can be done using Fourier or wavelet analysis. Note that these methods require a sufficient number of consecutive samples and should ideally cover an integer number of cycles to avoid spectral leakage, a condition that is often not met in practice. Typically, a visual inspection is sufficient to see the pattern.

A more detailed analysis is most interesting for helical sections. To orient the local helical structural element with respect to the remainder of the protein, we are interested in three parameters that characterize sequence-correlated values of $\Pi(O_2)$, $\Pi(NiEDDA)$, or ΔH^{-1}. These are (1) directionality (or phase), indicating which face of the helix points toward the solvent; (2) the absolute magnitude; and (3) the degree of anisotropy. The absolute magnitude depends on the concentration and diffusion of R_{ex}. The anisotropy reports the fractional accessibility of the helix surface. A single helix in solution will have zero anisotropy because all directions are equally accessible while a helix deeply embedded in the protein with only a narrow exposed strip will have a large anisotropy. These parameters can be obtained by converting the exposure measurements to vectors with the angle incremented according to the identified or assumed secondary structure (e.g., 100° for an α-helix). If there is an approximately even sampling of all directions, the vector sum points in the direction of highest exposure as measured by either ΔH^{-1} or Π (Altenbach et al., 1990). In the more general case where the data do not cover an integer number of cycles or if there are missing values, the vector sum will give a moderately distorted answer because certain directions are more heavily weighted while others are incompletely probed. It is more reliable to instead fit the vector endpoints to a circle, for example, using the method developed by Pratt (1987). The vector from the origin to the center of the fitted circle gives the direction and magnitude of highest accessibility. The length of this vector divided by the fitted best radius gives a measure of anisotropy (ranging from zero to one). Figure 8 shows a reanalysis of the 1990 oxygen collision data for bacteriorhodopsin residues 131–138 (Altenbach et al., 1990) by this method. Figure 8A shows the raw data and Fig. 8B the projection into the complex plane with a 100° angular increment corresponding to an α-helix (black to light gray arrows); the best fit circle (red (gray in the print version)) and the direction of highest accessibility (red (gray in the print version) arrow) are also indicated.

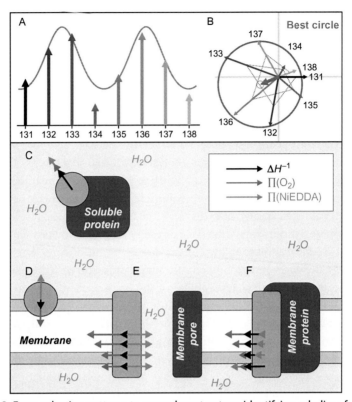

Figure 8 From solvation patterns to secondary structure. Identifying α-helices from the expected behavior of $\Pi(O_2)$, $\Pi(NiEDDA)$, and ΔH^{-1} for different situations that have been documented experimentally. (A) Oxygen accessibility of residues 131–138 in bacteriorhodopsin. A periodicity of 3.6 residues/turn is clearly visible. (B) Projection of the accessibility into the complex plane using a 100° angular increment between residues as indicative for an α-helix. The endpoints are fitted to a circle and the best fit circle shown in red (gray in the print version). The circle is also shown transformed as the red (gray in the print version) curve in (A). The exposure vector from the origin to the circle center points in the direction of highest accessibility and its magnitude relative to the fitted radius defines the anisotropy (0.47 in this case). The calculation is repeated for the inverse linewidth ΔH^{-1} and $\Pi(NiEEDA)$ data. (C) For a surface helix of a homogenously solvated protein, the exposure vectors for the various measures are all in phase (Isas et al., 2002). (D) For a helical segment lying parallel to the surface of a membrane (Altenbach, Froncisz, Hubbell, & Hyde, 1989a) or (E) lining a water-filled transmembrane pore (Oh et al., 1996), the helix is asymmetrically solvated by water and the fluid hydrophobic interior of the bilayer. These interesting situations can be identified by the out-of-phase helical periodicity of $\Pi(O_2)$ and $\Pi(NiEDDA)$ but a relatively constant ΔH^{-1} because all sites are surface located. (F) For a transmembrane helical segment in a compact folded protein, $\Pi(O_2)$ and $\Pi(NiEDDA)$ will vary periodically as shown in part (A), but there will be gradients in each along the direction of the membrane normal (see also discussion related to Fig. 9). This latter property allows one to distinguish between the membrane surface and pore-lining helical segments of (D) and (E), because no gradient will exist in the surface helix. *Panel (A): Data from Altenbach et al. (1990).*

Figure 8 shows the expected behavior of $\Pi(O_2)$, $\Pi(NiEDDA)$, and ΔH^{-1} for different situations that have been documented experimentally. For a surface helix of a protein in solution (homogeneous solvation), the exposure vectors for the various measures are all in phase. (Fig. 8C, see Isas et al., 2002 for an example). For a helical segment lying parallel to the surface of a membrane (Altenbach, Froncisz, Hyde, & Hubbell, 1989) or lining a water-filled transmembrane pore (Oh et al., 1996), the helix is asymmetrically solvated by water and the fluid interior of the bilayer. Due to the differential solubility of O_2 and NiEDDA in polar and nonpolar media, these situations can be identified by an out-of-phase helical periodicity of $\Pi(O_2)$ and $\Pi(NiEDDA)$ (Fig. 8D and E). For a transmembrane helical segment in a compact folded protein, $\Pi(O_2)$ and $\Pi(NiEDDA)$ will vary periodically as shown in part (A), but there will be gradients in each along the direction of the membrane normal (Fig. 8F), as discussed in the next section. This latter property allows one to distinguish between the membrane surface and pore-lining helical segments of (D) and (E), because no gradient will exist in the surface helix.

Given a known crystal structure, the solvent accessibility can be calculated from the coordinates and directly compared with the EPR data. An excellent example is provided by the soluble protein Annexin 12, where R1 scanning data revealed a periodicity in $\Pi(O_2)$, $\Pi(NiEDDA)$, and ΔH^{-1} with periods of 3.8 in helical regions consistent with the crystal structure (Isas et al., 2002). Similarly, accessibility measurements in the β-sheet protein cellular retinol-binding protein (CRBP) showed a sequence-dependent variation in $\Pi(O_2)$ and $\Pi(NiEDDA)$ with a periodicity of two that matches R1 mobility and is consistent with the local structure (Lietzow & Hubbell, 2004). Any significant differences in this comparison can indicate regions where the solution structure differs from the crystal structure.

Another important application of CW-based accessibility measurements has been the determination of topology of membrane proteins in lipid bilayers. In this regard, it is useful to define a parameter Φ that measures the ratio of collision frequency of O_2 to that of NiEDDA (or other hydrophilic paramagnetic metal complex) for a particular residue:

$$\Phi = \ln\left[\frac{\Delta P_{\frac{1}{2}}(O_2)}{\Delta P_{\frac{1}{2}}(NiEDDA)}\right] = \ln\left[\frac{\Pi(O_2)}{\Pi(NiEDDA)}\right] \tag{10}$$

Note that this ratio has the desirable property that all hardware and linewidth dependencies discussed in Section 4 cancel out as long as all data is acquired on the same instrument. For an R1 scan along a helical segment involving only surface sites, Φ is insensitive to local variations in structure and is thus approximately constant for a protein in homogeneous solution. However, it becomes a very useful parameter when studying membrane proteins. Due to the differential solubilities of O_2 and NiEDDA in the membrane interior and their dependence on depth from the aqueous interface, Φ has been shown to be a linear function of depth in the bilayer (Altenbach et al., 1994). For example, the results of a surface R1 scan along helix E in bacteriorhodopsin is shown in Fig. 9. Helix E has a very asymmetric membrane exposure pattern. Near residue 109 the exposed arc is much wider than that at residue 124, where the helix tucks between two neighboring helices. This asymmetry is clearly visible in the oxygen collision data (Fig. 9B, red (dark gray in the print version)), but is absent in the depth parameter (Fig. 9B, green (light gray in the print version)), which suggests exact symmetry between the helix halves.

The CW-based power saturation method in combination with nitroxide scanning has become widely used to determine secondary structure and

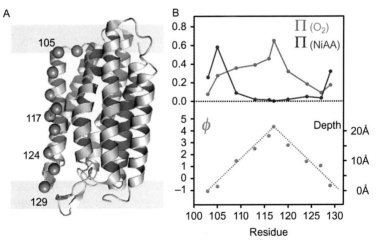

Figure 9 Determining membrane topology with CW power saturation. (A) Ribbon diagram of bacteriorhodopsin (PDB accession number 1FBB) showing the protein surface sites where R1 was introduced. (B) A plot showing the accessibilities of the indicated R1 residues to molecular oxygen and to NiAA. The log of the ratio of accessibility values (Φ) as a function of sequence is shown in green (light gray in the print version) and is a linear measure of depth. *Data used for panel (B) were taken from Altenbach et al. (1994).*

topology and for studying structural changes in a large library of soluble and membrane proteins with important biological roles such as bacteriorhodopsin (Fig. 9; Altenbach et al., 1994), rhodopsin (Altenbach, Cai, Khorana, & Hubbell, 1999; Altenbach, Klein-Seetharaman, Hwa, Khorana, & Hubbell, 1999), annexin (Isas et al., 2002), CRBP (Lietzow & Hubbell, 2004); myosin (Nelson, Blakely, Nesmelov, & Thomas, 2005), diphtheria toxin (Oh et al., 1996), KcsA (Perozo, Cortes, & Cuello, 1998), and colicin (Salwinski & Hubbell, 1999). It is noted that for correct interpretation of relative accessibilities in terms of topology of membrane proteins, the R1 sites should be engineered at solvent- and membrane-exposed sites in order to minimize complications arising from effects of local structure on accessibility values (Altenbach et al., 1994). In addition, although a pulsed saturation recovery method (Chapter 4.1, Yang, Bridges, Lerch, Altenbach & Hubbell, 2016) can be used as an alternative to CW power saturation for the determination nitroxide accessibility in a way that does not require additional measurements of a reference sample (DPPH), it does require the use of specialized instrumentation that is not widely available.

7. BACKBONE DYNAMICS

For situations where R1 is found at solvent-exposed surface sites and does not interact with other groups in the protein, the observed EPR spectra reflect only the internal motions of R1 and fluctuations of the protein backbone on the nanosecond timescale (assuming that rotational diffusion of the protein has been eliminated by one of the methods discussed in Section 3.2). The purely internal motion of R1 is characterized by $S \approx 0.5$, $\tau \approx 2$ ns, which is determined to a large extent by interactions within the R1 side chain itself and therefore is expected to be essentially the same at all surface sites (see Section 3.2). Fluctuations of the backbone on the ns timescale will add to this internal motion and lead to decreases in S and τ. Thus, variation in motion of R1 along a sequence reflects relative contributions from backbone dynamics. Values of S and τ can be obtained from spectral simulations, but as discussed in Section 3.2, the inverse central linewidth (ΔH^{-1}; Fig. 2C) of the EPR spectrum provides a simple alternative measure of R1 motional rate proportional to τ^{-1}. The variation in S is less than for τ, and τ or ΔH^{-1} are the most useful metrics, although the dynamic range of S can be increased with spin labels with a hindered internal motion (Columbus et al., 2001).

For convenience, ΔH^{-1} can be normalized to yield a "scaled mobility parameter" M_s (Columbus & Hubbell, 2002; Hubbell et al., 2000),

$$M_s = \frac{\left(\Delta H_{\exp}^{-1} - \Delta H_i^{-1}\right)}{\left(\Delta H_m^{-1} - \Delta H_i^{-1}\right)}, \qquad (11)$$

where the subscript exp denotes the linewidth for the site of interest, and i and m denote values corresponding to the most immobilized and mobile sites, respectively, observed under conditions where rotational diffusion of a protein does not contribute to the linewidth. Reasonable but not critical choices for these two ΔH values are 8.4 and 2.1 Gauss, respectively, as discussed in Columbus and Hubbell (2002).

In a typical application, R1 residues are placed, one at a time, at solvent-exposed sites along a selected helical segment and/or loop, and M_s is determined from the linewidth. If multiple components are present in a spectrum, ΔH^{-1} is dominated by the most mobile component, which generally reflects the influence of backbone motion. In the case of simulations, S and τ values are obtained for each component. The backbone contribution dominates for components with the smallest S and τ values; if they significantly exceed those for the reference state ($S \approx 0.5$, $\tau_i \approx 2$ ns), it can be assumed that interactions of R1 involve protein contacts.

The above strategy for monitoring backbone dynamics is a powerful means for identifying dynamically disordered sequences often implicated in recognition domains for protein–protein interactions. In addition, the method is sufficiently sensitive to reveal low-amplitude collective and local modes in stable helices. Examples of these two extreme cases are provided by the dynamically disordered helices in the DNA binding domain of the yeast transcription factor GCN4 and segmental backbone fluctuations in a stable helix of holomyoglobin.

Figure 10A shows solution-state models of free and DNA-bound GCN4, along with corresponding plots of M_s versus sequence position within the binding domain (residues 1–26) and in the stable leucine zipper (27–49); residues shown with circles identify solvent-exposed outer helix sites based on the crystal structure, and insets show example EPR spectra for R1 at sites in the two domains. For unbound GCN4 (purple (dark gray in the print version)), M_s values reveal a gradient of dynamics along the binding domain to the zipper, wherein the average motion is relatively constant; the gradient is illustrated in Fig. 10A as thickness of the helix ribbon. This

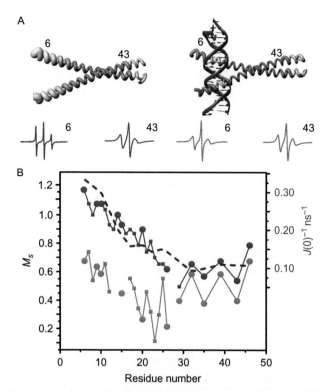

Figure 10 Monitoring the amplitudes and rates of backbone motions. (A) Ribbon diagram of GCN4 showing the dimer structure in solution and bound to DNA (black). The amplitude and rates of motions are proportional to the color and thickness of the tube. Representative EPR spectra are shown for the same two positions in each state. (B) The solution structure shows a strong gradient of backbone flexibility, as reported from the scaled mobility of EPR spectra (purple (dark gray in the print version)). Once bound to DNA, the dynamic fluctuations are suppressed (red (light gray in the print version)). Outside residues are outlined with larger circles. The dashed line shows results from NMR (see text for details).

result mirrors that obtained with spectral density mapping in NMR as shown by the plot of $J(0)^{-1}$ that monitors rate of motion on a ns timescale (Bracken, Carr, Cavanagh, & Palmer, 1999) (dashed line). Similar results are obtained with τ^{-1} determined from spectral simulations (Columbus & Hubbell, 2002) but the motion is of sufficiently large amplitude such that $S=0$ at all sites. NMR studies suggested that, the dynamic mode of the binding domain involves rupture of the backbone hydrogen bonds and transient population of nonhelical states (Bracken et al., 1999). Upon binding DNA (Fig. 10B, red (light gray in the print version)), the rate of motion is strongly damped but the trend in the binding domain remains, with the zipper region largely unchanged. This result is unique to SDSL and was not studied in NMR. The site-dependent modulation of M_s is due to tertiary structure and is discussed in Columbus & Hubbell (2004).

A second model, holomyoglobin, demonstrates segmental fluctuations of the protein backbone in a stable helix, a dynamic mode of backbone fluctuation unlike dynamic disorder. Figure 11A shows a ribbon model of holomyoglobin with solvent-exposed sites along helix H where R1 was substituted are designated by green (light gray in the print version) spheres. The corresponding EPR spectra can be simulated assuming a simple anisotropic motion (López et al., 2012). Figure 11B shows a plot of M_s versus sequence from 132 to 149, which reveals a clear increase in motion progressing from the N to C terminus. Interestingly, this trend is also observed in crystallographic thermal factors (Phillips, 1990). The dynamic mode detected by R1 is likely segmental fluctuations with a coherence length of one or two helical turns; the fluctuation could involve rocking modes about the helical symmetry axis, a motion that is not detected by the NMR S_{NH} order parameter (Columbus et al., 2001).

In addition to helix H, backbone motions in each of the holomyoglobin helices were sampled with 28 individual R1 residues located at solvent-exposed sites. M_s values from this dataset were used to generate the map of backbone flexibility shown by ribbon size and color in Fig. 11A. The data reveal enhanced fluctuations near helix termini and in the body of the short helix D. The amplitude and rate of the dynamic mode(s) detected by R1 is correlated to the fraction of buried surface on the helical segment to which R1 is attached, as shown by plots of S and τ^{-1} determined for the 28 sites by spectral simulation (Fig. 11C). This result suggests a simple picture in which the amplitude and rate of fluctuation are both decreased (increasing S, decreasing τ^{-1}) in proportion to the extent of contact with the core of the protein. Interestingly, extrapolation of S to 100% buried surface gives $S \approx 0.55$, close to the value estimated for R1 internal motion in a completely rigid protein (Section 3.2).

To investigate the suitability of CW EPR for mapping backbone dynamics in transmembrane domains, Lo and coworkers (Lo, Kroncke, Solomon, & Columbus, 2014) measured the mobility of 55 R1 sites of the membrane protein TM0026 and compared with the local backbone dynamics observed using NMR ^{15}N-relaxation data. The mobility of all sites

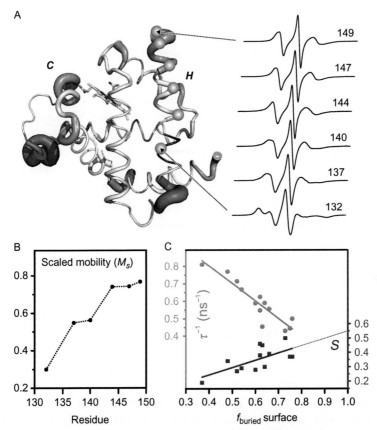

Figure 11 Monitoring the amplitude and rate of backbone fluctuations in flexible but well-ordered protein segments. (A) Cartoon representation of holomyoglobin indicating the ns mobility of the R1 side chain as judged by scaled mobility (M_s) values. The width and color of the backbone are proportional to M_s (gray = not determined; blue (dark gray in the print version) = $M_s \leq 0.3$; green (light gray in the print version) = $0.3 < M_s < 0.5$; yellow (light gray in the print version) = $0.5 \leq M_s \leq 0.7$; orange (light gray in the print version) = $0.7 < M_s < 0.9$; red (dark gray in the print version) = $M_s \geq 0.9$). Spectra of all surface sites along helix H reflect the gradient of backbone fluctuations. Green (light gray in the print version) spheres indicate the sites for the spectra shown on the right. (B) The scaled mobility values of the helix H spectra shown in (A). (C) A plot showing the correlation between tumbling rate and amplitude of R1 motion as a function of local fraction of surface buried observed for R1 sites in holomyoglobin (López, Oga, & Hubbell, 2012). *Panel (A): Adapted from López et al. (2012).*

investigated, as judged by M_s and the spectral second moment, showed a remarkable similarity with trends of backbone motions derived from NMR, strongly suggesting the ns backbone modes are reflected in the EPR spectra of R1-labeled membrane proteins.

8. IDENTIFYING SLOW CONFORMATIONAL EXCHANGE AND DETECTING CONFORMATIONAL CHANGES

The EPR spectrum of R1 at a solvent-exposed, noninteracting site on the surface of a helix corresponds to a single anisotropic mode of nitroxide motion (Figs. 2, 3, and 11). This is often the case when R1 is engineered at a solvent-exposed site based on a crystal structure model. However, in particular cases, the spectrum may have two resolved components corresponding to relatively "mobile" and "immobile" states of the nitroxide. The spectral resolution for X-band EPR is insufficient to resolve distinct components in the slow-motional regime, and the component corresponding to immobile states may be heterogeneous. Nevertheless, the spectra can be classified as "two-component" for the purposes discussed here. If the states are in an exchange equilibrium, the two spectral components will be well resolved without effects on the individual lineshapes as long as the exchange lifetime is long compared to the characteristic timescale of X-band CW EPR (0.1–100 ns). Exchange lifetimes short on this timescale will result in spectral averaging to a single component.

In principle, two-component spectra can arise from equilibrium between *two rotamers* of R1, one of which places the nitroxide in a position to interact with a nearby side chain, thereby immobilizing the nitroxide. Indeed, this origin of two-component spectra has been confirmed in a few cases—T4L 44R1 (Guo et al., 2008), T4L 115R1 (Guo et al., 2007), and T4L 119R1 (Bridges, Hideg, & Hubbell, 2010)—indicating that rotamer exchange lifetimes of R1 are longer than about 100 ns (Yang et al., 2016). Alternatively, two-component spectra can arise from equilibrium between *two conformational substates* if R1 is in a region where it has distinct interactions with the local environment in the substates (Fig. 12). The characteristic timescale of protein conformational exchange is μs–ms, much longer than that for X-band EPR, ensuring that the individual states will be resolved. Thus, the existence of multiple conformations at equilibrium can be revealed in the EPR spectrum as multiple components, but the lineshapes contain no information on exchange rate.

To enable SDSL as a tool for detecting conformational equilibria, it is necessary to distinguish the possible origins of two-component spectra as shown in Fig. 12. One CW strategy for this purpose is osmolyte perturbation. Stabilizing osmolytes such as sucrose and polyethylene glycol (PEG) shift protein conformational equilibria toward the least solvent-exposed

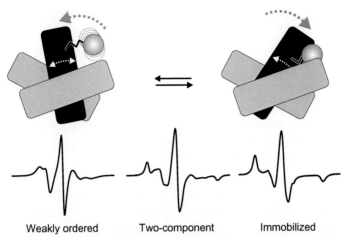

Figure 12 Slow conformational exchange giving rise to two-component CW EPR spectra. The protein is in equilibrium between two conformational states as indicated by the two position of the black helix. A suitably placed spin label will show different mobility in the two states. The two-component spectrum is a weighted sum of both spectra and the weights can be used to obtain the equilibrium constant.

state, but have little effect on rotameric equilibria of R1 (López et al., 2009). Thus, if osmolyte addition shifts the relative populations in a two-component spectrum, the origin of the two components is due to an equilibrium between distinct protein conformational states rather than rotameric equilibria of R1. Since the least solvent-exposed conformation of a protein is its most compact state, which increases opportunities for interaction of the R1 side chain, osmotic perturbation generally shifts spectral populations toward more immobilized states. For simple two-state exchange model, multicomponent spectra can be fit using the MOMD model to calculate equilibrium constants for the exchange, which can be used to determine the energy difference between two states in equilibrium (López et al, 2009).

To obtain observable osmolyte effects, relatively high concentrations of osmolyte are required. For sucrose, 30% w/w (≈ 1 M) is used; for PEG 400, similar concentrations are required (Cafiso, 2014). At these concentrations the osmolyte will generally change the solution viscosity and therefore alters the correlation time for protein rotational diffusion (τ_R) as well as the equilibrium between conformations. For small proteins (<60 kDa) where τ_R influences the EPR spectrum of R1, the effect of the osmolyte must be observed at constant viscosity. To this end 25% w/w Ficoll 70 can be used as a viscogen to match the effective viscosity of 30% w/w sucrose solution;

Ficoll 70 at this concentration does not alter solution osmolarity, though it does act as a molecular crowder (López et al., 2009). Thus, the pure osmolyte effect can be isolated in a comparison of the EPR spectra in 25% w/w Ficoll 70 and 30% w/w sucrose. In cases where one expects large molecular volume differences between conformational states, Ficoll 70 may show crowding effects in addition to viscosity changes. Such complications can be avoided by tethering the protein to a solid support (López et al, 2014). For large soluble proteins and membrane proteins, viscosity matching is not necessary and the effects of osmolyte can be directly observed.

An example of the osmolyte perturbation method for mapping sequence-specific conformational exchange is provided by myoglobin, where R1 was introduced one at a time at 38 solvent-exposed sites throughout the structure, sampling each of the helices. The protein was studied in the holo-, apo-, and molten globule states. For sites that exhibited multicomponent EPR spectra, thereby identifying putative regions in conformational exchange, spectra were compared in 25% w/w Ficoll 70 and 30% w/w sucrose. Figure 13 summarizes the results in ribbon models, with regions in conformational exchange identified by osmolyte perturbation designated in red (dark gray in the print version) for each state. The osmolyte perturbation method in combination with EPR has been applied to identify conformational exchange in several soluble (Chui et al., 2015) and membrane proteins with important biological roles (Cafiso, 2014; Fanucci, Lee, & Cafiso, 2003; Flores Jimenez, Do Cao, Kim, & Cafiso, 2010; Freed, Horanyi, Wiener, & Cafiso, 2010; López et al., 2009).

A major advantage of the osmolyte perturbation method is simplicity. A protein fold can be sampled with a small number of R1 sensors on the outer surface where structural perturbations are minimal and the response of the spin-labeled protein to osmolytes can be examined with a conventional EPR spectrometer. This method combined with lineshape analysis expands the capability of SDSL for facile mapping of protein flexibility from the picosecond to millisecond and longer timescales. It is noted that with the osmotic perturbation method, false negatives are possible in cases where the conformational exchange involves substates with equal solvent exposure (e.g., slow rotation of a helix about its own axis). For such cases, pulsed EPR-based strategies such as SR EPR (Bridges et al., 2010; Yang et al., 2016) and electron–electron double resonance (ELDOR) (Fleissner et al., 2011) provide alternative means to identify conformational equilibria.

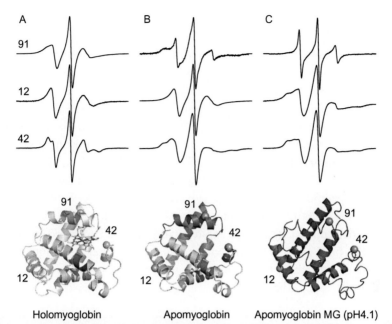

Figure 13 Characteristic spectra of myobglobin as a function of protein state. EPR spectra of residues H12R1, K42R1, and Q91R1 in the holo state (A), apo state (B), and molten globule state (C) are shown. The ribbon diagrams of myoglobin in each state are shown in the bottom panel. The spheres at the C_α atom identify the sites shown in the top panel. The red (dark gray in the print version) ribbon indicates regions of the molecule undergoing conformational exchange detected by osmolyte perturbation EPR. The thin tube representation of the backbone in the molten globule state indicates sequences sampling fully unfolded conformations.

9. OTHER APPLICATIONS OF CW EPR

The exquisite sensitivity of EPR spectral lineshapes to structural and dynamical changes in spin-labeled protein that can be triggered by a biomolecular interaction event makes CW EPR an attractive bioanalytical tool for qualitative and quantitative characterization of protein/protein and protein/ligand interactions. For this application, a spin label is introduced at a site that reports the binding event either via changes in local backbone dynamics, associated shifts in conformational equilibria, or by direct immobilization of the spin label due to close proximity to the interaction interface. The aforementioned strategy has been employed to determine the binding constant of a general anesthetic to an engineered cavity in T4 Lysozyme (López, Yang, Altenbach, & Hubbell, 2013) and to evaluate the relative affinities between spin-labeled mutants of the G_α subunit of the heterotrimeric G protein to the $G_{\beta\gamma}$ dimer (Van Eps, Oldham, Hamm, & Hubbell, 2006).

CW EPR can also be applied to monitor local protein unfolding and stability with site-specific resolution. This strategy relies on the sensitivity of the EPR spectra to changes in the amplitude of local backbone motions and in local packing density (Figs. 10 and 11). There are several studies that have reported the application of CW EPR spectroscopy to obtain thermodynamic information in spin-labeled proteins with site-specific resolution (Guo, 2003; Klug & Feix, 1998; Klug, Su, Liu, Klebba, & Feix, 1995). Additionally, a CW EPR approach can evaluate the kinetics of protein folding with site-specific resolution (Grigoryants, Veselov, & Scholes, 2000; Hubbell et al., 1996). For example, for iso-1-cytochrome c, the folding time constant determined by using a rapid mixing device combined with CW EPR detection was consistent with results from other techniques, thus validating the method for quantitative protein folding/unfolding studies (Grigoryants et al., 2000).

10. CONCLUSION AND FUTURE DIRECTIONS

Continuous-wave EPR spectroscopy is a powerful technique that can provide a wealth of structural and dynamical information in proteins with any degree of complexity under physiological conditions, from intrinsically disordered systems to well-ordered proteins that form high-molecular-weight complexes. One particular advantage of CW EPR is the ability to measure internal dynamics of proteins on the ps–ns timescale using relatively low protein amounts (50 pmol) and volumes (6 μL). Additionally, the intrinsic timescale of the CW EPR experiment enables the identification of protein regions with multiple conformations at equilibrium (lifetimes >1 μs), including partially unfolded states, uncomplicated by exchange events that sometimes lead to loss of information in the NMR timescale. Recent progress in the design of spin labels with different degrees of internal motions and of different reactivities has expanded the utility of the SDSL method and facilitated the application of the full complement of CW EPR-based techniques to measure structure, dynamics, stability, and binding in proteins that are difficult to study by other biophysical methods. Finally, immobilization of proteins onto solid supports for EPR applications along with the implementation of an EPR-based flow system should enable facile screening for detection and quantification of biomolecular interaction in soluble and membrane proteins; this application is currently under development.

REFERENCES

Altenbach, C., Cai, K., Khorana, H. G., & Hubbell, W. L. (1999). Structural features and light-dependent changes in the sequence 306–322 extending from helix VII to the palmitoylation sites in rhodopsin: A site-directed spin-labeling study. *Biochemistry*, *38*(25), 7931–7937. http://dx.doi.org/10.1021/bi9900121.

Altenbach, C., Froncisz, W., Hemker, R., Mchaourab, H., & Hubbell, W. L. (2005). Accessibility of nitroxide side chains: Absolute Heisenberg exchange rates from power saturation EPR. *Biophysical Journal*, *89*(3), 2103–2112. http://dx.doi.org/10.1529/biophysj.105.059063.

Altenbach, C., Froncisz, W., Hubbell, W. L., & Hyde, J. S. (1989). Conformation of spin-labeled melittin at membrane surfaces investigated by pulse saturation recovery and continuous wave power saturation electron paramagnetic resonance. *Biophysical Journal*, *56*, 1183–1191. http://dx.doi.org/10.1016%2FS0006-3495(89)82765-1.

Altenbach, C., Greenhalgh, D. A., Khorana, H. G., & Hubbell, W. L. (1994). A collision gradient method to determine the immersion depth of nitroxides in lipid bilayers: Application to spin-labeled mutants of bacteriorhodopsin. *Proceedings of the National Academy of Sciences of the United States of America*, *91*(5), 1667–1671. http://dx.doi.org/10.1073/pnas.91.5.1667.

Altenbach, C., Klein-Seetharaman, J., Hwa, J., Khorana, H. G., & Hubbell, W. L. (1999). Structural features and light-dependent changes in the sequence 59–75 connecting helices I and II in rhodopsin: A site-directed spin-labeling study. *Biochemistry*, *38*(25), 7945–7949. http://dx.doi.org/10.1021/bi990014l.

Altenbach, C., Kusnetzow, A. K., Ernst, O. P., Hofmann, K. P., & Hubbell, W. L. (2008). High-resolution distance mapping in rhodopsin reveals the pattern of helix movement due to activation. *Proceedings of the National Academy of Sciences of the United States of America*, *105*(21), 7439–7444. http://dx.doi.org/10.1073/pnas.0802515105.

Altenbach, C., Marti, T., Khorana, H. G., & Hubbell, W. L. (1990). Transmembrane protein structure: Spin labeling of bacteriorhodopsin mutants. *Science*, *248*, 1088–1092. http://dx.doi.org/10.1126/science.2160734.

Altenbach, C., Oh, K. J., Trabanino, R. J., Hideg, K., & Hubbell, W. L. (2001). Estimation of inter-residue distances in spin labeled proteins at physiological temperatures: Experimental strategies and practical limitations. *Biochemistry*, *40*(51), 15471–15482. http://dx.doi.org/10.1021/bi011544w.

Berliner, L., Grunwald, J., Hankovszky, H. O., & Hideg, K. (1982). A novel reversible thiol-specific spin label: Papain active site labeling and inhibition. *Analytical Biochemistry*, *119*, 450–455. http://dx.doi.org/10.1016/0003-2697(82)90612-1.

Beth, A. H., & Hustedt, E. J. (2005). Saturation transfer EPR, biomedical EPR, part B: Methodology, instrumentation, and dynamics. *Biological Magnetic, Resonance*, *24/B*, 369–407.

Boehr, D. D., Dyson, H. J., & Wright, P. E. (2006). An NMR perspective on enzyme dynamics. *Chemistry Review*, *106*(8), 3055–3079. http://dx.doi.org/10.1021/cr050312q.

Boehr, D. D., Nussinov, R., & Wright, P. E. (2009). The role of dynamic conformational ensembles in biomolecular recognition (vol. 5, pg 789, 2009). *Nature Chemical Biology*, *5*(12), 954. http://dx.doi.org/10.1038/nchembio1209-954d.

Bolen, D. W., & Rose, G. D. (2008). Structure and energetics of the hydrogen-bonded backbone in protein folding. *Annual Review of Biochemistry*, *77*, 339–362. http://dx.doi.org/10.1146/annurev.biochem.77.061306.131357. Volume publication date July 2008.

Borbat, P. P., Mchaourab, H. S., & Freed, J. H. (2002). Protein structure determination using long-distance constraints from double-quantum coherence ESR: Study of T4 lysozyme. *Journal of the American Chemical Society*, *124*(19), 5304–5314. http://dx.doi.org/10.1021/ja020040y.

Bordignon, E., Klare, J. P., Holterhues, J., Martell, S., Krasnaberski, A., Engelhard, M., et al. (2007). Analysis of light-induced conformational changes of natronomonas pharaonis sensory rhodopsin II by time resolved electron paramagnetic resonance spectroscopy. *Photochemistry and Photobiology*, *83*(2), 263–272. http://dx.doi.org/10.1562/2006-07-05-RA-960.

Bordignon, E., & Steinhoff, H.-J. (2007). Membrane protein structure and dynamics studied by site-directed spin-labeling ESR. In *ESR spectroscopy in membrane biophysics: Vol. 27.* (pp. 129–164). USA: Springer.

Bracken, C., Carr, P. A., Cavanagh, J., & Palmer, A. G., 3rd. (1999). Temperature dependence of intramolecular dynamics of the basic leucine zipper of GCN4: Implications for the entropy of association with DNA. *Journal of Molecular Biology*, *285*(5), 2133–2146. http://dx.doi.org/10.1006/jmbi.1998.2429.

Bridges, M. D., Hideg, K., & Hubbell, W. L. (2010). Resolving conformational and rotameric exchange in spin-labeled proteins using saturation recovery EPR. *Applied Magnetic Resonance*, *37*(1-4), 363. http://dx.doi.org/10.1007/s00723-009-0079-2.

Brown, L. J., Sale, K. L., Hills, R., Rouviere, C., Song, L., Zhang, X., et al. (2002). Structure of the inhibitory region of troponin by site directed spin labeling electron paramagnetic resonance. *Proceedings of the National Academy of Sciences of the United States of America*, *99*(20), 12765–12770. http://dx.doi.org/10.1073/pnas.202477399.

Budil, D. E., Lee, S., Saxena, S., & Freed, J. H. (1996). Nonlinear-least-squares analysis of slow-motion EPR spectra in one and two dimensions using a modified Levenberg–Marquardt algorithm. *Journal of Magnetic Resonance Series A*, *120*(2), 155–189. http://dx.doi.org/10.1006/jmra.1996.0113.

Cafiso, D. S. (2014). Identifying and quantitating conformational exchange in membrane proteins using site-directed spin labeling. *Accounts of Chemical Research*, *47*(10), 3102–3109. http://dx.doi.org/10.1021/ar500228s.

Chui, A. J., López, C. J., Brooks, E. K., Chua, K. C., Doupey, T. G., Foltz, G. N., et al. (2015). Multiple structural states exist throughout the helical nucleation sequence of the intrinsically disordered protein stathmin, as reported by EPR spectroscopy. *Biochemistry*, *54*, 1717–1728. http://dx.doi.org/10.1021/bi500894q.

Closs, G. L., Forbes, M. D. E., & Piotrowiak, P. (1992). Spin and reaction dynamics in flexible polymethylene biradicals as studied by EPR, NMR, optical spectroscopy, and magnetic field effects Measurements and mechanisms of scalar electron spin-spin coupling. *Journal of the American Chemical Society*, *114*(9), 3285–3294. http://dx.doi.org/10.1021/ja00035a020.

Columbus, L., & Hubbell, W. L. (2002). A new spin on protein dynamics. *Trends in Biochemical Sciences*, *27*(6), 288–295. http://dx.doi.org/10.1016/S0968-0004(02)02095-9.

Columbus, L., & Hubbell, W. L. (2004). Mapping backbone dynamics in solution with site-directed spin labeling: GCN4-58 bZip free and bound to DNA. *Biochemistry*, *43*(23), 7273–7287. http://dx.doi.org/10.1021/bi0497906.

Columbus, L., Kalai, T., Jeko, J., Hideg, K., & Hubbell, W. L. (2001). Molecular motion of spin labeled side chains in alpha-helices: Analysis by variation of side chain structure. *Biochemistry*, *40*(13), 3828–3846. http://dx.doi.org/10.1021/bi002645h.

Cooke, J. A., & Brown, L. J. (2011). Distance measurements by continuous wave EPR spectroscopy to monitor protein folding. *Methods in Molecular Biology*, *752*, 73–96. http://dx.doi.org/10.1007/978-1-60327-223-0_6.

Cunningham, T. F., McGoff, M. S., Sengupta, I., Jaroniec, C. P., Horne, W. S., & Saxena, S. (2012). High-resolution structure of a protein spin-label in a solvent-exposed beta-sheet and comparison with DEER spectroscopy. *Biochemistry*, *51*(32), 6350–6359. http://dx.doi.org/10.1021/bi300328w.

Fanucci, G. E., Lee, J. Y., & Cafiso, D. S. (2003). Spectroscopic evidence that osmolytes used in crystallization buffers inhibit a conformation change in a membrane protein. *Biochemistry*, *42*(45), 13106–13112. http://dx.doi.org/10.1021/bi035439t.

Farrens, D. L., Altenbach, C., Yang, K., Hubbell, W. L., & Khorana, H. G. (1996). Requirement of rigid-body motion of transmembrane helices for light activation of rhodopsin. *Science*, *274*(5288), 768–770. http://dx.doi.org/10.1126/science.274.5288.768.

Fawzi, N. L., Fleissner, M. R., Anthis, N. J., Kalai, T., Hideg, K., Hubbell, W. L., et al. (2011). A rigid disulfide-linked nitroxide side chain simplifies the quantitative analysis of PRE data. *Journal of Biomolecular NMR, 51*(1-2), 105–114. http://dx.doi.org/10.1007/s10858-011-9545-x.

Fiori, W. R., Miick, S. M., & Millhauser, G. L. (1993). Increasing sequence length favors alpha-helix over 310-helix in alanine-based peptides: Evidence for a length-dependent structural transition. *Biochemistry, 32*(45), 11957–11962. http://dx.doi.org/10.1021/bi00096a003.

Fleissner, M. R. (2007). *X-ray structures of nitroxide side chains in proteins: A basis for interpreting distance measurements and dynamic studies by electron paramagnetic resonance.* Thesis, Los Angeles: University of California.

Fleissner, M. R., Bridges, M. D., Brooks, E. K., Cascio, D., Kalai, T., Hideg, K., et al. (2011). Structure and dynamics of a conformationally constrained nitroxide side chain and applications in EPR spectroscopy. *Proceedings of the National Academy of Sciences of the United States of America, 108*(39), 16241–16246. http://dx.doi.org/10.1073/pnas.1111420108.

Fleissner, M. R., Cascio, D., & Hubbell, W. L. (2009). Structural origin of weakly ordered nitroxide motion in spin-labeled proteins. *Protein Science, 18*(5), 893–908. http://dx.doi.org/10.1002/pro.96.

Flores Jimenez, R. H., Do Cao, M. A., Kim, M., & Cafiso, D. S. (2010). Osmolytes modulate conformational exchange in solvent-exposed regions of membrane proteins. *Protein Science, 19*(2), 269–278. http://dx.doi.org/10.1002/pro.305.

Freed, J. H. (1976). Theory of slow tumbling ESR spectra for nitroxides. In L. J. Berliner (Ed.), *Spin labeling theory and applications* (pp. 53–132). New York: Academic Press.

Freed, D. M., Horanyi, P. S., Wiener, M. C., & Cafiso, D. S. (2010). Conformational exchange in a membrane transport protein is altered in protein crystals. *Biophysical Journal, 99*(5), 1604–1610. http://dx.doi.org/10.1016/j.bpj.2010.06.026.

Freed, D. M., Khan, A. K., Horanyi, P. S., & Cafiso, D. S. (2011). Molecular origin of electron paramagnetic resonance line shapes on beta-barrel membrane proteins: The local solvation environment modulates spin-label configuration. *Biochemistry, 50*(41), 8792–8803. http://dx.doi.org/10.1021/bi200971x.

Ghimire, H., Abu-Baker, S., Sahu, I. D., Zhou, A., Mayo, D. J., Lee, R. T., et al. (2012). Probing the helical tilt and dynamic properties of membrane-bound phospholamban in magnetically aligned bicelles using electron paramagnetic resonance spectroscopy. *Biochimica et Biophysica Acta, 1818*, 645–650. http://dx.doi.org/10.1016/j.bbamem.2011.11.030.

Ghimire, H., Hustedt, E. J., Sahu, I. D., Inbaraj, J. J., McCarrick, R., Mayo, D. J., et al. (2012). Distance measurements on a dual-labeled TOAC AChR M2δ peptide in mechanically aligned DMPC bilayers via dipolar broadening CW-EPR spectroscopy. *The Journal of Physical Chemistry B, 116*, 3866–3873. http://dx.doi.org/10.1021/jp212272d.

Griffith, O. H., & Jost, P. C. (1976). Lipid spin labels in biological membranes. In L. J. Berliner (Ed.), *Spin labeling theory and applications* (pp. 454–523). New York: Academic Press.

Grigoryants, V. M., Veselov, A. V., & Scholes, C. P. (2000). Variable velocity liquid flow EPR applied to submillisecond protein folding. *Biophysical Journal, 78*(5), 2702–2708. http://dx.doi.org/10.1016/s0006-3495(00)76814-7.

Guo, Z. (2003). *Correlation of spin label side-chain dynamics with protein structure studies of T4 lysozyme with site-directed mutagenesis and x-ray crystallography.* Thesis, Los Angeles: University of California.

Guo, Z., Cascio, D., Hideg, K., & Hubbell, W. L. (2008). Structural determinants of nitroxide motion in spin-labeled proteins: Solvent-exposed sites in helix B of T4 lysozyme. *Protein Science, 17*(2), 228–239. http://dx.doi.org/10.1110/ps.073174008.

Guo, Z., Cascio, D., Hideg, K., Kálái, T., & Hubbell, W. L. (2007). Structural determinants of nitroxide motion in spin-labeled proteins: Tertiary contact and solvent-inaccessible sites in helix g of T4 lysozyme. *Protein Science*, *16*, 1069–1086. http://dx.doi.org/10.1110/ps.062739107.

Henzler-Wildman, K., & Kern, D. (2007). Dynamic personalities of proteins. *Nature*, *450*(7172), 964–972. http://dx.doi.org/10.1038/nature06522.

Henzler-Wildman, K. A., Lei, M., Thai, V., Kerns, S. J., Karplus, M., & Kern, D. (2007). A hierarchy of timescales in protein dynamics is linked to enzyme catalysis. *Nature*, *450*(7171), 913–916. http://www.nature.com/nature/journal/v450/n7171/suppinfo/nature06407_S1.htm.

Hubbell, W. L., & Altenbach, C. (1994). Investigation of structure and dynamics in membrane proteins using site-directed spin labeling. *Current Opinion in Structural Biology*, *4*, 566–573. http://dx.doi.org/10.1016/S0959-440X(94)90219-4.

Hubbell, W. L., Altenbach, C., Hubbell, C. M., & Khorana, H. G. (2003). Rhodopsin structure, dynamics, and activation: A perspective from crystallography, site-directed spin labeling, sulfhydryl reactivity, and disulfide cross-linking. *Advances in Protein Chemistry*, *63*, 243–290. http://dx.doi.org/10.1016/S0065-3233(03)63010-X.

Hubbell, W. L., Cafiso, D. S., & Altenbach, C. (2000). Identifying conformational changes with site-directed spin labeling. *Nature Structural Biology*, *7*(9), 735–739. http://dx.doi.org/10.1038/78956.

Hubbell, W. L., Froncisz, W., & Hyde, J. S. (1987). Continuous and stopped flow EPR spectrometer based on a loop gap resonator. *The Review of Scientific Instruments*, *58*, 1879. http://dx.doi.org/10.1063/1.1139536.

Hubbell, W. L., Mchaourab, H. S., Altenbach, C., & Lietzow, M. A. (1996). Watching proteins move using site-directed spin labeling. *Structure*, *4*(7), 779–783. http://dx.doi.org/10.1016/S0969-2126(96)00085-8.

Hustedt, E. J., & Beth, A. H. (1999). Nitroxide spin-spin interactions: Applications to protein structure and dynamics. *Annual Review of Biophysics and Biomolecular Structure*, *28*, 129–153. http://dx.doi.org/10.1146/annurev.biophys.28.1.129.

Hyde, J. S., & Froncisz, W. (1989). Loop gap resonators. In A. J. Hoff (Ed.), *Advanced EPR: Applications in biology and medicine* (pp. 277–305). Amsterdam: Elsevier.

Hyde, J. S., & Thomas, D. D. (1980). Saturation-transfer spectroscopy. *Annual Review of Physical Chemistry*, *31*, 293–317. http://dx.doi.org/10.1146/annurev.pc.31.100180.001453.

Isas, J. M., Langen, R., Haigler, H. T., & Hubbell, W. L. (2002). Structure and dynamics of a helical hairpin and loop region in annexin 12: A site-directed spin labeling study. *Biochemistry*, *41*(5), 1464–1473. http://dx.doi.org/10.1021/bi011856z.

Jeschke, G., & Polyhach, Y. (2007). Distance measurements on spin-labelled biomacromolecules by pulsed electron paramagnetic resonance. *Physical Chemistry Chemical Physics*, *9*(16), 1895–1910. http://dx.doi.org/10.1039/b614920k.

Kim, C. S., Kweon, D. H., & Shin, Y. K. (2002). Membrane topologies of neuronal SNARE folding intermediates. *Biochemistry*, *41*(36), 10928–10933. http://dx.doi.org/10.1021/bi026266v.

Klare, J. P., & Steinhoff, H.-J. (2009). Spin labeling EPR. *Photosynthesis Research*, *102*, 377–390. http://dx.doi.org/10.1007/s11120-009-9490-7.

Klug, C. S., & Feix, J. B. (1998). Guanidine hydrochloride unfolding of a transmembrane β-strand in FepA using site-directed spin labeling. *Protein Science*, *7*(6), 1469–1476. http://dx.doi.org/10.1002/pro.5560070624.

Klug, C. S., Su, W., Liu, J., Klebba, P. E., & Feix, J. B. (1995). Denaturant unfolding of the ferric enterobactin receptor and ligand-induced stabilization studied by site-directed spin labeling. *Biochemistry*, *34*(43), 14230–14236. http://dx.doi.org/10.1021/bi00043a030.

Kroncke, B. M., Horanyi, P. S., & Columbus, L. (2010). Structural origins of nitroxide side chain dynamics on membrane protein alpha-helical sites. *Biochemistry*, *49*(47), 10045–10060. http://dx.doi.org/10.1021/bi101148w.

Lerch, M., Yang, Z., Altenbach, C., & Hubbell, W. L. (2015). High-pressure EPR and site-directed spin labeling for mapping molecular flexibility in proteins. *Methods in Enzymology, 563*.

Lietzow, M. A., & Hubbell, W. L. (2004). Motion of spin label side chains in cellular retinol-binding protein: Correlation with structure and nearest-neighbor interactions in an antiparallel beta-sheet. *Biochemistry, 43*(11), 3137–3151. http://dx.doi.org/10.1021/bi0360962.

Lin, Y., Nielsen, R., Murray, D., Hubbell, W. L., Mailer, C., Robinson, B. H., et al. (1998). Docking phospholipase A2 on membranes using electrostatic potential-modulated spin relaxation magnetic resonance. *Science, 279*, 1925–1929. http://dx.doi.org/10.1126/science.279.5358.1925.

Liu, Y. S., Sompornpisut, P., & Perozo, E. (2001). Structure of the KcsA channel intracellular gate in the open state. *Nature Structural Biology, 8*(10), 883–887. http://dx.doi.org/10.1038/nsb1001-883.

Lo, R. H., Kroncke, B. M., Solomon, T. L., & Columbus, L. (2014). Mapping membrane protein backbone dynamics: A comparison of site-directed spin labeling with NMR 15 N-relaxation measurements. *Biophysical Journal, 107*(7), 1697–1702. http://dx.doi.org/10.1016/j.bpj.2014.08.018.

López, C. J., Fleissner, M. R., Brooks, E. K., & Hubbell, W. L. (2014). Stationary-phase EPR for exploring protein structure, conformation, and dynamics in spin-labeled proteins. *Biochemistry, 53*(45), 7067–7075. http://dx.doi.org/10.1021/bi5011128.

López, C. J., Fleissner, M. R., Guo, Z., Kusnetzow, A. K., & Hubbell, W. L. (2009). Osmolyte perturbation reveals conformational equilibria in spin-labeled proteins. *Protein Science, 18*(8), 1637–1652. http://dx.doi.org/10.1002/pro.180.

López, C. J., Oga, S., & Hubbell, W. L. (2012). Mapping molecular flexibility of proteins with site-directed spin labeling: A case study of myoglobin. *Biochemistry, 51*(33), 6568–6583. http://dx.doi.org/10.1021/bi3005686.

López, C. J., Yang, Z., Altenbach, C., & Hubbell, W. L. (2013). Conformational selection and adaptation to ligand binding in T4 lysozyme cavity mutants. *Proceedings of the National Academy of Sciences of the United States of America, 110*(46), E4306–E4315. http://dx.doi.org/10.1073/pnas.1318754110.

Marsh, D. (2014). EPR moments for site-directed spin-labelling. *Journal of Magnetic Resonance, 248*, 60–66. http://dx.doi.org/10.1016/j.jmr.2014.09.006.

Mchaourab, H. S., Kalai, T., Hideg, K., & Hubbell, W. L. (1999). Motion of spin-labeled side chains in T4 lysozyme: Effect of side chain structure. *Biochemistry, 38*(10), 2947–2955. http://dx.doi.org/10.1021/bi9826310.

Mchaourab, H. S., Lietzow, M. A., Hideg, K., & Hubbell, W. L. (1996). Motion of spin-labeled side chains in T4 lysozyme. Correlation with protein structure and dynamics. *Biochemistry, 35*(24), 7692–7704. http://dx.doi.org/10.1021/bi960482k.

Mchaourab, H. S., Oh, K. J., Fang, C. J., & Hubbell, W. L. (1997). Conformation of T4 lysozyme in solution. Hinge-bending motion and the substrate-induced conformational transition studied by site-directed spin labeling. *Biochemistry, 36*(2), 307–316. http://dx.doi.org/10.1021/Bi962114m.

Mittermaier, A., & Kay, L. E. (2006). New tools provide new insights in NMR studies of protein dynamics. *Science, 312*(5771), 224–228. http://dx.doi.org/10.1126/science.1124964.

Nelson, W. D., Blakely, S. E., Nesmelov, Y. E., & Thomas, D. D. (2005). Site-directed spin labeling reveals a conformational switch in the phosphorylation domain of smooth muscle myosin. *Proceedings of the National Academy of Sciences of the United States of America, 102*(11), 4000–4005. http://dx.doi.org/10.1073/pnas.0401664102.

Oh, K. J., Altenbach, C., Collier, R. J., & Hubbell, W. L. (2000). Site-directed spin labeling of proteins, in bacterial toxins:Methods and protocols. *Methods in Molecular Biology, 145*, 147–169. http://dx.doi.org/10.1385/1-59259-052-7:147. New Jersey: Humana Press.

Oh, K. J., Zhan, H., Cui, C., Hideg, K., Collier, R. J., & Hubbell, W. L. (1996). Organization of diphtheria toxin T domain in bilayers: A site-directed spin labeling study. *Science, 273*(5276), 810–812. http://dx.doi.org/10.1126/science.273.5276.810.

Pake, G. E. (1948). Nuclear resonance absorption in hydrated crystals: Fine structure of the proton line. *The Journal of Chemical Physics, 16*, 327–336. http://dx.doi.org/10.1063/1.1746878.

Perozo, E., Cortes, D. M., & Cuello, L. G. (1998). Three-dimensional architecture and gating mechanism of a K+ channel studied by EPR spectroscopy. *Nature Structural Biology, 5*(6), 459–469. http://dx.doi.org/10.1038/nsb0698-459.

Phillips, G. N. (1990). Comparison of the dynamics of myoglobin in different crystal forms. *Biophysical Journal, 57*, 381–383. http://dx.doi.org/10.1016%2FS0006-3495(90)82540-6.

Pratt, V. (1987). Direct least-squares fitting of algebraic surfaces. In *SIGGRAPH '87 Proceedings of the 14th Annual Conference on Computer graphics and Interactive Techniques: Vol. 21* (pp. 145–152). http://dx.doi.org/10.1145/37401.37420.

Pyka, J., Ilnicki, J., Altenbach, C., Hubbell, W. L., & Froncisz, W. (2005). Accessibility and dynamics of nitroxide side chains in T4 lysozyme measured by saturation recovery EPR. *Biophysical Journal, 89*(3), 2059–2068. http://dx.doi.org/10.1529/biophysj.105.059055.

Rabenstein, M. D., & Shin, Y. K. (1995). Determination of the distance between two spin labels attached to a macromolecule. *Proceedings of the National Academy of Sciences, 92*(18), 8239–8243.

Sale, K., Sar, C., Sharp, K. A., Hideg, K., & Fajer, P. G. (2002). Structural determination of spin label immobilization and orientation: A monte carlo minimization approach. *Journal of Magnetic Resonance, 156*, 104–112. http://dx.doi.org/10.1006/jmre.2002.2529.

Salwinski, L., & Hubbell, W. L. (1999). Structure in the channel forming domain of colicin E1 bound to membranes: The 402-424 sequence. *Protein Science, 8*(3), 562–572. http://dx.doi.org/10.1110/ps.8.3.562.

Saxena, S., & Freed, J. H. (1997). Theory of double quantum two-dimensional electron spin resonance with application to distance measurements. *The Journal of Chemical Physics, 107*(5), 1317–1340. http://dx.doi.org/10.1063/1.474490.

Stagg, L., Zhang, S. Q., Cheung, M. S., & Wittung-Stafshede, P. (2007). Molecular crowding enhances native structure and stability of alpha/beta protein flavodoxin. *Proceedings of the National Academy of Sciences of the United States of America, 104*(48), 18976–18981. http://dx.doi.org/10.1073/pnas.0705127104. Epub 2007 Nov 16.

Toledo Warshaviak, D., Khramtsov, V. V., Cascio, D., Altenbach, C., & Hubbell, W. L. (2013). Structure and dynamics of an imidazoline nitroxide side chain with strongly hindered internal motion in proteins. *The Journal of Magnetic Resonance, 232*(0), 53–61. http://dx.doi.org/10.1016/j.jmr.2013.04.013.

Van Eps, N., Oldham, W. M., Hamm, H. E., & Hubbell, W. L. (2006). Structural and dynamical changes in an alpha-subunit of a heterotrimeric G protein along the activation pathway. *Proceedings of the National Academy of Sciences of the United States of America, 103*(44), 16194–16199. http://dx.doi.org/10.1073/pnas.0607972103.

Voet, D., & Voet, J. G. (2004). *Biochemistry* (3rd ed.). New York: John Wiley & Sons.

Warshaviak, D. T., Serbulea, L., Houk, K. N., & Hubbell, W. L. (2011). Conformational analysis of a nitroxide side chain in an alpha-helix with density functional theory. *The Journal of Physical Chemistry B, 115*(2), 397–405. http://dx.doi.org/10.1021/jp108871m.

Yang, Z., Bridges, M., Lerch, M., Altenbach, C., & Hubbell, W. L. (2016). Saturation recovery EPR and nitroxide spin labeling for exploring structure and dynamics in proteins. *Methods in Enzymology, 563*.

Yang, Z., Jimenez-Oses, G., López, C. J., Bridges, M. D., Houk, K. N., & Hubbell, W. L. (2014). Long-range distance measurements in proteins at physiological temperatures using saturation recovery EPR spectroscopy. *Journal of the American Chemical Society, 136*(43), 15356–15365. http://dx.doi.org/10.1021/ja5083206.

CHAPTER FOUR

Bifunctional Spin Labeling of Muscle Proteins: Accurate Rotational Dynamics, Orientation, and Distance by EPR

Andrew R. Thompson, Benjamin P. Binder, Jesse E. McCaffrey, Bengt Svensson, David D. Thomas[1]

Department of Biochemistry, Molecular Biology and Biophysics, University of Minnesota, Minneapolis, Minnesota, USA
[1]Corresponding author: e-mail address: DDT@ddt.umn.edu

Contents

1. Introduction	102
2. Methods of BSL Labeling	104
3. Rotational Dynamics	105
4. Orientation	108
4.1 Mechanically Aligned Systems	108
4.2 Magnetically Aligned Systems	111
5. Distance	113
5.1 Distance Measurements with DEER	113
5.2 The Problem of Orientation Selection	115
6. Discussion	116
6.1 Labeling Specificity and Protein Function	116
6.2 The BEER Technique	119
References	120

Abstract

While EPR allows for the characterization of protein structure and function due to its exquisite sensitivity to spin label dynamics, orientation, and distance, these measurements are often limited in sensitivity due to the use of labels that are attached via flexible monofunctional bonds, incurring additional disorder and nanosecond dynamics. In this chapter, we present methods for using a bifunctional spin label (BSL) to measure muscle protein structure and dynamics. We demonstrate that bifunctional attachment eliminates nanosecond internal rotation of the spin label, thereby allowing the accurate measurement of protein backbone rotational dynamics, including microsecond-to-millisecond motions by saturation transfer EPR. BSL also allows for accurate determination of helix orientation and disorder in mechanically and magnetically aligned systems, due to the label's stereospecific attachment. Similarly, labeling with a pair of BSL greatly enhances the resolution and accuracy of distance measurements measured by double electron–electron resonance (DEER). Finally, when BSL is applied to a protein with high helical content in an assembly with high orientational order (e.g., muscle fiber or membrane), two-probe DEER experiments can be combined with single-probe EPR experiments on an oriented sample in a process we call BEER, which has the potential for *ab initio* high-resolution structure determination.

1. INTRODUCTION

EPR has long been a mainstay technique in our study of muscle proteins due to its ability to make a diverse number of measurements for the characterization of protein structure and function including solvent accessibility (Surek & Thomas, 2008), nanosecond to millisecond dynamics (James, McCaffrey, Torgersen, Karim, & Thomas, 2012; Karim, Zhang, Howard, Torgersen, & Thomas, 2006; Thompson, Naber, Wilson, Cooke, & Thomas, 2008), orientation and disorder (Mello & Thomas, 2012; Moen, Thomas, & Klein, 2013), and distance measurement from 0.5 to >6 nm (Agafonov et al., 2009; Lin, Prochniewicz, James, Svensson, & Thomas, 2011; Moen, Klein, & Thomas, 2014). While complementary measurements are possible using fluorescence methods (Agafonov et al., 2009), spin labels offer distinct advantages over fluorescence probes due to their stability during measurement (i.e., they do not "bleach"), and their small size, typically on the order of a large amino acid side chain. In the measurement of distances, spin labels offer an additional advantage over fluorescent labels as they do not require distinct donor and acceptor probes.

EPR measurements have typically involved the use of monofunctionally attached labels targeted to cysteine residues, present either natively in the protein structure or introduced via protein mutation, a process commonly referred to as site-directed spin labeling (SDSL) (Altenbach, Marti, Khorana, & Hubbell, 1990). Due to the innate flexibility of such monofunctional attachments, the spectrum is influenced by both label and protein conformations, such that the sensitivity of EPR to changes in protein structure and function may be limited or masked entirely by spin label mobility (Fig. 1, top). While careful choice of spin label variety and labeling site may allow for restriction of spin label mobility, neither is a panacea for high-resolution measurement of dynamics, orientation, and distance.

Bifunctional spin labels, especially the label 3,4-bis-(methanethiosulfonylmethyl)-2,2,5,5-tetramethyl-2,5-dihydro-1H-pyrrol-1-

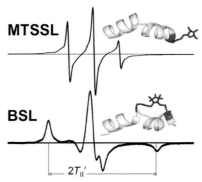

Figure 1 Conventional EPR spectra of the monofunctional spin label MTSSL versus the bifunctional spin label BSL attached to the myosin motor protein at positions 707 or 697/707, respectively. *Data from Thompson et al. (2008).*

yloxy spin label (HO-1944 or BSL) (Kalai, Balog, Jeko, & Hideg, 1999), eliminate these problems by addition of a second sulfhydryl linker, rigidly tethering the label to the protein backbone, thus eliminating the nanosecond rotational dynamics found in most monofunctionally attached labels (Fig. 1, bottom). When attached to a helix with cysteine residues engineered at i and $i + 4$ (or in some cases i and $i + 3$), BSL attains a stereospecifically defined orientation with respect to the helix axis (Fleissner et al., 2011), allowing for precise orientation measurements in oriented systems. The conformational restriction afforded by the second cysteine linker also allows for more precise measurement of distance distributions in dipolar EPR experiments, eliminating contributions from spin label flexibility in the measured distance distributions.

In this chapter, we demonstrate the power of BSL over traditional monofunctionally attached labels in the context of the characterization of muscle protein structural dynamics. We illustrate how the reduction of spin label mobility allows for measurement of microsecond-to-millisecond protein rotational dynamics, as explored by saturation transfer EPR (STEPR) studies of the myosin motor domain. We examine orientation measurements on several helices in the motor domain of myosin in mechanically aligned skinned muscle fibers, as well as on the transmembrane helix of phospholamban (PLB) in magnetically aligned lipid bicelles. In both cases, the precise coupling of BSL to the protein backbone allows for accurate measurement of helix orientation with respect to the symmetry axis of the ordered system, with resolution on the order of 1 degree. Finally, we show how distance measurements in myosin are further refined by BSL, and how such measurements can be used to constrain the computation of helix orientation and provide information about the relative motion of domains upon the addition of nucleotide, a process we call bifunctional electron–electron resonance (BEER).

2. METHODS OF BSL LABELING

With appropriate choices for labeling site and labeling conditions, attaching BSL to a protein is just as straightforward as attaching a conventional monofunctional spin label. BSL should be attached to cysteines on a helix at residues i and $i+4$ or i and $i+3$, or on a β-sheet at residues i and $i+2$ (Fleissner et al., 2011; Islam & Roux, 2015). The selection of specific residues for labeling is highly specific to the protein and its environment, and is beyond the scope of this chapter. In general, though, one should choose residues that avoid major steric hindrances by side chains and tertiary structure to ensure optimal labeling efficiency and minimize disruption of protein structure and function. When selecting helix labeling sites, one should aim to select residues at least a half turn away from the helix ends to limit the possibility that the addition of the nitroxide side chain will disrupt the native secondary structure. Additionally, labeling a helix at i and $i+4$ has been shown to be less disruptive of the native helical structure than at i and $i+3$ (Islam & Roux, 2015).

The general labeling procedure is as follows:

1. The target cysteines for labeling are reduced via the addition of a reducing agent such as dithiothreitol or tris(2-carboxyethyl)phosphine. This solution is kept stirring on ice for 1 h.
2. The reducing agent is removed by dialysis and/or by two sequential size-exclusion spin columns (preferred) such as a Pierce Zeba Spin desalting columns (Thermo Scientific).
3. The protein is labeled by the addition of 2–10 equivalents (spin labels/bifunctional labeling sites) of BSL (3,4-bis-(methanethiosulfonylmethyl)-2,2,5,5-tetramethyl-2,5-dihydro-1H-pyrrol-1-yloxy spin label, Toronto Research Chemicals) for 1–3 h on ice.
4. Excess unbound spin label is rapidly removed with two sequential spin columns. In this step, either centrifugal filters or spin columns may be used, but dialysis is not recommended because long equilibration times may contribute to a significant loss in labeling efficiency.
5. Labeled protein is either used immediately or stored in organic solvent such as trifluoroethanol at −20 °C or below to prevent the slow

accumulation of BSL released from the protein through disulfide exchange (see Section 6; Fava, Iliceto, & Camera, 1957).
6. Spin labeling efficiency is quantified by spin counting of labeled protein dissolved in pH neutral buffer or reconstituted in lipid or detergent, using a known standard such as TEMPO or TEMPOL (Eaton, Eaton, Barr, & Weber, 2010). Specific reconstitution conditions such as buffer configuration and spin label regeneration can be found in the literature, including Arata et al. (2003), Mello and Thomas (2012), Moen et al. (2013), and Thompson et al. (2008).

3. ROTATIONAL DYNAMICS

EPR is sensitive to rotational motion in the picosecond-to-millisecond range (Fig. 2). Despite their small size in comparison to fluorescent labels, spin labels can possess significant internal dynamics, which often manifest on the nanosecond timescale (Columbus, Kalai, Jeko, Hideg, & Hubbell, 2001). These motions are essentially indistinguishable from internal protein dynamics (detected by conventional EPR) (Goldman, Bruno, & Freed, 1972) and cause fast spectral diffusion which destroys the microsecond sensitivity of saturation transfer EPR (Squier & Thomas, 1986). Therefore, a rigidly bound label is essential for both fast and slow dynamics measurements.

Fortuitous placement of a monofunctional spin label on a protein can result in immobilization due to local steric and electrostatic interactions

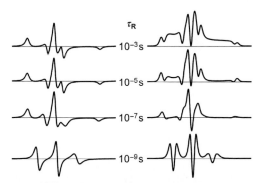

Figure 2 Sensitivity of conventional (left) and saturation transfer (right) EPR to isotropic rotational diffusion, shown by spectral simulations (Thomas & McConnell, 1974). τ_R is the rotational correlation time. Conventional EPR lacks sensitivity to dynamics for $\tau_R > 10^{-7}$ s, necessitating the use of saturation transfer EPR for these measurements.

(James et al., 2012; Thompson et al., 2008), but this is usually not the case. Often, labeling sites are at the far exterior of a protein, in order to avoid perturbations in internal structure or oligomeric interactions, and hence exert few local constraints on label dynamics. To reduce this ambiguity, it is desirable to use a spin label that makes two bonds with the protein. The synthetic spin-labeled amino acid 2,2,6,6-tetramethyl-piperidine-1-oxyl-4-amino-4-carboxyl (TOAC) spin label incorporates directly into the protein backbone via amide bonds, with two carbon atoms in the nitroxide-containing piperidine ring bonded directly to the α-carbon, and thus directly reports backbone dynamics (Karim, Kirby, Zhang, Nesmelov, & Thomas, 2004; Karim et al., 2006; Karim, Zhang, & Thomas, 2007). While compact and immobilized, TOAC requires incorporation via peptide synthesis and thus is not compatible with proteins with molecular weights above about 6 kDa.

BSL is an alternative to TOAC that also offers rigid attachment relative to the peptide backbone, but is compatible with SDSL by cysteine mutagenesis (Kalai et al., 1999). By reacting with two Cys residues, the probe's internal dynamics are restricted, resulting in the immobilization necessary to measure both fast and slow protein dynamics by conventional and saturation transfer EPR. Figure 3 shows a comparison between the monofunctional methanethiosulfonate spin label (MTSSL) and BSL attached to myosin at two different sites. At both sites, MTSSL demonstrates significant nanosecond rotational mobility as evidenced by narrow splittings in the spectrum. In contrast, BSL is significantly more immobilized, as indicated by the near rigid-limit splitting value. Figure 4 shows the conventional EPR spectra

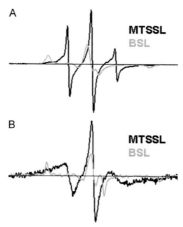

Figure 3 Conventional EPR spectra of MTSSL and BSL attached to myosin at several different locations. (A) 697 or 697/707, respectively. (B) 492 or 492/496, respectively.

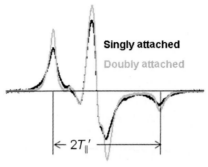

Figure 4 Singly and doubly attached BSL on the transmembrane helix of PLB at positions 32 and 32/36, respectively. Singly attached BSL ($2T_{\|}'=61.7$ G) is only slightly less immobilized than doubly attached BSL ($2T_{\|}'=62.2$ G).

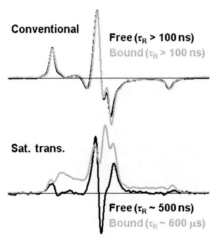

Figure 5 Conventional and saturation transfer EPR spectra of BSL-labeled myosin free in solution and bound to actin filaments. Rotational correlation times are determined from L''/L. Data adapted from Thompson et al. (2008).

of singly and doubly attached BSL on PLBs transmembrane helix. The outer splitting of singly attached BSL is slightly narrower, and the linewidth is slightly greater, indicating that it is only slightly more mobile than doubly attached BSL. This is consistent with previous studies that evaluated the effect of adding a bulky side chain to a monofunctional spin labels (Columbus et al., 2001).

Figure 5 shows the conventional and saturation transfer EPR spectra of BSL–myosin, both free in solution and bound to actin. On the conventional

EPR timescale (picoseconds to nanoseconds), the EPR spectra are quite similar and possess the characteristic powder lineshape, indicating strong label immobilization. In contrast, the saturation transfer EPR spectra are quite different, revealing a large decrease in myosin rotational mobility upon binding to actin, corresponding to an increase in rotational correlation time (τ_R) from 500 ns to 600 μs. Had BSL not been immobilized on myosin as shown in Fig. 5, the protein's microsecond global dynamics would not be detected, as the label's nanosecond dynamics would dominate the saturation transfer measurement (Wilcox, Parce, Thomas, & Lyles, 1990).

4. ORIENTATION
4.1 Mechanically Aligned Systems

The attachment of BSL is not only rigid on the conventional EPR time scale; it is also stereospecific with respect to the peptide backbone, permitting direct measurement of protein backbone orientation in a well-ordered, anisotropic system. The study of sarcomeric proteins stands to benefit greatly from this feature. Actin filaments within a muscle fiber can be oriented by positioning a fiber bundle within the cavity, thus setting up a biological scaffold upon which actin-binding proteins of interest may also be oriented (Fig. 6). Myosin is an excellent candidate for validating and exploiting this method, due to the high-binding affinity of the actomyosin complex and the abundance of mechanochemical coupling within its catalytic domain (CD) (Spudich, 2014; Thomas, Prochniewicz, & Roopnarine, 2002).

Figure 7C shows EPR spectra of myosin labeled with either MTSSL or BSL at equivalent sites on the C-terminal end of the relay helix. Skinned muscle fiber bundles were incubated with the spin-labeled protein and oriented with the actin symmetry axis either parallel or perpendicular to the applied magnetic field. Spectra of oriented MTSSL–myosin show poor orientational resolution, evidenced by virtually indistinguishable parallel and perpendicular lineshapes, while spectra of BSL show great disparity between fiber orientations, indicating a highly ordered spin label ensemble. When these decorated fiber bundles are minced, removing all spectral contributions from orientation (Fig. 7C), the MTSSL ensemble also reveals significant motional narrowing, while the BSL spectrum exhibits a characteristic powder lineshape, indicating negligible nanosecond motion.

The stereospecific attachment of BSL greatly simplifies the analysis of the resulting spectra and enables the derivation of label angular distributions. For the actomyosin data discussed above, the magnetic, hyperfine, and linewidth

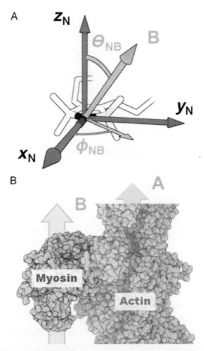

Figure 6 (A) Reference frame that defines the orientation dependence of nitroxide EPR (Earle & Budil, 2006). In the nitroxide frame, the applied magnetic field vector *B* forms the angles θ_{NB} and θ_{NB} relative to the nitroxide frame. (B) In oriented actomyosin samples, the actin filament forms the assembly axis vector *A*, which is usually aligned with *B*.

tensors can be determined from spectra of minced fibers, where there are no contributions from orientation. These values can then be fixed and applied to spectral simulations of the oriented samples, drastically reducing the parameter space of the subsequent fitting operation. With tensors thus predefined, the shape of the spectrum is entirely dependent on the ensemble spin label orientation, defined by θ_{NB} (axial orientation of the magnetic field in the nitroxide frame), ϕ_{NB} (azimuthal orientation in the nitroxide frame), and $\Delta\theta,\phi$ (width of the angular distribution) (Fig. 6A). The extremely high-anisotropic sensitivity of EPR grants exceptional resolution to these parameters in the spectrum, such that localized changes in the orientation of individual structural elements are easily resolved; an example of this is given in Fig. 7E, where a nucleotide-induced structural change is detected on myosin's relay helix.

In samples with the symmetry axis parallel to the magnetic field, these angular parameters also describe label orientation relative to the symmetry axis (actin, in this case), and therefore $\theta_{NB}=\theta_{NA}$ and $\phi_{NB}=\phi_{NA}$ in Fig. 6B.

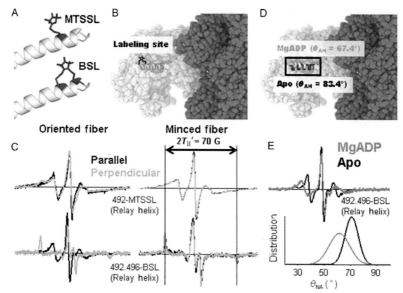

Figure 7 (A) MTSSL and BSL attached to the myosin relay helix. (B) Myosin bound to actin within an oriented muscle fiber. (C) Conventional EPR spectra of actomyosin complexes labeled with MTSSL and BSL within oriented muscle fibers demonstrate orientation sensitivity. Minced (randomly oriented) fiber samples show nanosecond dynamics of MTSSL but not BSL. (D and E) Addition of ADP reveals a previously undetected rotation of the relay helix within actin-bound myosin. (See the color plate.)

From there, powerful structural constraints for the system can be developed by taking advantage of the stereospecific nature of BSL's bifunctional attachment. For example, deriving the axial tilt of a myosin helix relative to actin requires two vectors in a common coordinate frame, one representing the actin vector, and the other representing the helix (Fig. 8). The actin vector can be accurately defined using the EPR-derived angle parameters (Fig. 8B), and the helix vector can be determined relative to the nitroxide frame (Fig. 8D) by geometric analysis of available crystal structures containing BSL (Fleissner et al., 2011). The helix tilt angle θ_{AH} is subsequently derived by finding the angle between these two vectors, thus generating a high-resolution structural constraint independent of the spin label itself (Fig. 8F).

Measurements obtained in this way for several sites across the myosin CD are in agreement with previous results from cryoelectron microscopy, in the absence of nucleotide (Holmes, Angert, Kull, Jahn, & Schroder, 2003). The real power of these derivations, though, lies in the ability to obtain constraints under a variety of biochemical conditions, potentially on the same protein sample. Figure 7D depicts the result of modeling the nucleotide-

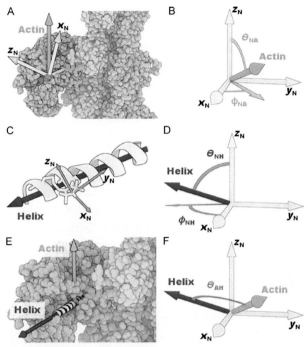

Figure 8 (A–D) Visualization of coordinate transformations used in the analysis of oriented muscle fiber EPR. Projection of the actin and helix vectors on the nitroxide frame yield the relative angle θ_{AH}, the axial tilt of a myosin helix relative to actin (E and F). (See the color plate.)

induced structural change on the relay helix, revealing a bending deformation that is similar to crystallographic results from myosin alone (Fisher et al., 1995; Smith & Rayment, 1996), but not previously observed for the actomyosin complex.

4.2 Magnetically Aligned Systems

In addition to sarcomeric proteins such as myosin, BSL has great potential for structural elucidation of membrane proteins through orientation measurements. Membrane-bound proteins often exhibit nanosecond backbone dynamics due to their fluid environment, resulting in additional spectral effects detected by conventional EPR that complicate the measurement of orientation. While this motion can be resolved through spectral simulation, rigid label attachment is necessary to decouple label and protein dynamics, allowing determination of orientation. Indeed, a rigidly bound label such as BSL is absolutely essential for this type of measurement.

To start, BSL-labeled membrane proteins must be reconstituted in an anisotropic environment to avoid orientational averaging. This can be accomplished through a mechanical bilayer such as a substrate-supported bilayer (Inbaraj, Laryukhin, & Lorigan, 2007), or a magnetically aligned bilayer such as bicelles (Cardon, Tiburu, & Lorigan, 2003). Bicelles offer a well-hydrated, homogeneous environment quite analogous to native vesicles, and isolated sarcoplasmic reticulum. They have seen extensive use in nuclear magnetic resonance (NMR) studies due to the high magnetic field available (De Angelis & Opella, 2007; Durr, Soong, & Ramamoorthy, 2013), but with lipid optimization and the addition of lanthanides, bicelles have seen increasing use in EPR experiments performed under physiological conditions and temperatures (Caporini, Padmanabhan, Cardon, & Lorigan, 2003; Cho, Dominick, & Spence, 2010; Garber, Lorigan, & Howard, 1999; Lu, Caporini, & Lorigan, 2004; McCaffrey, James, & Thomas, 2015).

Figure 9 shows EPR spectra of PLB with BSL at positions 32/36 in aligned bicelles. The substantial differences in the spectra shown in Fig. 9A indicate a narrow, well-defined orientation distribution. Fitting by spectral simulation determines a label tilt angle of $90° \pm 3°$. Figure 9B shows complementary molecular dynamics simulations of BSL on PLB's transmembrane helix, yielding an angle of 89° between the probe principal

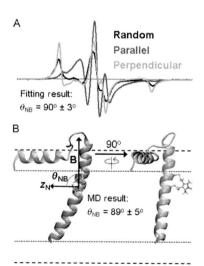

Figure 9 (A) EPR spectra of BSL attached at positions 32 and 36 on the transmembrane helix of PLB. Spectral analysis of this data yields a label tilt angle of $90° \pm 3°$. (B) Molecular dynamics simulation of BSL on PLB finds a label tilt angle of $89° \pm 5°$, consistent with the experimental result from (A).

axis and the membrane normal. These results are consistent with the analysis of Fig. 9A, as well as solid state NMR measurements of PLB (Traaseth, Buffy, Zamoon, & Veglia, 2006) along with the crystal structure of BSL as determined by Fleissner and coworkers (Fleissner et al., 2011).

Advanced EPR spectral simulation software such as NLSL (Budil, Lee, Saxena, & Freed, 1996; Khairy, Fajer, & Budil, 2002) and MultiComponent (Altenbach, Flitsch, Khorana, & Hubbell, 1989) allow partial macroscopic order microscopic disorder models that can accommodate nanosecond rotational motion and anisotropic orientational distributions characteristic of aligned bicelle EPR. Because of the complexity of analyzing spectra affected by both rotational motion and orientational anisotropy, acquiring spectra of control samples is recommended to determine magnetic and dynamics parameters, such as frozen samples and isotropic vesicle/bicelle samples (Zhang et al., 2010). These allow for better determination of label orientation, which affects the spectral splitting as does rotational motion.

Figure 10 shows a general scheme for analysis of EPR measurements on oriented samples. Magnetic tensors (electron g and hyperfine A) are determined from frozen or otherwise immobilized sample spectra, as they lack significant dynamics and orientational anisotropy. These tensor values are carried into the analysis of the randomly oriented sample (e.g., minced muscle fiber, vesicles, or isotropic bicelles) to determine dynamics parameters such as rotational correlation time (R or τ_R) and order parameter (S, realted to the simulated orienting potential coefficient c_{20}). These values are then carried into the analysis of the aligned sample spectra to determine label orientation relative to the magnetic field (diffusion angles α_D, β_D, and γ_D). To further restrict the fitting space during this step, multiple sample orientations (usually with the symmetry axis parallel and perpendicular to the applied field) can be globally analyzed. The final step (if applicable) uses previous atomic structures of the spin label in a similar environment to transform the label diffusion angles into helix angles that relate the helix orientation to the symmetry axis. A full explanation of EPR fitting parameters and notation in the context of the simulation program NLSL is available from Earle and Budil (2006).

5. DISTANCE

5.1 Distance Measurements with DEER

Distance measurements using EPR have the distinct advantage over the complementary fluorescence technique FRET in that there is no need to

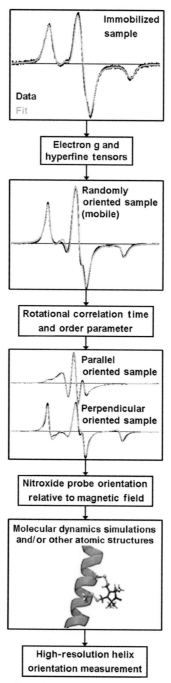

Figure 10 Typical workflow for EPR spectral analysis in oriented measurements.

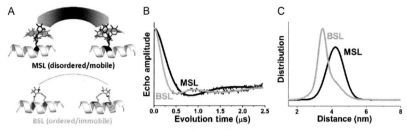

Figure 11 (A) The conformational flexibility of monofunctional labels introduces uncertainty in distance measurements between labels that is eliminated by bifunctional attachment. (B) DEER waveforms of MSL at position 498 and 639 on myosin, and BSL at positions 494/498 and 639/643 on myosin. (C) MSL reveals a single, broad distance, while BSL reveals two distinct distance populations.

introduce distinct donor and acceptor pairs in the system of interest (Kast, Espinoza-Fonseca, Yi, & Thomas, 2010); identical spin labels can be used and the distance is measured via the dipolar interaction. The two techniques most frequently used are continuous wave dipolar EPR, sensitive to distances between ~0.8 and 2 nm (Rabenstein & Shin, 1995), and double electron–electron resonance (DEER), which is capable of measuring distances from 2 to 6 nm in typical protein experiments (Pannier, Veit, Godt, Jeschke, & Spiess, 2000). DEER, in particular, is a powerful technique for the characterization of protein structure as the resultant waveforms are explicitly encoded with the distribution of distances between nearby labels (Jeschke, 2012).

The sensitivity of DEER to distance distributions is diminished by label flexibility, with monofunctionally attached labels typically producing broader, less defined distributions (Fig. 11C; Fleissner et al., 2011; Islam & Roux, 2015; Sahu et al., 2013). Label flexibility also reduces the reliability of detecting observed changes in distance distribution, as true structural changes may remain unresolved or, conversely, observed changes may be due merely to a change in label conformation. These ambiguities are eliminated through the use of the BSL.

5.2 The Problem of Orientation Selection

While the restricted label mobility afforded by BSL improves the colocalization of the label to the protein backbone, the rigid coupling introduces a new challenge to accurately analyzing data recorded by DEER, namely it introduces the effect of orientation selection (Marko et al., 2009). The traditional analysis of the DEER waveform assumes that there is no orientational dependence in the dipolar interaction between the interacting spin labels, which, in the case of BSL, is clearly not true. Fortunately, in the two most common frequency domains under which DEER is performed, X- and Q-band, orientation selection can be largely suppressed by following the standard protocol of pumping at the maximum of the EPR absorption spectrum and choosing nonselective pulses (Jeschke, 2012). Even under these conditions, though, the distribution width (but not center) extracted from the recorded waveforms is sensitive to choice of the position of the probe pulse, indicative of at least low levels of orientation selection, as shown in Fig. 12.

6. DISCUSSION

6.1 Labeling Specificity and Protein Function

While BSL possesses superior sensitivity to rotational dynamics, orientation, and distance, there are several technical points to consider when handling BSL and analyzing the data. First and foremost, labeling specificity is an important consideration for any spin label, but BSL in particular due to its two sulfhydryl groups. If the target protein contains native Cys residues, there is always a risk of nonspecific labeling that should be assessed with appropriate assays.

To evaluate nonspecific labeling, spin counting and mass spectrometry should be used to determine the label-to-protein ratio (Eaton et al., 2010; Karim et al., 2004). In addition to these routine measurements, conventional EPR on BSL-labeled proteins in oriented systems can also provide compelling evidence for nonspecific labeling. Indeed, measurements of dynamics alone are insufficient to identify nonspecific labeling, as singly attached BSL is often almost as immobilized as doubly attached BSL (Fig. 4). However, EPR spectra of oriented samples reveal the broader orientational distribution characteristic of singly attached spin labels (Fig. 13). In particular, singly attached labels tend to produce a powder-like spectrum, which is very distinct from the oriented components that typically arise from bifunctionally attached BSL. Thus, the presence of disordered spectral components from an oriented sample suggests nonspecific labeling.

The distinct behavior of singly and doubly attached BSL ultimately represents another advantage over monofunctional spin labels, as the doubly attached component of a spectrum can be identified and treated separately from any additional artifacts. Spectral resolution with BSL is usually sufficient to analyze both ordered and disordered components independently,

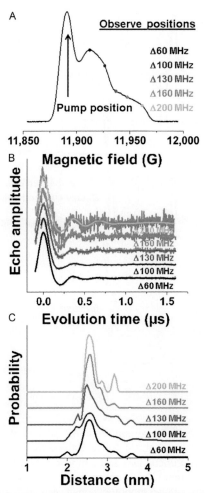

Figure 12 The presence of orientation selection was explored on myosin labeled at 192/196 and 639/643 with BSL via the acquisition of DEER at multiple pump-observe positions (A). The resultant DEER waveforms (B) were analyzed using LongDistances by Christian Altenbach, available for free download at http://www.chemistry.ucla.edu/directory/hubbell-wayne-l. (C) Changes in the distance distributions obtained from the waveforms in (B) are due to orientation selection.

pulling out accurate orientation parameters even when a disordered component is present. While orientation measurements with BSL are thus significantly less hindered by specificity concerns than traditional techniques, potential for nonspecific labeling should still be minimized as much as possible to avoid problems in DEER analysis and functional perturbation by additional modification. If nonspecific labeling is apparent, it can be

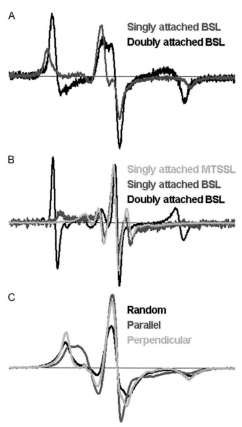

Figure 13 Orientational order is pronounced in BSL spectra of oriented systems when bifunctional attachment is present. (A) Conventional EPR spectra from myosin on oriented fibers, labeled monofunctionally and bifunctionally with BSL on the regulatory light chain at positions 122 and 122/126, respectively. (B) Conventional EPR spectra from myosin on oriented fibers, labeled monofunctionally with MTSSL or BSL at position 639, or bifunctionally with BSL at positions 639/643 on the myosin head. (C) Conventional EPR spectra from unaligned and aligned bicelles containing PLB labeled monofunctionally with BSL at position 32. See Fig. 9A for corresponding spectra with bifunctional BSL attachment.

addressed by careful substitution of native Cys with benign nonreactive analogs (such as Ser and Leu) (Shih, Gryczynski, Lakowicz, & Spudich, 2000).

In protein systems with engineered helical attachments points, BSL favors bifunctional attachment, presumably due to the rapid kinetics associated with disulfide bond formation. With nonideal cysteine orientations, monofunctional double labeling of the protein by BSL is possible even with steric restrictions, but this has been shown to be an uncommon occurrence

(Fleissner et al., 2011). While bifunctionally attached BSL is quite stable in practice, the release of label through disulfide exchange is apparent by the accumulation of unbound label over time. While we have not studied these effects in detail, potential mechanisms are the presence of reducing agent, or the formation of inter- or intra-disulfide bonds. As disulfide exchange is strongly temperature dependent, reversibility of BSL attachment can be minimized by maintaining samples at 4 °C or lower and is virtually eliminated by storage below 0 °C (Fava et al., 1957).

As with any approach that employs site-directed spectroscopy, it is essential to perform appropriate functional assays for the protein in question to determine whether the introduction of labeling sites and/or bifunctional labeling perturbs function. Small changes in enzymatic activity and/or substrate binding affinity are expected even with well-chosen labeling sites, but care should be taken to assess the extent of perturbation, and to move labeling sites if necessary. In addition to the considerations applicable to conventional labels, such as the potential for steric hindrance, bifunctional labeling can also hyperstabilize α-helices, and so should not be deployed on α-helical segments for which high flexibility or helical unfolding is key to function (Fleissner et al., 2011).

6.2 The BEER Technique

While the aforementioned techniques for measuring distance and orientation are powerful in their own right, BSL presents an unusual opportunity for parallel analysis, using the features of one method to constrain the other. Such analysis, here referred to collectively as BEER, represents a means to perform direct modeling of individual protein structural elements *de novo*, based on a small number of experiments. As discussed above, deployment of BSL in oriented systems can deliver accurate axial tilt angles for individual protein helices relative to the sample's symmetry axis. However, axial tilt alone is not enough to model each helix within a protein's tertiary structure, because the azimuthal angle with respect to the symmetry axis will always be degenerate. Furthermore, the intrinsic angular degeneracy of EPR allows for more than one potential helix tilt angle in most cases. These issues can often be resolved as long as a reference structure exists to contextualize the results, but for proteins without general established models, orientation measurements alone are not enough. In BEER, narrow intra-protein distance constraints obtained by DEER can significantly constrain the potential solutions given by orientation measurements, identifying the most likely conformer. Once the orientation of the given helices is thus established, the stereospecificity of BSL can be exploited to directly calculate the relative orientations of the labels used in DEER. This information may in turn be used to constrain the fitting of DEER waveforms by directly addressing orientation selection. BEER analysis may thus be performed iteratively to progressively constrain both results until an optimum solution is reached *de novo*, much like the model for analysis currently employed in structure determination by crystallography and NMR spectroscopy.

REFERENCES

Agafonov, R. V., Negrashov, I. V., Tkachev, Y. V., Blakely, S. E., Titus, M. A., Thomas, D. D., et al. (2009). Structural dynamics of the myosin relay helix by time-resolved EPR and FRET. *Proceedings of the National Academy of Sciences of the United States of America, 106*(51), 21625–21630.

Altenbach, C., Flitsch, S. L., Khorana, H. G., & Hubbell, W. L. (1989). Structural studies on transmembrane proteins. 2. Spin labeling of bacteriorhodopsin mutants at unique cysteines. *Biochemistry, 28*(19), 7806–7812. http://dx.doi.org/10.1021/Bi00445a042.

Altenbach, C., Marti, T., Khorana, H. G., & Hubbell, W. L. (1990). Transmembrane protein structure: Spin labeling of bacteriorhodopsin mutants. *Science, 248*(4959), 1088–1092.

Arata, T., Nakamura, M., Akahane, H., Aihara, T., Ueki, S., Sugata, K., et al. (2003). Orientation and motion of myosin light chain and troponin in reconstituted muscle fibers as detected by ESR with a new bifunctional spin label. *Advances in Experimental Medicine and Biology, 538*, 279–283. discussion 284.

Budil, D. E., Lee, S., Saxena, S., & Freed, J. H. (1996). Nonlinear-least-squares analysis of slow-motion EPR spectra in one and two dimensions using a modified Levenberg-Marquardt algorithm. *Journal of Magnetic Resonance, Series A, 120*(2), 155–189. http://dx.doi.org/10.1006/jmra.1996.0113.

Caporini, M. A., Padmanabhan, A., Cardon, T. B., & Lorigan, G. A. (2003). Investigating magnetically aligned phospholipid bilayers with various lanthanide ions for X-band spin-label EPR studies. *Biochimica et Biophysica Acta-Biomembranes, 1612*(1), 52–58. http://dx.doi.org/10.1016/S0005-2736(03)00085-3.

Cardon, T. B., Tiburu, E. K., & Lorigan, G. A. (2003). Magnetically aligned phospholipid bilayers in weak magnetic fields: Optimization, mechanism, and advantages for X-band EPR studies. *Journal of Magnetic Resonance, 161*(1), 77–90.

Cho, H. S., Dominick, J. L., & Spence, M. M. (2010). Lipid domains in bicelles containing unsaturated lipids and cholesterol. *Journal of Physical Chemistry B, 114*(28), 9238–9245. http://dx.doi.org/10.1021/Jp100276u.

Columbus, L., Kalai, T., Jeko, J., Hideg, K., & Hubbell, W. L. (2001). Molecular motion of spin labeled side chains in alpha-helices: Analysis by variation of side chain structure. *Biochemistry, 40*(13), 3828–3846.

De Angelis, A. A., & Opella, S. J. (2007). Bicelle samples for solid-state NMR of membrane proteins. *Nature Protocols, 2*(10), 2332–2338. http://dx.doi.org/10.1038/nprot.2007.329.

Durr, U. H. N., Soong, R., & Ramamoorthy, A. (2013). When detergent meets bilayer: Birth and coming of age of lipid bicelles. *Progress in Nuclear Magnetic Resonance Spectroscopy, 69*, 1–22. http://dx.doi.org/10.1016/j.pnmrs.2013.01.001.

Earle, K. A., & Budil, D. E. (2006). Calculating slow-motion ESR spectra of spin-labeled polymers. In S. Schlick (Ed.), *Advanced ESR methods in polymer research* (pp. 53–83). New York: John Wiley & Sons, Inc.

Eaton, G. R., Eaton, S. S., Barr, D. P., & Weber, R. T. (2010). *Quantitative EPR*. Vienna: Springer.

Fava, A., Iliceto, A., & Camera, E. (1957). Kinetics of the thiol-disulfide exchange. *Journal of the American Chemical Society, 79*(4), 833–838. http://dx.doi.org/10.1021/ja01561a014.

Fisher, A. J., Smith, C. A., Thoden, J., Smith, R., Sutoh, K., Holden, H. M., et al. (1995). Structural studies of myosin:nucleotide complexes: A revised model for the molecular basis of muscle contraction. *Biophysical Journal, 68*(Suppl. 4), 19S–26S. discussion 27S–28S.

Fleissner, M. R., Bridges, M. D., Brooks, E. K., Cascio, D., Kalai, T., Hideg, K., et al. (2011). Structure and dynamics of a conformationally constrained nitroxide side chain and applications in EPR spectroscopy. *Proceedings of the National Academy of Sciences of the United States of America, 108*(39), 16241–16246. http://dx.doi.org/10.1073/pnas.1111420108.

Garber, S. M., Lorigan, G. A., & Howard, K. P. (1999). Magnetically oriented phospholipid bilayers for spin label EPR studies. *Journal of the American Chemical Society, 121*(13), 3240–3241. http://dx.doi.org/10.1021/Ja984371f.

Goldman, S. A., Bruno, G. V., & Freed, J. H. (1972). Estimating slow-motional rotational correlation times for nitroxides by electron spin resonance. *The Journal of Physical Chemistry, 76*(13), 1858–1860. http://dx.doi.org/10.1021/j100657a013.

Holmes, K. C., Angert, I., Kull, F. J., Jahn, W., & Schroder, R. R. (2003). Electron cryomicroscopy shows how strong binding of myosin to actin releases nucleotide. *Nature, 425*(6956), 423–427.

Inbaraj, J. J., Laryukhin, M., & Lorigan, G. A. (2007). Determining the helical tilt angle of a transmembrane helix in mechanically aligned lipid bilayers using EPR spectroscopy. *Journal of the American Chemical Society, 129*(25), 7710–7711. http://dx.doi.org/10.1021/ja0715871.

Islam, S. M., & Roux, B. (2015). Simulating the distance distribution between spin-labels attached to proteins. *The Journal of Physical Chemistry. B, 119*(10), 3901–3911. http://dx.doi.org/10.1021/jp510745d.

James, Z. M., McCaffrey, J. E., Torgersen, K. D., Karim, C. B., & Thomas, D. D. (2012). Protein-protein interactions in calcium transport regulation probed by saturation transfer electron paramagnetic resonance. *Biophysical Journal, 103*(6), 1370–1378. http://dx.doi.org/10.1016/j.bpj.2012.08.032 (pii).

Jeschke, G. (2012). DEER distance measurements on proteins. *Annual Review of Physical Chemistry, 63*, 419–446. http://dx.doi.org/10.1146/annurev-physchem-032511-143716.

Kalai, T., Balog, M., Jeko, J., & Hideg, K. (1999). Synthesis and reactions of a symmetric paramagnetic pyrrolidine diene. *Synthesis, 6*, 973–980.

Karim, C. B., Kirby, T. L., Zhang, Z., Nesmelov, Y., & Thomas, D. D. (2004). Phospholamban structural dynamics in lipid bilayers probed by a spin label rigidly coupled to the peptide backbone. *Proceedings of the National Academy of Sciences of the United States of America, 101*(40), 14437–14442.

Karim, C. B., Zhang, Z., Howard, E. C., Torgersen, K. D., & Thomas, D. D. (2006). Phosphorylation-dependent conformational switch in spin-labeled phospholamban bound to SERCA. *Journal of Molecular Biology, 358*(4), 1032–1040.

Karim, C. B., Zhang, Z., & Thomas, D. D. (2007). Synthesis of TOAC spin-labeled proteins and reconstitution in lipid membranes. *Nature Protocols, 2*(1), 42–49.

Kast, D., Espinoza-Fonseca, L. M., Yi, C., & Thomas, D. D. (2010). Phosphorylation-induced structural changes in smooth muscle myosin regulatory light chain. *Proceedings of the National Academy of Sciences of the United States of America, 107*(18), 8207–8212.

Khairy, K., Fajer, P., & Budil, D. (2002). Simulation of spin label motion in EPR spectra of muscle fibers. *Biophysical Journal, 82*(1), 479a.

Lin, A. Y., Prochniewicz, E., James, Z., Svensson, B., & Thomas, D. D. (2011). Large-scale opening of utrophin's tandem CH domains upon actin binding, by an induced-fit mechanism. *Proceedings of the National Academy of Sciences of the United States of America, 108*(31), 12729–12733.

Lu, J. X., Caporini, M. A., & Lorigan, G. A. (2004). The effects of cholesterol on magnetically aligned phospholipid bilayers: A solid-state NMR and EPR spectroscopy study. *Journal of Magnetic Resonance, 168*(1), 18–30. http://dx.doi.org/10.1016/j.jmr.2004.01.013.

Marko, A., Margraf, D., Yu, H., Mu, Y., Stock, G., & Prisner, T. (2009). Molecular orientation studies by pulsed electron–electron double resonance experiments. *The Journal of Chemical Physics, 130*(6), 064102. http://dx.doi.org/10.1063/1.3073040.

McCaffrey, J. E., James, Z. M., & Thomas, D. D. (2015). Optimization of bicelle lipid composition and temperature for EPR spectroscopy of aligned membranes. *Journal of Magnetic Resonance, 250*, 71–75. http://dx.doi.org/10.1016/j.jmr.2014.09.026.

Mello, R. N., & Thomas, D. D. (2012). Three distinct actin-attached structural states of myosin in muscle fibers. *Biophysical Journal, 102*(5), 1088–1096. http://dx.doi.org/10.1016/j.bpj.2011.11.4027 (pii).

Moen, R. J., Klein, J. C., & Thomas, D. D. (2014). Electron paramagnetic resonance resolves effects of oxidative stress on muscle proteins. *Exercise and Sport Sciences Reviews, 42*(1), 30–36. http://dx.doi.org/10.1249/JES.0000000000000004.

Moen, R. J., Thomas, D. D., & Klein, J. C. (2013). Conformationally trapping the actin-binding cleft of myosin with a bifunctional spin label. *The Journal of Biological Chemistry, 288*(5), 3016–3024. http://dx.doi.org/10.1074/jbc.M112.428565.

Pannier, M., Veit, S., Godt, A., Jeschke, G., & Spiess, H. W. (2000). Dead-time free measurement of dipole-dipole interactions between electron spins. *Journal of Magnetic Resonance, 142*(2), 331–340. http://dx.doi.org/10.1006/jmre.1999.1944.

Rabenstein, M. D., & Shin, Y. K. (1995). Determination of the distance between 2 spin labels attached to a macromolecule. *Proceedings of the National Academy of Sciences of the United States of America, 92*(18), 8239–8243. http://dx.doi.org/10.1073/pnas.92.18.8239.

Sahu, I. D., McCarrick, R. M., Troxel, K. R., Zhang, R., Smith, H. J., Dunagan, M. M., et al. (2013). DEER EPR measurements for membrane protein structures via bifunctional spin labels and lipodisq nanoparticles. *Biochemistry, 52*(38), 6627–6632. http://dx.doi.org/10.1021/bi4009984.

Shih, W. M., Gryczynski, Z., Lakowicz, J. R., & Spudich, J. A. (2000). A FRET-based sensor reveals large ATP hydrolysis-induced conformational changes and three distinct states of the molecular motor myosin. *Cell, 102*(5), 683–694.

Smith, C. A., & Rayment, I. (1996). X-ray structure of the magnesium(II).ADP.vanadate complex of the dictyostelium discoideum myosin motor domain to 1.9 Å resolution. *Biochemistry, 35*(17), 5404–5417.

Spudich, J. A. (2014). Hypertrophic and dilated cardiomyopathy: Four decades of basic research on muscle lead to potential therapeutic approaches to these devastating genetic diseases. *Biophysical Journal, 106*(6), 1236–1249. http://dx.doi.org/10.1016/j.bpj.2014.02.011.

Squier, T. C., & Thomas, D. D. (1986). Methodology for increased precision in saturation transfer electron paramagnetic resonance studies of rotational dynamics. *Biophysical Journal, 49*(4), 921–935.

Surek, J. T., & Thomas, D. D. (2008). A paramagnetic molecular voltmeter. *Journal of Magnetic Resonance, 190*(1), 7–25.

Thomas, D. D., & McConnell, H. M. (1974). Calculation of paramagnetic resonance spectra sensitive to very slow rotational motion. *Chemical Physics Letters, 25*, 470–475.

Thomas, D. D., Prochniewicz, E., & Roopnarine, O. (2002). Changes in actin and myosin structural dynamics due to their weak and strong interactions. *Results and Problems in Cell Differentiation, 36,* 7–19.

Thompson, A. R., Naber, N., Wilson, C., Cooke, R., & Thomas, D. D. (2008). Structural dynamics of the actomyosin complex probed by a bifunctional spin label that cross-links SH1 and SH2. *Biophysical Journal, 95*(11), 5238–5246.

Traaseth, N. J., Buffy, J. J., Zamoon, J., & Veglia, G. (2006). Structural dynamics and topology of phospholamban in oriented lipid bilayers using multidimensional solid-state NMR. *Biochemistry, 45*(46), 13827–13834. http://dx.doi.org/10.1021/Bi0607610.

Wilcox, M., Parce, J., Thomas, M., & Lyles, D. (1990). A new bifunctional spin-label suitable for saturation-transfer EPR studies of protein rotational motion. *Biochemistry, 29*(24), 5734–5743.

Zhang, Z. W., Fleissner, M. R., Tipikin, D. S., Liang, Z. C., Moscicki, J. K., Earle, K. A., et al. (2010). Multifrequency electron spin resonance study of the dynamics of spin labeled T4 lysozyme. *Journal of Physical Chemistry B, 114*(16), 5503–5521. http://dx.doi.org/10.1021/Jp910606h.

CHAPTER FIVE

EPR Distance Measurements in Deuterated Proteins

Hassane El Mkami*, David G. Norman[†,1]
*School of Physics and Astronomy, University of St. Andrews, St. Andrews, United Kingdom
[†]Nucleic Acids Research Group, University of Dundee, Dundee, United Kingdom
[1]Corresponding author: e-mail address: d.g.norman@dundee.ac.uk

Contents

1. Introduction and Overview 126
   1.1 pEPR Spectroscopy 131
   1.2 PELDOR/DEER Experiment 132
   1.3 Relaxation and Its Impact on the PELDOR Experiment 135
   1.4 Effect of Deuteration on Relaxation and Distance Measurements 136
2. Deuteration 143
   2.1 Overview 143
   2.2 Methods for Producing Deuterated Proteins 144
   2.3 Measurement of Deuteration 144
3. The Pulse EPR Experiment 145
   3.1 T_m Measurements 146
   3.2 PELDOR/DEER Experiment Setup 146
4. Data Analysis 147
   4.1 T_m Data Fitting 147
   4.2 PELDOR Data Analysis 148
5. Concluding Remarks 149
Acknowledgments 150
References 150

Abstract

Pulsed electron double resonance technique, also known as double electron–electron resonance, jointly with site-directed spin labeling (SDSL) have been used extensively for studying structures and structural change. During the last decades, significant enhancements have been made by optimization of the experimental protocols, introducing new techniques for artifact suppression, and developing data analysis programs for extracting more reliable distance distributions. However, the distance determination by pulsed electron paramagnetic resonance is still facing some limitations, especially when studying spin-labeled proteins, due mainly to the fast relaxation time that imposes severe limitations on the maximum distances measurable and upon the sensitivity of such experiments. In the present work, we demonstrate the impact of the deuteration of the underlying protein, in addition to the solvent, on relaxation times, sensitivity, and on distance measurements.

1. INTRODUCTION AND OVERVIEW

Structural biology could be said to have started just over a hundred years ago with the discoveries in X-ray diffraction, followed later by the determination, by X-ray crystallography, of sperm whale myoglobin by John Kendrew et al. in 1958 (Kendrew, 1958; Kendrew et al., 1958). The determination of the structure of Hemoglobin by Perutz in 1963 followed shortly after (Perutz, 1963). In the last 40 years, there has been a huge increase in the number of biological structures determined, with currently well over 100,000 high-resolution structures deposited in the protein data bank. The explosion in structural information from crystallography is, arguably, at a peak in that the number of new folds determined has withered to almost nothing and it might be argued that now the most important task facing structural biology is to understand how proteins interact with each other. A range of new and sometimes old, biophysical techniques are being used, often in combination, to tease out information on protein complexes and interactions. Among these techniques that have had a renaissance in recent years is electron paramagnetic resonance (EPR). Although EPR is fundamentally quite an old technique, being first demonstrated by Zavoisky (1945), the development of pulsed microwave technology, coupled with spin labeling (Berliner & Grunwald, 1976), has opened up the possibility of investigating biomolecular structures in the solution state, in a way that brings a range of distinct differences and advantages from other techniques. As a technique, EPR and spin labeling have obvious similarities to nuclear magnetic resonance (NMR) as well as certain fluorescence methods, such as Förster resonance energy transfer (FRET). It has been amply demonstrated that EPR can measure distance and the distance distribution between spin labels, attached to biomolecules. The development in pulsed EPR (pEPR) extended the applicability of EPR to distance measurements (Jeschke, 2002; Milov, Salikhov, & Shirov, 1981; Schweiger & Jeschke, 2001). The limitations of this methodology relative to NMR are obvious in that essentially only a single distance is measured per experiment. The advantage is that the measurable distances can be very long in comparison to NMR. The distance range achievable has often been quoted as being between 18 and 80 Å. The long distances available from pEPR are obviously an attractive feature although the demands of sample preparation are somewhat onerous in relation to the quantity of data obtainable. FRET seems to be directly comparable to pEPR in that it has the potential to measure large distances from labeled biomolecules. One extreme advantage of FRET is the high sensitivity that allows experiments to be done at very low concentrations or indeed even at the single-molecule level (Ha et al., 1996). The disadvantages relative to EPR include, the often difficult task of, labeling the biomolecule or complex with two separate fluorophores and also with problems in interpreting the results, due partly to the orientation dependence of the Förster effect and to the generally much larger and often charged nature of the fluorophores. The distances measurable are often quoted at between 10 and 100 Å (Ha et al., 1996). The accuracy of FRET-derived distance measurement is somewhat diminished when the measurements are at any significant distance from the Förster distance. The distance determined from FRET, being related to the intensity of the detected signal, which is described by a sigmoidal relationship that leads to small changes when remote from the Förster distance. EPR on the other hand can measure a dipolar frequency, which is directly related to the distance between spin labels and that frequency changes in proportion to the inverse cube of the distance. At the extremes of measurable distance, the frequency becomes so small that the challenge becomes being able to capture enough time data to measure the dipolar frequency unambiguously. To measure increasingly longer distances, the major limiting factor becomes the relaxation time of the signal. Of the two major relaxation mechanisms, it is T_2 that is shorter than T_1 and so becomes limiting. Because distance measurements are made in the frozen state, the relaxation mechanism T_2 is usually referred to as T_m and describes the nonrecoverable dephasing of the electron spin echo (ESE) in the x–y plane. It has been commonly recognized that a proportion of the dephasing observed in spin-labeled protein solutions was caused by interactions with the protons of the solvent, usually water in which the protein was dissolved in. Replacement of the water with deuterated water led to a significant decrease in the rate of electron x–y dephasing due to the smaller magnetic moment of the deuteron. Use of deuterated glycerol as an agent to prevent ice formation was also widely adopted to again decrease the rate of dephasing. With both of these measures in place, data from a pulsed electron–electron double resonance (PELDOR or DEER) experiment could be gathered to several microseconds. It was empirically observed that even in the most favorable circumstances, an acquisition time in excess of 10 μs was not possible. In order to measure a distance with any reliability, one has to measure at least one dipolar-derived oscillation in the data. To measure not only a distance but also a distance distribution between spin labels, one has to be able to measure at least two oscillations (Jeschke, Bender, Paulsen, Zimmermann, & Godt, 2004). The wavelength of the dipolar oscillation seen in the PELDOR experiment can be plotted against the wavelength of the dipolar oscillation (Fig. 1) and this shows quite clearly that, if there is a practical limit for the acquisition of PELDOR data, of 10 μs, and if one requires to observe at least one full oscillation then the maximum distance measurable would be 80 Å.

The theoretical limit to the measurable distance between spin labels has been quoted in countless papers, which have used PELDOR for distance determination. The theoretical maximum distance had in practice never been achieved in biological molecules, with most reports describing distances in the 20–55 Å range. In 2009, we managed to achieve measurement of a couple of distances of 70 Å in the histone core octamer (Ward, Bowman, El Mkami, Owen-Hughes, & Norman, 2009). These data were collected to 8 μs and included just over one complete oscillation (see Fig. 2). These data exhibited persistent oscillations which continued to the end of the recorded data and although allowing determination of the distance,

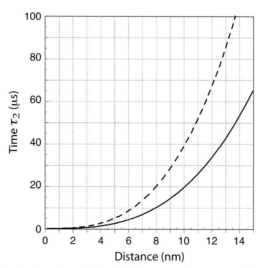

Figure 1 A graph showing the relationship between the measurable distance and the required time of the PELDOR experiment. The dotted line shows the relationship for one oscillation and the solid line shows the relationship for two full oscillations.

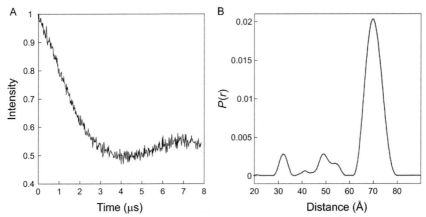

Figure 2 (A) Background-corrected PELDOR data for the histone octamer, labeled at position H3Q76C. (B) Tikhonov-derived distance distribution for H3Q76C.

made fitting of background decay somewhat problematic. The 70 Å distances were gathered at X-band and required experimental times if between 16 and 24 h. Practical restraints limited the possibility of making experiments over a significantly longer time period.

It had been noted previously that deuteration of the underlying protein might be expected to lead to increases of T_m over that achieved by simple solvent deuteration (Borbat & Freed, 2007). NMR spectroscopy of proteins has a related problem with T_2 that led to unacceptable peak widths in large proteins, thus limiting the size of protein that could be assigned. This had been ameliorated by using partial (or even complete) deuteration. Without protein deuteration NMR spectroscopy was largely limited to studying proteins of less than 100 amino acids in size. Partial deuteration of proteins increases T_2 and leads to narrower peaks enabling the maximum size in proteins studied and solved by NMR to increase substantially. Most NMR studies have used only partially deuterated proteins due to the requirement to retain some protons for measurement; however, it has also been demonstrated that backbone assignment and structure determination can be achieved by NMR in perdeuterated proteins (Koharudin, Bonvin, Kaptein, & Boelens, 2003; Venters, Metzler, Spicer, Mueller, & Farmer, 1995), relying only on the exchangeable protons that would be reconstituted once the protein was dissolved in aqueous buffer. Although molecular size was never a problem for EPR spectroscopy, due to either peak width or complexity, fast relaxation due to short T_m was a severe restriction to the dipolar coupling frequency that could be measured.

Although proteins can be expressed in deuterated or partially deuterated growth media, there are problems. It had previously been shown that *Escherichia coli* could be adapted to growth in deuterated growth media, by slowly increasing the level of D_2O in such media, although the growth rate and protein yields were always lower than those seen in a nondeuterated growth media. The availability of fully deuterated, rich growth media derived from algal hydrolysates had however been shown to give protein yields that were not significantly reduced over those in nondeuterated media (Cox, Chen, & Kabacoff, 1994). As a test of deuteration and also to increase the quality of the distance distributions, we decided to deuterate the nucleosome core particle (Ward et al., 2010). The core particle consists of eight histone proteins, two each of histones H3, H4, H2a, and H2b. The histones were expressed as inclusion bodies in *E. coli* and after purification and refolding they were assembled into the full histone octamer. Spin labeling was accomplished with (1-oxyl-2,2,5,5-tetramethylpyrroline-3-methyl) methanethiosulfonate (MTSSL) at position H3Q76C. The sample gave a measured T_m of 6 μs when protonated and 28 μs when fully deuterated.

The distance between spin labels at H3Q76C, on symmetry-related positions, was 70 Å. Data were initially collected to 25 μs, giving a result in which over two complete oscillations were recorded, with the oscillations essentially damping to zero by the end of the experiment. The other significant advantage obtained using a deuterated protein was that of sensitivity. An experiment taken to 7 μs showed a fourfold increase in sensitivity over one taken under identical conditions but on a nondeuterated protein. In practical terms, we were able to record PELDOR data, in 2 h that exhibited slightly better signal-to-noise ratio (s/n), when compared to data from a nondeuterated protein which took 15.5 h. A later study used nucleosome core histone octamers to examine the effect of segmental deuteration (El Mkami, Ward, Bowman, Owen-Hughes, & Norman, 2014). In this study, specific combinations of histones were deuterated and the rest protonated. This study showed that significant improvements in T_m were to be obtained even when the deuteration was somewhat remote from the site of the spin label. One could also say that it showed that deuteration of only the protein local to the spin label gave significant increases in T_m and sensitivity.

A number of alternative routes to the bacterial expression of deuterated protein were tested. Including a technique in which a solid form of the deuterated algal hydrolysate was used in an otherwise protonated growth media. Deuteration was also attempted, with some success using *E. coli* that had been grown in normal rich protonated media and then transferred to small volumes of deuterated media before induction of protein expression. We found that in general, simple substitution of growth media for fully deuterated rich media coupled with an extended growth period was the most successful approach in terms of both protein yield and isotopic substitution. If volumes of media were closely calculated, then sample production for pEPR could be done cost effectively. Assessment of deuteration was routinely made by mass spectroscopy; however, the calculated difference between protonated and partially protonated protein is quite small leading to some loss in accuracy. However, best estimates revealed close to 100% deuteration for samples grown in algae-derived, rich deuterated media.

1.1 pEPR Spectroscopy

EPR spectroscopy in its early days was a continuous wave (CW) technique used for a wide range of applications studying structure and dynamics properties of paramagnetic species. Reliable information derives mainly from knowing the EPR parameters that describe all the interactions between the unpaired electron and its surrounding nuclei with nonzero spins and other unpaired electrons. However, when the splitting resulting from interactions or couplings like hyperfine, super-hyperfine, and dipolar is smaller compared to the intrinsic linewidth, this leads to line broadening and therefore, most of the information is hidden underneath of the broad spectrum. Fortunately, this inhomogeneous broadening effect can be negated by the use of the ESE techniques. Several advanced ESE techniques have been developed, and depending on the type of interaction that needs to be resolved, a specific pulse sequence can be applied. For example, the pulsed electron nuclear double resonance (ENDOR) experiment has been used to identify the types of the nuclei that are coupled to a paramagnetic center and the hyperfine couplings (which give access to distances between the nuclei and the paramagnetic center); in this case, only the nuclear transitions are detected by the ENDOR pulse sequence. In a PELDOR (also known as DEER) experiment, the pulse sequence is mainly monitoring the effect of the dipolar coupling between electron spins.

The typical pulse EPR spectrometer consists, in addition to a standard CW EPR spectrometer, operating at high power, a pulse-forming unit, and a detection system that does not depend on field modulation. Extending such a pEPR spectrometer to be able to perform PELDOR requires only a second microwave source, with adjustable frequency, to be fed into the pulse-forming unit. The same cavity and high-power amplifier can be used as for a standard pulse EPR. The desired excitation bandwidth and hence pulse lengths are dictated by the type of pulse EPR experiments to be performed.

1.2 PELDOR/DEER Experiment

PELDOR is an (ESE) experiment for measuring dipole–dipole interactions between paramagnetic pairs (AB) in the nanometer range, where A and B are commonly either nitroxide–nitroxide, or nitroxide–metal or metal–metal centers. The initial three-pulse PELDOR (see Fig. 3A) introduced by Milov et al. (1981) was used widely but the method suffers from the inherent dead time that prevents reliable recording of the signal amplitude at time zero. This severely limits the usefulness of the method. A novel approach to overcome the dead-time problem proposed by Pannier et al. is a four-pulse dead-time free PELDOR (see Fig. 3B; Pannier, Veit, Godt, Jeschke, & Spiess, 2000). In both methods, two microwave frequencies are required to separately excite the coupled spins. The observer sequence (see Fig. 3A and B) consists of two parts: the first $(\pi/2)-\tau_1-(\pi)-\tau_1$ that

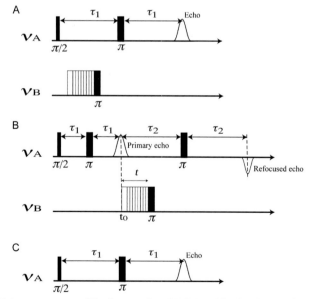

Figure 3 Pulse sequence: (A) three-pulse PELDOR, (B) dead-time free, four-pulse PELDOR, and (C) two-pulse Hahn echo. Where ν_A and ν_B correspond to the observer or pump frequencies, respectively. τ_1 and τ_2 represent the time delays.

creates a Hahn echo and the second τ_2–(π)–τ_2 that refocuses this echo. The four-pulse PELDOR signal is monitored by recording the amplitude of the refocused echo instead of the Hahn echo, as in the three-pulse version, which allows the maximum time domain signal (time zero) to be observed. The observer pulse sequence excites only A spins and refocuses all interactions that are experienced. The dipolar interaction, which is a part of these interactions, is manifested as a modulation of the ESE amplitude of the detected A spin, at the observer frequency ν_A, when the B spin is excited with an intense microwave pulse at the pump frequency ν_B, the B spin flips, resulting in a reversing of the sign of the dipolar coupling contribution to the local magnetic field experienced by spin A. This induces a phase lag of the Larmor frequency of the observer A spins that are no longer refocused and results in a decrease of the echo amplitude. The PELDOR trace is then obtained by incrementing the time 't' of the position of the pump pulse (see Fig. 3B) while the delay τ_1 and τ_2 are fixed. In the ideal situation when the A and B spins are excited selectively by the observer and pump pulse, respectively, the contribution of the dipolar coupling to the four-pulse PELDOR signal is described by

$$V(t, \tau_1, \tau_2) = 1 - \lambda_B (1 - \cos(\omega_{AB}(t - \tau_1)))$$

where λ_B is the inversion probability of the B spins by the pump pulse at time $t = \tau_1 + T$, and ω_{AB} is the dipole–dipole interaction frequency. In the case of weak coupling, when the distance between the partner spins is larger than 1.5 nm, the exchange coupling can be ignored and the frequency of the dipolar coupling is defined as

$$\omega_{AB} = \frac{g_A g_B \beta^2}{\hbar r_{AB}^3}(1 - 3\cos^2\theta)$$

r_{AB} is the distance between spins A and B, θ is the angle between the direction of the applied magnetic field and the vector r_{AB} connecting the spins in the pairs. g_A and g_B are the g tensors of the A and B spin labels.

PELDOR sensitivity relies on two mains aspects: the amplitude of the observed echo and the modulation depth of the PELDOR signal. The first depends mainly on the electron spin relaxation time T_m (the phase relaxation time which includes a wide variety of processes that contribute to the spin Hahn echo dephasing) and the second on the fraction of spins, inverted by the pump pulse, λ_B. The latter, also known as the modulation depth, depends on the resonator performances (Q factor, B_1, resonator bandwidth) and the experimental settings. In biological applications, the A and B electron spins are usually, chemically identical, nitroxide radicals. At X-band, the EPR spectrum (see Fig. 4) of the nitroxide is dominated by the hyperfine coupling ($\sim$100 MHz) with the ^{14}N (with nuclear spin $I=1$) nucleus along the molecular Z (g_{ZZ}) axis. The X (g_{XX}) and Y (g_{YY}) components of the hyperfine coupling are too weak to be resolved at this frequency. Therefore, the EPR spectrum consists of three lines corresponding to the nitrogen nuclear spin state ($M_I=0, \pm 1$). In a PELDOR experimental setup, B spins are chosen with a resonance frequency corresponding to the highest intensity point of the spectrum ($M_I=0$, see Fig. 4) and together with a short pump pulse result in maximizing the number of the inverted spins. This pump position excites all orientations of the molecules in the $M_I=0$ state and also those in $M_I=\pm 1$ state with the Z molecular axis perpendicular to the external magnetic field B_0. The observer pulse that excites the A spins is applied at a position with a resonance frequency corresponding to the molecules in the nitrogen nuclear spin state $M_I=-1$ (see Fig. 4) with the molecular Z axis parallel to B_0. The frequency offset $\Delta\nu = \nu_A - \nu_B$ is set in such a way (at least 65 MHz) as to keep the A and B spin excitations isolated. This condition is very important

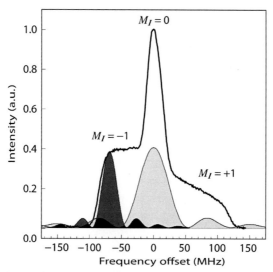

Figure 4 Echo-detected field-swept EPR spectrum of nitroxide biradicals recorded at 50 K. M_I ($I=-1,0,1$) is the nitrogen nuclear spin state. The black and gray areas correspond, respectively, to the simulated excitation profiles for 32 ns observer π pulse and 12 ns pump pulse.

since any overlapping excitations can reduce considerably, the modulation depth and in some cases leads to the nonobservation of dipolar coupling effects.

1.3 Relaxation and Its Impact on the PELDOR Experiment

The persistence of the ESE in the PELDOR experiment is one of the most crucial limitations to distance measurements. At a temperature of around 50 K, one of the predominant factors affecting persistence of an echo, and as such, the sensitivity and measurable distance between spin labels, is the spin echo dephasing time (T_m) (Huber et al., 2001; Lindgren et al., 1997). In a nondeuterated environment, short T_m, in the order of 2 μs, are usually observed, when studying spin-labeled proteins. A 2 μs T_m limits the measurement of distances, in the PELDOR experiment, to a maximum of 3–4 nm. T_m is affected by contributions from instantaneous and spectral diffusion as well as hyperfine interactions with surrounding nuclei. Unpaired electrons can show dipolar coupling to nuclear spins in the surrounding media and although individual nuclear spin flip-flop is slow, the large number of coupled nuclei in a typical protein makes these events highly probable and nuclear spin flip-flop in dipolar coupled electron–nuclei, change the precession frequency of the unpaired electron. Dipolar coupling is proportional to the magnetic moment, so proton spin diffusion is a more effective mechanism of dephasing electron spins than would be deuterium. It has become a common practice to use deuterated solvents and cryoprotectants, which has extended the T_m to around 5–6 μs. It has recently been demonstrated that total deuteration of the protein, containing a site-specific nitroxide spin-label pair, in addition of the solvent, extended T_m greatly giving a value of approximately 36 μs (Ward et al., 2010). The increase of T_m, upon deuteration of the underlying protein, is so large that it has quite significant implications to the measurement of distances by pEPR and to the sensitivity of pEPR experiments. The longest distance so far reported, measured by pEPR is 106 Å, measured in a deuterated protein system (Bowman et al., 2014). It is conceivable that with deuteration the maximum distance that could be measured might be as much as 140 Å. Recent work by Baber et al. has described an interesting problem in samples containing populations of spin label with different relaxation rates caused by different interactions with the underlying protein protons (Baber, Louis, & Clore, 2015). This paper showed that in a nondeuterated protein sample, the observed proportion of two distance distributions within a single sample could vary depending on the time (t) over which the data was gathered. In long t values the contribution from faster relaxing species is considerably attenuated relative to those with longer T_m. Although this potentially serious error could be corrected for by running a number of experiments and extrapolating values to $t = 0$, the problem essentially disappeared upon protein deuteration at high levels.

1.4 Effect of Deuteration on Relaxation and Distance Measurements

To demonstrate the impact of deuteration, we can show examples from three different protein systems: the histone core octamer (Ward, Bowman, et al., 2009), which was segmentally deuterated (Ward et al., 2010; Fig. 5); a fully deuterated histone chaperone Vps75 (Bowman et al., 2014; Fig. 6); and a pair of POTRA domains (Ward, Zoltner, et al., 2009; Fig. 7) with high levels of deuteration. In this context, deuteration refers to the nonfreely exchanging protons since all the experiments are conducted in deuterated aqueous buffer, the exchangeable protons are expected to be in deuteron form. In both cases, we have looked at the deuteration effects on the observed T_1 and T_m, providing a better understanding of the spatial relationship between the spin labels and protein protons and the extent and impact of spin diffusion.

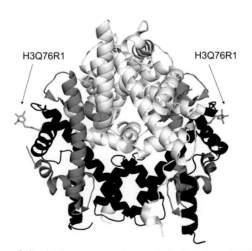

Figure 5 Structure of the histone core octamer (adapted from PDB code 1TZY). The positions of the spin labels H3Q76R1 are indicated. The H3 histones are shown as black ribbons, the H4 histones are indicated by medium gray ribbons, and the H2A/H2B histones are indicated by light gray ribbons.

Figure 6 Structure of the histone chaperone protein Vps75. Figure showing the structure of the Vps75 tetramer with each dimer shaded differently in gray and a dashed line indicating the distance between Rx-spin labels situated at the dyad axis of each dimer.

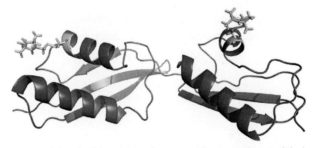

Figure 7 Structure of the double POTRA domain with nitroxide spin labels at positions 109 and 44 (Ward, Zoltner, et al., 2009).

1.4.1 Impact on Relaxation Time

The histone core octamer is composed of two copies of each, H3, H4, H2A, and H2B histones, making up of a central H3/H4 tetramer sandwiched between two H2A/H2B dimers. The multimeric form of this protein allowed the assembly of either all, a subset, or none of the histones in deuterated form. Segmentally deuteration of the octameric assemblies permitted investigation of the effect of the spatial distribution of the protons, of the underlying protein, on the the relaxation pathways.

We had labeled the nucleosome core octamer using MTSSL at the mutated position Q76C of histone H3, generating a symmetrical pair of labels with a spin–spin distance of 70 Å (Bowman, Ward, El-Mkami,

Owen-Hughes, & Norman, 2010; Ward, Bowman, et al., 2009). Measurement of T_m was made on constructs in which the constituent histones were deuterated, deuterated on the H3 histones, on H3 and H4, deuterated on H4 alone, or deuterated on all eight histones. The results of the echo decay curves (T_m) measured at 50 K for these constructs are shown in Fig. 8. The T_m value for the nondeuterated, octameric complex (in deuterated solvent) is 6.9 μs, which is at the high end of the reported T_m values for a spin label situated on the surface of a protein dissolved in deuterated buffer. Deuteration of H3 (the histone labeled) led to an approximate doubling of the T_m to 13.6 μs. The spin labels, at position H3Q76C, are quite close to parts of histone H4 and the effect of deuterating only histone H4 also had a large effect raising the T_m to 11.6 μs. The combined effect of deuterating both H3 and H4 led to a large jump in T_m to 31 μs. Deuteration of all histones in the octamer resulted in a final T_m value of 36 μs. It can be surmised that deuteration will be most efficient if the spin label is insulated, by deuteration, from the rest of the protein by a distance of approximately 25 Å.

As part of the study of the effects of protein deuteration, we also looked at the temperature dependence of the electron spin relaxation rate, $1/T_1$ and $1/T_m$. At temperatures below 50 K, $1/T_m$ is independent of temperature for

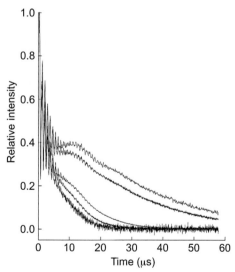

Figure 8 ESE decay curves for segmentally deuterated spin-labeled histone core octamer. The curves indicate the relaxation for the following constructs in order of increasing relaxation rate, fully deuterated, H3/H4 deuterated, H3 deuterated, H4 deuterated, and nondeuterated.

both deuterated and nondeuterated proteins, with constant values of 4.5 and 5.1 s^{-1}, respectively. This result suggests that T_m at low temperature is dominated by the nuclear spin diffusion due to the mutual spin flip-flop (Zecevic, Eaton, Eaton, & Lindgren, 1998). The slower rate for the fully deuterated system, being due to the fact, that the deuterium has approximately 6.51 times smaller magnetic moment. Above 50 K and up to 100 K, the phase memory relaxation rate for both samples (deuterated and nondeuterated) increases indicating that a thermally activated process arises, usually attributed to the rotation of the spin-label methyl groups effect (Nakagawa, Candelaria, Chik, Eaton, & Eaton, 1992; Rajca et al., 2010). In this study, the spin labels are nondeuterated and contain germinal methyl groups (protonate) and the temperature dependence of $1/T_m$ rate yielded activation energy of 1 kcal/mol which falls within the range of values reported for methyl group rotation in several nitroxyl labels (Nakagawa et al., 1992). Over this temperature range, the spin relaxation rate $1/T_1$ did not show any major difference between the deuterated and nondeuterated samples, which indicate that within this temperature range, the nuclear spins do not play a significant role in the spin–lattice relaxation mechanism. In this study, we have demonstrated that the relaxation effects of deuteration are manifested exclusively on the rate of spin dephasing, T_m. However, replacing protons with deuterons results in an increase in T_m of 5.5 times, and it is empirically shown that most of the effect, of deuteration on the rate of ESE dephasing, is due to nuclear–electron spin interactions within about 25 Å of the spin label. This observation has interesting implications to structural studies of protein complexes, in that even deuteration of parts of a complex can lead to significant increase in sensitivity and distances measurable.

In the case of the POTRA domains, in addition to deuteration of the protein, we also used a deuterated nitroxide spin label. Echo decay measurements were performed at W-band (Fig. 9). Although a large increase in T_m was seen upon protein deuteration, no added effect was observed when also using a deuterated spin label, this was not a surprising result; it having been previously predicted that methyl protons of the spin label would be decoupled from spin diffusion to the protein protons due to their large hyperfine coupling, so deuteration would not be expected to make any difference (Jeschke, 2007). On quick comparison between X- and W-band T_m data, one can see that the deuterium electron spin echo envelope modulation (ESEEM) modulations are strongly suppressed at W-band due to the modulation depth being scaled inversely with the square of the magnetic field (Figs. 8 and 9A). The removal of proton-driven ESE dephasing allows

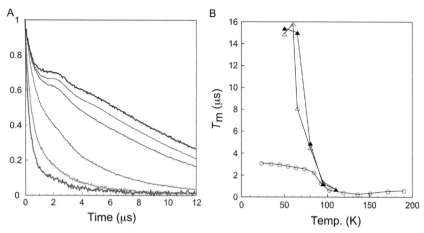

Figure 9 (A) ESE decay curves for double nitroxide-labeled, deuterated POTRA domains protein recorded at different temperatures. The curves indicate the relaxation at the following temperatures in order of increasing relaxation rate. 50, 60, 65, 80, 95, and 110 K. (B) Graph showing the relationship between T_m and temperature for double spin-labeled, nondeuterated, POTRA domains (circle); double spin-labeled, deuterated, POTRA domains (filled triangle); and double spin-labeled, deuterated, POTRA domains with deuterated nitroxide spin label (open triangle).

the easy observation of the electron dipole–dipole interaction. The latter is manifested as an oscillation that is superimposed on the echo decay signal. This effect is known as instantaneous diffusion and constitutes another dephasing process of the ESE. Instantaneous diffusion results from a simultaneous flip of dipolar coupled spins when the second microwave pulse is applied in a two-pulse echo experiment. The dephasing driven by this effect could lead to an oscillating echo signal, $V(\tau) \sim \cos(D\tau)$, if the spin pairs have a defined dipolar interaction, D, as it is the case with double-labeled systems.

One question that was addressed in the study of deuterated, double-labeled POTRA domains was that of the relationship between temperature and T_m. PELDOR experiments are routinely carried out at around 80 K, at which temperature the values for T_1 and T_m are optimized. Between 80 and 100 K, the T_m of a typical nitroxide spin label decreases sharply. It has been suggested that this effect is due to the flanking methyl groups beginning to rotate and so causing dephasing due to dynamic changes in the large hyperfine coupling to the methyl protons. Figure 9B shows a plot of the change in T_m over a range of temperatures for three different samples, nonprotein deuterated, protein deuterated, and protein deuterated with deuterated spin label. As can be clearly seen from this graph although deuteration of the protein causes long T_m values below about 70 K, compared to the nondeuterated sample, a rise in temperature causes the T_m value to decrease rapidly in the same manner as for the protonated protein, resulting in quite small T_m values above 100 K. Deuteration of the spin label surprisingly produced no change. One assumes that the hyperfine coupling with the methyl groups is so large that deuteration has no measureable effect in spite of a much smaller magnetic moment of the deuteron.

1.4.2 Impact on Sensitivity and Distance Measurements

The PELDOR experiment is based on the eventual detection of a modulated ESE. In order to examine the sensitivity enhancement upon protein deuteration, we recorded PELDOR data on two histone core octamer protein samples with identical concentrations using identical run conditions. One of the proteins was deuterated and the other, nondeuterated. A dipolar evolution time of 7 μs was used, which is a reasonable time at which a nondeuterated protein sample could normally be measured. Data was accumulated for 15.5 h (a typical X-band experimental run time) on the nondeuterated sample. The deuterated sample required only 2 h to accumulate data with the same signal to noise. This simple experiment clearly shows that the nondeuterated sample required about eight times more averaging to achieve a comparable quality of result. In order to make a more direct analytical comparison, we collected the data of the refocused spin echo using the standing four-pulse PELDOR sequence (in this sequence, the position of pump pulse is kept fixed and is used to optimize the evolution time window, the experiment is known as the standing-DEER experiment) (Jeschke et al., 2006) with a 7-μs delay between second and third pulses. This showed a fourfold increase in echo intensity resulting in increasing the signal-to-noise ratio from 14 to 55 for protonated and deuterated proteins, respectively (Ward et al., 2010). The implication of these experiments is that with protein deuteration extremely small concentration samples (a few μM) could be used to give acceptable result over a 16-h experiment. It should be noted that the dipolar oscillations are still strong at the echo detection time of 7 μs, in both samples.

Another example, demonstrating the impact of the deuteration, was shown in a study of the histone chaperone Vps75 (Bonangelino, Chavez, & Bonifacino, 2002; Selth & Svejstrup, 2007). Previous biophysical studies (Berndsen et al., 2008; Park, Sudhoff, Andrews, Stargell, & Luger, 2008; Tang, Meeth, Jiang, Luo, & Marmorstein, 2008) had revealed a homodimeric structure for Vps75 but there were indications that at a physiological salt concentration Vps75 formed tetramers. In order to test and characterize the tetrameric conformation, PELDOR was used in conjunction with the cross-linking spin-label Rx (Fleissner et al., 2011). The latter allowed the dimer form of the Vps75 to be singly labeled (cross-linking at the dyad axis of the homodimer). PELDOR measurement, in high salt, confirmed the presence of only dimeric forms due to the absence of spin-pair observation (only background decay observed). Lowering of the salt concentration led to the observation of dipolar coupling in the PELDOR signal. Because the maximum observation time was only 5 µs, one could surmise that there was a long distance present, but there was not enough data to reliably fit a background correction or make any accurate estimation of the distance (see Fig. 10). Upon deuteration of Vps75, it became possible to measure the dipolar evolution time up to 25 µs and a reliable distance of 78 Å was extracted (Fig. 10). This measurement not only confirmed the presence of a tetramer but also provided the information required to determine an accurate model of the tetramer structure (Bowman et al., 2014). Incidentally, the intrinsic rigidity, of the Rx label, has since permitted the measuring of spin-label orientations using analysis of W-band PELDOR data. Orientation measurements at a distance of 78 Å would be impossible without the added data available in a fully deuterated system.

Increasing sensitivity has been the aim of a number of studies and include a move to higher frequency (Ghimire, McCarrick, Budil, & Lorigan, 2009; Reginsson et al., 2012), alternative pulse EPR experiments (Borbat, Davis, Butcher, & Freed, 2004; Jeschke et al., 2004; Lovett, Lovett, & Harmer, 2012), and in the case of membrane proteins, the use of lipid minidisks

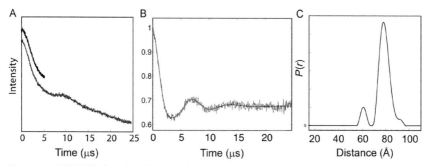

Figure 10 PELDOR data for histone chaperone Vps75 (Y35RX, low salt): (A) time traces for the nondeuterated (short trace) and deuterated (long trace) protein, absolute intensity for the short trace shifted for clarity. (B) Background-corrected time domain signal (data and fit). (C) Tikhonov-derived distance distribution.

(Sahu et al., 2013), non-spin-labeled protein dilution (Endeward, Butterwick, MacKinnon, & Prisner, 2009), and the use of more rigid spin labels. The large gains in echo persistence seen with protein deuteration provide significant gains in sensitivity and these should be additive to the other approaches mentioned.

2. DEUTERATION

2.1 Overview

Two other biophysical techniques, neutron diffraction and NMR, routinely use deuteration of proteins. Because the scattering of neutrons is dependent on interactions with the nucleus rather than electrons, as is the case with X-ray diffraction, it is a powerful method for determining the positions of hydrogen atoms. Deuterium nuclei exhibit significant differences in scattering length, in both the phase and magnitude, from hydrogen nuclei, and so deuteration has been used to highlight different components within large biological complexes. Different types of experiment require different extents of deuteration with neutron crystallography requiring complete protein deuteration and small-angle neutron scattering requiring only partial deuteration (around 70%) to achieve contrast between different components in a complex system. Protein NMR has always suffered from the problem of rapidly increasing linewidth with increasing molecular size, restricting the size range of proteins that could be studied. Linewidth in NMR is proportional to the inverse of T_2. Relaxation in solution NMR comes from a variety of mechanisms but a significant portion comes from dipole–dipole interactions and, a significant portion of that, can be eliminated by substituting protons for deuterons within the protein. Deuterons being spin 1 rather than spin 1/2 for protons and having a gyromagnetic ratio 6.5 times smaller. Because NMR techniques have used isotope enrichment to provide nuclei for direct or indirect detection, it has been a relatively small step to include partial deuteration into protein samples in order to slow T_2 relaxation and decrease linewidths. Perdeuteration of proteins (levels of >95%) increases both T_2 and T_1 and removes j-coupling from remaining protons. Random fractional deuteration of proteins retains some of the advantages of perdeuteration. Sophisticated tailoring of the type and extent of deuteration to the specific experiment required has an enormous impact on the size of protein approachable by NMR and the quality of the data obtainable.

2.2 Methods for Producing Deuterated Proteins

The methods for the production of deuterated proteins vary quite widely between applications. NMR uses a large range of increasingly sophisticated means to incorporate deuterium into protein samples as in perdeuteration, random fractional deuteration, and site-specific deuteration or protonation in a background of low or high deuteration. Neutron diffraction uses many of the techniques developed for NMR although mainly restricted to perdeuteration and random fractional deuteration. The fundamental problem in deuteration of proteins by endogenous expression of recombinant proteins in bacteria is the slow growth rate compared to that of normal, nondeuterated expression. It appears that the slower growth rate of bacteria in deuterated media is not due to a generalized isotope effect in slowing down metabolic processes but rather that the rate of growth is dictated by the effect of deuteration on a relatively small and rate limiting subset of biological processes. It has been common practice to attempt to adapt cells to deuteration by sequential increase in the percentage of deuterium in the growth media prior to large-scale growth and induction of protein expression. In this way minimal growth media in D_2O supplemented by a deuterated carbon source such as glucose or glycerol can be used successfully following deuterium adaption. An alternative approach to the production of perdeuterated proteins makes use of the availability of fully deuterated rich media derived from hydrolyzed algae grown in a fully deuterated media. Due to the more amenable growth conditions for algae, this can be an efficient way to proceed. Indeed, when a rich deuterated media is used then adaption to a deuterated environment is not required. Growth and expression times when using rich deuterated media are often slower and yield somewhat reduced. In many cases, the ease and efficiency of using deuterated rich media enable the routine expression of all but very difficult proteins. The main constraint on the use of rich media, being cost. If slightly lower levels of deuteration can be tolerated (lower T_m), then an alternative to using fully deuterated media is to use growth in nondeuterated minimal media, supplemented with added deuterated algal hydrolysate. If protein expression times are kept to a minimum, the presence of H_2O in the media does not vastly impact on the overall deuteration levels. The main bulk of the amino acids utilized, presumably coming from the algal hydrolysate without large amounts of deuterium–proton scrambling.

2.3 Measurement of Deuteration

Measurement of deuteration in proteins is generally carried out by mass spectrometry. A calculator spreadsheet for calculating percentage deuteration from amino acid composition and MS data is available from The Deuteration Laboratory, Institut Laue Langevin (https://www.ill.eu/sites/deuteration/Protocols/Mass%20spect%20calculation2.xls).

Calculation of percentage deuteration is complicated by the fact that many protons in a protein structure are exchangeable with water and the rate of exchange is hugely variable. Proteins isolated and purified in the normal way will have had plenty of time to exchange all the positions possible, with protons from the buffers that they are handled in. The calculation of theoretical mass of a deuterated protein must obviously take the exchangeable positions into account in the calculation. The mass difference between protonated and deuterated proteins is often only a few percent. Fractional deuteration will inevitably lead to a range of deuterated states. Highly accurate determination of the percentage deuteration can therefor be difficult.

3. THE PULSE EPR EXPERIMENT

All pulse experiments have been performed on an X-band (9.8 GHz) Bruker Elexsys E580 spectrometer, with a probe head supporting a dielectric resonator EN 4118X-MD4, equipped with a continuous helium-gas flow cryostat (Oxford CF-935) operating in 3.5–300 K temperature range. Pulses are amplified by a 1-kW pulsed traveling wave tube (TWT) amplifier. The resonator was overcoupled similarly for each experiment to a quality factor Q of approximately 100. This Q value can only be estimated if we do know the actual resonator bandwidth. Therefore, a nutation experiment is performed to map out the B1 profile versus the frequency in order to get the resonator bandwidth (for more details, on how to run such experiment, see the application notes provided by Bruker). The resonator bandwidth can be determined at different levels of strength of the overcoupling, which results in knowing the corresponding Q factor and therefore being able to reproduce the same Q value for similar experiments. However, we do not need necessarily to set up the same Q factor for all pulse experiments since some of them do not require a large resonator bandwidth as the PELDOR or DEER does.

The first experiment carried out usually is the field-swept echo, which generates an EPR spectrum analogous to the conventional CW EPR spectrum, but in the form of an absorption spectrum rather than the first derivative. This experiment is based on a detection sequence $\pi/2 - \tau - \pi - \tau$-echo in which the echo intensity is recorded, by integrating over the whole echo, as a function of the magnetic field. The typical values of $\pi/2$ range between 8 and 24 ns, at X-band, depending on resonators and available microwave power. The pulse separation τ is usually set at a value which is always larger than the dead-time (100 ns at X-band). The shot repetition time (SRT), which is the time between individual experiments, is an experimental parameter that needs to be judiciously adjusted and is usually set at three to five times T_1 to allow most of the magnetization to recover its equilibrium before starting the next experiment.

3.1 T_m Measurements

The T_m measurement is based on a detection sequence $\pi/2$–τ–π–τ–echo where the echo is monitored while the delay τ is increased in steps. There are two ways to perform this experiment; either one monitors the height of the echo (single point experiment) or one integrates a part of the echo symmetrically around its highest point. For the later experiment, it was suggested that the integration window should be as long as the longest π pulse in order to maximize the signal-to-noise ratio (Jeschke, 2007). A two-step phase cycle is required to remove receiver offsets. This can be performed using the PulseSPEL program provided by Bruker for T_2 measurements. The typical values used for our T_m measurements are $t\pi/2 = 16$ ns, $t\pi = 32$ ns, $\tau = 100$–400 ns, SRT = 4 ms, video bandwidth = 20 MHz, and step = 8–16 ns and the experiment is generally performed, for nitroxide spin label, at 50 K. The magnetic field is normally set at the high point of the field-swept echo. Usually, a short τ (100 ns) is recommended for such experiment since the data points times the step is shorter than the TWT gate (which is the time that the TWT can be turned on and is generally 10 μs). In this case both, pulses $\pi/2$ and π, are so close that they share the same TWT gate. When the time between the two pulses is over 10 μs (usually needed for measuring the longer T_m of deuterated proteins), a τ of at least 360 ns is used which allows each pulse to have its own TWT gate and therefore, there is no restriction on how far apart, the pulses are taken. This also can be used in PELDOR experiments when measuring distances over 6 nm or using an evolution time τ_2 which is over 10 μs. Alternatively, one can use the program provided by Bruker (2P_Split_TWT.exp), which controls the TWT gate splitting for experiments where the delay τ starts small and becomes large.

3.2 PELDOR/DEER Experiment Setup

Four-pulse experiments PELDOR/DEER are performed with the pulse sequence previously described. The typical value of the pump pulse with an MD4 resonator is around 16 ns. Usually, if the microwave power was optimized for a two-pulse Hahn echo with 16 and 32 ns as $\pi/2$ and π, the optimum pump pulse length is not less than 32 ns when using the stripline pulse-forming unit (SPFU) channels provided as defaults channels in the EPR spectrometer. This unit has four channels with fixed phases and one main attenuator. Unlike the SPFU channels, the optional microwave pulse-forming unit channels provide four channels with individual attenuators and phase shifters, which allow combining soft and hard pulses in the same experiment. This results in getting a shorter pump pulse even if the observer π pulse is set at 32 ns. PELDOR setup starts by recording a field-swept echo at the frequency corresponding to the center of the resonator dip. This allows the selection of the magnetic field position at which the pump pulse should be optimized and applied (usually, at highest sensitivity point of the absorption spectrum; see Fig. 4). Therefore, the pump pulse is set at this field position, while the observer pulses were set to an offset of 80 MHz which coincides with the low-field maximum of the spectrum (see Fig. 4). The observer pulses were set at 16 and 32 ns as $\pi/2$ and π, respectively. Since a deuterated solvent was used, the delay τ_1 (see Fig. 3B) was set to 380 ns, which corresponds to a blind spot of the deuterium modulation. Such choice is based on the fact that at this value of τ_1 most of the deuterium contribution via the ESEEM effect is suppressed. Proton modulation, which might also occur in the PELDOR data, can be suppressed by adding the signals of eight, four-pulse, PELDOR experiments. This corresponds to eight different τ_1 values that differ by increments of 8 ns. This loop is an option incorporated in the PELDOR running program. Two-step phase cycling is required to eliminate receiver offset, which avoid the latter being retained in the PELDOR data. In all experiments, the pump pulse position was set initially with a delay of 100 ns after the second observer pulse and the data points together with increment step were adjusted in a way that the pump pulse stops 100 ns before the last observer pulse. This avoids pulse overlapping that might introduce artifacts in PELDOR data. For longer evolution time τ_2 (>10 μs), the pump pulse position is set at 360 ns after the second observer π pulse in order to overcome the TWT limitation gate as mentioned in section T_m experiments.

4. DATA ANALYSIS
4.1 T_m Data Fitting

The phase relaxation times were determined using the Hahn echo sequence with variable interpulse, τ. The data was fitted with a stretched exponential:

$$y(\tau) = y_0 \exp\left[-\left(\frac{2\tau}{T_m}\right)^x\right]$$

where y_0 is the echo intensity extrapolated to time zero. T_m is the phase relaxation time. The parameter x depends upon the processes that dominate the echo dephasing (Eaton & Eaton, 1999). The strong ESEEM contribution, due to the hyperfine interaction with deuterium, to the echo signal makes fitting of the equation to the decay curve quite problematic. Several approaches have been suggested to get reliable fits although some discrepancies have been reported in the T_m values deriving from such approaches (Huber et al., 2001). At X-band, although the deuterium ESEEM signals obscure the first part of the phase memory decay curve, the decay is so long, when using a deuterated protein, that the first part of the decay curve can be safely ignored and the fit to the decay curve started from a point at which the ESEEM-derived oscillation have fully decayed. This results in easy fitting to the equation and so more accurate estimation of T_m. Unlike at X-band, W-band ESEEM-derived oscillation is not a major problem because as the modulation depth of the ESEEM contribution scales with the inverse square of the magnetic field, the deuterium nuclear modulations are significantly suppressed. However, in some cases another oscillation, attributed to electron–electron dipolar coupling, can be seen superimposed on the general echo decay. The presence of this dipolar coupling had been previously recognized (Kulik, Dzuba, Grigoryev, & Tsvetkov, 2001), although it is only with the extremely long decay curves and long distances that the oscillation becomes clearly visible in the decay curve, even in the presence of significant deuterium ESEEM signal.

4.2 PELDOR Data Analysis

The time domain signal of the PELDOR experiments consists of two contributions: the first results from interaction between two spins within the molecule of interest (known as intramolecular interaction) and the second from interaction between neighboring spins on different molecules (known as intermolecular interaction). The PELDOR signal is usually described as a product of these two contributions (Pannier et al., 2000):

$$V(t) = V(t)_{\text{inter}} \times V(t)_{\text{intra}}.$$

The PELDOR data analysis relies on the separation of these two contributions into a dipolar evolution function, $V(t)_{\text{inter}}$, and background decay usually described as monoexponential decay function. In simple cases, a first estimate of the distance can be achieved by processing the PELDOR data manually with the processing tools provided by Xepr Software (the same software that controls experiments). First, we fit the time trace of the PELDOR data with a monoexponential decay and by dividing the $V(t)$ by this decay function the modulated signal, $V(t)_{intra}$ can be obtained. The latter is apodized with a hamming function and zero filled. When Fourier transformed, the processed time trace yields a Pake pattern, frequency domain, with two singularities corresponding to $\nu_\perp$, at its highest point, and $\nu_\parallel$ at its edge. A distance can be then calculated by using the following expression:

$$r(\text{nm}) = \left(\frac{52.18}{|\nu_\perp|(\text{MHz})}\right)^{1/3}.$$

The most popular software for analysis of PELDOR data is DeerAnalysis (Jeschke et al., 2006), capable of fitting background decay in a variety of ways dependent on the nature of the sample, and using Tikhonov Regularization to derive a distance distribution. Several factors make processing of PELDOR data from deuterated proteins more satisfactory. First, the signal to noise is often better due to the sensitivity enhancement that has been described earlier. Second, the spin echo can be generated at such a time the oscillating signal can completely decay, this enables much better background fitting and correction compared to data in which the oscillations are still present at the longest time accumulated. Third, it is more likely that at least two complete oscillations can be measured allowing for not just accurate distance measurement but accurate distance distributions to be estimated (Jeschke, 2012), and finally, as recently reported (Baber et al., 2015), full deuteration abolishes artifacts sometimes seen in PELDOR data due to variation in the environmental effects on the T_m of the spin label.

5. CONCLUDING REMARKS

The importance of EPR to the understanding of molecular biology will depend upon a number of factors. Sensitivity is undoubtedly important as is the range of distances accessible and the quality of the measurements and analysis. The use of a deuterated matrix has become standard for PELDOR, enhancing all these factors; however, total deuteration of both the matrix and the substrate has now been demonstrated to provide significantly greater improvement. For many years, the limit to distance measurement by

PELDOR has been quoted as being 80 Å and even this theoretical limit has hardly, if ever, been closely approached in biological systems. Although the inverse cube relationship between dipolar coupling frequency and distance is going to impose a limit at some point, it is safe to speculate that the maximum distance measurable by PELDOR in a deuterated biological macromolecule will be in the region of 140 Å. Although the production of deuterated proteins can be a little financially and technically daunting, the increased sensitivity of a fully deuterated system partially compensates for that and the other factors described in this chapter make it a method that should be seriously considered for many studies.

ACKNOWLEDGMENTS
We would like to acknowledge the various contributions, of the following, to work described in this chapter, Graham Smith, Michael Stevens, Tom Owen-Hughes, Colin Hammond, Andrew Bowman, and Richard Ward.

REFERENCES
Baber, J. L., Louis, J. M., & Clore, G. M. (2015). Dependence of distance distributions derived from double electron–electron resonance pulsed EPR spectroscopy on pulse-sequence time. *Angewandte Chemie, International Edition, 127*, 1521–3757.
Berliner, L. J., & Grunwald, J. (1976). Immobilized bovine lactose synthetase—Method of topographical analysis of active-site. *Federation Proceedings, 35*(7), 1752.
Berndsen, C. E., Tsubota, T., Lindner, S. E., Lee, S., Holton, J. M., Kaufman, P. D., et al. (2008). Molecular functions of the histone acetyltransferase chaperone complex Rtt109-Vps75. *Nature Structural & Molecular Biology, 15*(9), 948–956.
Bonangelino, C. J., Chavez, E. M., & Bonifacino, J. S. (2002). Genomic screen for vacuolar protein sorting genes in *Saccharomyces cerevisiae*. *Molecular Biology of the Cell, 13*(7), 2486–2501.
Borbat, P. P., Davis, J. H., Butcher, S. E., & Freed, J. H. (2004). Measurement of large distances in biomolecules using double-quantum filtered refocused electron spin-echoes. *Journal of the American Chemical Society, 126*(25), 7746–7747.
Borbat, P. P., & Freed, J. H. (2007). Measuring distances by pulsed dipolar ESR spectroscopy. Spin-labeled histidine kinases. In *Two-component signaling systems (part B). Methods in enzymology: Vol. 423* (pp. 52–116). San Diego, California: Academic Press.
Bowman, A., Hammond, C. M., Stirling, A., Ward, R., Shang, W. F., El-Mkami, H., et al. (2014). The histone chaperones Vps75 and Nap1 form ring-like, tetrameric structures in solution. *Nucleic Acids Research, 42*(9), 6038–6051.
Bowman, A., Ward, R., El-Mkami, H., Owen-Hughes, T., & Norman, D. G. (2010). Probing the (H3-H4)(2) histone tetramer structure using pulsed EPR spectroscopy combined with site-directed spin labelling. *Nucleic Acids Research, 38*(2), 695–707.
Cox, J. C., Chen, H., & Kabacoff, C. (1994). Media for cell growth and method for making them. Patent US5324658 A.
Eaton, G. R., & Eaton, S. S. (1999). Solvent and temperature dependence of spin echo dephasing for chromium(V) and vanadyl complexes in glassy solution. *Journal of Magnetic Resonance, 136*(1), 63–68.

El Mkami, H., Ward, R., Bowman, A., Owen-Hughes, T., & Norman, D. G. (2014). The spatial effect of protein deuteration on nitroxide spin-label relaxation: Implications for EPR distance measurement. *Journal of Magnetic Resonance, 248*, 36–41.

Endeward, B., Butterwick, J. A., MacKinnon, R., & Prisner, T. F. (2009). Pulsed electron–electron double-resonance determination of spin-label distances and orientations on the tetrameric potassium Ion channel KcsA. *Journal of the American Chemical Society, 131*(42), 15246–15250.

Fleissner, M. R., Bridges, M. D., Brooks, E. K., Cascio, D., Kalai, T., Hideg, K., et al. (2011). Structure and dynamics of a conformationally constrained nitroxide side chain and applications in EPR spectroscopy. *Proceedings of the National Academy of Sciences of the United States of America, 108*(39), 16241–16246.

Ghimire, H., McCarrick, R. M., Budil, D. E., & Lorigan, G. A. (2009). Significantly improved sensitivity of Q-band PELDOR/DEER experiments relative to X-band is observed in measuring the intercoil distance of a leucine zipper motif peptide (GCN4-LZ). *Biochemistry, 48*(25), 5782–5784.

Ha, T., Enderle, T., Ogletree, D. F., Chemla, D. S., Selvin, P. R., & Weiss, S. (1996). Probing the interaction between two single molecules: Fluorescence resonance energy transfer between a single donor and a single acceptor. *Proceedings of the National Academy of Sciences of the United States of America, 93*(13), 6264–6268.

Huber, M., Lindgren, M., Hammarstrom, P., Martensson, L. G., Carlsson, U., Eaton, G. R., et al. (2001). Phase memory relaxation times of spin labels in human carbonic anhydrase II: Pulsed EPR to determine spin label location. *Biophysical Chemistry, 94*(3), 245–256.

Jeschke, G. (2002). Distance measurements in the nanometer range by pulse EPR. *Chemphyschem, 3*(11), 927–932.

Jeschke, G. (2007). Instrumentation and experimental setup. In M. A. Hemminga & L. J. Berliner (Eds.), *ESR spectroscopy in membrane biophysics: Vol. 27* (pp. 17–47). New York: Springer.

Jeschke, G. (2012). DEER distance measurements on proteins. *Annual Review of Physical Chemistry, 63*, 419–446.

Jeschke, G., Bender, A., Paulsen, H., Zimmermann, H., & Godt, A. (2004). Sensitivity enhancement in pulse EPR distance measurements. *Journal of Magnetic Resonance, 169*(1), 1–12.

Jeschke, G., Chechik, V., Ionita, P., Godt, A., Zimmermann, H., Banham, J., et al. (2006). DeerAnalysis2006—A comprehensive software package for analyzing pulsed ELDOR data. *Applied Magnetic Resonance, 30*(3–4), 473–498.

Kendrew, J. C. (1958). Architecture of a protein molecule. *Nature, 182*(4638), 764–767.

Kendrew, J. C., Bodo, G., Dintzis, H. M., Parrish, R. G., Wyckoff, H., & Phillips, D. C. (1958). A three-dimensional model of the myoglobin molecule obtained by X-ray analysis. *Nature, 181*(4610), 662–666.

Koharudin, L. M. I., Bonvin, A. M. J., Kaptein, R., & Boelens, R. (2003). Use of very long-distance NOEs in a fully deuterated protein: An approach for rapid protein fold determination. *Journal of Magnetic Resonance, 163*(2), 228–235.

Kulik, L. V., Dzuba, S. A., Grigoryev, I. A., & Tsvetkov, Y. D. (2001). Electron dipole-dipole interaction in ESEEM of nitroxide biradicals. *Chemical Physics Letters, 343*(3–4), 315–324.

Lindgren, M., Eaton, G. R., Eaton, S. S., Jonsson, B. H., Hammarstrom, P., Svensson, M., et al. (1997). Electron spin echo decay as a probe of aminoxyl environment in spin-labeled mutants of human carbonic anhydrase II. *Journal of the Chemical Society, Perkin Transactions 2*, (12), 2549–2554.

Lovett, J. E., Lovett, B. W., & Harmer, J. (2012). DEER-stitch: Combining three- and four-pulse DEER measurements for high sensitivity, deadtime free data. *Journal of Magnetic Resonance, 223*, 98–106.

Milov, A. D., Salikhov, K. M., & Shirov, M. D. (1981). Application of Eldor in electron-spin echo for paramagnetic center space distribution in solids. *Fizika Tverdogo Tela, 23*(4), 975–982.

Nakagawa, K., Candelaria, M. B., Chik, W. W. C., Eaton, S. S., & Eaton, G. R. (1992). Electron-spin relaxation-times of chromium(V). *Journal of Magnetic Resonance, 98*(1), 81–91.

Pannier, M., Veit, S., Godt, A., Jeschke, G., & Spiess, H. W. (2000). Dead-time free measurement of dipole-dipole interactions between electron spins. *Journal of Magnetic Resonance, 142*(2), 331–340.

Park, Y. J., Sudhoff, K. B., Andrews, A. J., Stargell, L. A., & Luger, K. (2008). Histone chaperone specificity in Rtt109 activation. *Nature Structural & Molecular Biology, 15*(9), 957–964.

Perutz, M. F. (1963). X-ray analysis of hemoglobin. *Science, 140*(3569), 863–869.

Rajca, A., Kathirvelu, V., Roy, S. K., Pink, M., Rajca, S., Sarkar, S., et al. (2010). A spirocyclohexyl nitroxide amino acid spin label for pulsed EPR spectroscopy distance measurements. *Chemistry: A European Journal, 16*(19), 5778–5782.

Reginsson, G. W., Hunter, R. I., Cruickshank, P. A. S., Bolton, D. R., Sigurdsson, S. T., Smith, G. M., et al. (2012). W-band PELDOR with 1 kW microwave power: Molecular geometry, flexibility and exchange coupling. *Journal of Magnetic Resonance, 216*, 175–182.

Sahu, I. D., McCarrick, R. M., Troxel, K. R., Zhang, R. F., Smith, H. J., Dunagan, M. M., et al. (2013). DEER EPR measurements for membrane protein structures via bifunctional spin labels and lipodisq nanoparticles. *Biochemistry, 52*(38), 6627–6632.

Schweiger, A., & Jeschke, G. (2001). *Principles of pulse electron paramagnetic resonance.* Oxford, UK; New York: Oxford University Press.

Selth, L., & Svejstrup, J. Q. (2007). Vps75, a new yeast member of the NAP histone chaperone family. *Journal of Biological Chemistry, 282*(17), 12358–12362.

Tang, Y., Meeth, K., Jiang, E., Luo, C., & Marmorstein, R. (2008). Structure of Vps75 and implications for histone chaperone function. *Proceedings of the National Academy of Sciences of the United States of America, 105*(34), 12206–12211.

Venters, R. A., Metzler, W. J., Spicer, L. D., Mueller, L., & Farmer, B. T. (1995). Use of H-1(N)-H-1(N) NOEs to determine protein global folds in perdeuterated proteins. *Journal of the American Chemical Society, 117*(37), 9592–9593.

Ward, R., Bowman, A., El Mkami, H., Owen-Hughes, T., & Norman, D. G. (2009). Long distance PELDOR measurements on the histone core particle. *Journal of the American Chemical Society, 131*(4), 1348–1349.

Ward, R., Bowman, A., Sozudogru, E., El-Mkami, H., Owen-Hughes, T., & Norman, D. G. (2010). EPR distance measurements in deuterated proteins. *Journal of Magnetic Resonance, 207*(1), 164–167.

Ward, R., Zoltner, M., Beer, L., El Mkami, H., Henderson, I. R., Palmer, T., et al. (2009). The orientation of a tandem POTRA domain pair, of the beta-barrel assembly protein BamA, determined by PELDOR spectroscopy. *Structure, 17*(9), 1187–1194.

Zavoisky, E. (1945). Spin-magnetic resonance in paramagnetics. *Journal of Physics USSR, 9*, 211–245.

Zecevic, A., Eaton, G. R., Eaton, S. S., & Lindgren, M. (1998). Dephasing of electron spin echoes for nitroxyl radicals in glassy solvents by non-methyl and methyl protons. *Molecular Physics, 95*(6), 1255–1263.

CHAPTER SIX

Spin labeling and Double Electron-Electron Resonance (DEER) to Deconstruct Conformational Ensembles of HIV Protease

Thomas M. Casey, Gail E. Fanucci[1]
Department of Chemistry, University of Florida, Gainesville, Florida, USA
[1]Corresponding author: e-mail address: gefanucci@gmail.com

Contents

1. Perspective—Conformational Sampling of HIV-1 Protease — 154
2. Site-Directed Spin-Labeling Electron Paramagnetic Resonance Spectroscopy — 156
   2.1 A Tool for Characterizing Functional Dynamics and Conformational Equilibria of HIV-1PR — 156
   2.2 Considerations for EPR Studies — 160
3. DEER Distance Profiles Reflect the Fractional Occupancies of HIV-1PR Conformational States — 165
   3.1 Gaussian Reconstruction of Tikhonov Regularization Distance Profiles — 165
   3.2 Establishing Confidence in Minor Components of DEER Distance Profiles — 167
   3.3 Statistical Analysis of DEER Distance Profiles — 168
4. "DEERconstruct," a Tool for Statistical Analysis of DEER Distance Profiles — 170
   4.1 Procedure — 170
   4.2 Case Studies — 175
References — 183

Abstract

An understanding of macromolecular conformational equilibrium in biological systems is oftentimes essential to understand function, dysfunction, and disease. For the past few years, our lab has been utilizing site-directed spin labeling (SDSL), coupled with electron paramagnetic resonance (EPR) spectroscopy, to characterize the conformational ensemble and ligand-induced conformational shifts of HIV-1 protease (HIV-1PR). The biomedical importance of characterizing the fractional occupancy of states within the conformational ensemble critically impacts our hypothesis of a conformational selection mechanism of drug-resistance evolution in HIV-1PR. The purpose of the following chapter is to give a timeline perspective of our SDSL EPR approach to characterizing conformational sampling of HIV-1PR. We provide detailed instructions for the procedure utilized in analyzing distance profiles for HIV-1PR obtained from pulsed electron–electron double resonance (PELDOR). Specifically, we employ a version of PELDOR known as double electron–electron resonance (DEER). Data are processed with the software package "DeerAnalysis" (http://www.epr.ethz.ch/software), which implements Tikhonov regularization (TKR), to generate a distance profile from electron spin-echo amplitude modulations. We assign meaning to resultant distance profiles based upon a conformational sampling model, which is described herein. The TKR distance profiles are reconstructed with a linear combination of Gaussian functions, which is then statistically analyzed. In general, DEER has proven powerful for observing structural ensembles in proteins and, more recently, nucleic acids. Our goal is to present our advances in order to aid readers in similar applications.

1. PERSPECTIVE—CONFORMATIONAL SAMPLING OF HIV-1 PROTEASE

Human Immunodeficiency Virus-1 protease (HIV-1PR) is an essential enzyme in the replicative lifecycle of HIV (Louis, Ishima, Torchia, & Weber, 2007; Robbins et al., 2010). As such, it is a critical target in the management of infection (Agniswamy et al., 2012; Joint United Nations Programme on HIV/AIDS, 2010; Martinez-Cajas & Wainberg, 2007; Robbins et al., 2010; Wlodawer & Vondrasek, 1998). HIV-1PR, shown in Fig. 1, is a symmetric homodimer where each monomer consists of 99 amino acids (Louis et al., 2007; Wlodawer & Gustchina, 2000). It is well known that HIV-1PR exhibits multiple conformational states during activity, which

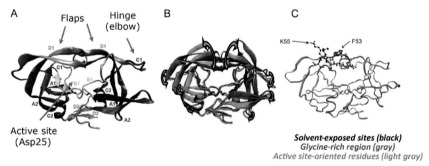

Figure 1 HIV-1 protease structure. (A) Ribbon diagram of HIV-1PR showing various structural elements as described by Wlodawer (Wlodawer & Gustchina, 2000). (B) Overlay of HIV-1PR conformations showing how regions of the protein move during opening and closing of the β-hairpin "flaps." (C) Ribbon diagram showing ball and stick representations of the spin-labeling sites K55 and F53, inward facing residues, and the glycine-rich tips of the flaps. *Panel (B): Structures are taken from MD simulations and were a gift from Carlos Simmerling.*

involves movement of two β-hairpin "flaps" (shown in Fig. 1) that mediate access to HIV-1PR's active site region (Ding, Layten, & Simmerling, 2008; Freedberg et al., 2002; Hornak, Okur, Rizzo, & Simmerling, 2006b; Ishima, Freedberg, Wang, Louis, & Torchia, 1999; Sadiq & De Fabritiis, 2010). Each state includes an ensemble of structures in which monomer orientations differ, primarily in the relative orientations of the flaps (Ding et al., 2008; Hornak, Okur, Rizzo, & Simmerling, 2006a; Ishima et al., 1999).

At present, our view of the conformational sampling landscape in HIV-1PR evokes a dynamic equilibrium between four distinct conformational states (Blackburn, Veloro, & Fanucci, 2009; Carter et al., 2014; de Vera, Smith, et al., 2013; de Vera, Blackburn, & Fanucci, 2012; Galiano, Bonora, & Fanucci, 2007; Galiano, Ding, et al., 2009; Huang et al., 2012, 2014; Kear, Blackburn, Veloro, Dunn, & Fanucci, 2009; Torbeev et al., 2009, 2011). The model landscape supposes an ensemble average of the four conformers sampled by the apoenzyme, where ligand binding or polymorphisms alter the relative stability, and hence, fractional occupancy of each conformer in the ensemble. These conformational states are referred to as closed, semi-open, curled/tucked, and wide open, where the last two are taken together as "open-like" states, as it is assumed that inhibitor can readily dissociate from the active site pocket (de Vera, Smith, et al., 2013; Huang et al., 2012, 2014). A comparison of ensemble conformers is shown in Fig. 2.

Numerous crystal structures of HIV-1PR show the flaps in either a semi-open or closed conformation (Hong, Zhang, Hartsuck, & Tang, 2000; Louis et al., 2007). Evidence for a large conformational opening of the flaps has been predicted from kinetic effects (Ermolieff, Lin, & Tang, 1997; Freedberg et al., 2002), nuclear magnetic resonance (NMR) studies

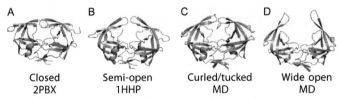

Figure 2 Conformational sampling model for HIV-1PR. Ribbon diagrams were made from coordinates obtained from X-ray data deposited within the protein data bank (PDB) or from MD simulations as indicated. The four flap conformers are termed (A) "closed," (B) "semi-open," (C) "curled/tucked," and (D) "wide open." Dynamic equilibrium between these four states correlates to substrate and inhibitor binding. Naturally occurring polymorphisms or drug-pressure selected mutations shift the equilibrium to alter enzymatic activity and substrate and inhibitor binding efficiencies. *MD coordinates were a gift from Adrian Roitberg.*

(Huang et al., 2012, 2013; Ishima & Louis, 2008), and molecular dynamics (MD) simulations (Freedberg et al., 2002; Sadiq & De Fabritiis, 2010; Wittayanarakul et al., 2005). Several X-ray structures have been solved in a more "open-like" state (Agniswamy et al., 2012; Coman et al., 2008; Martin et al., 2005; Robbins et al., 2010; Weber, Agniswamy, Fu, Shen, & Harrison, 2012), but the flaps are not as wide open as seen in snap-shots from MD simulations (Fig. 2; Ding et al., 2008; Hornak et al., 2006b). According to MD simulations of subtype B, HIV-1PR in the absence of substrate or inhibitor samples predominantly the semi-open state (60–85%), while the relative population of the wide-open state is consistently minor (<10%) (Ding et al., 2008; Hornak et al., 2006a, 2006b). The conformational equilibrium shifts to favor the closed-like states in the presence of substrate or inhibitor. Detailed NMR investigations also show that inhibitor binding alters the backbone dynamics of protease, particularly in the flap region. The glycine-rich tips of the flaps retain dynamics that occur on the nanosecond timescale, whereas backbone fluctuations elsewhere in the β-hairpin slow from the microsecond to millisecond regimes (Freedberg et al., 2002; Ishima & Torchia, 2003; Ishima et al., 1999; Katoh et al., 2003; Louis et al., 2007). Inhibitor binding also results in a shift of the conformational ensemble to a predominantly closed state (Ishima et al., 1999; Katoh et al., 2003; Louis et al., 2007).

Based on this notion of a dynamic equilibrium between closed, semi-open, and wide-open states, we aimed to utilize site-directed spin-labeling (SDSL) electron paramagnetic resonance (EPR) spectroscopy to characterize the "opening" and conformational dynamics of the flaps in the apoenzyme, along with ligand-induced conformational shifts to the closed state.

2. SITE-DIRECTED SPIN-LABELING ELECTRON PARAMAGNETIC RESONANCE SPECTROSCOPY

2.1 A Tool for Characterizing Functional Dynamics and Conformational Equilibria of HIV-1PR

Continuous-wave (CW) and pulsed EPR spectroscopy have become important tools for studying biomolecular structure and conformational dynamics (Fanucci & Cafiso, 2006; Hubbell, Cafiso, & Altenbach, 2000; Hubbell, Gross, Langen, & Lietzow, 1998; Hubbell, Lopez, Altenbach, & Yang, 2013; Hubbell, McHaourab, Altenbach, & Lietzow, 1996). In this method, unpaired electrons either exist or are incorporated into biomolecules at selected sites where the experimenter is interested in interrogating the local structure or dynamics of motion. Because the behavior(s) of unpaired electron(s) in externally applied static and oscillating magnetic fields is dependent on the spin probe's local environment, EPR spectra reflect structural and/or functional characteristics in that region. For our studies with HIV-1PR, site-directed mutagenesis strategies were employed, in which nitroxide-based spin labels were utilized as the source of unpaired electrons. Several reviews describe how the CW EPR spectrum for nitroxide spin probes reflects motion and the local environment (Casey et al., 2014; Hubbell et al., 1998, 2000; McHaourab, Lietzow, Hideg, & Hubbell, 1996; Zhang et al., 2010).

When applied to HIV-1PR, the X-band CW EPR line shapes for nitroxide spin labels incorporated at the chosen flap reporter site (K55C) did not reflect changes in structure or dynamics upon substrate or inhibitor binding (Galiano, Blackburn, Veloro, Bonora, & Fanucci, 2009; Galiano et al., 2007). Furthermore, even with different protease constructs, the X-band CW EPR line shapes are nearly identical (de Vera, Smith, et al., 2013; Kear et al., 2009). Figure 3 shows spectra obtained for the spin label at site K55C with and without inhibitor (overlain) for three different HIV-1PR constructs.

This finding, for us, was unusual considering conformational changes in spin-labeled macromolecules are usually accompanied by a change in the CW EPR line shape (Fanucci & Cafiso, 2006; Hubbell et al., 2000; Kim, Xu, Murray, & Cafiso, 2008). The conclusion drawn from these early studies was that the spin label dynamics at site K55C reflected mostly the internal motion of the spin label, which was largely decoupled from the backbone

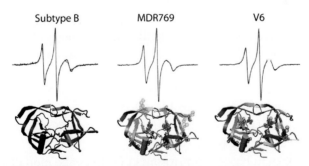

Figure 3 Selected CW EPR spectra for site K55R1 [R1 = Cys labeled with S-(1-oxyl-2,2,5, 5-tetramethyl-2,5-dihydro-1H-pyrrol-3-yl)methylmethanesulfonothioate (MTSL)] on three different HIV-1PR constructs in the absence and presence of inhibitor (scan width, 100 Gauss). The overlain spectra are nearly identical. Ribbon diagrams show the location of the amino acid substitutions relative to subtype B. (See the color plate.)

dynamics of the β-hairpin flap (Galiano, Ding, et al., 2009; Galiano et al., 2007). Unpublished data collected at the National Biomedical Research Center for Advanced ESR Technology using an EPR spectrometer operating at 140 GHz do, however, reveal differences in spectral line shapes for HIV-1PR constructs with and without inhibitors; thus demonstrating differential dynamics at these sites. Further analyses are underway that will elucidate the molecular origins of the motions that give rise to differences in the 140 GHz spectra.

Given the extensive NMR and X-ray crystallography data that exist for HIV-1PR, we were confident that inhibitor binding would induce a conformational change in this enzyme. Hence, we pursued an alternative approach in which it was assumed that changes in conformational sampling could be studied by observing corresponding changes in distances between points of interest in HIV-1PR, and that these distance changes would correlate with the conformational shifts. For this, we turned to a pulsed EPR technique known as double electron–electron resonance (DEER), which can be used to determine profiles of distances between two unpaired electrons in bio-macromolecules. The details of DEER theory have been described elsewhere (Borbat, Davis, Butcher, & Freed, 2004; Borbat, McHaourab, & Freed, 2002; Jeschke, 2002; Jeschke, Panek, Godt, Bender, & Paulsen, 2004; Jeschke & Polyhach, 2007; Pannier, Veit, Godt, Jeschke, & Spiess, 2000), and will not be covered here. Briefly, given HIV-1PR is a homodimer, a single reporter site (K55C) is used to attach two nitroxide spin labels within the protein, targeted to be within ∼2–6 nm of one another. Dipolar couplings between the unpaired electrons in the spin labels impart modulations on a stimulated electron spin–echo amplitude collected as a function of the time spacing between specific microwave pulses in the DEER pulse sequence. Encoded in the frequencies of the modulations are the distances between coupled electron spins. When a range of distances exists in a single sample, the time domain data are a product of the individual contributions, which can be transformed into a distance domain profile. Changes in conformational dynamics at spin-labeled sites corresponding to conformational shifts will be reflected in the distance profiles.

As can be seen in Fig. 4, comparisons of DEER echo modulation curves collected for HIV-1PR in the absence or presence of inhibitor (Ritonavir) clearly reflect changes in conformation (Blackburn et al., 2009; Galiano et al., 2007) and the corresponding distance profiles reflect our expected distances of 33 Å for the closed state and 36 Å for the semi-open state. We also

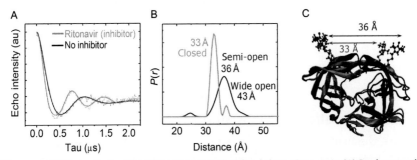

Figure 4 DEER results for HIV-1PR without and with inhibitor Ritonavir. (A) Background-corrected DEER echo modulation traces where solid lines are TKR fits of the data. (B) Corresponding distance profiles with distances and conformational states labeled accordingly. (C) Overlay of average crystallographic structure of HIV-1PR in the semi-open and closed flap conformations with MTSL appended at site K55C. Anticipated distances from modeled spin label rotamers are indicated.

proposed that the relatively small population in the uninhibited form (apo) at 43 Å corresponded to the wide-open state (Blackburn et al., 2009; Ding et al., 2008; Galiano et al., 2007). As we continued characterizing conformational sampling of other subtypes and effects of drug-pressure selected mutations, not only did we notice an increase in the fractional occupancy of the wide-open state (located at 43 Å in Fig. 4) but also the emergence of a population at a fourth distance within our ensemble (Galiano, Ding, et al., 2009; Kear et al., 2009). We currently propose that this population corresponds to an opening of the flaps in a less hindered side-to-side fashion (Carter et al., 2014; Huang et al., 2012, 2014).

Since our first investigation, we have used SDSL EPR and DEER to characterize conformational equilibria as a function of inhibitor binding (Blackburn et al., 2009; de Vera, Smith, et al., 2013; Galiano et al., 2007; Huang et al., 2012), naturally occurring polymorphisms (Huang et al., 2014; Kear et al., 2009), and drug-pressure selected mutations (Carter et al., 2014; de Vera, Blackburn, Galiano, & Fanucci, 2013; de Vera et al., 2012; Galiano, Ding, et al., 2009). Specifically, we have been successful in characterizing the expected shift to the closed conformation upon inhibitor binding as well as altered conformational landscapes induced by both natural polymorphisms and drug-pressure selected mutations. Importantly, we have also demonstrated a correlation between shifts in conformational sampling and the evolution of secondary mutations, thus providing a hypothesis for the mechanism by which mutations distal from the active site impact multidrug resistance without abolishing enzymatic activity (de Vera, Smith, et al., 2013). Our lab continues to utilize SDSL DEER to characterize conformational sampling in HIV-1PR with the goal of understanding the effects that natural polymorphisms and drug-pressure selected mutations have on conformational sampling, and the corresponding impacts on enzyme activity and drug resistance.

2.2 Considerations for EPR Studies

Before discussing the analyses of HIV-1PR DEER distance profiles in detail, we present our approach, beginning with our first experiments on HIV-1PR. When proposing SDSL EPR studies, it is essential to first explore the effects that the requirements of the experiment may have on the "native-like" system. Examples include the effects of glassing agents, cryoprotectants, sample freezing, and select mutations on conformational equilibria (Galiano, Ding, et al., 2009; Georgieva et al., 2012; Kim et al., 2008). In practice, it is ideal to repeat measurements. Typically, we do this by collecting data on similar samples prepared from two different protein batches (de Vera, Blackburn, et al., 2013; de Vera, Smith, et al., 2013).

2.2.1 Where to Incorporate the Spin Label

When deciding where to incorporate the spin label, we looked for an aqueous-exposed site. We assumed that the conformational distribution of the spin label in the aqueous-exposed region of the β-hairpin would generate fewer steric interactions, and hence, a more narrow spin label rotamer ensemble, resulting in less spin label distance heterogeneity. MD simulations for our chosen nitroxide validated our original assumption (Ding et al., 2008). Additionally, MD studies indicate that the rotomeric conformational sampling of the spin label MTSL incorporated at site K55C in HIV-1PR is not restricted by the S-H back-bonding described by Hubell as the "$\chi 4/\chi 5$ model" (Columbus, Kalai, Jeko, Hideg, & Hubbell, 2001). In fact, our MD data suggest that this interaction is present <10% of the time (unpublished data) and may account for the narrow distance profiles we obtain for MTSL in the inhibitor closed state.

For HIV-1PR, it was essential that we chose a site within the protein that could tolerate amino acid substitutions without changes in enzymatic activity. Given that at least 38 of 99 amino acid residues are modified as a result of drug-pressure selection or natural genetic drift (Martinez-Cajas, Pai, Klein, & Wainberg, 2009; Martinez-Cajas, Pant-Pai, Klein, & Wainberg, 2008; Martinez-Cajas & Wainberg, 2007), our choice of sites was limited.

Previous reports from other groups, where saturation mutagenesis was performed, showed limited sites in the flaps where mutations were readily tolerated (Shao et al., 1997). In Fig. 1C, potential target sites for spin labeling in the flaps are shown. In these regions, flexibility and dynamics are strongly correlated with function. Residues F53 and K55, for example, project into solution, making them ideal target sites. For site F53C, however, we encountered issues with protein precipitation after spin labeling. In addition, other reports showed that F53 forms important interactions with the flaps during opening and closing (Ermolieff et al., 1997; Hong et al., 2000; Saen-oon, Aruksakunwong, Wittayanarakul, Sompornpisut, & Hannongbua, 2007; Wittayanarakul et al., 2005). Site K55C, on the other hand, was found to tolerate a suite of spin labels and fluorophores without significantly impacting catalytic cleavage of a chromogenic substrate or causing precipitation during sample preparation (Blackburn et al., 2009).

Finally, the ideal site would lead to distance separations that are within the range of maximum sensitivity for DEER experiments (~30–50 Å). Because our original studies occurred before spin label rotamer modeling software, such as MMM (http://www.epr.ethz.ch/software), was readily available, we estimated anticipated distances of each conformational state using energy minimization with VMD (http://www.ks.uiuc.edu/Research/vmd). For an averaged semi-open structure (average of all structures in the database at that time) appended with the MTSL spin label, our simulations predicted an average distance of 36 Å. The same approach was utilized for the closed structure, where an average distance of 33 Å was predicted. Subsequently, through collaboration with Carlos Simmerling, full MD simulations were performed and the ensemble average of distances between spin labels was determined (Ding et al., 2008; Galiano, Ding, et al., 2009). The results, along with our VMD results, confirmed that the two K55 sites in the dimer had sampling distance ranges that are ideal for DEER investigations. Included in the results was the prediction of the aforementioned "wide-open" state, with anticipated distances centered near 43 Å. This finding was promising, as it suggested we should have access to the full range of dynamic conformers sampled by the HIV-1PR flaps.

As is the case with most spin-labeling studies, the naturally occurring cysteine (CYS) residues were initially targeted for mutagenesis. Given the rich history of HIV-1PR in structural biology, we looked to the literature and found that for various NMR and X-ray investigations it was common for the two naturally occurring CYS amino acids to be replaced with alanine (C67A and C95A) instead of the more common serine (SER) substitution (de Vera, Blackburn, et al., 2013). Although the structure of CYS is similar to SER, their hydrophobic/hydrophilic and hydrogen bonding properties are quite different. Because the naturally occurring CYS residues move into buried locations within the protein during conformational changes, the conservative mutation to SER is not well tolerated (de Vera, Blackburn, et al., 2013). As this example suggests, it can be erroneous to assume that SER is always the best choice for substitution of naturally occurring CYS residues.

2.2.2 Control Experiments for Validity of Conformational Sampling Results

2.2.2.1 Effects of Osmolytes on Conformational Sampling

Given our previous experience with using EPR for investigating the effects of osmolytes on conformational sampling of the vitamin B-12 transporter protein BtuB (a membrane protein) (Fanucci, Lee, & Cafiso, 2003; Kim, Xu, Fanucci, & Cafiso, 2006), as well as published work showing that osmolytes alter conformational sampling in T4 lysozyme (T4L) (Georgieva et al., 2012), we felt it important to assay the effects of cosolutes on the HIV-1PR conformational change. Figure 5 illustrates select results (Galiano, Blackburn, et al., 2009). These types of control experiments are essential before attempting to characterize conformational ensembles of any macromolecular system. This is especially true for DEER experiments as data are collected under cryogenic temperatures and glassing agents such as glycerol, sucrose, or Ficoll are used to improve uniformity in electron spin phase memory times (T_M) (de Vera, Blackburn, et al., 2013; Jeschke, 2002; Jeschke & Polyhach, 2007). However, this concern is eliminated in many membrane protein studies, as the lipids themselves act adequately as glassing agents and other cosolutes are not required.

The control studies on HIV-1PR show that glycerol and other polyethylene glycol-based polymers did alter the CW line shape for spin-labeled constructs. Specifically, the CW EPR spectra for the samples prepared with glycerol (Fig. 5) indicate changes in the spin label's rotational correlation time, whereas those of the sample prepared with Ficoll do not. However, DEER data collected with sucrose, glycerol, and Ficoll gave similar results, indicating that the changes seen in the CW spectra in the presence of glycerol do not arise from alterations in protein conformational ensembles. Additional studies further demonstrated that the glassing agents do not alter conformational sampling or shift the conformation to any given state (Galiano, Ding, et al., 2009). We rationalize these findings by considering that glycol-based

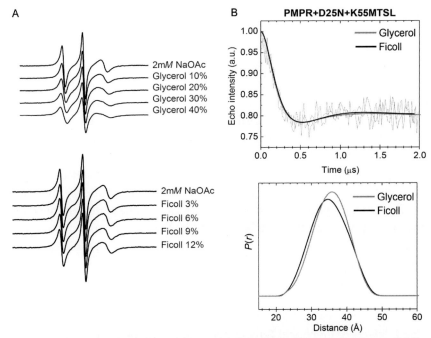

Figure 5 Effects of cosolutes on CW EPR spectra and DEER distance profiles. (A) Glycerol has a modest effect on the spectral line shape, whereas Ficoll does not perturb the spectra. (B) DEER showing that the Ficoll distance profile adequately matches that of glycerol. *Data are modified from Galiano, Blackburn, et al. (2009).*

cosolutes have specific surface interactions with HIV-1PR that impact surface hydration. As a result, the solvation changes in the spin-labeled region alter spin label dynamics without inducing the closed conformation that is observed upon addition of inhibitor or substrate (Galiano, Ding, et al., 2009; Fig. 4). For HIV-1PR, changes in surface hydration are not accompanied by changes in conformational sampling likely because the hydrophobic-exposed hydration surface area does not differ significantly in the various states (Fanucci et al., 2003; Kim et al., 2006). Based on these findings, HIV-1PR samples are prepared with glycerol (30% by volume).

In addition to ideal physical properties in solution, glycerol is typically less expensive than the other reagents and can readily be purchased in deuterated form. This is vital as the use of deuterated solvents and cosolutes is a common strategy for extending T_M to improve sensitivity in DEER experiments (Jeschke, 2002; Jeschke & Polyhach, 2007). As such, most of the HIV-1PR DEER experiments are performed on samples prepared in deuterated solvents and using deuterated cosolutes where possible (de Vera,

Blackburn, et al., 2013; de Vera, Smith, et al., 2013; de Vera et al., 2012; Huang et al., 2012).

It has recently been pointed out that solvent deuteration can lead to artifacts in situations where nonuniform solvent exposure of the spin labels leads to nonuniform T_M (El Mkami, Ward, Bowman, Owen-Hughes, & Norman, 2014). In addition, a recent study demonstrated that similar artifacts could be observed in DEER experiments performed at Q-Band (35 GHz; 1.3 T) due to the increase in the time over which the DEER pulse sequence is extended (Baber, Louis, & Clore, 2015). This has the effect of filtering frequency contributions from the time domain DEER signal according to the relationships between the total pulse sequence time and the T_M values leading to distorted distance profiles. The authors of the cited studies suggest the artifacts can be avoided by fully deuterating the biomolecule under investigation, in addition to deuterating the solvents and cosolutes. For DEER experiments at Q-Band, an alternative approach is to use a range of total pulse sequence times and analyze multiple distance profiles in parallel.

2.2.2.2 Effects of Cryogenic Temperatures on Ligand Binding

An additional concern was the effect of both temperature and cosolutes on ligand binding. In a detailed study, we utilized inhibitor titration effects on heteronuclear single-quantum coherence NMR spectra to characterize the time scale of ligand–protein interactions at near physiological temperatures in the absence of glycerol (Huang et al., 2013). The results were compared to DEER-determined inhibitor-induced conformational shifts where samples were prepared with glycerol added as a glassing agent (Huang et al., 2012). Similarities in findings from these two studies confirm the validity of the DEER approach to reliably report on structure and conformational sampling indicative of the native system. In addition, the results of the DEER study showed that for a drug resistant construct, the DEER-determined shifts to the closed state correlated with changes in drug IC_{50} values when compared to the consensus sequence (de Vera et al., 2012).

We have also explored the effects of freezing rate on the DEER conformational sampling ensemble of HIV-1PR. This was motivated by observations of Freed and coworkers in studies of T4L, in which the freezing rate affected spin label rotamer conformation (Georgieva et al., 2012). To date, we find no significant difference in ligand bound profiles between: (a) submersion of room temperature samples in isopentane suspended in liquid nitrogen ($\sim$1.5 s freezing time) or (b) equilibrating room temperature samples to 253 K followed by submersion in liquid nitrogen (~30–60 s total freezing time).

2.2.2.3 Effects of the Inactivating D25N Mutation

A final consideration regarding HIV-1PR was the presence or absence of the D25N substitution, which is typically used to prevent substrate or inhibitor processing and self-proteolysis and to enhance protein stability. It is well known that this mutation also alters inhibitor affinity (K_d) by nearly 100–1000-fold (de Vera, Smith, et al., 2013; Huang et al., 2012, 2014). However, since inhibitors bind in the nM range and the samples have enzyme concentrations of approx. 100–500 μM, which is still 10–$100 \times > K_d$, the change does not impact NMR and EPR investigations of inhibitor-induced conformational change. With the D25N mutation tolerated, we can discriminate between inhibitor binding modes that would not be detected in active enzyme. Resolution of the changes in relative inhibitor binding affinity also allows for interrogation of inhibitor-induced conformational shifts as a function of natural polymorphisms and drug-pressure selected mutations (de Vera, Smith, et al., 2013; Huang et al., 2012, 2014).

3. DEER DISTANCE PROFILES REFLECT THE FRACTIONAL OCCUPANCIES OF HIV-1PR CONFORMATIONAL STATES

3.1 Gaussian Reconstruction of Tikhonov Regularization Distance Profiles

In our studies, we observed differing degrees of conformational shifts to the closed state as a function of inhibitor binding (Fig. 6; Blackburn et al., 2009; Carter et al., 2014; de Vera et al., 2012; Galiano et al., 2007; Huang et al., 2012, 2014).

From these results, we developed the idea to regenerate the distance profiles as a linear combination of individual distance populations (Fig. 7). In an approach that is common in many electronic spectroscopies and computational methods, we used Gaussian-shaped distributions to model individual conformational states in the conformational landscape. In this approach, the relative percentage contribution of a given Gaussian distribution reflects of the fractional occupancy of the corresponding conformational state within the ensemble. We chose Gaussian-shaped functions based on the relative ease of use from a mathematical standpoint, as well as early observations that the MD-derived distances for HIV-1PR could be adequately fit with Gaussian

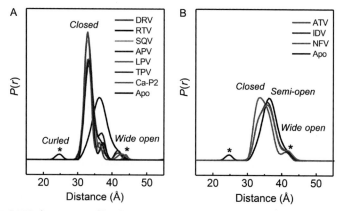

Figure 6 DEER distance profiles for subtype B HIV-1PR with various inhibitors. (A) These inhibitors shift the conformational sampling to predominantly the closed state at 33 Å, but residual population can be seen at 36 Å, which is the most probable distance in the apo protein, reflective of the semi-open state. (B) Distance profiles where the relative populations of the semi-open and closed states are roughly equal. *Data are taken from Blackburn et al. (2009).*

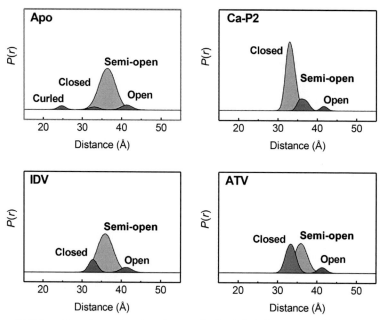

Figure 7 Select Gaussian reconstruction results for DEER distance profiles for subtype B HIV-1PR. "Apo" represents the protein alone, Ca-P2 is a nonhydrolyzable substrate mimic, IDV is the inhibitor Indinavir, and ATV is the inhibitor Atazanavir.

shapes (Ding et al., 2008; Galiano, Ding, et al., 2009). We have used this process to characterize the interactions of FDA-approved inhibitors with various HIV-1PR constructs, such as those containing natural polymorphisms (Huang et al., 2012, 2014; Kear et al., 2009) or drug-pressure-related mutations (Carter et al., 2014; de Vera, Smith, et al., 2013; de Vera et al., 2012). Shifts in fractional occupancies of populations in DEER distance profiles suggest conformational shifts that correlate well with IC_{50} values for DMR769 compared to consensus subtype B (de Vera et al., 2012). The concept of fractional occupancy has also been applied to characterize differences in the conformational sampling of apo HIV-1PR. The results demonstrate how the presence of natural polymorphisms or drug-pressure-selected mutations alters the conformational landscape (Kear et al., 2009). We have also provided a mechanism for how mutations distal from the active site can impact multidrug resistance while regaining enzymatic activity by showing a correlation between shifts in conformational sampling and the evolution of secondary mutations (de Vera et al., 2012). These findings suggest a mechanism whereby secondary mutations combine to stabilize open-like conformations at the expense of the closed state. For example, the semi-open conformation is found at greater than 65% fractional occupancy in all enzymes that have near wild-type activity. Constructs that exhibit multidrug resistance have an increase in the open-like populations with a concomitant decrease in the closed-like fractional occupancy, typically to <10%.

3.2 Establishing Confidence in Minor Components of DEER Distance Profiles

When characterizing HIV-1PR constructs with various drug-pressure selected mutations or natural polymorphisms we consistently observed a fourth distance in our ensemble that was not originally characterized in early MD simulations (denoted with * and labeled curled in Fig. 6) (Huang et al., 2012, 2014). We have come to describe this conformation as a "curled/tucked" state, where the flaps allow access to the active site, but where spin label distances are shortened, likely due to a more sideways opening of the flaps (Huang et al., 2012, 2014). Typically, the relative population of this novel state comprises less than 10% of the ensemble. We have also found that the relative population of the wide-open conformer varies from ~3–30% in various constructs (de Vera, Smith, et al., 2013; Kear et al., 2009). Given that both of these conformational states contribute to the "open-like" population and their mutual existence impacts how we think about motions in HIV-1PR and mechanisms of drug-resistance evolution, it is essential that we establish confidence in DEER distance profiles. Hence, we have developed error analysis procedures and population suppression algorithms for analyzing the distance profiles of HIV-1PR. The remaining sections of this chapter discuss our approach.

3.3 Statistical Analysis of DEER Distance Profiles

Given the significance of the emergence of minor populations in the DEER distance profiles in the context of our hypothesis that alterations to conformational sampling correlate with drug resistance, we set out to develop a method for translating the level of certainty in the time domain fit to a level of certainty in the minor components of the distance profiles. Scheme 1 illustrates our approach.

This process presumes the following: (1) The distance profile can be reconstructed as a linear combination of Gaussian functions whose relative areas reflect the fractional occupancies of the conformational states they represent and (2) the level of confidence in the time domain fit is directly related to the standard deviation of the time domain fit relative to the time domain data, which depends on the signal-to-noise ratio (SNR). In a separate review, we discussed the use of Tikhonov regularization (TKR) to fit the time domain data to generate distance profiles and detailed the process of regenerating the time domain fit from a linear superposition of Gaussian-shaped functions (de Vera, Blackburn, et al., 2013). It is known that certain processing steps or the tendency of TKR to fit noise components in time domain data with low SNR can each introduce minor artifacts in the distance profiles (Chiang, Borbat, & Freed, 2005; Jeschke, 2002; Jeschke et al., 2004). As mentioned above, solvent deuteration or increased pulse sequence time at Q-Band can lead to similar artifacts. The following procedure is designed to assess the likelihood that certain minor components of distance profiles are artifacts, as opposed to real representations of distances between spin labels. Briefly, we suppose that the distance profile obtained from fitting the time domain data (Scheme 1 "TKR distance profile") can be deconstructed into multiple components, which we call individual "populations," that each corresponds to a select conformational state within the ensemble. Next, we use Gaussian-shaped functions to represent each population and reconstruct our distance profiles as a linear combination of these functions (Scheme 1 "Gaussian reconstruction"). We then suppose that certain minor populations in the reconstructed distance profile could be artifacts. A minor population is typically chosen as a population comprising

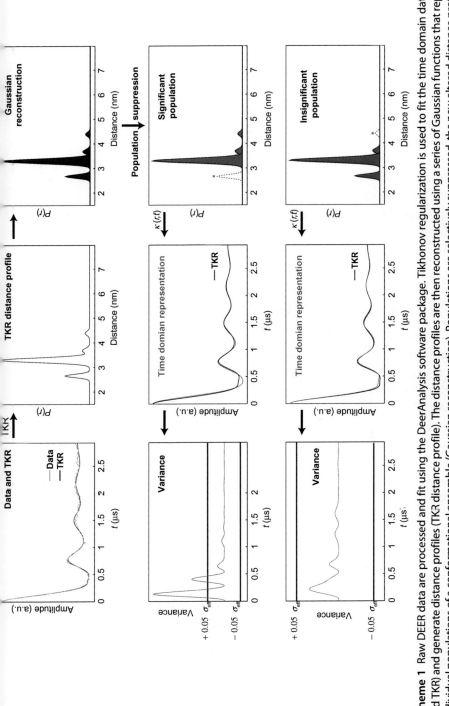

Scheme 1 Raw DEER data are processed and fit using the DeerAnalysis software package. Tikhonov regularization is used to fit the time domain data (data and TKR) and generate distance profiles (TKR distance profile). The distance profiles are then reconstructed using a series of Gaussian functions that represent individual populations of a conformational ensemble (Gaussian reconstruction). Populations are selectively suppressed, the new altered distance profiles are transformed back to time domain (time domain representation), and the variances between these and the time domain fit from TKR are calculated (variance). The variances are compared to a threshold related to the standard deviation between the time domain data and the TKR fit and individual populations are deemed insignificant contributions to the overall distance profile only if their suppression does not extend the variance beyond the threshold.

<10% of the area under the sum of all populations. To assess this notion, the linear combinations of Gaussian functions are modified to suppress populations that are user defined as minor or suspect components (Scheme 1 "significant" and "insignificant" populations). Upon suppression of a population, the new modified time domain representation is compared to the original time domain fit (Scheme 1 "time domain representation"). If the variance between the two curves falls within an interval set by the standard deviation in the original time domain fit relative to the time domain data (Scheme 1 "variance"), we conclude that the population contributes insignificantly to the error in the fit to the raw data and can be suppressed. Suppression proceeds in this manner for all possible combinations of the populations; excluding the combinations in which all or none of the populations are suppressed. The combination with the lowest number of populations, that results in a time domain representation that compares to the fit within the allowed variance interval, is chosen as the "best" solution.

4. "DEERconstruct," A TOOL FOR STATISTICAL ANALYSIS OF DEER DISTANCE PROFILES

The details of TKR methods, the best practices for processing and fitting raw DEER data to make distance profiles, and the process of regenerating time domain representations of the distance profiles have been reviewed previously (Chiang et al., 2005; de Vera, Blackburn, et al., 2013; Jeschke, 2002; Jeschke & Polyhach, 2007). For the remainder of this chapter, we will assume that the reader is at least moderately versed in common practices involved in DEER data analyses. A comprehensive tool called "DeerAnalysis" is available for processing DEER data using all of the necessary considerations (Jeschke et al., 2006). The following procedure will be presented in the context of a similar tool called "DEERconstruct," which is designed to work in tandem with DeerAnalysis to perform the entire statistical analysis procedure. DEERconstruct can be obtained either by contacting the authors of this chapter or by visiting Matlab Central File Exchange.

4.1 Procedure

We start by defining the data to be used in the procedure. First, we define "raw data" as the time domain signal after baseline correction and truncation (see aforementioned references for details regarding data processing steps). Next, we define "fit" as the time domain trace determined from analysis of the raw data. Although this is usually obtained using TKR, our procedure applies to distance profiles obtained from any fitting method. Finally, we define the "distance profile" as the plot of amplitude versus distance obtained by transformation of the fit from time domain to distance domain. These graphs are relative intensities ($P(r)$) plotted as a function of distance (r); reflecting the relative contributions to the raw data from spin pairs separated by each r value.

For the following sections, the reader may find it helpful to follow along with one of the demo data sets supplied with the DEERconstruct software.

4.1.1 Step 1—Processing Data and Generating Distance Profiles

We recommend that the DEER data be processed, fit, and transformed to a distance profile using DeerAnalysis (Jeschke et al., 2006). However, DEERconstruct will accept *.dat files that contain the raw data, fit data, and distance profile generated by any means. If using DeerAnalysis the saved results include files with the original data's filename appended with "_fit.dat" and "_distr.dat," which contain the time domain data (raw and fit) and distance profile, respectively. For the file containing the time domain data, the first column contains the time axis, the second column contains the raw data, and the third column contains the fit data. For the file containing the distance profile, the first column contains the distance axis and the second column contains the distance profile. If processing data by a means other than DeerAnalysis, the files should be formatted accordingly for use in DEERconstruct.

4.1.2 Step 2—Transformation of the Distance Profile to a Time Domain Representation

Just as a distance profile ($P(r)$) can be constructed from raw data ($R(t)$), $P(r)$ can be back-transformed to a time domain representation (Jeschke, Koch, Jonas, & Godt, 2002) (Scheme 1 "time domain representation") using the following two expressions:

$$E(t) = \int_{r_{min}}^{r_{max}} k(r, t) P(r) dr \quad (1)$$

$$k(r, t) = \int_{0}^{1} \cos\left(\frac{c}{r^3}(1 - 3\cos^2\theta)t\right) d\cos\theta. \quad (2)$$

In Eq. (1), $E(t)$ is the time domain representation of $P(r)$. Eq. (2) describes $k(r,t)$ where r are the distance points that form the x-axis of $P(r)$, t are the time points that form the x-axis of $E(t)$, θ are the angles relating the vectors connecting coupled electron spins to the direction of the external magnetic field (the dependence of $k(r,t)$ on θ is often neglected and an average is taken over all θ), and c is a constant proportional to the product of the g values for the coupled electron spins. When the data files are loaded, DEERconstruct performs three tasks and plots the results:

4.1.2.1 Task 1
The distance profile $P(r)$ is transformed to $E(t)$ using Eqs. (1) and (2).

4.1.2.2 Task 2
Two standard deviations (σ) are calculated using the expressions,

$$\sigma_{RF} = \sqrt{\frac{\sum_i \{R_i(t) - F_i(t)\}^2}{n}} \qquad (3)$$

$$\sigma_{FE} = \sqrt{\frac{\sum_i \{F_i(t) - E_i(t)\}^2}{n}} \qquad (4)$$

where n is number of data points (later updated to $n-N$, where N is set by the number of populations in the reconstructed distance profile), σ_{RF} compares $R(t)$ to the fit data ($F(t)$), and σ_{FE} compares $F(t)$ to $E(t)$.

4.1.2.3 Task 3
A variance threshold is set for qualifying/disqualifying modified $P(r)$ that will be generated in subsequent steps. The magnitude of σ_{RF} is defined by the SNR of $R(t)$, and thus directly reflects the confidence one can have in $F(t)$. The magnitude of σ_{FE} is defined by the mathematical precision of the calculation of $E(t)$ relative to $F(t)$. While σ_{FE} is usually orders of magnitude less than σ_{RF}, to be thorough we use an effective standard deviation, ($\sigma_{eff} = \sigma_{RF} + \sigma_{FE}$) to set the variance threshold at $\pm 0.05 * \sigma_{eff}$.

4.1.2.4 Optional Task 4
The user is also given the option to include the results of the "Validation" procedure offered by the DeerAnalysis software (Jeschke et al., 2006). This feature in DeerAnalysis establishes a level of uncertainty in $F(t)$ and represents this with standard deviations assigned to each point in $P(r)$. Similar standard deviation vectors obtained by a means other than DeerAnalysis can be arranged in the same format and incorporated in the same way. DEERconstruct uses these standard deviations to create a series of different profiles ($P_V(r)$) such that each $P_V(r)$ is a different variation of data points in $P(r)$, within the standard deviations determined by DeerAnalysis. Each $P_V(r)$ is transformed to time domain (Section 4.1.2.1) and compared to $F(t)$

(Eq. 4). If this option is used, the maximum value of σ, obtained by applying Eq. (4) to the full set of $P_V(r)$, is added to σ_{eff} and the variance threshold is updated accordingly.

4.1.3 Step 3—Gaussian Reconstruction of the Distance Profile

Two options are available in DEERconstruct for reconstructing $P(r)$ as a combination of individual Gaussian populations $(p_n(r))$ (Scheme 1 "Gaussian reconstruction"):

4.1.3.1 Option 1

The user defines the maximum height and corresponding r value (r_{max}) for each $p_n(r)$ visually using mouse clicks. The relative heights are controlled numerically by a weighting factor, η, such that $\eta = 1$ for the tallest $p_n(r)$ and η for the remaining $p_n(r)$ are fractions. A function then scans through values of full width at half maximum (FWHM) for each Gaussian-shaped $p_n(r)$ and chooses the appropriate FWHMs as corresponding to the minima in sums of residuals between each $p_n(r)$ and $P(r)$ in the regions encompassing each $p_n(r)$.

4.1.3.2 Option 2

The user enters the values for r_{max}, FWHM, and η manually.

Together, η and FWHM define the relative contribution of each $p_n(r)$ to the total area under $P(r)$. With either option for reconstructing $P(r)$ (here we will call $P'(r)$), DEERconstruct calculates the percentage that the area under each $p_n(r)$ contributes to the total area under $P'(r)$ and performs tasks similar to those defined above. To recap,

4.1.3.3 Task 1

$P'(r)$ is transformed to $E'(t)$ (Scheme 1 "time domain representation").

4.1.3.4 Task 2

A standard deviation $(\sigma_{E'E})$ is calculated for comparison of $E'(t)$ to $E(t)$.

4.1.3.5 Task 3

The variance thresholds are updated to include $\sigma_{E'E}$. If the reconstruction is acceptable, the magnitude of $\sigma_{E'E}$ is on the order of σ_{FE}. The updated σ_{eff} is defined as $(\sigma_{\text{eff}} = \sigma_{RF} + \sigma_{FE} + \sigma_{E'E})$.

The standard deviation $\sigma_{E'E}$ reflects the precision with which $P(r)$ can be reconstructed with a linear combination of Gaussian shapes. If $\sigma_{E'E}$

approaches σ_{RF}, $P'(r)$ needs to be entirely redefined or the present $P'(r)$ needs to be refined. For the latter, DEERconstruct allows you to attempt an optimization using least squares fitting of each individual $p_n(r)$ to $P(r)$ in their corresponding local regions. It is advisable to employ the optimization function in any case to achieve the closest possible representation of $P(r)$ before proceeding. However, the user may notice that such a procedure is inefficient when the $p_n(r)$ overlap significantly. For this reason, DEERconstruct offers several fitting options as well as the option to keep or discard the results of optimizations. Also included in the optimization function is the option to assume $p_n(r)$ are Lorentzian or pseudo-Vogtian shapes. In some cases, this can allow for a better graphical reconstruction of $P(r)$ but does not alter the procedures for manipulation of $P'(r)$ or the statistical analysis strategies.

With either option for defining $p_n(r)$, or for the optimized $p_n(r)$, the quality of $P'(r)$ can also be improved by manually adjusting the values for r_{max}, FWHM, or η in the data table. The percentage contributions to the overall $P'(r)$ are adjusted with each optimization or manual change. DEERconstruct will determine if the changes increased or decreased $\sigma_{E'E}$ and ask if the user would like to accept or reject the changes.

4.1.4 Step 4—Suppressions of Populations

For this procedure, our presumption is that some components of $P(r)$ may not be manifestations of real frequency components in $R(t)$, but instead a by-product of the imperfect fit between $F(t)$ and $R(t)$, which is usually a result of lower SNR. Assuming that $P'(r)$ is an acceptable reconstruction of $P(r)$, we can determine the effect each $p_n(r)$ has on $E'(t)$ by selectively removing them from $P'(r)$ to create a modified $P'(r)$ (here we call $P'_S(r)$) and monitoring the variance between $E'_S(t)$ and $E'(t)$. DEERconstruct offers three options for suppressing individual $p_n(r)$ (Scheme 1 "population suppression"):

4.1.4.1 Option 1
Manually remove the $p_n(r)$ from the data table.

4.1.4.2 Option 2
Use the button titled "Manually Suppress" and click at r_{max} to suppress the corresponding $p_n(r)$.

4.1.4.3 Option 3

Automatically test all combinations of suppression of $p_n(r)$ excluding the cases where all or none of the $p_n(r)$ are suppressed. This option is ideal for $P(r)$ having several components that appear to be artifacts where one can hypothesize that more than one $p_n(r)$ can be removed simultaneously.

4.1.5 Analyzing the Suppression Combinations

When a $p_n(r)$ is suppressed, DEERconstruct recalculates the percentage contributions of the areas under each remaining $p_n(r)$ to the total area under $P'_S(r)$, calculates $E'_S(t)$, calculates the variance ($v(t)$) between $E'(t)$ and $E'_S(t)$ as,

$$v(t) = \left\{ E'_i(t) - E'_{S,i}(t) \right\}^2 \qquad (5)$$

and calculates the standard deviation, $\sigma_{E'E'_S}$, that compares $E'(t)$ to $E'_S(t)$. If suppression of a $p_n(r)$ does not extend any portion of $v(t)$ beyond the thresholds ($\pm 0.05 * \sigma_{eff}$), then it is deemed statistically insignificant. In other words, it is classified as equally likely to be the result of the imperfect fit between $F(t)$ and $R(t)$ as it is to be the result of a distance between pairs of spin labels in the sample. In general, the "best" solution corresponds to the combination that contains the fewest $p_n(r)$ while remaining within the variance thresholds.

4.2 Case Studies

The following examples point out common situations encountered when using DEERconstruct.

4.2.1 Case 1—Overlapping Populations

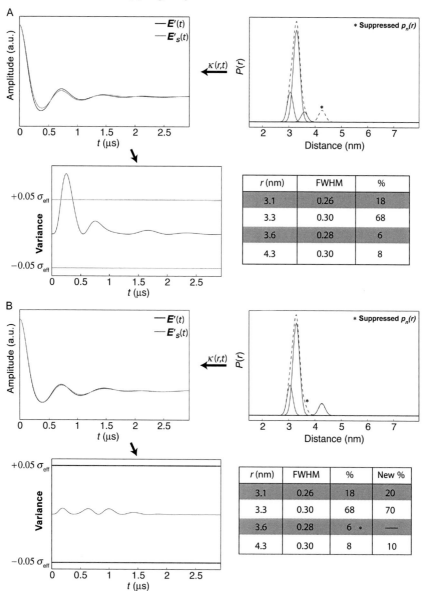

In this example, the distance profile is represented by a series of Gaussian functions, some of which overlap, and more than a single solution may exist.

The profile contains four populations, two of which each account for less than 10% of the total population. One of the two minor populations overlaps appreciably with the dominant feature, whereas the other minor population is well resolved. In Panel A, the variance plots show that suppression of the $p_n(r)$ at $r=4.3$ nm produces variance that exceeds the $\pm 0.05\sigma_{eff}$ threshold. This indicates that this $p_n(r)$ is a significant component of the data. In contrast, suppression of the overlapping $p_n(r)$ ($r=3.6$ nm) results in variance that is within threshold, and this $p_n(r)$ can be suppressed (Panel B). Before concluding that the "best" solution contains the three populations at $r=3.1, 3.3,$ and 4.3 nm, the populations must be suppressed in combinations. The best solution is obtained when the minimum number of populations that regenerate the TKR data also leads to a $E'_S(t)$ that is within the variance thresholds.

4.2.1.1 Combinations of Suppressions and Codependence of Variance

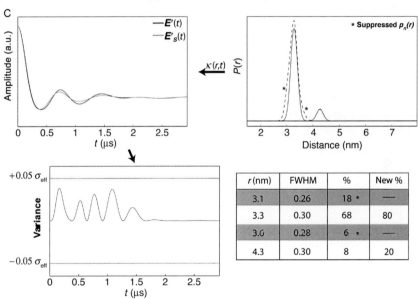

r (nm)	FWHM	%	New %
3.1	0.26	18 *	—
3.3	0.30	68	80
3.6	0.28	6 *	—
4.3	0.30	8	20

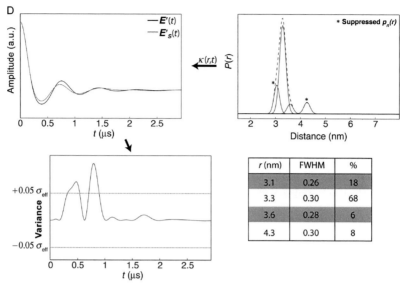

In some cases destructive interference between the variances introduced by suppressions of multiple populations damps the total variance such that individual populations cannot be suppressed within error, but combinations of $p_n(r)$ suppressions remain under the threshold. Panels C and D in Case 1 show this example. When populations at 3.1 and 3.6 nm are suppressed, the solution falls within the variance limit. However, when combining suppression of the populations at 3.1 and 4.3 nm, the variance exceeds the threshold. Interestingly, as shown in Panel E, suppressing only the $p_n(r)$ at 3.1 nm also extends the variance beyond the thresholds.

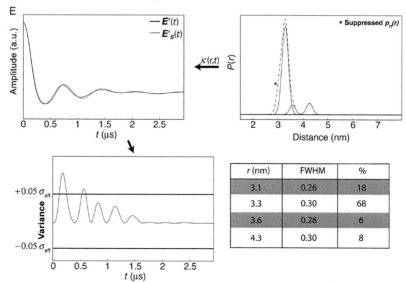

For this example, Panels B and C show solutions of $p_n(r)$ suppressions that are adequate solutions. Given that C contains only two populations, we would report this as the "best" solution.

4.2.2 Case 2—Well Resolved, Minor Components

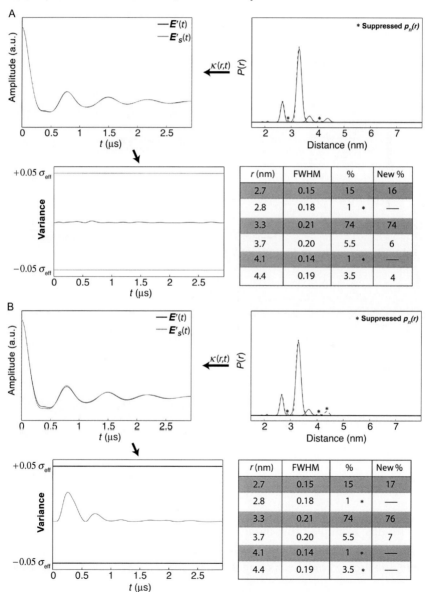

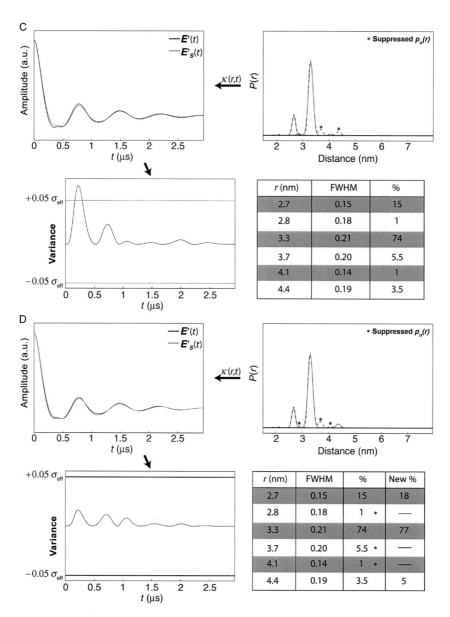

In this example, the distance profile is represented by a series of populations for which there is little overlap. For the specific case presented here, the profile contains six populations, where four represent <10% each.

Panels A and B in Case 2 demonstrate how combinations of small populations ($r = 2.8$, 3.7, 4.1, and 4.4 nm) can be suppressed without extending the variance in the time domain representation beyond the threshold. Panel C shows an example where a particular combination of populations cannot be suppressed ($r = 3.7$ and 4.4 nm). Panel D shows the "best" solution result for this example.

4.2.2.1 Effects of Signal-to-Noise Ratio

The effect of low SNR is to increase the variance thresholds, thus allowing for suppression of otherwise significant populations. Panel E in Case 2 shows how the variance threshold changes as a function of SNR (10 and 50; raw data not shown).

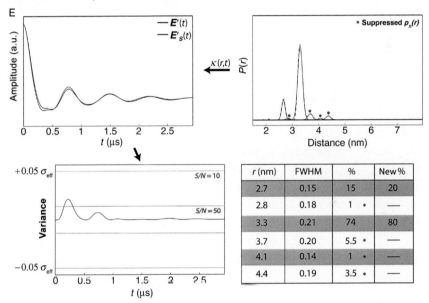

All four minor populations ($r = 2.8$, 3.7, 4.1, and 4.4 nm) can be suppressed when SNR = 10, but cannot be suppressed as a combination when SNR = 50. Hence, the presence of minor populations can be validated most accurately for data sets with high SNR.

4.2.3 Case 3—Profiles Extending Outside of Measurable Range

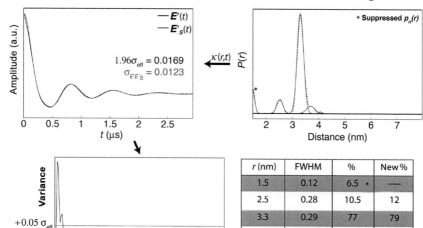

The data in this case consists of a set of five populations, one of which is centered at $r = 1.5$ nm; below the distance range generally accepted as measurable using DEER (Jeschke & Polyhach, 2007). This type of error usually arises from narrow artifacts near $t = 0$ in $R(t)$ resulting from improper zero-time approximation or noise near zero-time. When the $P'_S(r)$ with the population at 1.5 nm removed is transformed to $E'_S(t)$, the narrow, high amplitude, artifact near t_0 in $E(t)$ is missing, leading to exaggerated variance in this region. However, because the artifact is narrow, its contribution to $\sigma_{E'E'_S}$ (which is simply the square root of the normalized sum of $v(t)$) is comparatively less dramatic. In these situations, analysis of the closeness of the two time domain data using $\sigma_{E'E'_S}$ offers an alternative when components that are obviously artifacts dominate the variance analysis. As this example shows, $\sigma_{E'E'_S} = 0.0123$, which is less than $1.96\sigma_{eff} = 0.0169$ and hence, by this criteria, suppression of the population centered at $r = 1.5$ nm can be accepted as a solution. While this example demonstrates a special case, analyses of suppression combinations by maximum $v(t)$ or by $\sigma_{E'E'_S}$ generally lead to the same result.

REFERENCES

Agniswamy, J., Shen, C. H., Aniana, A., Sayer, J. M., Louis, J. M., & Weber, I. T. (2012). HIV-1 protease with 20 mutations exhibits extreme resistance to clinical inhibitors through coordinated structural rearrangements. *Biochemistry*, 51(13), 2819–2828. http://dx.doi.org/10.1021/bi2018317.

Baber, J. L., Louis, J. M., & Clore, G. M. (2015). Dependence of distance distributions derived from double electron–electron resonance pulsed EPR spectroscopy on pulse-sequence time. *Angewandte Chemie (International Ed. in English)*, 54(18), 5336–5339. http://dx.doi.org/10.1002/anie.201500640.

Blackburn, M. E., Veloro, A. M., & Fanucci, G. E. (2009). Monitoring inhibitor-induced conformational population shifts in HIV-1 protease by pulsed EPR spectroscopy. *Biochemistry*, 48(37), 8765–8767. http://dx.doi.org/10.1021/bi901201q.

Borbat, P. P., Davis, J. H., Butcher, S. E., & Freed, J. H. (2004). Measurement of large distances in biomolecules using double-quantum filtered refocused electron spin-echoes. *Journal of the American Chemical Society*, 126(25), 7746–7747. http://dx.doi.org/10.1021/ja049372o.

Borbat, P. P., McHaourab, H. S., & Freed, J. H. (2002). Protein structure determination using long-distance constraints from double-quantum coherence ESR: Study of T4 lysozyme. *Journal of the American Chemical Society*, 124(19), 5304–5314.

Carter, J. D., Gonzales, E. G., Huang, X., Smith, A. N., de Vera, I. M. S., D'Amore, P. W., et al. (2014). Effects of PRE and POST therapy drug-pressure selected mutations on HIV-1 protease conformational sampling. *FEBS Letters*, 588(17), 3123–3128. http://dx.doi.org/10.1016/J.Febslet.2014.06.051.

Casey, T. M., Liu, Z. L., Esquiaqui, J. M., Pirman, N. L., Milshteyn, E., & Fanucci, G. E. (2014). Continuous wave W- and D-Band EPR spectroscopy offer "sweet-spots" for characterizing conformational changes and dynamics in intrinsically disordered proteins. *Biochemical and Biophysical Research Communications*, 450(1), 723–728. http://dx.doi.org/10.1016/J.Bbrc.2014.06.045.

Chiang, Y. W., Borbat, P. P., & Freed, J. H. (2005). The determination of pair distance distributions by pulsed ESR using Tikhonov regularization. *Journal of Magnetic Resonance*, 172(2), 279–295. http://dx.doi.org/10.1016/J.Jmr.2004.10.012.

Columbus, L., Kalai, T., Jeko, J., Hideg, K., & Hubbell, W. L. (2001). Molecular motion of spin labeled side chains in alpha-helices: Analysis by variation of side chain structure. *Biochemistry*, 40(13), 3828–3846.

Coman, R. M., Robbins, A. H., Fernandez, M. A., Gilliland, C. T., Sochet, A. A., Goodenow, M. M., et al. (2008). The contribution of naturally occurring polymorphisms in altering the biochemical and structural characteristics of HIV-1 subtype C protease. *Biochemistry*, 47(2), 731–743. http://dx.doi.org/10.1021/Bi7018332.

de Vera, I. M. S., Blackburn, M. E., & Fanucci, G. E. (2012). Correlating conformational shift induction with altered inhibitor potency in a multidrug resistant HIV-1 protease variant. *Biochemistry*, 51(40), 7813–7815. http://dx.doi.org/10.1021/Bi301010z.

de Vera, I. M., Blackburn, M. E., Galiano, L., & Fanucci, G. E. (2013). Pulsed EPR distance measurements in soluble proteins by site-directed spin labeling (SDSL). *Current Protocols in Protein Science*. 74. http://dx.doi.org/10.1002/0471140864.ps1717s74. Unit 17.17.1–17.17.29.

de Vera, I. M., Smith, A. N., Dancel, M. C., Huang, X., Dunn, B. M., & Fanucci, G. E. (2013). Elucidating a relationship between conformational sampling and drug resistance in HIV-1 protease. *Biochemistry*, 52(19), 3278–3288. http://dx.doi.org/10.1021/bi400109d.

Ding, F., Layten, M., & Simmerling, C. (2008). Solution structure of HIV-1 protease flaps probed by comparison of molecular dynamics simulation ensembles and EPR experiments. *Journal of the American Chemical Society*, 130(23), 7184–7185. http://dx.doi.org/10.1021/ja800893d.

El Mkami, H., Ward, R., Bowman, A., Owen-Hughes, T., & Norman, D. G. (2014). The spatial effect of protein deuteration on nitroxide spin-label relaxation: Implications for EPR distance measurement. *Journal of Magnetic Resonance, 248*, 36–41. http://dx.doi.org/10.1016/j.jmr.2014.09.010.

Ermolieff, J., Lin, X., & Tang, J. (1997). Kinetic properties of saquinavir-resistant mutants of human immunodeficiency virus type 1 protease and their implications in drug resistance in vivo. *Biochemistry, 36*(40), 12364–12370. http://dx.doi.org/10.1021/bi971072e.

Fanucci, G. E., & Cafiso, D. S. (2006). Recent advances and applications of site-directed spin labeling. *Current Opinion in Structural Biology, 16*(5), 644–653. http://dx.doi.org/10.1016/j.sbi.2006.08.008.

Fanucci, G. E., Lee, J. Y., & Cafiso, D. S. (2003). Spectroscopic evidence that osmolytes used in crystallization buffers inhibit a conformation change in a membrane protein. *Biochemistry, 42*(45), 13106–13112. http://dx.doi.org/10.1021/bi035439t.

Freedberg, D. I., Ishima, R., Jacob, J., Wang, Y. X., Kustanovich, I., Louis, J. M., et al. (2002). Rapid structural fluctuations of the free HIV protease flaps in solution: Relationship to crystal structures and comparison with predictions of dynamics calculations. *Protein Science, 11*(2), 221–232. http://dx.doi.org/10.1110/ps.33202.

Galiano, L., Blackburn, M. E., Veloro, A. M., Bonora, M., & Fanucci, G. E. (2009). Solute effects on spin labels at an aqueous-exposed site in the flap region of HIV-1 protease. *The Journal of Physical Chemistry B, 113*(6), 1673–1680. http://dx.doi.org/10.1021/jp8057788.

Galiano, L., Bonora, M., & Fanucci, G. E. (2007). Interflap distances in HIV-1 protease determined by pulsed EPR measurements. *Journal of the American Chemical Society, 129*(36), 11004–11005. http://dx.doi.org/10.1021/ja073684k.

Galiano, L., Ding, F., Veloro, A. M., Blackburn, M. E., Simmerling, C., & Fanucci, G. E. (2009). Drug pressure selected mutations in HIV-1 protease alter flap conformations. *Journal of the American Chemical Society, 131*(2), 430–431. http://dx.doi.org/10.1021/ja807531v.

Georgieva, E. R., Roy, A. S., Grigoryants, V. M., Borbat, P. P., Earle, K. A., Scholes, C. P., et al. (2012). Effect of freezing conditions on distances and their distributions derived from Double Electron Electron Resonance (DEER): A study of doubly-spin-labeled T4 lysozyme. *Journal of Magnetic Resonance, 216*, 69–77. http://dx.doi.org/10.1016/j.jmr.2012.01.004.

Hong, L., Zhang, X. C., Hartsuck, J. A., & Tang, J. (2000). Crystal structure of an in vivo HIV-1 protease mutant in complex with saquinavir: Insights into the mechanisms of drug resistance. *Protein Science, 9*(10), 1898–1904. http://dx.doi.org/10.1110/ps.9.10.1898.

Hornak, V., Okur, A., Rizzo, R. C., & Simmerling, C. (2006a). HIV-1 protease flaps spontaneously close to the correct structure in simulations following manual placement of an inhibitor into the open state. *Journal of the American Chemical Society, 128*(9), 2812–2813. http://dx.doi.org/10.1021/ja058211x.

Hornak, V., Okur, A., Rizzo, R. C., & Simmerling, C. (2006b). HIV-1 protease flaps spontaneously open and reclose in molecular dynamics simulations. *Proceedings of the National Academy of Sciences of the United States of America, 103*(4), 915–920. http://dx.doi.org/10.1073/pnas.0508452103.

Huang, X., Britto, M. D., Kear-Scott, J. L., Boone, C. D., Rocca, J. R., Simmerling, C., et al. (2014). The role of select subtype polymorphisms on HIV-1 protease conformational sampling and dynamics. *The Journal of Biological Chemistry, 289*(24), 17203–17214. http://dx.doi.org/10.1074/jbc.M114.571836.

Huang, X., de Vera, I. M., Veloro, A. M., Blackburn, M. E., Kear, J. L., Carter, J. D., et al. (2012). Inhibitor-induced conformational shifts and ligand-exchange dynamics for HIV-1 protease measured by pulsed EPR and NMR spectroscopy. *The Journal of Physical Chemistry B, 116*(49), 14235–14244. http://dx.doi.org/10.1021/jp308207h.

Huang, X., de Vera, I. M., Veloro, A. M., Rocca, J. R., Simmerling, C., Dunn, B. M., et al. (2013). Backbone (1)H, (1)(3)C, and (1)(5)N chemical shift assignment for HIV-1 protease subtypes and multi-drug resistant variant MDR 769. *Biomolecular NMR Assignments, 7*(2), 199–202. http://dx.doi.org/10.1007/s12104-012-9409-7.

Hubbell, W. L., Cafiso, D. S., & Altenbach, C. (2000). Identifying conformational changes with site-directed spin labeling. *Nature Structural Biology, 7*(9), 735–739. http://dx.doi.org/10.1038/78956.

Hubbell, W. L., Gross, A., Langen, R., & Lietzow, M. A. (1998). Recent advances in site-directed spin labeling of proteins. *Current Opinion in Structural Biology, 8*(5), 649–656.

Hubbell, W. L., Lopez, C. J., Altenbach, C., & Yang, Z. (2013). Technological advances in site-directed spin labeling of proteins. *Current Opinion in Structural Biology, 23*(5), 725–733. http://dx.doi.org/10.1016/j.sbi.2013.06.008.

Hubbell, W. L., McHaourab, H. S., Altenbach, C., & Lietzow, M. A. (1996). Watching proteins move using site-directed spin labeling. *Structure, 4*(7), 779–783.

Ishima, R., Freedberg, D. I., Wang, Y. X., Louis, J. M., & Torchia, D. A. (1999). Flap opening and dimer-interface flexibility in the free and inhibitor-bound HIV protease, and their implications for function. *Structure, 7*(9), 1047–1055.

Ishima, R., & Louis, J. M. (2008). A diverse view of protein dynamics from NMR studies of HIV-1 protease flaps. *Proteins, 70*(4), 1408–1415. http://dx.doi.org/10.1002/prot.21632.

Ishima, R., & Torchia, D. A. (2003). Extending the range of amide proton relaxation dispersion experiments in proteins using a constant-time relaxation-compensated CPMG approach. *Journal of Biomolecular NMR, 25*(3), 243–248.

Jeschke, G. (2002). Distance measurements in the nanometer range by pulse EPR. *Chemphyschem, 3*(11), 927–932. http://dx.doi.org/10.1002/1439-7641(20021115)3:11<927::AID-CPHC927>3.0.CO;2-Q.

Jeschke, G., Chechik, V., Ionita, P., Godt, A., Zimmermann, H., Banham, J., et al. (2006). DeerAnalysis2006—A comprehensive software package for analyzing pulsed ELDOR data. *Applied Magnetic Resonance, 30*(3–4), 473–498. http://dx.doi.org/10.1007/Bf03166213.

Jeschke, G., Koch, A., Jonas, U., & Godt, A. (2002). Direct conversion of EPR dipolar time evolution data to distance distributions. *Journal of Magnetic Resonance, 155*(1), 72–82. http://dx.doi.org/10.1006/Jmre.2001.2498.

Jeschke, G., Panek, G., Godt, A., Bender, A., & Paulsen, H. (2004). Data analysis procedures for pulse ELDOR measurements of broad distance distributions. *Applied Magnetic Resonance, 26*(1–2), 223–244. http://dx.doi.org/10.1007/Bf03166574.

Jeschke, G., & Polyhach, Y. (2007). Distance measurements on spin-labelled biomacromolecules by pulsed electron paramagnetic resonance. *Physical Chemistry Chemical Physics, 9*(16), 1895–1910. http://dx.doi.org/10.1039/B614920k.

Joint United Nations Programme on HIV/AIDS (2010). *Global report: UNAIDS report on the global AIDS epidemic*. Geneva, Switzerland: Joint United Nations Programme on HIV/AIDS (pp. v).

Katoh, E., Louis, J. M., Yamazaki, T., Gronenborn, A. M., Torchia, D. A., & Ishima, R. (2003). A solution NMR study of the binding kinetics and the internal dynamics of an HIV-1 protease-substrate complex. *Protein Science, 12*(7), 1376–1385. http://dx.doi.org/10.1110/Ps.0300703.

Kear, J. L., Blackburn, M. E., Veloro, A. M., Dunn, B. M., & Fanucci, G. E. (2009). Subtype polymorphisms among HIV-1 protease variants confer altered flap conformations and flexibility. *Journal of the American Chemical Society, 131*(41), 14650–14651. http://dx.doi.org/10.1021/Ja907088a.

Kim, M., Xu, Q., Fanucci, G. E., & Cafiso, D. S. (2006). Solutes modify a conformational transition in a membrane transport protein. *Biophysical Journal, 90*(8), 2922–2929. http://dx.doi.org/10.1529/Biophysj.105.078246.

Kim, M., Xu, Q., Murray, D., & Cafiso, D. S. (2008). Solutes alter the conformation of the ligand binding loops in outer membrane transporters. *Biochemistry*, *47*(2), 670–679. http://dx.doi.org/10.1021/Bi7016415.

Louis, J. M., Ishima, R., Torchia, D. A., & Weber, I. T. (2007). HIV-1 protease: Structure, dynamics, and inhibition. *Advances in Pharmacology*, *55*, 261–298. http://dx.doi.org/10.1016/S1054-3589(07)55008-8.

Martin, P., Vickrey, J. F., Proteasa, G., Jimenez, Y. L., Wawrzak, Z., Winters, M. A., et al. (2005). "Wide-open" 1.3 Å structure of a multidrug-resistant HIV-1 protease as a drug target. *Structure*, *13*(12), 1887–1895. http://dx.doi.org/10.1016/j.str.2005.11.005.

Martinez-Cajas, J. L., Pai, N. P., Klein, M. B., & Wainberg, M. A. (2009). Differences in resistance mutations among HIV-1 non-subtype B infections: A systematic review of evidence (1996–2008). *Journal of the International AIDS Society*, *12*, 11. http://dx.doi.org/10.1186/1758-2652-12-11.

Martinez-Cajas, J. L., Pant-Pai, N., Klein, M. B., & Wainberg, M. A. (2008). Role of genetic diversity amongst HIV-1 non-B subtypes in drug resistance: A systematic review of virologic and biochemical evidence. *AIDS Reviews*, *10*(4), 212–223.

Martinez-Cajas, J. L., & Wainberg, M. A. (2007). Protease inhibitor resistance in HIV-infected patients: Molecular and clinical perspectives. *Antiviral Research*, *76*(3), 203–221. http://dx.doi.org/10.1016/j.antiviral.2007.06.010.

McHaourab, H. S., Lietzow, M. A., Hideg, K., & Hubbell, W. L. (1996). Motion of spin-labeled side chains in T4 lysozyme. Correlation with protein structure and dynamics. *Biochemistry*, *35*(24), 7692–7704. http://dx.doi.org/10.1021/bi960482k.

Pannier, M., Veit, S., Godt, A., Jeschke, G., & Spiess, H. W. (2000). Dead-time free measurement of dipole-dipole interactions between electron spins. *Journal of Magnetic Resonance*, *142*(2), 331–340. http://dx.doi.org/10.1006/jmre.1999.1944.

Robbins, A. H., Coman, R. M., Bracho-Sanchez, E., Fernandez, M. A., Gilliland, C. T., Li, M., et al. (2010). Structure of the unbound form of HIV-1 subtype A protease: Comparison with unbound forms of proteases from other HIV subtypes. *Acta Crystallographica Section D: Biological Crystallography*, *66*(Pt. 3), 233–242. http://dx.doi.org/10.1107/S0907444909054298.

Sadiq, S. K., & De Fabritiis, G. (2010). Explicit solvent dynamics and energetics of HIV-1 protease flap opening and closing. *Proteins*, *78*(14), 2873–2885. http://dx.doi.org/10.1002/prot.22806.

Saen-oon, S., Aruksakunwong, O., Wittayanarakul, K., Sompornpisut, P., & Hannongbua, S. (2007). Insight into analysis of interactions of saquinavir with HIV-1 protease in comparison between the wild-type and G48V and G48V/L90M mutants based on QM and QM/MM calculations. *Journal of Molecular Graphics & Modelling*, *26*(4), 720–727. http://dx.doi.org/10.1016/j.jmgm.2007.04.009.

Shao, W., Everitt, L., Manchester, M., Loeb, D. D., Hutchison, C. A., 3rd., & Swanstrom, R. (1997). Sequence requirements of the HIV-1 protease flap region determined by saturation mutagenesis and kinetic analysis of flap mutants. *Proceedings of the National Academy of Sciences of the United States of America*, *94*(6), 2243–2248.

Torbeev, V. Y., Raghuraman, H., Hamelberg, D., Tonelli, M., Westler, W. M., Perozo, E., et al. (2011). Protein conformational dynamics in the mechanism of HIV-1 protease catalysis. *Proceedings of the National Academy of Sciences of the United States of America*, *108*(52), 20982–20987. http://dx.doi.org/10.1073/pnas.1111202108.

Torbeev, V. Y., Raghuraman, H., Mandal, K., Senapati, S., Perozo, E., & Kent, S. B. (2009). Dynamics of "flap" structures in three HIV-1 protease/inhibitor complexes probed by total chemical synthesis and pulse-EPR spectroscopy. *Journal of the American Chemical Society*, *131*(3), 884–885. http://dx.doi.org/10.1021/ja806526z.

Weber, I. T., Agniswamy, J., Fu, G., Shen, C. H., & Harrison, R. W. (2012). Reaction intermediates discovered in crystal structures of enzymes. *Advances in Protein Chemistry and Structural Biology, 87*, 57–86. http://dx.doi.org/10.1016/B978-0-12-398312-1.00003-2.

Wittayanarakul, K., Aruksakunwong, O., Saen-oon, S., Chantratita, W., Parasuk, V., Sompornpisut, P., et al. (2005). Insights into saquinavir resistance in the G48V HIV-1 protease: Quantum calculations and molecular dynamic simulations. *Biophysical Journal, 88*(2), 867–879. http://dx.doi.org/10.1529/biophysj.104.046110.

Wlodawer, A., & Gustchina, A. (2000). Structural and biochemical studies of retroviral proteases. *Biochimica et Biophysica Acta, 1477*(1–2), 16–34.

Wlodawer, A., & Vondrasek, J. (1998). Inhibitors of HIV-1 protease: A major success of structure-assisted drug design. *Annual Review of Biophysics and Biomolecular Structure, 27*, 249–284. http://dx.doi.org/10.1146/annurev.biophys.27.1.249.

Zhang, Z., Fleissner, M. R., Tipikin, D. S., Liang, Z., Moscicki, J. K., Earle, K. A., et al. (2010). Multifrequency electron spin resonance study of the dynamics of spin labeled T4 lysozyme. *The Journal of Physical Chemistry B, 114*(16), 5503–5521. http://dx.doi.org/10.1021/jp910606h.

SECTION II

Spin Labeling Studies of Membrane and Membrane-Associated Proteins

CHAPTER SEVEN

Ionizable Nitroxides for Studying Local Electrostatic Properties of Lipid Bilayers and Protein Systems by EPR

Maxim A. Voinov, Alex I. Smirnov[1]
Department of Chemistry, North Carolina State University, Raleigh, North Carolina, USA
[1]Corresponding author: e-mail address: aismirno@ncsu.edu

Contents

1. Introduction	192
2. EPR Characterization of pH-Sensitive Thiol-Specific Nitroxide Labels for Mapping Local Protein Electrostatics	196
2.1 Calibration of Fast-Motion EPR Spectra of IMTSL- and IKMTSL-Labeled Thiols	199
2.2 Mapping of Local pH and Electrostatics of Peptide and Protein Systems with Methanethiosulfonate Derivatives of Ionizable Nitroxides	200
3. pH-Sensitive Spin-Labeled Lipids for Measuring Surface Electrostatics of Lipid Bilayers by EPR	202
3.1 Preparation of Spin-Labeled Lipid Vesicles	204
3.2 EPR Titration Experiments with Lipid Bilayers	204
3.3 Analysis of EPR Spectra in Titration Experiments: Least-Squares Decomposition of Experimental EPR Spectra Using a Two-Site Slow-Exchange Model	205
3.4 Interfacial pK_a and the Choice of an Electrically Neutral Reference Interface	207
3.5 Surface Charge and Potential Calculation Using the Gouy–Chapman Theory	210
4. Conclusions and Outlook	211
Acknowledgments	212
References	213

Abstract

Electrostatic interactions are known to play a major role in the myriad of biochemical and biophysical processes. Here, we describe biophysical methods to probe local electrostatic potentials of proteins and lipid bilayer systems that are based on an observation of reversible protonation of nitroxides by electron paramagnetic resonance (EPR). Two types of probes are described: (1) methanethiosulfonate derivatives of protonatable nitroxides for highly specific covalent modification of the cysteine's sulfhydryl groups and (2) spin-labeled phospholipids with a protonatable nitroxide tethered to the polar head group. The probes of both types report on their ionization state through changes in magnetic parameters and degree of rotational averaging, thus, allowing the electrostatic contribution to the interfacial pK_a of the nitroxide, and, therefore, the local electrostatic potential to be determined. Due to their small molecular volume, these probes cause a minimal perturbation to the protein or lipid system. Covalent attachment secures the position of the reporter nitroxides. Experimental procedures to characterize and calibrate these probes by EPR, and also the methods to analyze the EPR spectra by simulations are outlined. The ionizable nitroxide labels and the nitroxide-labeled phospholipids described so far cover an exceptionally wide range of ca. 2.5–7.0 pH units, making them suitable to study a broad range of biophysical phenomena, especially at the negatively charged lipid bilayer surfaces. The rationale for selecting proper electrostatically neutral interface for probe calibration, and examples of lipid bilayer surface potential studies, are also described.

1. INTRODUCTION

Electrostatic interactions are known to play one of the major roles in the myriad of cellular and molecular biology processes, ranging from protein folding to insertion of proteins, toxins, and viruses into membranes as well as other more complex events such as membrane fusion. Indeed, an essential feature of a living cell is its ability to maintain and control an electrical potential across the lipid bilayer membrane that separates the cell compartments from the rest of the world. Such a potential governs transport of electrically charged ions across the membrane, concentration of ions and other charged molecules at its surface, and, for charged residues of peripheral proteins, contributes to the free energy of binding to the membrane. The role of electrostatics is not limited to the membrane processes. Indeed, electrostatic forces are also directly involved in protein–protein association (Sheinerman, Norel, & Honig, 2000) and in structure and dynamics of DNA (Norberg & Nilsson, 2000).

Experimentally, local electrostatic interactions remain somewhat elusive parameters because of a relative scarcity of the spectroscopic methods capable of measuring these effects accurately and unambiguously. One of the methods for assessing local electric fields in proteins is based on the internal vibrational Stark effect of endogenous long-wavelength chromophores (Fried & Boxer, 2015; Lockhart & Kim, 1992, 1993; Park, Andrews, Hu, & Boxer, 1999; Steffen, Lao, & Boxer, 1994; Suydam & Boxer, 2003). Other methods include NMR (Cafiso, McLaughlin, McLaughlin, & Winiski, 1989; Crowell & Macdonald, 1999; Lindstrom,

Williamson, & Grobner, 2005), atomic force microscopy (Cuervo et al., 2014; Leonenko et al., 2007; Yang, Mayer, & Hafner, 2007), interaction force measurements (Marra, 1986), second harmonic generation (SHG) charge screening measurements (Troiano et al., 2015), fluorescent spectroscopy (Fernandez & Fromherz, 1977; Fromherz, 1989; Rottenberg, 1989), and spin probe electron paramagnetic resonance (EPR) (Barratt & Laggner, 1974; Bonnet, Roman, Fatome, & Berleur, 1990; Cafiso & Hubbell, 1981; Gaffney & Mich, 1976; Hecht, Honig, Shin, & Hubbell, 1995; Khramtsov, Marsh, Weiner, & Reznikov, 1992; Mehlhorn & Packer, 1979; Riske, Nascimento, Peric, Bales, & Lamy-Freund, 1999; Sankaram, Brophy, Jordi, & Marsh, 1990; Sanson, Ptak, Rigaud, & Gary-Bobo, 1976; Shin & Hubbell, 1992). All of these methods have some advantages as well as some limitations and drawbacks. For example, electrical potential of a lipid bilayer could be determined from measurements of a partition coefficient of a charged amphiphilic probe molecule between the lipid and the aqueous phases by analyzing continuous wave (CW) EPR spectra of the probe and an additional calibration (Cafiso & Hubbell, 1981; Gaffney & Mich, 1976; Mehlhorn & Packer, 1979). The partition coefficient is directly related to the Gibbs free energy of transferring an amphiphilic EPR probe molecule from aqueous to a membrane phase, but the location of the probe in the bilayer is largely unknown. One has also to be concerned about effects of the bilayer lipid composition on the probe's Gibbs energy as a source of additional errors.

The second group of EPR and NMR methods is based on measurements of molecular collisions between a specific site of a biomolecule, such as a protein, DNA, or a lipid, with other charged molecules. For example, local electrostatic potentials of lipid bilayer and DNA were determined from CW electron–electron double resonance measurements of the collision frequency of a freely diffusing, charged nitroxide with another nitroxide tag attached to a biomolecule in a site-directed manner (Hecht et al., 1995; Shin & Hubbell, 1992). While experimental results of the EPR collision exchange method demonstrated a remarkable agreement with the Debye–Hückel calculations for two small nitroxides (Surek & Thomas, 2008), in applications to larger molecules and membrane systems, this method is expected to suffer from the same uncertainty in the local diffusion constant as analogous NMR measurements of the site-specific relaxation enhancement upon collisions of exposed residues with charged paramagnetic relaxers (Cafiso et al., 1989; Likhtenshtein et al., 1999). Indeed, the latter complication has been noted upon completion of a detailed NMR study of electrostatically driven molecular collisions (Teng & Bryant, 2006). Specifically, it was concluded that "the three different combinations of the data sets do not yield internally consistent values for the electrostatic contribution to the intermolecular free energy" (Teng & Bryant, 2006).

The third group of methods for evaluating local electrostatic potential is based on observing reversible ionization of specific moieties by NMR, fluorescence, or EPR. The main advantage of NMR is in not relying on any exogenous probe molecules that are typically required for EPR and fluorescence measurements. Residue-specific pK_a can be measured directly by NMR for protein and RNA samples that are labeled with ^{13}C and ^{15}N, uniformly (Tollinger, Forman-Kay, & Kay, 2002) or selectively (Luptak, Ferre-D'Amare, Zhou, Zilm, & Doudna, 2001). Specialized NMR pulse sequences have been developed to measure the ^{13}C chemical shifts even for unfolded protein states where chemical shift overlap is the limiting resolution factor (Tollinger et al., 2002). However, for unambiguous pK_a determination, the dynamics of proton exchange has to be slow on the NMR time scale. The latter condition limits the number of systems that can be studied.

The use of the ionizable molecular probes and associated methods is well developed in fluorescence spectroscopy, mainly in applications to studies of electrostatic surface potentials of lipid bilayers. One of the most common fluorescent probes for such purposes is 4-alkyl-7-hydroxycoumarin. The alkyl tail of the probe inserts into the lipid bilayer and positions the ionizable OH-group just below the lipid phosphate moiety (Fromherz, 1989). As was noted by Fromherz, "a local perturbation of the interface by the chromophore is unavoidable, although the coumarin is a fluorescent probe of minimal size" (Fromherz, 1989). We also note that in lipid bilayers, the ionizable OH-group of coumarin is unavoidably positioned below the dipole of the phosphatidylcholine group. Such a location is somewhat deeper than one would wish to have for studying the bilayer surface potential. Also, a chromophore positioned right at the polar head region is expected to be sensitive to any changes in the lipid packing and, thus, would require additional calibrations for temperature and other effects.

Spin-labeling EPR represents a good alternative to fluorescence labeling of protein and membrane systems, primarily because of the complementary nature of the data on local structure and dynamics of biomolecules that this method provides (Klug & Feix, 2008). The distinguishing feature of the EPR method, relative to the fluorescence labeling, is a relatively small molecular volume of nitroxide-modified side chains that are comparable to phenylalanine (Feix & Klug, 2002), thus, introducing a minimal perturbation to the protein structure (Mchaourab, Lietzow, Hideg, & Hubbell, 1996). To this date, almost all site-directed spin-labeling (SDSL) EPR studies have been carried out with the nitroxide MTSL (S-(1-oxyl-2,2,5,5-tetramethyl-2,5-dihydro-1H-pyrrol-3-yl)methyl methanesulfonothioate), which has a methanethiosulfonate (MTS) moiety that is highly thiol-specific. While MTSL has been proven to be a very useful probe of molecular structure and dynamics, it cannot be used for local pH determination because the probe does not have any groups that are ionized within the useful biochemical pH range. In order to overcome this deficiency, we have introduced cysteine-specific labels (Smirnov, Ruuge, Reznikov, Voinov, & Grigor'ev, 2004; Voinov, Ruuge, Reznikov, Grigor'ev, & Smirnov, 2008) and lipid-mimicking EPR probes (Voinov, Kirilyuk, & Smirnov, 2009; Voinov, Rivera-Rivera, & Smirnov, 2013), that are based on derivatives of imidazoline and imidazolidine nitroxides. These nitroxides contain basic nitrogen functionalities in the heterocyclic ring, making their EPR spectra sensitive to pH changes near the probe pK_a. Specifically, the protonation of such functionality will result in acquiring a localized positive charge. Then, a component of the positive internal electric field directed along the N–O bond will partially stabilize the nitrogen p orbital, while destabilizing the p orbital of the oxygen nitrogen, leading to a partial shift of the spin density of the unpaired electron from N to O (Gulla & Budil, 2001). Decrease of the spin density on N will result in a smaller isotropic nitrogen hyperfine coupling constant, A_{iso}, and a larger isotropic g-factor, g_{iso}, that can be measured from the EPR spectra.

Imidazoline and imidazolidine nitroxides are considered to be the most suitable for biophysical applications because of the tunable pK_a, reversible pH effects, and high sensitivity of the EPR spectrum to pH changes (Khramtsov & Volodarsky, 2002). We note that while some of the imidazolidine nitroxides have been employed in studies of surface potentials and polarity of phospholipid bilayers (Khramtsov et al., 1992) and human serum albumin (Khramtsov et al., 1985, 1992) in the past, the probe attachment was not fully specific. Moreover, these earlier EPR studies were based solely on measurements of A_{iso}, which could be affected by experimental conditions other than the proton-exchange reactions.

In this chapter, we describe the use of nitroxides capable of reversible ionization by protonation to assess local proton concentration and electrostatic potentials of proteins and lipid bilayer systems by EPR. Two types of the electrostatic probes will be described. The first type of protonable nitroxides bears the MTS group that is highly specific for covalent modification of the cysteine's sulfhydryl groups. Such spin labels are very similar in magnetic parameters and chemical properties to conventional MTSL, making them suitable for studying local electrostatic properties of protein–lipid interfaces. The second type of EPR probes was designed as spin-labeled phospholipids (SLP), having a protonatable nitroxide tethered to the polar head group. The probes of both types report on their ionization state through changes in magnetic parameters and a degree of rotational averaging. We describe the methods of analysis of EPR spectra for evaluating the electrostatic contribution to the interfacial pK_a of the nitroxide, and, therefore, determining the local electrostatic potential. Finally, selective applications of these methods to study interfacial potentials are described. Potential problems and future directions are also outlined.

2. EPR CHARACTERIZATION OF pH-SENSITIVE THIOL-SPECIFIC NITROXIDE LABELS FOR MAPPING LOCAL PROTEIN ELECTROSTATICS

Figure 1 shows chemical structures of thiol-specific nitroxide spin labels, including conventional MTSL and two pH-sensitive nitroxides synthesized and characterized so far (Smirnov et al., 2004; Voinov et al., 2009). IMTSL has a tertiary amino group in the structure of the heterocycle, and IKMTSL is featured with an amidino group; the pK_a's of these functionalities determine the pK_a range that these probes are capable of reporting on. Note that the strong electron withdrawing effect of the nitroxide group decreases the intrinsic pK_a's of these groups by several units (Khramtsov et al., 1985; Khramtsov, Weiner, Grigoriev, & Volodarsky, 1982). Comparison of the chemical structure of IMTSL in its basic form

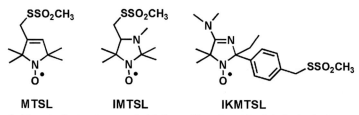

Figure 1 Chemical structures of thiol-specific nitroxide labels including conventional MTSL (S-(1-oxyl-2,2,5,5-tetramethyl-2,5-dihydro-1H-pyrrol-3-yl)methyl methanesulfonothioate) that is insensitive to pH changes and pH-sensitive IMTSL (S-(1-oxyl-2,2,3,5,5-pentamethylimidazolidin-4-yl)methyl methanesulfonothioate) and IKMTSL (S-4-(4-(dimethylamino)-2-ethyl-5,5-dimethyl-1-oxyl-2,5-dihydro-1H-imidazol-2-yl) benzyl methanesulfonothioate).

to that of MTSL (see Fig. 1) reveals many similarities, thus, raising expectations for similar physical properties, including solubility, while the same structure of the attachment group makes MTSL-labeling protocols applicable for site-directed protein chain modification with IMTSL. Uncharged IMTSL is expected to have little or no effects on the protein structure and function the same way as the structurally similar MTSL (Mchaourab et al., 1996). Akin to MTSL, the high specificity of the covalent attachment of pH-sensitive nitroxide tags to thiols was achieved by incorporating the MTS group into the structure. Chemical properties of both probes were found to be similar to MTSL. For example, it was reported that, above neutral pH, IMTSL would rapidly form biradicals, which are evidenced by characteristic five-line EPR spectra (not shown) (Voinov et al., 2008). Formation of the biradicals has also been observed for MTSL under similar conditions. It was speculated that at basic pH, some of the MTS groups are hydrolyzed to thiols, which would then react with the remaining MTSs, forming disulfide biradicals (Voinov et al., 2008). Thus, for achieving the optimal labeling it is desirable to avoid pH above the neutral value.

On the basis of the similarity of molecular sizes and geometry of the nitroxide rings of MTSL and the basic form of IMTSL, and the same length and the structure of the attachment tether, one would expect very similar rotational dynamics for the two spin labels and, therefore, similar EPR spectra. Indeed, akin to other nitroxides, EPR spectra of both IMTSL and IKMTSL, either free in solution or when covalently attached to small water-soluble peptides, fall into a fast motion regime, resulting in a well-resolved three-line EPR spectra similar to those shown in Fig. 2. The effect of pH on the EPR spectra of IMTSL and IKMTSL is associated with a chemical proton exchange between a radical $R^\bullet$ and its conjugated acid $R^\bullet H^+$:

$$R^\bullet + BH^+ \rightleftarrows R^\bullet H^+ + B \qquad (1)$$

Depending on the experimental conditions, this exchange could be fast or slow on the EPR time scale. Typically, pH-induced changes in EPR spectra of nitroxides in the fast motion limit are assessed from measurements of the isotropic nitrogen hyperfine coupling constant, A_{iso}. Assuming that for IMTSL the reaction (1) is diffusion controlled with a rate constant k_1 of $\approx 10^{10}\ M^{-1}\ s^{-1}$, as for other pH-sensitive nitroxides (Khramtsov et al., 1985), and that the difference in the resonant frequencies of the high-field nitrogen hyperfine components does not exceed $\Delta \nu \approx 4 \times 10^6\ s^{-1}$

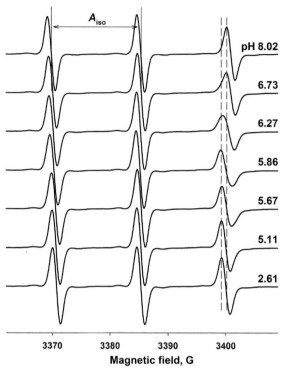

Figure 2 Representative X-band (9.5 GHz) EPR spectra of IKMTSL-2-mercaptoethanol adduct (IKMTSL-2me) measured at 17.0 °C in a series of 50 mM buffer solutions of various pH indicated next to the spectra. Vertical dashed lines mark approximate positions of the maximum of the high-field nitrogen hyperfine coupling components corresponding to the protonated $R^{\bullet}H^{+}$ and the nonprotonated $R^{\bullet}$ forms of IKMTSL-2me and are given as guides for an eye. Approximate magnitude of the isotropic nitrogen hyperfine coupling constant, A_{iso}, is shown by an arrow.

(i.e., ≈1.4 G), then the EPR spectra are in the fast exchange regime if (Khramtsov et al., 1985):

$$pK_a < \log(k_1/\Delta\nu) \approx 3.4 \qquad (2)$$

Thus, according to the estimate given by Eq. (2), the EPR spectra of the probes with pK_a close to the physiological range should yield resolved components corresponding to $R^{\bullet}$ and the $R^{\bullet}H^{+}$ nitroxide forms, thus, simplifying identification and characterization of the protonation phenomena. For probes with a low intrinsic pK_a, such as IMTSL, the slow exchange conditions could be achieved by carrying out experiments at high magnetic fields (i.e., by increasing $\Delta\nu$) (Smirnov et al., 2004; Voinov et al., 2008).

2.1 Calibration of Fast-Motion EPR Spectra of IMTSL- and IKMTSL-Labeled Thiols

Similar to fluorescent pH indicators, pH-sensitive nitroxides have to be calibrated to determine the intrinsic pK_a^0. However, pK_a of a spectroscopic probe could change upon reacting with the cysteine sulfhydryl group, if the chemical modification site is close to the moiety undergoing reversible protonation. This is indeed the case for IMTSL (Smirnov et al., 2004). Typically, such a calibration is carried out for adducts obtained by reacting the label with small unstructured cysteine-containing peptides (Smirnov et al., 2004) or 2-mercaptoethanol, to mimic the inductive effect of the mercaptoethyl phosphate group for labels attached to the polar head group of 1,2-dipalmitoyl-*sn*-glycero-3-phosphothioethanol (PTE) lipid (Voinov et al., 2009).

Figure 2 shows a typical series of X-band (9.5 GHz) EPR spectra obtained upon titration of an aqueous solution of a pH-sensitive nitroxide, specifically, IKMTSL-2-mercaptoethanol adduct (IKMTSL-2me, Fig. 5). All the spectra show three well-resolved nitrogen hyperfine coupling components indicating fast rotational motion of the nitroxide on the EPR time scale (Fig. 2). Similar spectra are expected for small spin-labeled peptides and unstructured protein domains. While for IKMTSL-2me no splitting of the high-field nitrogen hyperfine component has been observed, the spectra recorded at pH close to the label pK_a (such as at pH 5.86 and 6.27 in Fig. 2) exhibited some small asymmetry of the high-field nitrogen hyperfine coupling component. Such spectra were simulated using EWVoigt software (Smirnov & Belford, 1995; Smirnov, Smirnova, & Morse, 1995) as a superposition of two fast motion nitroxide components, whereas for the rest of the spectra simulations, a single component was sufficient. The magnitude of the isotropic nitrogen hyperfine coupling constant A_{iso} was determined directly from such fits. For the two-component spectra, the effective A_{iso} was averaged out proportionally to the weights of the individual components (Voinov et al., 2008). The observed decrease in A_{iso} upon lowering pH (Fig. 3) is associated with the protonation of the tertiary amine and, therefore, could be fitted to a modified Henderson–Hasselbalch equation:

$$A_{iso} = \frac{A_{iso}(R^{\bullet}H^+) \cdot 10^{(pH-pK_a)} + A_{iso}(R^{\bullet})}{1 + 10^{(pH-pK_a)}} \quad (3)$$

where $A_{iso}(R^{\bullet}H^+)$ and $A_{iso}(R^{\bullet})$ are the isotropic nitrogen hyperfine coupling constants for the fully protonated and nonprotonated forms of the nitroxide. These best-fit titration curves are shown as solid lines in Fig. 3 for

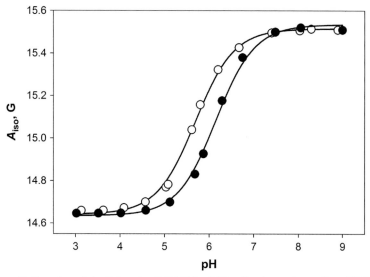

Figure 3 Experimental X-band (9.5 GHz) EPR titration data for IKMTSL-2-mercaptoethanol adduct (IKMTSL-2me) measured at 17.0 (•) and 48.0 °C (o) in 50 mM buffer solutions. Least-squares fits to the Henderson–Hasselbalch equation (3) are shown as solid lines with the parameters listed in Table 1.

IKMTSL-2me at 17.0 and 48.0 °C. The best-fit parameters are summarized in Table 1 together with other adducts of pH-sensitive nitroxides studied.

Table 1 demonstrates that for both IMTSL and IKMTSL, A_{iso} is a highly sensitive parameter for local pH measurements, with ΔA_{iso} ranging from ca. 0.87 to 1.41 G —such changes could be accurately measured from fast motion EPR spectra. For IKMTSL, the changes in the pK_a upon covalent attachments to sulfhydryl groups are insignificant, whereas the change for IMTSL is rather large (>1.5 pH units) because the chemical modification site for IKMTSL is further away from the protonatable group. Taken together, the two probes cover an exceptionally wide pH range from ca. 2.5 to 7.0 pH units.

2.2 Mapping of Local pH and Electrostatics of Peptide and Protein Systems with Methanethiosulfonate Derivatives of Ionizable Nitroxides

First and foremost, the ionizable nitroxides report on local proton concentration, measured as pH through changes in the EPR spectra recorded at pH close to the nitroxide pK_a. Thus, the initial EPR studies of peptide and

Table 1 Experimental pK_a and Isotropic Nitrogen Hyperfine Coupling Constants A_{iso} for the Fully Protonated ($R^·H^+$) and Nonprotonated ($R^·$) Forms of the Nitroxides Titrated in Aqueous Solutions Buffered at 50 mM and IMTSL-PTE Incorporated into Triton X-100 Micelles at 0.5 mol%

Nitroxide	T, °C	$A_{iso}(R^·H^+)$, G	$A_{iso}(R^·)$, G	pK_a
IMTSL	19.0	14.34 ± 0.04	15.75 ± 0.03	1.58 ± 0.03
IMTSL	37.0	14.36 ± 0.06	15.75 ± 0.04	1.54 ± 0.08
IMTSL-cys	19.0	14.52 ± 0.04	15.79 ± 0.04	3.29 ± 0.08
IMTSL-glu	19.0	14.51 ± 0.01	15.82 ± 0.02	3.17 ± 0.03
IMTSL-2me	19.0	14.58 ± 0.01	15.86 ± 0.01	3.33 ± 0.03
IKMTSL	17.0	14.56 ± 0.04	15.33 ± 0.04	5.68 ± 0.01
IKMTSL-cys	17.0	14.58 ± 0.02	15.48 ± 0.02	5.92 ± 0.04
IKMTSL-glu	17.0	14.62 ± 0.01	15.46 ± 0.02	6.15 ± 0.03
IKMTSL-2me	17.0	14.64 ± 0.01	15.53 ± 0.01	6.16 ± 0.03
IKMTSL-2me	48.0	14.65 ± 0.01	15.52 ± 0.01	5.70 ± 0.02
IMTSL-PTE, Triton X-100	23.0	14.19 ± 0.01	15.19 ± 0.01	2.52 ± 0.01
IMTSL-PTE, Triton X-100	48.0	14.25 ± 0.01	15.13 ± 0.01	2.39 ± 0.03

Full chemical names of IMTSL and IKMTSL adducts are given in the text.

protein systems with MTS derivatives of ionizable nitroxides were focused on measurements of local proton concentrations.

The first demonstrations of the method have been carried out using a series of IMTSL-labeled biomolecules including amino acid cysteine, short peptides (glutathione and P11 peptide (residues 925–933 of the B1 chain of the basement membrane specific glycoprotein, laminin)) (Smirnov et al., 2004), bacteriorhodopsin (Mobius et al., 2005) and iso-1-cytochrome c from the yeast *Saccharomyces cerevisiae* (Voinov et al., 2008). It was shown that, for IMTSL and IMTSL-labeled amino acid cysteine and a tripeptide glutathione, an asymmetric charge acquired upon protonation and located chiefly on the N_3 atom has a large effect on the nitroxide isotropic hyperfine coupling constant, $\Delta A_{iso} \approx 1.2$–1.3 G (Table 1), and also the isotropic g-factor, $\Delta g_{iso} \approx 0.00026$ (Voinov et al., 2008). The latter shift can be easily measured in high-field/high-frequency (HF) EPR experiments at 95 GHz (W-band) and above (Smirnov et al., 2004; Voinov et al., 2008). The charge asymmetry with respect to the nitroxide ring provides a basis for differentiating the

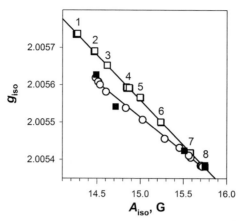

Figure 4 Isotropic magnetic parameters g_{iso} versus A_{iso} obtained from room temperature solution W-band (95 GHz) EPR spectra. Open squares: IMTSL in a series of protic and aprotic solvents and their mixtures: 1, toluene; 2, acetonitrile; 3, acetone; 4, isopropanol; 5, ethanol; 6, water/ethanol mixture (3:7, v/v); 7, water/ethanol mixture (7:3, v/v); 8, water (buffered to pH 6.0). Titration parameters for: IMTSL (open circles) and IMTSL-P11 (filled squares). Linear regressions are shown as solid lines and discussed in the text. Reproduced with permission from Voinov et al. (2008).

protonation and solvent polarity effects on magnetic parameters of IMTSL. Specifically, W-band EPR experiments with IMTSL in a series of solvents indicated that while g_{iso} correlates linearly with A_{iso}, the correlations are different for the neutral and the charged forms of the nitroxide. This was related to the effects of solvent on the spin density at the oxygen atom of the N—O$^{\bullet}$ group and on the excitation energy of the oxygen lone-pair orbital (Voinov et al., 2008). The observed difference in g_{iso} vs. A_{iso} correlation has been utilized in a W-band EPR study of the IMTSL-labeled P11 peptide to demonstrate that the changes in g_{iso} and A_{iso} are related to the proton-exchange reaction of the spin label, but not to the local polarity changes (Fig. 4) (Smirnov et al., 2004). Thus, enhanced g-factor resolution of HF EPR and g_{iso} vs. A_{iso} correlations, such as the one shown in Fig. 4, could be employed for mapping local polarity and proton concentration of protein surfaces.

3. pH-SENSITIVE SPIN-LABELED LIPIDS FOR MEASURING SURFACE ELECTROSTATICS OF LIPID BILAYERS BY EPR

Figure 5 shows chemical structures of SLP with pH-reporting nitroxides IMTSL or IKMTSL covalently attached to the lipid's polar

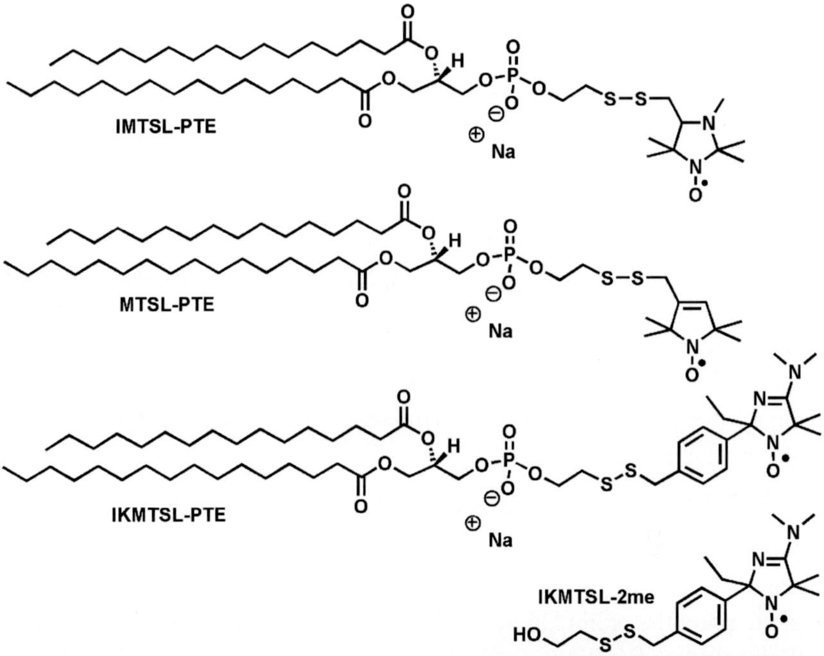

Figure 5 Chemical structures of phospholipid-based nitroxide electrostatic EPR probes IMTSL-PTE ((S)-2,3-bis(palmitoyloxy)propyl 2-(((1-oxyl-2,2,3,5,5-pentamethylimidazolidin-4-yl)methyl)disulfanyl)ethyl phosphate) and IKMTSL-PTE ((S)-2,3-bis(palmitoyloxy)propyl 2-((4-(4-(dimethylamino)-2-ethyl-1-oxyl-5,5-dimethyl-2,5-dihydro-1H-imidazol-2-yl)benzyl)disulfanyl)ethyl phosphate) are compared with nonionizable analogue of IMTSL-PTE–MTSL-PTE ((S)-2,3-bis(palmitoyloxy)propyl 2-(((1-oxyl-2,2,5,5-tetramethyl-2,5-dihydro-1H-pyrrol-3-yl)methyl)disulfanyl)ethyl phosphate), and of the adduct of IKMTSL with 2-mercaptoethanol (IKMTSL-2me).

head group. The synthesis of these lipids has been described by Voinov et al. (2009) and is based on a covalent modification of a synthetic phospholipid 1,2-dipalmitoyl-sn-glycero-3-phosphothioethanol (PTE) with either IMTSL or IKMTSL. Molecular probes of this type mimic the molecular structure of phospholipids as closely as possible, in both the acyl chain and the polar head regions. Due to their lipid-like nature, these probes do not partition between the lipid and aqueous phases of a lipid bilayer membrane but instead become its integral part. The chemical structure of pH-sensitive phospholipids ensures that the nitroxide moiety is positioned at the lipid bilayer interface. Perturbations to the lipid bilayer, including the phosphate group region, are expected to be minimal because of the compact volume of the nitroxide reporter group, especially for IMTSL-PTE.

3.1 Preparation of Spin-Labeled Lipid Vesicles

Both IMTSL-PTE and IKMTSL-PTE are headgroup-modified phospholipids and, therefore, have very low solubility in water. For these reasons, mixing of these EPR probes with other lipids should be done as chloroform solutions, although other organic solvents could be also used. Typically, doping lipid bilayers with 0.5–2.0 mol% of IMTSL-PTE or IKMTSL-PTE is sufficient for obtaining an easily detectable EPR signal without compromising properties of lipid bilayers. For the preparation of multilamellar lipid vesicles (MLVs), the readers could follow the standard procedures (Voinov et al., 2013). Homogeneously sized unilamellar or plurilamellar vesicles could be then prepared from MLVs by extrusion, utilizing a filter with a desired pore diameter (Mayer, Hope, & Cullis, 1986).

3.2 EPR Titration Experiments with Lipid Bilayers

Similar to the EPR experiments with peptides and proteins labeled with ionizable nitroxides described in Section 2, measurements of the surface electrostatic potential of lipid bilayers involve recording and analyzing a series of EPR spectra as a function of pH. However, one should pay attention to two additional aspects. Firstly, the phase transition temperatures of lipid bilayers are known to be pH dependent (Fernandez, Gonzalezmartinez, & Calderon, 1986) and, therefore, one should properly choose the temperature of the experiment so that the lipid bilayer phase would remain the same over the entire titration range. Secondly, lipid bilayers represent a barrier to charged molecules, including protons. Thus, the titration should be carried out in a way to ensure uniform pH in the sample. The latter could be achieved by either incorporating a gramicidin A channel into lipid bilayers to provide for proton permeability (Woldman et al., 2009) or by subjecting the lipid preparation to a few freeze–thaw cycles (Voinov et al., 2013). The former approach is imperative for titration of unilamellar vesicles.

Finally, equilibration of lipid bilayers with electrolytes and buffers and the consequent biophysical measurements should always be carried out at the same temperature, or at least within an interval of temperatures corresponding to the same bilayer phase. Indeed, results of an EPR titration of DMPG (1,2-dimyristoyl-*sn*-glycero-3-(phospho-*rac*-(1-glycerol))) bilayer doped with IMTSL-PTE at 48 °C were found to be significantly different for the MLV samples that were pH equilibrated at 40 and 20 °C (Voinov et al., 2013). A likely reason for such a discrepancy is a difference in the binding constants of counterions and/or protons to the lipid bilayers in different phase states. Thus, the bulk pH and electrolyte concentration of the aqueous phase of a sample containing a high fraction of DMPG (typically, 10–20 vol%), equilibrated at 20 °C when the bilayer is in the gel phase, would not necessarily remain the same after heating the same sample to 48 °C and changing the bilayer phase to fluid.

3.3 Analysis of EPR Spectra in Titration Experiments: Least-Squares Decomposition of Experimental EPR Spectra Using a Two-Site Slow-Exchange Model

Figure 6 shows typical results of EPR titration experiments of lipid bilayers doped with headgroup-labeled phospholipids. The left panel of Fig. 6 demonstrates nearly identical EPR spectra for DMPG MLVs, doped with 1 mol% of the nonionizable MTSL-PTE reference probe, from pH 2.0 to 7.0. The minor changes below pH 2.0 are likely caused by a bilayer reorganization upon protonation of the lipid phosphatidyl groups that occurs at $pK_a \approx 2$–3 (Watts, Harlos, Maschke, & Marsh, 1978). Excluding these minor effects, the EPR spectra of MTSL-PTE —a lipid labeled with a nonionizable nitroxide —appear to be essentially insensitive to pH changes even in the range of protonation of the phosphatidyl moiety.

In contrast, EPR spectra of DMPG MLVs doped with 1 mol% of ionizable IMTSL-PTE display significant changes (Fig. 6, right panel) that were also found to be reversible (Voinov et al., 2013). Because the nonionizable MTSL-PTE reveals no changes in local microviscosity (Fig. 6, left), the observed changes in rotational dynamics of IMTSL-PTE must

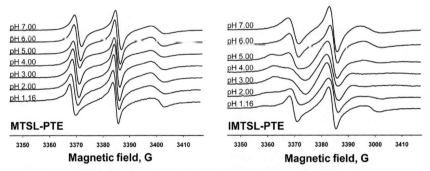

Figure 6 Representative X-band (9.5 GHz) EPR spectra from pH titration experiments of DMPG MLVs doped with 1 mol% of either pH-sensitive IMTSL-PTE (right) or control MTSL-PTE (left). Spectra were acquired at 17 °C when the bilayer is in a gel phase. Reproduced with permission from Voinov et al. (2013).

be attributable to the appearance of an ionized fraction of the nitroxide and electrostatic interactions of the protonated nitroxide with the negatively charged bilayer interface (Voinov et al., 2013). This ionized fraction is responsible for the appearance of a low-field shoulder at pH <5.0, that is characteristic of a slower and a more restricted nitroxide tumbling. We note here that while the magnetic parameters of the nitroxide are also affected by protonation, it is the effect of the electrostatic interactions on the nitroxide tumbling rate that causes the largest and most easily detectable changes in the EPR spectra shown in Fig. 6 (right panel).

For the purpose of EPR titration experiments, we are only interested in determining a fraction of the protonated species, f_{RH^+}, rather than extracting the whole set of nitroxide motion parameters from the experimental spectra. Therefore, in order to reduce the number of adjustable parameters and to increase the accuracy of f_{RH^+} measurement, the following simplified slow chemical exchange model was found to be useful. Specifically, an experimental EPR spectrum at an intermediate pH, $I(B)$, is assumed to be a superposition of the spectra from the neutral and ionized species, $F_R(B)$ and $F_{RH^+}(B)$, respectively:

$$I(B) = a \cdot F_R(B) + b \cdot F_{RH^+}(B) \qquad (4)$$

For the purpose of our decomposition procedure $F_R(B)$ and $F_{RH^+}(B)$ spectra could be measured experimentally and then used for a least-squares decomposition to derive the coefficients a and b. The fraction $f = D_R/(D_R + D_{RH^+})$ of the nonprotonated form of the nitroxide is then calculated from the double integrals D_R and D_{RH^+} of the corresponding $F_R(B)$ and $F_{RH^+}(B)$ spectra. However, we note that some of the experimental EPR spectra may contain an admixture of an out-of-phase EPR signal and also be misaligned in the magnetic field due to a shift in the resonator frequency. To account for these experimental factors, we have employed a convolution filtering that allows for correcting of the phase distortion by digital convolution with a Lorentzian function of the desired phase (Smirnov, 2008). Specifically, the phase shifts $\Delta\varphi$ and $\Delta\varphi_+$, magnetic field parameters ΔB and ΔB_+, as well as the intensity coefficients a and b were adjusted during the least-squares Levenberg–Marquardt optimization to minimize the following expression:

$$I(B) \otimes L(0, 0, \Delta B_{pp}) - \{a \cdot f_R(B) \otimes L(\Delta B_R, \Delta\varphi_R, \Delta B_{pp}) \\ + b \cdot f_{RH^+}(B) \otimes L(\Delta B_{RH^+}, \Delta\varphi_{RH^+}, \Delta B_{pp})\} \to 0 \qquad (5)$$

where the operator $\otimes$ stands for the convolution integral and $L(\Delta B_i, \Delta \phi_i, \Delta B_{pp})$ are the amplitude-normalized Lorentzian functions with the same peak-to-peak width of ΔB_{pp} but different phase shifts from the pure absorption of $\Delta \varphi_i$, and magnetic field shifts ΔB_i.

This form of the spectral fitting allows for adjusting the phase misalignment and the field positions by varying parameters of the corresponding $L(\Delta B_i, \Delta \phi_i, \Delta B_{pp})$ functions that are calculated analytically while $F_R(B)$ and $F_{RH^+}(B)$ are measured experimentally. This approach is also advantageous for implementing the Levenberg–Marquardt optimization that requires calculating partial derivative functions with respect to each of the adjustable parameters: the partial derivatives of $L(\Delta B_i, \Delta \phi_i, \Delta B_{pp})$ functions are calculated analytically and only the last step requires digital convolution using the Fast Fourier Transform algorithm. The fast and accurate calculations of the partial derivative functions ensure the rapid fit convergence. Typically, the parameter ΔB_{pp} (which should be the same for all the three Lorentzian functions in the expression (5)) was set to be equal to three digitizing intervals of an EPR spectrum.

Note that this decomposition method requires experimental reference spectra for the protonated and nonprotonated nitroxide forms. Because pH corresponding to such spectra is not known *a priori*, the EPR spectra are usually measured at increments of about 0.5 pH units and then compared. If no measurable changes are observed in the two consecutive spectra, either could be taken as a reference.

Figure 7 illustrates a least-squares decomposition of an experimental two-component X-band EPR spectrum from 1 mol% IMTSL-PTE in 65 nm unilamellar DMPG vesicles equilibrated at 17 °C with 50 mM buffer at pH 4.88 and measured at 17 °C. The least-squares decomposition into nonprotonated and protonated components is shown as B and C, respectively. The residual of the fit—the difference between the experimental and two simulated spectra (Fig. 7E)—demonstrates that the two-site slow exchange model and the decomposition procedure through convolution filtering work exceptionally well.

3.4 Interfacial pK_a and the Choice of an Electrically Neutral Reference Interface

For an ionizable group of a molecule located at a charged interface, such as pH-sensitive nitroxides tethered to the lipid head groups, the observed interfacial pK_a^i contains contributions arising from changes in the Gibbs free energy, ΔG_{pol}, upon transferring the probe from the bulk water onto the

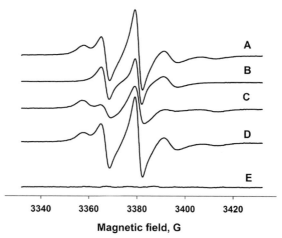

Figure 7 (A) Experimental X-band (9.5 GHz) EPR spectrum of 65 nm unilamellar IMTSL-PTE-doped POPG vesicles (1:100 mol%) equilibrated at 17 °C with 50 mM buffer at pH 4.88 and measured at 17 °C shown after a digital convolution with a Lorentzian function with peak-to-peak linewidth of just $\Delta B_{pp} = 0.146$ G (or three digitizing intervals). Least-squares decomposition of the spectrum (A) into component corresponding to nonprotonated (B) and protonated (C) forms of the nitroxide. (D) Simulated composite spectrum. (E) Residual of the fit, i.e., the difference between the experimental and the sum of the simulated spectra. Panels B and C are the actual reference spectra convolved with the same $\Delta B_{pp} = 0.146$ G and scaled by amplitude.

interface with a different local electric permittivity, ε_i, and the term ΔG_{el} attributed to the local electric potential, Ψ, that affects the equilibrium between charged and uncharged species. Then the observed interfacial pK_a^i is given by:

$$pK_a^i = pK_a^0 + \Delta pK_a^{el} + \Delta pK_a^{pol}, \quad (6)$$

where pK_a^0 is an intrinsic pK_a of the probe observed in pure water, and ΔpK_a^{pol} and ΔpK_a^{el} are the electrostatic and the polarity contributions, respectively (Fernandez & Fromherz, 1977; Fromherz, 1989). The surface electrostatic potential Ψ is related to ΔpK_a^{el} as:

$$\Delta pK_a^{el} = -e\Psi/\ln(10)kT, \quad (7)$$

where e is the elementary charge, k is the Boltzmann constant, and T is the absolute temperature.

The electrostatic pK_a shift, ΔpK_a^{el}, can be derived from Eq. (6) if other contributions to pK_a^i are otherwise determined. While pK_a^0 can be obtained from the experimental EPR titration in water of nitroxides or model compounds (i.e., IMTSL-2-mercaptoethanol adduct that models IMTSL-PTE, which has limited water solubility; Voinov et al., 2009), the value of $\Delta p K_a^{pol}$ could be affected by the position of the spectroscopic reporter group with respect to the interface, as well as by specific chemical moieties in the immediate vicinity of the probe. Experimentally, this problem could be resolved by measuring pK_a^i for a probe incorporated into a model system with a chemically similar but uncharged interface, so that $\Psi = 0$ and $\Delta p K_a^{el} = 0$. Early studies of fluorescent pH indicators employed micelles composed of the nonionic detergent polyoxyethylene isooctyl phenyl ether (Triton X-100), as a neutral (uncharged) reference interface (Fernandez & Fromherz, 1977; Fromherz, 1989). More recently, slightly more polar nonionic amphiphilic alkylglucoside surfactants have also been suggested and tested (Whiddon, Bunton, & Soderman, 2003). One disadvantage of the alkylglucoside micelles is that the measured $\Delta p K_a^{pol}$ value could be affected by specific interactions of the hydroxyl groups of the surfactant with the molecular probe and the surrounding water molecules (Drummond, Warr, Grieser, Ninham, & Evans, 1985). Other authors employed zwitterionic DMPC vesicles as an "uncharged" reference for deriving the electrostatic surface potential of charged lipid bilayers (Barbosa, Fornes, Curi, Procopio, & Ito, 2000; Khramtsov et al., 1992; Kleinschmidt & Marsh, 1997). We note here that while the vesicles formed from zwitterionic DMPC exhibit zero net charge, there is a charge separation within the lipid polar head group. The latter leads to an electrical potential across the polar head region of the lipid (or detergent) monolayer and may produce a proton (ion) concentration gradient between the interface and the bulk water. Moreover, in addition to formal charges localized at the polar head groups of the lipid molecules, a significant density of molecular dipoles is thought to exist between the head group and the hydrophobic region of the membrane (Brockman, 1994). These combined dipole moments may noticeably affect the local surface potential especially for bilayers formed from zwitterionic lipids even though these molecules formally have net zero charge (Belaya, Levadny, & Pink, 1994a, 1994b). For these and other reasons, the use of zwitterionic DPPC and DSPC references for $\Delta p K_a^{pol}$ was noted to be inappropriate (Lukac, 1983). Furthermore, the experimental pK_a^i of IMTSL-PTE in DMPC MLVs reported by Voinov et al. (2013) provided another argument that DMPC vesicles may be unsuitable as electrically neutral references for measurement of bilayer electrostatic potential. To conclude, despite numerous literature suggestions for better reference systems, Triton X-100 micelles are thought to be the most suitable choice for the neutral (uncharged) reference interface, as first described more than 30 years ago (Fernandez & Fromherz, 1977; Fromherz, 1989).

Thus, Triton X-100 micelles appear as the best choice for the reference interface (Voinov et al., 2009, 2013). The pK_a^i of IMTSL-PTE in Triton X-100 micelles was shown to have only a slight temperature dependence (see also Table 1), yielding very similar values of 2.52 ± 0.01 and 2.39 ± 0.03 units of pH at 23.0 and 48.0 °C, respectively (Voinov et al., 2009). Using the intrinsic $pK_a^0 = 3.33 \pm 0.03$ of IMTSL-PTE, the ΔpK_a^{pol} values were determined to be -0.81 ± 0.03 (23.0 °C) and -0.94 ± 0.04 (48.0 °C). These values of ΔpK_a^{pol} demonstrate that the pH-sensitive nitroxide of IMTSL-PTE experiences a significantly less polar environment at the interface of the Triton X-100 micelles vs. that of pure water. Voinov et al. (2009) also reported a series of EPR titrations of IMTSL-2-mercaptoethanol adduct in mixed water/isopropyl alcohol solution that allowed for calibrating the polarity-induced pK_a shifts, ΔpK_a^{pol}, vs. bulk solvent dielectric permittivity, ε. These calibration data allowed for estimating the local dielectric constant, ε_{eff}, experienced by the reporter nitroxide of the IMTSL-PTE lipid incorporated into the nonionic Triton X-100 micelles as 60 ± 5 and 57 ± 5 at 23 and 48 °C, respectively (Voinov et al., 2009).

To conclude, once the reference interface is chosen and ΔpK_a^{pol} is determined, then ΔpK_a^{el} and the corresponding local electrical potential Ψ can be determined from Eqs. (6) and (7). For the lipid bilayer interface, one can use experimental data for IMTSL-PTE in Triton X-100 micelles, which is summarized in Table 1, to determine ΔpK_a^{pol}. Other ionizable EPR probes can be calibrated in the same way.

3.5 Surface Charge and Potential Calculation Using the Gouy-Chapman Theory

The experimental measurements of the bilayer surface potential described in the preceding section could be compared with the widely accepted Gouy-Chapman (GC) theory. According to the GC theory, the surface potential, Ψ_{GC}, of a lipid bilayer is given by (Schwarz & Beschiaschvili, 1989):

$$\Psi_{GC} = \frac{2k_B T}{e_0} a \sinh\left(\frac{\lambda_D e_0 \sigma}{2\varepsilon_0 \varepsilon k_B T}\right), \quad (8)$$

where σ is the lipid surface charge density, λ_D is the Debye screening length, k_B is the Boltzmann constant, e_0 is the elementary electric charge, ε_0 is the permittivity of vacuum, ε is the dielectric constant of the medium ($\varepsilon = 78$ for water), and T is the absolute temperature.

The Debye screening length is given by:

$$\lambda_D = \sqrt{\frac{\varepsilon_0 \varepsilon k_B T}{2000 e_0^2 N_A C_{el}}}, \qquad (9)$$

where N_A is the Avogadro's number and C_{el} is the bulk molar electrolyte concentration.

The surface charge density can be estimated as:

$$\sigma = -\frac{e_0 \alpha}{A_L}, \qquad (10)$$

where A_L is the surface area per ionizable group of a lipid and α is the degree of dissociation of the phosphatidyl group of a lipid. If the pK_a of this group is lower than the range of the pH values used in the experiments, then $\alpha \approx 1$.

Experimentally determined surface electrostatic potentials can also be used to estimate σ of lipid bilayers by using the GC theory. For a monovalent electrolyte, σ is given by:

$$\sigma = \sqrt{8000 k_B T \varepsilon \varepsilon_0 C_{el} N_A} \sinh\left(\frac{e_0 \Psi}{2 k_B T}\right), \qquad (11)$$

4. CONCLUSIONS AND OUTLOOK

In this chapter, we described the use of nitroxides capable of reversible protonation to probe local proton concentration and electrostatic potentials of proteins and lipid bilayer systems by EPR. These probes report on their ionization state through changes in magnetic parameters and the degree of rotational averaging. We have also outlined experimental procedures to characterize and calibrate these probes by EPR, and also the methods to analyze the EPR spectra by least-squares simulations. We have summarized EPR and ionization parameters for IMTSL and IKMTSL and their derivatives, including phospholipids. Together, these two probes cover an exceptionally wide pH range from ca. 2.5 to 7.0 pH units, making them suitable for study of a broad range of proton and charge transfer-related phenomena, especially at the negatively charged lipid bilayer surfaces. We have also described the rationale for selecting a proper electrostatically neutral interface for calibrating such probes, and described an example of studying the surface potential of a lipid bilayer. The methods are general and can be applied to characterize other ionizable nitroxides synthesized in the future.

The authors believe that the use of ionizable nitroxides in biophysical EPR is largely unexplored. The method is expected to shed new light on the role of electrostatic interactions in protein function—from the operation of ion channels, to assembly of membrane proteins, and to the charge separation in membranes of photoactive proteins. The method is also uniquely capable of mapping local dielectric constant along the membrane protein–lipid bilayer interface as a function of the basic thermodynamic parameters, such as temperature and ion concentration. Thus, the fundamental questions regarding effects of protein sequence and lipid composition on the dielectric properties of the interface could be answered by this method. Another promising direction lies in studies of local electrostatics of oligonucleotides and oligonucleotide–protein complexes.

However, several challenges remain. Firstly, at the moment of this writing, none of the ionizable nitroxide labels or headgroup-labeled lipids described in this chapter are available commercially. This could be changed rapidly as the related synthetic methods described in the literature are not protected by US Patents, to the best of the authors' knowledge. Secondly, new probes with higher intrinsic pK_a would be beneficial for probing the local dielectric environment inside lipid bilayers, where the local proton concentration is a few orders of magnitude lower than in bulk solution. One may also wish to develop ionizable nitroxides in the form of unnatural amino acids, and that are less bulky than IKMTSL. Finally, but not lastly, additional method development is needed to evaluate the polarity contribution, ΔpK_a^{pol}, at different protein sites. In this respect, HF EPR experiments at low temperatures could be very useful, because magnetic parameters of the nitroxides (both g-matrix and A-tensor), determined from the rigid limit HF EPR with a high accuracy, are known to report on local polarity (Smirnova et al., 2007; Smirnova & Smirnov, 2007). However, one should proceed with a caution, because the effective polarity of frozen solvent glasses in the vicinity of dipolar solutes could be altered by a partial ordering of the solute molecules (Bublitz & Boxer, 1998).

ACKNOWLEDGMENTS

This work was supported by grant no. DE-FG-02-02ER153 (to A.I.S.) from the U.S. Department of Energy. EPR instrumentation was supported by grants from the National Institutes of Health (no. RR023614), the National Science Foundation (no. CHE-0840501), and NCBC (no. 2009-IDG-1015).

REFERENCES

Barbosa, M. P., Fornes, J. A., Curi, R., Procopio, J., & Ito, A. S. (2000). Incorporation of arachidonic and palmitic acids in large unilamellar vesicles. A comparison of electrical surface parameters. *Physical Chemistry Chemical Physics*, *2*(20), 4779–4783.

Barratt, M. D., & Laggner, P. (1974). The pH-dependence of ESR spectra from nitroxide probes in lecithin dispersions. *Biochimica et Biophysica Acta (BBA)—Biomembranes*, *363*(1), 127–133. http://dx.doi.org/10.1016/0005-2736(74)90011-X.

Belaya, M., Levadny, V., & Pink, D. A. (1994a). Electric double-layer near soft permeable interfaces. 1. Local electrostatics. *Langmuir*, *10*(6), 2010–2014. http://dx.doi.org/10.1021/la00018a061.

Belaya, M., Levadny, V., & Pink, D. A. (1994b). Electric double-layer near soft permeable interfaces. 2. Nonlocal theory. *Langmuir*, *10*(6), 2015–2024. http://dx.doi.org/10.1021/la00018a062.

Bonnet, P. A., Roman, V., Fatome, M., & Berleur, F. (1990). Carboxylic-acid or primary amine titration at the lipid-water interface—On the role of electric charges and phospholipid acyl chain composition—A spin labeling experiment. *Chemistry and Physics of Lipids*, *55*(2), 133–143. http://dx.doi.org/10.1016/0009-3084(90)90074-2.

Brockman, H. (1994). Dipole potential of lipid membranes [Review]. *Chemistry and Physics of Lipids*, *73*(1–2), 57–79. http://dx.doi.org/10.1016/0009-3084(94)90174-0.

Bublitz, G. U., & Boxer, S. G. (1998). Effective polarity of frozen solvent glasses in the vicinity of dipolar solutes. *Journal of the American Chemical Society*, *120*(16), 3988–3992. http://dx.doi.org/10.1021/ja971665c.

Cafiso, D. S., & Hubbell, W. L. (1981). EPR determination of membrane potentials. *Annual Review of Biophysics and Bioengineering*, *10*, 217–244.

Cafiso, D., McLaughlin, A., McLaughlin, S., & Winiski, A. (1989). Measuring electrostatic potentials adjacent to membranes. In B. F. Sidney Fleischer (Ed.), *Methods in enzymology: Vol. 171* (pp. 342–364). New York, NY: Academic Press.

Crowell, K. J., & Macdonald, P. M. (1999). Surface charge response of the phosphatidylcholine head group in bilayered micelles from phosphorus and deuterium nuclear magnetic resonance. *Biochimica et Biophysica Acta (BBA)—Biomembranes*, *1416*(1-2), 21–30. http://dx.doi.org/10.1016/s0005-2736(98)00206-5.

Cuervo, A., Dans, P. D., Carrascosa, J. L., Orozco, M., Gomila, G., & Fumagalli, L. (2014). Direct measurement of the dielectric polarization properties of DNA. *Proceedings of the National Academy of Sciences*, *111*(35), E3624–E3630. http://dx.doi.org/10.1073/pnas.1405702111.

Drummond, C. J., Warr, G. G., Grieser, F., Ninham, B. W., & Evans, D. F. (1985). Surface-properties and micellar interfacial microenvironment of n-dodecyl beta-D-maltoside. *Journal of Physical Chemistry*, *89*(10), 2103–2109. http://dx.doi.org/10.1021/j100256a060.

Feix, J., & Klug, C. (2002). Site-directed spin labeling of membrane proteins and peptide-membrane interactions. In L. Berliner (Ed.), *Biological magnetic resonance: Vol. 14* (pp. 251–281). New York, NY: Springer.

Fernandez, M. S., & Fromherz, P. (1977). Lipoid pH indicators as probes of electrical potential and polarity in micelles. *The Journal of Physical Chemistry*, *81*(18), 1755–1761. http://dx.doi.org/10.1021/j100533a009.

Fernandez, M. S., Gonzalezmartinez, M. T., & Calderon, E. (1986). The effect of pH on the phase-transition temperature of dipalmitoylphosphatidylcholine-palmitic acid liposomes. *Biochimica et Biophysica Acta (BBA)—Biomembranes*, *863*(2), 156–164. http://dx.doi.org/10.1016/0005-2736(86)90255-5.

Fried, S. D., & Boxer, S. G. (2015). Measuring electric fields and noncovalent interactions using the vibrational stark effect. *Accounts of Chemical Research*, *48*(4), 998–1006. http://dx.doi.org/10.1021/ar500464j.

Fromherz, P. (1989). Lipid coumarin dye as a probe of interfacial electrical potential in biomembranes. *Methods in Enzymology*, *171*, 376–387.
Gaffney, B. J., & Mich, R. J. (1976). New measurement of surface-charge in model and biological lipid membranes. *Journal of the American Chemical Society*, *98*(10), 3044–3045. http://dx.doi.org/10.1021/ja00426a076.
Gulla, A. F., & Budil, D. E. (2001). Orientation dependence of electric field effects on the g factor of nitroxides measured by 220 GHz EPR. *Journal of Physical Chemistry B*, *105*(33), 8056–8063. http://dx.doi.org/10.1021/jp0109224.
Hecht, J. L., Honig, B., Shin, Y. K., & Hubbell, W. L. (1995). Electrostatic potentials near-the-surface of DNA—Comparing theory and experiment. *Journal of Physical Chemistry*, *99*(19), 7782–7786. http://dx.doi.org/10.1021/j100019a067.
Khramtsov, V. V., Marsh, D., Weiner, L., & Reznikov, V. A. (1992). The application of pH-sensitive spin labels to studies of surface-potential and polarity of phospholipid membranes and proteins. *Biochimica et Biophysica Acta (BBA)—Biomembranes*, *1104*(2), 317–324. http://dx.doi.org/10.1016/0005-2736(92)90046-o.
Khramtsov, V., & Volodarsky, L. (2002). Use of imidazoline nitroxides in studies of chemical reactions ESR measurements of the concentration and reactivity of protons, thiols, and nitric oxide. In L. Berliner (Ed.), *Biological magnetic resonance: Vol. 14* (pp. 109–180). New York, NY: Springer.
Khramtsov, V. V., Weiner, L. M., Eremenko, S. I., Belchenko, O. I., Schastnev, P. V., Grigor'ev, I. A., et al. (1985). Proton-exchange in stable nitroxyl radicals of the imidazoline and imidazolidine series. *Journal of Magnetic Resonance*, *61*(3), 397–408. http://dx.doi.org/10.1016/0022-2364(85)90180-5.
Khramtsov, V. V., Weiner, L. M., Grigoriev, I. A., & Volodarsky, L. B. (1982). Proton-exchange in stable nitroxyl radicals—Electron paramagnetic resonance study of the pH of aqueous solutions. *Chemical Physics Letters*, *91*(1), 69–72. http://dx.doi.org/10.1016/0009-2614(82)87035-8.
Kleinschmidt, J. H., & Marsh, D. (1997). Spin-label electron spin resonance studies on the interactions of lysine peptides with phospholipid membranes. *Biophysical Journal*, *73*(5), 2546–2555.
Klug, C. S., & Feix, J. B. (2008). Methods and applications of site-directed spin labeling EPR spectroscopy. In L. J. Berliner (Ed.), *Spin labeling. The next millennium. Biological magnetic resonance: Vol. 14* (pp. 251–281). New York, NY: Plenum Press.
Leonenko, Z., Gill, S., Baoukina, S., Monticelli, L., Doehner, J., Gunasekara, L., et al. (2007). An elevated level of cholesterol impairs self-assembly of pulmonary surfactant into a functional film. *Biophysical Journal*, *93*(2), 674–683. http://dx.doi.org/10.1529/biophysj.107.106310.
Likhtenshtein, G. I., Adin, I., Novoselsky, A., Shames, A., Vaisbuch, I., & Glaser, R. (1999). NMR studies of electrostatic potential distribution around biologically important molecules. *Biophysical Journal*, *77*(1), 443–453. http://dx.doi.org/10.1016/s0006-3495(99)76902-x.
Lindstrom, F., Williamson, P. T. F., & Grobner, G. (2005). Molecular insight into the electrostatic membrane surface potential by N-14/P-31 MAS NMR spectroscopy: Nociceptin-lipid association. *Journal of the American Chemical Society*, *127*(18), 6610–6616. http://dx.doi.org/10.1021/ja042325b.
Lockhart, D. J., & Kim, P. S. (1992). Internal stark-effect measurement of the electric-field at the amino terminus of an alpha helix. *Science*, *257*(5072), 947–951. http://dx.doi.org/10.1126/science.1502559.
Lockhart, D. J., & Kim, P. S. (1993). Electrostatic screening of charge and dipole interactions with the helix backbone. *Science*, *260*(5105), 198–202. http://dx.doi.org/10.1126/science.8469972.

Lukac, S. (1983). Surface-potential at surfactant and phospholipid-vesicles as determined by amphiphilic pH indicators. *Journal of Physical Chemistry*, *87*(24), 5045–5050. http://dx.doi.org/10.1021/j150642a053.

Luptak, A., Ferre-D'Amare, A. R., Zhou, K. H., Zilm, K. W., & Doudna, J. A. (2001). Direct pK(a) measurement of the active-site cytosine in a genomic hepatitis delta virus ribozyme. *Journal of the American Chemical Society*, *123*(35), 8447–8452.

Marra, J. (1986). Direct measurement of the interaction between phosphatidylglycerol bilayers in aqueous-electrolyte solutions. *Biophysical Journal*, *50*(5), 815–825.

Mayer, L. D., Hope, M. J., & Cullis, P. R. (1986). Vesicles of variable sizes produced by a rapid extrusion procedure. *Biochimica et Biophysica Acta (BBA)—Biomembranes*, *858*(1), 161–168. http://dx.doi.org/10.1016/0005-2736(86)90302-0.

Mchaourab, H. S., Lietzow, M. A., Hideg, K., & Hubbell, W. L. (1996). Motion of spin-labeled side chains in T4 lysozyme, correlation with protein structure and dynamics. *Biochemistry*, *35*(24), 7692–7704. http://dx.doi.org/10.1021/bi960482k.

Mehlhorn, R. J., & Packer, L. (1979). Membrane surface potential measurements with amphiphilic spin labels. *Methods in Enzymology*, *56*, 515–526.

Mobius, K., Savitsky, A., Wegener, C., Rato, M., Fuchs, M., Schnegg, A., et al. (2005). Combining high-field EPR with site-directed spin labeling reveals unique information on proteins in action. *Magnetic Resonance in Chemistry*, *43*, S4–S19. http://dx.doi.org/10.1002/mrc.1690.

Norberg, J., & Nilsson, L. (2000). On the truncation of long-range electrostatic interactions in DNA. *Biophysical Journal*, *79*, 1537–1553. http://dx.doi.org/10.1016/s0006-3495(00)76405-8.

Park, E. S., Andrews, S. S., Hu, R. B., & Boxer, S. G. (1999). Vibrational stark spectroscopy in proteins: A probe and calibration for electrostatic fields. *Journal of Physical Chemistry B*, *103*(45), 9813–9817. http://dx.doi.org/10.1021/jp992329g.

Riske, K. A., Nascimento, O. R., Peric, M., Bales, B. L., & Lamy-Freund, M. T. (1999). Probing DMPG vesicle surface with a cationic aqueous soluble spin label. *Biochimica et Biophysica Acta (BBA)—Biomembranes*, *1418*(1), 133–146. http://dx.doi.org/10.1016/s0005-2736(99)00019-x.

Rottenberg, H. (1989). Determination of surface-potential of biological membranes. *Methods in Enzymology*, *171*, 364–375.

Sankaram, M. B., Brophy, P. J., Jordi, W., & Marsh, D. (1990). Fatty-acid pH titration and the selectivity of interaction with extrinsic proteins in dimyristoylphosphatidylglycerol dispersions—Spin label ESR studies. *Biochimica et Biophysica Acta (BBA)—Biomembranes*, *1021*(1), 63–69. http://dx.doi.org/10.1016/0005-2736(90)90385-2.

Sanson, A., Ptak, M., Rigaud, J. L., & Gary-Bobo, C. M. (1976). An ESR study of the anchoring of spin-labeled stearic acid in lecithin multilayers. *Chemistry and Physics of Lipids*, *17*(4), 435–444. http://dx.doi.org/10.1016/0009-3084(76)90045-1.

Schwarz, G., & Beschiaschvili, G. (1989). Thermodynamic and kinetic-studies on the association of melittin with a phospholipid bilayer. *Biochimica et Biophysica Acta (BBA)—Biomembranes*, *979*(1), 82–90. http://dx.doi.org/10.1016/0005-2736(89)90526-9.

Sheinerman, F. B., Norel, R., & Honig, B. (2000). Electrostatic aspects of protein-protein interactions [Review]. *Current Opinion in Structural Biology*, *10*(2), 153–159. http://dx.doi.org/10.1016/s0959-440x(00)00065-8.

Shin, Y. K., & Hubbell, W. L. (1992). Determination of electrostatic potentials at biological interfaces using electron electron double resonance [Article]. *Biophysical Journal*, *61*(6), 1443–1453.

Smirnov, A. I. (2008). Post-processing of IEPR spectra by convolution filtering: Calculation of a harmonics' series and automatic separation of fast-motion components from spin-label EPR spectra. *Journal of Magnetic Resonance*, *190*(1), 154–159. http://dx.doi.org/10.1016/j.jmr.2007.10.006.

Smirnov, A. I., & Belford, R. L. (1995). Rapid quantitation from inhomogeneously broadened EPR-spectra by a fast convolution algorithm. *Journal of Magnetic Resonance, Series A*, *113*(1), 65–73. http://dx.doi.org/10.1006/jmra.1995.1057.

Smirnov, A. I., Ruuge, A., Reznikov, V. A., Voinov, M. A., & Grigor'ev, I. A. (2004). Site-directed electrostatic measurements with a thiol-specific pH-sensitive nitroxide: Differentiating local pK and polarity effects by high-field EPR. *Journal of the American Chemical Society*, *126*(29), 8872–8873. http://dx.doi.org/10.1021/ja048801f.

Smirnov, A. I., Smirnova, T. I., & Morse, P. D. (1995). Very high-frequency electron-paramagnetic-resonance of 2,2,6,6-tetramethyl-1-piperidinyloxy in 1,2-dipalmitoyl-sn-glycero-3-phosphatidylcholine liposomes—Partitioning and molecular-dynamics. *Biophysical Journal*, *68*(6), 2350–2360.

Smirnova, T. I., Chadwick, T. G., Voinov, M. A., Poluektov, O., van Tol, J., Ozarowski, A., et al. (2007). Local polarity and hydrogen bonding inside the Sec14p phospholipid-binding cavity: High-field multi-frequency electron paramagnetic resonance studies. *Biophysical Journal*, *92*(10), 3686–3695. http://dx.doi.org/10.1529/biophysj.106.097899.

Smirnova, T., & Smirnov, A. (2007). High-field ESR spectroscopy in membrane and protein biophysics. *ESR spectroscopy in membrane biophysics: Vol. 27* (pp. 165–251). New York, NY: Springer.

Steffen, M. A., Lao, K. Q., & Boxer, S. G. (1994). Dielectric asymmetry in the photosynthetic reaction center. *Science*, *264*(5160), 810–816. http://dx.doi.org/10.1126/science.264.5160.810.

Surek, J. T., & Thomas, D. D. (2008). A paramagnetic molecular voltmeter. *Journal of Magnetic Resonance*, *190*(1), 7–25. http://dx.doi.org/10.1016/j.jmr.2007.09.020.

Suydam, I. T., & Boxer, S. G. (2003). Vibrational Stark effects calibrate the sensitivity of vibrational probes for electric fields in proteins. *Biochemistry*, *42*(41), 12050–12055. http://dx.doi.org/10.1021/bi0352926.

Teng, C. L., & Bryant, R. G. (2006). Spin relaxation measurements of electrostatic bias in intermolecular exploration. *Journal of Magnetic Resonance*, *179*(2), 199–205. http://dx.doi.org/10.1016/j.jmr.2005.12.001.

Tollinger, M., Forman-Kay, J. D., & Kay, L. E. (2002). Measurement of side-chain carboxyl pK(a) values of glutamate and aspartate residues in an unfolded protein by multinuclear NMR spectroscopy. *Journal of the American Chemical Society*, *124*(20), 5714–5717.

Troiano, J. M., Olenick, L. L., Kuech, T. R., Melby, E. S., Hu, D. H., Lohse, S. E., et al. (2015). Direct probes of 4 nm diameter gold nanoparticles interacting with supported lipid bilayers. *Journal of Physical Chemistry C*, *119*(1), 534–546. http://dx.doi.org/10.1021/jp512107z.

Voinov, M. A., Kirilyuk, I. A., & Smirnov, A. I. (2009). Spin-labeled pH-sensitive phospholipids for interfacial pK(a) determination: Synthesis and characterization in aqueous and micellar solutions. *Journal of Physical Chemistry B*, *113*(11), 3453–3460. http://dx.doi.org/10.1021/jp810993s.

Voinov, M. A., Rivera-Rivera, I., & Smirnov, A. I. (2013). Surface electrostatics of lipid bilayers by EPR of a pH-sensitive spin-labeled lipid. *Biophysical Journal*, *104*(1), 106–116. http://dx.doi.org/10.1016/j.bpj.2012.11.3806.

Voinov, M. A., Ruuge, A., Reznikov, V. A., Grigor'ev, I. A., & Smirnov, A. I. (2008). Mapping local protein electrostatics by EPR of pH-sensitive thiol-specific nitroxide. *Biochemistry*, *47*(20), 5626–5637. http://dx.doi.org/10.1021/bi800272f.

Watts, A., Harlos, K., Maschke, W., & Marsh, D. (1978). Control of structure and fluidity of phosphatidylglycerol bilayers by pH titration. *Biochimica et Biophysica Acta (BBA)—Biomembranes*, *510*(1), 63–74. http://dx.doi.org/10.1016/0005-2736(78)90130-x.

Whiddon, C. R., Bunton, C. A., & Soderman, O. (2003). Titration of fatty acids in sugar-derived (APG) surfactants: A C-13 NMR study of the effect of headgroup size, chain length, and concentration on fatty acid pK_a at a nonionic micellar interface. *Journal of Physical Chemistry B, 107*(4), 1001–1005. http://dx.doi.org/10.1021/jp0263875.

Woldman, Y. Y., Semenov, S. V., Bobko, A. A., Kirilyuk, I. A., Polienko, J. F., Voinov, M. A., et al. (2009). Design of liposome-based pH sensitive nanoSPIN probes: Nano-sized particles with incorporated nitroxides. *Analyst, 134*(5), 904–910. http://dx.doi.org/10.1039/b818184e.

Yang, Y., Mayer, K. M., & Hafner, J. H. (2007). Quantitative membrane electrostatics with the atomic force microscope. *Biophysical Journal, 92*(6), 1966–1974. http://dx.doi.org/10.1529/biophysj.106.093328.

CHAPTER EIGHT

Peptide–Membrane Interactions by Spin-Labeling EPR

Tatyana I. Smirnova[1], Alex I. Smirnov
Department of Chemistry, North Carolina State University, Raleigh, North Carolina, USA
[1]Corresponding author: e-mail address: tismirno@ncsu.edu

Contents

1. Introduction — 220
2. Peptide Labeling with EPR Active Probes and Preparation of Membrane Mimetic Systems — 221
   2.1 Spin Labeling of Peptides by Covalent Modification of Natural Amino Acid Side Chains — 221
   2.2 Spin Labeling by Inserting Unnatural Amino Acid via Peptide Synthesis — 225
   2.3 Membrane Mimetic: Liposomes Preparation — 226
3. EPR Measurements of Membrane Peptide Binding — 227
   3.1 Experimental Considerations — 227
   3.2 Analysis of Peptide Binding by a Signal Amplitude Method — 230
   3.3 Analysis of Peptide Binding by Spectral Simulations — 230
4. Analysis of Binding Isotherms — 233
5. Topology of Membrane-Associated Peptides: Paramagnetic Relaxers Accessibility EPR Experiments — 236
   5.1 Accessibility Measurements by EPR — 237
   5.2 Experimental Considerations — 239
6. Detecting Membrane-Induced Aggregation and Agglomeration of Peptides — 242
   6.1 CW EPR Method: A Diamagnetic Dilution Experiment — 242
   6.2 Distance Measurements and Spin Counting by Pulsed Electron-Electron Double Resonance — 249
7. Conclusions and Outlook — 250
Acknowledgments — 251
References — 251

Abstract

Site-directed spin labeling (SDSL) in combination with electron paramagnetic resonance (EPR) spectroscopy is a well-established method that has recently grown in popularity as an experimental technique, with multiple applications in protein and peptide science. The growth is driven by development of labeling strategies, as well as by considerable technical advances in the field, that are paralleled by an increased availability of EPR instrumentation. While the method requires an introduction of a paramagnetic probe at a well-defined position in a peptide sequence, it has been shown to be minimally destructive to the peptide structure and energetics of the peptide–membrane interactions.

In this chapter, we describe basic approaches for using SDSL EPR spectroscopy to study interactions between small peptides and biological membranes or membrane mimetic systems. We focus on experimental approaches to quantify peptide–membrane binding, topology of bound peptides, and characterize peptide aggregation. Sample preparation protocols including spin-labeling methods and preparation of membrane mimetic systems are also described.

1. INTRODUCTION

Biological activity of many peptides arises from the targeted interactions with cellular membranes. Examples of naturally occurring peptides that bind to biological membranes include bacterial toxins and antimicrobial peptides. These naturally occurring peptides inspired an intensive search for additional natural and synthetic peptides with enhanced cell surface binding properties for more efficient drug targeting, including cancer therapeutics (Aina, Sroka, Chen, & Lam, 2002; Boohaker, Lee, Vishnubhotla, Perez, & Khaled, 2012; Torchilin, 2008). The rational design of such membrane-active peptides, and the needs for understanding the peptide biological functions, stimulated further research on structural changes in the peptide upon membrane binding and elucidating the mode of peptide interactions with the membranes. Currently, a number of experimental biophysical methods are available to investigate the peptide membrane interactions, including not only the binding coefficients but also the orientation and immersion depth of peptides in biological membranes or membrane-mimetic systems. Those methods mainly include solution and solid-state nuclear magnetic resonance, spin-labeling electron paramagnetic resonance (EPR), fluorescence spectroscopy, infrared- and oriented circular dichroism spectroscopy, and small-angle neutron scattering. Spin-labeling EPR spectroscopy offers an informative and convenient way to probe interactions of small peptides with biological membranes and to determine the secondary structure and peptide location and orientation in membrane systems. Although spin-labeling EPR approaches discussed here are fully applicable for studies of membrane proteins and complexes, this chapter is focused primarily on experimental methods and data interpretation for binding and association of small peptides with biological membranes and their mimetics.

2. PEPTIDE LABELING WITH EPR ACTIVE PROBES AND PREPARATION OF MEMBRANE MIMETIC SYSTEMS

Natural and synthetic peptides are EPR silent (unless chelated to paramagnetic metal ion(s)) and, therefore, labeling of peptides by covalent modification of natural side chains represents the necessary step in the sample preparation. In this chapter, we restrict our discussion to labeling with stable organic free radicals and, particularly, nitroxides. From a chemical perspective, one can introduce a new unnatural chemical moiety (a nitroxide) through (a) peptide synthesis using an unnatural amino acid, (b) covalent modification of a natural side chain, and (c) covalent modification of an unnatural amino acid. Currently, the first two approaches are more common and will be discussed in further detail.

2.1 Spin Labeling of Peptides by Covalent Modification of Natural Amino Acid Side Chains

Peptides are generally more stable than proteins and, therefore, are more amenable to chemical modifications including covalent attachment of nitroxides. While site-directed spin labeling (SDSL) of proteins typically relies on an incorporation of a unique amino acid (cysteine) and the highly specific reactivity of its sulfhydryl group with a methanethiosulfonate-derivatized nitroxide, one can easily employ other chemistries for peptide labeling. For example, the terminal hydroxyl at the peptide C-terminus can be labeled using a modification of the mild one-pot esterification method by Hassner and Alexanian (1978) as was demonstrated by Dzikovski et al. for spin labeling of a pentadecapeptide antibiotic gramicidin A (Dzikovski, Borbat, & Freed, 2004). Amino groups of a peptide can be readily modified by a reaction with succinimidyl-2,2,5,5-tetramethyl-3-pyrroline-1-oxyl-3-carboxylate (SSL, Fig. 1, compound **6**). If a peptide has several amino groups, the multiple products of the reaction carried out without excess of the labeling reagent can be separated by HPLC, yielding selectively labeled peptide adducts (e.g., see Altenbach & Hubbell, 1988).

The most common method of a site-specific incorporation of a nitroxide into a peptide, however, is based on covalent modification of the Cys sulfhydryl group. Similar to the protein labeling, the essential prerequisite of the sample preparation involves site-directed mutations of the peptide sequence to replace the unwanted native cysteines with another amino acid (typically

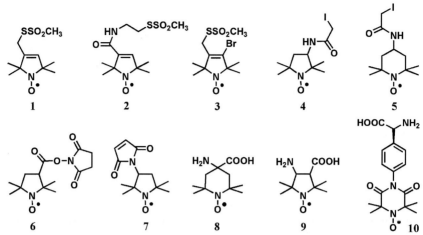

Figure 1 Chemical structures of the representative spin labels: Compound **1**, (1-oxyl-2,2,5,5-tetramethylpyrroline-3-methyl)methanethiosulfonate [MTSL]; Compound **2**, (1-oxyl-2,2,5,5-tetramethylpyrroline-3-yl)carbamidoethyl methanethiosulfonate; Compound **3**, 4-bromo-(1-oxyl-2,2,5,5-tetramethylpyrroline-3-methyl) methanethiosulfonate; Compound **4**, 3-(2-iodoacetamido)-2,2,5,5-tetramethyl-1-pyrrolidinyl-1-oxyl [3-(2-iodo acetamido)-PROXYL]; Compound **5**, 4-(2-iodoacetamido)-2,2,6,6-tetramethylpiperidine 1-oxyl [4-(2-iodoacetamido)-TEMPO]; Compound **6**, 1-oxyl-2,2,5,5-tetramethyl-1-pyrrolidinyl-3-carboxylate N-hydroxysuccinimide ester; Compound **7**, 3-maleimido-2,2,5,5-tetramethyl-1-pyrrolidinyl-1-oxyl [3-maleimido-PROXYL]; Compound **8**, 4-amino-4-carboxy-2,2,6,6-tetramethylpiperidine 1-oxyl [TOAC]; Compound **9**, 3-amino-1-oxyl-2,2,5,5-tetramethylpyrrolidine-4-carboxylic acid [POAC]; and Compound **10**, 4-(3,3,5,5-tetramethyl-2,6-dioxo-4-oxylpiperazin-1-yl)-l-phenylglycine [TOPP].

Ser) and the native amino acid with Cys at desired site(s). Then peptides of the desired sequences are prepared synthetically or expressed *in vitro*. In order to attain more complete structural information, one may replace the native amino acids in the peptide sequence by cysteine in a systematic way to obtain a library of structurally similar spin-labeled (SL) peptides. Such a systematic approach is known as a "nitroxide scan" of a peptide or a protein domain.

A number of sulfhydryl-reactive nitroxide labels are available commercially. The structures of the most common labels, (1-oxyl-2,2,5,5-tetramethylpyrroline-3-methyl)methanethiosulfonate (MTSL, Fig. 1, compound **1**) (Berliner, Grunwald, Hankovszky, & Hideg, 1982), 3-maleimido-2,2,5,5-tetramethyl-1-pyrrolidinyl-1-oxyl (maleimido-proxyl, MSL, Fig. 1, compound **7**) (Griffith & McConnel, 1966), and 3-(2-iodoacetamido)-2,2,5,5-tetramethyl-1-pyrrolidinyl-1-oxyl (iodoacetamido-proxyl, IAP, Fig. 1, compound **4**) (Ogawa & McConnel, 1967; Ogawa, McConnel, & Horwitz, 1968) and the structures of the resulting attachment linkers are shown in Figs. 1 and 2. MTSL (sometimes abbreviated as MTSSL) is the label that is used the most often. The label was first described by Berliner et al. as a reagent to quantify accessible thiols in proteins because of the reversibility of the covalent modification of sulfhydryl groups (Berliner et al., 1982). Nowadays, the label is primarily used in SDSL EPR because of its high specificity to Cys. The side chain formed upon the labeling reaction is comparable in molecular volume to phenylalanine or tryptophan, and is commonly abbreviated in literature as *R1*. A number of methanethiosulfonate spin labels with varied length and structure of the tether were synthesized including commercially available (1-oxyl-2,2,5,5-tetramethylpyrroline-3-yl)carbamidoethyl methane-thiosulfonate (MTS-4-oxyl, Fig. 1, compound **2**) and 4-bromo-(1-oxyl-2,2,5,5-tetramethylpyrroline-3-methyl) methanethiosulfonate (4-bromo-MTSL, Fig. 1, compound **3**) (Pistolesi, Pogni, & Feix, 2007).

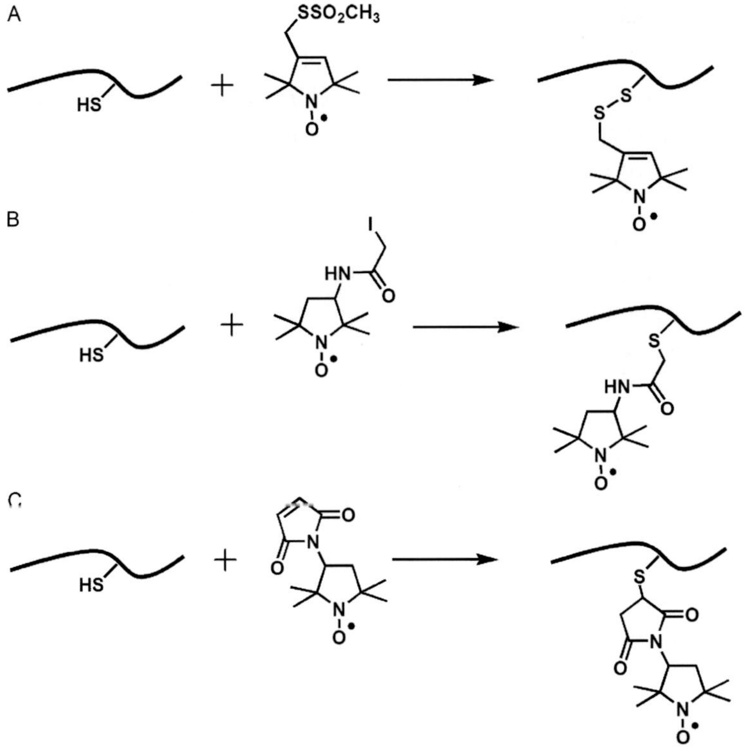

Figure 2 Structures of the nitroxide side chains produced by reactions of the sulfhydryl group of a cysteine residue with common spin labels. (A) (1-oxyl-2,2,5,5-tetramethylpyrroline-3-methyl)methanethiosulfonate [MTSL], **1**; (B) 3-(2-iodoacetamido)-2,2,5,5-tetramethyl-1-pyrrolidinyl-1-oxyl [3-(2-iodoacetamido)-PROXYL], **4**; and (C) 3-maleimido-2,2,5,5-tetramethyl-1-pyrrolidinyl-1-oxyl [3-maleimido-PROXYL], **7**.

While experimental protocols for covalent modifications of the native amino acids are well developed and easily reproducible in a lab, the method has some disadvantages. The main disadvantage is in a large conformational space accessible to the reporting NO group. The distance between C_β and NO group varies between 4 and 8 Å, depending on the label and its conformation, introducing an uncertainty in the position with respect to the peptide backbone and, therefore, affecting data interpretation. A recently introduced label with two MTS attachment groups reduces the uncertainty in the nitroxide position and is advantageous for distance measurements in proteins. This attachment, however, also can make the peptide structure potentially more rigid and requires two Cys mutations per one spin label attached (Fleissner et al., 2011).

2.1.1 Spin Labeling of Peptides by MTSL

For labeling a peptide with MTSL, a stock solution of the label is prepared at 100–200 mM concentration in acetonitrile; however, ethanol or DMSO can also be used as solvents. The stock solution should be kept at or below $-20\ °C$ and protected from light. The label from the stock solution is added to a peptide solution at 5- to 10-fold molar excess, gently mixed, and then incubated at room temperature from 1 to 3 h, often followed by labeling overnight at 4 °C. With small water-soluble peptides, we were able to achieve >95% cysteine modification by labeling at room temperature for just 1 h. The unreacted spin label is then separated by HPLC or other methods depending on the nature of the peptide. For some sample preparations, a detachment of MTSL tag is observed upon storage, especially at pH above 8.0. The X-band (9 GHz) EPR spectrum of a free nitroxide label consists of three sharp lines corresponding to each of the three ^{14}N hyperfine transitions. The lines have almost equal peak-to-peak intensities expected for a nitroxide in the fast-motional narrowing regime. The EPR signal from free label would overlap with the signal from the labeled peptide in solution and, thus, complicate, for example, the analysis of the membrane-binding experiments. The signal from the free label can be difficult to accurately account for, especially in case of small, highly dynamic peptides because of very similar EPR spectra at X-band. This consideration highlights the need for a careful separation of the free label from the labeled peptide.

2.1.2 Spin Labeling of Peptides by Maleimido- and Iodoacetamido-Derivatives of Nitroxides

MSL and IAP nitroxide tags have some advantages over the methanethiosulfonate label for EPR experiments carried out under mild reducing conditions that can lead to the cleavage of the disulfide bond formed by the latter label. Reaction with maleimides is highly specific to Cys at pH 6–7.5. At pH above 8, however, the maleimide label can react with the primary amines and modify the lysine side chain or the amino group of the N-terminus (Brewer & Riehm, 1967; Steinhoff, Dombrowsky, Karim, & Schneiderhahn, 1991).

It should be noted that all common reducing agents such as tris(2-carboxyethyl)phosphine (TCEP), dithiothreitol (DTT), or mercaptoethanol that are widely used in peptide and protein preparations to prevent Cys oxidation would have to be removed from the peptide solution prior to the reaction with a spin label. Both DTT and TCEP interfere with the labeling by maleimide and methanethiosulfonate reactants, and, to lesser extent, with iodoacetamido derivatives, and the inhibition is more pronounced with DTT (Getz, Xiao, Chakrabarty, Cooke, & Selvin, 1999; Shafer, Inman, & Lees, 2000). In addition, the presence of reducing agents, even at relatively low concentrations, results in a loss of EPR signal. TCEP was shown to react with nitroxides more slowly than DTT; however, those measurements were carried out at 0 °C (Getz et al., 1999), whereas the majority of EPR studies of biomolecules are conducted at physiological temperatures, resulting in faster nitroxide reduction.

2.2 Spin Labeling by Inserting Unnatural Amino Acid via Peptide Synthesis

A number of paramagnetic unnatural amino acids were reported in the literature (Balog et al., 2003, 2004; Kalai, Schindler, Balog, Fogassy, & Hideg, 2008; Tominaga et al., 2001; Wright et al., 2007). The most commonly used one is 2,2,6,6-tetramethyl-N-oxyl-4-amino-4-carboxylic acid (TOAC, Fig. 1, compound **8**) (Rassat & Rey, 1967; Schreier, Bozelli, Marín, Vieira, & Nakaie, 2012; Shafer, Nakaie, Deupi, Bennett, & Voss, 2008; Thomas et al., 2005). Typically, TOAC is inserted into a peptide via solid-phase peptide synthesis. Being incorporated into peptides through a peptide bond, TOAC is an achiral and very rigid cyclic molecule with only one degree of freedom—the conformation of the six-member ring. Because of the rigidity and a close position of the nitroxide moiety to the peptide backbone, TOAC has been very useful in studies of backbone dynamics, peptide secondary structure, and the orientation of peptides with respect to the membrane normal (Anderson et al., 1999; Ghimire et al., 2012; Marsh, Jost, Peggion, & Toniolo, 2007; McNulty & Millhauser, 2000; Newstadt, Mayo, Inbaraj, Subbaraman, & Lorigan, 2009; Sahu et al., 2014; Shafer et al., 2008). However, one cannot exclude a possibility of a distortion of the α-helical structure by the presence of this structurally rigid unnatural amino acid (Elsasser, Monien, Haehnel, & Bittl, 2005). Another disadvantage of TOAC is in a relatively low coupling yield during the peptide synthesis that is caused by a steric hindrance. An alternative beta-amino acid, POAC (2,2,5,5-tetramethylpyrrolidine-*N*-oxyl-3-amino-4-carboxylic acid, Fig. 1, compound **9**), was shown to provide higher coupling yields compared to TOAC (Tominaga et al., 2001). Incorporation of another chiral amino acid, TOPP (4-(3,3,5,5-tetramethyl-2,6-dioxo-4-oxylpiperazin-1-yl)-l-phenylglycine, Fig. 1, compound **10**), into peptide had been shown to not perturb the peptide structure in solutions (Stoller, Sicoli, Baranova, Bennati, & Diederichsen, 2011), thus, providing another promising tool for structural studies of peptides and proteins.

2.3 Membrane Mimetic: Liposomes Preparation

While both multilamellar and unilamellar vesicles have been widely used in studies of peptide–bilayer interactions, unilamellar liposomes are generally preferred because of a capability of providing control over the available binding surface and, to some extent, the bilayer curvature. Traditionally, unilamellar vesicles with diameters up to 100 nm are classified as small unilamellar vesicles (SUVs), from 100 nm to few μm as large unilamellar vesicles (LUVs), and larger vesicles, typically with an average diameter of 100 μm, are called giant unilamellar vesicles (GUVs). The use of GUVs in EPR is impractical because of a low average lipid concentration in the sample.

A typical liposome preparation protocol involves making stock solutions of lipids in chloroform or other suitable solvents and mixing the solutions to achieve the desired bilayer composition. Organic solvents are then removed at room temperature using a round-bottom flask connected to a rotary evaporator to form a thin lipid film on the flask wall. If the solution volume is small, the solvents can be removed under a flow of dry nitrogen gas. The residual organic solvent should be removed from the lipid film by a vacuum pump operated for a few hours or overnight. The lipid film is then hydrated with a buffer by vortexing at a temperature above the gel–liquid crystal phase transition temperature, T_m, of the lipid bilayer. Hydrated lipids are then freeze-thaw cycled by dipping the sample vial into liquid nitrogen and warming it up to a temperature above T_m, up to 10×. This procedure results in a formation of multilamellar vesicles (MLVs).

SUVs with diameters in the range of 15–50 nm can be formed by high-power sonication of the MLVs, until the aqueous solution becomes clear (Huang, 1969; Yamaguchi, Nomura, Matsuoka, & Koda, 2009). Depending on the lipid and vesicle diameter, the outer leaflet of the SUV bilayer contains about 60% of lipid molecules (Michaels, Horwitz, & Klein, 1973). SUVs are not as stable as the LUVs and, therefore, should be used in EPR or other experiments as soon as they are prepared, because even charged vesicles may fuse to form larger aggregates.

SUVs with diameters greater than 40–50 nm and LUVs are typically prepared by extrusion of MLVs through two stacked polycarbonate filters of a specific pore diameter using an extruder maintained at a temperature above T_m (Traikia, Warschawski, Recouvreur, Cartaud, & Devaux, 2000). Commercial liposome extruders are available from Avanti Polar Lipids, Inc. (Alabaster, AL) and Northern Lipids, Inc. (Burnaby, BC, Canada). Typically, up to 10 passages through the filters are carried out to ensure the best achievable size distribution that is verified by dynamic light scattering. In our experiments with DOPC, DOPC/DOPG, DOPC/DOPS mixed liposomes, an extrusion through 100 nm pores produces LUVs with mean diameters from 120 to 135 nm, depending on the lipid composition. The final lipid concentration after the extrusion can be determined by microdetermination of phosphorus (Chen, Toribara, & Warner, 1956; Itaya & Ui, 1966). LUVs should be stored under argon at temperatures above the T_m of the lipids to avoid rapid vesicle fusion (Larrabee, 1979). Under such conditions, LUVs have an excellent storage stability of up to a few days (Lasic, 1988; Lasic & Martin, 1990), 1:1 leaflet lipid stoichiometry, and the lipid lateral packing density close to that of eukaryotes' membranes. All these features make LUVs an excellent model system for studies of membrane–peptide interactions.

3. EPR MEASUREMENTS OF MEMBRANE PEPTIDE BINDING

3.1 Experimental Considerations

The general methodology for quantifying binding of SL peptide to lipid bilayer membranes is well established. Detection of the peptide binding relies on sensitivity of the nitroxide EPR spectrum to rotational diffusion of the probe. Typically, peptide-binding experiments are carried out using conventional X-band (9.5 GHz) spectrometers. Continuous wave (CW) X-band EPR spectra of nitroxides reflect both the rotational anisotropy and correlation time of the label motion at the ps to ns timescale, making this method suitable for characterization of protein secondary structure and dynamics (Hubbell, Gross, Langen, & Lietzow, 1998; Hubbell, Mchaourab, Altenbach, & Lietzow, 1996; Mchaourab, Lietzow, Hideg, & Hubbell, 1996). X-band CW EPR spectra from SL peptides in aqueous solutions fall into a fast-motional regime with characteristic rotational correlation time of the label in the sub-ns range, regardless of whether the peptide is unstructured or possesses secondary structure. Upon membrane binding, the rotational motion of the peptide and the label become restricted, resulting in an EPR spectrum that is easily distinguishable from the one corresponding to the free peptide in the aqueous phase.

Binding titrations are typically carried out by adding lipids to about 100–200 µL sample of a peptide solution. A small volume of peptide/lipid solution, typically about 5 µL, is drawn into a quartz or a glass capillary with inner diameter (i.d.) of 0.5–1.0 mm (or less) that is placed into an EPR resonator. After recording the spectra, the sample is unloaded back into the bulk peptide/lipid mixture, a new portion of lipids is added and a new sample is drawn for EPR spectroscopy. Alternatively, a quartz capillary with plunger that allows for drawing in and pushing out the sample without removal of the capillary from the resonator can be used (Victor & Cafiso, 2001). If the amount of the available peptide and lipids is not an issue, the samples for each titration point can be prepared separately with a given concentration of the peptide and a varied concentration of the lipids.

Temperature stabilization is essential in the peptide-binding experiments, as its variations would not only affect the peptide-binding equilibrium to the lipid bilayer but also the line shape of the EPR spectra. The latter might cause errors in determination of the fractions of free and bound peptide. In our NCSU laboratory for measurements using a Varian Century Series X-band EPR spectrometer (Palo Alto, CA), the temperature is stabilized better than $\pm 0.01\ °C$ using an in-house-built variable temperature accessory comprising from a digitally controlled circulator bath that pumps fluid through high-efficiency aluminum radiators attached to an EPR resonator (Alaouie & Smirnov, 2006). Variable temperature systems capable of temperature stabilization in the physiologically relevant range are commercially available from Bruker.

Peptide concentration in solution should be kept below the critical micelle concentration (CMC). In order to determine the CMC of a peptide, the amplitude of the EPR signal, typically the peak-to-peak amplitude of the central nitrogen hyperfine component, $m_I = 0$, is measured as a function of the peptide concentration. Below the CMC, when the peptide is monomeric, the dependence is linear with zero intercept. Above the CMC, the peak-to-peak amplitude of the sharp, three-line spectra becomes essentially concentration independent. Upon micellization of the peptide, spin labels get in close contact, providing conditions for strong exchange and dipole–dipole interaction between the electron spins. As a result, the EPR signal from the peptide in a micelle will appear as a very broad single line that would not contribute significantly to the intensity of the central peak of the sharp free component.

Figure 3 illustrates typical changes in X-band CW EPR spectra from a small unstructured peptide observed upon binding to LUVs. NOD3 peptide with amino acid sequence, KKKKKKKFFC(R1)F, was labeled with a threefold excess of MTSL for 4 h at room temperature and separated from the unreacted spin label by a gradient HPLC. EPR spectra from 50 µM NOD3-R1 in a buffer (100 mM MOPS containing 100 mM KCl at pH 7.0), and in the presence of LUVs formed from a mixture of 1-hexadecanoyl-2-(9Z-octadecenoyl)-sn-glycero-3-phosphocholine (POPC) and 1-hexadecanoyl-2-(9Z-octadecenoyl)-sn-glycero-3-phospho-L-serine (POPS) in 24:1 molar ratio (total lipid concentration of 8 mg/mL) are shown in Fig. 3.

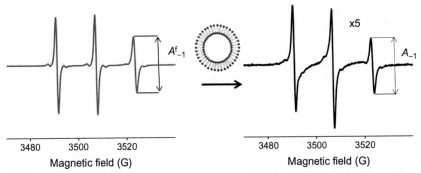

Figure 3 Changes in room temperature X-band (9 GHz) CW EPR spectra from 50 µM MTSL-labeled NOD3 peptide upon binding to POPC:POPG (24:1 molar ratio) LUVs. Left: a freely tumbling peptide in an aqueous buffer. Right: EPR spectrum of the peptide in the presence of POPC:POPG LUVs is shown at fivefold amplification. The peak-to-peak intensities, A_{-1}, of the high field ($m_I = -1$) nitrogen hyperfine components are shown by arrows.

3.2 Analysis of Peptide Binding by a Signal Amplitude Method

For each EPR spectrum obtained in the course of a titration, the fraction of the membrane-bound peptide can be determined from measuring A_{-1}, the peak-to-peak amplitude of the high-field, $m_I = -1$, nitrogen hyperfine component (Cafiso & Hubbell, 1981; Mchaourab, Hyde, & Feix, 1994). The fraction of the bound peptide, f_b, can be calculated using the relation:

$$f_b = \frac{A^f_{-1} - A_{-1}}{A^f_{-1} - A^b_{-1}} \quad (1)$$

where A^f_{-1} and A_{-1} are the peak-to-peak amplitudes of the high-field line for the peptide free in buffer and in presence of LUVs, respectively, and A^b_{-1} is the amplitude of the high-field line when the peptide is fully bound to the membrane (Fig. 3). For weakly binding peptides, the condition of the complete binding may be difficult to achieve in the titration experiment as there is a practical limit of how much lipids can be added. For the latter cases, A^b_{-1} can be treated as an adjustable parameter during the binding isotherm analysis. If the EPR spectrum from the bound peptide is much broader than the one from the free peptide in solution (for example, the spectrum from the bound peptide falls in the slow-motional regime), then the bound peptide contributes little to the overall peak-to-peak EPR amplitude and the contribution of the A^b_{-1} term can be neglected. As an example, in a binding experiment, A^b_{-1} was estimated to be less than 3% of A^f_{-1} (Bhargava & Feix, 2004).

In experiments when lipids are titrated into a sample with the given amount of the peptide, the total concentration of the peptide may be considered constant during the titration. Then the peak-to-peak amplitude of the high-field nitrogen hyperfine component can be measured directly from the experimental EPR spectra, and has to be corrected only for the signal gain setting of the spectrometer if changed from sample to sample. In titrations when the labeled peptide is added to lipids, CW EPR spectra should be intensity normalized to the same double integral before measuring the signal amplitude. The concentrations of the free and the bound peptides are calculated from f_b and the known total concentration of the peptide in the sample.

3.3 Analysis of Peptide Binding by Spectral Simulations

Experimental EPR spectra from peptide–liposome mixtures can be decomposed into the individual components, corresponding to the free peptide in solution and bound to the lipids, by using spectral simulation. Typically, the EPR spectrum from the free peptide can be modeled as a superposition of Voigt line shapes (a convolution integral of Lorentzian and Gaussian functions) with different homogeneous linewidth assigned to each of the three nitrogen hyperfine components. The simulations could be further improved by accounting for additional unresolved hyperfine interactions. For example, the quality of the fit of EPR spectra of NOD3-R1 peptide improved significantly by including a hyperfine interaction with one unique proton and satellite lines due to natural abundance ^{13}C. A fitting program based on a fast convolution algorithm with a Levenberg–Marquardt optimization method (Smirnov, 2008; Smirnov & Belford, 1995; Smirnova et al., 1997) was employed to extract ^{14}N, ^{1}H, and ^{13}C isotropic hyperfine coupling parameters, as well as Lorentzian and Gaussian contributions to the linewidth of three nitroxide components. The experimental spectrum from the peptide in the presence of LUVs was modeled as a superposition of two spectra: a free peptide in the aqueous phase and a slower moving bound peptide. The line shape was modeled as described above with an explicit unresolved hyperfine splitting on one ^{1}H and ^{13}C nuclei as determined from simulations of the free peptide spectrum. All the spectral parameters including a small admixture of the dispersion contribution were adjustable for both centers, except for the probability of ^{13}C and the unique hydrogen hyperfine. Figure 4 shows an example of an experimental EPR spectrum and the least-squares simulated components corresponding to the free and the bound peptides. The fraction of the bound peptide can be calculated from the double integrals of the simulated spectra available from the fit as:

$$f_b = \frac{I_b}{I_f + I_b} \qquad (2)$$

where I_b and I_f are the double-integrated intensities of the bound and the free components, respectively. Figure 5 shows experimentally measured binding isotherms for an interaction of NOD3 peptide with LUVs prepared from pure POPC lipids and from POPC lipids doped with the anionic lipid POPG at 4 mol%. The fraction of the peptide bound and the concentration of the free peptide were calculated based on the spectral simulations, as described above.

When an EPR spectrum of the bound form of a peptide does not fall into the fast-motion regime it cannot be simulated by a Voigt function. This is more typical for binding of basic peptides to liposomes with a high content

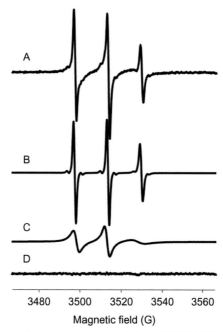

Figure 4 (A) Experimental room temperature X-band (9 GHz) EPR spectrum from 50 μM NOD3-R1 in the presence of 8 weight% LUVs (POPC:POPG in 24:1 molar ratio) LUVs. Simulated spectra corresponding to the free peptide in solution and bound to LUVs are shown as (B) and (C), respectively; (D) is the fit residual—a difference between the experimental and the sum of the two simulated components.

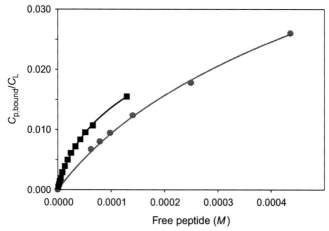

Figure 5 Binding of NOD3-R1 peptide to LUVs composed of POPC (red (gray in the print version) circles) and POPC:POPS lipids in 24:1 molar ratio (black squares). A degree of binding, $c_{p,\,bound}/c_L$ (mole of bound peptide per mole of lipid in the outer leaflet), is plotted versus concentration of the free peptide in solution. The solid curves are the best least-squares fits obtained using a model described by Eqs. (7)–(9).

of anionic lipids, and/or for TOAC-labeled peptides. While one can employ slow-motion models for EPR spectra, such simulations would significantly increase the computation burden, especially when an entire binding series of EPR spectra has to be simulated. An easier and equally effective approach for determination of the peptide-binding constant involves modeling of the experimental spectrum in the presence of lipids, $F(B)$, as a superposition of experimentally detected spectra, $F_f(B)$ from the free peptide and $F_b(B)$ from the bound form.

$$I(B) = a \cdot F_f(B) + b \cdot F_b(B) \qquad (3)$$

and the fraction of bound form is calculated as:

$$f_b = \frac{b \cdot I_b}{a \cdot I_f + b \cdot I_b} \qquad (4)$$

where a and b are weight coefficients, and I_b and I_f are double-integrated intensities of the experimental spectra $F_f(B)$ and $F_b(B)$, respectively.

If a peptide binds only weakly, obtaining an experimental EPR spectrum from the bound form without any contributions from the free peptide may be problematic. Then the free peptide component could be subtracted manually by adjusting the amplitude of the free peptide signal till the residual spectrum corresponding to the bound form would not contain any sharp components as confirmed by a visual inspection. Spectra manipulation options in *Xepr* acquisition and data processing suite for the Bruker Biospin ELEXSYS series of EPR spectrometers (Billerica, MA) allow for an easy subtraction of the experimental spectra and evaluation of the double-integral intensities. Alternatively, a convolution-based fitting of filtered EPR spectra can be applied for an automatic least-squares separation of fast and broad slow-motion nitroxide components without the explicit simulation of the slow-motion spectrum (Smirnov, 2008).

4. ANALYSIS OF BINDING ISOTHERMS

When discussing peptide–membrane interactions, the term "binding" is used rather loosely, typically referring to interactions by a physical adsorption and partitioning. Specific peptide–phospholipid interactions, when a peptide forms a well-defined stoichiometric peptide–lipid complex, are uncommon (Breukink et al., 1999; Choung, Kobayashi, Inoue, et al., 1988; Choung, Kobayashi, Takemoto, Ishitsuka, & Inoue, 1988) and more particular for large membrane proteins. A review of EPR methods for studying lipid–protein interactions is given by Marsh (2010).

Once the fraction of the bound peptide, f_b, is determined, concentrations of the free and the bound forms of the peptide can be calculated from the known total amount of the labeled peptide added to the sample. In a partitioning model, no assumptions on the peptide–lipid binding stoichiometry are made. The affinity of the peptide to the lipid membrane is described by a partition constant, K_p:

$$x_b = \frac{c_{p,\text{bound}}}{a \cdot c_L} = K_p c_{p,m} \quad (5)$$

where $c_{p,m}$ is the concentration of the peptide immediately above the membrane surface, $c_{p,\text{bound}}$ is the concentration of the peptide bound, a is the fraction of the lipids in the outer leaflet accessible for binding ($a = 0.5$ for LUVs), and c_L is the total lipid concentration. In a simple partitioning model, the concentration of the peptide at the membrane surface is the same as the bulk concentration of the free peptide in the aqueous phase, $c_{p,f}$, and the partition coefficient is obtained from the linear part of the plot of x_b versus concentration of the free peptide. The partitioning model is suitable for the binding driven by hydrophobic interactions at a low peptide-to-lipid ratio. The linear relation is anticipated, for example, for interactions of uncharged peptides with membranes at nonsaturating conditions.

The Langmuir adsorption isotherm is often employed to describe peptide–membrane interaction. If an assumption that the binding site is made up of n noninteracting and energetically equivalent lipid molecules is valid, then the Langmuir binding isotherm can be written as

$$\frac{\theta}{1-\theta} = K c_{p,f} \quad (6)$$

where $\theta = \frac{n \cdot c_{p,b}}{a \cdot c_L}$ is the mole fraction of the occupied binding sites. The Eq. (6) is equivalent to the ligand-binding model when specific binding between the peptide and n equivalent lipids, forming PL_n complex, is assumed.

Electrostatic forces can significantly contribute to the peptide–lipid interactions resulting in deviations from the binding behavior predicted by partitioning or the Langmuir models. For an electrostatically driven peptide binding, the Eq. (5) is still valid; however, the contribution of the electrostatic interaction would change the effective concentration of the peptide immediately above the membrane surface, $c_{p,m}$, compared to that in the bulk solution (Seelig, 2004; Wieprecht & Seelig, 2002). Effective $c_{p,m}$ depends on the peptide charge and the membrane surface electrostatic potential, ψ_0, and can be described using the Boltzmann distribution:

$$c_{p,m} = c_{p,f} \exp\left(-\frac{z_p e_0 \psi_0}{kT}\right) \quad (7)$$

where z_p is the electric charge associated with the peptide, e_0 is the elementary charge, T is the temperature, and k is the Boltzmann constant. The corresponding surface potential, ψ_0, can be determined by solving the Gouy–Chapman equation (Aveyard & Haydon, 1973; Mathias, McLaughlin, Baldo, & Manivannan, 1988; McLaughlin, 1977, 1989):

$$\sigma^2 = 2000\varepsilon_0 \kappa RT \sum_i c_i \left(\exp\left\{-\frac{z_i e_0 \psi_0}{kT}\right\} - 1\right) \quad (8)$$

where σ is the two-dimensional density of electric charge on the surface, ε_0 is the permittivity of the free space, κ is the dielectric constant of the solution, R is the universal gas constant, F is the Faraday constant, c_i is the concentration of the i-th electrolyte in the bulk aqueous phase, and z_i is the signed valency of the i-th species. The use of Eq. (8) requires an evaluation of the surface charge density, σ, for the given lipid bilayer system, as well as taking into account changes in σ from binding of cationic peptides. Surface charge density can be calculated from the known fraction of the charged lipids and literature data for the average area per lipid, or measured experimentally as described by Voinov, Rivera-Rivera, and Smirnov (2013). The presence of cations, such as Na^+ or K^+, that may specifically bind to anionic lipids, could reduce the effective membrane surface charge. This effect can be accounted for by modeling the ion–membrane interaction using the Langmuir binding isotherm. To describe binding of a cationic peptide with an electric charge z to a membrane containing x_L mole fraction of an anionic lipid with a charge z_L, in the presence of salt NaCl, we assume that:

$$\sigma = \frac{e}{A}(z_L x_L (x_{Na} - 1) + z x_b); \quad x_{Na} = \frac{K_{Na} c_{Na}}{1 + K_{Na} c_{Na}} \quad (9)$$

where A is the surface area per lipid, c_{Na} and K_{Na} are concentration of cations and its binding constant to the anionic lipid, respectively, and e is the elementary charge (Seelig, 2004; Wieprecht & Seelig, 2002). Binding of a charged peptide to an electrically neutral lipid membrane generates a positive electric charge at the membrane surface. From the extent of binding, x_b, the surface charge density can be calculated assuming the first term in Eq. (9) to be zero. Equation (9) can be modified to include a correction factor to take into account the possible penetration of peptide into the lipid membrane and the resulting increase in the membrane surface area (Kuchinka & Seelig, 1989); however, at a low peptide-to-lipid ratio this effect could be neglected.

Combining Eqs. (7) and (8), and calculating σ from the experimental values of x_b, a numerical solution for K, ψ_0, and $c_{p,m}$ for each of the experimental data pairs of x_b and $c_{p,f}$ can be found. The partition coefficient, K_p, describes a hydrophobic binding or an adsorption of the peptide and should be independent of the peptide concentration for the given lipid system, as long as the model is applicable. Figure 5 shows experimental binding isotherms of NOD3 peptide to neutral POPC LUVs and to POPC LUVs containing 4 mol% of negatively charged POPG lipid and the best fit of the experimental data to the electrostatic binding model (Eqs. 5, 7–9). During the fit, K_p and the peptide charge, z, were used as adjustable parameters. As expected, the cationic NOD3 peptide binds stronger to LUVs containing anionic lipids than to zwitterionic DOPC lipids. Least-squares fitting of the binding isotherm for DOPC:DOPG liposomes, however, led to z estimated to be +2.5. This was unexpected as the peptide contains seven lysine residues. It is possible that not all the lysine side chains are protonated because of close proximity to each other.

5. TOPOLOGY OF MEMBRANE-ASSOCIATED PEPTIDES: PARAMAGNETIC RELAXERS ACCESSIBILITY EPR EXPERIMENTS

Determination of the topology of membrane-associated peptides relies on measurements of molecular accessibility of the SL side chains to paramagnetic relaxation agents with differential solubility in water and in lipid environments. The method is well established and was reviewed on multiple occasions (Fanucci & Cafiso, 2006; Klug & Feix, 2008). In such experiments, molecular oxygen is served as a nonpolar relaxer that partitions preferably into a lipid phase with a gradient of an increasing concentration toward the center of the membrane. Ni(II) ethylenediaminediacetate (NiEDDA) is a neutral solute that, while partitioning primarily into water, would also penetrate into the lipid headgroup region and slightly into the hydrophobic section of the bilayer. Another commonly employed relaxer is a negatively charged chromium oxalate (Crox) complex that cannot penetrate into the bilayer (Altenbach, Froncisz, Hubbell, & Hyde, 1988; Altenbach, Froncisz, Hyde, & Hubbell, 1989). Experimental data indicate that the Heisenberg exchange is the dominant mechanism for magnetic interactions of nitroxide spin labels with both molecular oxygen and transition metal ions, as the spin-lattice relaxation time of these paramagnetic relaxers is much shorter than the spin-lattice relaxation time of the nitroxide. The contribution from the dipole–dipole interaction can be neglected in most cases. Effect of the Heisenberg exchange on the nitroxide spin-lattice relaxation time T_{1e} is directly proportional to the nitroxide-relaxer collisional rate and, therefore, can be used as a measure of the nitroxide accessibility to the relaxer. Effect of the relaxer on T_{1e} of a nitroxide label can be measured by either a power saturation CW EPR experiment or a direct pulsed saturation recovery (SR) method.

5.1 Accessibility Measurements by EPR

In the power saturation experiment, the peak-to-peak amplitude of the central line of the first derivative nitroxide EPR signal, A_{pp}, is measured as a function of the incident microwave power and then fitted to the following function:

$$A_{pp} = IP^{0.5}\left[1 + \left(2^{\frac{1}{\varepsilon}} - 1\right)P/P_{1/2}\right]^{-\varepsilon} \quad (10)$$

where I is a scaling factor, $P_{1/2}$ is the microwave power required to reduce the resonance amplitude to half of its maximum unsaturated value, and ε is a measure of the homogeneity of the saturation of the resonance line, $\varepsilon = 1.5$ for the homogeneous and 0.5 for the inhomogeneous saturation limits. From the least-squares fitting of the experimental CW EPR "roll-over" curves, the $P_{1/2}$ parameter is obtained. Then, in order to measure the relaxer accessibility, values for $P_{1/2}$ are determined for the sample equilibrated with N_2, air (20.95% O_2), and N_2 but in the presence of a certain concentration of a hydrophobic relaxer such as, for example, NiEDDA. The collision rate with a fast-relaxing agent is directly related to the difference in $P_{1/2}$ in the presence and absence of the relaxer:

$$\Delta P_{1/2} = P_{1/2}(\text{relaxer}) - P_{1/2}(N_2) \propto \frac{2W_{ex}}{T_{2e}^*} \quad (11)$$

The accessibility parameter, Π^{relaxer}, for a given relaxer is determined as (Farahbakhsh, Altenbach, & Hubbell, 1992):

$$\Pi^{\text{relaxer}} = \frac{\Delta P'_{1/2}(\text{relaxer})}{P'_{1/2}(\text{DPPH})}$$
$$= \frac{P_{1/2}(\text{relaxer})/\Delta H_{\text{p-p}}(\text{relaxer}) - P_{1/2}(N_2)/\Delta H_{\text{p-p}}(N_2)}{P_{1/2}(\text{DPPH})/\Delta H_{\text{p-p}}(\text{DPPH})} \quad (12)$$

where $\Delta H_{\text{p-p}}$ is the peak-to-peak linewidth of the central nitrogen hyperfine component of the nitroxide EPR spectrum and $\Delta H_{\text{p-p}}(\text{DPPH})$ is the peak-to-peak linewidth of the solid DPPH (2,2-diphenyl-1-picrylhydrazyl) reference sample. Division by the peak-to-peak linewidth makes $\Delta P'_{1/2}$ independent upon T^*_{2e}, and normalization by the data for the reference compound DPPH compensates for variations in the EPR resonator quality factor Q.

The location of the spin label within a bilayer membrane and at the water–membrane interface can be characterized by the depth parameter, Φ, defined as:

$$\Phi = \ln\left[\frac{\Pi^{\text{Oxy}}}{\Pi^{\text{NiEDDA}}}\right] \quad (13)$$

The parameter Φ is directly related to the difference in the standard state chemical potentials of the reagents at any depth, independent of viscosity or steric constraints imposed by the environment and also independent of the EPR line shape.

The calibration curve for the depth parameter as a function of the label position can be easily obtained using lipid bilayers doped with n-doxyl PC (1-palmitoyl-2-stearoyl-(n-doxyl)-sn-glycero-3-phosphocholine) lipids labeled at a position n along the acyl chain. n-Doxyl PC lipids with $n=5$, 7, 10, 12, 14, and 16 and Tempo PC, a lipid molecule with a nitroxide tethered to the lipid headgroup, are all commercially available from Avanti Polar Lipids, Inc. The dependence of Φ upon the depth is approximately linear for labels within the bilayer interior, as measured using SL lipids (Qin & Cafiso, 1996), an SL membrane protein (Altenbach, Greenhalgh, Khorana, & Hubbell, 1993), and a transmembrane peptide WALP (Nielsen, Che, Gelb, & Robinson, 2005). Although the gradients in concentration of both oxygen and metal ions level off quickly in the aqueous phase, the calibration curve was extended to the aqueous phase using a membrane-bound SL protein (Frazier, Wisner, Falke, & Cafiso, 2000), allowing for determination of distances up to about 5 Å in the aqueous side of the membrane–solution interface (Victor & Cafiso, 2001).

Alternatively, molecular accessibility parameters can be measured using pulsed SR experiment, which allows for direct measurement of the electronic spin-lattice relaxation time T_{1e} (Eaton & Eaton, 2005; Pyka, Ilnicki, Altenbach, Hubbell, & Froncisz, 2005). Changes in the spin-lattice electronic relaxation rates, ΔR_1, for oxygen and NiEDDA can be used as the time domain analogues of the accessibility parameter:

$$\Delta R_1 = \left[\frac{1}{T_{1e}(\text{relaxer})} - \frac{1}{T_{1e}(N_2)}\right] \propto [\text{relaxer}] \quad (14)$$

where $T_{1e}(N_2)$ is the nitroxide spin-lattice relaxation time measured for the nitrogen-equilibrated sample, and $T_{1e}(\text{relaxer})$ is the effective relaxation time in the presence of a relaxation agent, and [relaxer] is the concentration of the relaxer. The depth parameter is defined as:

$$\Phi = \ln\left[\frac{\Delta R_1^{O_2}}{\Delta R_1^{\text{NiEDDA}}}\right] \quad (15)$$

Experimental SR traces are typically least squares fitted to a single-exponential decay to extract the relaxation rate $R_{1e} = \frac{1}{T_{1e}}$. However, for SL systems with two distinctive states, the SR data should be analyzed in terms of a double-exponential decay (Nielsen et al., 2005; Pyka et al., 2005; Yin, Feix, & Hyde, 1990). The main advantage of the SR method is in the ability to resolve multiple nitroxide populations with different molecular accessibilities to the relaxer molecules (Eaton & Eaton, 2005; Kusumi, Subczynski, & Hyde, 1982; Pyka et al., 2005; Subczynski, Widomska, Wisniewska, & Kusumi, 2007). The use of the SR method in peptide binding and other biophysical EPR experiments, however, is limited by the need for specialized instrumentation. Although commercial spectrometer systems capable of SR EPR are available from Bruker BioSpin (Billerica, MA), the installations are not as common as CW EPR spectrometers.

5.2 Experimental Considerations

CW EPR power saturation method has been the method of choice for most spin-labeling accessibility studies as it could be performed using a standard CW spectrometer typically equipped with a loop-gap resonator to achieve the maximum microwave power on the sample required for at least a partial signal saturation. Accessibility measurements require a strict control of molecular oxygen concentration in the sample that is typically achieved by drawing a solution into gas-permeable TPX (polymethylpentene or PMP) capillaries (available from L&M EPR Supplies, Milwaukee, MI, or from Bruker, Karlsruhe, Germany) and continuously flushing nitrogen gas through the resonator space. Alternatively, liquid samples can be drawn into a disposable gas-permeable thin-wall PTFE (polytetrafluoroethylene) capillary, 0.81 mm i.d., 0.86 mm outer diameter (o.d.) tubing is available from Zeus Industrial Products, Inc. (Orangeburg, SC). Such a thin and flexible tubing allows for the ends of the capillary to be closed by crimping that will stay in place when the capillary is inserted into an open-ended quartz tube (3 mm i.d.). The quartz tube is then placed inside a variable temperature dewar of the X-band EPR or/and EPR resonator that is continuously flushed with a gas mixture with a controlled oxygen concentration (Smirnov, Clarkson, & Belford, 1996).

The optimal concentration of NiEDDA in EPR accessibility experiments varies depending on the location of the nitroxide label within the membrane. Typically, 20 mM concentration is sufficient while twice as high concentrations of NiEDDA may be needed for peptides with the nitroxide chains deeply embedded into a bilayer or located at sterically protected sites. The latter are expected for aggregated membrane peptides. For nitroxide labels exposed to the aqueous phase, concentrations as low as 3 mM NiEDDA may be sufficient to measure changes in the nitroxide relaxation properties. Accessibility data can be easily normalized to 20 mM NiEDDA concentration for the comparison purposes.

Figures 6 and 7 demonstrate an application of the depth parameter approach to elucidate the transmembrane location of STM23, a 23-amino-acid peptide containing a highly hydrophobic 16-amino-acid segment. Figure 6 shows experimental SR traces for the peptide labeled with MTSL at residue 4 (STM23-4R1) and incorporated into LMVs composed of 1,2-dipalmitoleoyl-*sn*-glycero-3-phosphoethanolamine (DPoPE) and 1-palmitoyl-2-oleoyl-*sn*-glycero-3-phosphocholine (POPC) lipid in 2:3 molar ratio. Two samples were prepared, one using the buffer of choice and the second one using a buffer containing 20 mM NiEDDA. SR traces were obtained for a nitrogen-equilibrated sample prepared with 20 mM NiEDDA, and for air- and nitrogen-equilibrated samples without the metal complex. The corresponding fits with a single-exponential decay function are superimposed on experimental curves. The depth parameter as a function of the label position is presented in Fig. 7 and was calculated using

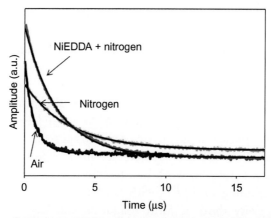

Figure 6 Effects of paramagnetic broadening agents on experimental X-band (9 GHz) room temperature saturation recovery traces for a MTSL-labeled peptide (STM23-4R1) bound to DPoPE/POPC LUVs with position of the nitroxide-labeled chain approximately 8 Å below the lipid phosphate groups: green (light gray in the print version), nitrogen-equilibrated sample; blue (dark gray in the print version), air-equilibrated sample; red (gray in the print version), nitrogen-equilibrated sample containing 20 mM NiEDDA. Least-squares fits to single-exponential decays are shown as solid black lines.

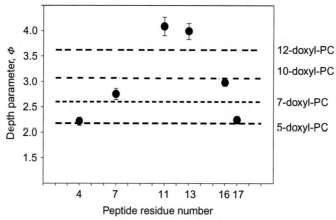

Figure 7 The depth parameter Φ as a function of the nitroxide label position along the STM23 peptide incorporated into LUVs consisting of DPoPE and DOPC lipids in 2:3 molar ratio. The horizontal lines represent the experimental depth parameters measured for the LMVs of the same composition doped with n-doxyl PC ($n=5$, 7, 10, and 12) at 1 mol%. The depth parameter clearly indicates a transmembrane insertion of the peptide.

Eqs. (14) and (15). The membrane depth calibration data were obtained using LMVs of the same composition labeled with n-doxyl PC ($n=5, 7, 10$, and 12) in the presence of an unlabeled STM23, and are shown in Fig. 7 as horizontal lines. The data demonstrate a transmembrane insertion of the STM23 peptide into DPoPE/POPC bilayers.

Dependence of the depth parameter upon the label position in the peptide sequence can provide information on the secondary structure of the peptide bound to phospholipid bilayers. Such an approach requires measurements of the EPR accessibility parameter for an extended set of consecutive SL sites and it is known to work best for the surface-associated peptides. For a peptide forming an α-helix with ≈3.6 residues per turn, a spin label positioned at every third or forth residue would have about the same accessibility to a specific paramagnetic relaxer. That is, the surface-exposed residues would exhibit a higher accessibility to a water-soluble relaxer, like NiEDDA. The corresponding sites on the other side of the helix would be buried deeper into the bilayer, showing a lower accessibility to NiEDDA, but a higher accessibility to oxygen. The plot of the corresponding depth parameter versus the residue number would then reveal the characteristic periodicity of 3.6 for an α-helical and 2 for a β-sheet structure. Highly illustrative examples of this approach can be found in the following references (Klug, Su, & Feix, 1997; Sato & Feix, 2006; Turner, Braide, Mills, Fanucci, & Long, 2014).

6. DETECTING MEMBRANE-INDUCED AGGREGATION AND AGGLOMERATION OF PEPTIDES

6.1 CW EPR Method: A Diamagnetic Dilution Experiment

Spin-labeling EPR provides an easy and informative way to detect formation of peptide–peptide clusters or aggregation that sometimes is promoted by peptide–membrane interactions. The approach is based on measuring a broadening of a CW EPR signal from the labeled peptide due to magnetic interactions between the labels brought into close proximity by aggregation. At a low peptide-to-lipid ratio, and in the absence of peptide–peptide interactions, peptides are distributed randomly through the membrane, providing a sufficiently long distance between the nitroxide labels and allowing to neglect spin–spin magnetic interactions. The EPR spectra for such peptides are determined by the label dynamics. If two or more labels are located at distances <25 Å, the CW EPR signal is broadened and has lower peak-to-peak amplitude than the spectrum from the same concentration of the noninteracting spins. Typically, peptide aggregates are more tightly packed, causing a slower rotational dynamics of the nitroxide and an additional spectral broadening. Then, the onset of the aggregation or cluster formation is detected as a decrease in the amplitude of the CW EPR signal, normalized for the given number of spins. Thus, the peptide aggregation or cluster formation can be readily detected by comparing the EPR signal from the cluster sample with the signal from a diamagnetically "diluted" sample taken under the same conditions (Altenbach & Hubbell, 1988; Margittai & Langen, 2006).

First, let us consider the types of magnetic interactions responsible for the spectral broadening. For two nitroxide labels separated by a distance in the range from 8–10 to 25 Å, only the dipole–dipole interaction contributes to the observed broadening (Altenbach, Oh, Trabanino, Hideg, & Hubbell, 2001; Hubbell, Cafiso, & Altenbach, 2000) while at distances below 8 Å, the Heisenberg spin exchange starts to play a measurable role (Molin, Salikhov, & Zamaraev, 1980). At a very close contact between the labels, the spin exchange results in a collapse of the nitrogen hyperfine spectral features, resulting in an observation of a single-line spectrum. Although such spectra are atypical for a membrane-associated peptide aggregation, exchange-narrowed single-line spectra were reported for SL amyloids (Chen, Margittai, Chen, & Langen, 2007; Cobb, Sönnichsen, Mchaourab, & Surewicz, 2007) and tau filaments (Margittai & Langen, 2004). The broadening effect due to the dipole–dipole interactions is detectable at both cryogenic and physiological temperatures, although in the latter case the dipole–dipole interactions can be partially averaged by translational motion of spin labels or a rotation of the dipolar vector with respect to the magnetic field.

In a typical experiment, a diamagnetically diluted sample is prepared by mixing labeled and unlabeled peptides at various ratios. Alternatively, a sample could be labeled with a mixture of MTSL and its diamagnetic analog methyl methanethiosulfonate (MMTS, available from Toronto Research Chemicals Inc. (Toronto, ON, Canada; Gross, Columbus, Hideg, Altenbach, & Hubbell, 1999) mixed at different ratios and, therefore, producing samples with the required magnetic dilution. The use of the diamagnetic analog ensures that the steric interactions due to the presence of the nitroxide labels are preserved in the diamagnetically diluted sample. Preserving such interactions could be of significance, especially if information on local dynamics of the peptide in an aggregate or a cluster is of interest. Figure 8 schematically shows a reduction in the number of the close-contact

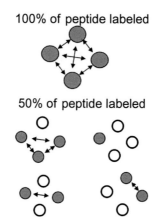

Figure 8 A cartoon illustrating a reduction in the number of close-contact spin pairs (indicated by arrows between filled circles depicting spin-labeled peptides) in aggregated peptide samples upon a diamagnetic dilution with EPR-silent peptides (open circles).

spin pairs in an aggregated sample upon the diamagnetic dilution. Figure 9A depicts an example of spectral changes observed upon diamagnetic dilution in the case of an aggregation of STM25, a 25-amino-acid peptide containing a highly hydrophobic transmembrane segment of 18 amino acids. STM25 was labeled at various positions, n, with MTSL (STM25-nR1) and incorporated in LMVs composed of POPC and POPE (1-palmitoyl-2-oleoyl-sn-glycero-3-phosphoethanolamine) lipids in the 1:2 mole ratio and containing varying amounts of cholesterol. The spectrum shown in dashed line was recorded for a sample containing 100%-labeled STM25-11R1 peptide while the spectrum in solid line is from a magnetically diluted sample containing the peptide at 25 mol% labeling. In both samples, the peptide-to-lipid ratio was 1:100. Figure 9B shows analogous spectra from the same peptide incorporated in POPC/POPE LUVs containing 30 mol% of cholesterol. The aggregation-induced broadening is readily observable in both cases as a decrease in the amplitude of these intensity-normalized spectra.

In order to estimate the broadening effect, we have applied a one-linewidth-parameter fitting model that was previously developed for fitting inhomogeneously broadened EPR spectra by Smirnov and Belford (1995), and used by us to extract relaxer-induced broadening and changes in linewidth due to dynamic effects (Smirnova, Smirnov, Clarkson, & Belford, 1995a,1995b). A similar approach was described by Mchaourab et al. to extract the spectral broadening from pairwise spin–spin interactions

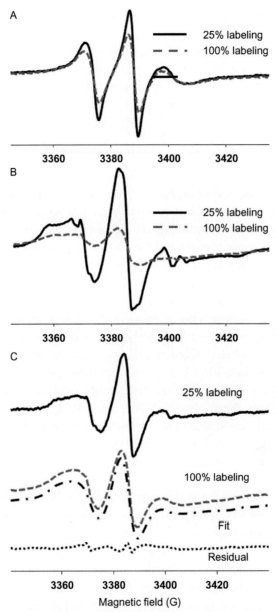

Figure 9 Intensity normalized room temperature CW X-band (9 GHz) EPR spectra of STM25-11R1 in POPE:POPC 1:2 LMVs in the presence of 10 mol% (A) and 30 mol% (B) cholesterol and STM25-11R1 in POPE:POPC 1:2 LMVs in presence of 25 mol% (C): 100% labeled peptide (red, light gray in the print version dashed line) and 25% labeled peptide (black solid line). C: The spectrum from 100% labeled peptide was least squares fitted to a model assuming a homogenous (Lorentzian) broadening due to magnetic interactions. The result of the fit is shown as a dash-dot line. The fit residual, *i.e.*, difference between the experimental and the simulated spectra, is shown at the bottom as a dotted line.

(Mchaourab, Oh, Fang, & Hubbell, 1997). This model is based on simulating the CW EPR spectrum from the aggregated sample by using a convolution integral given by Eq. (16), in which $F_0(B')$ is the spectrum in the absence of magnetic interactions (the infinitely diluted sample), $F(B)$ is the predicted spectrum from an aggregated sample, and $m(B)$ is the Lorentzian broadening, measured as the $\delta(\Delta B_L^{p-p})$ peak-to-peak linewidth:

$$F(B) = \int_{-\infty}^{+\infty} F_0(B')m(B-B')dB' \qquad (16)$$

In practice, the spectrum $F_0(B')$ from the "infinitely diamagnetically diluted sample" could be difficult to obtain experimentally because of the vanishingly low signal intensity. Thus, the signal-to-noise consideration would typically dictate a compromise value for the diamagnetic dilution. In our analysis, we took an experimental EPR spectrum from the sample containing a peptide with 25 mol% of labeling as an approximation for the "infinitely diluted" spectrum $F_0(B')$. In this case, the broadening is measured relative to the one already present in the diamagnetically diluted sample. The approach is illustrated in Fig. 9C, which displays the experimental spectrum for a sample containing 100%-labeled STM25-11R1 peptide incorporated in POPC/POPE LUVs containing 25 mol% of cholesterol shown in dashed line, spectrum from a magnetically diluted sample containing the peptide at 25 mol% labeling shown in solid line, and the result of the fit using Eq. (16) in dash-dotted line. The fit residual, i.e., difference between the experimental and the simulated spectra, is shown at the bottom in dotted line. The aggregation broadening was determined to be 0.31 G for the fully labeled STM25-11R1 in a lipid bilayer containing 10 mol% of cholesterol, and 1.6 and 4.7 G in a membrane containing 25 and 30 mol% of cholesterol correspondently. The broadening can be used as an empirical parameter to characterize the extent of the aggregation as the larger broadening corresponds to shorter distances between the labels. Figure 10 displays an aggregation-induced broadening for STM25-11R1 peptide in POPC/POPE LMVs as a function of the cholesterol content. A clear onset of the aggregation for STM25-11R1 peptide is observed at a cholesterol concentration above 20 mol %. An interesting dependence of the broadening is observed for labels located at different positions along the peptide chain.

If the peptide–peptide interactions are highly specific, some information on the mode of the peptide–peptide interaction can be elucidated by comparing the broadening effects for the labels located at different positions of

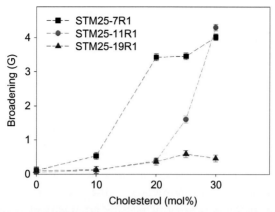

Figure 10 Lorentzian broadening for room temperature X-band (9 GHz) EPR spectra of STM25-nR1 peptide fully labeled with the MTSL side chain (R1) at positions $n=7, 11$, and 19 as a function of cholesterol content in POPE:POPC (1:2 molar ratio) LMVs. The samples of identical lipid and peptide composition but diamagnetically diluted with unlabeled STM25 to 25% of STM25-nR1 were used to obtain the reference unbroadened EPR spectra. See text for further details.

the peptide chain. This approach is demonstrated in Fig. 10 by comparing the broadening observed for the labels located at the positions 7, 11, and 19 of the STM25 peptide. Although the aggregation effect is clearly detected by the labels at the positions 7 and 11 at 25 mol% of cholesterol, the relative broadening is much higher for the position 7 than for the position 11. Unexpectedly, the difference disappears in the samples with 30 mol% of cholesterol. The observed broadening is, however, much smaller for the peptide labeled at the position 19, even for the sample with the highest cholesterol content. This broadening trend can be explained by considering that 7R1 and 11R1 are located in a highly hydrophobic region of the peptide, while 19R1 is close to the hydrophilic tail of the peptide containing five consecutive lysine residues that are exposed to the aqueous phase. The presence of positive charges on the lysine side chain prevents the aggregation of the hydrophilic tail. Combined with the accessibility measurements that point to STM25 being located at the membrane surface in the bilayer preparations with cholesterol content above 20 mol%, the broadening data suggest a "star-like" arrangement of the peptide at the membrane surface: the hydrophobic tails of the peptide aggregate to minimize the contact with the aqueous phase forming the center of the "star," while the hydrophilic tails show a large separation forming the "arms." It is not clear, however, how many peptides participate in the formation of an individual aggregate.

Quantitative interpretation of the observed broadening effects in terms of spin–spin distances, and the aggregate structure, can be difficult to obtain, because this would require testing specific models of the peptide arrangements, in order to agree with the observed spectral broadening. The approach is based on considering a specific peptide arrangement, for example, dimer, trimer, hexamer, etc., and calculating possible spin–spin pairwise distances, discarding ones above 2.5 nm, as the contribution of these pairs in the CW EPR spectrum broadening can be neglected. Next, the changes in EPR spectrum from noninteracting spins (taken at the high diamagnetic dilution limit) due to the predicted distance distributions can be calculated and compared with the experimentally observed spectrum from a nondiluted sample. If the geometry of peptide arrangement is such that only one distance contributes significantly to the broadening, then the methods for determination of distances between the spins in spin pairs are readily available (Altenbach et al., 2001; Banham et al., 2008; Hubbell et al., 2000; Rabenstein & Shin, 1995; Steinhoff et al., 1997) to test the validity of the proposed peptide–peptide arrangement. For example, the convolution approach was employed to extract effective distance and distance distribution to characterize the membrane-induced aggregation of alamethicin in bicelles (Bortolus, De Zotti, Formaggio, & Maniero, 2013). Rigid limit CW EPR linewidth broadening as measured for frozen samples ($T = 78.5$ K) was used to estimate spin–spin distances for membrane-imbedded TOAC-labeled trichogin GA IV, a lipopeptide with antibiotic activity, assuming formation of a dimer in a model membrane (Syryamina et al., 2012).

Another way to quantify magnetic interactions is by the second moment analysis of the EPR spectra that is based on the famous work of Van Vleck (1948). The second moment of the broadening function, μ_2, is obtained from the intensity-normalized and baseline-corrected broadened, $F_0(B)$, and nonbroadened, $F_0(B')$, spectra (Abragam, 1961):

$$\mu_2 = \sum (B - \bar{B}_d)^2 F_0(B) - \sum (B' - \bar{B})^2 F_0(B') \qquad (17)$$

where B is the applied magnetic field, and $\bar{B}$ is the magnetic field corresponding to the mean of the spectrum. The second moment of μ_2 can be used as an empirical parameter to characterize the extent of the spin–spin interactions, similar to the Lorentzian broadening in the example above, or be related to the pairwise distance between nitroxide labels.

The μ_2 of the powder sample of like spins (case of a strong dipolar coupling) can be calculated using the Van Vleck (1948) formula, assuming a well-defined distance r, resulting in:

$$\mu_2 = \frac{a}{r^6} \quad (18)$$

where $a = 1.56 \cdot 10^{-60}$ T^2 m^6. Equation (18) can be solved for the interspin distance, r:

$$r(\text{nm}) = \frac{2.32}{(\mu_2 \cdot 10^8)^{1/6}} \quad (19)$$

This approach has been employed to investigate an aggregation of a model transmembrane WALP23 peptide (Scarpelli et al., 2009) as a function of the lipid composition. The main source of error in this analysis comes from the polynomial baseline correction errors in the integration of the first-derivative EPR spectra.

6.2 Distance Measurements and Spin Counting by Pulsed Electron-Electron Double Resonance

Essential information on number of spins in a peptide aggregate can be obtained using advanced EPR pulse methods. Pulsed electron–electron double resonance (PELDOR, also known as DEER, or double electron–electron resonance) is widely used to measure the dipolar electron–electron coupling, in order to determine distances and distance distributions in the nanometer range, from 1.5 to about 8 nm, as well as to provide information on the mutual orientation of the interacting spins. In addition to providing distance information, PELDOR can be used to count the number of SL monomers in a complex. In this article, we will not discuss the theoretical basis or experimental details of this approach. Readers are referred to excellent reviews on this topic (Borbat & Freed, 2014; Jeschke, 2012, 2014). In brief, the approach is based on a relationship between the depth of the modulation, Δ, of the PELDOR signal and the number of spins in the cluster, N:

$$N = 1 - \frac{\ln(1 - \Delta)}{\ln(1 - \lambda \cdot f)} \quad (20)$$

where λ is the B spin inversion efficiency, a parameter defined by the experimental conditions, and f is the efficiency of the labeling (Jeschke, 2012;

Milov, Maryasov, & Tsvetkov, 1998; Milov, Ponomarev, & Tsvetkov, 1984). To obtain the number of spins in the cluster, an experimental evaluation of λ is required. The latter can be readily achieved by measurements on a model biradical system. Several studies using model bi-, tri-, and tetraradicals with known labeling efficiency reported the overall error of PELDOR spin counting of less than 5% (Bode et al., 2007). The PELDOR spin-counting method was successfully applied to study protein oligomerization (Hilger et al., 2005) and self-assembly and aggregation of peptides (Milov et al., 2009, 2000; Salnikov et al., 2006).

We note that PELDOR measurement requires highly specialized pulse EPR spectrometers, operated by a well-trained staff and this currently limits a widespread application of this very useful and informative technique. Commercial spectrometers operating at X-, Q-, and W-band frequencies offered by Bruker BioSpin and a number of home-built spectrometers show excellent technical specifications. PELDOR experiments typically require 50 μM protein concentration and about 30 μL of sample volume for measurements using commercial X-band EPR spectrometers. Q-band spectrometers offer higher sensitivity and require much smaller samples (down to 5 μL) (Ghimire, McCarrick, Budil, & Lorigan, 2009); however, these spectrometers are even less common than the one operating at X-band EPR frequency.

7. CONCLUSIONS AND OUTLOOK

EPR, in combination with spin-labeling methods, offers researchers access to a wide range of structural and functional information on peptide–membrane interactions beyond the data on the binding constant. Although the methods do require highly specialized instrumentation, the experimental protocols are well developed and the instrumentation is generally accessible to broad groups of researchers from academia as well as industry. Continuous progress in EPR instrumentation, especially the recent introduction of portable, low-cost CW EPR spectrometers from Bruker BioSpin and Active Spectrum (Foster City, CA) to the market, opens up further opportunities for even wider utilization of this method in academic research, student training, and industrial applications. Further development of spin-labeling methods and synthesis of new, specific spin labels hold a promise to extend the number of biological systems amenable to such studies, and to expand the information content of the method.

ACKNOWLEDGMENTS

A part of this work was supported by National Science Foundation Grants MCB-0451510 and MCB-0843632 (to T.I.S). A part of this work including development of simulation software and analysis of membrane-binding data for charged peptides was supported by Grant No. DE-FG-02-02ER153 (to A.I.S.) from the U.S. Department of Energy. EPR instrumentation was supported by grants from the National Institutes of Health (No. RR023614), the National Science Foundation (No. CHE-0840501), and NCBC (No. 2009-IDG-1015). The authors are thankful to Dr. Maxim A. Voinov (NCSU) for his expert help in preparing some of the figures and numerous discussions and suggestions.

REFERENCES

Abragam, A. (1961). *The principles of nuclear magnetism*. London: Oxford University Press.
Aina, O. H., Sroka, T. C., Chen, M. L., & Lam, K. S. (2002). Therapeutic cancer targeting peptides. *Biopolymers*, 66(3), 184–199. http://dx.doi.org/10.1002/bip.10257.
Alaouie, A. M., & Smirnov, A. I. (2006). Ultra-stable temperature control in EPR experiments: Thermodynamics of gel-to-liquid phase transition in spin-labeled phospholipid bilayers and bilayer perturbations by spin labels. *Journal of Magnetic Resonance*, 182(2), 229–238. http://dx.doi.org/10.1016/j.jmr.2006.07.002.
Altenbach, C., Froncisz, W., Hubbell, W., & Hyde, J. (1988). The orientation of membrane-bound, spin-labeled melittin as determined by electron-paramagnetic-res saturation recovery measurements. *Biophysical Journal*, 53(2), A94.
Altenbach, C., Froncisz, W., Hyde, J. S., & Hubbell, W. L. (1989). Conformation of spin-labeled melittin at membrane surfaces investigated by pulse saturation recovery and continuous wave power saturation electron-paramagnetic resonance. *Biophysical Journal*, 56(6), 1183–1191.
Altenbach, C., Greenhalgh, D. A., Khorana, H. G., & Hubbell, W. L. (1993). E depth measurements of nitroxides in membrane bilayers using spin labeled mutants of bacteriorhodopsin. *Biophysical Journal*, 64(2), A51.
Altenbach, C., & Hubbell, W. L. (1988). The aggregation state of spin-labeled melittin in solution and bound to phospholipid-membranes—Evidence that membrane-bound melittin is monomeric. *Proteins*, 3(4), 230–242. http://dx.doi.org/10.1002/prot.340030404.
Altenbach, C., Oh, K. J., Trabanino, R. J., Hideg, K., & Hubbell, W. L. (2001). Estimation of inter-residue distances in spin labeled proteins at physiological temperatures: Experimental strategies and practical limitations. *Biochemistry*, 40(51), 15471–15482. http://dx.doi.org/10.1021/bi011544w.
Anderson, D. J., Hanson, P., McNulty, J., Millhauser, G., Monaco, V., Formaggio, F., et al. (1999). Solution structures of TOAC-labeled trichogin GA IV peptides from allowed (g approximate to 2) and half-field electron spin resonance. *Journal of the American Chemical Society*, 121(29), 6919–6927. http://dx.doi.org/10.1021/ja984255c.
Aveyard, R., & Haydon, D. A. (1973). *An introduction to the principles of surface chemistry*. London: Cambridge University Press.
Balog, M., Kalai, T., Jeko, J., Berente, Z., Steinhoff, H. J., Engelhard, M., et al. (2003). Synthesis of new conformationally rigid paramagnetic alpha-amino acids. *Tetrahedron Letters*, 44(51), 9213–9217. http://dx.doi.org/10.1016/j.tetlet.2003.10.020.
Balog, M. R., Kalai, T. K., Jeko, J., Steinhoff, H. J., Engelhard, M., & Hideg, K. (2004). Synthesis of new 2,2,5,5-tetramethyl-2,5-dihydro-1H-pyrrol-1-yloxyl radicals and 2-substituted-2,5,5-trimethylpyrrolidin-1-yloxyl radicals based alpha-amino acids. *Synlett*, 2004(14), 2591–2593. http://dx.doi.org/10.1055/s-2004-834806.

Banham, J. E., Baker, C. M., Ceola, S., Day, I. J., Grant, G. H., Groenen, E. J. J., et al. (2008). Distance measurements in the borderline region of applicability of CW EPR and DEER: A model study on a homologous series of spin-labelled peptides. *Journal of Magnetic Resonance, 191*(2), 202–218. http://dx.doi.org/10.1016/j.jmr.2007.11.023.

Berliner, L. J., Grunwald, J., Hankovszky, H. O., & Hideg, K. (1982). A novel reversible thiol-specific spin label: Papain active site labeling and inhibition. *Analytical Biochemistry, 119*(2), 450–455. http://dx.doi.org/10.1016/0003-2697(82)90612-1.

Bhargava, K., & Feix, J. B. (2004). Membrane binding, structure, and localization of cecropin-mellitin hybrid peptides: A site-directed spin-labeling study. *Biophysical Journal, 86*(1), 329–336. http://dx.doi.org/10.1016/s0006-3495(04)74108-9.

Bode, B. E., Margraf, D., Plackmeyer, J., Duerner, G., Prisner, T. F., & Schiemann, O. (2007). Counting the monomers in nanometer-sized oligomers by pulsed electron–electron double resonance. *Journal of the American Chemical Society, 129*(21), 6736–6745. http://dx.doi.org/10.1021/ja065787t.

Boohaker, R. J., Lee, M. W., Vishnubhotla, P., Perez, J. M., & Khaled, A. R. (2012). The use of therapeutic peptides to target and to kill cancer cells. *Current Medicinal Chemistry, 19*(22), 3794–3804 (Review).

Borbat, P. P., & Freed, J. H. (2014). Pulse dipolar electron spin resonance: Distance measurements. In C. R. Timmel & J. R. Harmer (Eds.), *Structural information from spin-labels and intrinsic paramagnetic centres in the biosciences: Vol. 152* (pp. 1–82). Heidelberg: Springer.

Bortolus, M., De Zotti, M., Formaggio, F., & Maniero, A. L. (2013). Alamethicin in bicelles: Orientation, aggregation, and bilayer modification as a function of peptide concentration. *Biochimica et Biophysica Acta, 1828*(11), 2620–2627. http://dx.doi.org/10.1016/j.bbamem.2013.07.007.

Breukink, E., Wiedemann, I., van Kraaij, C., Kuipers, O. P., Sahl, H. G., & de Kruijff, B. (1999). Use of the cell wall precursor lipid II by a pore-forming peptide antibiotic. *Science, 286*(5448), 2361–2364. http://dx.doi.org/10.1126/science.286.5448.2361.

Brewer, C. F., & Riehm, J. P. (1967). Evidence for possible nonspecific reactions between N-ethylmaleimide and proteins. *Analytical Biochemistry, 18*(2), 248–255. http://dx.doi.org/10.1016/0003-2697(67)90007-3.

Cafiso, D. S., & Hubbell, W. L. (1981). Electron-paramagnetic-res determination of membrane-potentials. *Annual Review of Biophysics and Bioengineering, 10*, 217–244. http://dx.doi.org/10.1146/annurev.bb.10.060181.001245.

Chen, M., Margittai, M., Chen, J., & Langen, R. (2007). Investigation of alpha-synuclein fibril structure by site-directed spin labeling. *Journal of Biological Chemistry, 282*(34), 24970–24979. http://dx.doi.org/10.1074/jbc.M700368200.

Chen, P. S., Toribara, T. Y., & Warner, H. (1956). Microdetermination of phosphorus. *Analytical Chemistry, 28*(11), 1756–1758. http://dx.doi.org/10.1021/ac60119a033.

Choung, S. Y., Kobayashi, T., Inoue, J., Takemoto, K., Ishitsuka, H., & Inoue, K. (1988). Hemolytic-activity of a cyclic peptide ro09-0198 isolated from Streptoverticillium. *Biochimica et Biophysica Acta, 940*(2), 171–179. http://dx.doi.org/10.1016/0005-2736(88)90192-7.

Choung, S. Y., Kobayashi, T., Takemoto, K., Ishitsuka, H., & Inoue, K. (1988). Interaction of a cyclic peptide, RO09-0198, with phosphatidylethanolamine in liposomal membranes. *Biochimica et Biophysica Acta, 940*(2), 180–187. http://dx.doi.org/10.1016/0005-2736(88)90193-9.

Cobb, N. J., Sönnichsen, F. D., Mchaourab, H., & Surewicz, W. K. (2007). Molecular architecture of human prion protein amyloid: A parallel, in-register β-structure. *Proceedings of the National Academy of Sciences of the United States of America, 104*(48), 18946–18951. http://dx.doi.org/10.1073/pnas.0706522104.

Dzikovski, B. G., Borbat, P. P., & Freed, J. H. (2004). Spin-labeled gramicidin A: Channel formation and dissociation. *Biophysical Journal, 87*(5), 3504–3517. http://dx.doi.org/10.1529/biophysj.104.044305.

Eaton, S., & Eaton, G. (2005). Saturation recovery EPR. In S. Eaton, G. Eaton, & L. Berliner (Eds.), *Biomedical EPR, Part B: Methodology, instrumentation, and dynamics: Vol. 24/B*. (pp. 3–18). New York, NY: Springer US.

Elsasser, C., Monien, B., Haehnel, W., & Bittl, R. (2005). Orientation of spin labels in de novo peptides. *Magnetic Resonance in Chemistry*, 43, S26–S33. http://dx.doi.org/10.1002/mrc.1692.

Fanucci, G. E., & Cafiso, D. S. (2006). Recent advances and applications of site-directed spin labeling. *Current Opinion in Structural Biology*, 16(5), 644–653. http://dx.doi.org/10.1016/j.sbi.2006.08.008.

Farahbakhsh, Z. T., Altenbach, C., & Hubbell, W. L. (1992). Spin labeled cysteines as sensors for protein lipid interaction and conformation in rhodopsin. *Photochemistry and Photobiology*, 56(6), 1019–1033. http://dx.doi.org/10.1111/j.1751-1097.1992.tb09725.x.

Fleissner, M. R., Bridges, M. D., Brooks, E. K., Cascio, D., Kalai, T., Hideg, K., et al. (2011). Structure and dynamics of a conformationally constrained nitroxide side chain and applications in EPR spectroscopy. *Proceedings of the National Academy of Sciences of the United States of America*, 108(39), 16241–16246. http://dx.doi.org/10.1073/pnas.1111420108.

Frazier, A., Wisner, M., Falke, J., & Cafiso, D. (2000). Determination of calcium-induced structural changes and membrane binding orientation of cPLA2-C2 domain. *Biophysical Journal*, 78(1), 415A.

Getz, E. B., Xiao, M., Chakrabarty, T., Cooke, R., & Selvin, P. R. (1999). A comparison between the sulfhydryl reductants tris(2-carboxyethyl)phosphine and dithiothreitol for use in protein biochemistry. *Analytical Biochemistry*, 273(1), 73–80. http://dx.doi.org/10.1006/abio.1999.4203.

Ghimire, H., Hustedt, E. J., Sahu, I. D., Inbaraj, J. J., McCarrick, R., Mayo, D. J., et al. (2012). Distance measurements on a dual-labeled TOAC AChR M2δ peptide in mechanically aligned DMPC bilayers via dipolar broadening CW-EPR spectroscopy. *The Journal of Physical Chemistry. B*, 116(12), 3866–3873. http://dx.doi.org/10.1021/jp212272d.

Ghimire, H., McCarrick, R. M., Budil, D. E., & Lorigan, G. A. (2009). Significantly improved sensitivity of Q-band PELDOR/DEER experiments relative to X-band is observed in measuring the intercoil distance of a leucine zipper motif peptide (GCN4-LZ). *Biochemistry*, 48(25), 5782–5784. http://dx.doi.org/10.1021/bi900781u.

Griffith, O. H., & McConnel, H. M. (1966). A nitroxide-maleimide spin label. *Proceedings of the National Academy of Sciences of the United States of America*, 55(1), 8–11. http://dx.doi.org/10.1073/pnas.55.1.8.

Gross, A., Columbus, L., Hideg, K., Altenbach, C., & Hubbell, W. L. (1999). Structure of the KcsA potassium channel from Streptomyces lividans: A site-directed spin labeling study of the second transmembrane segment. *Biochemistry*, 38(32), 10324–10335. http://dx.doi.org/10.1021/bi990856k.

Hassner, A., & Alexanian, V. (1978). Synthetic methods. 12. Direct room-temperature esterification of carboxylic-acids. *Tetrahedron Letters*, 19(46), 4475–4478.

Hilger, D., Jung, H., Padan, E., Wegener, C., Vogel, K. P., Steinhoff, H. J., et al. (2005). Assessing oligomerization of membrane proteins by four-pulse DEER: pH-dependent dimerization of NhaA Na+/H+ antiporter of E-coli. *Biophysical Journal*, 89(2), 1328–1338. http://dx.doi.org/10.1529/biophysj.105.062232.

Huang, C. H. (1969). Studies on phosphatidylcholine vesicles. Formation and physical characteristics. *Biochemistry*, 8(1), 344–352. http://dx.doi.org/10.1021/bi00829a048.

Hubbell, W. L., Cafiso, D. S., & Altenbach, C. (2000). Identifying conformational changes with site-directed spin labeling. *Nature Structural Biology*, 7(9), 735–739. http://dx.doi.org/10.1038/78956.

Hubbell, W. L., Gross, A., Langen, R., & Lietzow, M. A. (1998). Recent advances in site-directed spin labeling of proteins. *Current Opinion in Structural Biology*, 8(5), 649–656. http://dx.doi.org/10.1016/s0959-440x(98)80158-9.

Hubbell, W. L., Mchaourab, H. S., Altenbach, C., & Lietzow, M. A. (1996). Watching proteins move using site-directed spin labeling. *Structure, 4*(7), 779–783. http://dx.doi.org/10.1016/s0969-2126(96)00085-8.

Itaya, K., & Ui, M. (1966). A new micromethod for the colorimetric determination of inorganic phosphate. *Clinica Chimica Acta, 14*, 361–366.

Jeschke, G. (2012). DEER distance measurements on proteins. In M. A. Johnson & T. J. Martinez (Eds.), *Annual review of physical chemistry: Vol. 63* (pp. 419–446). Palo Alto, CA: Annual Reviews.

Jeschke, G. (2014). Interpretation of dipolar EPR data in terms of protein structure. In C. R. Timmel & J. R. Harmer (Eds.), *Structural information from spin-labels and intrinsic paramagnetic centres in the biosciences: Vol. 152* (pp. 83–120). Heidelberg: Springer.

Kalai, T., Schindler, J., Balog, M., Fogassy, E., & Hideg, K. (2008). Synthesis and resolution of new paramagnetic alpha-amino acids. *Tetrahedron, 64*(6), 1094–1100. http://dx.doi.org/10.1016/j.tet.2007.11.020.

Klug, C. S., & Feix, J. B. (2008). Methods and applications of site-directed spin labeling EPR spectroscopy. In J. J. Correia & H. W. Detrich (Eds.), *Biophysical tools for biologists: Vol. 1 in vitro techniques: Vol. 84* (pp. 617–658). San Diego: Academic Press.

Klug, C. S., Su, W. Y., & Feix, J. B. (1997). Mapping of the residues involved in a proposed beta-strand located in the ferric enterobactin receptor FepA using site-directed spin-labeling. *Biochemistry, 36*(42), 13027–13033. http://dx.doi.org/10.1021/bi971232m.

Kuchinka, E., & Seelig, J. (1989). Interaction of melittin with phosphatidylcholine membranes—Binding isotherm and lipid headgroup conformation. *Biochemistry, 28*(10), 4216–4221. http://dx.doi.org/10.1021/bi00436a014.

Kusumi, A., Subczynski, W. K., & Hyde, J. S. (1982). Oxygen-transport parameter in membranes as deduced by saturation recovery measurements of spin-lattice relaxation-times of spin labels. *Proceedings of the National Academy of Sciences of the United States of America, 79*(6), 1854–1858. http://dx.doi.org/10.1073/pnas.79.6.1854.

Larrabee, A. L. (1979). Time-dependent changes in the size distribution of distearoylphosphatidylcholine vesicles. *Biochemistry, 18*(15), 3321–3326. http://dx.doi.org/10.1021/bi00582a019.

Lasic, D. D. (1988). The mechanism of vesicle formation. *Biochemical Journal, 256*(1), 1–11.

Lasic, D. D., & Martin, F. J. (1990). On the mechanism of vesicle formation. *Journal of Membrane Science, 50*(2), 215–222. http://dx.doi.org/10.1016/s0376-7388(00)80317-8.

Margittai, M., & Langen, R. (2004). Template-assisted filament growth by parallel stacking of tau. *Proceedings of the National Academy of Sciences of the United States of America, 101*(28), 10278–10283. http://dx.doi.org/10.1073/pnas.0401911101.

Margittai, M., & Langen, R. (2006). Spin labeling analysis of amyloids and other protein aggregates. In I. Kheterpal & R. Wetzel (Eds.), *Amyloid, prions, and other protein aggregates, pt C: Vol. 413* (pp. 122–139). Amsterdam: Elsevier.

Marsh, D. (2010). Electron spin resonance in membrane research: Protein-lipid interactions from challenging beginnings to state of the art. *European Biophysics Journal, 39*(4), 513–525. http://dx.doi.org/10.1007/s00249-009-0512-3.

Marsh, D., Jost, M., Peggion, C., & Toniolo, C. (2007). TOAC spin labels in the backbone of alamethicin: EPR studies in lipid membranes. *Biophysical Journal, 92*(2), 473–481. http://dx.doi.org/10.1529/biophysj.106.092775.

Mathias, R. T., McLaughlin, S., Baldo, G., & Manivannan, K. (1988). The electrostatic potential due to a single fixed charge at a membrane-solution interface. *Biophysical Journal, 53*(2), A128.

Mchaourab, H. S., Hyde, J. S., & Feix, J. B. (1994). Binding and state of aggregation of spin-labeled cecropin ad in phospholipid-bilayers—Effects of surface-charge and fatty acyl-chain length. *Biochemistry, 33*(21), 6691–6699. http://dx.doi.org/10.1021/bi00187a040.

Mchaourab, H. S., Lietzow, M. A., Hideg, K., & Hubbell, W. L. (1996). Motion of spin-labeled side chains in T4 lysozyme, correlation with protein structure and dynamics. *Biochemistry, 35*(24), 7692–7704. http://dx.doi.org/10.1021/bi960482k.

Mchaourab, H. S., Oh, K. J., Fang, C. J., & Hubbell, W. L. (1997). Conformation of T4 lysozyme in solution. Hinge-bending motion and the substrate-induced conformational transition studied by site-directed spin labeling. *Biochemistry, 36*(2), 307–316. http://dx.doi.org/10.1021/bi962114m.

McLaughlin, S. (1977). Electrostatic potentials at membrane solution interfaces. In *Current Topics in Membranes and Transport: Vol. 9* (pp. 71–144). San Diego: Academic Press.

McLaughlin, S. (1989). The electrostatic properties of membranes. *Annual Review of Biophysics and Biophysical Chemistry, 18*, 113–136. http://dx.doi.org/10.1146/annurev.biophys.18.1.113.

McNulty, J. C., & Millhauser, G. L. (2000). TOAC—The rigid nitroxide side chain. In L. Berliner, G. Eaton, & S. Eaton (Eds.), *Distance Measurements in Biological Systems by EPR: Vol. 19.* (pp. 277–307). New York, NY: Springer.

Michaels, D. M., Horwitz, A. F., & Klein, M. P. (1973). Transbilayer asymmetry and surface homogeneity of mixed phospholipids in cosonicated vesicles. *Biochemistry, 12*(14), 2637–2645. http://dx.doi.org/10.1021/bi00738a014.

Milov, A. D., Maryasov, A. G., & Tsvetkov, Y. D. (1998). Pulsed electron double resonance (PELDOR) and its applications in free-radicals research. *Applied Magnetic Resonance, 15*(1), 107–143.

Milov, A. D., Ponomarev, A. B., & Tsvetkov, Y. D. (1984). Electron electron double-resonance in electron-spin echo—Model biradical systems and the sensitized photolysis of decalin. *Chemical Physics Letters, 110*(1), 67–72. http://dx.doi.org/10.1016/0009-2614(84)80148-7.

Milov, A. D., Samoilova, R. I., Tsvetkov, Y. D., De Zotti, M., Formaggio, F., Toniolo, C., et al. (2009). Structure of self-aggregated alamethicin in ePC membranes detected by pulsed electron-electron double resonance and electron spin echo envelope modulation spectroscopies. *Biophysical Journal, 96*(8), 3197–3209. http://dx.doi.org/10.1016/j.bpj.2009.01.026.

Milov, A. D., Tsvetkov, Y. D., Formaggio, F., Crisma, M., Toniolo, C., & Raap, J. (2000). Self-assembling properties of membrane-modifying peptides studied by PELDOR and CW-ESR spectroscopies. *Journal of the American Chemical Society, 122*(16), 3843–3848. http://dx.doi.org/10.1021/ja993870t.

Molin, Y. N., Salikhov, K. M., & Zamaraev, K. I. (1980). *Spin exchange*. Berlin: Springer.

Newstadt, J. P., Mayo, D. J., Inbaraj, J. J., Subbaraman, N., & Lorigan, G. A. (2009). Determining the helical tilt of membrane peptides using electron paramagnetic resonance spectroscopy. *Journal of Magnetic Resonance, 198*(1), 1–7. http://dx.doi.org/10.1016/j.jmr.2008.12.007.

Nielsen, R. D., Che, K. P., Gelb, M. H., & Robinson, B. H. (2005). A ruler for determining the position of proteins in membranes. *Journal of the American Chemical Society, 127*(17), 6430–6442. http://dx.doi.org/10.1021/ja042782s.

Ogawa, S., & McConnel, H. M. (1967). Spin-label study of hemoglobin conformations in solution. *Proceedings of the National Academy of Sciences of the United States of America, 58*(1), 19–26. http://dx.doi.org/10.1073/pnas.58.1.19.

Ogawa, S., McConnel, H. M., & Horwitz, A. (1968). Overlapping conformation changes in spin-labeled hemoglobin. *Proceedings of the National Academy of Sciences of the United States of America, 61*(2), 401–405. http://dx.doi.org/10.1073/pnas.61.2.401.

Pistolesi, S., Pogni, R., & Feix, J. B. (2007). Membrane insertion and bilayer perturbation by antimicrobial peptide CM15. *Biophysical Journal, 93*(5), 1651–1660. http://dx.doi.org/10.1529/biophysj.107.104034.

Pyka, J., Ilnicki, J., Altenbach, C., Hubbell, W. L., & Froncisz, W. (2005). Accessibility and dynamics of nitroxide side chains in T4 lysozyme measured by saturation recovery EPR. *Biophysical Journal, 89*(3), 2059–2068. http://dx.doi.org/10.1529/biophysj.105.059055.

Qin, Z. H., & Cafiso, D. S. (1996). Membrane structure of protein kinase C and calmodulin binding domain of myristoylated alanine rich C kinase substrate determined by site-directed spin labeling. *Biochemistry, 35*(9), 2917–2925. http://dx.doi.org/10.1021/bi9521452.

Rabenstein, M. D., & Shin, Y. K. (1995). Determination of the distance between 2 spin labels attached to a macromolecule. *Proceedings of the National Academy of Sciences of the United States of America, 92*(18), 8239–8243. http://dx.doi.org/10.1073/pnas.92.18.8239.

Rassat, A., & Rey, P. (1967). Nitroxides. 23. Preparation of amino-acid free radicals and their complex salts. *Bulletin de la Société Chimique de France, 3*, 815–818.

Sahu, I. D., Hustedt, E. J., Ghimire, H., Inbaraj, J. J., McCarrick, R. M., & Lorigan, G. A. (2014). CW dipolar broadening EPR spectroscopy and mechanically aligned bilayers used to measure distance and relative orientation between two TOAC spin labels on an antimicrobial peptide. *Journal of Magnetic Resonance, 249*, 72–79. http://dx.doi.org/10.1016/j.jmr.2014.09.020.

Salnikov, E. S., Erilov, D. A., Milov, A. D., Tsvetkov, Y. D., Peggion, C., Formaggio, F., et al. (2006). Location and aggregation of the spin-labeled peptide trichogin GA IV in a phospholipid membrane as revealed by pulsed EPR. *Biophysical Journal, 91*(4), 1532–1540. http://dx.doi.org/10.1529/biophysj.105.075887.

Sato, H., & Feix, J. B. (2006). Peptide–membrane interactions and mechanisms of membrane destruction by amphipathic α-helical antimicrobial peptides. *Biochimica et Biophysica Acta, 1758*(9), 1245–1256. http://dx.doi.org/10.1016/j.bbamem.2006.02.021.

Scarpelli, F., Drescher, M., Rutters-Meijneke, T., Holt, A., Rijkers, D. T. S., Killian, J. A., et al. (2009). Aggregation of transmembrane peptides studied by spin-label EPR. *Journal of Physical Chemistry B, 113*(36), 12257–12264. http://dx.doi.org/10.1021/jp901371h.

Schreier, S., Bozelli, J. C., Marín, N., Vieira, R. F. F., & Nakaie, C. R. (2012). The spin label amino acid TOAC and its uses in studies of peptides: Chemical, physicochemical, spectroscopic, and conformational aspects. *Biophysical Reviews, 4*(1), 45–66. http://dx.doi.org/10.1007/s12551-011-0064-5.

Seelig, J. (2004). Thermodynamics of lipid-peptide interactions. *Biochimica et Biophysica Acta, 1666*(1-2), 40–50. http://dx.doi.org/10.1016/j.bbamem.2004.08.004.

Shafer, D. E., Inman, J. K., & Lees, A. (2000). Reaction of tris(2-carboxyethyl)phosphine (TCEP) with maleimide and alpha-haloacyl groups: Anomalous elution of TCEP by gel filtration. *Analytical Biochemistry, 282*(1), 161–164. http://dx.doi.org/10.1006/abio.2000.4609.

Shafer, A. M., Nakaie, C. R., Deupi, X., Bennett, V. J., & Voss, J. C. (2008). Characterization of a conformationally sensitive TOAC spin-labeled substance P. *Peptides, 29*(11), 1919–1929. http://dx.doi.org/10.1016/j.peptides.2008.08.002.

Smirnov, A. I. (2008). Post-processing of IEPR spectra by convolution filtering: Calculation of a harmonics' series and automatic separation of fast-motion components from spin-label EPR spectra. *Journal of Magnetic Resonance, 190*(1), 154–159. http://dx.doi.org/10.1016/j.jmr.2007.10.006.

Smirnov, A. I., & Belford, R. L. (1995). Rapid quantitation from inhomogeneously broadened EPR-spectra by a fast convolution algorithm. *Journal of Magnetic Resonance, Series A, 113*(1), 65–73. http://dx.doi.org/10.1006/jmra.1995.1057.

Smirnov, A. I., Clarkson, R. B., & Belford, R. L. (1996). EPR linewidth (T-2) method to measure oxygen permeability of phospholipid bilayers and its use to study the effect of low ethanol concentrations. *Journal of Magnetic Resonance. Series B, 111*(2), 149–157. http://dx.doi.org/10.1006/jmrb.1996.0073.

Smirnova, T. I., Smirnov, A. I., Clarkson, R. B., & Belford, R. L. (1995a). W-band (95 Ghz) EPR spectroscopy of nitroxide radicals with complex proton hyperfine-structure: Fast motion. *Journal of Physical Chemistry, 99*(22), 9008–9016. http://dx.doi.org/10.1021/J100022a011.

Smirnova, T. I., Smirnov, A. I., Clarkson, R. B., & Belford, R. L. (1995b). Accuracy of oxygen measurements in T-2 (line-width) Epr oximetry. *Magnetic Resonance in Medicine, 33*(6), 801–810. http://dx.doi.org/10.1002/mrm.1910330610.

Smirnova, T. I., Smirnov, A. I., Clarkson, R. B., Belford, R. L., Kotake, Y., & Janzen, E. G. (1997). High-frequency (95 GHz) EPR spectroscopy to characterize spin adducts. *Journal of Physical Chemistry B, 101*(19), 3877–3885. http://dx.doi.org/10.1021/Jp963066i.

Steinhoff, H. J., Dombrowsky, O., Karim, C., & Schneiderhahn, C. (1991). 2-Dimensional diffusion of small molecules on protein surfaces—An epr study of the restricted translational diffusion of protein-bound spin labels. *European Biophysics Journal, 20*(5), 293–303. http://dx.doi.org/10.1007/bf00450565.

Steinhoff, H. J., Radzwill, N., Thevis, W., Lenz, V., Brandenburg, D., Antson, A., et al. (1997). Determination of interspin distances between spin labels attached to insulin: Comparison of electron paramagnetic resonance data with the x-ray structure. *Biophysical Journal, 73*(6), 3287–3298.

Stoller, S., Sicoli, G., Baranova, T. Y., Bennati, M., & Diederichsen, U. (2011). TOPP: A novel nitroxide-labeled amino acid for EPR distance measurements. *Angewandte Chemie-International Edition, 50*(41), 9743–9746. http://dx.doi.org/10.1002/anie.201103315.

Subczynski, W., Widomska, J., Wisniewska, A., & Kusumi, A. (2007). Saturation-recovery electron paramagnetic resonance discrimination by oxygen transport (DOT) method for characterizing membrane domains. In T. McIntosh (Ed.), *Lipid rafts: Vol. 398.* (pp. 143–157). Totowa, NJ: Humana Press.

Syryamina, V. N., De Zotti, M., Peggion, C., Formaggio, F., Toniolo, C., Raap, J., et al. (2012). A molecular view on the role of cholesterol upon membrane insertion, aggregation, and water accessibility of the antibiotic lipopeptide trichogin GA IV as revealed by EPR. *Journal of Physical Chemistry B, 116*(19), 5653–5660. http://dx.doi.org/10.1021/jp301660a.

Thomas, L., Scheidt, H. A., Bettio, A., Huster, D., Beck-Sickinger, A. G., Arnold, K., et al. (2005). Membrane interaction of neuropeptide Y detected by EPR and NMR spectroscopy. *Biochimica et Biophysica Acta, 1714*(2), 103–113. http://dx.doi.org/10.1016/j.bbamem.2005.06.012.

Tominaga, M., Barbosa, S. R., Poletti, E. F., Zukerman-Schpector, J., Marchetto, R., Schreier, S., et al. (2001). Fmoc-POAC: (9-fluorenylmethyloxycarbonyl)-2,2,5,5-tetramethylpyrrolidine-N-oxyl-3-amino-4 carboxylic acid: A novel protected spin labeled beta-amino acid for peptide and protein chemistry. *Chemical & Pharmaceutical Bulletin, 49*(8), 1027–1029. http://dx.doi.org/10.1248/cpb.49.1027.

Torchilin, V. P. (2008). Tat peptide-mediated intracellular delivery of pharmaceutical nanocarriers. *Advanced Drug Delivery Reviews, 60*(4–5), 548–558. http://dx.doi.org/10.1016/j.addr.2007.10.008.

Traikia, M., Warschawski, D. E., Recouvreur, M., Cartaud, J., & Devaux, P. F. (2000). Formation of unilamellar vesicles by repetitive freeze-thaw cycles: Characterization by electron microscopy and P-31-nuclear magnetic resonance. *European Biophysics Journal, 29*(3), 184–195. http://dx.doi.org/10.1007/s002490000077.

Turner, A. L., Braide, O., Mills, F. D., Fanucci, G. E., & Long, J. R. (2014). Residue specific partitioning of KL4 into phospholipid bilayers. *Biochimica et Biophysica Acta, 1838*(12), 3212–3219. http://dx.doi.org/10.1016/j.bbamem.2014.09.006.

Van Vleck, J. H. (1948). The dipolar broadening of magnetic resonance lines in crystals. *Physical Review, 74*(9), 1168–1183. http://dx.doi.org/10.1103/PhysRev.74.1168.

Victor, K. G., & Cafiso, D. S. (2001). Location and dynamics of basic peptides at the membrane interface: Electron paramagnetic resonance spectroscopy of tetramethyl-piperidine-N-oxyl-4-amino-4-carboxylic acid-labeled peptides. *Biophysical Journal*, *81*(4), 2241–2250.

Voinov, M. A., Rivera-Rivera, I., & Smirnov, A. I. (2013). Surface electrostatics of lipid bilayers by EPR of a pH-sensitive spin-labeled lipid. *Biophysical Journal*, *104*(1), 106–116. http://dx.doi.org/10.1016/j.bpj.2012.11.3806.

Wieprecht, T., & Seelig, J. (2002). Isothermal titration calorimetry for studying interactions between peptides and lipid membranes. In T. J. M. Sidney & A. Simon (Eds.), *Current topics in membranes: Vol. 52* (pp. 31–56). San Diego, CA: Academic Press.

Wright, K., Sarciaux, M., de Castries, A., Wakselman, M., Mazaleyrat, J. P., Toffoletti, A., et al. (2007). Synthesis of enantiomerically pure cis- and trans-4-amino-1-oxyl-2,2,6,6-tetramethylpiperidine-3-carboxylic acid: A spin-labelled, cyclic, chiral beta-amino acid, and 3D-Structural analysis of a doubly spin-labelled beta-hexapeptide. *European Journal of Organic Chemistry*, *2007*(19), 3133–3144. http://dx.doi.org/10.1002/ejoc.200700153.

Yamaguchi, T., Nomura, M., Matsuoka, T., & Koda, S. (2009). Effects of frequency and power of ultrasound on the size reduction of liposome. *Chemistry and Physics of Lipids*, *160*(1), 58–62. http://dx.doi.org/10.1016/j.chemphyslip.2009.04.002.

Yin, J. J., Feix, J. B., & Hyde, J. S. (1990). Mapping of collision frequencies for stearic-acid spin labels by saturation-recovery electron-paramagnetic resonance. *Biophysical Journal*, *58*(3), 713–720.

CHAPTER NINE

Structural Characterization of Membrane-Curving Proteins: Site-Directed Spin Labeling, EPR, and Computational Refinement

Mark R. Ambroso*, Ian S. Haworth[†], Ralf Langen*,[1]

*Department of Biochemistry and Molecular Biology, Zilkha Neurogenetic Institute, University of Southern California, Los Angeles, California, USA
[†]Department of Pharmacology and Pharmaceutical Sciences, University of Southern California, Los Angeles, California, USA
[1]Corresponding author: e-mail address: langen@usc.edu

Contents

1. Introduction	260
1.1 Membrane-Curving Proteins	260
2. SDSL, EPR, and Computational Refinement Are a Powerful Combination for Studying Membrane-Curving Proteins Bound to Membranes of Defined Curvature	261
3. Sample Preparation Methodology for SDSL	263
3.1 Site-Directed Spin Labeling	264
3.2 Protein Purification and Spin Labeling	266
3.3 Protein–Lipid Complex Preparation	267
4. EPR Measurements	270
4.1 Mobility Measurements	270
4.2 Accessibility	272
4.3 Distances	274
5. Computational Refinement	277
5.1 Construction of a Starting Structure	277
5.2 Simulated Annealing	280
5.3 Reconstruction of the Protein–Lipid Assembly	283
6. Outlook	284
Acknowledgment	285
References	285

Abstract

Endocytosis and other membrane remodeling processes require the coordinated generation of different membrane shapes. Proteins capable of manipulating lipid bilayers mediate these events using mechanisms that are not fully understood. Progress is limited by the small number of structures solved for proteins bound to different membrane shapes and tools capable of resolving such information. However, recent studies have shown site-directed spin labeling (SDSL) in combination with electron paramagnetic resonance (EPR) to be capable of obtaining high-resolution structural information for proteins bound to different membrane shapes. This technique can be applied to proteins with no known structure or proteins with structures known in solution. By refining the data obtained by EPR with computational modeling, 3D structures or structural models of membrane-bound proteins can be generated. In this chapter, we highlight the basic considerations and steps required to investigate the structures of membrane-bound proteins using SDSL, EPR, and computational refinement.

1. INTRODUCTION

Membranes act as partitions to establish cellular compartments, and the ability to regulate and remodel membrane shape is vital for cellular function. Highly coordinated membrane remodeling events are observed in essential processes including endo- and exocytosis, intracellular membrane trafficking, organelle shape, viral budding, phagocytosis, autophagy, and cell movement (Doherty & McMahon, 2009). In synaptic endocytosis, cells must regulate the uptake of synaptic vesicles carrying neurotransmitters on millisecond time scales (Watanabe et al., 2013). This process involves the careful generation of pre-endocytotic invaginations, a spherical intermediate connected to the membrane by a cylindrical tube, and scission events in order to release a vesicle into the cytosol. Events like this involve a series of multiple different membrane structures, indicating that cells are capable of orchestrating specific membrane shapes with precise regulation and direction. In fact, the disruption of these faculties results in the generation of neurological diseases and muscular myopathies (De Camilli et al., 1993; Nicot et al., 2007). The field of membrane curvature is still young and the precise mechanisms that underlie how cells regulate this process are still not fully understood. This is due in large part to the insufficiency of methods for studying the structure of proteins bound to membranes. Site-directed spin labeling (SDSL) together with electron paramagnetic resonance (EPR) is one of the few approaches capable of resolving how proteins interact with membranes.

1.1 Membrane-Curving Proteins

It has become increasingly clear that protein–membrane interactions are the primary driving force behind most if not all membrane events (Doherty & McMahon, 2009; Drin & Antonny, 2010; Farsad & Camilli, 2003; McMahon & Gallop, 2005; Qualmann, Koch, & Kessels, 2011). Entire protein families of membrane-curving proteins have been discovered and observed to remodel membranes *in vitro* and/or *in vivo*, including BIN/Amphiphysin/Rvs (BAR) (Ambroso, Hegde, & Langen, 2014; Boucrot et al., 2012; Farsad et al., 2001; Gallop et al., 2006; Isas, Ambroso, Hegde, Langen, & Langen, 2015; Jao et al., 2010; Meinecke et al., 2013; Peter et al., 2004), Eps15-homology (Daumke et al., 2007; Shah et al., 2014), epsin (Boucrot et al., 2012; Lai et al., 2012), synaptotagmins (Martens, Kozlov, & McMahon, 2007), and synucleins (Varkey et al., 2010; Westphal & Chandra, 2013). Most of these proteins are capable of generating multiple membrane shapes, such as small vesicles or lipid tubes.

Several mechanisms have been proposed to explain how proteins bend membranes, including membrane crowding (Stachowiak et al., 2012), bilayer coupling, membrane insertions of amphipathic helices or loop regions (wedging) (Gallop et al., 2006; Jao et al., 2010), and scaffolding (Ambroso et al., 2014; Boucrot et al., 2012; Gallop et al., 2006). For review, see Drin and Antonny (2010), Farsad and Camilli (2003), and Qualmann et al. (2011). Of these mechanisms, wedging and scaffolding are the most thoroughly studied (Fig. 1B). Amphipathic helices are defined as helical segments whose properties allow the helix to associate on one side with polar or charged lipid headgroups and on the other with hydrophobic acyl chains. As an amphipathic helix inserts into a leaflet, it begins to take up space and wedge apart nearby lipids. Proteins can also bind membranes by acting as a scaffold and force them to conform to their intrinsic shape. While some proteins utilize a single mechanism to bend membranes, others have the ability to use a combination of mechanisms. Furthermore, recent studies have shown that the structures of membrane-bound N-BAR proteins, amphiphysin, and endophilin depend on the shape they generate (Fig. 1B; Ambroso et al., 2014; Isas et al., 2015; Jao, Hegde, Chen, Haworth, & Langen, 2008; Varkey et al., 2010). Therefore, to understand how proteins generate different membrane shapes, it is imperative to be able to study their three-dimensional structures on distinct membrane shapes.

2. SDSL, EPR, AND COMPUTATIONAL REFINEMENT ARE A POWERFUL COMBINATION FOR STUDYING MEMBRANE-CURVING PROTEINS BOUND TO MEMBRANES OF DEFINED CURVATURE

While cell biology-based approaches have identified a number of protein families responsible for regulating the geometry of membranes, the structural mechanisms underlying how a protein curves membranes is unclear.

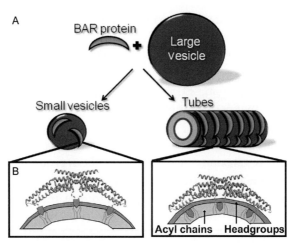

Figure 1 Proteins can use different structures and mechanisms to generate distinct membrane shapes. (A) BAR proteins are banana-shaped proteins capable of inducing both small vesicles and membrane tubes when incubated with large spherical vesicles. (B) A SDSL and EPR study using approximately 30 spin-labeled sites of BAR protein endophilin A1 revealed that this protein uses different structures and mechanisms to induce small vesicles or membrane tubes (Ambroso et al., 2014). Accessibility measurements revealed that the BAR domain (crystal structure: PDB ID 2C08) only makes contact with the membrane when the protein is bound to tubes. Using accessibility measurements, the amphipathic helices (red (gray in the print version) wedges) were observed to insert into the lipid membrane on vesicles and tubes, but at different depths. Simulations have suggested that the force a protein has on a lipid (blue (gray in the print version) acyl chains) bilayer is dependent on the depth to which it penetrates (Campelo, McMahon, & Kozlov, 2008). *The figure was adapted from PNAS (Ambroso et al., 2014).*

Crystallizing proteins in the presence of lipids is difficult, and it is even more difficult to examine different membrane shapes (i.e., tubes vs. vesicles) in crystals. Analysis by solution NMR typically requires the use of detergents or membrane-mimicking systems such as nanolipoprotein particles or bicelles, instead of larger lipid assemblies with different, physiologically relevant membrane shapes (Raschle, Hiller, Etzkorn, & Wagner, 2010). EPR with SDSL, however, is a powerful approach for studying the 3D structure of proteins bound to tubes, vesicles, or other membrane shapes.

SDSL relies on the modification of an amino acid side chain through covalent conjugation with a small paramagnetic spin label containing a single unpaired electron. Many spin labels are commercially available, but the most commonly utilized labels are thiol-reactive conjugates that can be linked to proteins via cysteine residues. Of these, the best-characterized spin label,

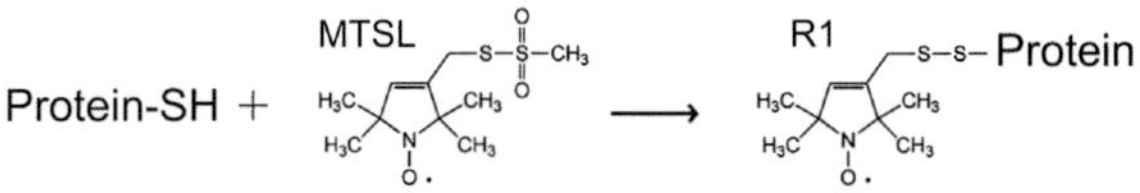

Figure 2 Spin labeling of proteins through conjugation with a cysteine side chain. The spin label MTSL, which contains an unpaired electron in its nitroxide moiety, can be conjugated to a target protein through reaction with an activated thiol side chain.

[1-oxy-2,2,5,5-tetramethylpyrroline-3-methyl]-methane-thiosulfonate (MTSL), forms the side chain R1 (Fig. 2). There is also an array of synthesized lipids containing spin labels at specific locations along their acyl chains, lipid phosphates, and lipid headgroups. Spin-labeled lipids can be used to measure lateral lipid mobility (Bartucci et al., 2006), membrane protein interactions (Marsh & Horváth, 1998), and to calibrate the accessibility of the membrane (Altenbach, Greenhalgh, Khorana, & Hubbell, 1994; Frazier, Roller, Havelka, Hinderliter, & Cafiso, 2003). The ability to gather information from both the protein and the lipid environment is a major reason why EPR with SDSL is such a powerful tool for studying membrane protein structure.

By measuring differences in spin label mobility of a single spin-labeled protein in solution or bound to membranes, changes in protein structure can be determined (Hubbell, Gross, Langen, & Lietzow, 1998). Whether a residue is membrane bound can be determined by measuring its immersion depth (Altenbach et al., 1994). Spin-labeled double-cysteine mutants can be used to measure intramolecular distances between sites of the same protein (Jeschke & Polyhach, 2007). For proteins with crystal structures, distance measurements can be used to determine if the membrane-bound protein has a similar structure. Finally, distances between singly labeled proteins can be measured to determine quaternary interactions. Using computational refinement based on mobility, immersion depth, and distance measurements from different spin-labeled sites, 3D structures or structural models of proteins bound to membranes with distinct membrane shapes can be elucidated (Gallop et al., 2006; Isas et al., 2015; Jao et al., 2008, 2010; Lai et al., 2012; Shah et al., 2014).

3. SAMPLE PREPARATION METHODOLOGY FOR SDSL

Prior to analysis by EPR, spin-labeled proteins need to be prepared and the lipid-binding conditions need to be optimized to achieve the desired membrane shape and curvature. An overview of considerations for generating samples to be analyzed by EPR is provided here.

3.1 Site-Directed Spin Labeling

3.1.1 Basic Considerations for SDSL

Practical considerations are important in deciding whether a target protein is amenable to study by SDSL and EPR. The approach relies on the introduction of one or more spin labels, whose unpaired electron EPR properties can then be measured. SDSL commonly requires multiple different cysteine analogs of the same protein for different regions of a protein of interest. Therefore, the generation of multiple distinct cysteine mutant protein constructs is most practically done by utilizing a recombinant expression system or by peptide synthesis. In order to generate constructs containing a single cysteine, it is often required to mutate native cysteines to other residues, commonly to an alanine or a serine. Proteins with many cysteines or proteins, whose structure or function relies on the presence of native cysteines, are not ideal targets for this approach. Using target protein-specific functional assays and biophysical methods, one should also verify that cysteine mutagenesis does not significantly alter the endogenous structure or function of the protein.

EPR with SDSL is used to measure the local environment of a number of different R1 side chains. However, not every residue in a protein needs to be mutated and spin labeled to obtain structural information by EPR. To limit the number of cysteine constructs that need to be generated, SDSL can be strategically applied to proteins of known or unknown structure using different strategies. Figure 3 outlines a general strategy to investigate structures of proteins with previously defined structures similar to what was done for endophilin (Jao et al., 2010) or for *de novo* structure analysis as was done for synuclein (Jao et al., 2008). In the case of endophilin, EPR measurements of approximately 25 spin-labeled sites in regions in the crystal structure or regions unresolved by crystallography were used to generate 3D structural models of vesicle-bound endophilin A1. For our study on α-synuclein, which has no crystal structure, *de novo* 3D structural models were built using over 73 single and 17 double spin-labeled derivatives to define its structure bound to vesicles.

3.1.2 Protocol for the Generation of Cysteine Mutant Constructs

The following protocol describes the typical procedure for generating a single-cysteine mutant DNA construct for recombinant expression in *Escherichia coli*.

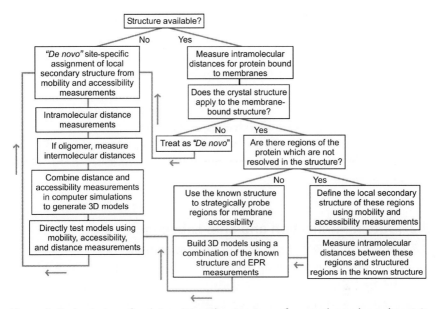

Figure 3 Basic strategy for determining the structure of a membrane-bound protein using SDSL, EPR, and computational refinement. The experimental strategy is typically divided into two parts: (1) solving *de novo* structures which have little to no previous characterization and (2) resolving the membrane-bound structure of proteins with known solution structures. Regions that are selected to be examined for local secondary structure information are mutated to cysteines, one or two at a time, spin labeled, and subjected to mobility (Section 4.1), accessibility (Section 4.2), or distance measurements (Section 4.3). Structural attributes determined form EPR measurements are used to computationally generate 3D models (Section5). These models can be directly tested using additional EPR measurements.

(1) Clone cDNA encoding the protein of interest into a recombinant expression vector using appropriate restriction sites. The plasmid should have selective antibiotic resistance and a T7 RNA polymerase promoter such as the pET (Agilent) systems.[1]

(2) Design DNA primers aligning to the region of the cDNA which has been selected to contain a mutation to cysteine. Replace the codon of a predetermined site with a cysteine codon, TGC or TGT. Follow

---

[1] An additional consideration is to use a vector system containing protein tags or conjugates which can expedite the purification of the expressed protein. Polyhistidine and glutathione *S*-transferase tags can eliminate several purification steps by binding to Ni^{2+} or glutathione sepharose beads, respectively. Proteins can also be expressed in conjunction with proteins such as thioredoxin to increase their solubility. Protein tags can be removed postexpression by engineering protease cleavage sites between the tag and the protein of interest.

standard QuikChange (Agilent) site-directed mutagenesis protocols to mutagenize the parent cDNA with the designed primers.
(3) Transform the mutagenized cDNA into competent *E. coli* and allow colonies to grow overnight on an LB-agar plate with the appropriate antibiotic.
(4) Pick a colony from the LB-agar plate and expand it overnight in 50 mL of LB media at 37 °C with shaking at 225 rpm.
(5) Purify the DNA using standard mini-prep kits (Qiagen) and verify the integrity of the cysteine mutant construct using DNA sequencing.

3.2 Protein Purification and Spin Labeling
3.2.1 Basic Considerations for Protein Purification

Common protein purification techniques can be used as long as special consideration is taken to prevent oxidation of the cysteine residues. Oxidation prevents the downstream conjugation of thiol-reactive spin labels with cysteine side chains. Adding reducing agents such as dithiothreitol or tris(2-carboxyethyl)phosphine to a final concentration of 1–10 mM in buffers used during purification prevents cysteine oxidation. These agents should be removed using gel filtration immediately prior to spin labeling the purified protein. Additional considerations should be made to ensure that the resulting proteins are highly pure and free from contamination of cysteine-containing proteins. As a control, a cysteine-less protein can be subjected to the same spin-labeling procedure as that applied to the cysteine-containing analogs and its EPR spectra examined to detect contaminating signals.

It is important to optimize the expression and purification process of the target protein as these steps will need to be repeated for each cysteine mutant analog. In our experience, the same purification protocol can be used for all cysteine mutant analogs of the same protein. An additional consideration is whether introduced mutations affect protein stability or structure. Circular dichroism can be used to verify that no significant changes in secondary structure were induced by the introduction of the spin labels. Light scattering can be used to monitor protein aggregation.

Protocols describing how to purify amphiphysin (Isas et al., 2015), α-synuclein (Jao et al., 2008; Varkey et al., 2010), EHD2 (Daumke et al., 2007; Shah et al., 2014), and epsin (Lai et al., 2012) can be found in their respective references.

3.2.2 Spin-Labeling Protocol

For proteins, spin labeling with MTSL is the most commonly used method of labeling. Immediately prior to spin labeling, reducing agents must be removed, often by passing protein samples through a PD-10 (GE) gel filtration column. Stocks of spin label MTSL are typically made at 40 mg/mL in acetonitrile, split into small aliquots, and stored by freezing at $-20\,°C$. Adding a 5- to 10-fold molar excess of MTSL to protein ensures close to quantitative labeling of nearly all but the most buried of sites. To facilitate the reactivity of the thiol-based reaction, labeling should be performed around pH 7. At pH greater than 8, the label can detach from the protein and produce several spin-labeled side products. Spin labeling becomes slower at lower pH and acidic pH lower than 6 can lead to incomplete labeling. Excess spin label can be removed using PD-10 columns.

(1) Use gel filtration (PD-10) or dialysis to exchange the protein of interest into a buffer devoid of reducing agents and with a pH of $\sim$7.0.
(2) Add spin label in a 5- to 10-fold molar excess and incubate overnight at 4 °C or 1 h at room temperature.
(3) Remove excess spin label from spin-labeled protein using gel filtration (PD-10) or dialysis.
(4) Confirm the presence of spin label on the target protein by recording an EPR spectrum (see Section 4.1) of a small aliquot of the protein.

3.3 Protein–Lipid Complex Preparation

3.3.1 Basic Considerations for Sample Preparation

Because EPR methods used with SDSL measure the bulk properties of all spin labels, it is advisable to work with homogenous samples. For example, a sample, which contains vesicle and tube-bound protein, will give EPR signals from both states. In order to gain membrane-shape context-specific structural information, it is typically required to optimize sample homogeneity. Conditions also need to be highly reproducible because typically measurements need to be repeated for a large number of different spin-labeled analogs. This becomes an especially difficult task in the case of metastable protein–lipid complexes. It is also desirable to find conditions which allow the complex to be isolated from contaminants or unbound protein. This can be accomplished by either optimizing the homogeneity of the samples to maximize signal or by using biochemical methods to purify the desired complex. Sample optimization is typically achieved in our laboratory by varying several parameters, including the protein to lipid ratio, lipid composition, vesicle dimensions, ionic strength, and incubation time. Specific lipids known to be important for membrane-mediated processes, such as $PI(4,5)P_2$ or other inositols, can also be tested for their ability to augment membrane binding as well as remodeling into specific membrane shapes. The starting size of lipid vesicles can be modulated by extruding the aqueously suspended vesicles through polycarbonate membranes containing pores of specific diameters (Avanti Polar Lipids, AL). A matrix of conditions created by varying the above conditions is screened using negative-stain transmission electron microscopy (TEM) (see Section 3.2.2).

Due to their physiological relevance, lipid extracts from biological samples tend to be the first type of lipid used for optimization of a protein–lipid complex. However, such extracts suffer from batch-to-batch variation, requiring reoptimization of conditions for each new batch. In order to increase sample consistency, synthetic lipids can be utilized. The exact lipid composition used is highly dependent on the protein being studied as well as the desired membrane shape.

Once samples have been optimized for homogeneity, it is also important to evaluate sample stability overtime. In our experience, membrane tubes formed using α-synuclein can be stable for days but those formed by EHD2 tend to be more transient. Although most EPR measurements are short in duration and often only require 10–20 min, it is important to ensure that the sample quality does not degrade within the time of measurement. Sample stability can be conveniently tested using negative-stain TEM.

Despite differences in experimental systems, all lipid mixtures are initially prepared using the same methods. Lipid stocks are mixed to the desired molar ratios in chloroform or methanol, dried under a nitrogen gas stream, and desiccated overnight to remove any residual trace of organic solvent. The dried lipid films are resuspended in aqueous buffer. Protein is then added and samples are screened by negative-stain TEM. Unless otherwise specified, the separation of unbound protein from protein bound to small vesicles (20–30 nm) is performed using centrifugation at $120,000 \times g$ for 25 min in an ultra centrifuge (Beckman Coulter, Inc., Brea, CA). For separating unbound protein from protein bound to larger vesicles or membrane tubes, samples are spun at $16,000 \times g$ for 20 min. Care should be taken to ensure that samples survive centrifugation. Again, negative-stain TEM is an appropriate tool for testing this.

3.3.2 Protocol for Optimizing Protein–Lipid Complexes

It is imperative that each different protein–lipid complex be optimized for sample homogeneity and stability prior to EPR measurements. For the sake of simplicity, we will outline a general protocol for the optimization of a protein–lipid complex using negative-stain TEM to screen samples. While negative-stain TEM works well for screening shapes, certain artifacts can arise due to the nature of using a stain. Verification of membrane shapes can be done using cryoelectron microscopy, which may mitigate the artifacts. Examples of conditions that were found for optimizing endophilin A1 bound to membrane tubes are provided in the notes.

(1) Perform sample optimization by creating a large number of samples in a matrix of varied lipid compositions,[2] protein-to-lipid molar ratios, starting vesicle size, and ionic strengths.[3]

(2) Aliquot droplets ($\sim$10 μL) of the prepared samples onto a clean plastic surface and suspend carbon-coated formvar copper grids (Electron Microscopy Sciences) on top for 10 min.

(3) During the incubation of the copper grids on the samples, aliquot a 10 μL droplet of 1% uranyl acetate onto the plastic surface. Carefully remove excess sample from the carbon grids using Whatman filter paper and suspend the grid, sample side down, onto one of the droplets of 1% uranyl acetate. Incubate the grid on top of the uranyl acetate for 1–2 min, and dry the grid on Whatman filter paper.

(4) Examine the samples for the presence of the desired protein–lipid complex using TEM.[4]

(5) Prior to analysis by EPR, separate tube-bound from unbound protein by centrifugation of samples that are determined to contain the desired protein–lipid complex.

---

[2] In the case of endophilin A1, optimization of conditions while binding total brain lipids resulted in only $\sim$10% of the sample being converted into membrane tubes. Optimization was then performed using synthetic vesicle preparations and by varying the anionic lipid type, concentration, starting vesicle size, ionic strength, and protein-to-lipid molar ratio. As determined by qualitative analysis using negative-stain TEM, increasing concentrations of the anionic lipid DOPG to around 66% of total lipid resulted in increasing amounts of tubulation.

[3] Optimal tubulation was observed for vesicles that were not extruded to smaller diameters and at protein-to-lipid molar ratios around 1:70, such that decreasing and increasing ratios result in lower levels of tubulation. Increasing the concentration of NaCl beyond 150 mM resulted in decreased levels of membrane tubes. Under optimized protein-to-lipid ratios and NaCl conditions, i.e., dried lipids films of 2:1 mass ratio of DOPG:DOPE resuspended in 20 mM Hepes (pH 7.4) to a final concentration of 4 mg/mL, a mixture of 2:1 DOPG:DOPE produced near 100% tubulation. Under these conditions, we found the protein and lipid incubation time to be a negligible consideration.

[4] Specimen observation is performed in our laboratory on a JEOL 1400 transmission electron microscope accelerated to 100 kV. Endophilin A1-induced tubes were observed by negative-stain and cryoelectron microscopy to be on average 40–50 nm in diameter and exhibited a highly repetitive protein coat.

(6) Remove the supernatant and scan the pellet to determine the quality of the EPR signal. Optimize samples so that the final product produces enough signal for mobility (10–100 µM), accessibility (20–100 µM), or double electron–electron resonance (DEER) measurements (50–100 µM).

4. EPR MEASUREMENTS

Once samples have been prepared and optimized, EPR measurements can be made to reveal information about a spin label's mobility and local environment, accessibility to polarity-sensitive reagents, or proximity to neighboring spin labels.

4.1 Mobility Measurements

4.1.1 Basic Considerations for Mobility Measurements

The local structure and dynamics of R1 are reflected in the line shape of its continuous wave (CW) X-band EPR spectrum, which is recorded as a first derivative of the absorption spectrum to improve the signal-to-noise ratio. The nitrogen (predominantly ^{14}N) of the nitroxide moiety has three different nuclear spin states (magnetic moments). These nuclear spin states differently interact with the unpaired electron leading to hyperfine splitting, resulting in a three-line EPR spectrum. A highly mobile R1 gives rise to relatively sharp and narrow lines whereas low mobility causes line broadening. Quantification of the line broadening is typically done through measuring the inverse of the central line width (ΔH_0^{-1}) or scaled mobility (M_s) (Hubbell, Cafiso, & Altenbach, 2000). In fact, these two mobility parameters are capable of distinguishing whether an R1 is in buried, helix surface, loop, or membrane contacting positions (Isas et al., 2003; Margittai, Fasshauer, Jahn, & Langen, 2003; Mchaourab, Lietzow, Hideg, & Hubbell, 1996). Systematic studies combining X-ray crystallography (Fleissner, Cascio, & Hubbell, 2009; Guo, Cascio, Hideg, & Hubbell, 2008; Hubbell, Altenbach, Hubbell, & Khorana, 2003; Langen, Oh, Cascio, & Hubbell, 2000), mutagenesis (Columbus, Kálai, Jekö, Hideg, & Hubbell, 2001), and spectral simulations (Mchaourab et al., 1996) have further correlated local structure to R1 mobility in α-helical proteins. Comparison of EPR spectra before and after the addition of lipid membranes can additionally elucidate lipid-dependent structural changes. This principle has been utilized in combination with accessibility measurements to determine which residues on the concave surface of the BAR domain contact the membrane in the cases of amphiphysin and endophilin (Ambroso et al., 2014; Isas et al., 2015; Jao et al., 2010), identify global structural changes in annexins (Isas et al., 2003; Langen, Isas, Luecke, Haigler, & Hubbell, 1998), and resolve the structuring of amphipathic helices in the membrane as in the cases of α-synuclein (Jao et al., 2008) and the N-terminal helices of N-BAR proteins (Ambroso et al., 2014; Gallop et al., 2006; Isas et al., 2015; Jao et al., 2010).

Analysis of a spin-labeled protein binding to membranes should involve a comparison of protein in the absence and presence of lipids. Major changes in CW spectra between these two states can be used to determine what regions of the protein undergo conformational changes upon interaction with membranes. This information can help to identify membrane-binding sites, but this assignment needs to be further verified using accessibility and membrane immersion depth measurements (see Section 4.2).

CW EPR spectra can also reveal tertiary and quaternary structural information. Spin labels in close proximity undergo spin–spin interaction (either dipolar or exchange), which is reflected in their EPR spectra. Spin–spin interactions can be used to determine inter- as well as intramolecular interactions as in the case of oligomerized endophilin A1 on membrane vesicles (Jao et al., 2010). A more detailed description of spin–spin interactions and coupling is provided in Section 4.3. Dilution of spin-labeled protein with Cys-less protein or protein labeled with a diamagnetic moiety can be added to reduce signals arising from coupling. Conformational changes in the protein–lipid complex overtime can also be observed in mobility spectra. This method can be utilized to examine sample stability overtime.

4.1.2 Protocol for Obtaining Mobility Parameters

(1) Incubate spin-labeled protein and lipids under the optimized conditions discussed in Section 3.3.2. Final spin-labeled protein concentrations should be at least 10 μM.
(2) Load quartz capillaries[5] (VitroCom Inc., NJ) with sample, seal the capillary, and insert into the resonator ensuring that the sample volume is in the observation cell.
(3) Tune the resonator with the sample in place by following the instructions from the manual of the spectrometer.[5] Set the appropriate scan width, attenuation, modulation, and any other parameters highlighted in the manual.

---

[5] EPR measurements are conducted in our laboratory using X-band Bruker EMX spectrometers. CW EPR spectra are acquired by fitting the spectrometer with an ER4119HS resonator.

(4) Based on the desired signal-to-noise ratio, scan the sample 1–20 times and save the spectrum file.
(5) Determine the mobility parameter (Hubbell et al., 2000) by measuring the central line width and plot the inverse by residue number.

4.2 Accessibility
4.2.1 Basic Considerations for Accessibility Measurements
Like mobility measurements, accessibility measurements can reveal detailed structural information of membrane-bound proteins. These measurements involve the addition of paramagnetic colliders, which affect the relaxation properties of R1 in a concentration-dependent manner. Relaxation properties can be measured using specific resonators through (a) pulse EPR or (b) power saturation. Power saturation measurements are comparatively simpler and require only CW EPR spectra. Thus, they are typically the method of choice. By comparing the relaxation of R1 in the presence of a paramagnetic collider that prefers the aqueous solution, Ni(II) ethylenediaminediacetate (NiEDDA), versus that in the presence of one that prefers the hydrophobic lipid bilayer, O_2, membrane immersion depth can be measured (Altenbach et al., 1994). This is accomplished by scanning the central absorption line at increasing attenuations and plotting its amplitude by microwave power in the presence of both reagents. The fit of each plot yields to an experimental Π value as described in Altenbach et al. (1994). These values are commonly expressed in relation to one another using the depth parameter Φ [$\Phi = \ln(\Pi_{O_2}/\Pi_{NiEDDA})$]. Secondary structural information can be elucidated by systematically scanning a protein sequence and determining periodicity. Membrane-bound amphipathic helices have been shown to have periodic accessibilities to O_2 similar to the periodicity of an α-helix (~3.6 amino acids per turn) (Hubbell et al., 1998) and this approach has also been used for membrane curvature inducing or sensing proteins (Ambroso et al., 2014; Jao et al., 2008, 2010; Lai et al., 2012). Alternatively, a membrane-penetrating N-terminal domain of EHD2 lacked this periodicity and was determined to bind the membrane in a non-helical "loop" (Shah et al., 2014).

The depth parameter, Φ, measured for an R1 moiety can be calibrated in terms of membrane immersion depth by using lipids containing spin labels along their acyl chains and headgroups. 1-Palmitoyl-2-doxyl-stearoyl-*sn*-glycero-3-phosphocholine (Avanti Polar Lipids, Alabaster, AL) spin labeled at positions 5, 7, 10, and 12 on the acyl chains as well as the tempo-labeled

(1-palmitoyl-2-oleoyl-*sn*-glycero-3-phospho(tempo)choline) derivative containing a spin label in the headgroup region can be doped into the lipid composition being used in an experiment, incubated with unlabeled protein, and measured for their respective accessibilities (Frazier et al., 2003). As the spin labels are positioned at known depths, the depth parameter Φ can be calibrated for a given lipid bilayer from the lipid headgroups to the bottom of the acyl chains, from a range of 5 Å above to 20 Å below the lipid phosphates.

This information has been used to determine how N-BAR proteins endophilin and amphiphysin insert their amphipathic N-terminal helices at different depths depending on the membrane shape (Ambroso et al., 2014; Isas et al., 2015; Fig. 1). It is therefore possible to obtain structural information for protein domains embedded in the bilayer as well as domains which are more remote from the membrane. In interpreting results from depth measurements, it is important to keep in mind that highly buried spin-labeled sites can have low accessibilities to both O_2 and NiEDDA. This occluded volume effect (Isas, Langen, Haigler, & Hubbell, 2002) can create positive Φ values even though a given site may not be in direct contact with the membrane. These false-positives can be identified by finding sites that have abnormally low accessibilities to both reagents. For an example, see Isas et al. (2015).

4.2.2 Accessibility Measurement Protocol

(1) Add spin-labeled protein to lipids to form the optimized protein–lipid complex (described in Section 3.3.2) with final spin label concentrations of 20–100 μ*M*.
(2) Load the sample using a pipette into a gas-permeable TPX capillary.[6] Oftentimes, pellets of protein–lipid complexes must be diluted with small volumes of buffer to reduce viscosity prior to being loaded into capillaries. Seal the capillary.
(3) Load a TPX capillary into the TPX sample holder and insert into the resonator.[7] Typically, only 3–5 μL of sample is required.
(4) Tune the resonator and set experimental parameters as described in the manual provided by the resonator's vendor.

---

[6] TPX or other gas-permeable capillaries (ER 221, Bruker) are used for power saturations experiments to allow the samples to equilibrate with O_2 or N_2.
[7] Power saturation experiments require efficient resonators capable of generating strong magnetic fields. Accessibility measurements are performed in our laboratory using a Bruker dielectric or loop gap resonators.

(5) Power saturation measurements in the presence of O_2 use the oxygen present in air. Thus, samples are merely equilibrated with air. The power saturation data are obtained by scanning the central line at increasing microwave powers. To evaluate the effect of O_2, the power saturation experiments are performed in the absence of O_2 by applying a steady stream of N_2 to evacuate O_2 from the sample.[6] After an equilibration time of >10 min, the power saturation experiments are repeated. Similarly, accessibility measurements by power saturation with NiEDDA also require the evacuation of O_2 using a N_2 stream.
(6) Plot the peak-to-peak amplitude of the central line width as a function of incident microwave power (Altenbach et al., 1994). Use the fit of this plot to calculate Π_{O_2} and Π_{NiEDDA} as described in Altenbach et al. (1994).
(7) Calculate the depth parameter Φ $[\Phi = \ln(\Pi_{O_2}/\Pi_{NiEDDA})]$.
(8) Repeat the measurements using samples prepared with unlabeled protein and vesicles containing around a 1:1000 molar ratio of spin-labeled lipid. Lipids containing a spin label either at the 5, 7, 10, or 12 doxyl, as well as the lipid headgroup,[8] should be individually doped into different stocks of the lipid mixture being used in the experimental system. Accessibilities to O_2 and NiEDDA are measured and converted into the depth parameter Φ as described in Section 4.2.1.
(9) The depth parameter Φ obtained from each labeled lipid is plotted as a function of the immersion depth of the spin-labeled site. This calibration curve can then be used to convert Φ into membrane immersion depth in Å for measurements made for spin-labeled protein.

4.3 Distances

4.3.1 Basic Considerations for Distance Measurements

Using DEER and pulse EPR, long range distances can be measured in ranges of 15–80 Å for ideal systems but around 15–60 Å for typical protein samples. Shorter distances (<20 Å) can be measured by CW EPR (Hubbell et al., 2000, 1998). Both CW and pulse EPR measure dipolar interactions, but in the case of very close spin labels (especially when multiple labels come together) CW EPR can resolve spin exchange interactions as well (Margittai & Langen, 2008). Longer distances between spin labels result in decreased dipolar coupling making detection using CW EPR challenging at longer distances (>20 Å). DEER measurements do not suffer from this

---

[8] The depths of spin labels on spin-labeled lipids have been defined previously (Altenbach et al., 1994; Bretscher, Buchaklian, & Klug, 2008; Dalton, McIntyre, & Fleischer, 1987).

limitation and allow for an enhanced distance range. Intramolecular distance information can be obtained from doubly labeled proteins. This information can be used to generate new distance constraints for obtaining new structural insights or for testing the validity of existing crystallographic information (Hubbell et al., 1998; Jao et al., 2008, 2010; Shah et al., 2014). Alternatively, distances between singly labeled proteins can be used to determine quaternary structures and interactions between proteins (Jao et al., 2008; Margittai & Langen, 2008).

4.3.2 Application of Distance Measurements

Distance measurements allow for 3D structural information with respect to secondary, tertiary, and quaternary structures by measuring intra- or intermolecular distances. By strategically determining inter- or intramolecular distances, a 3D mapping and computational refinement of a protein's structure can be accomplished. One common strategy is to make doubly labeled proteins in which only one site is varied and one stays the same. This allows for the triangulation of the constant site and allows for structures to be built in reference to a single location; for example, this method was used to locate the N-terminus of EHD2 to a hydrophobic pocket in its G domain (Shah et al., 2014). Distances between singly labeled proteins can also provide detailed quaternary structural information. For example, to determine the structure of the inserted helices relative to one another when the N-BAR homodimer is bound to membranes, distances between the single spin-labeled sites along the helices were measured (Jao et al., 2010). Increasing distances from the start to end of the helices indicated that these regions extend away from one another in an antiparallel conformation.

4.3.3 Distance Measurement Protocol
(A) Measuring distances using CW EPR
  (1) Add spin-labeled protein to lipids to form the optimized protein–lipid complex described in Section 3.3.2. To reveal spin–spin interactions, prepare multiple samples for scanning by CW EPR. For both single- and double-cysteine mutants, one sample should contain proteins fully labeled with paramagnetic spin label. The second sample should contain a mix of proteins fully labeled with paramagnetic (i.e., R1) and diamagnetic spin label[9] typically referred to as R_1'. By diluting the paramagnetic spin label these

---

[9] Our studies have typically used a 4:1 molar ratio of protein fully labeled with diamagnetic label to protein fully labeled with paramagnetic spin label. Similarly, diamagnetic and paramagnetic labels can be mixed in a 4:1 molar ratio and used to label protein.

samples can be used to provide reference spectra of noninteracting spins, and when combining with the fully labeled spectra exposes intermolecular signals. Intramolecular spin–spin interactions can be diluted out for experiments using double-cysteine mutant derivatives by labeling with a mixture of paramagnetic and diamagnetic spin label.[9] Load the samples into quartz capillaries.

(2) EPR distance measurements are typically done in the frozen state[10] and may require the use of a cryoprotectant.[11] Toward this end, the sample cell of the resonator needs to be cooled to maintain a constant temperature. We typically obtain spectra at 233 K using a Bruker temperature controller (ER4131VT).

(3) Insert the sample-loaded capillary into the resonator and allow the sample to equilibrate in temperature.

(4) Tune the sample according to the resonator's manual and scan the sample. Initial scans should be recorded at large gauss scan widths (~200) to ensure that all spectral broadening is recorded in the spectral scan.

(5) Distances are determined through simulation of dipolar broadening from spectra containing the dipolar interaction with spectra in which this dipolar interaction has been diluted. We use software, as described previously (Altenbach, Oh, Trabanino, Hideg, & Hubbell, 2001).

(B) Measuring distances using DEER
(1) Prepare samples as described in Section 3.3.2 and test their stability under conditions of the experiment.

(2) Add a cryoprotectant[11] to protect the sample during the freezing process. Load the sample into a quartz capillary and use gentle centrifugation (~$500 \times g$) to move the sample to the bottom of the capillary.

(3) Prepare a bath of butanol[12] suspended in liquid N_2. Freeze the sample by submerging the capillary into the chilled butanol solution and either immediately load the sample into the resonator[13]

---

[10] For some large and slowly tumbling samples with slowly rotating interspin vector, measurements can be done at room temperature (Altenbach et al., 2001).
[11] In our laboratory, we typically use glycerol or sucrose up to 25%.
[12] Butanol is used to reduce the Leidenfrost effect.
[13] DEER measurements are performed on an Elexsys E580 X-band pulse EPR spectrometer (Bruker) fitted with a 3-mm split ring (MS-3) resonator using a continuous-flow cryostat and a temperature controller (Oxford Instruments).

or store the sample at −80 °C. Most samples are stable for a year or more.
(4) Following the protocols outlined in the spectrometer's manual, set up the DEER experiments. For examples of specific settings used in DEER experiments, see Jao et al. (2008).
(5) A number of programs are available for the analysis of DEER data. We use the DeerAnalysis2013 program (Jeschke & Polyhach, 2007) which can be run in MATLAB.

5. COMPUTATIONAL REFINEMENT

5.1 Construction of a Starting Structure

Transformation of DEER data and depth constraints into three-dimensional structure requires a computational method. These data are amenable to treatment with simulated annealing molecular dynamics (SAMD). In broad terms, the approach is similar to that used for nOe data from NMR experiments. However, there are specific challenges in performing these calculations with DEER distances to determine a complete or partial structure of a protein embedded in a lipid layer.

The first task is building a starting protein conformation that is appropriately spin labeled and located in a defined position relative to a curved lipid surface and the center of an imaginary vesicle. This is achieved through a computer algorithm that permits construction of a curved lipid surface, protein construction, and spin labeling of defined amino acid positions. Input of DEER distance constraints and depth parameters from EPR experiments is included in this step for generation of a set of constraints for subsequent SAMD calculations.

An example of the setup for calculations performed on α-synuclein (Jao et al., 2008) on a lipid vesicle is shown in Fig. 4. In this example, the entire protein is a single α-helix, but the same approach can be used for more complex structures (Jao et al., 2010), as indicated in the protocols of Section 6. The initial structure is produced in the context of a lipid surface and with a defined center of the lipid vesicle (Fig. 4A and B). The lipid surface is not present in the SAMD calculation, but it is useful to construct this surface at this stage to ensure that the protein is correctly positioned. The complete protein (Fig. 4C) includes labels at all sites used in DEER experiments and for evaluation of lipid depth. In the example shown, 26 of 81 amino acid positions are labeled. The initial pdb file also contains a bar (Fig. 4D,

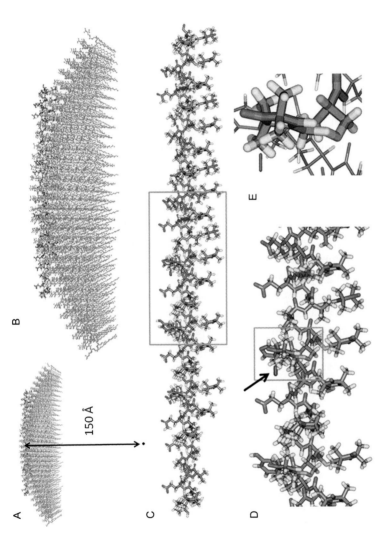

Figure 4 Starting structure for α-synuclein for use in SAMD calculations. (A) Position of a straight α-helix with respect to the surface of a lipid vesicle of radius 150 Å. The center of the vesicle is used for implementation of depth constraints in the calculation. (B) A closer view of the α-helix and lipid surface. (C) The 81-amino acid α-helix including 26 labels. (D) A closer view of the central part of the α-helix (blue box in C, showing several spin labels in more detail. The purple bar (indicated by an arrow) indicates the "surface" of the lipid vesicle. (E) A detailed view of a single label (blue box in D). This and all other labels are added in a "m,m" conformation for χ_1 and χ_2. This conformation is favored for an MTSL-labeled amino acid in an α-helix due to the interaction between Sδ (lighter yellow atom in the label) and Hα (green atom). (See the color plate.)

arrow) that shows the position of the "surface" of the lipid vesicle with respect to the central amino acid of the α-helix. The labels are added in a defined conformation in which χ_1 and χ_2 are 'm,m' (gauche(−), gauche(−)) for consistency with X-ray data for MTSL labels on α-helices (Guo et al., 2008).

The following is a typical protocol with general considerations required for the construction of a starting structure. An input file (Scheme 1) for construction of the molecular assembly shown in Fig. 4A is included to illustrate the protocol. More details of these inputs and information on executables are provided at https://ihlab.hsc.usc.edu/research/plc/.

(1) Definition of the lipid vesicle (orange (light gray in the print version) section in Scheme 1). Parameters include the vesicle radius (150.0 Å in the example shown), the angle between lipid axes (3.6°), two

```
Structure building for simulated annealing
  1 -1  0  0   1   0 -1   1
Lipid vesicle
  150.0    3.6 0.6 1.7   65
   2   1   1   0   0
    SER 11   0 13   0    40 S13 -1
    CHO 11   0 13   0    60 C13  0
     7.5 180.0 22 43
   100.0 180.0 22 43 0
Peptide (26 labels)
SKXKEGVVAAAEKXKQGXAEXXGKXKEGXLYXXSKXKEGXVHXXATVXEKXKEQXT
NXXGAXVTGXTAXXQKXVE*
GXGSXX
   0   0   0   0   0   1
  -1    4.0    90.0     90.0     90.0     0.0    240.0
SSlabel_mm90.zmat      26
   12    25    0
    3    18 25.
   55    73 26.
    3       12.8
  ...
   81        8.2
```

DEER distances		DEPTH distances			
3	18 25.	3	12.8	48	11.7
3	33 48.	18	11.3	51	9.5
14	44 49.	21	8.0	55	12.3
18	48 44.	22	12.2	58	9.0
29	59 42.	25	12.8	59	11.1
33	48 23.	29	13.9	62	11.2
33	59 37.	32	10.4	66	12.0
33	62 41.	33	11.6	69	10.6
36	59 36.	36	11.5	70	7.3
40	59 29.	40	7.9	73	13.1
48	77 42.	43	9.8	77	10.9
55	73 26.	44	10.9	80	6.8
				81	8.2

Scheme 1 Input for construction of an α-helix for use in SAMD calculations based on DEER-measured distances and depth data. The file is broken into title lines (light blue (gray in the print version)) and input for the lipid surface (orange (light gray in the print version)), protein (red (dark gray in the print version)), and constraint data (green (gray in the print version)). The inset shows detailed constraint data for DEER distance and EPR-determined lipid depths.

parameters for adjustment of lipid positions (generally unchanged for this size vesicle), and the number of "rings" of lipids in the vesicle (65: the vesicle is formed from concentric rings of lipids). The types of lipids and the extent to which the lipid vesicle is output are defined on the next four lines (more details given in Section 5.3).

(2) Definition of the protein sequence and position (red (dark gray in the print version) section in Scheme 1). The sequence of the protein is given in single-letter amino acid nomenclature with "X" indicating a label position. A new line (indicated by *) is required every 75 amino acids. The next line is important because this line defines the geometry of the protein with respect to the surface of the lipid vesicle. In Scheme 1, the helix axis passing through the central amino acid of the α-helix is positioned 4 Å below the vesicle surface [note that this protocol can also be applied to a partially known protein structure, with the sequence read in replaced by an uploaded pdb file (see https://ihlab.hsc.usc.edu/research/plc/)].

(3) Definition of the label geometry and the DEER and depth data (green (gray in the print version) section in Scheme 1). The label geometry can be specified through an external file (SSlabel_mm90.zmat, indicated to be used at 26 positions based on the "X" entries in the protein sequence) [this is defined differently if a protein structure is uploaded, but the same principle is used]. Data for DEER distances (12 in Scheme 1) and depth (25 in Scheme 1) are converted into a constraint file for simulated annealing (currently AMBER format constraints).

5.2 Simulated Annealing

We have generally run simulated annealing calculations using the AMBER force field, but the principles outlined here should be applicable for other platforms. Constraints (Table 1) generated in Section 5.1 (protocol step 3) are used to limit the calculations. These constraints include the DEER distances, lipid depth data, and several constraints on the protein conformation and protein–label contacts. The exact ranges used depend on the quality of the experimental data, and thus the values in Table 1 for the DEER and depth data should be used for guidance only. We also found it necessary to restrain the α-helix conformation in the SAMD calculations, using relatively loose constraints for backbone torsions and

Table 1 Restraints Used in SAMD Calculations

Geometry Element	Structural Element	Value	Range
Interlabel distance	Labels DEER pairs	Exptl. N (Å)	±3 Å
Label-vesicle center distance	Labels with depth data	Exptl. N (Å)	$N-1$ to $N+3$ Å
ϕ Torsion angle	All amino acids	−57°	±20°
ψ Torsion angle	All amino acids	−47°	±20°
ω Torsion angle	All amino acids	180°	±3°
Backbone H-bond distance	All amino acids	2.15 Å	1.3 Å
Hα–Sδ distance	All labels	2.8 Å	±1 Å

intramolecular hydrogen bonds (Table 1). We also include constraints to maintain the Hα–Sδ contact between the protein backbone and the label side chain (see Section 5.1).

Our general approach to the SAMD calculations is to obtain stable simulations over 10 cycles of heating from 0 to >1000 K and cooling back to 0 K. The molecule used in the calculation has labels substituted at all positions for which DEER distances or depth data are available. If sufficient data have been collected, it is helpful to leave out some data so that these can be used as a post-calculation check to determine consistency of the unused data with the derived structures. Distances between spin labels are defined based on the N atom of the nitroxide groups of each label. A dummy atom is used to represent the center of the vesicle, so that the depth data can be included using distances from the labels to the dummy atom. We also include charged amino acids (Lys, Arg, Asp, and Glu) as neutral side chains (although structurally still containing NH_3, COO, etc., in the side chain, but with an overall neutral residue). This is because we found that inclusion of charged residues in gas-phase SAMD calculations results in formation of salt bridges that is unlikely to occur in an aqueous environment.

Despite the relatively small number of constraints in the SAMD calculations, we observed significant distortion of the protein at higher temperature, as shown in Fig. 5A. The α-helix shown in green in Fig. 5A is the structure obtained at the end of a SAMD cycle, while the red and blue helices are snapshots from the high-temperature phase of the same cycle.

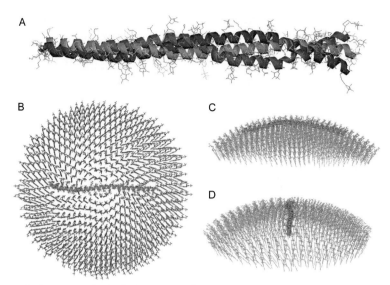

Figure 5 Structures formed during simulated annealing and reconstruction of the protein–lipid surface. (A) Deviation of structures at low temperature (green) and higher temperature (red and blue). (B–D) Reconstructed protein–lipid surface from three perspectives, with the helix shown in green. (See the color plate.)

The overall helical structure is maintained, but the orientation is variable. It is important to note that the green helix in Fig. 5A is representative of structures formed at the end of multiple SAMD cycles (Jao et al., 2008). We have attempted to release the protein further from the constraints in Table 1, but to date these simulations have been unsuccessful. There is clearly room for improvement of the parameters and methodology for these simulations.

The following is a typical protocol for the SAMD calculations. The suggested conditions are from our simulations of α-synuclein, but the conditions are general and should be applicable to similar systems.

(1) Energy minimize the starting structure.
(2) Perform 10 SAMD cycles of 30 ps each, including a heating phase from 0 to 1200 K in 4 ps using a step of 0.002 ps, during which the force constants for the restraints are increased from 0.1 to 10.0; maintenance of the temperature at 1200 K for a further 6 ps; and then cooling to 0 K over 20 ps.
(3) Use a time step of 0.002 ps, a distance-dependent dielectric of 4, and a cut off of 10.0 Å. The "atom" representing the center of the imaginary lipid vesicle is constrained to the origin with a large force constant.

(4) Collect structures every 1 ps for evaluation. Most importantly, collect the structure at the end of each cycle. Identify geometry parameters that can be used to evaluate fluctuation from structures at the end of the cycle. Evaluate the consistency of this parameter among low-temperature structures. Also, evaluate the RMSD among these structures.

5.3 Reconstruction of the Protein–Lipid Assembly

The protein structures that emerge from the SAMD calculations carry spin labels that need to be replaced by the original amino acids. These structures are also positioned in space with respect to an imaginary lipid surface (see Section 5.1). Thus, it is necessary to replace the labels with the original amino acid side chains and to reconstruct the protein–lipid system so that the location of the protein with respect to the lipid surface can be visualized.

Replacement of the labels is straightforward and can be achieved using many programs. The Cβ atom of the label should be retained in the converted structure, so that the amino acid side chain position mimics that of the label as closely as possible. There is clearly error in this process and this reduces the accuracy of the final structure. However, the orientation of the amino acid side chain is correct and structural conclusions can be drawn based on this orientation (Jao et al., 2008, 2010).

Reconstruction of the lipid surface is particularly valuable for understanding the structural context of the derived structure (Fig. 5B–D). In our approach, lipids are built around the protein, using simple clash avoidance in construction of the lipids (Fig. 5B). The depth of the protein in the lipid (as defined by the depth parameters) emerges in this reconstruction (Fig. 5C and D). In addition to visualization, these structures are of value as starting points for further simulations.

The following is a typical protocol for rebuilding the protein–lipid system. An input file (Scheme 2) is shown for construction of the protein–lipid assembly shown in Fig. 5B–D. The algorithm used for this reconstruction has considerable flexibility for building of curved lipid surfaces around proteins. However, here, we focus on visualization of the lipid surface around a protein structure derived from a SAMD simulation of α-synuclein. The protocol is generally applicable to any protein.

(1) Upload a protein structure from the SAMD calculation as a pdb file (red (dark gray in the print version) section in Scheme 2).

```
Reconstruction of protein and lipid surface
  0 -1   0   1   0   0   0
Protein_SAMD.pdb
Lipid vesicle
150.0    3.6 0.6 1.7   65
  2   0   1   1   0
    SER 11   0 13   0   40 S13 -1
    CHO 11   0 13   0   60 C13  0
100.0 180.0 22 43 1
```

Scheme 2 Input for reconstruction of a protein–lipid system using structures obtained from SAMD calculations based on DEER-measured distances and depth data. The file is broken into title lines (light blue (gray in the print version)) and input for the lipid surface (orange (light gray in the print version)) and protein (red (dark gray in the print version)).

(2) Rebuild the lipid surface around the protein (orange (light gray in the print version) section in Scheme 2). Line 1 is similar to the description in Section 5.1. The following lines define the details of the lipids. Lipids can be added with serine (SER) or choline (CHO) headgroups (more options are available, see https://ihlab.hsc.usc.edu/research/plc/), and the lipid hydrocarbon chains may be of any carbon length (11 and 13 are used in Scheme 2) and may contain a double bond (not activated in Scheme 2). The percentage of lipids in the vesicle can be defined (40% serine, 60% choline in Scheme 2). The last line indicates that lipids will be output if they are within 100 Å of the protein and in rings (see Section 6.1) 22–43.

(3) A pdb file in standard format is produced with the protein inserted into the lipid surface. This file can become large if a large expanse of lipid surface is selected for output.

6. OUTLOOK

In this chapter, we have discussed how SDSL, EPR, and computational refinement can be used to determine the structure and mechanisms of proteins bound to different membrane shapes. The ability to gather structural information on proteins bound to distinct membrane curvatures in real-time establishes EPR with SDSL as one of the most effective tools for studying membrane-curving proteins. These data are highly amenable for use in generating 3D structures by computational refinement and have been used to verify crystallographic data. Due to the limited number of methods which can generate similar information, SDSL, EPR, and computational refinement are currently at the forefront of techniques for future studies on membrane-interacting proteins.

ACKNOWLEDGMENT
This work was supported by National Institutes of Health Grant GM063915 (to R.L.).

REFERENCES
Altenbach, C., Greenhalgh, D. A., Khorana, H. G., & Hubbell, W. L. (1994). A collision gradient method to determine the immersion depth of nitroxides in lipid bilayers: Application to spin-labeled mutants of bacteriorhodopsin. *Proceedings of the National Academy of Sciences of the United States of America*, *91*(5), 1667–1671.

Altenbach, C., Oh, K. J., Trabanino, R. J., Hideg, K., & Hubbell, W. L. (2001). Estimation of inter-residue distances in spin labeled proteins at physiological temperatures: Experimental strategies and practical limitations. *Biochemistry*, *40*(51), 15471–15482.

Ambroso, M. R., Hegde, B. G., & Langen, R. (2014). Endophilin A1 induces different membrane shapes using a conformational switch that is regulated by phosphorylation. *Proceedings of the National Academy of Sciences of the United States of America*, *111*(19), 6982–6987. http://dx.doi.org/10.1073/pnas.1402233111.

Bartucci, R., Erilov, D. A., Guzzi, R., Sportelli, L., Dzuba, S. A., & Marsh, D. (2006). Time-resolved electron spin resonance studies of spin-labelled lipids in membranes. *Chemistry and Physics of Lipids*, *141*(1–2), 142–157. http://dx.doi.org/10.1016/j.chemphyslip.2006.02.009.

Boucrot, E., Pick, A., Çamdere, G., Liska, N., Evergren, E., McMahon, H. T., et al. (2012). Membrane fission is promoted by insertion of amphipathic helices and is restricted by crescent BAR domains. *Cell*, *149*(1), 124–136. http://dx.doi.org/10.1016/j.cell.2012.01.047.

Bretscher, L. E., Buchaklian, A. H., & Klug, C. S. (2008). Spin-labeled lipid A. *Analytical Biochemistry*, *382*(2), 129–131. http://dx.doi.org/10.1016/j.ab.2008.08.001.

Campelo, F., McMahon, H. T., & Kozlov, M. M. (2008). The hydrophobic insertion mechanism of membrane curvature generation by proteins. *Biophysical Journal*, *95*(5), 2325–2339. http://dx.doi.org/10.1529/biophysj.108.133173.

Columbus, L., Kálai, T., Jekö, J., Hideg, K., & Hubbell, W. L. (2001). Molecular motion of spin labeled side chains in alpha-helices: Analysis by variation of side chain structure. *Biochemistry*, *40*(13), 3828–3846.

Dalton, L. A., McIntyre, J. O., & Fleischer, S. (1987). Distance estimate of the active center of N-beta-hydroxybutyrate dehydrogenase from the membrane surface. *Biochemistry*, *26*(8), 2117–2130.

Daumke, O., Lundmark, R., Vallis, Y., Martens, S., Butler, P. J. G., & McMahon, H. T. (2007). Architectural and mechanistic insights into an EHD ATPase involved in membrane remodelling. *Nature*, *449*(7164), 923–927. http://dx.doi.org/10.1038/nature06173.

De Camilli, P., Thomas, A., Cofiell, R., Folli, F., Lichte, B., Piccolo, G., et al. (1993). The synaptic vesicle-associated protein amphiphysin is the 128-kD autoantigen of Stiff-Man syndrome with breast cancer. *The Journal of Experimental Medicine*, *178*(6), 2219–2223.

Doherty, G. J., & McMahon, H. T. (2009). Mechanisms of endocytosis. *Annual Review of Biochemistry*, *78*, 857–902. http://dx.doi.org/10.1146/annurev.biochem.78.081307.110540.

Drin, G., & Antonny, B. (2010). Amphipathic helices and membrane curvature. *FEBS Letters*, *584*(9), 1840–1847. http://dx.doi.org/10.1016/j.febslet.2009.10.022.

Farsad, K., & Camilli, P. D. (2003). Mechanisms of membrane deformation. *Current Opinion in Cell Biology*, *15*(4), 372–381. http://dx.doi.org/10.1016/S0955-0674(03)00073-5.

Farsad, K., Ringstad, N., Takei, K., Floyd, S. R., Rose, K., & De Camilli, P. (2001). Generation of high curvature membranes mediated by direct endophilin bilayer interactions. *The Journal of Cell Biology*, *155*(2), 193–200. http://dx.doi.org/10.1083/jcb.200107075.

Fleissner, M. R., Cascio, D., & Hubbell, W. L. (2009). Structural origin of weakly ordered nitroxide motion in spin-labeled proteins. *Protein Science: A Publication of the Protein Society*, *18*(5), 893–908. http://dx.doi.org/10.1002/pro.96.

Frazier, A. A., Roller, C. R., Havelka, J. J., Hinderliter, A., & Cafiso, D. S. (2003). Membrane-bound orientation and position of the synaptotagmin I C2A domain by site-directed spin labeling. *Biochemistry*, *42*(1), 96–105. http://dx.doi.org/10.1021/bi0268145.

Gallop, J. L., Jao, C. C., Kent, H. M., Butler, P. J. G., Evans, P. R., Langen, R., et al. (2006). Mechanism of endophilin N-BAR domain-mediated membrane curvature. *The EMBO Journal*, *25*(12), 2898–2910. http://dx.doi.org/10.1038/sj.emboj.7601174.

Guo, Z., Cascio, D., Hideg, K., & Hubbell, W. L. (2008). Structural determinants of nitroxide motion in spin-labeled proteins: Solvent-exposed sites in helix B of T4 lysozyme. *Protein Science: A Publication of the Protein Society*, *17*(2), 228–239. http://dx.doi.org/10.1110/ps.073174008.

Hubbell, W. L., Altenbach, C., Hubbell, C. M., & Khorana, H. G. (2003). Rhodopsin structure, dynamics, and activation: A perspective from crystallography, site-directed spin labeling, sulfhydryl reactivity, and disulfide cross-linking. *Advances in Protein Chemistry*, *63*, 243–290.

Hubbell, W. L., Cafiso, D. S., & Altenbach, C. (2000). Identifying conformational changes with site-directed spin labeling. *Nature Structural Biology*, *7*(9), 735–739. http://dx.doi.org/10.1038/78956.

Hubbell, W. L., Gross, A., Langen, R., & Lietzow, M. A. (1998). Recent advances in site-directed spin labeling of proteins. *Current Opinion in Structural Biology*, *8*(5), 649–656.

Isas, J. M., Ambroso, M. R., Hegde, P. B., Langen, J., & Langen, R. (2015). Tubulation by amphiphysin requires concentration-dependent switching from wedging to scaffolding. *Structure*, *23*(5), 873–881. http://dx.doi.org/10.1016/j.str.2015.02.014.

Isas, J. M., Langen, R., Haigler, H. T., & Hubbell, W. L. (2002). Structure and dynamics of a helical hairpin and loop region in annexin 12: A site-directed spin labeling study. *Biochemistry*, *41*(5), 1464–1473. http://dx.doi.org/10.1021/bi011856z.

Isas, J. M., Patel, D. R., Jao, C., Jayasinghe, S., Cartailler, J.-P., Haigler, H. T., et al. (2003). Global structural changes in annexin 12. The roles of phospholipid, Ca^{2+}, and pH. *The Journal of Biological Chemistry*, *278*(32), 30227–30234. http://dx.doi.org/10.1074/jbc.M301228200.

Jao, C. C., Hegde, B. G., Chen, J., Haworth, I. S., & Langen, R. (2008). Structure of membrane-bound alpha-synuclein from site-directed spin labeling and computational refinement. *Proceedings of the National Academy of Sciences of the United States of America*, *105*(50), 19666–19671. http://dx.doi.org/10.1073/pnas.0807826105.

Jao, C. C., Hegde, B. G., Gallop, J. L., Hegde, P. B., McMahon, H. T., Haworth, I. S., et al. (2010). Roles of amphipathic helices and the bin/amphiphysin/rvs (BAR) domain of endophilin in membrane curvature generation. *The Journal of Biological Chemistry*, *285*(26), 20164–20170. http://dx.doi.org/10.1074/jbc.M110.127811.

Jeschke, G., & Polyhach, Y. (2007). Distance measurements on spin-labelled biomacromolecules by pulsed electron paramagnetic resonance. *Physical Chemistry Chemical Physics (PCCP)*, *9*(16), 1895–1910. http://dx.doi.org/10.1039/b614920k.

Lai, C.-L., Jao, C. C., Lyman, E., Gallop, J. L., Peter, B. J., McMahon, H. T., et al. (2012). Membrane binding and self-association of the epsin N-terminal homology domain. *Journal of Molecular Biology*, *423*(5), 800–817. http://dx.doi.org/10.1016/j.jmb.2012.08.010.

Langen, R., Isas, J. M., Luecke, H., Haigler, H. T., & Hubbell, W. L. (1998). Membrane-mediated assembly of annexins studied by site-directed spin labeling. *Journal of Biological Chemistry, 273*(35), 22453–22457. http://dx.doi.org/10.1074/jbc.273.35.22453.

Langen, R., Oh, K. J., Cascio, D., & Hubbell, W. L. (2000). Crystal structures of spin labeled T4 lysozyme mutants: Implications for the interpretation of EPR spectra in terms of structure. *Biochemistry, 39*(29), 8396–8405.

Margittai, M., Fasshauer, D., Jahn, R., & Langen, R. (2003). The Habc domain and the SNARE core complex are connected by a highly flexible linker. *Biochemistry, 42*(14), 4009–4014. http://dx.doi.org/10.1021/bi027437z.

Margittai, M., & Langen, R. (2008). Fibrils with parallel in-register structure constitute a major class of amyloid fibrils: Molecular insights from electron paramagnetic resonance spectroscopy. *Quarterly Reviews of Biophysics, 41*(3-4), 265–297. http://dx.doi.org/10.1017/S0033583508004733.

Marsh, D., & Horváth, L. I. (1998). Structure, dynamics and composition of the lipid-protein interface. Perspectives from spin-labelling. *Biochimica et Biophysica Acta, 1376*(3), 267–296.

Martens, S., Kozlov, M. M., & McMahon, H. T. (2007). How synaptotagmin promotes membrane fusion. *Science, 316*(5828), 1205–1208. http://dx.doi.org/10.1126/science.1142614.

Mchaourab, H. S., Lietzow, M. A., Hideg, K., & Hubbell, W. L. (1996). Motion of spin-labeled side chains in T4 lysozyme. Correlation with protein structure and dynamics. *Biochemistry, 35*(24), 7692–7704. http://dx.doi.org/10.1021/bi960482k.

McMahon, H. T., & Gallop, J. L. (2005). Membrane curvature and mechanisms of dynamic cell membrane remodelling. *Nature, 438*(7068), 590–596. http://dx.doi.org/10.1038/nature04396.

Meinecke, M., Boucrot, E., Camdere, G., Hon, W.-C., Mittal, R., & McMahon, H. T. (2013). Cooperative recruitment of dynamin and BIN/amphiphysin/Rvs (BAR) domain-containing proteins leads to GTP-dependent membrane scission. *The Journal of Biological Chemistry, 288*(9), 6651–6661. http://dx.doi.org/10.1074/jbc.M112.444869.

Nicot, A.-S., Toussaint, A., Tosch, V., Kretz, C., Wallgren-Pettersson, C., Iwarsson, E., et al. (2007). Mutations in amphiphysin 2 (BIN1) disrupt interaction with dynamin 2 and cause autosomal recessive centronuclear myopathy. *Nature Genetics, 39*(9), 1134–1139. http://dx.doi.org/10.1038/ng2086.

Peter, B. J., Kent, H. M., Mills, I. G., Vallis, Y., Butler, P. J. G., Evans, P. R., et al. (2004). BAR domains as sensors of membrane curvature: The amphiphysin BAR structure. *Science, 303*(5657), 495–499. http://dx.doi.org/10.1126/science.1092586.

Qualmann, B., Koch, D., & Kessels, M. M. (2011). Let's go bananas: Revisiting the endocytic BAR code. *The EMBO Journal, 30*(17), 3501–3515. http://dx.doi.org/10.1038/emboj.2011.266.

Raschle, T., Hiller, S., Etzkorn, M., & Wagner, G. (2010). Nonmicellar systems for solution NMR spectroscopy of membrane proteins. *Current Opinion in Structural Biology, 20*(4), 471–479. http://dx.doi.org/10.1016/j.sbi.2010.05.006.

Shah, C., Hegde, B. G., Morén, B., Behrmann, E., Mielke, T., Moenke, G., et al. (2014). Structural insights into membrane interaction and caveolar targeting of dynamin-like EHD2. *Structure, 22*(3), 409–420. http://dx.doi.org/10.1016/j.str.2013.12.015.

Stachowiak, J. C., Schmid, E. M., Ryan, C. J., Ann, H. S., Sasaki, D. Y., Sherman, M. B., et al. (2012). Membrane bending by protein-protein crowding. *Nature Cell Biology, 14*(9), 944–949. http://dx.doi.org/10.1038/ncb2561.

Varkey, J., Isas, J. M., Mizuno, N., Jensen, M. B., Bhatia, V. K., Jao, C. C., et al. (2010). Membrane curvature induction and tubulation are common features of synucleins and apolipoproteins. *The Journal of Biological Chemistry, 285*(42), 32486–32493. http://dx.doi.org/10.1074/jbc.M110.139576.

Watanabe, S., Rost, B. R., Camacho-Pérez, M., Davis, M. W., Söhl-Kielczynski, B., Rosenmund, C., et al. (2013). Ultrafast endocytosis at mouse hippocampal synapses. *Nature*, *504*(7479), 242–247. http://dx.doi.org/10.1038/nature12809.

Westphal, C. H., & Chandra, S. S. (2013). Monomeric synucleins generate membrane curvature. *The Journal of Biological Chemistry*, *288*(3), 1829–1840. http://dx.doi.org/10.1074/jbc.M112.418871.

CHAPTER TEN

Determining the Secondary Structure of Membrane Proteins and Peptides Via Electron Spin Echo Envelope Modulation (ESEEM) Spectroscopy

Lishan Liu, Daniel J. Mayo, Indra D. Sahu, Andy Zhou, Rongfu Zhang, Robert M. McCarrick, Gary A. Lorigan[1]

Department of Chemistry and Biochemistry, Miami University, Oxford, Ohio, USA
[1]Corresponding author: e-mail address: lorigag@miamioh.edu

Contents

1. Introduction — 290
　1.1 Membrane Protein Secondary Structure — 290
　1.2 ESEEM Spectroscopy — 293
2. Integration of Membrane Peptides into Lipid Bilayer — 294
3. ESEEM Spectroscopy on Model Peptides in a Lipid Bilayer — 298
　3.1 ESEEM Principles — 298
　3.2 ESEEM Experimental Setup and Data Analysis — 301
4. Development of ESEEM Secondary Structure Determination Approach — 302
　4.1 Determine α-Helical Secondary Structure of Membrane Peptides — 302
　4.2 Distinguishing α-Helices from β-Strands — 304
5. Summary and Future Direction — 308
Acknowledgments — 308
References — 309

Abstract

Revealing detailed structural and dynamic information of membrane embedded or associated proteins is challenging due to their hydrophobic nature which makes NMR and X-ray crystallographic studies challenging or impossible. Electron paramagnetic resonance (EPR) has emerged as a powerful technique to provide essential structural and dynamic information for membrane proteins with no size limitations in membrane systems which mimic their natural lipid bilayer environment. Therefore, tremendous efforts have been devoted toward the development and application of EPR spectroscopic techniques to study the structure of biological systems such as membrane proteins and peptides.

This chapter introduces a novel approach established and developed in the Lorigan lab to investigate membrane protein and peptide local secondary structures utilizing the pulsed EPR technique electron spin echo envelope modulation (ESEEM) spectroscopy. Detailed sample preparation strategies in model membrane protein systems and the experimental setup are described. Also, the ability of this approach to identify local secondary structure of membrane proteins and peptides with unprecedented efficiency is demonstrated in model systems. Finally, applications and further developments of this ESEEM approach for probing larger size membrane proteins produced by overexpression systems are discussed.

1. INTRODUCTION

Membrane-associated and embedded proteins comprise 30% of sequenced genes (Landreh & Robinson, 2015; Moraes, Evans, Sanchez-Weatherby, Newstead, & Stewart, 2014). They are responsible for the exchange of signals and physical materials across the membranes and play vital roles in different aspects of cellular activities (Baker, 2010b; Congreve & Marshall, 2010). Mutations or misfolding of membrane proteins are associated with numerous human dysfunctions, disorders, and diseases (Cheung & Deber, 2008; Conn, Ulloa-Aguirre, Ito, & Janovick, 2007). Currently, half of all the FDA approved drugs target membrane proteins (von Heijne, 2007). Detailed structural and dynamic information for membrane proteins are vital for elucidating protein functions, intermolecular interactions, and regulations. Better structural knowledge of membrane protein systems is also crucial to our understanding of the basic mechanisms of disease pathways and benefit novel clinical therapy development (Rask-Andersen, Almén, & Schiöth, 2011; Shukla, Vaitiekunas, & Cotter, 2012). Despite the abundance and importance of membrane proteins, there is very limited knowledge about structure, function, and dynamics of these complicated biological systems (Das, Park, & Opella, 2015; Kang, Lee, & Drew, 2013).

1.1 Membrane Protein Secondary Structure

The majority of membrane proteins structural motifs fall into two categories: membrane-spanning or surface-associated α-helix or α-helical bundles and β-barrels (Chothia, Levitt, & Richardson, 1977; McLuskey, Roszak, Zhu, & Isaacs, 2010; White & Wimley, 1999). It has been shown previously that the local secondary structure affects membrane proteins packing and interactions with its lipid environment (Kurochkina, 2010). Generally, better knowledge about secondary structure, particularly site-specific secondary structure, is useful toward the understanding of the function, dynamics, and interactions of membrane proteins (Kubota, Lacroix, Bezanilla, & Correa, 2014; Yu & Lorigan, 2014). Also, the formation and transition of secondary structural components are crucial for a variety of cellular processes ranging from protein folding and refolding to the amyloid deposits in various neurodegenerative disorders such as Alzheimer's disease, Huntington's disease, and Parkinson's syndrome (Gross, 2000).

While enormous efforts have been placed on accessing membrane protein structural information over the past two decades, membrane proteins are inherently difficult to study (Baker, 2010a; Kang et al., 2013). Traditional structural biology techniques such as NMR and X-ray crystallography have revealed an increasing number of atomic level 3D structures of proteins. However, only a small portion of those are membrane proteins (Garman, 2014; Harris, 2014; Wang & Ladizhansky, 2014). In addition to these traditional biophysical techniques, the structural biology community has also benefited greatly from other structural approaches to tackle challenging biological systems (Bahar, Lezon, Bakan, & Shrivastava, 2010; Cowieson, Kobe, & Martin, 2008; Feng, Pan, & Zhang, 2011). Biophysical and biochemical techniques such as mass spectrometry, IR, Raman spectroscopy, fluorescence resonance energy transfer spectroscopy, chemical cross-linking, and computational modeling have all been utilized successfully to provide valuable information about structure, dynamics, and interactions of membrane proteins (Chattopadhyay & Haldar, 2014; King et al., 2008; Ladokhin, 2014; Tang & Clore, 2006).

There are several established biophysical techniques that are used to study the secondary structure of membrane proteins. Circular dichroism (CD) is an excellent tool for rapid determination of the secondary structure and folding properties of proteins (Greenfield, 2006; Whitmore & Wallace, 2008). CD spectroscopy detects the differential absorption of left- and right-handed circular polarized light that can be used to determine the global secondary structure of a protein. CD has the advantage that it can measure samples containing 20 μg or less of proteins in physiological buffers in a short period of time. However, it only yields the overall secondary structure of the entire complex and does not provide the specific secondary structure of different segments of the protein. Solid-state NMR spectroscopy can be utilized to determine local secondary structures based on the backbone chemical shift assignment and dipolar couplings

(Fritzsching, Yang, Schmidt-Rohr, & Hong, 2013). However, it requires milligram scales of isotope-labeled protein or peptide samples and days to weeks of data-acquisition time while still suffering from low sensitivity. Other methods such as FT-Raman spectroscopy, ATR FT-IR, and continuous-wave EPR dipolar wave analysis also can provide secondary structure information (Carbonaro & Nucara, 2010; Roach, Simpson, & JiJi, 2012). Data obtained by these methods are sometimes ambiguous and often require extensive data analysis.

Electron paramagnetic resonance (EPR) is a powerful and sensitive biophysical technique for studying chemical and biological systems with unpaired electron spins. It was first observed over a half century ago and has been particularly useful in characterizing organic radicals, metal complexes, and biomolecules with paramagnetic centers (Brückner, 2010; Goldfarb, 2006). However, with the development of site-directed spin-labeling (SDSL) techniques to target biological systems, there has been a significant increase in the application of EPR spectroscopy to study protein structure and dynamics (Alexander, Bortolus, Al-Mestarihi, Mchaourab, & Meiler, 2008; Altenbach, Flitsch, Khorana, & Hubbell, 1989; Fanucci & Cafiso, 2006; Hirst, Alexander, McHaourab, & Meiler, 2011; Hubbell, Gross, Langen, & Lietzow, 1998; Hubbell, López, Altenbach, & Yang, 2013; Sahu, McCarrick, & Lorigan, 2013; Sahu, McCarrick, Troxel, et al., 2013). SDSL EPR is sensitive to dynamics on the picoseconds to microsecond timescales, which cover a wide range of motions in biological and molecular systems (Barnes, Liang, Mchaourab, Freed, & Hubbell, 1999; Casey et al., 2014; Nesmelov, 2014). Also, the topology of a membrane protein can be explored with respect to the lipid bilayer with SDSL coupled with CW-EPR spectroscopy. Adding relaxation enhancers such as chelated nickel and oxygen can alter the electron spin–lattice and spin–spin relaxation rates and distinguish between solvent-exposed regions and residues buried in the membrane (Altenbach, Greenhalgh, Khorana, & Hubbell, 1994; Huang et al., 2015; van Wonderen et al., 2014). Utilizing different experimental approaches, EPR spectroscopy can also access distance information between different spin labels from several angstroms up to 10 nm (Baber, Louis, & Clore, 2015; Edwards et al., 2013; Sahu, Hustedt, et al., 2014; Sahu, McCarrick, Troxel, et al., 2013). Pulsed EPR techniques such as double electron–electron resonance provide important structural information on membrane proteins (Baber et al., 2015; Sahu, Kroncke, et al., 2014).

This work describes a novel approach established and developed in our lab to investigate membrane protein and peptide secondary structure utilizing the pulsed EPR technique electron spin echo envelope modulation (ESEEM). ESEEM spectroscopy coupled with SDSL can provide valuable local secondary structural information (α-helix and β-strand) of membrane proteins and peptides in a lipid bilayer with short data-acquisition times and straightforward data analysis.

1.2 ESEEM Spectroscopy

ESEEM spectroscopy has been widely utilized to study the electronic environment of paramagnetic metal centers and metalloproteins and provide valuable information on metalloenzyme mechanisms, metal binding, substrate binding, and the ligand coordination sphere (Cieslak, Focia, & Gross, 2010; Deligiannakis, Boussac, & Rutherford, 1995; Hernández-Guzmán et al., 2013; Warncke, 2005). SDSL and ESEEM spectroscopy have been used to study the supermolecular structure of biological systems, the penetration depth of water into the membrane and in KcsA K$^+$ channels, localization of proteins or lipids in membranes, and protein folding (Carmieli et al., 2006; Cieslak et al., 2010) (Bartucci, Guzzi, Esmann, & Marsh, 2014; Dzuba & Raap, 2013; Matalon, Faingold, Eisenstein, Shai, & Goldfarb, 2013).

In this chapter, a novel ESEEM approach developed in the Lorigan lab is discussed. By using SDSL coupled with ESEEM spectroscopy, the secondary structure of membrane peptides and proteins can be determined by detecting ^{2}H modulation between a ^{2}H-labeled amino acid and a nearby spin-labeled cysteine residue (Liu et al., 2012; Mayo et al., 2011; Zhou et al., 2012). A cysteine-mutated nitroxide spin label (MTSL) is positioned strategically at 1, 2, 3, and 4 residues away from an amino acid (i) with a deuterated side chain (denoted as i + 1 to i + 4). The characteristic periodicity of the α-helix or β-strand structure has unique patterns in the individual ESEEM spectra. A typical α-helical periodicity consists of 3.6 amino acids per turn. The distance from the beginning to the end of the turn in the α-helix is 5.4 Å. Taking this into account every three or four residues in an α-helical segment should have a minimum distance between the side chain residues, assuming that the helix is straight. The second predominant secondary structure, β-strand, is an extended stretch of polypeptides chain with every other two amino acid side chains approximately 6 Å apart. ESEEM spectroscopy can detect dipolar interactions between a nitroxide spin label and a ^{2}H nucleus out to a maximum of approximately 8 Å. For an α-helical structure, one set of ESEEM data from different samples should show a pattern in which ^{2}H modulation can be detected for i + 3 and i + 4 samples, but not for the i + 2 samples because they are outside the 8 Å detection limit. However, for peptides or protein segments adopting an extended structure such as a β-strand, the ESEEM spectra should show exactly the opposite results. In this case, ^{2}H modulation would be detected for the i + 2 sample, but not for the i + 3 and i + 4 samples.

This novel pulsed EPR ESEEM secondary structure approach is advantageous because it has no protein or protein-complex size limitations and is very sensitive when compared to NMR spectroscopy. Moreover, this approach can provide direct local secondary structural information qualitatively without complicated data analysis. Generally, each set of ESEEM experiments requires small amounts and concentrations of labeled protein sample (~25 μL and ~100 μM). Also, the ESEEM data acquisition is fairly fast when compared to NMR and only takes about an hour. With selective isotopic labeling, this approach can be adopted in an overexpression system and, therefore, can be applied to larger proteins and protein complexes. In Section 2, detailed experimental procedures for sample preparation, spectrometer setup, and data analysis are described.

2. INTEGRATION OF MEMBRANE PEPTIDES INTO LIPID BILAYER

In order to demonstrate the ability of this novel pulsed EPR approach for determining secondary structural components, well-characterized model peptides with known α-helical or β-sheet secondary structures were chosen to prove the concept (Opella et al., 1999; Zerella, Chen, Evans, Raine, & Williams, 2000). The nicotinic acetylcholine receptor M2δ segment, which is a well-studied 23 amino acid residue transmembrane peptide was selected to represent a model α-helix in a lipid bilayer (PDB entry: 1EQ8). The 17-amino acid residue of Ubiquitin was chosen to represent a model β-strand (PDB entry: 1E0Q). In this section, sample preparation with model peptides in a membrane mimetic system is described.

1. *Solid-phase peptide synthesis (SPPS)*: SPPS is the standard method for synthesizing peptides and small proteins in the lab. It allows for the synthesis of natural peptides and small proteins which are difficult to express, as well as the incorporation of unnatural or isotope-labeled amino acids and the synthesis of D-amino acid proteins. Also, unlike ribosomal protein synthesis, SPPS can proceed in both C-terminal and N-terminal fashions. SPPS has the ability to synthesize peptides with up to 70 amino

acids and can potentially make proteins and peptides with 150+ amino acids with the help of chemical ligation (Chandrudu, Simerska, & Toth, 2013; Raibaut, El Mahdi, & Melnyk, 2015). The two most commonly used forms of SPPS are Fmoc and Boc, which have different protecting groups used on the C-terminal or N-terminal residues of each amino acid block. Table 1 shows the wild-type and ESEEM experimental construct sequences for the two model peptides AChR M2δ (α-helix) and Ubiquitin 17 (β-sheet). Four different peptides were designed by positioning the ^{2}H-labeled amino acid at position i and the cysteine (X) at four successive positions (i+1 to i+4). Both the M2δ and Ubiquitin peptide constructs were synthesized on a CEM microwave-assisted peptide synthesizer using Fmoc protection chemistry. Low loading (0.2 mmol/g) and high swelling rate solid supports were chosen to increase the yield of these relatively hydrophobic peptide sequences. A solution of ^{2}H-labeled amino acid such as d_{10} Leu or d_8 Val (Isotec) dissolved in N-methyl-2-pyrrolidone was used as the ^{2}H probe and incorporated into peptides at a designated position (i). The peptides were cleaved, deprotected, and isolated from their resin support in an acidic environment. The cleavage and deprotection cocktail was designed and optimized according to the sequence, length, and protection groups used on the amino acid side chain (Góngora-Benítez, Tulla-Puche, & Albericio, 2013; Mäde, Els-Heindl, & Beck-Sickinger, 2014). Most cocktails are TFA-based and the amino acid composition of the peptide dictates the final concentration of TFA, type of scavengers used, and reaction times.

2. *Peptide purification and validation*: After the peptides were cleaved from their solid support, the cleavage cocktail was evaporated by N_2 gas flow or via rotary evaporation until the peptide precipitation started to appear. Methyl *tert*-butyl ether was added to assist the precipitation of peptide and wash off residual TFA. The crude peptide was dried under a vacuum overnight followed by purification via reverse-phase HPLC with a C4 preparation column using a linear gradient of 5–95% solvent B (90% acetonitrile). This gradient is usually sufficient to purify typical peptides from SPPS. However, high impurity and multiple major truncations from a bad synthesis can increase the difficulty of the separation. In those cases, a mobile-phase gradient and component can be adjusted to achieve a better separation. The HPLC fraction of the target peptide was collected and lyophilized to a solid powder for further steps and storage. Matrix-assisted laser desorption/ionization time-of-flight

Table 1 Peptide Constructs for AChR M2δ and Ubiquitin Peptides

	AChR M2δ	Ubiquitin Peptide
Wild type	NH$_2$-EKMSTAISVLLAQAVFLLLTSQR-COOH	NH$_2$-MQIFVKTLDGKTITLEV-COOH
i+1	NH$_2$-EKMSTAISV**X**iAQAVFLLLTSQR-COOH	NH$_2$-MQIFVK**X**iDGKTITLEV-COOH
i+2	NH$_2$-EKMSTAIS**X**LiAQAVFLLLTSQR-COOH	NH$_2$-MQIFV**X**TiDGKTITLEV-COOH
i+3	NH$_2$-EKMSTAI**X**VLiAQAVFLLLTSQR-COOH	NH$_2$-MQIF**X**KTiDGKTITLEV-COOH
i+4	NH$_2$-EKMSTA**X**SVLiAQAVFLLLTSQR-COOH	NH$_2$-MQI**X**VKTiDGKTITLEV-COOH

Wild-type and experimental constructs of AChR M2δ (α-helix) and Ubiquitin peptide (β-sheet) were listed in this table. **i** stands for positions where ^{2}H-labeled d$_{10}$ Leu was placed. **X** makes positions where amino acid is replaced by Cys for MTSL incorporation.

(MALDI-TOF) mass spectroscopy was used to confirm the molecular weight and purity of the peptides after HPLC purification.
3. *Attachment of spin label*: Purified peptides were labeled with S-(2,2,5,5-tetramethyl-2,5-dihydro-1H-pyrrol-3-yl)methyl methanesulfonothioate (MTSL) at 10× molar excess in DMSO for 12 h. Reversed-phase HPLC was used to remove the excess MTSL (Gorka et al., 2012; Zhao et al., 2012). For small proteins and peptides, chromatography is more efficient when compared to dialysis for the removal of excess MTSL. HPLC fractions of the targeted peptides were lyophilized and stored in −20 °C for further processing. MALDI-TOF was utilized to confirm the molecular weight, purity, and labeling efficiency of the target peptides qualitatively. A series of tempo solutions with standardized concentrations were prepared and a spin concentration calibration curve was generated. The concentrations of the spin-labeled peptide samples were directly calculated from the CW-EPR spectra. Spin-labeling efficiency was determined by comparing the spin concentration obtained from CW-EPR data with the protein or peptide concentration.
4. *Integration of synthetic peptide into lipid bilayers*: MTSL-labeled M2δ peptides were integrated into DMPC/DHPC (3.5/1) bicelles at a 1:1000 peptide to lipid molar ratio. Both spin-labeled peptide and lipids were dissolved in chloroform in a pear-shape flask. N_2 gas was applied to evaporate the solvent, while the flask was slowly rotated to form a uniform film of lipid and peptide mixture along the wall of the flask. This lipid/peptide film was dried under vacuum overnight to remove any remaining solvent. 200 µL HEPES buffer at pH 7.4 was added to rehydrate the lipid/peptide film followed by a combination of vortex, freeze–thaw, and sonication steps until the bicelle sample turned clear. For these experiments, bicelles were used as a membrane mimic system and yielded high-quality ESEEM data. The final spin label concentrations of the peptides were ~100 µM. Comparable ESEEM data could be obtained with micelles, vesicles, and lipodisq nanoparticles (data not published). The Ubiquitin peptide was dissolved in an aqueous buffer using a previously published protocol (Zerella et al., 2000). CW-EPR spectra of bicelle samples were taken to verify the incorporation of peptide and successful removal of free spin label.
5. *Peptide secondary structure validation*: M2δ bicelles and aqueous Ubiquitin peptide samples with concentrations ranging from 0.01 to 0.1 mg/mL were analyzed using a Jasco J-810 spectrometer over a wavelength range of 190–250 nm. The CD spectrum of i + 3 Ubiquitin construct in Fig. 1

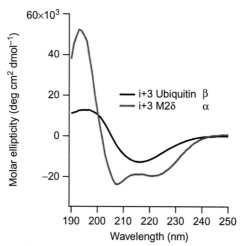

Figure 1 CD spectra of i+3 AChR M2δ and Ubiquitin peptide constructs in DMPC/DHPC bicelles with a lipid protein ratio of 200:1 at 298 K and pH 7. AChR M2δ spectrum (red line (dark gray in the print version)) shows two negative bands at 208 nm and 222 nm which indicate α-helical structure. The Ubiquitin peptide spectrum (black line) shows a large and broad negative band centered at 218 nm indicating a β-sheet secondary structure.

shows a large and broad negative band centered at 218 nm indicating a β-sheet secondary structure. The M2δ CD spectrum indicates α-helical secondary structure through the two negative bands at 222 nm and 208 nm. After baseline subtraction, CD data were analyzed with DichroWeb and showed pure α-helical content for M2δ and β-sheet for the Ubiquitin peptide after spin label incorporation.

The strategies described above can be applied to other synthetic proteins and peptides with minor modifications. Utilizing site-directed mutagenesis and selective isotope labeling, this ESEEM approach has been demonstrated that it can be adapted to overexpressed and reconstituted proteins and peptides (data not shown).

3. ESEEM SPECTROSCOPY ON MODEL PEPTIDES IN A LIPID BILAYER

3.1 ESEEM Principles

Pulsed EPR techniques such as ESEEM involve the application of a series of short time-dependent microwave pulses at the appropriate frequency to an electron spin system in a constant external magnetic field. The corresponding magnetization of the electron spins can be measured in the form of an emitted microwave signal, which provides information about the local environment of the electron spin system. The standard two-pulse spin echo or "Hahn echo" sequence is shown in Fig. 2A. In the two-pulse experiment, two microwave pulses separated by a time interval τ are applied to the electron spin system at a microwave frequency and magnetic field, which satisfies the magnetic resonance condition of the electron spin system. The first $\pi/2$ pulse rotates the electron spin magnetization by 90°, thus creating a short free induction signal. However, this signal rapidly decays due to the rapid spin dephasing resulting from inherent inhomogeneous line broadening. Thus, instead of directly observing the free induction decay as for NMR, a second π pulse is applied, which flips the magnetization of the spin system 180° in such a way that the spins refocus at the moment τ after the second pulse and create an electron spin echo that can be easily measured. Generally, in a two-pulse EPR experiment, the intensity of the echo is measured after the instrumental dead time as a function of increasing τ which generates the original time domain signal (Schweiger & Jeschke, 2001). The intensity of this echo decreases with increasing τ due to transverse spin relaxation. Nuclear spins that are coupled to the electron spin modulate the amplitude of the electron spin echo periodically as a function of τ. This phenomenon is called ESEEM. This periodic modulation results from the nuclear spin precession of nuclei in close vicinity of the unpaired electron spin. Basically, ESEEM spectroscopy indirectly observes NMR transitions through an electron spin coupled to a nearby NMR active nucleus (Deligiannakis & Rutherford, 2001; Hoffman, 2003).

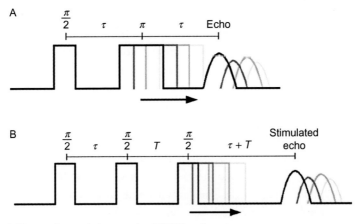

Figure 2 Two-pulse and three-pulse ESEEM pulse sequence.

In the three-pulse ESEEM sequence (Fig. 2B), three $\pi/2$ pulses are used to monitor the modulated echo as function of time. Instead of refocusing the spin, it creates a polarization grading with the first two pulses. The third $\pi/2$ pulse flips the stored polarization back to the transverse plane for measurement. All three pulses together generated a stimulated echo at the time of τ after the third pulse. Since only the longitudinal magnetization created by the second pulse contributes to the formation of the stimulated echo signal, the signal decay depends on the corresponding spin–lattice relaxation time T_1. By utilizing the three-pulse technique, ESEEM data can be obtained much further out in time and dramatically increases the resolution of the frequency domain data following Fourier transformation (FT) (Kevan & Schwartz, 1979).

For three-pulse ESEEM, the intensity of the stimulated echo is measured as a function of the evolution time T. However, the value of the fixed τ leads to artifacts within the ESEEM data, which are known as "τ-dependent blind spots" (Stoll, Calle, Mitrikas, & Schweiger, 2005). A nuclear spin with a Larmor frequency ω_n is suppressed at the appropriate τ value ($\tau = 2\pi n/\omega_n$). Therefore, for ESEEM experiments with multiple NMR active nuclei involved, different τ values should be examined to optimize the spectra. Proton modulation can be suppressed and deuterium modulation can be maximized with this approach.

Another shortcoming of the three-pulse ESEEM is that it generates more than the stimulated echo: it also generates one refocused echo and three primary echoes. For certain values of T and τ, those echoes can overlap with the stimulated echo and cause spectrum distortion. However, four-step phase cycling can be utilized to remove those unwanted echoes and prevent distortion of the spectrum.

The modulation depths of both two- and three-pulse ESEEM experiments have been discussed thoroughly in previous publications and reviews (Schweiger & Jeschke, 2001). It is affected by the static field B_0, g tensor of the electron spin and gyromagnetic ratio of the nuclei. For the same type of nucleus in a constant magnetic field, the modulation depth is proportional to r^{-6}, where r is the distance between the NMR active nucleus and the center of the unpaired electron spin density. Thus, it suffers from a rapid decay as the distance between the electron spin and nuclei spins increases. Typically, ESEEM can only detect dipolar interactions between a NMR active nucleus and a spin label through a dipolar coupling within a short of distance. For deuterium coupled to a nitroxide spin label, the detection limitation is about

8 Å. In the case of several nuclei, the ESEEM signal can be calculated as a product of the expression for each individual nuclear spin. In the next session, the analysis and interpretation of our ESEEM data will be discussed in detail.

3.2 ESEEM Experimental Setup and Data Analysis

In order to determine local secondary structural information from this ESEEM approach, a set of ESEEM spectra from different spin-labeled position need to be compared. Therefore, it requires consistency with sample preparation, experimental setup, data collection, and analysis.

1. *Experimental setup*: ESEEM data were collected at X-band on a Bruker ELEXSYS E580 spectrometer equipped with an MS3 split ring resonator. The three-pulse ESEEM sequence was chosen to maximize the low-frequency modulation of ^{2}H. The measurements were conducted at 80 K at a microwave frequency of 9.269 GHz with 16 ns $\pi/2$ pulse widths. A starting T of 368 ns and 512 points in 12 ns increments were used for all samples. A τ value of 200 ns was chosen to suppress proton modulation. 30% glycerol was added to rehydrolyzed bicelle sample as a cryoprotectant to prevent water crystallization during the freezing process. 40 μL of sample with a final concentration of 100 μmol was pipetted into a 3 mm ESEEM tube and fast frozen in liquid N_2. 30 scans with four-step phase cycling were used to obtain the required signal-to-noise ratio.

2. *Data analysis and interpretation*: The original ESEEM time domain data consists of two components: the unmodulated decay and the modulation of nuclei at the corresponding Larmor frequency. The time domain data were fit to a two-component exponential decay. The maximum value of the exponential fit was scaled to 1 and the same factor was applied to the time domain data. The exponential fit was then subtracted from the time domain data and yielded a scaled ESEEM spectrum with modulation about 0. A cross-term averaged FT was performed to the resulting spectrum to generate the corresponding frequency domain with minimized dead time artifacts (Tarabek, Bonifacić, & Beckert, 2006). The analysis can be performed use Matlab and our ESEEM data-processing package at http://epr.muohio.edu/user-resources/oaeprl-plotting-package. Any modulation presented from every weakly coupled nucleus will show a peak at its corresponding Larmor frequency. The maximum intensity of the deuterium peak at 2.3 MHz was measured in an arbitrary unit and peak intensity was recorded for further analysis.

4. DEVELOPMENT OF ESEEM SECONDARY STRUCTURE DETERMINATION APPROACH

Here, we demonstrate the ability and efficiency of this ESEEM approach to probe the secondary structure of membrane proteins and peptides in a model lipid bilayer system.

4.1 Determine α-Helical Secondary Structure of Membrane Peptides

In order to map out the α-helical content of a segment of the M2δ peptide, ^{2}H-labeled d_8 Val was positioned at Val9 (i) and the MTSL spin label (X) was strategically placed at four successive positions (i + 1 to i + 4). Figure 3 shows the normalized ESEEM spectra for ^{2}H-labeled d_8 Val9 AChR M2δ (EKMSTAISiLL**X**QAVFLLLTSQR) including both the time domain and frequency domain data (Fig. 3, red (light gray in the print version) and blue (dark gray in the print version)). A control sample was prepared such that Val9 was not ^{2}H-labeled (Fig. 3, black) at position i + 3 (EKMSTAISVLL**X**QAVFLLLTSQR). The spectra compared the three-pulse ESEEM data of the M2δ peptide with ^{2}H-labeled d_8 Val and a SL three residues away (i + 3) at two different τ values with a non-deuterated sample

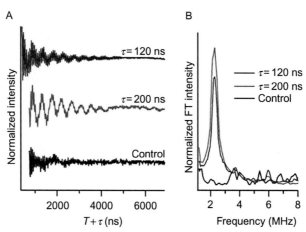

Figure 3 Three-pulse ESEEM spectra of AChR M2δ with ^{2}H-labeled d_8 Val9 in DMPC/DHPC bicelles. (A) Time-domain data for i+3 AChR M2δ with ^{2}H-labeled d_8 Val9 at τ=120 ns (blue (dark gray in the print version)), τ=200 ns (red (light gray in the print version)) and non-deuterated control sample at τ=200 ns (black). (B) Corresponding frequency domain data for i+3 AChR M2δ with ^{2}H-labeled d_8 Val9 at τ=120 ns (blue (dark gray in the print version)), τ=200 ns (red (light gray in the print version)) and nondeuterated control sample at τ=200 ns (black).

as a control. Figure 3A reveals an obvious low-frequency ^{2}H modulation in the time domain data for both τ values in the ^{2}H-labeled Val sample when compared to the control sample, which had normal protonated Val instead of ^{2}H-labeled Val. The corresponding Fourier transform frequency domain data in Fig. 3B revealed a large well-resolved peak centered at the ^{2}H-Larmor frequency of 2.3 MHz originating from weakly coupled ^{2}H nuclei. ^{2}H modulation is not detected in the control sample at any τ values. Also, the ESEEM data showed that both τ values provided high-quality time domain and FT ESEEM data. However, it was obvious that when ^{2}H modulation was optimized, the proton modulation was also effectively suppressed with a τ equal to 200 ns. The optimal τ values can vary depending on the field and frequency under which the experiment was performed, as well as the type of isotopic label used in the experiment (Kevan & Schwartz, 1979).

Figure 4 shows the three-pulse ESEEM data for ^{2}H-labeled d$_8$ Val9 M2δ peptides at all four successive positions (i + 1 to i + 4) in bicelles. ^{2}H modulation

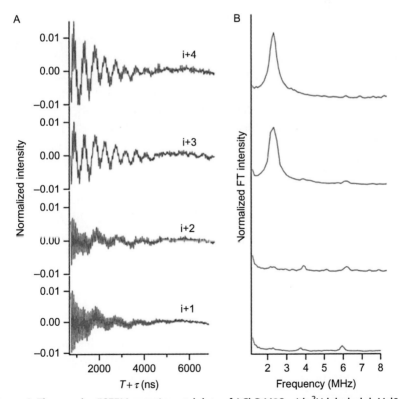

Figure 4 Three-pulse ESEEM experimental data of AChR M2δ with ^{2}H-labeled d$_8$ Val9 in DMPC/DHPC bicelles at $\tau=200$ ns for the i+1 to i+4 in (A) time domain and (B) frequency domain.

was observed in the time domain spectrum and a corresponding peak centered at the ^{2}H-Larmor frequency in the frequency domain data for the i + 3 and i + 4 ^{2}H-labled d$_8$ Val9 M2δ samples. However, no ^{2}H modulation was detected for the i + 1 or i + 2 positions. These ESEEM spectral pattern of large ^{2}H peaks observed at i + 3 and i + 4 positions were consistent with the structural characteristic that there are 3.6 amino acids per turn for an α-helix. The ^{2}H-labeled Val side chain and the spin labels are located on the same side of the helix when they are three or four amino acid residues away. These data clearly showed the utility of this technique for determining the α-helical secondary structure of membrane peptides and proteins.

This approach was further explored and expanded to utilize ^{2}H-labeled d$_{10}$ Leu as a ^{2}H-labeled probe for this novel ESEEM approach. The three-pulse ESEEM data for the ^{2}H-labeled d$_{10}$ Leu11 (i + 1 through i + 4) M2δ peptides are shown in Fig. 5. For ^{2}H-labeled d$_{10}$ Leu11 M2δ peptides, the ^{2}H modulation was observed in the time domain for i + 1, i + 3, and i + 4 samples. Also, a ^{2}H peak was clearly observed at the ^{2}H-Larmor frequency in the frequency domain data. However, there was no ^{2}H modulation for the ^{2}H-labeled d$_{10}$ Leu11 i + 2 M2δ sample. Despite the longer side chain of Leu when compared to Val, ESEEM spectra still revealed a similar pattern for this α-helix. Moreover, Fig. 6A shows the comparison of the ESEEM frequency domain data between ^{2}H-labeled d$_{10}$ Leu10 and ^{2}H-labeled d$_8$ Val9 peptides at the i + 4 position. This frequency domain data revealed a dramatic signal enhancement when ^{2}H-labeled d$_{10}$ Leu was used instead of ^{2}H-labeled d$_8$ Val. Distances and conformations provided by molecular dynamic simulations also supported this result (Liu et al., 2012). Figure 6B indicates that the additional C–C bonds in the Leu side chain brought the deuterons closer to the N–O nitroxide bond of the spin label when compared to the Val side chain. Therefore, the MTSL had a higher probability of being able to detect the ^{2}H nuclei at a closer distance resulting in a significant increase in ^{2}H peak intensity for Leu at this position, when compared to Val.

4.2 Distinguishing α-Helices from β-Strands

In addition to demonstrating this novel ESEEM approach's ability to identify an α-helix, this exact labeling paradigm can be applied to an ideal β-sheet peptide to distinguish these two most predominant secondary structures.

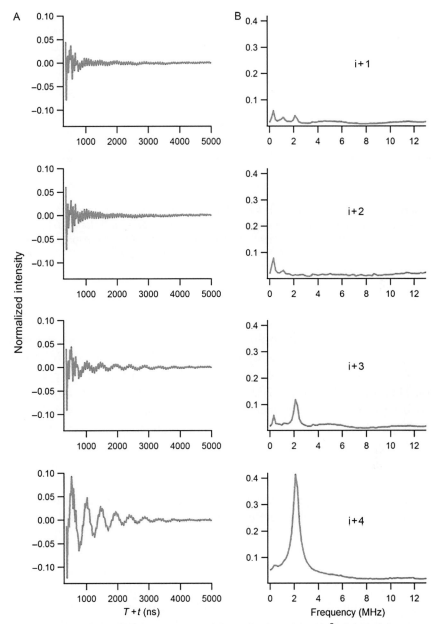

Figure 5 Three-pulse ESEEM experimental data of AChR M2δ with ^{2}H-labeled d_{10} Leu11 in DMPC/DHPC bicelles at $\tau = 200$ ns for the i+1 to i+4 in (A) time domain and (B) frequency domain.

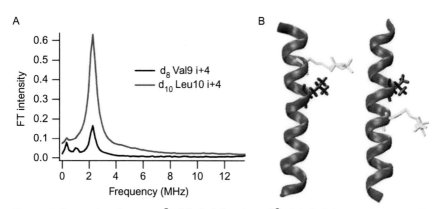

Figure 6 Comparison between ^{2}H-labeled d_8 Val9 and ^{2}H-labeled d_{10}Leu10 M2δ bicelle samples. (A) Normalized FT frequency domain modulation data for ^{2}H-labeled d_{10}Leu10 i+4 (black) and ^{2}H-labeled d_8 Val9 i+4 (red (dark gray in the print version)). (B) Likely conformation of M2δ with and ^{2}H-labeled d_{10}Leu10 (left) and M2δ with ^{2}H-labeled d_8 Val9 (right) both with MTSL at i+4 from MD simulation.

Figure 7 compares three-pulse ESEEM data for samples with the spin label in the i+2 and i+3 positions for both ^{2}H-labeled d_{10} Leu17 M2δ and ^{2}H-labeled d_{10} Leu8 Ubiquitin peptide constructs. In Fig. 7A, the presence of ^{2}H modulation in the i+2 Ubiquitin time domain spectrum indicated a weak dipolar coupling between deuterium nuclei and the spin label. Also, a corresponding FT peak at the ^{2}H-Larmor frequency was revealed in the frequency domain which indicated that the distance between the ^{2}H nuclei on Leu and the spin labels must be within the ∼8 Å detection limit. The absence of ^{2}H modulation in the i+3 Ubiquitin sample implied that the ^{2}H-SL distances are greater than 8 Å. This ESEEM spectra pattern was caused by the extended structure of the β-sheet in which the ^{2}H nucleus is closer to the i+2 position than the i+3 position. Also, the MTSL and the d_{10} Leu side chain pointed toward the same side of the β-sheet in i+2 constructs, while they pointed to opposite sides of the β-sheet in i+3 constructs. Figure 7B shows the corresponding ESEEM data for the M2δ peptide in a bicelle. Conversely, the M2δ i+2 spectrum did not contain any ^{2}H modulation, whereas the M2δ i+3 spectrum clearly shows ^{2}H modulation and a ^{2}H FT peak. The MTSL and d_{10} Leu side chain pointed toward opposite sides of the helix in i+2 constructs and wrap around to point toward the same side of the helix in the i+3 construct which was

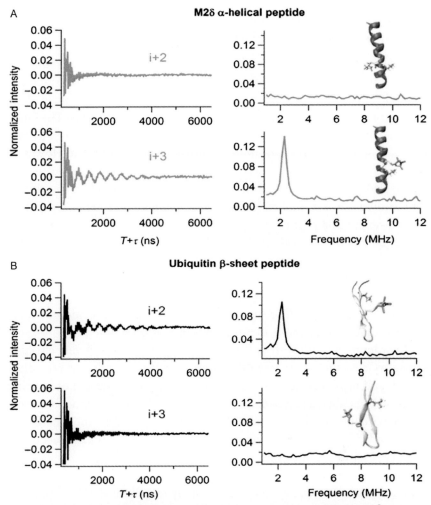

Figure 7 Three-pulse ESEEM experimental data. (A) Ubiquitin peptide with ^{2}H-labeled d_{10} Leu8 in buffer at $\tau = 200$ ns for the i+2 and i+3 in time domain (left) and frequency domain (right). (B) AChR M2δ with ^{2}H-labeled d_{10} Leu11 in DMPC/DHPC bicelles at $\tau = 200$ ns for the i+2 and i+3 in time domain (left) and frequency domain (right).

similar to the illustrations shown in Fig. 6B. The complementary results of the i+2/i+3 spectra for an α-helices and β-sheets obtained utilizing this novel ESEEM approach demonstrated the establishment of a simple qualitative method for determining site-specific secondary structure of any given protein system.

5. SUMMARY AND FUTURE DIRECTION

Here, we established and developed a novel approach to probe the secondary structure of membrane proteins and peptides in lipid bilayers. Results showed that this approach can be used to identify α-helical and β-sheet secondary structures with multiple isotopically labeled amino acids. The modulation depth with different probes varied according to the side chain length, side chain flexibility, backbone motion, and the local environment. However, the $i+x$ pattern for each secondary structure is similar regardless of which amino acid probe was used. Thus, this ESEEM approach should be valuable for identifying an α-helical region, as well as distinguishing between α-helical and a β-strand in a peptide or protein. This efficient ESEEM spectroscopic technique does not provide the same high-resolution structural information obtained from NMR spectroscopy or X-ray crystallography, but does provide very important qualitative secondary structural information for membrane protein systems or other biological systems where those techniques are not applicable. For SDSL EPR researchers, this approach will provide additional spectroscopic tools to probe the structures of biological systems.

In order to fully establish and expand the application of this very effective secondary structure approach, other amino acids with different side chain lengths and flexibility need to be examined. Also, the potential of utilizing other isotope-labeled amino acids (such as ^{13}C, ^{15}N, or ^{19}F) could be investigated with this novel technique. This powerful method has the potential to be extended to detecting random coils and less predominant secondary structure such as 3_{10} and π helices. Also, it can be adapted to detect secondary structural transitions and local conformational changes.

Further studies will be conducted to apply this ESEEM SDSL approach to larger integral membrane proteins, which are overexpressed in bacteria and then reconstituted into lipid bilayers. Cys residues can be introduced to desired positions through site-directed mutagenesis. Also, selective amino acid isotopic labeling can be used to ^{2}H label certain amino acids such as Leu or Val. However, overexpression conditions need to be optimized to prevent scrambling of the isotope-labeled amino acids.

ACKNOWLEDGMENTS

This work was generously supported by National Institutes of Health Grant R01 GM108026 and by the National Science Foundation Grant CHE-1011909. The pulsed EPR spectrometer was purchased through the NSF and the Ohio Board of Regents Grants (MRI-0722403).

REFERENCES

Alexander, N., Bortolus, M., Al-Mestarihi, A., Mchaourab, H., & Meiler, J. (2008). De novo high-resolution protein structure determination from sparse spin-labeling EPR data. *Structure, 16*(2), 181–195.

Altenbach, C., Flitsch, S. L., Khorana, H. G., & Hubbell, W. L. (1989). Structural studies on transmembrane proteins. 2. Spin labeling of bacteriorhodopsin mutants at unique cysteines. *Biochemistry, 28*(19), 7806–7812.

Altenbach, C., Greenhalgh, D. A., Khorana, H. G., & Hubbell, W. L. (1994). A collision gradient method to determine the immersion depth of nitroxides in lipid bilayers: Application to spin-labeled mutants of bacteriorhodopsin. *Proceedings of the National Academy of Sciences of the United States of America, 91*(5), 1667–1671.

Baber, J. L., Louis, J. M., & Clore, G. M. (2015). Dependence of distance distributions derived from double electron–electron resonance pulsed EPR spectroscopy on pulse-sequence time. *Angewandte Chemie (International Ed. in English), 54*, 5336–5339.

Bahar, I., Lezon, T. R., Bakan, A., & Shrivastava, I. H. (2010). Normal mode analysis of biomolecular structures: Functional mechanisms of membrane proteins. *Chemical Reviews, 110*(3), 1463–1497.

Baker, M. (2010a). Making membrane proteins for structures: A trillion tiny tweaks. *Nature Methods, 7*(6), 429–434.

Baker, M. (2010b). Structural biology: The gatekeepers revealed. *Nature, 465*(7299), 823–826.

Barnes, J. P., Liang, Z., Mchaourab, H. S., Freed, J. H., & Hubbell, W. L. (1999). A multifrequency electron spin resonance study of T4 lysozyme dynamics. *Biophysical Journal, 76*(6), 3298–3306.

Bartucci, R., Guzzi, R., Esmann, M., & Marsh, D. (2014). Water penetration profile at the protein-lipid interface in Na, K-ATPase membranes. *Biophysical Journal, 107*(6), 1375–1382.

Brückner, A. (2010). In situ electron paramagnetic resonance: A unique tool for analyzing structure-reactivity relationships in heterogeneous catalysis. *Chemical Society Reviews, 39*(12), 4673–4684.

Carbonaro, M., & Nucara, A. (2010). Secondary structure of food proteins by Fourier transform spectroscopy in the mid-infrared region. *Amino Acids, 38*(3), 679–690.

Carmieli, R., Papo, N., Zimmermann, H., Potapov, A., Shai, Y., & Goldfarb, D. (2006). Utilizing ESEEM spectroscopy to locate the position of specific regions of membrane-active peptides within model membranes. *Biophysical Journal, 90*(2), 492–505.

Casey, T. M., Liu, Z., Esquiaqui, J. M., Pirman, N. L., Milshteyn, E., & Fanucci, G. E. (2014). Continuous wave W- and D-band EPR spectroscopy offer "sweet-spots" for characterizing conformational changes and dynamics in intrinsically disordered proteins. *Biochemical and Biophysical Research Communications, 450*(1), 723–728.

Chandrudu, S., Simerska, P., & Toth, I. (2013). Chemical methods for peptide and protein production. *Molecules, 18*(4), 4373–4388.

Chattopadhyay, A., & Haldar, S. (2014). Dynamic insight into protein structure utilizing red edge excitation shift. *Accounts of Chemical Research, 47*(1), 12–19.

Cheung, J. C., & Deber, C. M. (2008). Misfolding of the cystic fibrosis transmembrane conductance regulator and disease. *Biochemistry, 47*(6), 1465–1473.

Chothia, C., Levitt, M., & Richardson, D. (1977). Structure of proteins: Packing of alpha-helices and pleated sheets. *Proceedings of the National Academy of Sciences of the United States of America, 74*(10), 4130–4134.

Cieslak, J. A., Focia, P. J., & Gross, A. (2010). Electron spin-echo envelope modulation (ESEEM) reveals water and phosphate interactions with the KcsA potassium channel. *Biochemistry, 49*(7), 1486–1494.

Congreve, M., & Marshall, F. (2010). The impact of GPCR structures on pharmacology and structure-based drug design. *British Journal of Pharmacology, 159*(5), 986–996.

Conn, P. M., Ulloa-Aguirre, A., Ito, J., & Janovick, J. A. (2007). G protein-coupled receptor trafficking in health and disease: Lessons learned to prepare for therapeutic mutant rescue in vivo. *Pharmacological Reviews, 59*(3), 225–250.

Cowieson, N. P., Kobe, B., & Martin, J. L. (2008). United we stand: Combining structural methods. *Current Opinion in Structural Biology, 18*(5), 617–622.

Das, B. B., Park, S. H., & Opella, S. J. (2015). Membrane protein structure from rotational diffusion. *Biochimica et Biophysica Acta, 1848*(1 Pt. B), 229–245.

Deligiannakis, Y., Boussac, A., & Rutherford, A. W. (1995). ESEEM study of the plastoquinone anion radical (QA--) in 14N- and 15N-labeled photosystem II treated with CN. *Biochemistry, 34*(49), 16030–16038.

Deligiannakis, Y., & Rutherford, A. W. (2001). Electron spin echo envelope modulation spectroscopy in photosystem I. *Biochimica et Biophysica Acta, 1507*(1–3), 226–246.

Dzuba, S. A., & Raap, J. (2013). Spin-echo electron paramagnetic resonance (EPR) spectroscopy of a pore-forming (Lipo)peptaibol in model and bacterial membranes. *Chemistry & Biodiversity, 10*(5), 864–875.

Edwards, D. T., Ma, Z., Meade, T. J., Goldfarb, D., Han, S., & Sherwin, M. S. (2013). Extending the distance range accessed with continuous wave EPR with Gd3+ spin probes at high magnetic fields. *Physical Chemistry Chemical Physics, 15*(27), 11313–11326.

Fanucci, G. E., & Cafiso, D. S. (2006). Recent advances and applications of site-directed spin labeling. *Current Opinion in Structural Biology, 16*(5), 644–653.

Feng, W., Pan, L., & Zhang, M. (2011). Combination of NMR spectroscopy and X-ray crystallography offers unique advantages for elucidation of the structural basis of protein complex assembly. *Science China. Life Sciences, 54*(2), 101–111.

Fritzsching, K. J., Yang, Y., Schmidt-Rohr, K., & Hong, M. (2013). Practical use of chemical shift databases for protein solid-state NMR: 2D chemical shift maps and amino-acid assignment with secondary-structure information. *Journal of Biomolecular NMR, 56*(2), 155–167.

Garman, E. F. (2014). Developments in x-ray crystallographic structure determination of biological macromolecules. *Science, 343*(6175), 1102–1108.

Goldfarb, D. (2006). High field ENDOR as a characterization tool for functional sites in microporous materials. *Physical Chemistry Chemical Physics, 8*(20), 2325–2343.

Góngora-Benítez, M., Tulla-Puche, J., & Albericio, F. (2013). Handles for Fmoc solid-phase synthesis of protected peptides. *ACS Combinatorial Science, 15*(5), 217–228.

Gorka, J., Rohmer, M., Bornemann, S., Papasotiriou, D. G., Baeumlisberger, D., Arrey, T. N., et al. (2012). Perfusion reversed-phase high-performance liquid chromatography for protein separation from detergent-containing solutions: An alternative to gel-based approaches. *Analytical Biochemistry, 424*(2), 97–107.

Greenfield, N. J. (2006). Using circular dichroism spectra to estimate protein secondary structure. *Nature Protocols, 1*(6), 2876–2890.

Gross, M. (2000). Proteins that convert from alpha helix to beta sheet: Implications for folding and disease. *Current Protein & Peptide Science, 1*(4), 339–347.

Harris, J. R. (2014). Transmission electron microscopy in molecular structural biology: A historical survey. *Archives of Biochemistry and Biophysics*, In press.

Hernández-Guzmán, J., Sun, L., Mehta, A. K., Dong, J., Lynn, D. G., & Warncke, K. (2013). Copper(II)-bis-histidine coordination structure in a fibrillar amyloid β-peptide fragment and model complexes revealed by electron spin echo envelope modulation spectroscopy. *Chembiochem, 14*(14), 1762–1771.

Hirst, S. J., Alexander, N., McHaourab, H. S., & Meiler, J. (2011). RosettaEPR: An integrated tool for protein structure determination from sparse EPR data. *Journal of Structural Biology, 173*(3), 506–514.

Hoffman, B. M. (2003). Electron-nuclear double resonance spectroscopy (and electron spin-echo envelope modulation spectroscopy) in bioinorganic chemistry. *Proceedings of the National Academy of Sciences of the United States of America, 100*(7), 3575–3578.

Huang, S., Green, B., Thompson, M., Chen, R., Thomaston, J., DeGrado, W. F., et al. (2015). C-terminal juxtamembrane region of full-length M2 protein forms a membrane surface associated amphipathic helix. *Protein Science, 24*(3), 426–429.

Hubbell, W. L., Gross, A., Langen, R., & Lietzow, M. A. (1998). Recent advances in site-directed spin labeling of proteins. *Current Opinion in Structural Biology, 8*(5), 649–656.

Hubbell, W. L., López, C. J., Altenbach, C., & Yang, Z. (2013). Technological advances in site-directed spin labeling of proteins. *Current Opinion in Structural Biology, 23*(5), 725–733.

Kang, H. J., Lee, C., & Drew, D. (2013). Breaking the barriers in membrane protein crystallography. *The International Journal of Biochemistry & Cell Biology, 45*(3), 636–644.

Kevan, L., & Schwartz, R. N. (Eds.), (1979). *Time domain electron spin resonance*. New York: Wiley-Interscience.

King, G. J., Jones, A., Kobe, B., Huber, T., Mouradov, D., Hume, D. A., et al. (2008). Identification of disulfide-containing chemical cross-links in proteins using MALDI-TOF/TOF-mass spectrometry. *Analytical Chemistry, 80*(13), 5036–5043.

Kubota, T., Lacroix, J. J., Bezanilla, F., & Correa, A. M. (2014). Probing α-3(10) transitions in a voltage-sensing S4 helix. *Biophysical Journal, 107*(5), 1117–1128.

Kurochkina, N. (2010). Helix-helix interactions and their impact on protein motifs and assemblies. *Journal of Theoretical Biology, 264*(2), 585–592.

Ladokhin, A. S. (2014). Measuring membrane penetration with depth-dependent fluorescence quenching: Distribution analysis is coming of age. *Biochimica et Biophysica Acta, 1838*(9), 2289–2295.

Landreh, M., & Robinson, C. V. (2015). A new window into the molecular physiology of membrane proteins. *The Journal of Physiology, 593*(2), 355–362.

Liu, L., Sahu, I. D., Mayo, D. J., McCarrick, R. M., Troxel, K., Zhou, A., et al. (2012). Enhancement of electron spin echo envelope modulation spectroscopic methods to investigate the secondary structure of membrane proteins. *The Journal of Physical Chemistry B, 116*(36), 11041–11045.

Mäde, V., Els-Heindl, S., & Beck-Sickinger, A. G. (2014). Automated solid-phase peptide synthesis to obtain therapeutic peptides. *Beilstein Journal of Organic Chemistry, 10*, 1197–1212.

Matalon, E., Faingold, O., Eisenstein, M., Shai, Y., & Goldfarb, D. (2013). The topology, in model membranes, of the core peptide derived from the T-cell receptor transmembrane domain. *Chembiochem, 14*(14), 1867–1875.

Mayo, D., Zhou, A., Sahu, I., McCarrick, R., Walton, P., Ring, A., et al. (2011). Probing the structure of membrane proteins with electron spin echo envelope modulation spectroscopy. *Protein Science, 20*(7), 1100–1104.

McLuskey, K., Roszak, A. W., Zhu, Y., & Isaacs, N. W. (2010). Crystal structures of all-alpha type membrane proteins. *European Biophysics Journal, 39*(5), 723–755.

Moraes, I., Evans, G., Sanchez-Weatherby, J., Newstead, S., & Stewart, P. D. (2014). Membrane protein structure determination—The next generation. *Biochimica et Biophysica Acta, 1838*(1 Pt. A), 78–87.

Nesmelov, Y. E. (2014). Protein structural dynamics revealed by site-directed spin labeling and multifrequency EPR. *Methods in Molecular Biology, 1084*, 63–79.

Opella, S. J., Marassi, F. M., Gesell, J. J., Valente, A. P., Kim, Y., Oblatt-Montal, M., et al. (1999). Structures of the M2 channel-lining segments from nicotinic acetylcholine and NMDA receptors by NMR spectroscopy. *Nature Structural Biology, 6*(4), 374–379.

Raibaut, L., El Mahdi, O., & Melnyk, O. (2015). Solid phase protein chemical synthesis. *Topics in Current Chemistry, 363*, 103–154.
Rask-Andersen, M., Almén, M. S., & Schiöth, H. B. (2011). Trends in the exploitation of novel drug targets. *Nature Reviews Drug Discovery, 10*(8), 579–590.
Roach, C. A., Simpson, J. V., & JiJi, R. D. (2012). Evolution of quantitative methods in protein secondary structure determination via deep-ultraviolet resonance Raman spectroscopy. *Analyst, 137*(3), 555–562.
Sahu, I. D., Hustedt, E. J., Ghimire, H., Inbaraj, J. J., McCarrick, R. M., & Lorigan, G. A. (2014). CW dipolar broadening EPR spectroscopy and mechanically aligned bilayers used to measure distance and relative orientation between two TOAC spin labels on an antimicrobial peptide. *Journal of Magnetic Resonance, 249C*, 72–79.
Sahu, I. D., Kroncke, B. M., Zhang, R., Dunagan, M. M., Smith, H. J., Craig, A., et al. (2014). Structural investigation of the transmembrane domain of KCNE1 in proteoliposomes. *Biochemistry, 53*(40), 6392–6401.
Sahu, I. D., McCarrick, R. M., & Lorigan, G. A. (2013). Use of electron paramagnetic resonance to solve biochemical problems. *Biochemistry, 52*(35), 5967–5984.
Sahu, I. D., McCarrick, R. M., Troxel, K. R., Zhang, R., Smith, H. J., Dunagan, M. M., et al. (2013). DEER EPR measurements for membrane protein structures via bifunctional spin labels and lipodisq nanoparticles. *Biochemistry, 52*(38), 6627–6632.
Schweiger, A., & Jeschke, G. (2001). *Principles of pulse electron paramagnetic resonance*. Oxford: Oxford University Press.
Shukla, H. D., Vaitiekunas, P., & Cotter, R. J. (2012). Advances in membrane proteomics and cancer biomarker discovery: Current status and future perspective. *Proteomics, 12*(19–20), 3085–3104.
Stoll, S., Calle, C., Mitrikas, G., & Schweiger, A. (2005). Peak suppression in ESEEM spectra of multinuclear spin systems. *Journal of Magnetic Resonance, 177*(1), 93–101.
Tang, C., & Clore, G. M. (2006). A simple and reliable approach to docking protein-protein complexes from very sparse NOE-derived intermolecular distance restraints. *Journal of Biomolecular NMR, 36*(1), 37–44.
Tarabek, P., Bonifačić, M., & Beckert, D. (2006). Time-resolved FT EPR and optical spectroscopy study on photooxidation of aliphatic alpha-amino acids in aqueous solutions; electron transfer from amino vs carboxylate functional group. *The Journal of Physical Chemistry. A, 110*(22), 7293–7302.
van Wonderen, J. H., McMahon, R. M., O'Mara, M. L., McDevitt, C. A., Thomson, A. J., Kerr, I. D., et al. (2014). The central cavity of ABCB1 undergoes alternating access during ATP hydrolysis. *The FEBS Journal, 281*(9), 2190–2201.
von Heijne, G. (2007). The membrane protein universe: What's out there and why bother? *Journal of Internal Medicine, 261*(6), 543–557.
Wang, S., & Ladizhansky, V. (2014). Recent advances in magic angle spinning solid state NMR of membrane proteins. *Progress in Nuclear Magnetic Resonance Spectroscopy, 82*, 1–26.
Warncke, K. (2005). Characterization of the product radical structure in the Co(II)-product radical pair state of coenzyme B12-dependent ethanolamine deaminase by using three-pulse 2H ESEEM spectroscopy. *Biochemistry, 44*(9), 3184–3193.
White, S. H., & Wimley, W. C. (1999). Membrane protein folding and stability: Physical principles. *Annual Review of Biophysics and Biomolecular Structure, 28*, 319–365.
Whitmore, L., & Wallace, B. A. (2008). Protein secondary structure analyses from circular dichroism spectroscopy: Methods and reference databases. *Biopolymers, 89*(5), 392–400.
Yu, X., & Lorigan, G. A. (2014). Secondary structure, backbone dynamics, and structural topology of phospholamban and its phosphorylated and Arg9Cys-mutated forms in phospholipid bilayers utilizing 13C and 15N solid-state NMR spectroscopy. *The Journal of Physical Chemistry B, 118*(8), 2124–2133.

Zerella, R., Chen, P. Y., Evans, P. A., Raine, A., & Williams, D. H. (2000). Structural characterization of a mutant peptide derived from ubiquitin: Implications for protein folding. *Protein Science*, *9*(11), 2142–2150.

Zhao, J., Guo, L., Zeng, H., Yang, X., Yuan, J., Shi, H., et al. (2012). Purification and characterization of a novel antimicrobial peptide from Brevibacillus laterosporus strain A60. *Peptides*, *33*(2), 206–211.

Zhou, A., Abu-Baker, S., Sahu, I. D., Liu, L., McCarrick, R. M., Dabney-Smith, C., et al. (2012). Determining α-helical and β-sheet secondary structures via pulsed electron spin resonance spectroscopy. *Biochemistry*, *51*(38), 7417–7419.

CHAPTER ELEVEN

Spin Labeling Studies of Transmembrane Signaling and Transport: Applications to Phototaxis, ABC Transporters and Symporters

Johann P. Klare, Heinz-Jürgen Steinhoff[1]

Physics Department, University of Osnabrück, Barbarastr. 7, Osnabrück, Germany
[1]Corresponding author: e-mail address: hsteinho@uni-osnabrueck.de

Contents

1. Introduction	316
1.1 Information from EPR Spectroscopy on Spin-Labeled Proteins	317
2. Applications to Membrane Proteins	322
2.1 Sample Preparation	324
2.2 Instrumentation	326
2.3 SRII/HtrII: Static and Time-Resolved Detection of Signaling Events	326
2.4 HisQMP: Conformational Changes Studied by DEER Distance Measurements	331
2.5 PutP: A Spin Labeling Site Scan Reveals the Structure of eL4	336
3. Outlook	340
Acknowledgments	341
References	341

Abstract

Membrane proteins still represent a major challenge for structural biologists. This chapter will focus on the application of continuous wave and pulsed EPR spectroscopy on spin-labeled membrane proteins. Site-directed spin labeling EPR spectroscopy has evolved as a powerful tool to study the structure and dynamics of proteins, especially membrane proteins, as this method works largely independently of the size and complexity of the biological system under investigation. This chapter describes applications of this technique to three different systems: the archaeal photoreceptor/-transducer complex SRII/HtrII as an example for transmembrane signaling and two transport systems, the histidine ATP-binding cassette transporter HisQMP, and the sodium–proline symporter PutP.

1. INTRODUCTION

Pioneered by Wayne L. Hubbell and coworkers about 25 years ago (Altenbach, Flitsch, Khorana, & Hubbell, 1989; Altenbach, Marti, Khorana, & Hubbell, 1990) site-directed spin labeling (SDSL) in combination with electron paramagnetic resonance (EPR) spectroscopy has emerged as a powerful tool to investigate the structure and conformational dynamics of proteins and nucleic acids. This technique is applicable to soluble molecules as well as to membrane proteins, the latter either being in detergent, in nanodiscs, or embedded in a lipid bilayer. Advantageously, the size and complexity of the system is almost arbitrary, and most experiments can be performed under conditions close to the physiological state of the system under investigation (for reviews see, e.g., Bordignon & Steinhoff, 2007; Hubbell, Gross, Langen, & Lietzow, 1998; Hubbell, Mchaourab, Altenbach, & Lietzow, 1996; Klare, 2013; Klare & Steinhoff, 2009; Klug & Feix, 2008).

The most commonly used SDSL approach utilizes the reactivity of the sulfhydryl group of cysteine residues engineered into the protein under investigation by site-directed mutagenesis. Consequently, this approach usually requires that the protein possesses accessible cysteine residues only at the desired sites, and that native cysteines at other sites are not accessible or can be replaced by serines or alanines without impairment of the protein's structure or function. Among the various spin labels available the (1-oxyl-2,2,5,5-tetramethylpyrroline-3-methyl) methanethiosulfonate spin label (MTSSL) (Berliner, Grunwald, Hankovszky, & Hideg, 1982) is most often used in SDSL studies due to its sulfhydryl specificity and its small molecular volume comparable to a tryptophan side chain (Fig. 1A and B).

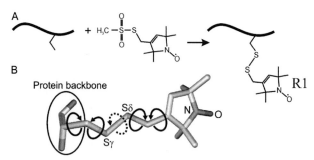

Figure 1 *Site-directed spin labeling* (A) using the (1-oxyl-2,2,5,5-tetramethylpyrroline-3-methyl) methanethiosulfonate spin label (MTSSL). (B) Conformational degrees of freedom for the R1 side chain.

This spin label is bound to the protein via a disulfide bond, and the resulting spin label side chain is commonly abbreviated as R1. The linker between the nitroxide ring and the protein backbone renders the R1 side chain flexible (Fig. 1B), minimizing disturbances of the protein structure. Moreover, the dynamic properties of R1 as revealed from analysis of room temperature EPR spectra can provide detailed structural information as described below. Besides MTSSL, a variety of spin label compounds are commercially available, comprising longer or sterically more demanding linkers. Alternatively, maleimide- or iodoacetamide-functionalized spin label compounds can be used if the disulfide linkage appears labile under the required experimental conditions.

Although being beyond the scope of this chapter, it should be mentioned that in the past years a wider range of orthogonal coupling schemes has been established, including, for example, the highly specific introduction of the spin label side chain via "click chemistry" (Kolb, Finn, & Sharpless, 2001) or the direct incorporation of spin-labeled amino acids during protein synthesis. The latter, genetically encoded, labeling methods lack the above mentioned restrictions of cysteine replacement mutagenesis (Cornish et al., 1994; Schmidt, Borbas, Drescher, & Summerer, 2014; Tyagi & Lemke, 2013), enabling SDSL EPR studies also on proteins where native cysteines cannot be removed without structural and functional impairment. Furthermore, these highly specific labeling reactions can be used for *in vivo* studies, where the complex cellular environment usually precludes cysteine labeling approaches. A detailed review about recent developments in orthogonal labeling techniques can be found in Ramil and Lin (2013).

1.1 Information from EPR Spectroscopy on Spin-Labeled Proteins

1.1.1 Spin Label Mobility

The flexible linker of the R1 spin label side chain renders its reorientational motion strongly dependent on its immediate environment, i.e., neighboring side chains and adjacent secondary structure elements. Thus, it can report about the local structure where the label is attached to the protein. The shape of room temperature EPR spectra reflects the restriction of the reorientational motion of the label and the influence of its dynamics on the spectral shape has been extensively reviewed (Berliner, 1976, 1979; Berliner and Reuben, 1989). Furthermore, its relationship to protein structure has been systematically investigated with T4 lysozyme (Columbus, Kalai, Jekö, Hideg, & Hubbell, 2001; Columbus & Hubbell, 2002;

Mchaourab, Lietzow, Hideg, & Hubbell, 1996). Spin-labeled sites exposed to the bulk water exhibit reorientational correlation times of the label side chain in the ns range, and the resulting EPR spectra are characterized by small line widths of the center lines, ΔH_0, and small apparent hyperfine splitting, i.e., the distance between the outer lines (Fig. 2A). Restriction of the mobility of the spin label side chain by interaction with neighboring side chains or backbone atoms is reflected in increased line widths and hyperfine splittings. R1 side chains buried in the interior of a protein exhibit completely restricted reorientational motion and a so-called EPR powder spectrum is observed. A plot of the inverse center line width, ΔH_0^{-1}, versus residue number reveals secondary structure elements through periodic variation of the spin label mobility, allowing the assignment of α-helices, β-strands, or random structures. Moreover, regions accommodating buried, surface-exposed, or loop residues can be identified from the correlation between the spectral breadth characterized by the inverse of its second moment, $\langle H^2 \rangle^{-1}$, and the inverse of the central line width, based on the X-ray structures of T4 lysozyme and annexin 12 (Hubbell et al., 1996; Isas, Langen, Haigler, & Hubbell, 2002; Mchaourab et al., 1996).

Proteins in their native state often exhibit several conformational substates playing an important role in their function (Cooper, 1976; Frauenfelder, Parak, & Young, 1988), for example, corresponding to

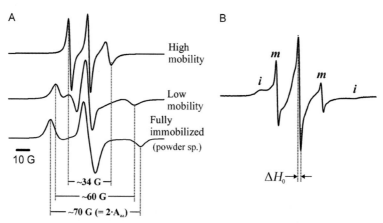

Figure 2 *Spin label side chain mobility.* (A) Room temperature EPR spectra recorded at X-band (~9.5 GHz, ~0.34 mT) of protein-bound MTSSL (R1) with high mobility, low mobility, and fully immobilized (in a frozen sample). (B) Multicomponent spectrum with separated spectral intensities corresponding to mobile (*m*) and immobile (*i*) components. The peak-to-peak distance ΔH_0 of the central resonance line is indicated.

different signaling states or open and closed states of a ligand-binding pocket. Due to their lifetimes in the μs to ms regime, these substates can be recognized in the cw spectra if they are characterized by different spin label side chain mobility (Fig. 2B). Such spectra are called multicomponent spectra and permit investigation of these conformational substates by, for example, osmolyte perturbation (Lopez, Fleissner, Guo, Kusnetzow, & Hubbell, 2009), saturation recovery (Bridges, Hideg, & Hubbell, 2010), and high-pressure EPR (McCoy & Hubbell, 2011).

For a more quantitative interpretation of the experimental data in terms of dynamic mechanisms and local tertiary interaction, simulations of the EPR spectra can be performed either employing dynamic models of the spin label side chain (Barnes, Liang, Mchaourab, Freed, & Hubbell, 1999; Borbat, Costa-Filho, Earle, Moscicki, & Freed, 2001; Freed, 1976) or on the basis of molecular dynamics (MD) simulations (Beier & Steinhoff, 2006; Budil, Sale, Khairy, & Fajer, 2006; DeSensi, Rangel, Beth, Lybrand, & Hustedt, 2008; Oganesyan, 2007; Sezer, Freed, & Roux, 2008; Steinhoff & Hubbell, 1996; Steinhoff, Müller, Beier, & Pfeiffer, 2000). Especially the latter approach provides a direct link between the molecular structure on atomic level and the EPR spectral line shape, thus allowing verification, refinement, or even suggestion of structural models of spin-labeled proteins based on EPR spectral data.

1.1.2 Spin Label Accessibility

The location of the spin label side chain with respect to the protein/water/membrane boundaries can be determined from its accessibility toward paramagnetic probes, which selectively partition in the different environments. Accessibility of the R1 side chain is defined by its collision (Heisenberg exchange) frequency, W_{ex}, with a paramagnetic exchange reagent diffusing in its environment. Metal ion complexes, for example Ni(II)ethylenediamine diacetate (NiEDDA) or chromium oxalate, quantify the accessibility from the bulk water phase, whereas molecular oxygen or hydrophobic organic radicals accumulate in the hydrophobic part of the lipid bilayer. The concentration gradients along the membrane normal for these probes can be used to characterize the immersion depth of the spin label side chain with respect to the membrane/water interface, calculated from the accessibilities to O_2 and NiEDDA (immersion depth parameter $\varphi = \ln(W_{ex}(O_2))/\ln(W_{ex}(\text{NiEaDDA}))$ (Altenbach, Greenhalgh, Khorana, & Hubbell, 1994; Hubbell & Altenbach, 1994; Marsh, Dzikovski, & Livshits,

2006). Also here, periodic variation of the spin label accessibility can allow the assignment of secondary structure elements.

The Heisenberg exchange mechanism affects the relaxation properties of the spin label attached at the investigated sites, and several methods can be applied to determine the exchange frequency W_{ex}. Most commonly, exchange rates for nitroxide side chains in proteins are measured using cw power saturation. Here, the EPR signal amplitude is monitored as a function of the incident microwave power in the absence and presence of the paramagnetic quencher. From the saturation behavior of the spin label a dimensionless "accessibility parameter," π, can be extracted, which is proportional to the exchange frequency W_{ex} (Altenbach, Flitsch, Khorana, & Hubbell, 1989; Altenbach, Froncisz, Hemker, Mchaourab, & Hubbell, 2005; Farahbakhsh, Altenbach, & Hubbell, 1992). Alternatively, W_{ex} can be determined directly from the change of the spin–lattice relaxation time, T_{1e}, in the presence of the quencher using pulse EPR methods (Altenbach, Froncisz, Hyde, & Hubbell, 1989; Altenbach et al., 2005; Feix et al., 1984; Mchaourab & Hyde, 1993; Nielsen et al., 2004; Pyka, Ilnicki, Altenbach, Hubbell, & Froncisz, 2005; Yin, Pasenkiewicz-Gierula, & Hyde, 1987). In addition, the water accessibility of the spin label side chain can be determined by electron spin echo envelope modulation (Urban & Steinhoff, 2013; Volkov, Dockter, Bund, Paulsen, & Jeschke, 2009).

1.1.3 Polarity of the Spin Label Environment

Changes of the polarity and proticity of the spin label microenvironment can provide insights into conformational changes taking place, for example, at membrane–protein–water interfaces, if transport proteins shuttle between conformational states with different accessibilities of the solute transport pathway to both sides of the membrane. Both, polarity and proticity, are reflected in the hyperfine tensor component A_{zz} and the g tensor component g_{xx}. A polar environment shifts A_{zz} to higher values and g_{xx} is decreased. The A_{zz} component can be determined from cw X-band ESR spectra of spin-labeled proteins in frozen samples. For determination of the g tensor components and their variation high-field EPR techniques have to be used that provide enhanced Zeeman resolution (Huber & Törring, 1995; Steinhoff, Savitsky, et al., 2000). In surface-exposed regular secondary structure elements, e.g., α-helices, the water density, and hence A_{zz} and g_{xx}, are a periodic function of residue number. These data can be used similarly to the mobility and accessibility data to obtain structural and topological information. The hyperfine tensor component A_{zz} (cf. Fig. 2A) as a measure for the polarity is determined from a line shape analysis using, e.g., the program Dipfit (Steinhoff et al. (1997)).

The abovementioned information obtainable from cw EPR spectra recorded at room temperature and from frozen samples will be exemplified with a spin labeling site scan, revealing the structure of the extracellular loop 4 (eL4) of the sodium–proline symporter PutP, in the second part of this chapter.

1.1.4 Time-Resolved cw EPR Measurements

Conformational changes of proteins induced, for example, by light excitation or ligand binding, can be followed in time by monitoring spectral changes caused by variations of the spin label side chain mobility with down to 100 μs time resolution using conventional EPR instrumentation and detection schemes (Altenbach, Kusnetzow, Ernst, Hofmann, & Hubbell, 2008; Bordignon et al., 2007; Farahbakhsh, Hideg, & Hubbell, 1993; Farrens, Altenbach, Yang, Hubbell, & Khorana, 1996; Klose et al., 2014, Steinhoff et al., 1994, Thorgeirsson et al., 1997). The transient EPR signal after activation, monitored at a fixed magnetic field, can be recorded using an analog-to-digital converter read out by a standard PC, if none of the available spectrometers offers built-in functions for recording transient EPR signals. Most suitable for time-resolved detection of spin label mobility changes appears to be the center EPR line, as the amplitude of this line strongly depends on the reorientational correlation time of the label side chain. Nevertheless, constant magnetic fields are required, because small variations can lead to shifts of the spectrum, moving the often sharp center peak out of the detection position. Less problematic are time-resolved measurements in the low-field region of the spectrum, where spectral components characteristic for mobile and immobilized label side chains are often well separated.

1.1.5 Distance Measurements

If two spin label side chains are introduced into a protein or two molecules carrying single spin labels are in a stable macromolecular complex, their distance can be determined through quantification of their spin–spin interaction, and thus providing valuable structural information that can be used to determine the protein's topology and/or orientation in macromolecular complexes. If the molecules of interest can be trapped in intermediate states of their catalytic cycle, e.g., conformational changes coupled to function can be identified and quantified. In EPR spectroscopy, two basic approaches are used to determine the dipolar spin–spin interaction. The static dipolar interaction in an unordered immobilized sample, i.e., in frozen or highly viscous solution, leads to considerable broadening of the cw EPR spectrum if the interspin distance is less than about 2 nm. This interspin distance can be quantified by a detailed line shape analysis using spectra convolution or deconvolution techniques (Altenbach, Oh, Trabanino, Hideg, & Hubbell, 2001; Rabenstein & Shin, 1995; Steinhoff, Dombrowsky, Karim, & Schneiderhahn, 1991; Steinhoff et al., 1997).

Distances in the range from 1.5 up to 8 nm (Borbat & Freed, 1999; Pannier, Veit, Godt, Jeschke, & Spiess, 2000), or even >10 nm, if fully deuterated samples are used (Bowman et al., 2014), can be measured using pulse EPR methods. Two major protocols are successfully applied, 4-pulse double electron–electron resonance (DEER), also called pulse electron double resonance (PELDOR), and double quantum coherence (DQC). DEER experiments can be performed on commercially available pulse EPR spectrometers at X- ($\sim$9 GHz) or Q-band ($\sim$35 GHz) frequencies. For experimental details, see Section 2.2 and the example of HisQMP$_2$ in the second part of this chapter. DQC experiments require very high microwave power (up to 2 kW) as the whole EPR spectrum is excited with a single microwave pulse, and can only be performed on specialized spectrometers. A comprehensive description of the method(s) is beyond the scope of this chapter and can be found, for example, in Schiemann & Prisner (2007), where both approaches are compared.

Combining cw and pulse EPR techniques provide means to determine interspin distances in the range from 1 to 8 nm, and thereby covering the most important distance regime necessary for structural investigations on proteins and protein complexes. Notably, when combining cw and pulse EPR distance data borderline effects in the region from 1.6 to 1.9 nm have to be taken into account (Banham et al., 2008; Grote et al., 2008).

2. APPLICATIONS TO MEMBRANE PROTEINS

The ability to investigate membrane proteins in their native environment, i.e., in a lipid bilayer, or in membrane mimics like nanodiscs, lipodisqs, or micelles/bicelles, renders SDSL EPR a valuable tool to study signaling and transport processes across biological membranes. The method has been, for example, successfully applied to study conformational changes in course of the transport cycle of different membrane transporters, mainly based on the analyzes of distance measurements. Grote et al. used cw and pulse EPR spectroscopy (Grote et al., 2008, 2009) for the analysis of conformational changes of the nucleotide-binding (NBD) subunits, MalK$_2$, of the maltose ABC transporter MalFGK$_2$, revealing a "tweezers-like" model of closure and reopening of the NBDs during the catalytic cycle. Another study on the homotrimeric Na$^+$-coupled aspartate transporter Glt$_{Ph}$ showed that the transporting domains sample multiple conformations, consistent with large-scale movements during the transport cycle, and that substrate-binding influences the occupancies of the different conformational states. Interestingly, this study also revealed that the membrane environment favored conformations different from those observed in detergent micelles (Hänelt et al., 2013). Other examples for investigations on active and passive transport systems include studies on the structural mechanism of gating of the mechanosensitive channel of small conductance (MscS), where mainly mobility and accessibility data have been used to suggest a structural mechanism for transition from the closed to the open channel state (Vasquez, Sotomayor, Cordero-Morales, Schulten, & Perozo, 2008; Vasquez, Sotomayor, Marien-Cortez, et al., 2008), and mechanistic studies on the potassium ion channel KcsA both solubilized in detergent and reconstituted in lipids (Cordero-Morales et al., 2006; Endeward, Butterwick, MacKinnon, & Prisner, 2009; Gross, Columbus, Hideg, Altenbach, & Hubbell, 1999). Applications of SDSL EPR on membrane-bound signaling systems mainly comprise visual rhodopsin/transducin/arrestin (Altenbach et al., 2008; Farrens et al., 1996; Kim et al., 2012; Van Eps et al., 2010) and bacterial and archaeal chemo- (Airola et al., 2013; Bhatnagar et al., 2010; Ottemann, Thorgeirsson, Kolodziej, Shin, & Koshland, 1998; Park et al., 2006) and phototaxis systems (reviewed in Klare, Bordignon, Engelhard, & Steinhoff, 2011).

In the following section, first some general aspects concerning protein mutagenesis for spin labeling, labeling of purified proteins, and the instrumentation required to perform the different EPR experiments are discussed. Afterwards, three examples from our laboratory are described in more detail:

1. Light-induced conformational changes in the first HAMP domain of the SRII/HtrII photoreceptor/transducer complex from *Natronomonas pharaonis*.
2. The conformational dynamics of the histidine ABC transporter HisQMP$_2$ from *Salmonella enterica serovar Typhimurium* in three states of the ATP hydrolysis cycle.
3. A complete spin labeling site scan of eL4 of the *Escherichia coli* Na$^+$/proline symporter PutP revealing its secondary structure.

2.1 Sample Preparation
2.1.1 Mutagenesis
The most common technique for spin labeling of proteins is the introduction of a cysteine residue at the desired site using standard polymerase chain reaction methods, for example, using commercially available kits like QuikChange from Stratagene. To assure specific labeling, all native cysteines, which are accessible for spin labeling, have to be replaced by appropriate amino acids like alanine or serine. It is important to note that for all mutations and modifications (i.e., attachment of the spin label side chain) the protein's structure and functionality should be tested using appropriate assays (e.g., transport or nucleotide hydrolysis assays) to ensure that the information obtained from the labeled proteins is of biological relevance. Although proteins carrying multiple native cysteine side chains are in general less eligible for spin labeling studies, a number of examples show that often neither removal of several cysteines nor introduction of the spin label side chain has severe effects on the proteins' functionality. For example, in human guanylate binding protein 1 all seven cysteine residues could be removed and one or two R1 side chains could be introduced without significantly affecting GTP hydrolysis or nucleotide-induced oligomerization of the protein (Vöpel et al., 2014).

The choice of the labeling positions naturally depends on the question addressed, but some general points should be taken into account. Effective labeling requires good accessibility of the cysteine residue, and thus rendering surface-exposed sites more suitable for labeling than buried positions. Furthermore, at surface-exposed sites mutations can be expected to have less influence on the protein's fold and/or functionality. If structural information is available from X-ray crystallography, NMR or homology modeling, freely available software packages like MMM (Polyhach, Bordignon, & Jeschke, 2011) or MTSSLWizard (Hagelueken, Ward, Naismith, & Schiemann, 2012) can be used for the selection of labeling positions based on their putative accessibility for the label reagent and for the expected range for distance measurements between two labels. For the choice of the labeling position(s), the type of amino acid replaced by cysteine should also be considered. In general, removing aromatic or charged residues can be expected to have more influence on the proteins properties than, for example, replacing serine or alanine. The latter are also the most conservative replacements for cysteines, but it has to be kept in mind that the resulting R1 side chain is larger.

2.1.2 Protein Purification and Spin Labeling

Effective coupling of the spin label side chain requires the cysteine to be present in its sulfhydryl (-SH) form. Especially, oxygen can oxidize the cysteine and consequently reducing agents, like dithiothreitol (DTT), should be present during the last purification steps or used as a treatment step prior to labeling, but have to be removed completely for incubation with the label reagent. The use of a protective atmosphere (Ar, N_2) for these steps can help to increase labeling efficiencies. High purity of the protein is another requirement, as contaminations with other cysteine-containing proteins have to be avoided.

The purification of membrane proteins usually involves solubilization of the molecules using detergents. Labeling can then be carried out either in solution or when the protein is bound, for example, via an affinity tag to a chromatography matrix. If applicable, the latter approach offers significant advantages, as cysteine reduction, removal of the reducing agent, incubation with the spin label, and removal of unbound label are integrated in the protein purification procedure, thus minimizing additional treatment steps that can affect protein stability or reduce the final amount of labeled protein. In our standard protocol for labeling, protein solutions (10–200 μM) are incubated with 1 mM spin label from a 100 mM MTSSL (Enzo Life Sciences or Toronto Research Chemicals) stock solution in DMSO or acetonitrile for 1–16 h at room temperature or 4 °C, depending on the accessibility of the cysteine and the conditions necessary to keep the protein stable. Unbound label is removed via gel filtration columns, by washing in centrifugal filter units or dialysis. To assure the integrity of the disulfide bond between the protein and the label in all subsequent treatment steps, the presence of reducing agents should be avoided.

Labeling efficiencies are determined by double integration of the (first derivate) room temperature EPR spectra, and comparison to reference samples of known spin label concentration and the protein concentration determined by colorimetric or spectroscopic (UV/Vis) assays. In general labeling efficiencies should be maximized to be close to completeness. High labeling efficiencies are less crucial for mobility, accessibility, or polarity investigations, but should be as high as possible or at least be accurately known for interspin distance measurements, as spectral broadening in low temperature cw spectra and modulation depth in pulse EPR experiments strongly depend on the labeling efficiency.

2.2 Instrumentation

In our laboratory, we use X-band (~9.5 GHz) cw EPR spectrometers with dielectric microwave resonators (Bruker Flexline MD-5, 0.5 mW microwave power, 0.15 mT B-field modulation) for ambient temperature cw EPR, a super-high Q resonator (Bruker Biospin GmbH, Rheinstetten, Germany) for cw EPR on frozen samples (160 K, in a continuous flow cryostat Oxford ESR900 controlled by an Intelligent Temperature Controller (ITC 4, Oxford Instruments)), and a loop-gap resonator (Hubbell, Froncisz, & Hyde, 1987; Pfeiffer, Rink, Gerwert, Oesterhelt, & Steinhoff, 1999) for accessibility measurements. Pulse EPR experiments at X-band are carried out with a Bruker Elexsys 580 spectrometer equipped with a split-ring resonator (Bruker Flexline ER 4118X-MS3) in a continuous flow helium cryostat (CF935; Oxford Instruments) controlled by an Oxford Intelligent Temperature Controller ITC 503S adjusted to 50 K. Samples at a protein concentration of 100–250 μM are loaded into 0.9-mm ID glass capillaries for mobility measurements (sample volume 10–20 μl) or in gas permeable TPX (polymethylpenten) capillaries (Rototec Spintec GmbH, Biebesheim, Germany, sample volume ~7 μl) for accessibility measurements. For pulse EPR experiments 40–50 μl sample are loaded into 3 mm OD quartz capillaries. Notably, DEER distance measurements can be carried out with higher signal-to-noise and shorter accumulation times at higher microwave frequencies (Q-band, 35 GHz) (Polyhach et al., 2012).

2.3 SRII/HtrII: Static and Time-Resolved Detection of Signaling Events

In the phototaxis system of *N. pharaonis*, being homologous to the enterobacterial chemotaxis system, light activation of the photoreceptor SRII in the membrane leads to activation of its cognate transducer HtrII. The extended coiled-coil signaling protein HtrII protrudes ~20 nm into the cytoplasm (Fig. 3A), where it modulates the activity of the histidine kinase CheA being part of a two-component signaling cascade.

2.3.1 Sample Preparation

C-terminally His-tagged SRII (wt) and transducer constructs truncated at position 157 (HtrII$_{157}$) carrying cysteine mutations for labeling are seperately expressed in *E.coli*. The membrane fraction of the cells is solubilized with 2% dodecylmaltoside as detergent. The proteins are purified via a

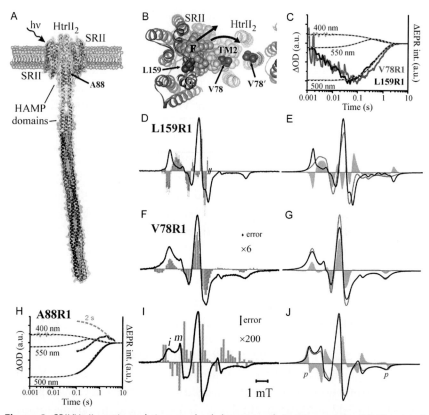

Figure 3 SRII/HtrII: static and time-resolved detection of signaling events. (A) Structural model of the SRII/HtrII complex in the membrane. The periplasmic side is on the top. (B) View from the cytplasmic side. (C) Transient optical absorbance changes determined at 400, 500, and 550 nm (gray) with multiexponential fits (black, dotted), and EPR signal changes for SRII-L159R1/HtrII (black, bold) and SRII/HtrII-V78R1 (dark gray, bold) upon light excitation by a laser flash. (D) Experimental light - dark difference spectrum (gray bars, with error bars) and RT cw EPR spectrum for SRII-L159R1. Note, that the difference spectrum has not been determined for the high-field region of the EPR spectrum (marked by //). (E) Simulated difference spectrum (gray shaded area) determined from simulations of the RT cw spectrum (black) and of a spectrum of a mobilized spin label side chain (gray). (F and G) Corresponding experimental and simulated spectra for SRII/HtrII-V78R1. Here, a decreased dipolar interaction was simulated. (H) Transient optical absorbance changes and EPR differences upon light excitation by a LED at 500 nm for SRII/HtrII$_{157}$-A88R1 reconstituted into purple membrane lipids with global triexponential fits (black, dashed). For the EPR transients the fits are decomposed in light (black) and heat (dashed gray) induced contributions. (I) cw EPR spectrum for SRII/HtrII$_{157}$-A88R1 (black, bold) with mobile (*m*) and immobile (*i*) components and transient EPR difference spectrum (gray). (J) Spectra simulations, which reveal an increased population of the immobile component by ~10% and a polarity change for A88R1 (gray). The difference spectra in panels F and I are scaled up for visibility.

single Ni-affinity purification step. Spin labeling of the transducer constructs with MTSSL is carried out on the Ni-NTA column during purification. After a 1-h incubation period with 1 mM DTT, the reducing reagent is washed out and exchanged for a 1 mM MTS spin label solution in degassed buffer followed by incubation over night at 4 °C and washing out of unbound excess spin label prior to elution. The obtained labeling efficiencies depend on labeling position and are typically 75–100%.

2.3.2 Transient Absorption Measurements

Transient absorption measurements are carried out with a 50-W halogen lamp equipped with an infrared cutoff filter (KG-2) and either 400, 500, or 550 nm interference filters for illumination of the sample-filled quartz cuvette. The sample holder is temperature controlled to 298 K. The transmitted light passes through a monochromator and is detected by a photodiode. Sample activation perpendicular to the transmission beam is carried out with a flashlight equipped with a 475-nm edge filter, flash duration is 80 μs. The amplified signal is recorded with an analog-to-digital converter connected to a standard PC.

2.3.3 Static and Time-Resolved EPR Measurements

Room temperature cw EPR spectra are recorded on a home-built X-band EPR spectrometer equipped with a dielectric resonator (microwave power: 1.0 mW, B-field modulation amplitude: 0.35 mT), and the sample temperature is stabilized to 297 ± 1 K by a temperature-controlled gas stream through the resonator. 15 μl sample (~100 μM) are loaded into 0.9-mm ID glass capillaries.

For the time-resolved cw EPR measurements signaling is induced by illuminating the samples inside the resonator. Different types of light source can be used for activation, partly depending on the type of experiment. We use pulsed (20 ns pulse duration) laser sources of appropriate wavelength (λ_{max} for NpSRII = 498 nm) with energies per light pulse between 1 and 1.5 mJ (Holterhues et al., 2011) or flashlights (high-energy xenon flash lamp (bulb supplied by ILC Technologies) 80 J energy/pulse with 30–80 μs flash duration) (Rink, Riesle, Oesterhelt, Gerwert, & Steinhoff, 2000; Steinhoff et al., 1994) for transient experiments with high (μs–ms) time resolution. Comparative experiments with laser and flash sources show full reproducibility of the transients independent of the light source used (Klose et al., 2014). If lower time resolution (~100 ms) and partial accumulation of the activated signaling state are sufficient difference spectra can be obtained by switching of a constant light source, e.g., a white light emitting diode (3 W, Edixeon A, EDEW-3LA3-E3) equipped with an infrared cutoff filter (KG-2) and a 500-nm edge filter to reduce heating of the sample (Klose et al., 2014). The transient EPR signal upon illumination is recorded using an analog-to-digital converter read out by a standard PC. Typically, 500–10,000 time traces at 20–100 B-field positions covering the EPR spectrum are averaged to achieve sufficient signal-to-noise ratio.

Figure 3B shows the location of residues chosen for SDSL, being diagnostic for light-induced conformational changes in the transmembrane region of the SRII/HtrII complex. Spin label side chain L159R1 is buried between SRII helices F and G and reports on relative motions of the two helices, and V78R1 in the HtrII dimer interface is sensitive to conformational changes of the four helix bundle. Transient changes of the EPR signal are observed for both positions (Fig. 3C), indicating light-induced conformational changes occurring 2–3 ms after light activation that correspond to a spectrally silent transition (M1 → M2) in the SRII photocycle (Klare et al., 2004). The back reaction of SRII coincides with recovery of the ground state optical absorbance at 500 nm, whereas HtrII appears decoupled and returns back to its original conformation delayed by about 200 ms. Difference spectra are calculated as the amplitude spectra of exponential functions used to fit the EPR transients covering the EPR spectrum (Fig. 3D and F). A more quantitative analysis using spectra simulations based on the solution of the stochastic Liouville equation (Budil, Lee, Saxena, & Freed, 1996), carried out with the software "multicomponent" from Chr. Altenbach (https://sites.google.com/site/altenbach/labview-programs/epr-programs/multicomponent) is depicted in Fig. 3C and G. For L159R1 simulation of a light-induced increase of the spin label mobility yields a difference spectrum similar to the experimentally observed one. For V78R1 a simulated difference spectrum considering a change in dipolar broadening agrees with the experimental difference spectrum. The observed transient mobilization of L159R1 indicates a light-induced conformation of SRII with the cytoplasmic moieties of TM helices F and G being more separate. The investigations of additional single spin-labeled SRII variants and interspin distance measurements between the SRII TM helices allowed to conclude a light-induced outward movement of the cytoplasmic side of helix F (Wegener, Chizhov, Engelhard, & Steinhoff, 2000). The decreased dipolar coupling between HtrII positions V78R1 and V78R1' upon light excitation (together with the unchanged coupling between L82R1 and L82R1') (Wegener, Klare, Engelhard, & Steinhoff, 2001) is in agreement with the assumption of a rotary and/or piston-like motion of HtrII TM2 as a result of the SRII helix F tilt (see arrows in Fig. 3B).

Despite the provisions described above, sample heating due to the extended illumination period occurs when using the LED source, leading for example to transient shifts of a thermodynamic equilibrium between R1 side chain orientations of different mobility, changes of the protein dynamics or to a decrease of the sample viscosity. For spin label side chains like SRII-L159R1 and HtrII-V78R1 exhibiting strong light-induced EPR spectral changes, these effects on the R1 side chain dynamics can be neglected. Nevertheless, in case that the light-induced spectral changes are small, exemplified in Fig. 3H–J for a spin label attached to transducer position A88 in the first HAMP domain, the transient EPR data are composed of contributions originating from light- and heat-induced spectral changes, that are distinguishable according to their different relaxation times. Time constants for heating and cooling of the sample can be determined by control experiments carried out with unbound spin label mixed with wild-type protein complexes (carrying no cysteine residues). The contributions are separated using biexponential fitting of the transient data sets (Fig. 3H). Difference spectra are calculated as the amplitude spectrum of the exponential functions with the functional relevant time constants. The difference spectrum for HtrII-A88R1 (Fig. 3I) exhibits a transiently decreased mobile (m) spectral component, while the immobile (i) component is increased upon light activation. An analysis using spectra simulations reveals an additional slight increase in the local polarity as evident from the agreement between experimental and simulated difference spectra, where a polarity change is included (p) for this mutant (Fig. 3I and J).

Notably, the origin of the transient EPR spectral changes (increase or decrease of the mobility, dipolar interaction, or environmental polarity) can easily be identified by the shape of the difference spectra. Summarizing, time-resolved EPR experiments provide means to follow propagation of the light-induced signal to the HAMP domain and to characterize the kinetic coupling of HtrII to SRII: Light activation of the retinal chromophore triggers an outward movement of helix F in SRII with formation of the M2 photocycle intermediate, and thereby inducing a rotary/piston-like movement of the tightly coupled TM2 helix of HtrII. This conformational change, as revealed from transient EPR data recorded for a comprehensive set of mutants (Klose et al., 2014), leads to a shift of the thermodynamic equilibrium between states of the downstream HAMP domain characterized by mobile (dynamic HAMP) and immobile (compact HAMP) spectral components toward the more compact state (reviewed e.g., in Klare et al., 2011).

2.4 HisQMP: Conformational Changes Studied by DEER Distance Measurements

ATP-binding cassette (ABC) transporters permit the ATP-driven translocation of solutes across biological membranes in organisms from bacteria to man and play an important role, for example, in the immune system. In prokaryotes, they mediate the uptake of nutrients and vitamins or the export of toxic compounds or virulence factors. ABC transporters minimally consist of four protein modules: two transmembrane domains (TMDs) forming the translocation path and two nucleotide-binding domains (NBDs). In prokaryotes, a periplasmic substrate-binding protein (SBP) is required as an additional component. Biochemical, biophysical and structural data for one of the best studied transporters, the maltose transporter MalEFGK$_2$, support an "alternate access" model for type I ABC importers, in which the transporter resides in the inward-facing (resting) state in the absence of substrate. In the presence of ATP the loaded SBP promotes tight association of the NBDs and the transport path opens to the extracellular side (outward-facing), which in turn induces substrate release from the SBP to the binding site in MALFG. Finally, ATP hydrolysis allows release of the substrate to the cytoplasm, and dissociation of phosphate and ADP followed by ATP binding restores the resting state. The *Salmonella* histidine ABC transporter (cf. Fig. 5A) is composed of the TMDs HisQ and HisM and a homodimer of the NBDs, HisP$_2$. DEER interspin distance measurements on HisQMP$_2$ incorporated into liposomes in different states of the transport cycle (in the absence of nucleotides, in the ATP-bound, and in the posthydrolysis state) reveal relative motions of the domains during the transport cycle (Sippach et al., 2014).

2.4.1 Sample Preparation

Cysteine residues are introduced using the Stratagene QuikChange Lightning kit on a plasmid carrying a cysteine-less variant of HisQMP. *E. coli* BL21-T1® (DE3) pLysS cells carrying the respective plasmids are grown in TB medium at 30 °C and shaking to OD$_{650nm}$ = 0.8–1. Induction with 0.1 m*M* IPTG is carried out at 22 °C for 4 h. After harvesting, the cell pellet is resuspended in 50 m*M* Tris/HCl (pH 7.5), 10% glycerol, 0.1 m*M* phenylmethylsulfonyl fluoride (PMSF), and the cells are broken in a high-pressure cell disrupter (Model Basic Z, Constant Systems Ltd.). The membrane fraction is collected by ultracentrifugation (200.000 × *g*, 90 min, 4 °C) and solubilized with 1.1% *n*-dodecyl-β-maltoside (DDM) for 1 h at 4 °C. HisQMP$_2$ is purified from the supernatant after ultracentrifugation *via* metal-affinity chromatography (TALON resin, Clontech, equilibrated in 50 mM Tris/HCl (pH 7.5), 20% glycerol, 0.05% DDM, 100 mM NaCl, 4 mM β-mercaptoethanol, and 0.1 mM PMSF). The protein is eluted with 100 mM imidazole, that is removed from the samples using PD-10 columns equilibrated with 50 mM MOPS/KOH (pH 7.5), 20% glycerol, 0.05% DDM, 100 mM NaCl, and stored at −80 °C.

Spin labeling is carried out with 10-fold molar excess of MTSSL for 4 h or overnight (depending on the label position) at 4 °C under gentle stirring. Unbound label is removed using PD-10 columns and the protein is concentrated using Centrifugal Filter Units (Amicon® Ultra-4, YM10, 10 kDa cut-off, Millipore) to a final concentration of ∼3 mg/ml. The labeling efficiencies determined by cw EPR and absorption spectroscopy are usually between 75% and 95%.

Reconstitution into liposomes (*E. coli* total Lipid Extract, Avanti Polar Lipids, Inc.) is carried out by mixing proteins and preformed lipid vesicles in a 5:1 ratio (w/w). The detergent is removed using detergent-adsorbing beads (Bio-Beads SM-2 Adsorbent, BioRad) by overnight incubation at 4 °C. The beads are replaced once with fresh beads to ensure complete removal of the detergent. The resulting proteoliposomes are collected by ultracentrifugation and resuspended in 50 mM MOPS/KOH (pH 7.5) to a final complex concentration of 100 μM.

Proteoliposomes are assayed for ATPase activity, revealing for both wild-type and the cysteine-less variant ∼threefold ATPase stimulation in the presence of HisJ/His. Introduction of R1 side chains on two of the four subunits (Q/M or P/P′) reduced stimulation to 1.6 to 2.4-fold, depending on the label positions, and thus suggesting functional coupling of ATP hydrolysis to substrate transport.

Different steps of the transport cycle are mimicked supplementing the samples with 100 μM HisJ/1 mM histidine in the presence of 10 mM ATP/1 mM EDTA (added during reconstitution) or 2 mM ATP/3 mM MgCl$_2$. Mg^{2+}-ions permeabilize liposomes, thereby allowing access of ATP also to the lumen of the vesicles.

2.4.2 DEER Interspin Distance Measurements

DEER interspin distance measurements are performed at X-band using the four-pulse DEER sequence (Fig. 4; Pannier et al., 2000):

$$\frac{\pi}{2}(v_{obs}) - \tau_1 - \pi(v_{obs}) - t' - \pi(v_{pump}) - (\tau_1 + \tau_2 - t') - \pi(v_{obs}) - \tau_2 - \text{echo}$$

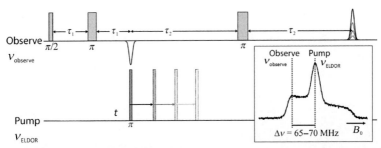

Figure 4 *Pulse sequence of the DEER experiment.* The inset shows the positions of the observer and pump frequencies on an EPR absorption spectrum (usually detected with a (Hahn) echo-detected field sweep).

with observer pulse (v_{obs}) lengths of 16 ns for $\pi/2$ and 32 ns for π pulses and a pump pulse (v_{pump}) length of 12 ns. A two-step phase cycling ($+\infty, -\infty$) is performed on $\pi/2(v_{obs})$. Time t' is varied with fixed values for τ_1 and τ_2. The dipolar evolution time is given by $t = t' - \tau_1$. Data are analyzed only for $t > 0$. The resonator is overcoupled. The pump frequency v_{pump} is set to the center of the resonator dip (coinciding with the maximum of the EPR spectrum, see inset in Fig. 4). The observer frequency v_{obs} is set 65–70 MHz higher, at the low-field local maximum of the EPR spectrum. Proton modulation ($v_H \approx 14$–15 MHz at X-band magnetic fields (≈ 0.34 T) for nitroxide spin labels, $g \sim 2$) is averaged by adding traces at eight different τ_1 values, starting at $\tau_{1,0} = 200$ ns with increments $\Delta\tau_1 = 8$ ns. Data points are collected in 8 ns time steps. The total measurement time for each sample is ~ 24 h.

2.4.3 DEER Data Analysis

Analysis of the data is carried out with the freely available software package DeerAnalysis2011 (Jeschke et al., 2006). Figure 5 depicts the course of data analysis for spin label pairs HisQ-197R1/HisM-104R1, HisP-101R1/HisP'-101R1 and HisP-153R1/HisP'-153R1 (Fig. 5A). The raw data traces (Fig. 5B) are the product of the contributions from spin–spin interactions in the HisQMP$_2$ complex and of interactions with the other complexes distributed in the sample (background), the latter appearing as an exponential decay of the signal. For soluble proteins homogeneously distributed in solution, a three-dimensional background can be assumed, but for membrane proteins in proteoliposomes the dimensionality of the background depends on the protein density in the liposomes and on the vesicle size. In extreme cases, where membrane proteins form a two-dimensional lattice in large liposomes, a 2D background might describe the situation best. Fitting of the

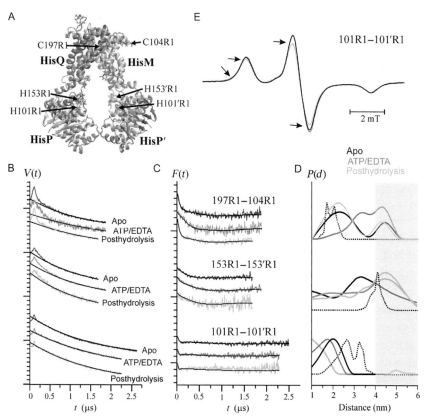

Figure 5 *Conformational changes of HisQMP studied by distance measurements.* (A) Model of the HisQMP$_2$ complex with spin label pairs indicated. (B) DEER dipolar evolution data V(t) and background fits. (C) DEER form factors F(t) after background correction of the traces shown in panel B and fits obtained by Tikhonov regularization. (D) Resulting interspin distance distributions. The calculated distance distributions for the spin label pairs obtained with a rotamer library approach from the model shown in panel A are depicted as dotted lines. (E) Superimposed continuous wave EPR spectra of 101R1–101'R1 for the apo state (black), the ATP bound state (grey), and the post hydrolysis state (light grey). The spectra are normalized to the same spin number. Positions in the spectrum where changes are observed are marked by arrows.

background dimensionality (e.g., using the DeerAnalysis software; Jeschke et al., 2006) often yields values between 2 and 3 for proteoliposomes, reflecting different extents of inhomogeneity of the long-range spin label distribution. In the present case, fitting of an exponential representing a three-dimensional background of interacting spins reasonably well accounts for the background behavior for samples showing clear oscillations (197R1–104R1). If such oscillations are not obvious (153R1–153'R1), what is a well known problem for DEER on membrane proteins reconstituted in liposomes, fitting of the exponential background decay is less reliable and may lead to intracomplex distance contributions shifted into the background or *vice versa*, thus distorting the resulting distance distribution.

Background correction leads to the form factors shown in Fig. 5C together with fits to the data obtained by Thikonov regularization. The resulting distance distributions are depicted in Fig. 5D. Regularization methods are usually the best choice for DEER data analysis, as translation of the dipolar evolution data into an interspin distance distribution is an ill-posed problem (Jeschke & Polyhach, 2007). Their advantage is that no assumptions have to be made about the shape of the distance distribution. Nevertheless, in some cases, e.g., if clear oscillations are absent in the data traces, indicating very broad distance distributions, fitting with one or multiple Gaussians describing the interspin distance distribution can prevent overinterpretation of the data in terms of the distribution shape.

In order to enable comparison between the EPR data and a structural model for HisQMP$_2$ based on the crystal structure of the *E. coli* methionine transporter MetNI (Fig. 5A), *in silico* spin labeling can be carried out. Here, simulation of the expected interspin distances was obtained based on a rotamer library approach implemented in the software MMM (Polyhach et al., 2011). This kind of simulation technique provides spin label side chain rotamers which are populated in the given structure and predicts the resulting interspin distances, here shown overlayed to the experimental distance distribution in Fig. 5D.

Clearly different DEER traces, and thus distinct distance distributions are observed for the pair 197R1–104R1 at the periplasmic side of HisQM for the three states investigated. The major interspin distance of 2.2 nm for the apo-state coincides reasonably well with the prediction from the structural model and is shifted to 3.2 nm in the ATP-bound state and back to 2.1 nm upon ATP hydrolysis. Obviously, nucleotide-dependent opening and closing of the periplasmic side of the transporter takes place, reflected in average distance changes of more than 1 nm, allowing the conclusion that conformational transitions of the NBDs are communicated to the periplasmic side of the TMDs (Sippach et al., 2014).

The pair 101R1–101'R1 in the NBDs reveals distance distributions with significant contributions below 2 nm for all three states investigated. This is the borderline range between pulse and cw EPR analysis and cw EPR spectra ($T = 160$ K) are recorded on the same samples. Increased dipolar broadening upon hydrolysis is recognizable when comparing the cw spectra (Fig 5E, indicated by arrows). A line shape analysis by fitting of simulated spectra to the experimental ones using the program DIPFIT (Steinhoff et al., 1997), yields an average distance of 1.4 nm for the posthydrolysis state. Remarkably, the experimental distances are shorter than predicted from the structural model for $HisQMP_2$, indicating that the NBDs in solution are more closely associated. As the data for this label pair closely resemble those observed for the homologous positions 83 and 83′ in $MalK_2$ (Grote et al., 2008), the experiments provide strong evidence that the NBDs in $HisQMP_2$ undergo similar conformational transitions as $MalK_2$ in $MalFGK_2$ (Sippach et al. (2014).

The DEER traces observed for the label pair 153R1–153′R1 do not reveal clear oscillations. Furthermore, neither the raw data nor the form factors reveal significant differences, and Tikhonov regularizations yield very broad distance distributions which cannot unambiguously be attributed to spin–spin interaction within the NBD dimer. Similar contributions are observed at longer distances for 197R1–104R1. Thus, interactions between $HisQMP_2$ complexes in the membrane are likely to contribute. Since from the data of the spin label pair 153R1–153′R1 inter- and intramolecular interactions cannot be clearly separated, no further conclusion should be drawn from this experiment.

2.5 PutP: A Spin Labeling Site Scan Reveals the Structure of eL4

The Na^+/proline symporter PutP belongs to the so-called LeuT-fold structural family which features ten core transmembrane domains connected by extra- and intracellular loops. These loops have been discussed to play a crucial role in the gating function in the alternating access model of secondary active transport processes (Forrest & Rudnick, 2009; Jardetzky, 1966), which is strongly supported by crystal structures of LeuT-type transporters in different conformations and by EPR studies (Claxton et al., 2010). Accordingly, eL4, working as a (thick) gate, regulates access to the substrate binding sites. According to structural data eL4 of LeuT exhibits a helix-turn-helix fold. For PutP a homology model exists (Fig. 6A) based on the structure of the Na^+/glucose transporter VSGLT, but this model reveals only one short α-helical stretch in the middle of the 30 aa loop that appears to be largely disordered, putatively due to inaccuracies of the homology model in this region.

Based on mobility, polarity, and accessibility profiles, the secondary structure of eL4 in PutP reconstituted into liposomes can be revealed by a spin labeling site scan of the corresponding region as shown in the following (Raba et al., 2014).

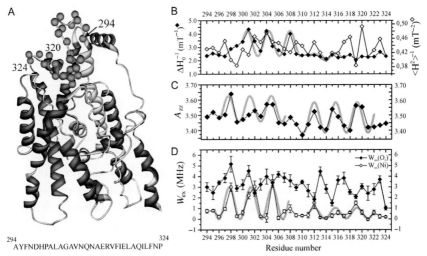

294
AYFNDHPALAGAVNQNAERVFIELAQILFNP
324

Figure 6 *Spin labeling site scan of PutP extracellular loop 4* (A) PutP homology model based on the crystal structure of the Na⁺/galactose transporter vSGLT (Olkhova, Raba, Bracher, Hilger, & Jung, 2011) in ribbon representation. Positions in eL4 connecting cTMs 7 and 8 are marked by spheres at the positions of the respective Cα atoms. The sequence of eL4 is given at the bottom. (B) Mobility parameter, ΔH_0^{-1}, the estimated error is ±0.01 mT^{-1} due to uncertainties of the line widths, and mobility parameter second moment, $\langle H^2 \rangle^{-1}$; the estimated error is ±0.01 mT^{-2} due to the strong influence of the baseline quality of the EPR spectra. (C) Polarity parameter, A_{zz}. (D) Collision frequencies W_{ex} for O_2 and NiEDDA (20 mM). All parameters are plotted versus residue number. Error bars represent standard deviations of the best-fit W_{ex} parameter calculated from ten saturation data points.

2.5.1 Sample Preparation

PutP variants carrying single cysteines, based on a functional cysteine-less variant (PutP(ΔCys)), 5 native cysteines replaced with Ala (Cys-12, -141, and -281) or Ser (Cys-344 and -349), are overexpressed in *E. coli* WG170 cells and prepared from isolated membranes by solubilization and purified via Ni–NTA affinity chromatography (Jung, Tebbe, Schmid, & Jung, 1998).

Spin labeling with MTSSL (Toronto Research Chemicals) is carried out with the protein bound to the Ni–NTA matrix. 1 mM MTSSL in buffer (50 mM KPi, pH8.0, 300 mM KCl, 10 mM imidazole, 10% glycerol (v/v), 0.04% β-D-dodecylmaltoside (w/v)) is applied to the column and incubated for 3 h at 4 °C. Unreacted label is removed by washing the column with the same buffer, and the labeled protein is eluted with buffer containing 200 mM Imidazol.

Information on the functionality of the resulting single Cys PutP variants is obtained from a ^{14}C-proline transport assay, revealing eleven mutants with activities >80%, twelve mutants with intermediate activities of 30% to 80% and eight positions (Leu302, Phe314, Glu311, Glu316, Gln319, Phe322, Asn323, and Pro324) with transport activities <30% compared to PutP (ΔCys). All PutP variants bind ^{3}H-proline with $\geq$60% of the binding activity of PutP(ΔCys) (Raba et al., 2014).

Reconstitution into preformed and detergent-destabilized (as determined by changes in absorbance at 540 nm) liposomes composed of *E. coli* lipids (Polar lipid extract; Avanti Polar Lipids) is carried out at a lipid/protein ratio of 20:1 (w/w). Detergent is removed by repeated use of detergent-adsorbing beads (Bio-Beads SM-2 Adsorbent, BioRad). Finally, the proteoliposomes are washed twice with 50 mM Tris/HCl, pH 7.5 and stored in liquid nitrogen until use.

2.5.2 EPR Experiments

For instrumentation see Section 2.2. Accessibilities are determined for O_2 and NiEDDA by EPR power saturation (Altenbach et al., 2005) on a home-built X-band machine equipped with a loop-gap resonator on 6,3 µl sample solution in gas permeable TPX (polymethylpenten) capillaries (Rototec Spintec GmbH). The B-field modulation is set to 0.15 mT and the applied microwave power is varied between 0.1 and 50 mW. For reference measurements, the sample is deoxygenated in a nitrogen gas flux, revealing the saturation behavior in the absence of any collision reagents. Accessibilities for molecular oxygen (hydrophobic) are determined using a 100% O_2 flux. Water-soluble NiEDDA is used with a final concentration of 20 mM, and during the measurements a nitrogen gas flux is applied. Prior to each experiment the sample was fluxed for at least 30 min to ensure saturation of the samples with the respective gas. Heisenberg exchange frequencies, W_{ex}, are calculated based on the relation

$$W_{ex} \propto \Delta P'_{1/2} = \frac{\Delta P_{1/2}}{\Delta H_0},$$

where $\Delta P_{1/2}$ is the difference between the microwave powers in the absence and in the presence of the collision reagent at which the spectral amplitude reaches half of the theoretical value in the absence of saturation (Farahbakhsh et al., 1992), and are calculated here using a homemade computer program. The $\Delta P'_{1/2}$ values are divided by a resonator specific proportionality factor α (here: 1.87 MHz^{-1}), determined from calibration measurements using different NiEDDA concentrations, to obtain W_{ex} rates according to Altenbach et al. (2005).

2.5.3 Data Analysis

All cw EPR spectra reveal the presence of at least two spectral components characterized by different mobilities ($i=$ immobile and $m=$ mobile, see spectrum in Fig. 2B). This can result either from local structural constraints leading to different stable spin label side chain conformations or from different protein conformations in equilibrium (Hubbell et al., 1996; Langen, Oh, Cascio, & Hubbell, 2000). Here, the spectra for all label positions are multicomponent spectra, rendering it unlikely that similar local structural constraints exist for all label positions, indicating that eL4 is involved in a protein conformational equilibrium at room temperature.

The spin label side chain mobility is quantified by the reciprocal width of the central resonance line (ΔH_0^{-1}) and analyzed as a function of sequence position (Fig. 6B). The resulting mobility profile reveals two distinct subdomains, one with mobile R1 side chains ($\Delta H_0^{-1} = \sim 2.8$–$4.3$ mT^{-1}) and the other showing intermediate mobilities ($\Delta H_0^{-1} = 2.0$–2.8 mT^{-1}). The latter values indicate R1 side chains with tertiary contacts. In addition, the inverse second moment, $\langle H^2 \rangle^{-1}$, is calculated:

$$\langle H^2 \rangle = \frac{\int (B - \langle H \rangle)^2 S(B) dB}{\int S(B) dB} \text{ with } \langle H \rangle = \frac{\int B S(B) dB}{\int S(B) dB}$$

(Fig. 6B). In contrast to ΔH_0^{-1} being strongly biased by mobile spectral components, $\langle H^2 \rangle^{-1}$ is dominated by the immobile spectral components.

The hyperfine tensor element A_{zz} is taken as a measure for the polarity of the spin label environment, providing a polarity profile (Fig. 6C) for eL4. This indicates that the N-terminal part (294–305) is located in a more polar environment ($\bar{A}_{zz} = 3.52$ mT) compared to the C-terminal part (306–324, $\bar{A}_{zz} = 3.46$ mT). The accessibility profiles (W_{ex} vs. residue number) for O_2 and NiEDDA are shown in Fig. 6D and reveal the same two-subdomain organization of eL4 observed in the mobility and polarity profiles.

2.5.4 Secondary Structure Determination

In the ΔH_0^{-1} profile (Fig. 6B), the more mobile region (298–308) displays a clear periodical pattern with a periodicity of ~ 3.6 (indicated in the figure), being characteristic for a surface-exposed α-helix. A similar pattern, even though less pronounced, is also visible in the range from residues 313 to 317, where an α-helical secondary structure was proposed from the homology model (cf. Fig. 6A; Olkhova et al., 2011).

The $\langle H^2 \rangle^{-1}$ profile (Fig. 6B) shows this periodical pattern for the whole eL4 sequence, suggesting that both eL4 conformations reflected in the mobile and immobile spectral components exhibit a helical secondary structure.

The same periodicity and phase are also observed in the polarity and accessibility profiles (indicated by gray harmonic graphs in the figures). Consequently, the combined results from the mobility, polarity, and accessibility analyzes provide strong evidence that eL4 is composed of two α-helices connected by a short loop, and that the C-terminal helix is more closely associated to other protein components than the N-terminal one, thus allowing to refine the present homology model (Raba et al. (2014)).

3. OUTLOOK

SDSL in combination with EPR spectroscopy is a powerful tool for the investigation of the structural and dynamic properties of biomolecules, in particular of membrane proteins involved in signaling and transport, and the number of SDSL EPR studies on such systems is steadily increasing. Combined use of the methods described above in conjunction with computational methods has been shown to provide detailed insights into the structure and function of a number of different systems, including those described in this chapter and other examples like the maltose transporter (reviewed in Bordignon, Grote, & Schneider, 2011; Klare & Steinhoff, 2013) or signaling complexes like rhodopsin and its heterotrimeric G protein (reviewed by Farrens, 2010). Numerous other systems might be good candidates for investigations using SDSL EPR.

In the past several new techniques, for example, osmolyte perturbation (Lopez et al., 2009) and high-pressure EPR (McCoy & Hubbell, 2011) have been developed that extend the possible use of SDSL EPR for membrane protein studies by providing deeper insights into protein and spin label conformational equilibria. Future developments, like the design of new spin labels for *in vivo* applications, improvements of spectrometer sensitivity by optimized hardware and new detection schemes will further enhance the applicability of this technique in membrane protein research.

ACKNOWLEDGMENTS

The studies described in this chapter have been funded by the Deutsche Forschungsgemeinschaft: SFB-944-P10, STE 640/10, and STE640/11.

REFERENCES

Airola, M. V., Sukomon, N., Samanta, D., Borbat, P. P., Freed, J. H., Watts, K. J., et al. (2013). HAMP domain conformers that propagate opposite signals in bacterial chemoreceptors. *PLoS Biology*, *11*, e1001479.

Altenbach, C., Flitsch, S. L., Khorana, H. G., & Hubbell, W. L. (1989). Structural studies on transmembrane proteins. 2. Spin labeling of bacteriorhodopsin mutants at unique cysteines. *Biochemistry*, *28*, 7806–7812.

Altenbach, C., Froncisz, W., Hemker, R., Mchaourab, H., & Hubbell, W. L. (2005). Accessibility of nitroxide side chains: Absolute Heisenberg exchange rates from power saturation EPR. *Biophysical Journal*, *89*, 2103–2112.

Altenbach, C., Froncisz, W., Hyde, J. S., & Hubbell, W. L. (1989). Conformation of spin-labeled melittin at membrane surfaces investigated by pulse saturation recovery and continuous wave power saturation electron paramagnetic resonance. *Biophysical Journal*, *56*, 1183–1191.

Altenbach, C., Greenhalgh, D. A., Khorana, H. G., & Hubbell, W. L. (1994). A collision gradient method to determine the immersion depth of nitroxides in lipid bilayers: Application to spin-labeled mutants of bacteriorhodopsin. *Proceedings of the National Academy of Sciences of the United States of America*, *91*, 1667–1671.

Altenbach, C., Kusnetzow, A. K., Ernst, O. P., Hofmann, K. P., & Hubbell, W. L. (2008). High-resolution distance mapping in rhodopsin reveals the pattern of helix movement due to activation. *Proceedings of the National Academy of Sciences of the United States of America*, *105*, 7439–7444.

Altenbach, C., Marti, T., Khorana, H. G., & Hubbell, W. L. (1990). Transmembrane protein structure: Spin labeling of bacteriorhodopsin mutants. *Science*, *248*, 1088–1092.

Altenbach, C., Oh, K. J., Trabanino, R. J., Hideg, K., & Hubbell, W. L. (2001). Estimation of inter-residue distances in spin labeled proteins at physiological temperatures: Experimental strategies and practical limitations. *Biochemistry*, *40*, 15471–15482.

Banham, J. E., Baker, C. M., Ceola, S., Day, I. J., Grant, G. H., Groenen, E. J. J., et al. (2008). Distance measurements in the borderline region of applicability of CW EPR and DEER: A model study on a homologous series of spin-labelled peptides. *Journal of Magnetic Resonance*, *191*, 202–218.

Barnes, J. P., Liang, Z., Mchaourab, H. S., Freed, J. H., & Hubbell, W. L. (1999). A multifrequency electron spin resonance study of T4 lysozyme dynamics. *Biophysical Journal*, *76*, 3298–3306.

Beier, C., & Steinhoff, H.-J. (2006). A structure-based simulation approach for electron paramagnetic resonance spectra using molecular and stochastic dynamics simulations. *Biophysical Journal*, *91*, 2647–2664.

Berliner, L. J. (Ed.). (1976). *Spin labeling: Theory and applications*. New York: Academic Press.

Berliner, L. J. (Ed.). (1979). *Spin labeling II: Theory and applications*. New York: Academic Press.

Berliner, L. J., & Reuben, J. (Eds.). (1989). *Spin labeling theory and applications, Biological magnetic resonance: Vol. 8*. New York: Plenum Press.

Berliner, L. J., Grunwald, J., Hankovszky, H. O., & Hideg, K. (1982). A novel reversible thiol-specific spin label: Papain active site labeling and inhibition. *Analytical Biochemistry*, *119*, 450–455.

Bhatnagar, J., Borbat, P. P., Pollard, A. M., Bilwes, A. M., Freed, J. H., & Crane, B. R. (2010). Structure of the ternary complex formed by a chemotaxis receptor signaling domain, the CheA histidine kinase, and the coupling protein CheW as determined by pulsed dipolar ESR spectroscopy. *Biochemistry, 49,* 3824–3841.

Borbat, P. P., Costa-Filho, A. J., Earle, K. A., Moscicki, J. K., & Freed, J. H. (2001). Electron spin resonance in studies of membranes and proteins. *Science, 291,* 266–269.

Borbat, P. P., & Freed, J. H. (1999). Multiple-quantum ESR and distance measurements. *Chemical Physics Letters, 313,* 145–154.

Bordignon, E., Grote, M., & Schneider, E. (2011). The maltose ATP-binding cassette transporter in the 21st century—Towards a structural dynamic perspective on its mode of action. *Molecular Microbiology, 77,* 1354–1366.

Bordignon, E., Klare, J. P., Holterhues, J., Martell, S., Krasnaberski, A., Engelhard, M., et al. (2007). Analysis of light-induced conformational changes of natronomonas pharaonis sensory rhodopsin II by time resolved electron paramagnetic resonance spectroscopy. *Photochemistry and Photobiology, 83,* 263–272.

Bordignon, E., & Steinhoff, H.-J. (2007). Membrane protein structure and dynamics studied by site-directed spin labeling ESR. In M. A. Hemminga, & L. J. Berliner (Eds.), *ESR spectroscopy in membrane biophysics* (pp. 129–164). New York, NY: Springer Science and Business Media.

Bowman, A., Hammond, C. M., Stirling, A., Ward, R., Shang, W., El-Mkami, H., et al. (2014). The histone chaperones Vps75 and Nap1 form ring-like, tetrameric structures in solution. *Nucleic Acids Research, 42,* 6038–6051.

Bridges, M. D., Hideg, K., & Hubbell, W. L. (2010). Resolving conformational and rotameric exchange in spin-labeled proteins using saturation recovery EPR. *Applied Magnetic Resonance, 37,* 363–390.

Budil, D. E., Lee, S., Saxena, S., & Freed, J. H. (1996). Nonlinear-least-squares analysis of slow-motion EPR spectra in One and Two dimensions using a modified Levenberg-Marquardt algorithm. *Journal of Magnetic Resonance A, 120,* 155–189.

Budil, D. E., Sale, K. L., Khairy, K., & Fajer, P. G. (2006). Calculating slow-motional electron paramagnetic resonance spectra from molecular dynamics using a diffusion operator approach. *The Journal of Physical Chemistry. A, 110,* 3703–3713.

Claxton, D. P., Quick, M., Shi, L., de Carvalho, F. D., Weinstein, H., Javitch, J. A., et al. (2010). Ion/substrate-dependent conformational dynamics of a bacterial homolog of neurotransmitter: Sodium symporters. *Nature Structural and Molecular Biology, 17,* 822–829.

Columbus, L., & Hubbell, W. L. (2002). A new spin on protein dynamics. *Trends in Biochemical Sciences, 27,* 288–295.

Columbus, L., Kalai, T., Jekö, J., Hideg, K., & Hubbell, W. L. (2001). Molecular motion of spin labeled side chains in α-helices: Analysis by variation of side chain structure. *Biochemistry, 40,* 3828–3846.

Cooper, A. (1976). Thermodynamic fluctuations in protein molecules. *Proceedings of the National Academy of Sciences of the United States of America, 73,* 2740–2741.

Cordero-Morales, J. F., Cuello, L. G., Zhao, Y., Jogini, V., Cortes, D. M., Roux, B., et al. (2006). Molecular determinants of gating at the potassium-channel selectivity filter. *Nature Structural and Molecular Biology, 13,* 311–318.

Cornish, V. W., Benson, D. R., Altenbach, C., Hideg, K., Hubbell, W. L., & Schultz, P. G. (1994). Site-Specific incorporation of biophysical probes into proteins. *Proceedings of the National Academy of Sciences of the United States of America, 91,* 2910–2914.

DeSensi, S. C., Rangel, D. P., Beth, A. H., Lybrand, T. P., & Hustedt, E. J. (2008). Simulation of nitroxide electron paramagnetic resonance spectra from Brownian trajectories and molecular dynamics simulations. *Biophysical Journal, 94,* 3798–3809.

Endeward, B., Butterwick, J. A., MacKinnon, R., & Prisner, T. F. (2009). Pulsed electron-electron double-resonance determination of spin-label distances and orientations on the tetrameric potassium Ion channel KcsA. *Journal of the American Chemical Society, 131*, 15246–15250.

Farahbakhsh, Z. T., Altenbach, C., & Hubbell, W. L. (1992). Spin labeled cysteines as sensors for protein-lipid interaction and conformation in rhodopsin. *Photochemistry and Photobiology, 56*, 1019–1033.

Farahbakhsh, Z. T., Hideg, K., & Hubbell, W. L. (1993). Photoactivated conformational changes in rhodopsin: A time-resolved spin label study. *Science, 262*, 1416–1419.

Farrens, D. L. (2010). What site-directed labeling studies tell us about the mechanism of rhodopsin activation and G-protein binding. *Photochemical and Photobiological Sciences, 9*, 1466–1474.

Farrens, D. L., Altenbach, C., Yang, K., Hubbell, W. L., & Khorana, H. G. (1996). Requirement of rigid-body motion of transmembrane helices for light activation of rhodopsin. *Science, 274*, 768–770.

Feix, J. B., Popp, C. A., Venkataramu, S. D., Beth, A. H., Park, J. H., & Hyde, J. S. (1984). An electron-electron double-resonance study of interactions between [14 N]- and [15 N]stearic acid spin-label pairs: Lateral diffusion and vertical fluctuations in dimyristoylphosphatidylcholine. *Biochemistry, 23*, 2293–2299.

Forrest, L. R., & Rudnick, G. (2009). The rocking bundle: A mechanism for ion-coupled solute flux by symmetrical transporters. *Physiology, 24*, 377–386.

Frauenfelder, H., Parak, F. G., & Young, R. D. (1988). Conformational substates in proteins. *Annual Review in Biophysics and Biophysical Chemistry, 17*, 451–479.

Freed, J. H. (1976). Theory of slow tumbling ESR spectra for nitroxides. In L. J. Berliner (Ed.), *Spin labeling: Theory and applications* (pp. 53–132). New York, NY: Academic Press.

Gross, A., Columbus, L., Hideg, K., Altenbach, C., & Hubbell, W. L. (1999). Structure of the KcsA potassium channel from streptomyces lividans: A site-directed spin labeling study of the second transmembrane segment. *Biochemistry, 38*, 10324–10335.

Grote, M., Bordignon, E., Polyhach, Y., Jeschke, G., Steinhoff, H. J., & Schneider, E. (2008). A comparative EPR study of the nucleotide-binding domains' catalytic cycle in the assembled maltose ABC-importer. *Biophysical Journal, 95*, 2924–2938.

Grote, M., Polyhach, Y., Jeschke, G., Steinhoff, H.-J., Schneider, E., & Bordignon, E. (2009). Transmembrane signaling in the maltose ABC transporter MALFGK2-E: The periplasmic MalF-P2 loop communicates substrate availability to the ATP-bound MalK dimer. *The Journal of Biological Chemistry, 284*, 17521–17526.

Hagelueken, G., Ward, R., Naismith, J., & Schiemann, O. (2012). MtsslWizard: In silico spin-labeling and generation of distance distributions in PyMOL. *Applied Magnetic Resonance, 42*, 377–391.

Hänelt, I., Wunnicke, D., Bordignon, E., Steinhoff, H.-J., & Slotboom, D. J. (2013). Conformational heterogeneity of the aspartate transporter Glt_{Ph}. *Nature Structural and Molecular Biology, 20*, 210–214.

Holterhues, J., Bordignon, E., Klose, D., Rickert, C., Klare, J. P., Martell, S., et al. (2011). The signal transfer from the receptor NpSRII to the transducer NpHtrII is not hampered by the D75N mutation. *Biophysical Journal, 100*, 2275–2282.

Hubbell, W. L., & Altenbach, C. (1994). Investigation of structure and dynamics in membrane proteins using site-directed spin labeling. *Current Opinion in Structural Biology, 4*, 566–573.

Hubbell, W. L., Froncisz, W., & Hyde, J. S. (1987). Continuous and stopped flow electron-paramagnetic-Res spectrometer based on a loop Gap resonator. *The Review of Scientific Instruments, 58*, 1879–1886.

Hubbell, W. L., Gross, A., Langen, R., & Lietzow, M. A. (1998). Recent advances in site-directed spin labeling of proteins. *Current Opinion in Structural Biology, 8*, 649–656.

Hubbell, W. L., Mchaourab, H. S., Altenbach, C., & Lietzow, M. A. (1996). Watching proteins move using site-directed spin labeling. *Structure, 4*, 779–783.

Huber, M., & Törring, J. T. (1995). High-field EPR on the primary electron donor cation radical in single crystals of heterodimer mutant reaction centers of photosynthetic bacteria—First characterization of the G-tensor. *Chemical Physics, 194*, 379–385.

Isas, J. M., Langen, R., Haigler, H. T., & Hubbell, W. L. (2002). Structure and dynamics of a helical hairpin and loop region in annexin 12: A site-directed spin labeling study. *Biochemistry, 41*, 1464–1473.

Jardetzky, O. (1966). Simple allosteric model for membrane pumps. *Nature, 211*, 969–970.

Jeschke, G., Chechik, V., Ionita, P., Godt, A., Zimmermann, H., et al. (2006). DeerAnalysis2006—A comprehensive software package for analyzing pulsed ELDOR data. *Applied Magnetic Resonance, 30*, 473–498.

Jeschke, G., & Polyhach, Y. (2007). Distance measurements on spin-labelled biomacromolecules by pulsed electron paramagnetic resonance. *Physical Chemistry Chemical Physics, 9*, 1895–1910.

Jung, H., Tebbe, S., Schmid, R., & Jung, K. (1998). Unidirectional reconstitution and characterization of purified Na+/proline transporter of *Escherichia coli*. *Biochemistry, 37*, 11083–11088.

Kim, M., Vishnivetskiy, S. A., Van Eps, N., Alexander, N. S., Cleghorn, W. M., et al. (2012). Conformation of receptor-bound visual arrestin. *Proceedings of the National Academy of Sciences of the United States of America, 109*, 18407–18412.

Klare, J. P. (2013). Site-directed spin labeling EPR spectroscopy in protein research. *Biological Chemistry, 394*, 1281–1300.

Klare, J. P., Bordignon, E., Engelhard, M., & Steinhoff, H.-J. (2011). Transmembrane signal transduction in archaeal phototaxis: The sensory rhodopsin II-transducer complex studied by electron paramagnetic resonance spectroscopy. *European Journal of Cell Biology, 90*, 731–739.

Klare, J. P., Gordeliy, V. I., Labahn, J., Büldt, G., Steinhoff, H.-J., & Engelhard, M. (2004). The archaeal sensory rhodopsin II-transducer complex: A model for transmembrane signal transfer. *FEBS Letters, 564*, 219–224.

Klare, J. P., & Steinhoff, H.-J. (2009). Spin labeling EPR. *Photosynthesis Research, 102*, 377–390.

Klare, J. P., & Steinhoff, H.-J. (2013). Structural information from spin-labelled membrane-bound proteins. *Structure and Bonding, 152*, 205–248.

Klose, D., Voskoboynikova, N., Orban-Glaß, I., Rickert, C., Engelhard, M., Klare, J. P., et al. (2014). Light-induced switching of HAMP domain conformation and dynamics revealed by time-resolved EPR spectroscopy. *FEBS Letters, 588*, 3970–3976.

Klug, C. S., & Feix, J. B. (2008). Methods and applications of site-directed spin labeling EPR spectroscopy. In J. J. Correia, & H. W. Detrich (Eds.), *Methods in cell biology. Biophysical tools for biologists, volume One: In vitro techniques* (pp. 617–658). New York, NY: Academic Press.

Kolb, H., Finn, M. G., & Sharpless, K. B. (2001). Click chemistry: Diverse chemical function from a Few good reactions. *Angewandte Chemie International Edition, 40*, 2004–2021.

Langen, R., Oh, K. J., Cascio, D., & Hubbell, W. L. (2000). Crystal structures of spin labeled T4 lysozyme mutants: Implications for the interpretation of EPR spectra in terms of structure. *Biochemistry, 39*, 8396–8405.

Lopez, C. J., Fleissner, M. R., Guo, Z., Kusnetzow, A. N., & Hubbell, W. L. (2009). Osmolyte perturbation reveals conformational equilibria in spin-labeled proteins. *Protein Science, 18*, 1637–1652.

Marsh, D., Dzikovski, B. G., & Livshits, V. A. (2006). Oxygen profiles in membranes. *Biophysical Journal, 90*, L49–L51.

McCoy, J., & Hubbell, W. L. (2011). High-pressure EPR reveals conformational equilibria and volu-metric properties of spin-labeled proteins. *Proceedings of the National Academy of Sciences of the United States of America, 108*, 1331–1336.

Mchaourab, H. S., & Hyde, J. S. (1993). Dependence of the multiple-quantum EPR signal on the spin-lattice relaxation time. Effect of oxygen in spin-labeled membranes. *Journal of Magnetic Resonance, Series B, 101*, 178–184.

Mchaourab, H. S., Lietzow, M. A., Hideg, K., & Hubbell, W. L. (1996). Motion of spin-labeled side chains in T4 lysozyme. Correlation with protein structure and dynamics. *Biochemistry, 35*, 7692–7704.

Nielsen, R. D., Canaan, S., Gladden, J. A., Gelb, M. H., Mailer, C., & Robinson, B. H. (2004). Comparing continuous wave progressive saturation EPR and time domain saturation recovery EPR over the entire motional range of nitroxide spin labels. *Journal of Magnetic Resonance, 169*, 129–163.

Oganesyan, V. S. (2007). A novel approach to the simulation of nitroxide spin label EPR spectra from a single truncated dynamical trajectory. *Journal of Magnetic Resonance, 188*, 196–205.

Olkhova, E., Raba, M., Bracher, S., Hilger, D., & Jung, H. (2011). Homology model of the Na^+/proline transporter PutP of *Escherichia coli* and its functional implications. *Journal of Molecular Biology, 406*, 59–74.

Ottemann, K. M., Thorgeirsson, T. E., Kolodziej, A. F., Shin, Y.-K., & Koshland, D. E., Jr. (1998). Direct measurement of small ligand-induced conformational changes in the aspartate chemoreceptor using EPR. *Biochemistry, 37*, 7062–7069.

Pannier, M., Veit, S., Godt, A., Jeschke, G., & Spiess, H. W. (2000). Dead-time free measurement of dipole-dipole interactions between electron spins. *Journal of Magnetic Resonance, 142*, 331–340.

Park, S. Y., Borbat, P. P., Gonzalez-Bonet, G., Bhatnagar, J., Pollard, A. M., et al. (2006). Reconstruction of the chemotaxis receptor-kinase assembly. *Nature Structural and Molecular Biology, 13*, 400–407.

Pfeiffer, M., Rink, T., Gerwert, K., Oesterhelt, D., & STeinhoff, H.-J. (1999). Site-directed spin-labeling reveals the orientation of the amino acid side-chains in the E-F loop of bacteriorhodopsin. *Journal of Molecular Biology, 187*, 163–171.

Polyhach, Y., Bordignon, E., & Jeschke, G. (2011). Rotamer libraries of spin labelled cysteines for protein studies. *Physical Chemistry Chemical Physics, 13*, 2356–2366.

Polyhach, Y., Bordignon, E., Tschaggelar, R., Gandra, S., Dodt, A., & Jeschke, G. (2012). High sensitivity and versatility of the DEER experiment on nitroxide radical pairs at Q-band frequencies. *Physical Chemistry Chemical Physics, 14*, 10762–10773.

Pyka, J., Ilnicki, J., Altenbach, C., Hubbell, W. L., & Froncisz, W. (2005). Accessibility and dynamics of nitroxide side chains in T4 lysozyme measured by saturation recovery EPR. *Biophysical Journal, 89*, 2059–2068.

Raba, M., Dunkel, S., Hilger, D., Lipiszko, K., Polyhach, Y., Jeschke, G., et al. (2014). Extracellular loop 4 of the proline transporter PutP controls the periplasmic entrance to ligand binding sites. *Structure, 22*, 769–780.

Rabenstein, M. D., & Shin, Y. K. (1995). Determination of the distance between 2 spin labels attached to a macromolecule. *Proceedings of the National Academy of Sciences of the United States of America, 92*, 8239–8243.

Ramil, C. P., & Lin, Q. (2013). Bioorthogonal chemistry: Strategies and recent developments. *Chemical Communications, 49*, 11007–11022.

Rink, T., Riesle, J., Oesterhelt, D., Gerwert, K., & Steinhoff, H.-J. (2000). Spin-labeling studies of the conformational changes in the vicinity of D36, D38, T46, and E161 of bacteriorhodopsin during the photocycle. *Biophysical Journal, 73*, 983–993.

Schiemann, O., & Prisner, T. F. (2007). Long-range distance determinations in biomacromolecules by EPR spectroscopy. *Quarterly Reviews of Biophysics, 40*, 1–53.

Schmidt, M. J., Borbas, J., Drescher, M., & Summerer, D. (2014). A genetically encoded spin label for electron paramagnetic resonance distance measurements. *Journal of the American Chemical Society, 136*, 1238–1241.

Sezer, D., Freed, J. H., & Roux, B. (2008). Simulating electron spin resonance spectra of nitroxide spin labels from molecular dynamics and stochastic trajectories. *The Journal of Chemical Physics, 128*, 165106–165116.

Sippach, M., Weidlich, D., Klose, D., Abe, C., Klare, J. P., Schneider, E., et al. (2014). Conformational changes of the histidine ATP-binding cassette transporter studied by double electron-electron resonance spectroscopy. *Biochimica et Biophysica Acta, 1838*, 1760–1768.

Steinhoff, H.-J., Dombrowsky, O., Karim, C., & Schneiderhahn, C. (1991). Two dimensional diffusion of small molecules on protein surfaces: An EPR study of the restricted translational diffusion of protein-bound spin labels. *European Biophysics Journal, 20*, 293–303.

Steinhoff, H.-J., & Hubbell, W. L. (1996). Calculation of electron paramagnetic resonance spectra from Brownian dynamics trajectories: Application to nitroxide side chains in proteins. *Biophysical Journal, 71*, 2201–2212.

Steinhoff, H.-J., Mollaaghababa, R., Altenbach, C., Hideg, K., Krebs, M. P., Khorana, H. G., et al. (1994). Time-resolved detection of structural changes during the photocycle of spin-labeled bacteriorhodopsin. *Science, 266*, 105–107.

Steinhoff, H.-J., Müller, M., Beier, C., & Pfeiffer, M. (2000). Molecular dynamics simulation and EPR spectroscopy of nitroxide side chains in bacteriorhodopsin. *Journal of Molecular Liquids, 84*, 17–27.

Steinhoff, H.-J., Radzwill, N., Thevis, W., Lenz, V., Brandenburg, D., Antson, A., et al. (1997). Determination of interspin distances between spin labels attached to insulin: Comparison of electron paramagnetic resonance data with the X- ray structure. *Biophysical Journal, 73*, 3287–3298.

Steinhoff, H. J., Savitsky, A., Wegener, C., Pfeiffer, M., Plato, M., & Möbius, K. (2000). High-field EPR studies of the structure and conformational changes of site-directed spin labeled bacteriorhodopsin. *Biochimica et Biophysica Acta, 1457*, 253–262.

Thorgeirsson, T. E., Xiao, W., Brown, L. S., Needleman, R., Lanyi, J. K., & Shin, Y.-K. (1997). Transient channel-opening in bacteriorhodopsin: An EPR study. *Journal of Molecular Biology, 273*, 951–957.

Tyagi, S., & Lemke, E. A. (2013). Chapter 9—Genetically encoded click chemistry for single-molecule FRET of proteins. In P. M. Conn (Ed.), *Laboratory methods in cell biology imaging: Vol. 113* (pp. 169–187). New York, NY: Academic Press.

Urban, L., & Steinhoff, H.-J. (2013). Hydrogen bonding to the nitroxide of protein bound spin labels. *Molecular Physics, 111*, 2873–2881.

Van Eps, N., Anderson, L. L., Kisselev, O. G., Baranski, T. J., Hubbell, W. L., & Marshall, G. R. (2010). Electron paramagnetic resonance studies of functionally active, nitroxide spin-labeled peptide analogues of the C-terminus of a G-protein α subunit. *Biochemistry, 49*, 6877–6886.

Vasquez, V., Sotomayor, M., Cordero-Morales, J., Schulten, K., & Perozo, E. (2008). A structural mechanism for MscS gating in lipid bilayers. *Science, 321*, 1210–1214.

Vasquez, V., Sotomayor, M., Marien-Cortez, D., Roux, B., Schulten, K., & Perozo, E. (2008). Three-dimensional architecture of membrane-embedded MscS in the closed conformation. *Journal of Molecular Biology, 378*, 55–70.

Volkov, A., Dockter, C., Bund, T., Paulsen, H., & Jeschke, G. (2009). Pulsed EPR determination of water accessibility to spin-labeled amino acid residues in LHCIIb. *Biophysical Journal, 96*, 1124–1141.

Vöpel, T., Hengstenberg, C. S., Peulen, T. O., Ajaj, Y., Seidel, C. A. M., Herrmann, C., et al. (2014). Triphosphate induced dimerization of human guanylate binding protein 1 involves association of the C-terminal helices—A joint DEER and FRET study. *Biochemistry, 53,* 4590–4600.

Wegener, A. A., Chizhov, I., Engelhard, M., & Steinhoff, H.-J. (2000). Time-resolved detection of transient movement of helix F in spin-labelled pharaonis sensory rhodopsin II. *Journal of Molecular Biology, 301,* 881–891.

Wegener, A. A., Klare, J. P., Engelhard, M., & Steinhoff, H.-J. (2001). Structural insights into the early steps of receptor-transducer signal transfer in archaeal phototaxis. *The EMBO Journal, 20,* 5312–5319.

Yin, J. J., Pasenkiewicz-Gierula, M., & Hyde, J. S. (1987). Lateral diffusion of lipids in membranes by pulse saturation recovery electron spin resonance. *Proceedings of the National Academy of Sciences of the United States of America, 84,* 964–968.

CHAPTER TWELVE

Navigating Membrane Protein Structure, Dynamics, and Energy Landscapes Using Spin Labeling and EPR Spectroscopy

Derek P. Claxton[1], Kelli Kazmier, Smriti Mishra, Hassane S. Mchaourab[1]

Department of Molecular Physiology and Biophysics, Vanderbilt University School of Medicine, Nashville, Tennessee, USA
[1]Corresponding authors: e-mail address: derek.p.claxton@vanderbilt.edu; hassane.mchaourab@vanderbilt.edu

Contents

1. Introduction	350
2. An EPR Primer	354
2.1 Strategy of Site-Directed Spin Labeling	354
2.2 The EPR Tool Kit	356
2.3 Spin Labels as Molecular Spies of Local Structure	357
3. Principles of DEER Spectroscopy to Uncover Conformational Dynamics	361
3.1 The Distance Distribution	361
3.2 Factors Influencing DEER Data Analysis and Interpretation	363
3.3 From Distance Distributions to Structure	367
3.4 Connecting Structure and Thermodynamics to Elucidate Mechanism	370
4. Practical Considerations in Sample Preparation	372
4.1 Criteria for Selection of Labeling Sites	372
4.2 General Principles for Avoiding Erroneous Interpretation of EPR Data	373
4.3 Labeling and Handling of Purified Membrane Proteins in Detergent	374
4.4 Lipid Environments	377
5. Perspective	379
Acknowledgments	381
References	381

Abstract

A detailed understanding of the functional mechanism of a protein entails the characterization of its energy landscape. Achieving this ambitious goal requires the integration of multiple approaches including determination of high-resolution crystal structures, uncovering conformational sampling under distinct biochemical conditions, characterizing the kinetics and thermodynamics of transitions between functional intermediates using spectroscopic techniques, and interpreting and harmonizing the data into novel computational models. With increasing sophistication in solution-based and ensemble-oriented biophysical approaches such as electron paramagnetic resonance (EPR) spectroscopy, atomic resolution structural information can be directly linked to conformational sampling in solution. Here, we detail how recent methodological and technological advances in EPR spectroscopy have contributed to the elucidation of membrane protein mechanisms. Furthermore, we aim to assist investigators interested in pursuing EPR studies by providing an introduction to the technique, a primer on experimental design, and a description of the practical considerations of the method toward generating high quality data.

1. INTRODUCTION

Structural biology is at the cusp of a fundamental transition in its focus. As the number of structures in the protein data bank (PDB) surpassed 100,000 protein structures, it has shown a steady, but perhaps predictable, decline in the number of novel folds added each year (Levitt, 2007). Within this vast trove of predominantly atomic resolution structures, we may already have access to the overwhelming majority of protein folds sampled by nature (Sillitoe et al., 2015). Structural redundancy has emerged as the rule of protein evolution rather than the exception. Recurrent folds have been found in functionally distinct protein superfamilies suggesting that elements of this conserved architecture are central to the mechanism (Forrest, Kramer, & Ziegler, 2011; ter Beek, Guskov, & Slotboom, 2014). However, it has become increasingly apparent that although proteins adopt similar folds, they can have different conformational dynamics necessitated by the evolution of the underlying mechanisms presumably in adaptation to their specific functional contexts (Bhabha et al., 2013; Faham et al., 2008; Kazmier, Sharma, Islam, Roux, & McHaourab, 2014; Krishnamurthy, Piscitelli, & Gouaux, 2009; Ma et al., 2012; Perez, Koshy, Yildiz, & Ziegler, 2012). We contend that revealing mechanistic commonalities and differences and the underlying interplay between sequence, structure and dynamics will catalyze a transition in focus for structural biology from the collection of static structures to the characterization of energy landscapes. This will necessitate uncovering intermediate protein states and the pathways of the transitions between them by combining atomistic models with spectroscopic, biochemical, thermodynamic, and kinetic studies with a central role for computational biology in guiding experimental investigation and integrating data from diverse techniques.

This vision is particularly pertinent to membrane proteins which are implicated in a spectrum of diseases and represent 50% of pharmaceutical drug targets (Nigam, 2015; Overington, Al-Lazikani, & Hopkins, 2006; Yildirim, Goh, Cusick, Barabasi, & Vidal, 2007). Membrane proteins are often involved in cellular signaling and signal transduction pathways enabling cells to respond to their environments and carry out regulatory processes vital to the physiology of the organism (Lin, Yee, Kim, & Giacomini, 2015). Despite their clinical importance, membrane proteins have posed significantly greater challenges for structural analysis due to their large size, intrinsic dynamic properties, and the inherent complications arising from the necessity of detergent solubilization and formation of stable crystal contacts. While various developments in stabilization and conformational selection have extended the reach of X-ray crystallography into the realm of membrane proteins, such as hyperthermophilic target selection (Wiener, 2004; Yamashita, Singh, Kawate, Jin, & Gouaux, 2005), detergent optimization (Sonoda et al., 2011), antibody chaperones (Griffin & Lawson, 2011; Pardon et al., 2014), and mutagenic thermostabilization (Penmatsa, Wang, & Gouaux, 2013; Serrano-Vega, Magnani, Shibata, & Tate, 2008), the degree to which these modifications alter the energy landscapes of these protein remains largely undetermined.

Thus, membrane proteins pose two distinct challenges in the transition from static structures to mechanism. The first challenge is that membrane protein crystal structures, while they are generally accurate representations of the tertiary fold, are often crystallized under conditions that may obscure the position of these structures in the functional energy landscape (Cross, Sharma, Yi, & Zhou, 2011; Cuello, Cortes, & Perozo, 2004; Freed, Horanyi, Wiener, & Cafiso, 2010; Kazmier, Sharma, Quick, et al., 2014). Therefore, these structures require rigorous validation to assign their mechanistic identities. The second challenge arises from the fact that mechanistic descriptions of membrane proteins, like all dynamic proteins, require an understanding of conformational sampling under biochemical conditions that mimic the *in vivo* context. For example, how are the relative populations of intermediate states altered in response to ligand or drug binding, or energy input? It is only with this comprehensive view that we can move toward an understanding of the energy landscape that underlies the full mechanistic description of function.

Describing energy landscapes requires methodologies to measure thermodynamics (ΔG) and kinetics ($\Delta G^{\ddagger}$) of conformational changes at a resolution adequate to enable comparisons between solution conformations and atomic-scale models. Spectroscopic approaches like nuclear magnetic resonance (NMR), electron paramagnetic resonance (EPR), and fluorescence resonance energy transfer (FRET) are well suited for this task with access to different timescales and amplitudes of structural changes (Fig. 1). Liquid-state NMR describes solution structure and heterogeneity of proteins at high resolution and monitors conformational changes from the local backbone to the domain levels with high sensitivity (Mittermaier & Kay, 2009). Unfortunately, most membrane proteins are not amenable to NMR analysis primarily due to size restrictions. Single-molecule (SM) FRET has the distinct advantage of monitoring the kinetics of conformational change at the level of individual molecules (Akyuz, Altman, Blanchard, & Boudker, 2013; Akyuz et al., 2015; Zhao et al., 2011, 2010), which is key for identifying transition-state free energies and rate-limiting steps. The application of SMFRET requires relatively large probes which limit their placement in the sequence and may compromise the identification of the nature and magnitude of conformational changes. Extracting distances from FRET efficiencies is nontrivial, and therefore data-driven

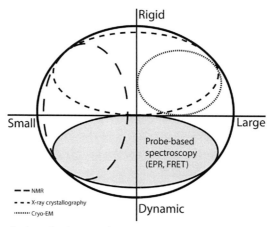

Figure 1 Biophysical methods to study protein structure and dynamics. Whereas X-ray crystallography is the most robust method to determine high-resolution structures of small and large proteins, cryoelectron microscopy is best suited for large proteins and protein complexes. Despite its utility to investigate dynamic properties, current molecular size constraints limit the applicability of liquid-state NMR to <50,000 MW. In contrast, EPR and fluorescence spectroscopies can interrogate dynamic processes regardless of size or complexity. The application of these probe-based methods to proteins of known structure amplifies the interpretation of data toward understanding mechanism.

computational modeling of intermediates using FRET data remains challenging (Brunger, Strop, Vrljic, Chu, & Weninger, 2011).

EPR analysis of spin-labeled proteins enables direct observation of triggered movements in domains and secondary structural elements as well as equilibrium fluctuations of these units arising from the isomerization of the protein between conformations (Fig. 2; McHaourab, Steed, & Kazmier, 2011). Although EPR suffers from low throughput and its structural resolution is moderate, it is no longer unusual to see EPR datasets to describe conformational intermediates, assign mechanistic identity, and

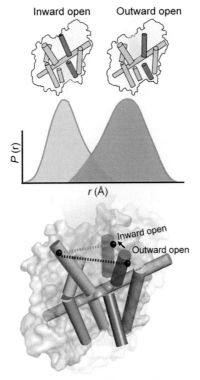

Figure 2 EPR methods report on the ensemble of conformations in solution. In a DEER experiment, each molecule in solution reports a characteristic interprobe distance consistent with its conformation. The distance distribution reports these distances as a function of their frequency within the ensemble. Thus, discrete conformations undergoing equilibrium fluctuations in solution (inward facing and outward facing) at ambient temperatures are represented as distinct distance populations in the DEER distance distribution (yellow and green) in the solid state. Therefore, individual conformations can be described using distance parameters generated from ensemble-based measurements. (See the color plate.)

discover novel intermediate states in conjunction with computational modeling (Cuello et al., 2010; Durr et al., 2014; Freed, Lukasik, Sikora, Mokdad, & Cafiso, 2013; Georgieva, Borbat, Ginter, Freed, & Boudker, 2013; Hanelt, Wunnicke, Bordignon, Steinhoff, & Slotboom, 2013; Jao, Hegde, Chen, Haworth, & Langen, 2008; Kazmier, Sharma, Islam, et al., 2014; Masureel et al., 2014; Wen, Verhalen, Wilkens, McHaourab, & Tajkhorshid, 2013). EPR has a number of advantages that are well suited for the investigation of membrane proteins. In addition to the lack of size limitations on protein targets, experiments can be conducted in a variety of conditions including detergent micelles, proteoliposomes, and nanodiscs, which yield unique insights into conformational dynamics. Furthermore, limited quantities of protein are sufficient for experimentation in EPR due to relatively high signal-to-noise ratios compared to NMR. Importantly, spin labeling at appropriate sites produces limited structural and functional perturbations due to small probe size.

With the emerging recognition of the need for a multifaceted view of protein structure and dynamics, structural biology and biochemistry laboratories are becoming more interested in adding EPR spectroscopy to complement other approaches. The goal of this chapter is to provide information on the practical aspects of conducting EPR measurements and interpreting EPR data from the perspective of a structural biologist. We will also discuss how EPR data can be effectively incorporated with data from other approaches to generate mechanistic descriptions of protein function.

2. AN EPR PRIMER

2.1 Strategy of Site-Directed Spin Labeling

Due to the rarity of stable unpaired electrons in nature, applications of EPR have historically been limited to biological systems that naturally incorporate EPR active transition metals, such as photosynthetic reaction centers (Britt et al., 2004; Calvo, Passeggi, Isaacson, Okamura, & Feher, 1990), organic radicals including biradical and triplet-state molecules (Weil & Bolton, 2007), and oxidation/reduction reactions (Bhattacharjee et al., 2011; Zielonka et al., 2014). For these reasons, development of methodologies that allow site-specific incorporation of spin labels bearing stable unpaired electrons into protein systems was a highly sought advancement in the field. The introduction of sulfhydryl-specific spin probes combined with site-directed cysteine mutagenesis ushered the ability to selectively attach spin labels at essentially any site along the polypetide chain. This was coined as site-directed spin labeling (SDSL; Fig. 3A) (Altenbach, Marti, Khorana, & Hubbell, 1990; Hubbell & Altenbach, 1994).

SDSL requires initial mutagenesis to remove endogenous, labile cysteine residues followed by reintroduction of cysteines only at selected sites of interest typically as single or double mutants. Sites for cysteine replacement are selected to avoid structural perturbation and are typically located on the protein surface at nonconserved residues. Importantly, native residue

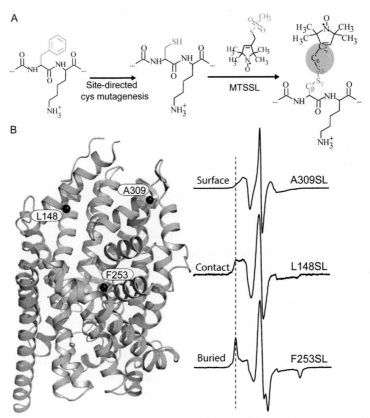

Figure 3 Site-directed spin labeling and correlation of the EPR spectrum with local structure. (A) Targeted cysteine mutagenesis introduces a sulfhydryl moiety for the attachment of a nitroxide spin label, such as MTSSL. Rotational isomerization of MTSSL predominantly around the bonds highlighted in gray is reflected in the EPR spectral lineshape. (B) The degree of rotational freedom of the label is determined by the local packing environment. Fast rotational correlation times ($\sim$1 ns) correspond to spin labels attached to surface-exposed sites. Tertiary contact interactions or buried sites that restrict spin label motion reduce the rate and amplitude of isomerization leading to broadening of the lineshape. The dashed line emphasizes the progressive appearance of a slow-motion component associated with restricted rotation.

substitution and spin labeling have been found to have little effect on protein structural and functional properties at these locations (Mchaourab, Lietzow, Hideg, & Hubbell, 1996). Spin probes are covalently attached to the protein backbone through thiol reactive functional groups including methanethiosulfonate, maleimide, and iodoacetamide moieties (Klare & Steinhoff, 2009).

The most commonly used spin label is the methanethiosulfonate nitroxide label ((1-Oxyl-2,2,5,5-tetramethylpyrroline-3-methyl) methanethiosulfonate, or MTSSL; Fig. 3A) (Berliner, Grunwald, Hankovszky, & Hideg, 1982). In this molecule, the free electron is located in a π-like orbital along the N–O bond. The radical is stable due to steric shielding provided by the proximal set of dimethyl groups of the pyrrole ring. Unlike maleimide and iodoacetamide derivatives, the thiol moiety of MTSSL is highly specific to cysteine modification creating a disulfide linkage that can be easily cleaved with reducing agents for control experiments. Furthermore, MTSSL is theoretically well characterized (Columbus & Hubbell, 2002; Columbus, Kalai, Jeko, Hideg, & Hubbell, 2001; Mchaourab, Kalai, Hideg, & Hubbell, 1999; Mchaourab et al., 1996) and possesses the molecular flexibility to label most sites. Because the protein is modified, it is imperative to establish the structural and functional integrity of mutant proteins through activity assays. From this information, an accurate assessment of the effect of the mutation and labeling can be made, and spin-labeled mutants in which perturbations that are judged to be too severe can be removed from the dataset.

2.2 The EPR Tool Kit

EPR analysis of spin-labeled proteins yields a number of parameters that describe the local environment of the label as well as its distance from a second site specifically introduced paramagnetic center (Fig. 4). The structural interpretation of these parameters has been established in model systems, and a number of reviews present a detailed description of their applications (Hubbell, Cafiso, & Altenbach, 2000; Hubbell, Lopez, Altenbach, & Yang, 2013; Hubbell, McHaourab, Altenbach, & Lietzow, 1996; Klare, 2013; McHaourab et al., 2011). Briefly, the local steric environment of the spin label determines to a large extent its dynamics as reflected in the lineshape of the continuous-wave (cw-)EPR spectrum. Solvent exposure is a function of the topological location of the spin label, whether buried in the hydrophobic core or at the interface of helices, or in direct contact with solvent. Accessibility and mobility are parameters which reflect the local structure. In contrast, distance-dependent dipolar interactions report

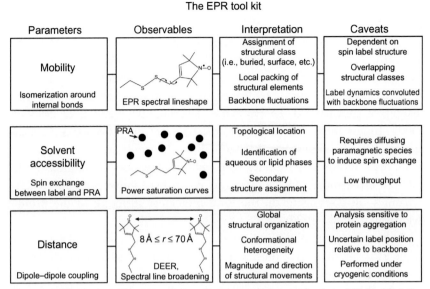

Figure 4 EPR spectroscopy at a glance. Summary of the methods in EPR highlighting structural interpretation and caveats for each method.

a more global perspective enabling the determination of the spatial relationships between secondary structures or domains. The selection of which parameter to measure depends on the questions to be addressed but, more often than not, informative spin labeling analysis requires the integration of all three parameters. Systematic and exhaustive investigations are critical to permit an unequivocal interpretation of the data.

2.3 Spin Labels as Molecular Spies of Local Structure

The EPR lineshape uncovers essential features of the local environment by describing the properties of spin label motion. The rotation of the spin label about internal bonds (Fig. 3) and local dynamic fluctuations of the backbone to which the nitroxide is attached contribute to the overall "mobility" observed in the EPR lineshape (Hubbell et al., 1996). For large macromolecules (>50,000 MW), such as membrane proteins, the overall rate of protein tumbling is too slow ($\tau_c \cong 10^{-8}$ s or slower depending on solution viscosity) to have an effect on the EPR spectrum (10^{-11}–10^{-9} s timescale). The lineshape displays a range of dynamic motion depending on the degree of steric interaction experienced by the spin label due to side and main chain atoms. A detailed motional analysis of MTSSL suggested that intrinsic spin label rotation is largely limited to the C_ε–C_ζ and the C_ε–S_δ bonds (Figs. 3A and 4; Mchaourab et al., 1996). Rotation about the disulfide linkage is restricted by a sufficiently large energy barrier (≥ 7 kcal mol^{-1}) (Fraser, Boussard, Saunders, & Lambert, 1971; Jiao, Barfield, Combariza, & Hruby, 1992). Rotation about the S_γ–C_β is severely hindered by interaction of the S_γ atom with the C_α hydrogen, as was supported by crystallographic analysis of spin-labeled T4 lysozyme (Langen, Oh, Cascio, & Hubbell, 2000). Importantly, the chemical structure of the nitroxide side chain impacts the EPR lineshape (Mchaourab et al., 1999) and provides a way to increase motional sensitivity to backbone dynamics (Columbus et al., 2001). Furthermore, Mchaourab and coworkers showed that spin label dynamics are contingent on the local molecular structure of the protein (Mchaourab et al., 1996).

Although deconvolution of the dynamic modes requires full simulation of the EPR lineshape (Columbus et al., 2001; Columbus & Hubbell, 2002), in most cases, a phenomenological analysis of the EPR spectrum can qualitatively differentiate between spin labels attached to buried, surface or loop sites and those in tertiary interaction (Fig. 3B). Mchaourab et al. (1996) and Columbus et al. (2001) established the use of the inverse line width from the central resonance line as a measure of spin label mobility. A remarkable correlation of this parameter with the structural class of the spin label, i.e., buried, exposed, or in tertiary contact, has been demonstrated. In favorable cases, it is possible to extract a measure of the flexibility of the backbone from analysis of the EPR lineshape. Although these calibration studies were carried out on the water-soluble protein T4 lysozyme, the parameterization of the lineshape analysis can be extended to membrane proteins in general. It was noted however that spin label interactions with local side chains are modified at sites located on membrane-exposed α-helices relative to α-helices on water-soluble proteins (Kroncke, Horanyi, & Columbus, 2010).

Solvent accessibility is a complementary parameter to the spectroscopic signature of local structure reported in the EPR spectrum. Although pulsed-based methods have been developed (Subczynski, Mainali, Camenisch, Froncisz, & Hyde, 2011), spin label solvent accessibility (Π) is commonly determined by monitoring the peak-to-peak EPR central line amplitude as a function of increasing microwave power in the presence and absence of a small, paramagnetic fast-relaxing agent (PRA; Fig. 5) (Altenbach, Froncisz, Hemker, McHaourab, & Hubbell, 2005). These power saturation experiments produce multipoint curves in which their shapes are

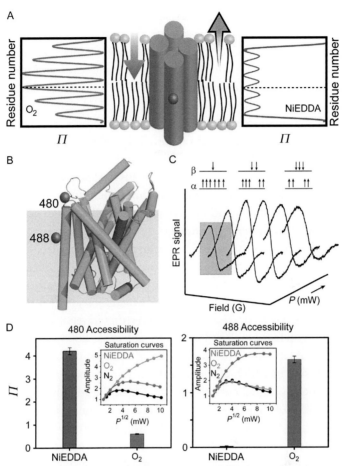

Figure 5 Spin-label solvent accessibility and the correlation with local environment. (A) The differential solubility of fast-relaxing PRAs (NiEDDA and O_2) allows the determination of spin label environment. Nitroxide scanning of an α-helix that is asymmetrically solvated between aqueous and hydrocarbon milieu will report the gradient of oxygen accessibility toward the center of the bilayer in accordance with helix periodicity. The dotted line highlights the site of expected maximum in O_2 accessibility. The NiEDDA accessibility profile, which probes water exposure, is 180° out-of-phase with the O_2 profile. (B) Two spin-labeled sites are shown on a model of LeuT (PDB 2A65), which are used to probe the membrane–water interface. An approximate position for this interface is outlined by an orange box. (C) Power saturation experiments showing the reduction in signal intensity as a function of microwave power. (D) The high NiEDDA accessibility at site 480 in LeuT relative to O_2 and N_2 determined from power saturation curves (inset) suggests a water-exposed position of the spin label. In contrast, the high O_2 accessibility at site 488 indicates that the spin label samples the lipid bilayer. (See the color plate.)

informed by the collisional frequency of the nitroxide with a PRA. Due to slow nitroxide relaxation times ($T_1 \approx 1$ μs), the EPR signal intensity decreases at high powers as a result of equalized spin state populations (saturation) in the absence of a PRA (Fig. 5C). Direct collision of the nitroxide with a PRA shifts the saturation curve to the right by means of enhanced T_1 relaxation rates incurred through Heisenberg spin exchange (Fig. 5D). Curves generated in the absence and presence of PRAs are characterized by a $P_{1/2}$ value, which is the microwave power required to achieve half of the unsaturated EPR signal amplitude and used to determine the dimensionless Π parameter relative to a standard sample. Although limited by the PRA diffusion, this method is advantageous over other accessibility methods such as cysteine alkylation, which is inferred from chemical reactivity profiles that are dependent upon thiol acid dissociation, steric constraints, and the charge or size of the modifying reagent (Zhu & Casey, 2007). Importantly, the spin label accessibility parameter reflects the steady-state exposure as opposed to the reactivity-based methods which could represent rare excursions of protein to the trapped state.

Paramagnetic Ni(II)ethylenediaminediacetic acid (NiEDDA) and molecular oxygen (O_2) possess ideal characteristics to serve as PRAs, including small size and fast T_1 relaxation (Altenbach et al., 2005). Especially valuable for membrane protein studies, these compounds display disparate solubility profiles in which O_2 demonstrates a finite concentration in water, but a concentration gradient into the center of the membrane bilayer. By contrast, NiEDDA is almost exclusively water soluble and diffuses into the membrane–water interface. Differential accessibility to the aqueous and lipid phase defines the membrane–water interface, topological organization, and independently assigns secondary structure (Fig. 5A; Altenbach, Greenhalgh, Khorana, & Hubbell, 1994; Hubbell et al., 1996).

Mobility and accessibility can be used as a readout of conformational changes. However, characterization of the underlying structural arrangements can be problematic owing to the local nature of these parameters. They are most informative when used in conjunction with long-range distance measurements which are more conducive to assessing the nature and amplitude of movements. In our laboratory, these parameters were utilized to outline the ion and substrate-dependent formation and collapse of a water-permeable pathway within the protein interior of the Na^+-coupled leucine transporter, LeuT (Claxton et al., 2010) and to map the ATP-dependent structural transition of the lipid flippase MsbA which drives alternating access to a hydrated central cavity (Dong, Yang, & McHaourab, 2005).

Distance information can be extracted from pairs of spin labels by exploiting the distance dependence of dipolar interactions and utilized to establish the spatial relationships of structural elements within or between proteins in solution. The energy of the dipolar interaction is inversely proportional to the cube of the distance (r^3). When $r < 20$ Å, spin–spin coupling significantly alters the EPR lineshape through broadening of the spectrum. The strength of the interaction can be assessed qualitatively from the degree of line broadening or measured directly by a variety of lineshape analysis techniques to extract distance information (Hustedt et al., 2006; McHaourab, Oh, Fang, & Hubbell, 1997; Rabenstein & Shin, 1995). For $r > 20$ Å, the intrinsic width of the absorption lines in the EPR spectrum (6–8 G) obscures the line broadening due to dipolar interactions (3–5 G). The introduction of pulse methods, particularly double electron–electron resonance (DEER, or PELDOR), allows detection of distances up to 70 Å under favorable conditions (Borbat, McHaourab, & Freed, 2002; Jeschke & Polyhach, 2007).

DEER spectroscopy employs a four-pulse sequence to selectively interrogate the dipolar interaction between spin labels (Fig. 6A; Pannier, Veit, Godt, Jeschke, & Spiess, 2000). In this experiment, the first set of pulses generates a spin echo, which contains the dipolar information. Extraction of this information is achieved by a second pulse at a slightly different frequency which modulates the dipolar interaction and is reported by a change in the spin echo intensity. The second pulse is varied along specified time intervals within a defined data collection window, which leads to oscillations in the intensity of the spin echo decay. The period of the oscillatory frequency (or frequencies) that describes this time-dependent decay reflects the distance between probes.

3. PRINCIPLES OF DEER SPECTROSCOPY TO UNCOVER CONFORMATIONAL DYNAMICS

3.1 The Distance Distribution

Data analysis of the dipolar interaction between spin labels yields a probability distribution, $P(r)$, of distances defined by an average distance (r_{av}) and width or standard deviation (σ) (Fig. 6B). The shape of the distribution, whether unimodal or multimodal, is informed by the collective dynamic processes associated with protein motion ranging from spin label side chain isomerization to backbone fluctuations, as well as the ensemble of distinct protein conformers. Because DEER experiments are carried out in a solid (glassy) phase, dynamic equilibria at ambient temperatures lead to broad

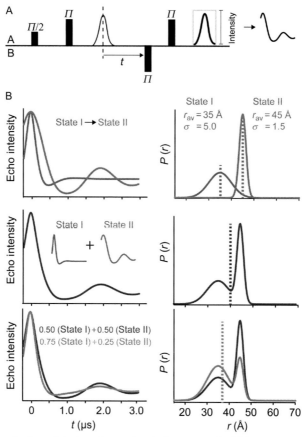

Figure 6 Monitoring global conformational changes with DEER spectroscopy. (A) The four-pulse protocol for DEER spectroscopy is designed to interrogate distance-dependent dipolar interactions between spin A and spin B. The inversion Π pulse on spin B modulates the echo decay of spin A as a function of time t, and the frequency of the resulting oscillation is inversely proportional to the average distance. The decay rate of the spin echo modulation is informed by the distribution of distances in the sample. (B) A simulated conformational change between two states of discrete energies as shown in the spin echo decay (top panels) is manifested by distinct r_{av} and σ in the unimodal distance distribution. The middle panels illustrate an equilibrium between two states, which is the sum of contributions from each conformation. For simplicity, each conformation is equally populated. In the bottom panels, a shift in the equilibrium (induced by ligand binding, for instance) toward an increase in population of the short distance component is visualized in the spin echo decay. The dotted lines in $P(r)$ show the position of r_{av}.

or multicomponent distance distributions. Thus, not only can transitions between distinct states be detected from the change in average distance but also shifts in preexisting conformational equilibria are manifested in the width and shape of the distribution.

As shown in the upper panels of Fig. 6B, $P(r)$ with a specific r_{av} and σ arises from a spin echo decay characterized by the oscillation frequency and rate of signal decay. Transitions between discrete energetic states of a protein, as reported by unimodal distributions, are reflected by changes in r_{av}. In an ensemble undergoing equilibrium transitions between conformers of different energies, the DEER signal reflects a composite of multiple frequencies each with oscillation amplitude related to the relative population of the associated distance component. In the context of unresolved multimodal distributions (Fig. 6B, middle and lower panels), r_{av} is inadequate in the global description of protein movement since it does not correspond to a particular distance component and hence it does not describe a specific conformation of the protein. Instead, the relative populations of the distance components uncover conformational preference in the molecular ensemble under a defined set of conditions. A shift in the ensemble toward another equilibrium conformer under a new set of conditions (i.e., addition of ligands) will lead to a change in r_{av}, but more importantly the change in the relative amplitude of the distance component implies a change in the energetic preference of the protein.

The result of DEER experiments between numerous nitroxide pairs is a web of distances that can be used to evaluate crystal structures, describe structural features of intermediate states or used as constraints into computational structure determination. Variations in r_{av} and σ reveal unique properties of structural states induced by different biochemical conditions, such as changes in absolute distance between conformers or increased or decreased conformational sampling, often rationalizing site-specific changes in mobility and accessibility. This has been demonstrated exceptionally well in MsbA where systematic analysis of distance changes correlated with the pattern of spin label accessibility under conditions that define the power stroke of its transport mechanism (Fig. 7; Borbat et al., 2007; Dong et al., 2005; Zou, Bortolus, & McHaourab, 2009; Zou & McHaourab, 2009).

3.2 Factors Influencing DEER Data Analysis and Interpretation

A number of experimental factors contribute to the generation of a reliable $P(r)$ from DEER measurements. The spin echo dephasing time, T_m, defines

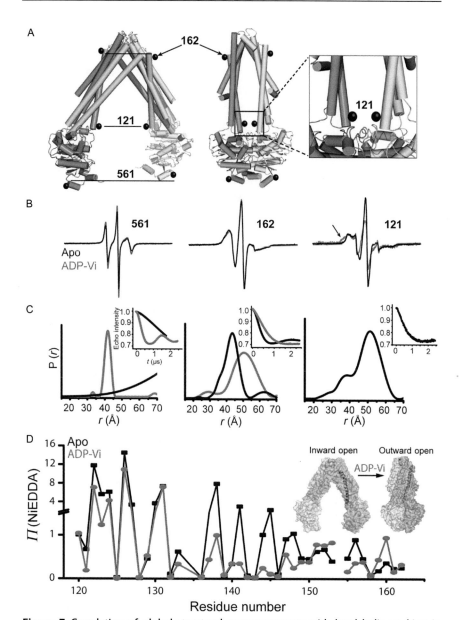

Figure 7 Correlation of global structural rearrangements with local helix packing in MsbA. (A) Model of the MsbA homodimer in the open, Apo (PDB 3B5W) and the closed, AMP-PNP-bound (PDB 3B60) states showing symmetry-related sites for spin label incorporation. Individual monomers are identified by the color scheme. (B) EPR spectra of spin labels at these positions and the corresponding distance distributions (C) in the Apo and ADP-Vi-bound states (trapped posthydrolysis). Labels at 561 and 162 show opposite distance changes between states, consistent with rigid body movement of

the time interval that the dipolar interaction can be observed and sets limits to the detection time of the experiment, which in turn imposes upper boundaries on the distance range and determines confidence levels in the width of the distribution. Notably, the T_m of MTSSL (and other nitroxides) is too short at room temperature to capture the spin echo decay. Since T_m is primarily a function of T_2 relaxation mechanisms, echo coherence can be improved by decreasing the absolute temperature. For MTSSL, a complete characterization of the echo decay requires lowering the temperature substantially, typically between 50 and 80 K. Samples for DEER experiments that are performed under cryogenic conditions are supplemented with a cryoprotectant such as 20–25% (v/v) glycerol. Below 80 K, T_m can also be increased by deuteration of the sample buffer and glycerol which reduces proton-mediated T_2 relaxation (Jeschke & Polyhach, 2007).

The low-temperature acquisition of DEER data potentially hampers the interpretation of resulting distance distributions. Because the short T_m of nitroxides at ambient temperatures is a consequence of rotation of the dimethyl moieties of the ring, alteration of the nitroxide design could conceivably remove this source of spin echo dephasing and increase the accessible temperature range. This elegant approach has recently been shown effective in measuring distances between sites on T4 lysozyme (Meyer et al., 2015; Yang et al., 2012). In one case, exchange of the dimethyl groups for spirocyclohexyl moieties in combination with protein immobilization in a trehalose glass matrix permitted accurate distance measurements of ~30 Å at room temperature (Meyer et al., 2015).

In addition to T_m, the impact of intermolecular dipolar coupling on the DEER signal cannot be overstated. The oscillating spin echo decay that characterizes the DEER spectrum is a product of an intramolecular term arising from dipolar coupling between spin labels within the same protein and an intermolecular term, referred to as the background, which describes the randomized spatial distribution of neighboring proteins. The background component is therefore directly proportional to the spin-labeled protein concentration. During processing, this background must be removed to isolate the intramolecular term. Without explicit

---

helices in an alternating access mechanism. Although separated by ~50 Å in the Apo state (C), spin labels at site 121 are within 20 Å in the ADP-Vi-bound state as indicated by broadening of the EPR lineshape (arrow in B). (D) Formation of a closed conformation on the intracellular side according to distance analysis is consistent with changes in the NiEDDA accessibility profile of transmembrane helix 3 induced by ADP-Vi. (See the color plate.)

determination, the background is generally assumed *a priori* to be a stretched exponential decay that dampens the intramolecular dipolar interaction and effectively reduces the distance range and sensitivity of DEER measurements (Brandon, Beth, & Hustedt, 2012; Jeschke et al., 2007). At sufficiently high spin concentrations (above 200 μM), the background signal can dominate the DEER signal by forming a steep sloping baseline that is difficult to remove. Furthermore, an ill-defined background as a result of a short data collection window will introduce artifacts in the distance distribution, such as the appearance of an artificial long-distance component or alteration of σ. Defining the background often introduces a tradeoff, since longer collection windows result in a decrease in signal-to-noise as a practical consequence of T_m, necessitating more signal averaging. To practically account for these limiting factors, we suggest designing spin label pairs not exceeding 50 Å based on an estimate from the C_α–C_β projection or simulations. This conservative approach recognizes that spin labels can add more than 10 Å to the predicted distance (based on the C_α–C_β projection) due to the length of the nitroxide linker and relative orientation between nitroxide rotamers.

In the absence of a unique spin label orientation relative to the backbone, the inherent flexibility of the probe sets a lower boundary on σ. Thus, interpretation of σ requires a quantitative understanding of rotamer populations. Although still being evaluated, the development and application of rotamer libraries derived from spin-labeled T4 lysozyme crystal structures and molecular dynamics (MD) simulations appears to be a valid approach to predict distance distributions which account for spin label flexibility at specific sites (Polyhach, Bordignon, & Jeschke, 2011). Experimentally, we have found that σ varies as an approximate Gaussian function from 1 to 5 Å for surface-exposed sites (McHaourab et al., 2011). Thus, broad distance distributions indicated by σ values larger than this likely reflect dynamic fluctuations of the protein backbone.

Since the spin echo decay is a convolution of $P(r)$ with the ensemble average of dipolar coupling, isolating $P(r)$ from DEER data is an ill-posed mathematical problem. As a result, a number of approaches have emerged that impose additional constraints in order to obtain a tractable $P(r)$. The DeerAnalysis suite developed by Jeschke employs Tikhonov regularization, which is contingent on the identification of an appropriate regularization parameter defined by the L-curve criterion (Chiang, Borbat, & Freed, 2005; Jeschke et al., 2006; Jeschke, Wegener, Nietschke, Jung, & Steinhoff, 2004). This approach generally assumes *a priori* knowledge of the intermolecular background. Although "model free," the regularization enforces a degree of smoothness onto $P(r)$ resulting in Gaussian-like distributions. Another approach to identify $P(r)$ utilizes a parameterized model to fit the DEER signal and describes all distance components as Gaussian in shape (Fajer, Brown, & Song, 2007). Recently, a method for direct fitting of DEER data without *a priori* background correction has been developed which demonstrates the capacity to analyze multiple datasets simultaneously to identify global changes in $P(r)$ (Brandon et al., 2012; Mishra et al., 2014) and see "A Straightforward Approach to the Analysis of DEER Data" by Stein, Beth, and Hustedt (this volume).

3.3 From Distance Distributions to Structure

The first step in interpretation of $P(r)$ is to link medium-resolution EPR data to the available high-resolution structural models including crystal structures or homology models. To accomplish this task, computationally calculated distributions are generated using MD simulations of the distance between spin labels (Islam, Stein, McHaourab, & Roux, 2013). These simulations can be directly compared to experimentally derived $P(r)$ to assign distance populations to structural models. Overlap between experimental and simulated distributions suggests that the structure or model is sampled in solution. Disagreements that manifest consistently across the dataset can often be used to suggest specific differences between solution intermediates and structural models, though this approach requires caution and rigorous validation. While explicit rendering of spin labels is possible for a small number of mutants (Alexander, Bortolus, Al-Mestarihi, McHaourab, & Meiler, 2008), it becomes computationally expensive at the level of global EPR datasets that commonly include 20–100 mutants. Therefore, programs have been developed that leverage rotamer libraries to approximate spin label flexibility and simplify the underlying calculations. Two of the most successful approaches have been Jeschke's MMM (multiscale modeling of macromolecules) program (Polyhach et al., 2011) and Roux's dummy label approach (Islam et al., 2013; Roux & Islam, 2013). We recently compared DEER data and available crystal structures for LeuT and the Na^+/hydantoin transporter Mhp1 using MMM and the dummy label approach (Kazmier, Sharma, Islam, et al., 2014; Kazmier, Sharma, Quick, et al., 2014). We were able to conclude that, in general, the Mhp1 conformations identified in crystal structures were consistent with those we described in solution by EPR spectroscopy. For LeuT, we concluded that many of the crystal structures were consistent with an outward-facing solution conformation. Furthermore, we proposed and later experimentally verified that a putative inward-facing conformation, crystallized from a heavily mutated construct, was not a major conformation in solution, arguing for a reevaluation of the proposed mechanism (Kazmier, Sharma, Quick, et al., 2014).

Once an accurate model of the protein has been established, relatively few mutants are required to identify the motifs that undergo conformational changes, describe the magnitude and directionality of these transitions, and evaluate established crystallographic or computational mechanistic models. In brief, multicomponent distance distributions are most commonly associated with structural motifs undergoing equilibrium fluctuations which can be shifted by changes in biochemical conditions (e.g., ligand binding). To detect conformational changes, however, transitions must result in distance changes between spin labels. Thus, a unimodal distribution or static equilibrium does not itself exclude the possibility of dynamics, as different conformations can report similar or identical distances. Therefore, identifying dynamic motifs requires triangulation, with at least two distance measurements per site (Fig. 8). To characterize the movement of motifs in detail, the data density should be increased to three or four distances per site, producing quadrangles and pyramids, respectively. The quality of such investigations is judged primarily through internal consistency; mutants that cannot be made to fit the established lattice of triangles also often display deviations from WT in functional assays. Such mutants are then removed from the composite structural analysis but may be important for identifying functionally important residues.

If it becomes clear that available crystal structures need refinement or that conformations sampled in solution (as reflected in the EPR dataset) represent novel states not yet observed, then the data density requirements increase significantly. At this stage, it is advisable to conduct pyramid analysis for each secondary structural element predicted to function as part of a dynamic motif. This can require a substantial commitment of time and resources. However, this resolution is required if intermediate structures are to be described in detail, and accurate models are to be generated in conjunction with computational modeling. The long-term quest to transcend a qualitative description of conformational changes to structural models that capture the spatial information encoded in the EPR data has been recently achieved. Although fold determination from spin labeling is not a central focus of this review, we note that new methods have been recently developed that have demonstrated the feasibility of the approach: RosettaEPR focuses primarily on *de novo* modeling of proteins of unknown structures; whereas an MD method (Hirst, Alexander, McHaourab, & Meiler, 2011), restrained

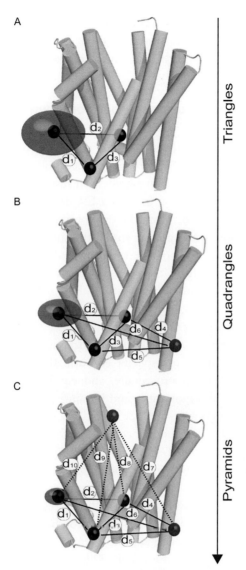

Figure 8 EPR distance measurements and structure elucidation. Analysis of protein structure using EPR distance measurements requires triangulation of spin label positions. (A) Triangles represent the least dense labeling strategy that can identify whether a motif is undergoing conformational reorganization. (B) More dense strategies, like quadrangles, narrow the possible space that spin label can occupy thereby providing a more detailed description of protein structure and more informative restraints for modeling. (C) To effectively identify positions in three dimensions, a pyramid scheme is required.

ensemble molecular dynamics, emphasizes modeling intermediate states starting from high-resolution structures (Islam et al., 2013). These methods are now accessible to users through direct contacts with the Meiler and Roux laboratories, respectively.

3.4 Connecting Structure and Thermodynamics to Elucidate Mechanism

As described previously, in the solid-phase conditions under which DEER data are collected, each molecule will possess a conformation that is reported by the distance between the spin labels. Each distance is represented by a specific frequency of oscillation in the DEER decay and leads to a distinct distance component in $P(r)$. Thus, the distribution represents a snapshot of the energy landscape of the protein under the specific set of experimental conditions. Therefore, it is possible to estimate the change in free energy between sampled states from their relative populations (Georgieva et al., 2013). With optimized sample preparation, these values tend to be consistent across mutants, although small perturbations in these equilibria due to label incorporation are not unexpected.

A mechanistic analysis requires an EPR dataset in which each distance population has been linked to a specific conformation, all major conformations have been described structurally, and equilibrium information has been generated for all important ligand binding conditions. In LeuT, we compared multicomponent distance distributions in apo (ligand-free), Na^+-bound, and Na^+/Leu-bound conditions for spin label pairs on the intracellular and extracellular sides of the transporter (Fig. 9; Kazmier, Sharma, Quick, et al., 2014). In the apo condition, LeuT favored a closed conformation on the extracellular side and an open conformation on the intracellular side. This inward-facing state represented 50–60% of the ensemble under apo conditions. Upon saturation binding of Na^+, the equilibrium shifted to favor the open conformation on the extracellular side and the closed conformation on the inside. This outward-facing intermediate represented a majority (60–70%) of the molecular ensemble in the Na^+-bound state. In contrast, addition of Leu drove an equilibrium shift toward a closed conformation favored on both sides of the transporter (70–90%), which was evidence of a doubly occluded intermediate. Notably, the population ratios in all conditions implied that conformational sampling of the extracellular and intracellular sides is governed by unique equilibrium constants, suggesting a degree of structural independence in the functional cycle. The composite analysis underscores the role of conformational dynamics

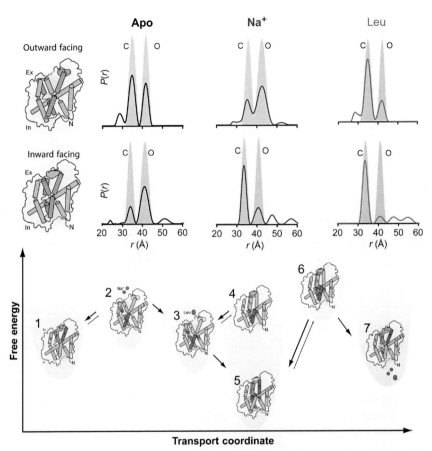

Figure 9 EPR reveals equilibria that can be used to describe energy landscapes and mechanism. LeuT is a Na^+-coupled amino acid transporter. We conducted DEER experiments that monitored the conformational transitions on the extracellular and intracellular sides, shown here for helix 6/intracellular loop 3 (orange). We observed conformational equilibria between inward-facing, outward-facing, and occluded conformations associated with apo (ligand-free, black), Na^+-bound (blue), and Na^+/Leu-bound (red) conditions. These were used along with structural characterization of intermediate states (numerically identified) and a biochemical description of transport to produce a novel description of alternating access in LeuT. (See the color plate.)

evident in distance distributions to identify intermediate states sampled in solution and to assign their relative positions in the energy landscape.

Through linking global EPR data to structural and biochemical descriptions, we generated a description of the mechanistic cycle of LeuT (Fig. 9). In our EPR-derived model, the apo equilibrium favors a previously uncharacterized inward-facing conformer (1), but a minor outward-facing state (2) must be sampled to bind ligand from the extracellular side. Upon Na^+ binding, the outward-facing conformation (3) becomes favored so that the majority of Na^+-bound LeuT is primed to bind the cotransported Leu substrate. The Na^+-bound equilibrium also suggests the presence of a minor inward-facing population (4). Upon binding of Leu, a new equilibrium is adopted between an occluded conformation (5) and rarely sampled inward-facing conformation (6). Upon sampling of this conformation (6), ion and substrate will dissociate due to the presence of cellular electrochemical gradients (7). Thus, we described a mechanistic cycle of alternating access in LeuT that introduced a novel ligand-coupling mechanism and previously unidentified conformational intermediates.

4. PRACTICAL CONSIDERATIONS IN SAMPLE PREPARATION

4.1 Criteria for Selection of Labeling Sites

Although not absolutely required, the availability of crystal structures greatly improves the experimental design and interpretation of EPR data by providing a high-resolution reference to generate mechanistic hypotheses and rationalizes findings concerning protein dynamics, ligand binding, or protein–protein interactions. These structures are initially used as a guide for appropriate site selection of single and double mutants, the goal of which is to avoid sites that would interfere with structural integrity or function. In general, mutation of conserved residues, which are often buried within the protein core or contribute to ligand binding interactions, should be avoided. Due to the relatively low throughput of the approach, it is critical to design experiments that will monitor predicted conformational movements and evaluate specific mechanistic models. Secondary structural elements that form or contribute to ion or substrate binding sites, or those anticipated to participate in pathways relevant to ligand binding are ideal candidates for labeling. Incorporation of biochemical or computational studies in conjunction with crystal structures provides an additional reference for locating appropriate sites for spin labeling (Claxton et al., 2010). If more than one crystal structure is available, identification of domains or helices that undergo the largest displacements between structures are likely to yield the greatest information content for data interpretation. Subsequent incorporation of background mutations that interfere with functional activity (disruption of ion or substrate binding for instance) may uncover novel roles for these structural elements in a functional cycle. If such mutations have been used to stabilize and trap the membrane protein in a specific conformation for subsequent crystallization experiments, elucidating their effects is critical to evaluating the mechanistic relevance of the crystal structures (Kazmier, Sharma, Quick, et al., 2014).

Generating a model of the protein and its conformational changes requires implementation of a spin label network that fingerprints the 3D fold. Initially, experiments are conducted to generate a system of overlapping triangles within a plane parallel to membrane (Fig. 8). This triangulation of spin-labeled pairs defines the relative spatial distribution between nitroxide centers in two dimensions and is often an information-rich approach when dealing with topological restriction associated with membrane-embedded proteins. However, a more sophisticated matrix of distance pyramids between labeled sites is required to outline the 3D fold (Fig. 8). Although pyramids substantially increase the number of DEER measurements, the results will distinguish between many types of conformational movements, including horizontal and vertical translation, rotation, or tilt and decrease the error in the resulting model. The impact of these movements on local packing interactions can be determined using site-specific changes in mobility and accessibility.

In order to obtain optimal information content, spin labeling sites for DEER studies should be reserved to exposed sites in secondary structural elements (Kazmier, Alexander, Meiler, & McHaourab, 2011). Labeling of unstructured loops increases conformational entropy of the spin label leading to featureless time-dependent spin echo decays and broad distance distributions that are not informative for computational modeling and may obscure relevant structural movements. If possible, the mobility of spin-labeled sites reported in the EPR spectrum should be insensitive to the biochemical conditions of the DEER experiment as this information can be used to exclude changes in the rotameric freedom of the label as the origin of distance changes. Changes in the spin label mobility may indicate selection of specific nitroxide rotamer populations that may induce shifts in the distance distributions, therefore confounding the interpretation of conformational changes.

4.2 General Principles for Avoiding Erroneous Interpretation of EPR Data

The most significant and potentially dooming source of experimental error is found in the quality of sample preparation. Isolation of stable and functionally optimized protein devoid of unattached spin label, impurities, degradation products, and aggregation should be viewed as a prerequisite for EPR analysis. Not only does compromised sample integrity impact experimental design and throughput, but it also invariably leads to compounded errors in data analysis and subsequent structural interpretation. The intrinsic spin label sensitivity to the local environment will report altered packing arrangements and tertiary contacts relative to the structure or the model as a consequence of direct or indirect structural perturbations, which then may obfuscate pertinent conformational changes associated with function. Furthermore, the relative thermodynamic relationship between biochemical intermediates can be perturbed even under mildly destabilizing conditions. A number of factors should be considered to control the potential for aggregation induced by compromised membrane protein stability. Site-directed mutagenesis designed to interrogate putative transport mechanisms can generate a reduction in protein expression and/or increase heterogeneity by promoting aggregation. Although classified as a "negative" result, such observations may help to identify previously unknown key residues associated with activity. Separate from other buffer components, detergent selection for extraction from the native membrane environment and subsequent purification should be carefully considered for maintaining global homogeneity and functional activity (Roy, 2015). Harsh detergents, even those that are routinely used in crystallization, should be avoided. In our experience, the mild nonionic dodecyl-β-D-maltoside (DDM) efficiently solubilizes a variety of proteins without altering structural or functional integrity. However, it should be noted that removal of functionally relevant biomolecules through the purification process, including lipids and substrates, can reduce membrane protein stability or compromise function. In such cases, purification in the presence of lipids or substrates may be necessary.

4.3 Labeling and Handling of Purified Membrane Proteins in Detergent

We express proteins of interest fused with an N- or C-terminal polyhistidine tag to facilitate purification by standard methods using metal affinity chromatography with either Ni^{2+} or Co^{2+} resin. The histidine tag is not removed by proteolytic cleavage unless it interferes with analysis. Samples are labeled immediately after elution from the resin with two consecutive rounds of sulfhydryl-reactive MTSSL (dissolved in dimethylformamide at 100–200 mM) for 2 h each round at room temperature with 10- to

20-fold molar excess reagent per labeling site. We note that reducing the temperature to 4 °C may be necessary if cloudiness in the solution appears during the reaction or if protein stability dictates purification at low temperatures. To further drive label attachment, another round of spin label is added and the solution placed on ice overnight. Generally, reactivity can be site and protein specific. Due to the nature of the reaction, labeling efficiency will theoretically increase above pH 7. However, MTSSL forms dimers under these conditions, effectively reducing the pool of reactive species. Since affinity purification is typically performed above pH 7 to increase sample binding to the resin, titration with a weak acid (i.e., 1 M Mes hydrate) can be used to decrease sample pH to neutral for labeling purposes. Any precipitation that forms during overnight incubation is removed by centrifugation, syringe filtration (0.45 µm pore size), or both.

To ensure good quality and consistent sample preparation, we subsequently employ size exclusion chromatography to remove unreacted spin label and assess global structural properties and purity. This step is especially important to reduce contamination from sample aggregation that can confound global structural analysis by introducing artificial distance components (Fig. 10). Determination of labeling efficiency is a necessary means of evaluating the tractability of a sample for EPR analysis and to establish a correlation between labeling and functional activity. Labeling efficiency can be approximated by quantifying the absolute number of spins in the sample through double integration of the cw-EPR spectrum and comparing this value to the concentration of available labeling sites. For

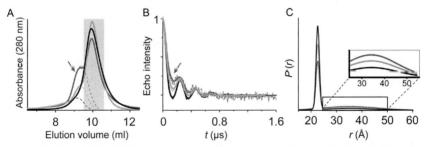

Figure 10 Impact of protein aggregates on the spin echo decay and the resulting distance distribution. (A) Preparative size exclusion chromatography demonstrating different levels of protein aggregation for the same mutant as indicated by the leading shoulder. The traces were normalized by area. The peak fractions pooled for DEER analysis are highlighted by a gray rectangle. Changes in the intermolecular background of the DEER experiment tracked with the presence of aggregated species (B), introducing a long-distance artifact in the distance distributions (C).

singly labeled proteins, the concentration of labeling sites is equal to the protein concentration.

Recently, a systematic analysis of ligand binding in DDM-purified LeuT indicated that binding of radiolabeled substrate was significantly impaired upon an increase in DDM-to-protein ratio with a concomitant loss of phospholipid content (Quick, Shi, Zehnpfennig, Weinstein, & Javitch, 2012). Further binding studies and subsequent computational experiments correlated the increase in DDM concentration with detergent occupation of the second substrate binding site (Khelashvili et al., 2013), an observation consonant with functional impairment imposed through binding of the crystallization detergent octyl-glucoside in a similar location (Quick et al., 2009). At a practical level, these studies emphasize the need for monitoring detergent-induced effects on membrane protein function by optimizing and maintaining an appropriate detergent-to-protein ratio.

If sample limitations require concentrating the protein prior to EPR analysis, centrifugal filters can be used with caution. In such cases, one must choose suitable filtration units when concentrating membrane protein samples by taking account of the molecular weight of the protein and the physicochemical properties of the detergent. For instance, the aggregation number (which depends on buffer composition) and molecular weight of DDM monomers suggests a micelle molecular weight of approximately 50,000 on average. LeuT, which is a 60,000 molecular weight monomer in solution, would be expected to form proteomicelles with DDM that are greater than a molecular weight of 100,000. Indeed, DDM concentration increased when LeuT proteomicelles were subjected to concentration in centrifugal filters with a molecular weight pore size of 50,000 (Quick et al., 2012). In preparation of LeuT in DDM micelles, we conservatively estimate a transporter loss of 20% when using a 100,000 molecular weight cut off.

Additionally, subjecting membrane proteins to concentrating procedures also introduces the possibility of *decreasing* the detergent-to-protein ratio. Overconcentrating samples using ideal centrifugal filters can shift the ratio through loss of empty micelles and detergent monomers in equilibrium with proteomicelles. The loss of detergent promotes formation of proteomicelles with more than one protein per micelle, or nonspecific multimerization. We find that this phenomenon rationalizes the appearance of a broad range of distance components as a result of increased intermolecular dipolar interactions between neighboring proteins in DEER experiments. The population distribution of these components will reflect the relative orientation between individual proteins in the micelle, potentially confounding distance information associated with unique structural intermediates. Although this behavior is expected to be dependent on detergent and protein properties, we suggest that final protein concentrations not exceed 100 μM in samples prepared for DEER analysis to reduce this effect.

4.4 Lipid Environments

We have found that EPR analysis of membrane proteins in nondestabilizing detergents provides an appropriate starting point for assessing conformational dynamics. In general, EPR spectra of membrane-exposed sites demonstrate similar lineshapes in detergent and lipid environments (Claxton et al., 2010; Dong et al., 2005). Additionally, local and global changes in structure have been shown to be comparable (Claxton et al., 2010). However, reconstitution into more native-like lipid environments adds a new dimension to investigate the role of specific lipids on conformational equilibria and to map changes in the packing and orientation of structural elements within a membrane (Fig. 11). Although liposomes are the

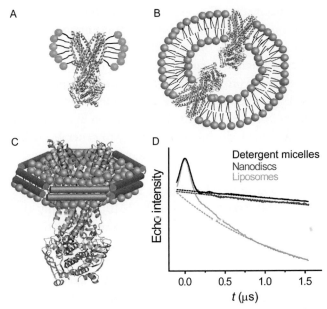

Figure 11 Detergent and lipid environments and the consequence on the DEER signal background. Solvation of membrane protein in (A) detergent micelles, (B) liposomes, and (C) nanodiscs. Liposome reconstitution often introduces more than one protein copy per liposome. As a result, the higher effective spin concentration increases the contribution of the background decay in the DEER signal relative to proteomicelles and nanodiscs (D). (See the color plate.)

traditional platform for such experiments, reconstitution into nanodiscs is a viable and promising alternative. Nanodiscs are discoidal phospholipid bilayers in which the size and water solubility are determined by an annulus of amphipathic membrane scaffold protein (MSP) from an engineered fragment of apolipoprotein A-1 (Bayburt & Sligar, 2010). Like liposomes, the phospholipid content can be controlled to reflect the conditions for optimal activity of the protein.

The use of nanodiscs for DEER sample preparation and subsequent analysis using Q-band (34 GHz) pulse spectrometers yielded an order of magnitude greater sensitivity despite reduced sample requirements. The origin of this striking result lies at the convergence of the intrinsic increase in signal-to-noise achieved with higher microwave frequencies relative to X-band (9.5 GHz) and the reduced contribution of background intermolecular dipolar coupling in the nanodisc environment (Ghimire, McCarrick, Budil, & Lorigan, 2009; Zou & McHaourab, 2010). In proteoliposomes, multiple proteins can incorporate into a single liposome, which compresses the bulk spatial distribution of spins from three dimensions to approximately two (Fig. 11; Hilger et al., 2005). This change in protein distribution increases the effective spin concentration leading to steep intermolecular backgrounds that are difficult to remove during processing and reduce the measurable distance range. In contrast, a one-to-one incorporation of membrane protein into nanodics can be achieved, maintaining the bulk spatial distribution in three dimensions. Given this benefit, we encourage the use of nanodiscs for EPR studies in lipids. The primary disadvantage of nanodiscs is the inherent inability to generate gradients that are often necessary for functional studies of transporters.

Our detailed protocols for reconstitution into either liposomes (Claxton et al., 2010; Zou et al., 2009; Zou & McHaourab, 2009) or nanodiscs (Mishra et al., 2014; Zou & Mchaourab, 2010) are readily available elsewhere, and we make additional suggestions here. For both environments, efficient reconstitution requires careful manipulation of component molar ratios. Furthermore, optimization of nanodisc reconstitution is dependent on the target protein and the use of the appropriate size MSP. In our hands, using the MSP1E3D1 construct (Bayburt & Sligar, 2010), the appropriate molar ratios for nanodisc reconstitution appear to correlate with molecular weight of the membrane protein. We have found the following molar ratios to be good initial conditions for smaller proteins (50–60,000 MW, e.g., LeuT): protein-to-MSP, 1:5; MSP-to-lipid, 1:60; lipid-to-DDM, 1:5. Further decreasing the MSP-to-lipid ratio to ~1:120 was necessary for larger proteins (over 100,000 MW), such as MsbA and BmrCD. Following reconstitution, performing size exclusion chromatography on an HPLC equipped with a dual UV/fluorescence detector will help to resolve the membrane protein–nanodisc complex from unincorporated target protein, lipids, and MSP, which tend to run in the void volume of a Superdex200 10/300 column. Subsequent SDS-PAGE will confirm the presence of the transporter and scaffold protein. Stoichiometry between MSP and membrane protein can be corroborated by combining densitometry measurements from SDS-PAGE for the scaffold protein with the absolute spin concentration from the protein.

5. PERSPECTIVE

Following the seminal work of Hubbell and coworkers on Rhodopsin, the application of spin labeling and EPR spectroscopy has expanded to include virtually every class of membrane protein. While an exhaustive account is beyond the scope of this chapter, we note some of the major contributions and the unique insight that emerged from them. SDSL studies of Rhodopsin preceded the determination of its crystal structures, revealing local dynamics and structures in transmembrane helices (Hubbell, Altenbach, Hubbell, & Khorana, 2003; Palczewski et al., 2000; Salom et al., 2006). Importantly, distance measurements between spin labels were instrumental in uncovering the movement of helix 6 upon photoexcitation (Altenbach, Kusnetzow, Ernst, Hofmann, & Hubbell, 2008), which proved to be a conserved functional feature in GPCRs (Choe et al., 2011).

One of the most extensive spin labeling studies targeted MsbA combining nitroxide scanning of five of its transmembrane segments and the determination of at least 60 distances in the context of known but mechanistically suspect crystal structures (Borbat et al., 2007; Dong et al., 2005; Zou et al., 2009; Zou & McHaourab, 2009). The dataset revealed that ATP energy input changes the hydration of a transmembrane chamber, evident in changes in local accessibility, and dynamics of spin labels, concomitant with large amplitude distance changes between the two halves of the transporter (Fig. 7). These dramatic conformational changes were interpreted as reflecting the ATP-coupled transition from inward-facing to outward-facing states. The degree to which this blueprint of conformational changes is conserved across ABC transporters is being explored in other members of the family (Mishra et al., 2014).

Similar questions regarding mechanistic commonalities and divergence that focus on the patterns of ion- and substrate-dependent conformational changes have been posed in ion-coupled transporters of the LeuT-fold class. Here, crystal structures have left the question of the structural mechanics and the ion dependence of the inward- to outward-facing transition clouded with ambiguity (Krishnamurthy & Gouaux, 2012; Shimamura et al., 2010; Weyand et al., 2008; Yamashita et al., 2005). Spin labeling studies of LeuT and Mhp1, two members of this fold class, have illuminated aspects of the mechanistic divergence between these transporters (Kazmier, Sharma, Islam, et al., 2014). Finally, spin labeling investigation of an ion-coupled multidrug transporter from the major facilitator superfamily is a recent example of extensive DEER analysis in the context of a homology model (Masureel et al., 2014). Despite the lack of a high-resolution structures, the authors were able to identify a protonation sensitive switch critical for the antiporter mechanism of this transporter.

A large body of work from the Perozo lab illustrates the application of spin labeling and EPR spectroscopy to ion channels. Extensive analysis of accessibility, mobility, and short-range distances in conjunction with modeling enabled insight into the conformational changes associated with gating of multiple classes of ion channels (Cordero-Morales et al., 2007; Cortes, Cuello, & Perozo, 2001; Perozo, Cortes, & Cuello, 1999; Perozo, Cortes, Sompornpisut, Kloda, & Martinac, 2002; Vasquez, Sotomayor, Cordero-Morales, Schulten, & Perozo, 2008). A methodologically novel contribution was the combination of EPR spectroscopy and crystallography to define aspects of channel function such as inactivation (Cuello et al., 2010).

One remarkable example of the application of spin labeling is the studies of Cafiso and coworkers on the *E. coli* vitamin B_{12} outer membrane transporter BtuB. In a collection of experiments, EPR spectroscopy was used to show that a short, yet conserved N-terminal segment (the Ton box) that couples BtuB to the inner membrane protein TonB undergoes a conformational equilibrium between folded and unfolded states (Fanucci, Coggshall, et al., 2003; Xu, Ellena, Kim, & Cafiso, 2006). Importantly, crystal lattice interactions and osmolytes in the crystallization milieu were found to inhibit this structural transition, shifting the free energy difference between the unfolded and folded forms of the Ton box by ~ 3 kcal mol^{-1} (Fanucci, Lee, & Cafiso, 2003; Freed et al., 2010).

While these examples and others unequivocally highlight the contribution of spin labeling to understanding the mechanisms of all classes of membrane proteins, the next frontier is the integration of EPR parameters with other techniques. There are early examples of the combined use of crystallography cryoEM and EPR to characterize the conformational changes in ligand-gated ion channels (Durr et al., 2014). In combination with computational tools, such multipronged investigations hold the promise for elucidation of energy landscapes.

ACKNOWLEDGMENTS
The authors wish to thank Dr. Richard A Stein and other members of the Mchaourab laboratory for helpful comments during the preparation of this chapter. Research in the Mchaourab laboratory is funded by NIH grants U54-GM087519 and R01-GM077659.

REFERENCES
Akyuz, N., Altman, R. B., Blanchard, S. C., & Boudker, O. (2013). Transport dynamics in a glutamate transporter homologue. *Nature, 502*(7469), 114–118.

Akyuz, N., Georgieva, E. R., Zhou, Z., Stolzenberg, S., Cuendet, M. A., Khelashvili, G., et al. (2015). Transport domain unlocking sets the uptake rate of an aspartate transporter. *Nature, 518*(7537), 68–73.

Alexander, N., Bortolus, M., Al-Mestarihi, A., McHaourab, H., & Meiler, J. (2008). De novo high-resolution protein structure determination from sparse spin-labeling EPR data. *Structure, 16*(2), 181–195.

Altenbach, C., Froncisz, W., Hemker, R., McHaourab, H., & Hubbell, W. L. (2005). Accessibility of nitroxide side chains: Absolute Heisenberg exchange rates from power saturation EPR. *Biophysical Journal, 89*(3), 2103–2112.

Altenbach, C., Greenhalgh, D. A., Khorana, H. G., & Hubbell, W. L. (1994). A collision gradient method to determine the immersion depth of nitroxides in lipid bilayers: Application to spin-labeled mutants of bacteriorhodopsin. *Proceedings of the National Academy of Sciences of the United States of America, 91*(5), 1667–1671.

Altenbach, C., Kusnetzow, A. K., Ernst, O. P., Hofmann, K. P., & Hubbell, W. L. (2008). High-resolution distance mapping in rhodopsin reveals the pattern of helix movement due to activation. *Proceedings of the National Academy of Sciences of the United States of America, 105*(21), 7439–7444.

Altenbach, C., Marti, T., Khorana, H. G., & Hubbell, W. L. (1990). Transmembrane protein structure: Spin labeling of bacteriorhodopsin mutants. *Science, 248*(4959), 1088–1092.

Bayburt, T. H., & Sligar, S. G. (2010). Membrane protein assembly into Nanodiscs. *FEBS Letters, 584*(9), 1721–1727.

Berliner, L. J., Grunwald, J., Hankovszky, H. O., & Hideg, K. (1982). A novel reversible thiol-specific spin label: Papain active site labeling and inhibition. *Analytical Biochemistry, 119*(2), 450–455.

Bhabha, G., Ekiert, D. C., Jennewein, M., Zmasek, C. M., Tuttle, L. M., Kroon, G., et al. (2013). Divergent evolution of protein conformational dynamics in dihydrofolate reductase. *Nature Structural & Molecular Biology, 20*(11), 1243–1249.

Bhattacharjee, S., Deterding, L. J., Chatterjee, S., Jiang, J., Ehrenshaft, M., Lardinois, O., et al. (2011). Site-specific radical formation in DNA induced by Cu(II)-H(2)O(2) oxidizing system, using ESR, immuno-spin trapping, LC-MS, and MS/MS. *Free Radical Biology & Medicine, 50*(11), 1536–1545.

Borbat, P. P., McHaourab, H. S., & Freed, J. H. (2002). Protein structure determination using long-distance constraints from double-quantum coherence ESR: Study of T4 lysozyme. *Journal of the American Chemical Society, 124*(19), 5304–5314.

Borbat, P. P., Surendhran, K., Bortolus, M., Zou, P., Freed, J. H., & McHaourab, H. S. (2007). Conformational motion of the ABC transporter MsbA induced by ATP hydrolysis. *PLoS Biology, 5*(10). e271.

Brandon, S., Beth, A. H., & Hustedt, E. J. (2012). The global analysis of DEER data. *Journal of Magnetic Resonance, 218*, 93–104.

Britt, R. D., Campbell, K. A., Peloquin, J. M., Gilchrist, M. L., Aznar, C. P., Dicus, M. M., et al. (2004). Recent pulsed EPR studies of the photosystem II oxygen-evolving complex: Implications as to water oxidation mechanisms. *Biochimica et Biophysica Acta, 1655*(1–3), 158–171.

Brunger, A. T., Strop, P., Vrljic, M., Chu, S., & Weninger, K. R. (2011). Three-dimensional molecular modeling with single molecule FRET. *Journal of Structural Biology, 173*(3), 497–505.

Calvo, R., Passeggi, M. C., Isaacson, R. A., Okamura, M. Y., & Feher, G. (1990). Electron paramagnetic resonance investigation of photosynthetic reaction centers from *Rhodobacter sphaeroides* R-26 in which $Fe2+$ was replaced by $Cu2+$. Determination of hyperfine interactions and exchange and dipole-dipole interactions between $Cu2+$ and QA. *Biophysical Journal, 58*(1), 149–165.

Chiang, Y. W., Borbat, P. P., & Freed, J. H. (2005). The determination of pair distance distributions by pulsed ESR using Tikhonov regularization. *Journal of Magnetic Resonance, 172*(2), 279–295.

Choe, H. W., Kim, Y. J., Park, J. H., Morizumi, T., Pai, E. F., Krauss, N., et al. (2011). Crystal structure of metarhodopsin II. *Nature, 471*(7340), 651–655.

Claxton, D. P., Quick, M., Shi, L., de Carvalho, F. D., Weinstein, H., Javitch, J. A., et al. (2010). Ion/substrate-dependent conformational dynamics of a bacterial homolog of neurotransmitter:sodium symporters. *Nature Structural & Molecular Biology, 17*(7), 822–829.

Columbus, L., & Hubbell, W. L. (2002). A new spin on protein dynamics. *Trends in Biochemical Sciences, 27*(6), 288–295.

Columbus, L., Kalai, T., Jeko, J., Hideg, K., & Hubbell, W. L. (2001). Molecular motion of spin labeled side chains in alpha-helices: Analysis by variation of side chain structure. *Biochemistry, 40*(13), 3828–3846.

Cordero-Morales, J. F., Jogini, V., Lewis, A., Vasquez, V., Cortes, D. M., Roux, B., et al. (2007). Molecular driving forces determining potassium channel slow inactivation. *Nature Structural & Molecular Biology, 14*(11), 1062–1069.

Cortes, D. M., Cuello, L. G., & Perozo, E. (2001). Molecular architecture of full-length KcsA: Role of cytoplasmic domains in ion permeation and activation gating. *The Journal of General Physiology, 117*(2), 165–180.

Cross, T. A., Sharma, M., Yi, M., & Zhou, H. X. (2011). Influence of solubilizing environments on membrane protein structures. *Trends in Biochemical Sciences, 36*(2), 117–125.

Cuello, L. G., Cortes, D. M., & Perozo, E. (2004). Molecular architecture of the KvAP voltage-dependent $K+$ channel in a lipid bilayer. *Science, 306*(5695), 491–495.

Cuello, L. G., Jogini, V., Cortes, D. M., Pan, A. C., Gagnon, D. G., Dalmas, O., et al. (2010). Structural basis for the coupling between activation and inactivation gates in $K(+)$ channels. *Nature, 466*(7303), 272–275.

Dong, J., Yang, G., & McHaourab, H. S. (2005). Structural basis of energy transduction in the transport cycle of MsbA. *Science, 308*(5724), 1023–1028.

Durr, K. L., Chen, L., Stein, R. A., De Zorzi, R., Folea, I. M., Walz, T., et al. (2014). Structure and dynamics of AMPA receptor GluA2 in resting, pre-open, and desensitized states. *Cell, 158*(4), 778–792.

Faham, S., Watanabe, A., Besserer, G. M., Cascio, D., Specht, A., Hirayama, B. A., et al. (2008). The crystal structure of a sodium galactose transporter reveals mechanistic insights into $Na+$/sugar symport. *Science, 321*(5890), 810–814.

Fajer, P. G., Brown, L., & Song, L. (2007). Practical pulsed dipolar ESR (DEER). In *Vol. 27. Biological magnetic resonance: ESR spectroscopy in membrane biophysics* (pp. 95–128). New York: Kluwer Academic/Plenum Publishers.

Fanucci, G. E., Coggshall, K. A., Cadieux, N., Kim, M., Kadner, R. J., & Cafiso, D. S. (2003). Substrate-induced conformational changes of the periplasmic N-terminus of an outer-membrane transporter by site-directed spin labeling. *Biochemistry, 42*(6), 1391–1400.

Fanucci, G. E., Lee, J. Y., & Cafiso, D. S. (2003). Spectroscopic evidence that osmolytes used in crystallization buffers inhibit a conformation change in a membrane protein. *Biochemistry, 42*(45), 13106–13112.

Forrest, L. R., Kramer, R., & Ziegler, C. (2011). The structural basis of secondary active transport mechanisms. *Biochimica et Biophysica Acta, 1807*(2), 167–188.

Fraser, R. R., Boussard, G., Saunders, J. K., & Lambert, J. B. (1971). Barriers to rotation about the sulfur-sulfur bond in acyclic disulfides. *Journal of the American Chemical Society, 93*(15), 3822–3823.

Freed, D. M., Horanyi, P. S., Wiener, M. C., & Cafiso, D. S. (2010). Conformational exchange in a membrane transport protein is altered in protein crystals. *Biophysical Journal, 99*(5), 1604–1610.

Freed, D. M., Lukasik, S. M., Sikora, A., Mokdad, A., & Cafiso, D. S. (2013). Monomeric TonB and the Ton box are required for the formation of a high-affinity transporter-TonB complex. *Biochemistry, 52*(15), 2638–2648.

Georgieva, E. R., Borbat, P. P., Ginter, C., Freed, J. H., & Boudker, O. (2013). Conformational ensemble of the sodium-coupled aspartate transporter. *Nature Structural & Molecular Biology, 20*(2), 215–221.

Ghimire, H., McCarrick, R. M., Budil, D. E., & Lorigan, G. A. (2009). Significantly improved sensitivity of Q-band PELDOR/DEER experiments relative to X-band is observed in measuring the intercoil distance of a leucine zipper motif peptide (GCN4-LZ). *Biochemistry, 48*(25), 5782–5784.

Griffin, L., & Lawson, A. (2011). Antibody fragments as tools in crystallography. *Clinical and Experimental Immunology, 165*(3), 285–291.

Hanelt, I., Wunnicke, D., Bordignon, E., Steinhoff, H. J., & Slotboom, D. J. (2013). Conformational heterogeneity of the aspartate transporter Glt(Ph). *Nature Structural & Molecular Biology, 20*(2), 210–214.

Hilger, D., Jung, H., Padan, E., Wegener, C., Vogel, K. P., Steinhoff, H. J., et al. (2005). Assessing oligomerization of membrane proteins by four-pulse DEER: pH-dependent dimerization of NhaA Na+/H+ antiporter of *E. coli*. *Biophysical Journal, 89*(2), 1328–1338.

Hirst, S. J., Alexander, N., McHaourab, H. S., & Meiler, J. (2011). RosettaEPR: An integrated tool for protein structure determination from sparse EPR data. *Journal of Structural Biology, 173*(3), 506–514.

Hubbell, W. L., & Altenbach, C. (1994). Investigation of structure and dynamics in membrane proteins using site-directed spin labeling. *Current Opinion in Structural Biology, 4*(4), 566–573.

Hubbell, W. L., Altenbach, C., Hubbell, C. M., & Khorana, H. G. (2003). Rhodopsin structure, dynamics, and activation: A perspective from crystallography, site-directed spin labeling, sulfhydryl reactivity, and disulfide cross-linking. *Advances in Protein Chemistry, 63*, 243–290.

Hubbell, W. L., Cafiso, D. S., & Altenbach, C. (2000). Identifying conformational changes with site-directed spin labeling. *Nature Structural Biology, 7*(9), 735–739.

Hubbell, W. L., Lopez, C. J., Altenbach, C., & Yang, Z. (2013). Technological advances in site-directed spin labeling of proteins. *Current Opinion in Structural Biology, 23*(5), 725–733.

Hubbell, W. L., McHaourab, H. S., Altenbach, C., & Lietzow, M. A. (1996). Watching proteins move using site-directed spin labeling. *Structure, 4*(7), 779–783.

Hustedt, E. J., Stein, R. A., Sethaphong, L., Brandon, S., Zhou, Z., & Desensi, S. C. (2006). Dipolar coupling between nitroxide spin labels: The development and application of a tether-in-a-cone model. *Biophysical Journal, 90*(1), 340–356.

Islam, S. M., Stein, R. A., McHaourab, H. S., & Roux, B. (2013). Structural refinement from restrained-ensemble simulations based on EPR/DEER data: Application to T4 lysozyme. *The Journal of Physical Chemistry. B, 117*(17), 4740–4754.

Jao, C. C., Hegde, B. G., Chen, J., Haworth, I. S., & Langen, R. (2008). Structure of membrane-bound alpha-synuclein from site-directed spin labeling and computational refinement. *Proceedings of the National Academy of Sciences of the United States of America, 105*(50), 19666–19671.

Jeschke, G., Chechik, V., Ionita, P., Godt, A., Zimmermann, H., Banham, J., et al. (2006). DeerAnalysis2006—A comprehensive software package for analyzing pulsed ELDOR data. *Applied Magnetic Resonance, 30*, 473–498.

Jeschke, G., & Polyhach, Y. (2007). Distance measurements on spin-labelled biomacromolecules by pulsed electron paramagnetic resonance. *Physical Chemistry Chemical Physics, 9*(16), 1895–1910.

Jeschke, G., Wegener, C., Nietschke, M., Jung, H., & Steinhoff, H. J. (2004). Interresidual distance determination by four-pulse double electron–electron resonance in an integral membrane protein: The Na+/proline transporter PutP of *Escherichia coli*. *Biophysical Journal, 86*(4), 2551–2557.

Jiao, D., Barfield, M., Combariza, J. E., & Hruby, V. J. (1992). Ab initio molecular orbital studies of the rotational barriers and the sulfur-33 and carbon-13 chemical shieldings for dimethyl disulfide. *Journal of the American Chemical Society, 114*(10), 3639–3643.

Kazmier, K., Alexander, N. S., Meiler, J., & McHaourab, H. S. (2011). Algorithm for selection of optimized EPR distance restraints for de novo protein structure determination. *Journal of Structural Biology, 173*(3), 549–557.

Kazmier, K., Sharma, S., Islam, S. M., Roux, B., & McHaourab, H. S. (2014). Conformational cycle and ion-coupling mechanism of the Na+/hydantoin transporter Mhp1. *Proceedings of the National Academy of Sciences of the United States of America, 111*(41), 14752–14757.

Kazmier, K., Sharma, S., Quick, M., Islam, S. M., Roux, B., Weinstein, H., et al. (2014). Conformational dynamics of ligand-dependent alternating access in LeuT. *Nature Structural & Molecular Biology, 21*(5), 472–479.

Khelashvili, G., LeVine, M. V., Shi, L., Quick, M., Javitch, J. A., & Weinstein, H. (2013). The membrane protein LeuT in micellar systems: Aggregation dynamics and detergent binding to the S2 site. *Journal of the American Chemical Society, 135*(38), 14266–14275.

Klare, J. P. (2013). Site-directed spin labeling EPR spectroscopy in protein research. *Biological Chemistry, 394*(10), 1281–1300.

Klare, J. P., & Steinhoff, H. J. (2009). Spin labeling EPR. *Photosynthesis Research, 102*(2–3), 377–390.

Krishnamurthy, H., & Gouaux, E. (2012). X-ray structures of LeuT in substrate-free outward-open and apo inward-open states. *Nature, 481*(7382), 469–474.

Krishnamurthy, H., Piscitelli, C. L., & Gouaux, E. (2009). Unlocking the molecular secrets of sodium-coupled transporters. *Nature, 459*(7245), 347–355.

Kroncke, B. M., Horanyi, P. S., & Columbus, L. (2010). Structural origins of nitroxide side chain dynamics on membrane protein alpha-helical sites. *Biochemistry, 49*(47), 10045–10060.

Langen, R., Oh, K. J., Cascio, D., & Hubbell, W. L. (2000). Crystal structures of spin labeled T4 lysozyme mutants: Implications for the interpretation of EPR spectra in terms of structure. *Biochemistry, 39*(29), 8396–8405.

Levitt, M. (2007). Growth of novel protein structural data. *Proceedings of the National Academy of Sciences of the United States of America, 104*(9), 3183–3188.

Lin, L., Yee, S. W., Kim, R. B., & Giacomini, K. M. (2015). SLC transporters as therapeutic targets: Emerging opportunities. *Nature Reviews. Drug Discovery, 14*(8), 543–560.

Ma, D., Lu, P., Yan, C., Fan, C., Yin, P., Wang, J., et al. (2012). Structure and mechanism of a glutamate-GABA antiporter. *Nature, 483*(7391), 632–636.

Masureel, M., Martens, C., Stein, R. A., Mishra, S., Ruysschaert, J. M., McHaourab, H. S., et al. (2014). Protonation drives the conformational switch in the multidrug transporter LmrP. *Nature Chemical Biology, 10*(2), 149–155.

Mchaourab, H. S., Kalai, T., Hideg, K., & Hubbell, W. L. (1999). Motion of spin-labeled side chains in T4 lysozyme: Effect of side chain structure. *Biochemistry, 38*(10), 2947–2955.

Mchaourab, H. S., Lietzow, M. A., Hideg, K., & Hubbell, W. L. (1996). Motion of spin-labeled side chains in T4 lysozyme. Correlation with protein structure and dynamics. *Biochemistry, 35*(24), 7692–7704.

McHaourab, H. S., Oh, K. J., Fang, C. J., & Hubbell, W. L. (1997). Conformation of T4 lysozyme in solution. Hinge-bending motion and the substrate-induced conformational transition studied by site-directed spin labeling. *Biochemistry, 36*(2), 307–316.

McHaourab, H. S., Steed, P. R., & Kazmier, K. (2011). Toward the fourth dimension of membrane protein structure: Insight into dynamics from spin-labeling EPR spectroscopy. *Structure, 19*(11), 1549–1561.

Meyer, V., Swanson, M. A., Clouston, L. J., Boratynski, P. J., Stein, R. A., McHaourab, H. S., et al. (2015). Room-temperature distance measurements of immobilized spin-labeled protein by DEER/PELDOR. *Biophysical Journal, 108*(5), 1213–1219.

Mishra, S., Verhalen, B., Stein, R. A., Wen, P. C., Tajkhorshid, E., & McHaourab, H. S. (2014). Conformational dynamics of the nucleotide binding domains and the power stroke of a heterodimeric ABC transporter. *eLife, 3*. e02740.

Mittermaier, A. K., & Kay, L. E. (2009). Observing biological dynamics at atomic resolution using NMR. *Trends in Biochemical Sciences, 34*(12), 601–611.

Nigam, S. K. (2015). What do drug transporters really do? *Nature Reviews. Drug Discovery, 14*(1), 29–44.

Overington, J. P., Al-Lazikani, B., & Hopkins, A. L. (2006). How many drug targets are there? *Nature Reviews. Drug Discovery, 5*(12), 993–996.

Palczewski, K., Kumasaka, T., Hori, T., Behnke, C. A., Motoshima, H., Fox, B. A., et al. (2000). Crystal structure of rhodopsin: A G protein-coupled receptor. *Science, 289*(5480), 739–745.

Pannier, M., Veit, S., Godt, A., Jeschke, G., & Spiess, H. W. (2000). Dead-time free measurement of dipole-dipole interactions between electron spins. *Journal of Magnetic Resonance, 142*(2), 331–340.

Pardon, E., Laeremans, T., Triest, S., Rasmussen, S. G., Wohlkonig, A., Ruf, A., et al. (2014). A general protocol for the generation of nanobodies for structural biology. *Nature Protocols, 9*(3), 674–693.

Penmatsa, A., Wang, K. H., & Gouaux, E. (2013). X-ray structure of dopamine transporter elucidates antidepressant mechanism. *Nature, 503*(7474), 85–90.

Perez, C., Koshy, C., Yildiz, O., & Ziegler, C. (2012). Alternating-access mechanism in conformationally asymmetric trimers of the betaine transporter BetP. *Nature, 490*(7418), 126–130.

Perozo, E., Cortes, D. M., & Cuello, L. G. (1999). Structural rearrangements underlying K^+-channel activation gating. *Science, 285*(5424), 73–78.

Perozo, E., Cortes, D. M., Sompornpisut, P., Kloda, A., & Martinac, B. (2002). Open channel structure of MscL and the gating mechanism of mechanosensitive channels. *Nature, 418*(6901), 942–948.

Polyhach, Y., Bordignon, E., & Jeschke, G. (2011). Rotamer libraries of spin labelled cysteines for protein studies. *Physical Chemistry Chemical Physics, 13*(6), 2356–2366.
Quick, M., Shi, L., Zehnpfennig, B., Weinstein, H., & Javitch, J. A. (2012). Experimental conditions can obscure the second high-affinity site in LeuT. *Nature Structural & Molecular Biology, 19*(2), 207–211.
Quick, M., Winther, A. M., Shi, L., Nissen, P., Weinstein, H., & Javitch, J. A. (2009). Binding of an octylglucoside detergent molecule in the second substrate (S2) site of LeuT establishes an inhibitor-bound conformation. *Proceedings of the National Academy of Sciences of the United States of America, 106*(14), 5563–5568.
Rabenstein, M. D., & Shin, Y. K. (1995). Determination of the distance between two spin labels attached to a macromolecule. *Proceedings of the National Academy of Sciences of the United States of America, 92*(18), 8239–8243.
Roux, B., & Islam, S. M. (2013). Restrained-ensemble molecular dynamics simulations based on distance histograms from double electron–electron resonance spectroscopy. *The Journal of Physical Chemistry. B, 117*(17), 4733–4739.
Roy, A. (2015). Membrane preparation and solubilization. *Methods in Enzymology, 557*, 45–56.
Salom, D., Lodowski, D. T., Stenkamp, R. E., Le Trong, I., Golczak, M., Jastrzebska, B., et al. (2006). Crystal structure of a photoactivated deprotonated intermediate of rhodopsin. *Proceedings of the National Academy of Sciences of the United States of America, 103*(44), 16123–16128.
Serrano-Vega, M. J., Magnani, F., Shibata, Y., & Tate, C. G. (2008). Conformational thermostabilization of the beta1-adrenergic receptor in a detergent-resistant form. *Proceedings of the National Academy of Sciences of the United States of America, 105*(3), 877–882.
Shimamura, T., Weyand, S., Beckstein, O., Rutherford, N. G., Hadden, J. M., Sharples, D., et al. (2010). Molecular basis of alternating access membrane transport by the sodium-hydantoin transporter Mhp1. *Science, 328*(5977), 470–473.
Sillitoe, I., Lewis, T. E., Cuff, A., Das, S., Ashford, P., Dawson, N. L., et al. (2015). CATH: Comprehensive structural and functional annotations for genome sequences. *Nucleic Acids Research, 43*(Database issue), D376–D381.
Sonoda, Y., Newstead, S., Hu, N. J., Alguel, Y., Nji, E., Beis, K., et al. (2011). Benchmarking membrane protein detergent stability for improving throughput of high-resolution X-ray structures. *Structure, 19*(1), 17–25.
Subczynski, W. K., Mainali, L., Camenisch, T. G., Froncisz, W., & Hyde, J. S. (2011). Spin-label oximetry at Q- and W-band. *Journal of Magnetic Resonance, 209*(2), 142–148.
ter Beek, J., Guskov, A., & Slotboom, D. J. (2014). Structural diversity of ABC transporters. *The Journal of General Physiology, 143*(4), 419–435.
Vasquez, V., Sotomayor, M., Cordero-Morales, J., Schulten, K., & Perozo, E. (2008). A structural mechanism for MscS gating in lipid bilayers. *Science, 321*(5893), 1210–1214.
Weil, J. A., & Bolton, J. R. (2007). *Electron paramagnetic resonance: Elementary theory and practical applications* (2nd ed.). Hoboken, NJ: Wiley-Interscience.
Wen, P. C., Verhalen, B., Wilkens, S., McHaourab, H. S., & Tajkhorshid, E. (2013). On the origin of large flexibility of P-glycoprotein in the inward-facing state. *The Journal of Biological Chemistry, 288*(26), 19211–19220.
Weyand, S., Shimamura, T., Yajima, S., Suzuki, S., Mirza, O., Krusong, K., et al. (2008). Structure and molecular mechanism of a nucleobase-cation-symport-1 family transporter. *Science, 322*(5902), 709–713.
Wiener, M. C. (2004). A pedestrian guide to membrane protein crystallization. *Methods, 34*(3), 364–372.
Xu, Q., Ellena, J. F., Kim, M., & Cafiso, D. S. (2006). Substrate-dependent unfolding of the energy coupling motif of a membrane transport protein determined by double electron–electron resonance. *Biochemistry, 45*(36), 10847–10854.

Yamashita, A., Singh, S. K., Kawate, T., Jin, Y., & Gouaux, E. (2005). Crystal structure of a bacterial homologue of Na^+/Cl^--dependent neurotransmitter transporters. *Nature*, *437*(7056), 215–223.

Yang, Z., Liu, Y., Borbat, P., Zweier, J. L., Freed, J. H., & Hubbell, W. L. (2012). Pulsed ESR dipolar spectroscopy for distance measurements in immobilized spin labeled proteins in liquid solution. *Journal of the American Chemical Society*, *134*(24), 9950–9952.

Yildirim, M. A., Goh, K. I., Cusick, M. E., Barabasi, A. L., & Vidal, M. (2007). Drug-target network. *Nature Biotechnology*, *25*(10), 1119–1126.

Zhao, Y., Terry, D. S., Shi, L., Quick, M., Weinstein, H., Blanchard, S. C., et al. (2011). Substrate-modulated gating dynamics in a Na^+-coupled neurotransmitter transporter homologue. *Nature*, *474*(7349), 109–113.

Zhao, Y., Terry, D., Shi, L., Weinstein, H., Blanchard, S. C., & Javitch, J. A. (2010). Single-molecule dynamics of gating in a neurotransmitter transporter homologue. *Nature*, *465*(7295), 188–193.

Zhu, Q., & Casey, J. R. (2007). Topology of transmembrane proteins by scanning cysteine accessibility mutagenesis methodology. *Methods*, *41*(4), 439–450.

Zielonka, J., Cheng, G., Zielonka, M., Ganesh, T., Sun, A., Joseph, J., et al. (2014). High-throughput assays for superoxide and hydrogen peroxide: Design of a screening workflow to identify inhibitors of NADPH oxidases. *The Journal of Biological Chemistry*, *289*(23), 16176–16189.

Zou, P., Bortolus, M., & McHaourab, H. S. (2009). Conformational cycle of the ABC transporter MsbA in liposomes: Detailed analysis using double electron–electron resonance spectroscopy. *Journal of Molecular Biology*, *393*(3), 586–597.

Zou, P., & McHaourab, H. S. (2009). Alternating access of the putative substrate-binding chamber in the ABC transporter MsbA. *Journal of Molecular Biology*, *393*(3), 574–585.

Zou, P., & McHaourab, H. S. (2010). Increased sensitivity and extended range of distance measurements in spin-labeled membrane proteins: Q-band double electron–electron resonance and nanoscale bilayers. *Biophysical Journal*, *98*(6), L18–L20.

CHAPTER THIRTEEN

Spin Labeling of Potassium Channels

Dylan Burdette, Adrian Gross[1]

Department of Biochemistry and Molecular Biology, Rosalind Franklin University of Medicine and Science, North Chicago, Illinois, USA
[1]Corresponding author: e-mail address: adrian.gross@rosalindfranklin.edu

Contents

1. Introduction	390
2. General Considerations for Spin Labeling Studies of Potassium Channels	391
2.1 Selection of Target DNA	391
2.2 Toxicity to the Host Organism	391
2.3 Flask Versus Fermentor Growth	392
2.4 Labeling Reactions and Reagents	393
2.5 Quality Control of the Labeling Reaction	394
2.6 Reconstitution	395
3. Examples of Labeling Methods	395
3.1 Labeling of the Prokaryotic KcsA Potassium Channel	395
3.2 Labeling of the Prokaryotic KvAP Potassium Channel	397
References	399

Abstract

Potassium channels are the ion channels most extensively studied by structural techniques. Whereas high-resolution crystal structures have provided key insights into the molecular architecture of these channels, spin labeling studies have helped to unveil the dynamic structural aspects underlying their function. From a practical standpoint, the popularity of spin labeling studies of potassium channels lies in their small size and relative ease of overexpression. The inherent fourfold symmetry of most potassium channels has also greatly facilitated spin labeling studies. This chapter focuses on the overexpression, purification, spin labeling, and subsequent reconstitution of modified potassium channels. It will discuss the general methods used to produce a suitable spin-labeled potassium channel sample and highlight some of the common pitfalls that can occur along the way. At the end of the chapter, we provide detailed methods to produce spin-labeled samples of KcsA and KvAP, the two most commonly studied potassium channels.

1. INTRODUCTION

The technique of site-directed spin labeling has been widely applied to ion channels. The large majority of these applications have targeted potassium channels (Chakrapani, Cuello, Cortes, & Perozo, 2008; Cieslak, Focia, & Gross, 2010; Cortes, Cuello, & Perozo, 2001; Cortes & Perozo, 1997; Cuello, Cortes, & Perozo, 2004, Cuello, Romero, Cortes, & Perozo, 1998; Gross, Columbus, Hideg, Altenbach, & Hubbell, 1999; Gross & Hubbell, 2002; Perozo, Cortes, & Cuello, 1998, 1999, Perozo, Kloda, Cortes, & Martinac, 2001; Vamvouka, Cieslak, Van Eps, Hubbell, & Gross, 2008; Vasquez et al., 2008). Potassium channels are similarly overrepresented in databases of solved crystal structures. This is due to the relative ease with which potassium channels can be obtained in the required quantities and purity. In addition, the relatively short protein sequence of potassium channels compared to sodium and calcium channels allows for easier manipulation at the DNA level: fewer endogenous cysteine residues to mutate and fewer mutants to investigate. The small size of potassium channels is a direct consequence of their inherent fourfold (or in some cases, twofold) symmetry. This symmetry not only provides additional advantages for electron paramagnetic resonance (EPR) studies but also leads to a few disadvantages. For instance, EPR experiments are greatly simplified in that only a single site needs to be mutated and labeled at any time. Symmetry then accounts for the presence of multiple (usually four) spin labels per protein. This is useful for distance determinations. In addition, the signal intensity of the sample is four times higher than it would be for an asymmetric target under comparable conditions. Symmetry also imposes strong and useful structural constraints for the interpretation of distance measurement data (Vamvouka et al., 2008). On the other hand, distance measurements are made more complicated by the presence of two distinct distances arising from a single mutant (or even eight distances from a double mutant). In addition, symmetry can lead to undesirable perturbations at sites near the symmetry axis, as the presence of symmetry mates can perturb the immediate environment of the nitroxide. Overall, the advantages of small size and symmetry clearly outweigh the disadvantages for most EPR studies, as is evident from numerous published studies.

This chapter focuses on the overexpression, purification, spin labeling, and subsequent reconstitution of potassium channels. Experimental details about EPR techniques and applications follow those of other membrane proteins and are described in this volume and elsewhere (Columbus & Hubbell, 2002; Fanucci & Cafiso, 2006; Hubbell, Cafiso, & Altenbach, 2000, Hubbell, Gross, Langen, & Lietzow, 1998; Hubbell, Mchaourab, Altenbach, & Lietzow, 1996; Mchaourab, Steed, & Kazmier, 2011; Oh, Altenbach, Collier, & Hubbell, 2000; Perozo, Cuello, Cortes, Liu, & Sompornpisut, 2002).

2. GENERAL CONSIDERATIONS FOR SPIN LABELING STUDIES OF POTASSIUM CHANNELS

2.1 Selection of Target DNA

The first step in any EPR study is the construction of an appropriate DNA sequence for the expression vector. Some early studies on potassium channels utilized sequences present in native organisms (Ruta, Jiang, Lee, Chen, & MacKinnon, 2003; Schrempf et al., 1995). Toward this end, genomic DNA is either obtained directly from appropriate sources or produced by culturing the chosen organism under suitable conditions. The target DNA is then amplified by PCR and subcloned into an overexpression vector. Alternatively, the target DNA can be obtained through direct DNA synthesis techniques. The latter approach has the advantage that the DNA sequence can be easily optimized for high-level expression in *Escherichia coli* or yeast and fitted with useful additional flanking sequences to either enhance expression levels and/or facilitate protein detection and purification. The progressively lower cost of DNA synthesis has made this approach the current method of choice. Once the appropriate DNA sequence has been introduced into the host organism, the protein can be overexpressed and purified.

2.2 Toxicity to the Host Organism

Potassium channels flux potassium ions across cell membranes and are therefore intimately involved in the setting of the membrane potential of the host organism. KcsA, the first potassium channel studied with EPR techniques, has the substantial practical advantage that the open probability of the channel under physiological conditions is essentially zero (Cuello et al., 1998; Schrempf et al., 1995). This means that potentially deleterious ion fluxes across the membrane of the host organism are largely avoided. As a result, plasmid maintenance and overexpression procedures are straightforward and high protein yields (>10 mg/l culture) can typically be obtained. For other potassium channels of interest, however, an intrinsic basic open probability can lead to ionic fluxes of consequence, changes of membrane potential, and hence toxic effects on the host organism. Overexpression of the voltage-dependent channel KvAP in *E. coli* for instance often leads to substantial cell death during the production phase. A useful method to prevent these toxic effects is the addition of barium chloride to the growth medium at the time of induction (Ruta et al., 2003). Barium competes with potassium for binding at ion binding sites at the selectivity filter of the channel. The high binding affinity and low ion permeability of barium efficiently blocks channel function during protein production (Neyton & Miller, 1988a,1988b). However, the presence of barium in the resulting cell pellet interferes with purification methods based on poly-histidine binding to metal chelate affinity matrices. To avoid this problem, the added barium must be carefully removed from the cell suspension prior to any contact with these affinity matrices. This can be achieved by simple washing procedures or, more effectively, by taking the cells through an intermediate wash buffer containing EDTA. Since EDTA not only chelates barium but may also strip metal ions (usually nickel or cobalt) off affinity matrices, this chelator must also be carefully removed before metal affinity purification.

Toxicity to the host organism also leads to challenges in vector stability as adaptations resulting in lower protein expression and hence less toxicity are advantageous to the transformed organism and can accumulate temporally. As a result, growth and expression procedures may lose their effectiveness over time and may ultimately lead to low and insufficient protein expression levels. In these circumstances, close attention needs to be paid to minimizing protein expression during the DNA production steps as well as optimizing the growth procedures prior to protein induction. Retransforming the overexpression vector is usually effective in restoring protein yields to their previously observed levels.

2.3 Flask Versus Fermentor Growth

Growing host organisms in flasks is simple, cheap, and fast. Nitroxide scanning studies in particular require relatively small amounts of many different protein variants. It is advantageous to ensure, however, that the wild-type or cysteine-free base variant produces substantially more protein than required since nitroxide scanning invariably leads to mutants that express at noticeably lower level than the base construct. When protein material is required for pulsed EPR experiments or if yields in flasks are too low, overexpression in a small (~2 l) fermentation vessel is advisable. The fermentor allows for tighter control of growth and induction conditions, such as temperature, pH, oxygen tension, and nutrient levels, which can lead to substantially higher cell densities and protein yields.

2.4 Labeling Reactions and Reagents

The most common labeling reaction for site-directed spin labeling studies utilizes cysteine residues on the target and a thiol-specific labeling reagent, typically a methanethiosulfonate derivative. The nitroxide reagent MTSL (Fig. 1A) has established itself as the reagent of choice for most EPR studies. A cross-linking reagent (Fig. 1B) has also been used successfully (Fleissner et al., 2011; Li, Wanderling, Sompornpisut, & Perozo, 2014). The homobifunctional reagent substantially reduces the accessible volume of the unpaired electron to the singly attached MTSL reagent. This facilitates the determination of protein conformational changes through distance measurements using pulsed EPR methods. However, since the labeling reaction of this particular reagent seems to be more easily reversible, it is important to ensure that both disulfides have formed and that this labeling state remains

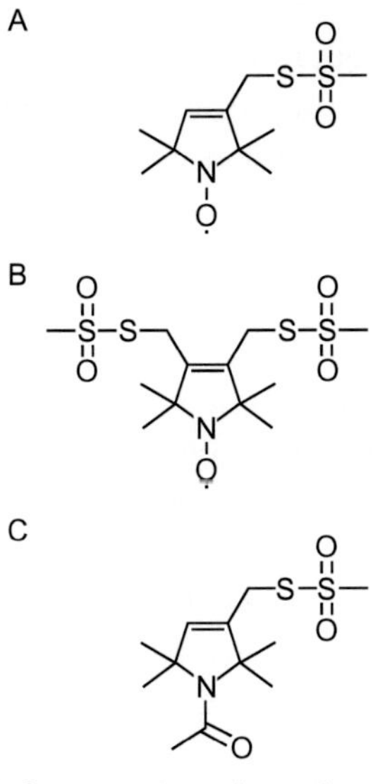

Figure 1 Common spin labels for potassium channel studies. (A) MTSL: ([1-oxy-2,2,5,5 tetramethyl-Δ3-pyrroline-methyl] methanethiosulfonate, (B) homobifunctional MTSL: Trans-3,4-Bis[[(methylsulfonyl)thio]methyl]-2,2,5,5-tetramethylpyrrolidin-1-yloxyl, and (C) isosteric diamagnetic analog of MTSL: (1-acetyl-2,2,5,5 tetramethyl-Δ3-pyrroline-methyl) methanethiosulfonate.

predominant at the time of the experiment. The most straightforward technique to evaluate the state of labeling is mass spectroscopy (see below).

Figure 1C depicts a diamagnetic reagent that is nearly isosteric to paramagnetic MTSL (Gross et al., 1999). This reagent is very effective for spin dilution experiments in which the protein sample is reacted with a mixture of paramagnetic and diamagnetic reagents. The identical labeling chemistry and similar overall size lead to a fully labeled sample with a reduced number of nitroxides per channel. Assuming random and equal reactivity of the two reagents, the resulting channel populations follow a binomial distribution. While spin dilution experiments can also be conducted by underlabeling the sample, i.e., by reacting a known quantity of spin label with an excess of protein, the isosteric reagent has the advantage that the available protein sample is optimally used and that the environment surrounding the nitroxide remains approximately isosteric.

2.5 Quality Control of the Labeling Reaction

Not all sites on potassium channels can be spin-labeled in a quantitative manner (Gross et al., 1999; Kelly & Gross, 2003). Some sites are simply inaccessible to the spin label. In addition, symmetry leads to crowding effects and steric hindrance at sites near the symmetry axis. These sites do not permit complete labeling. Some spin label products are unstable and the nitroxide moiety has a tendency to dissociate over time. The resulting non-stoichiometric labeling is of particular concern for distance measurements and appropriate quality control may be required (Gross et al., 1999).

2.5.1 Quantification of Spin Labeling

The extent of labeling can be measured by counting the number of spins and the number of protein molecules in a given sample. Spins are counted by integration of the EPR spectrum with respect to a reference standard. This can be achieved by integrating the spectrum derived from the labeled protein or by releasing the nitroxide from the protein by breaking the disulfide bond without reducing the nitroxide moiety. The accuracy of the first method varies with the spectral properties at a given site. The accuracy is higher at mobile sites with sharp spectral features and lower at immobile sites with broad features. The second method is advantageous in that all spectra derive from the same free nitroxide species (Oh et al., 2006). In addition, the resulting sharp spectral features increase the sensitivity of the method. The number of protein molecules can be determined using a variety of well-established spectroscopic analytical assays to determine protein concentrations (e.g., Bradford or BCA protein assays). Alternatively, the number of proteins can be determined by amino acid analysis (Gross et al., 1999).

2.5.2 Mass Spectrometry

The extent of labeling is directly observable from a mass spectrum by integrating the respective peaks for the expected masses of the labeled and unlabeled species. Both electrospray and MALDI ionization techniques have been successfully used on potassium channels (Doyle et al., 1998; Kelly & Gross, 2003; le Coutre et al., 2000). The underlying assumption is that the probability of ionization of the channel is not measurably altered by the addition of a spin label.

2.6 Reconstitution

Although potassium channels can be studied by EPR techniques in detergent micelles, it is usually desirable to study them embedded in a lipid membrane to more closely mimic native conditions. Toward this end, detergent-solubilized protein is mixed with detergent-solubilized lipids and the detergent is subsequently removed. Detergent removal can be achieved by diluting the detergent via size exclusion chromatography using the method of Green and Belll (1984) or by binding to hydrophobic beads (e.g., Bio-Beads SM; Biorad). Both of these techniques lead to multilamellar liposomes with a wide size distribution. If unilamellar liposomes of uniform size are required, liposomes are first fused by freeze-thawing, followed by extrusion through a filter. Suitable polycarbonate filters typically have a pore size of 100 nm.

3. EXAMPLES OF LABELING METHODS

3.1 Labeling of the Prokaryotic KcsA Potassium Channel

3.1.1 Procedure

A synthetic gene encoding KcsA is cloned into the pQE60 (Qiagen) vector in-frame with a carboxyterminal hexa-histidine tag and the vector is then transformed into *E. coli* using standard protocols. The cells are grown in flasks in Luria broth at 37 °C. At an OD_{600} of 0.8–1.2, protein expression is induced by the addition of 1 mM IPTG. The cells are cultured for 2 h at 37 °C or overnight at room temperature. The cells are pelleted by centrifugation, washed in lysis buffer, and pelleted again. At this stage, the cell pellet can be frozen at −80 °C for later use. The cells are resuspended in lysis buffer and disrupted by either sonication or French pressing. Suitable protease inhibitors (leupeptin, aprotinin, and phenylmethanesulfonylfluoride) to prevent degradation are added before, during, and after cell disruption. At this stage, KcsA can be solubilized either from a cell membrane pellet (after isolation of cell membranes by an initial low speed centrifugation step ($500-1000 \times g$) to pellet unbroken cells, cell walls, etc. and a second hard ultracentrifugation ($400,000 \times g$) to pellet cell membranes) or directly from the lysis solution. The first approach is more economical, as it reduces the working volume and the amount of detergent required, but more laborious and ultimately less efficient in terms of overall protein yield. KcsA can be solubilized in a functional state by many different detergents (Cortes & Perozo, 1997). Decylmaltoside has been used in crystallization studies of KcsA (Doyle et al., 1998; Zhou, Morais-Cabral, Kaufman, & MacKinnon, 2001) and is therefore a sensible choice. In addition, the relatively high critical micelle concentration of decylmaltoside facilitates the downstream reconstitution of the channel into liposomes. A high initial concentration of decylmaltoside (40 mM) is used to solubilize KcsA. For all subsequent steps, the concentration can be lowered to 5 mM. The solubilizing solution is cleared by centrifugation and the supernatant is mixed with a metal affinity resin (such as Ni-NTA from Qiagen or Talon from Clontech) equilibrated in wash buffer. Unbound material is removed by gentle centrifugation to prevent breakage of resin ($<1000 \times g$) and removal of the supernatant or by placing the resin into a small gravity chromatography column. The resin is washed with lysis buffer to further remove impurities. At this stage, KcsA can either be eluted from the column and spin-labeled in solution or the spin-labeling reaction can be performed while KcsA is immobilized on the affinity resin. On-resin labeling has the advantage that excess label can be removed more effectively after the labeling reaction. Note that the removal of unreacted spin label is a critical step in the labeling procedure of a membrane protein (Gross et al., 1999). While desalting on Sephadex G-50 effectively removes excess reagent for soluble proteins, this procedure is frequently not sufficient with membrane proteins. Great care needs to be taken to achieve as low a background signal as possible. An effective strategy to minimize background signal is on-resin labeling with subsequent removal of the excess label by washing while the channel is immobilized, followed by elution with 500 mM imidazole and size exclusion chromatography of the spin-labeled KcsA on Superdex 200 (GE Helthcare) in Superdex buffer.

Labeled KcsA in a solution containing 5 mM decylmaltoside is then mixed at a molar lipid: protein ratio of $\sim$1000: 1 with 4:1

1-palmitoyl-2-oleoyl-sn-glycero-3-phosphocholine (POPC): 1-palmitoyl-2-oleoyl-sn-glycero-3-[phospho-rac-(1-glycerol)] (POPG) (Avanti Polar) and reconstituted on Sephadex G-50 (GE Healthcare) by the method of Green and Bell (1984). The resulting proteoliposomes are then pelleted by ultracentrifugation (20 h at 400,000 × g at 20 °C) and resuspended with liposome suspension buffer and loaded into EPR tubes for analysis. The samples can be flash-frozen in liquid nitrogen and stored at −80 °C until measurement. Due to the proton sensitivity of KcsA, the gating state of the channel can be affected by varying the pH of the liposome resuspension buffer (Hulse et al., 2014; Liu, Sompornpisut, & Perozo, 2001).

3.1.2 Solutions Required
Lysis buffer: 20 mM Tris (pH 8.0), 100 mM KCl
Wash buffer: 20 mM Tris (pH 8.0), 100 mM KCl, 5 mM decylmaltoside
Superdex buffer: 50 mM Tris (pH 7.5), 150 mM KCl, 5 mM decylmaltoside
Liposome resuspension buffer: 50 mM Tris (pH 7.5), 150 mM KCl, 25% (v/v) glycerol

3.2 Labeling of the Prokaryotic KvAP Potassium Channel
3.2.1 Procedure
A synthetic DNA sequence coding for KvAP is obtained and subcloned into the pQE60 expression vector (Qiagen). *E. coli* are transformed and grown in a 2-l fermentor (Bioflow 110; New Brunswick Scientific) in K12 growth media supplemented with K12 trace metals solution (10 ml/l), magnesium sulfate solution (2 ml/l), and thiamine solution (500 μl/l). Throughout fermentation, ampicillin solution is added at a rate of 350 μl/min. When the OD$_{600}$ of the culture reaches 2–3, the culture is supplemented with additional K12 trace metal solution (5 ml/l), magnesium sulfate solution (1 ml/l), thiamine solution (250 μl/l), and glucose solution is fed at the rate of 1.7 ml/min. When the OD$_{600}$ reaches 11–13, additional ampicillin is added, and expression of KvAP is induced with 0.6 mM IPTG in the presence of 10 mM BaCl$_2$ (Ruta et al., 2003). After induction at 37 °C for 3 h, the cells are collected by centrifugation. The cell debris is removed, and the remaining cells are resuspended in 80 ml chilled 0.5 M EDTA (pH 8). Cells are washed thrice in 2 l of chilled lysis buffer. At this point, the cell pellet can be frozen at −80 °C for future use. The cells are resuspended in lysis buffer and disrupted by sonication. Suitable protease inhibitors (leupeptin, aprotinin, and phenylmethanesulfonylfluoride) are added before, during,

and after cell disruption. KvAP is solubilized with 40 mM decylmaltoside in lysis buffer, and the solution is spun to clear cell debris (Cuello et al., 2004; Jiang et al., 2003; Ruta et al., 2003; Vamvouka et al., 2008). KvAP is maintained in solution with 5 mM decylmaltoside in all subsequent steps. The supernatant is mixed with metal affinity resin (Ni-NTA from Qiagen or Talon from Clontech) equilibrated with wash buffer. Unbound material is removed via gentle centrifugation (<1000 × g) or by placing the resin in a small chromatography column and removing excess solution by gravity. As with KcsA, KvAP can be eluted and spin-labeled in solution or while immobilized on the resin. If the protein is to be embedded into liposomes, on-resin spin labeling is recommended to decrease background signal. After washing the resin with lysis buffer, KvAP is spin-labeled overnight with a 10-fold molar excess of MTSL or a 1:3 mixture (Gross et al., 1999) of MTSL and its isosteric diamagnetic analog (both Toronto Research), extensively rinsed with wash buffer on a small gravity chromatography column to remove excess spin label, and eluted with 500 mM imidazole. KvAP is further purified by gel filtration on a Superdex 200 column in Superdex buffer. Labeled KvAP in a solution containing 5 mM decylmaltoside is reconstituted into liposomes using Bio-beads SM (BioRad) (1 g beads/5 ml solution) into POPC:POPG (4:1) (Avanti Polar) at a molar lipid:protein ratio of ~1000:1. Proteoliposomes are concentrated by ultracentrifugation (20 h at 400,000 × g), resuspended with liposome resuspension buffer, and loaded into EPR tubes. The samples are either measured immediately or flash-frozen in liquid nitrogen and stored at −80 °C until measurement. Under these experimental conditions, KvAP is expected to be in an open, inactivated conformation (Li et al., 2014; Ruta et al., 2003).

3.2.2 Solutions Required

K12 growth media: yeast extract (5 g/l), KH_2PO_4 (4 g/l), K_2HPO_4 (6 g/l), $(NH_4)_2SO_4$ (10 g/l)

K12 trace metals solution: NaCl (5 g/l), $ZnSO_4 \cdot 7H_2O$ (1 g/l), $MnCl_2 \cdot 4H_2O$ (4 g/l), $FeCl_3 \cdot 6H_2O$ (4.75 g/l), $CuSO_2 \cdot 5H_2O$ (0.4 g/l), H_3BO_3 (0.575 g/l), $Na_2MoO_4 \cdot 2H_2O$ (0.5 g/l), 6N H_2SO_4 (12.5 ml/l)

Magnesium sulfate solution: 50% (w/v) $MgSO_4$ (500 g/l)

Thiamine solution: 0.2% (w/v) thiamine (2 g/l)

Ampicillin solution: ampicillin (0.08 g/l)

Glucose feed solution: 50% (w/v) glucose (500 g/l)

Lysis buffer: 20 mM Tris (pH 8.0), 100 mM KCl

Wash buffer: 20 mM Tris (pH 8.0), 100 mM KCl, 5 mM decylmaltoside
Superdex buffer: 50 mM Tris (pH 7.5), 150 mM KCl, 5 mM decylmaltoside
Liposome resuspension buffer: 50 mM Tris (pH 7.5), 150 mM KCl, 25% (v/v) glycerol

REFERENCES

Chakrapani, S., Cuello, L. G., Cortes, D. M., & Perozo, E. (2008). Structural dynamics of an isolated voltage-sensor domain in a lipid bilayer. *Structure, 16,* 398–409.
Cieslak, J. A., Focia, P. J., & Gross, A. (2010). Electron spin-echo envelope modulation (ESEEM) reveals water and phosphate interactions with the KcsA potassium channel. *Biochemistry, 49,* 1486–1494.
Columbus, L., & Hubbell, W. L. (2002). A new spin on protein dynamics. *Trends in Biochemical Sciences, 27,* 288–295.
Cortes, D. M., Cuello, L. G., & Perozo, E. (2001). Molecular architecture of full-length KcsA: Role of cytoplasmic domains in ion permeation and activation gating. *The Journal of General Physiology, 117*(2), 165–180.
Cortes, D. M., & Perozo, E. (1997). Structural dynamics of the Streptomyces lividans K+ channel (SKC1): Oligomeric stoichiometry and stability. *Biochemistry, 36,* 10343–10352.
Cuello, L. G., Cortes, D. M., & Perozo, E. (2004). Molecular architecture of the KvAP voltage-dependent K+ channel in a lipid bilayer. *Science, 306,* 491–495.
Cuello, L. G., Romero, J. G., Cortes, D. M., & Perozo, E. (1998). pH-dependent gating in the Streptomyces lividans K+ channel. *Biochemistry, 37,* 3229–3236.
Doyle, D. A., Cabral, J. M., Pfuetzner, R. A., Kuo, A., Gulbis, J. M., Cohen, S. L., et al. (1998). The structure of the potassium channel: Molecular basis of K+ conduction and selectivity. *Science, 280,* 69–77.
Fanucci, G. E., & Cafiso, D. S. (2006). Recent advances and applications of site-directed spin labeling. *Current Opinion in Structural Biology, 16,* 644–653.
Fleissner, M. R., Bridges, M. D., Brooks, E. K., Cascio, D., Kalai, T., Hideg, K., et al. (2011). Structure and dynamics of a conformationally constrained nitroxide side chain and applications in EPR spectroscopy. *Proceedings of the National Academy of Sciences of the United States of America, 108*(39), 16241–16246. http://dx.doi.org/10.1073/pnas.1111420108.
Green, P. R., & Bell, R. M. (1984). Asymmetric reconstitution of homogeneous Escherichia coli sn-glycerol-3-phosphate acyltransferase into phospholipid vesicles. *The Journal of Biological Chemistry, 259,* 14688–14694.
Gross, A., Columbus, L., Hideg, K., Altenbach, C., & Hubbell, W. L. (1999). Structure of the KcsA potassium channel from Streptomyces lividans: A site-directed spin labeling study of the second transmembrane segment. *Biochemistry, 38,* 10324–10335.
Gross, A., & Hubbell, W. L. (2002). Identification of protein side chains near the membrane-aqueous interface: A site-directed spin labeling study of KcsA. *Biochemistry, 41,* 1123–1128.
Hubbell, W. L., Cafiso, D. S., & Altenbach, C. (2000). Identifying conformational changes with site-directed spin labeling. *Nature Structural Biology, 7*(9), 735–739.
Hubbell, W. L., Gross, A., Langen, R., & Lietzow, M. A. (1998). Recent advances in site-directed spin labeling of proteins. *Current Opinion in Structural Biology, 8,* 649–656.
Hubbell, W. L., Mchaourab, H. S., Altenbach, C., & Lietzow, M. A. (1996). Watching proteins move using site-directed spin labeling. *Structure, 4,* 779–783.
Hulse, R. E., Sachleben, J. R., Wen, P. C., Moradi, M., Tajkhorshid, E., & Perozo, E. (2014). Conformational dynamics at the inner gate of KcsA during activation. *Biochemistry, 53*(16), 2557–2559. http://dx.doi.org/10.1021/bi500168u.

Jiang, Y., Lee, A., Chen, J., Ruta, V., Cadene, M., Chait, B. T., et al. (2003). X-ray structure of a voltage-dependent K^+ channel. *Nature*, *423*, 33–41.

Kelly, B. L., & Gross, A. (2003). Potassium channel gating observed with site-directed mass tagging. *Nature Structural Biology*, *10*(4), 280–284.

le Coutre, J., Whitelegge, J. P., Gross, A., Turk, E., Wright, E. M., Kaback, H. R., et al. (2000). Proteomics on full-length membrane proteins using mass spectrometry. *Biochemistry*, *39*, 4237–4242.

Li, Q., Wanderling, S., Sompornpisut, P., & Perozo, E. (2014). Structural basis of lipid-driven conformational transitions in the KvAP voltage-sensing domain. *Nature Structural & Molecular Biology*, *21*(2), 160–166. http://dx.doi.org/10.1038/nsmb.2747.

Liu, Y. S., Sompornpisut, P., & Perozo, E. (2001). Structure of the KcsA channel intracellular gate in the open state. *Nature Structural Biology*, *8*(10), 883–887.

Mchaourab, H. S., Steed, P. R., & Kazmier, K. (2011). Toward the fourth dimension of membrane protein structure: Insight into dynamics from spin-labeling EPR spectroscopy. *Structure*, *19*(11), 1549–1561. http://dx.doi.org/10.1016/j.str.2011.10.009.

Neyton, J., & Miller, C. (1988a). Discrete Ba^{2+} block as a probe of ion occupancy and pore structure in the high-conductance Ca^{2+}-activated K^+ channel. *The Journal of General Physiology*, *92*, 569–586.

Neyton, J., & Miller, C. (1988b). Potassium blocks barium permeation through a calcium-activated potassium channel. *The Journal of General Physiology*, *92*, 549–567.

Oh, K. J., Altenbach, C., Collier, R. J., & Hubbell, W. L. (2000). Site-directed spin labeling of proteins. Applications to diphtheria toxin. *Methods in Molecular Biology*, *145*, 147–169.

Oh, K. J., Barbuto, S., Pitter, K., Morash, J., Walensky, L. D., & Korsmeyer, S. J. (2006). A membrane-targeted BID BCL-2 homology 3 peptide is sufficient for high potency activation of BAX in vitro. *The Journal of Biological Chemistry*, *281*(48), 36999–37008. http://dx.doi.org/10.1074/jbc.M602341200.

Perozo, E., Cortes, D. M., & Cuello, L. G. (1998). Three-dimensional architecture and gating mechanism of a K+ channel studied by EPR spectroscopy. *Nature Structural Biology*, *5*, 459–469.

Perozo, E., Cortes, D. M., & Cuello, L. G. (1999). Structural rearrangements underlying K+ channel activation gating. *Science*, *285*, 73–78.

Perozo, E., Cuello, L. G., Cortes, D. M., Liu, Y. S., & Sompornpisut, P. (2002). EPR approaches to ion channel structure and function. *Novartis Foundation Symposium*, *245*, 146–158 (discussion 158–164, 165–148).

Perozo, E., Kloda, A., Cortes, D. M., & Martinac, B. (2001). Site-directed spin-labeling analysis of reconstituted MscL in the closed state. *The Journal of General Physiology*, *118*(2), 193–206.

Ruta, V., Jiang, Y., Lee, A., Chen, J., & MacKinnon, R. (2003). Functional analysis of an archaebacterial voltage-dependent K+ channel. *Nature*, *422*, 180–185.

Schrempf, H., Schmidt, O., Kummerlen, R., Hinnah, S., Muller, D., Betzler, M., et al. (1995). A prokaryotic potassium ion channel with two predicted transmembrane segments from Streptomyces lividans. *The EMBO Journal*, *14*, 5170–5178.

Vamvouka, M., Cieslak, J., Van Eps, N., Hubbell, W., & Gross, A. (2008). The structure of the lipid-embedded potassium channel voltage sensor determined by double-electron–electron resonance spectroscopy. *Protein Science*, *17*, 506–517.

Vasquez, V., Sotomayor, M., Cortes, D. M., Roux, B., Schulten, K., & Perozo, E. (2008). Three-dimensional architecture of membrane-embedded MscS in the closed conformation. *Journal of Molecular Biology*, *378*, 55–70.

Zhou, Y., Morais-Cabral, J. H., Kaufman, A., & MacKinnon, R. (2001). Chemistry of ion coordination and hydration revealed by a K+ channel-Fab complex at 2.0 Å resolution. *Nature*, *414*(6859), 43–48.

SECTION III

Spin Labeling Studies of Nucleic Acids

CHAPTER FOURTEEN

Advanced EPR Methods for Studying Conformational Dynamics of Nucleic Acids

B. Endeward*, A. Marko*, V.P. Denysenkov*, S.Th. Sigurdsson[†], T.F. Prisner*,[1]

*Institute of Physical and Theoretical Chemistry and Center of Biomolecular Magnetic Resonance, Goethe University Frankfurt am Main, Frankfurt am Main, Germany
[†]Department of Chemistry, Science Institute, University of Iceland, Reykjavık, Iceland
[1]Corresponding author: e-mail address: prisner@chemie.uni-frankfurt.de

Contents

1. Introduction	404
2. Spin Labels	407
3. Theory for Orientation-Selective PELDOR	408
4. Experimental Procedure for Multifrequency/Multifield PELDOR	410
4.1 X-Band PELDOR (Low Magnetic Field)	410
4.2 Q-Band PELDOR (Medium Magnetic Field)	412
4.3 G-Band PELDOR (High Magnetic Field)	412
5. Analysis of Multifrequency/Multifield PELDOR Data	415
6. Summary and Outlook	418
Acknowledgments	421
References	421

Abstract

Pulsed electron paramagnetic resonance (EPR) spectroscopy has become an important tool for structural characterization of biomolecules allowing measurement of the distances between two paramagnetic spin labels attached to a biomolecule in the 2–8 nm range. In this chapter, we will focus on applications of this approach to investigate tertiary structure elements as well as conformational dynamics of nucleic acid molecules. Both aspects take advantage of using specific spin labels that are rigidly attached to the nucleobases, as they allow obtaining not only the distance but also the relative orientation between both nitroxide moieties with high accuracy. Thus, not only the distance but additionally the three Euler angles between both the nitroxide axis systems and the two polar angles of the interconnecting vector with respect to the nitroxide axis systems can be extracted from a single pair of spin labels. To extract all these parameters independently and unambiguously, a set of multifrequency/multifield pulsed EPR experiments have to be performed. We will describe the experimental procedure as well as newly developed spin labels, which are helpful to disentangle all these parameters, and tools which we have developed to analyze such data sets. The procedures and analyses will be illustrated by examples from our laboratory.

1. INTRODUCTION

Understanding the function of biomolecules relies on the knowledge of their structure and dynamics on an atomistic level. X-ray crystallography is the most successful technique to obtain structures of biological macromoleculos, but requires crystals, a nonnatural environment for biomolecules. Also, due to this ordered crystallized state, limited information about conformational flexibility or dynamics of the molecules is available. Spectroscopic methods, like nuclear magnetic resonance or FRET (Förster resonance energy transfer), on the other hand, enable to gain information on structure and dynamics of the molecules in solution. PELDOR (pulsed electron–electron double resonance) (Milov, Ponomarev, & Tsvetkov, 1984; Milov, Salikov, & Shirov, 1981), also called DEER (double electron–electron resonance) (Larsen, Halkides, & Singel, 1993), is a method which can determine distances in the 2–8 nm range, similar to FRET.

PELDOR relies on the magnetic dipole interaction between two paramagnetic spin labels covalently attached to the macromolecule (Hubbell, Cafiso, & Altenbach, 2000). Pulsed EPR (electron paramagnetic resonance) techniques allow measurement of these weak interactions between the two spin labels, even if the strength of their interaction is much weaker than the interaction of the unpaired spin with the external magnetic field and surrounding nuclear spins. For a distance of 2 nm, the interaction strength in frequency units accounts to 7 MHz. The PELDOR time trace signal is modulated by this frequency and can be detected easily by Fourier transform of the signal or by other mathematical transformations (Chiang, Borbat, & Freed, 2005; Jeschke et al., 2006; Jeschke, Koch, Jonas, & Godt, 2002). For nitroxide spin labels, the time traces can usually be recorded for up to 10 μs at a temperature of 50 K. This results in a frequency resolution of about 100 kHz, corresponding to a very high distance accuracy of about 20 pm. Unfortunately, the internal flexibility of the covalently attached spin label itself blurs this high resolution. For the most commonly used nitroxide spin label MTSSL ((1-oxyl-2,2,5,5-tetramethylpyrroline-3-methyl) methanethiosulfonate) (Altenbach, Marti, Khorana, & Hubbell, 1990), which can be covalently attached to cysteine amino acids residues, the distance distribution resulting from the rotameric freedom of the spin label itself limits the accuracy of the distance to about 300 pm (Fajer, Li, Yang, & Fajer, 2007; Jeschke, 2012; Sale, Song, Liu, Perozo, & Fajer, 2005).

Intrinsic paramagnetic cofactors, as amino acid radicals, metal ions, and iron sulfur centers exhibit a much higher accuracy in the distance determination, because they are usually rigidly incorporated into the protein environment (Becker & Saxena, 2005; Bennati et al., 2003; Biglino, Schmidt, Reijerse, & Lubitz, 2006; Denysenkov, Prisner, Stubbe, & Bennati, 2006; Kay, Elsässer, Bittl, Farrell, & Thorpe, 2006; Roessler et al., 2010; Van Amsterdam, Ubbink, Canters, & Huber, 2003). However, additional effort has to be undertaken to extract the distance information for such centers. This is caused by the broad anisotropic spectra, which inhibit an equal excitation of all random orientations of the interconnecting vector R by the microwave pulses for metal centers and clusters, and because of delocalized spin densities in coordinated metal ions and spin projection factors in FeS clusters (Bode, Plackmeyer, Bolte, Prisner, & Schiemann, 2009; Elsaesser, Brecht, & Bittl, 2005; Riplinger et al., 2009). Here, the orientation of the anisotropic hyperfine and g-tensors between both paramagnetic centers and with respect to the distance vector R also effects the dipolar oscillations of the PELDOR time traces and have, therefore, to be taken into account. On the other hand, if the orientation of these tensors within the molecular frame is known, this offers valuable additional angular information. Thus, instead of only determining spheres with a given distance between the two paramagnetic centers, all relevant angles between the two molecules can be determined, yielding complete structural information if the location of the paramagnetic cofactor in the protein environment is known. Several independent PELDOR experiments with distinct pump and detection frequencies have to be performed to extract this information unambiguously. Because the orientation between both paramagnetic centers cannot be assumed to be randomly distributed, Tikhonov regularization (Chiang et al., 2005; Jeschke et al., 2006, 2002) cannot be used to obtain the distance distribution function directly from the PELDOR time trace. Instead, more elaborate simulation methods have to be used to disentangle distances and angular information (Denysenkov et al., 2006; Margraf, Bode, Marko, Schiemann, & Prisner, 2007; Marko et al., 2010, 2009; Marko & Prisner, 2013; Polyhach, Godt, Bauer, & Jeschke, 2007; Schiemann, Cekan, Margraf, Prisner, & Sigurdsson, 2009).

The benefit of this additional angular information for structural biology was first demonstrated on a ribonucleotide reductase dimer. PELDOR measurements at X-band frequencies allowed determination of the distance between the two tyrosyl radicals of the dimer (Bennati et al., 2005, 2003). At high magnetic fields (6.4 T), the anisotropy of the tyrosyl g-tensor is fully resolved, which allowed the measurement of the dipolar interaction for all different orientations of the tyrosyl radicals with respect to the external magnetic field (Denysenkov, Biglino, Lubitz, Prisner, & Bennati, 2008; Denysenkov et al., 2006). From this set of measurements, the orientation of the two tyrosyl radicals with respect to the interconnecting R vector could be extracted, enabling full determination of the dimer structure just from a single pair of tyrosyl radicals.

As explained above, this orientation information is usually lost for covalently attached MTSSL spin labels, due to the large rotameric mobility of the spin label. Only in rare cases does interaction of the nitroxide spin labels with amino acid side groups or bound lipid molecules restrict the geometry of the labels (Abé et al., 2012; Endeward, Butterwick, MacKinnon, & Prisner, 2009; Sezer, Freed, & Roux, 2009). PELDOR experiments performed at different detection frequencies can easily detect such cases and indicate that more elaborate analysis of the PELDOR time traces is required to quantitatively interpret the data. Some spin labels, like the TOAC spin label (Karim, Kirby, Zhang, Nesmelov, & Thomas, 2004; McNulty & Millhauser, 2000) or newly developed spin labels for proteins (Hubbell, López, Altenbach, & Yang, 2013; Sajid, Jeschke, Wiebcke, & Godt, 2009), are more rigid and have the potential to allow determination of more precise distances and angular information in proteins.

Rigid spin labels already exist for nucleic acids, either intercalated into abasic sites of duplex DNA (Shelke & Sigurdsson, 2010) or as paramagnetic variants of the nucleobases cytosine or uracil. For the rigid spin labels, the orientation of the nitroxide moiety is locked to the structure of double-stranded DNA or RNA helices. This allows one to relate the obtained angular information directly to the tertiary structure of the oligonucleotide molecule. In this chapter, we describe spin labels for nucleic acids molecules with different degree of rotameric freedom and explain how the multi-frequency/multifield experiments, necessary to disentangle the distance and orientation information, has to be performed. In addition, we show how the data can be quantitatively analyzed to yield information about tertiary structure and conformational flexibility of the DNA and RNA molecules.

2. SPIN LABELS

Approximately 25 years ago, Spaltenstein, Robinson, and Hopkins (1989) developed a site-directed spin-labeling strategy for nucleic acids, where the spin label 2,2,5,5-tetramethylpyrroline-1-yloxy-3-acetylene (TPA), linked to the 5-position of uracil, was incorporated into DNA by solid-phase synthesis. This spin label was less flexible than other reported labels (Edwards & Sigurdsson, 2007; Piton et al., 2005; Qin et al., 2007; Qin, Hideg, Feigon, & Hubbell, 2003; Ramos & Varani, 1998; Schiemann et al., 2004, 2007; Sigurdsson & Eckstein, 1996; Strube, Schiemann, MacMillan, Prisner, & Engels, 2001), but still had a relatively large conformational freedom through rotation of the single bonds flanking the acetylene. Since this time, several other spin-labeling approaches have been reported (Shelke & Sigurdsson, 2012) some of which have less conformational flexibility than TPA. In particular, the rigid label Ç (c-spin) for DNA (Barhate, Cekan, Massey, & Sigurdsson, 2007) and the corresponding derivative for RNA, Çm (Höbartner, Sicoli, Wachowius, Gophane, & Sigurdsson, 2012; Fig. 1B) have an isoindoline nitroxide fused to a nucleobase, thus eliminating dynamics of the spin label independent of the nucleobase. These rigid labels were shown to be nonperturbing to DNA (Edwards et al., 2011) and RNA duplexes (Höbartner et al., 2012).

Due to the synthetic effort required to prepare Ç and Çm, the benzimidazole and benzoxazole derivatives ^{Im}U or ^{Ox}U were prepared (Fig. 1C; Gophane & Sigurdsson, 2013). Although rotation is possible around the single bond connecting the spin label to the base, making these labels less rigid than Ç, PELDOR experiments showed that due to an intramolecular hydrogen bond in ^{Im}U, the rotation around the single bond is strongly

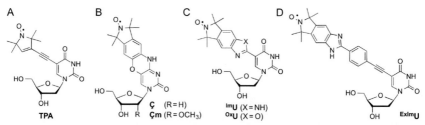

Figure 1 Semi-rigid and rigid spin labels. (A) The TPA spin label. (B) The rigid spin labels Ç and Çm. (C) The isoindoline-based spin labels ^{Im}U and ^{Ox}U. (D) The extended imidazole-based ^{ExIm}U spin label, which freely rotates around N–O axis (x-axis) of the nitroxide.

reduced (Gophane, Endeward, Prisner, & Sigurdsson, 2014). When the benzimidazole spin label was projected away from the nucleic acid with a longer and flexible linker, the orientation selection that was possible with ^{Im}U vanished. However, in this so-called conformationally unambiguous spin label, a rotation around the N–O axis (x-axis) of the spin label did not compromise the fixed position of the spin label relative to the nucleic acid. Therefore, the distance distribution that can be collected by one PELDOR time trace allows estimation of the distance as well as the flexibility of the nucleic acid. In this chapter, we will only concentrate on EPR applications of nucleic acid structures that contain rigid spin labels.

3. THEORY FOR ORIENTATION-SELECTIVE PELDOR

PELDOR is a two-frequency experiment which is performed on an ensemble of spin pairs in frozen disordered solutions (Milov et al., 1984, 1981). The dead-time-free four-pulse DEER sequence is commonly used (Fig. 2; Martin et al., 1998; Pannier, Veit, Godt, Jeschke, & Spiess, 2000). In order to observe the dipolar interaction between the two spins, the Larmor frequency of the first spin (A-spin) has to coincide with the detection frequency ν_A, and the Larmor frequency of the second spin (B-spin) has to be equal to the pump pulse frequency ν_B. For nitroxide spin labels, this condition is only fulfilled for a small part of the ensemble of spin-labeled molecules that have a special orientation with respect to the static magnetic field B_0, as explained below.

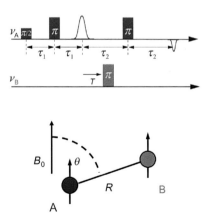

Figure 2 The dead-time-free four-pulse DEER sequence used for the experiments and a graphical representation of the parameters describing the magnetic dipolar coupling between both paramagnetic centers.

The A-spins which are excited by the three detection pulses, separated by the times τ_1 and τ_2, give rise to the refocused echo at $2(\tau_1 + \tau_2)$. The pump pulse inverts the B-spins that are resonant with the pump frequency. If a spin pair contains one A- and one B-spin, then the inversion of the B-spin changes the magnetic field at the position of the A-spin. This leads to a change of the A-spin Larmor frequency by the value ω_d determined by the strength of the magnetic dipolar interaction between the spin pair. The dipolar frequency

$$\omega_d(R, \theta) = \frac{D}{R^3}(1 - 3\cos^2\theta)$$

depends on the length of the distance vector R and its orientation θ with respect to the magnetic field B_0 (Fig. 2). $D = \mu_0 \mu_B^2 g_A g_B / (4\pi\hbar)$ is the dipolar interaction constant, wherein μ_0 is the magnetic vacuum permeability, μ_B is the Bohr magneton, and $\hbar$ is the reduced Planck constant. For nitroxide spin labels, the g-values g_A and g_B correspond to 2.006, resulting in a value of the dipolar constant $D = 2\pi$ 52.04 MHz nm^3. The application of the pump pulse leads to a phase shift and a modulation of the refocused echo intensity. This phase shift, and therefore modulation of the intensity of the echo, depends on the time T between the A-spin Hahn echo and the pump pulse (Marko, Denysenkov, & Prisner, 2013). In nitroxide spin ensembles, only a fraction of the B-spin is flipped in each spin pair. Therefore, the refocused echo magnitude consists of a nonmodulated part $1-\lambda$ and a modulated part $\lambda \cos(\omega_d T)$, where λ is the probability to flip B-spins by the pump pulse.

Nitroxide spin labels in frozen solution have an inhomogeneous linewidth, which is dominated by the anisotropy of the nitrogen hyperfine interaction A (at X-band frequencies) or by the anisotropy of the g-tensor (at G-band frequencies). This inhomogeneous linewidth is larger than the spectral width of the mw-pulses, which allows excitation of different subensembles of the spin system by the pump and detection pulses (Fig. 3). Because the nitrogen hyperfine tensor A and the g-tensor are known (Möbius et al., 2005) in the molecular axis system, the specific orientations excited by pump and detection pulses can easily be calculated if the microwave field strength B_1 and the microwave frequency are known.

If the relative orientation of the nitroxides in the doubly labeled nucleic acid molecules of the ensemble is fixed, then the efficiency to excite simultaneously one spin by the detection pulse sequence and invert one spin by

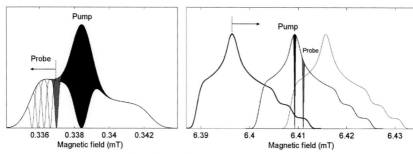

Figure 3 EPR absorption line shape for nitroxides at X-band frequencies (left) and at G-band frequencies (right). Typical pump (black) and detection (gray) pulse excitation profiles and positions are shown in the figures. Whereas at X-band frequencies the frequency offset between pump and detection pulses is varied, the magnetic field position is changed at G-band frequencies to obtain two-dimensional data sets.

the pump pulse strongly depends on the orientation of the nucleic acid molecule with respect to the external magnetic field B_0. The total normalized signal from an ensemble of such identical rigid biradicals in a powder sample is described by

$$S(T) = e^{-\gamma T}\left(1 - \int_0^{\pi/2} \lambda(\theta)(\cos(\omega_\mathrm{d}(r,\theta)T) - 1)\sin\theta\,\mathrm{d}\theta\right)$$

Here, $\lambda(\theta)$ is the orientation-dependent B-spin flip probability density (Larsen & Singel, 1993; Milov, Maryasov, & Tsvetkov, 1998; Milov & Tsvetkov, 1997). The shape of this function is determined by the detection and pump pulses frequencies, relative orientation of nitroxides in double-labeled molecules, and the values of g- and hyperfine interaction tensors. The exponential factor corresponds to the intermolecular magnetic spin interaction.

4. EXPERIMENTAL PROCEDURE FOR MULTIFREQUENCY/MULTIFIELD PELDOR

4.1 X-Band PELDOR (Low Magnetic Field)

At X-band frequency, the edges of the EPR spectra of nitroxides are governed by the large nitrogen hyperfine coupling along the molecular z-axis (out of plane axis). The contribution of the g-tensor anisotropy, as well as the hyperfine components along the x- (N–O bond direction) and y-axis, build up the intensity in the center of the spectrum (Margraf et al., 2007). Therefore, in the center of the nitroxide EPR spectrum almost all orientation of the nitroxide with respect to the external magnetic field contribute, whereas the low- and high-field edges of the spectrum mainly relate to nitroxides, which have an alignment of the nitroxide z-axis parallel to the magnetic field. Thus, differently oriented subensembles of nuclear acid molecules can be selected by choosing different pump and detection frequencies. Taking advantage of these properties in the PELDOR experiment is called orientation selection (Margraf et al., 2007; Polyhach et al., 2007).

Typically, PELDOR experiments at 0.3 T magnetic field are performed by setting the pump pulse frequency resonant with spins in the center of the nitroxide spectra to reach a maximum inversion efficiency λ. The detection pulse sequence is set resonant with spins at the low-field edge of the nitroxide spectra to minimize spectral overlap between the pump and detection pulses. This simple picture already indicates some restrictions for performing orientation-selective measurements, where the detection frequency is varied (Fig. 3). A short pump pulse is chosen (typically 12 ns with an excitation bandwidth of 80 MHz) to excite a broad frequency range in the center of the nitroxide spectra resulting in a modulation depth parameter λ of approximately 0.43. To reach optimum B_1 field strengths, the frequency ν_B is adjusted to the microwave resonator frequency. The orientation selectivity is achieved by longer and more selective detection frequency pulses (typically 32 ns with a bandwidth of 30 MHz). Such pulses can be easily achieved off-resonant from the microwave cavity dip but have to be adjusted for each different offset. Experiments are performed with detection frequency offsets with respect to the pump pulse frequency, ranging from 40 to 100 MHz (up to the edge of the nitroxide spectrum). For the lowest offsets, a strong overlap between pump and detection pulse excitation occurs, which reduces the observable echo signal intensity and makes a quantitative prediction of the modulation depth parameter λ more difficult. Nevertheless, for quantitative analysis of the orientation selection, it is also important to acquire data sets where orientations of the magnetic field close to the x- and y-axis of the nitroxides are detected. Of course, such multifrequency PELDOR experiments require more measurement time to achieve a good signal-to-noise ratio for all different detection frequencies. On the other hand, if only distance information is of interest, they can all be summed up, giving almost a similar signal-to-noise ratio as a classical 1D-PELDOR measurement. To average out the orientation selection, a data set with detection frequencies ranging from 30 to 90 MHz is required

(Prisner, Marko, & Sigurdsson, 2015). Further reducing the microwave power of pump and detection pulses increases the selectivity, which can be specifically beneficial for long distances. Of course, this also reduces the modulation depth further and, therefore, the signal-to-noise ratio of the dipolar oscillations. Thus, a compromise depending on the parameters of the specific system investigated has to be found. Gaussian or adiabatic pulses with better excitation profiles could improve the experiments for small pump–detection offsets.

4.2 Q-Band PELDOR (Medium Magnetic Field)

Q-band experiments offer a strongly improved signal-to-noise ratio even without a high power microwave amplifier (Gromov et al., 2001; Polyhach et al., 2012). For orientation-selective experiments, the situation is somewhat more complicated compared to X-band frequencies (where the hyperfine anisotropy is dominant) or G-band frequencies (where the g-tensor anisotropy defines the width of the spectrum). At Q-band frequencies, the contribution of the g-tensor and A-tensor anisotropies has similar sizes, making the orientation selection less intuitive and the analysis more complicated. Nevertheless, orientation selection can be observed at Q-band frequencies, as demonstrated on a double-labeled double-stranded DNA molecule in Fig. 4. Whereas the semi-rigid spin label ^{Im}U (Fig. 1; DNA spin labeled at two positions 8 nucleobases apart from each other) shows different PELDOR time traces for different offsets between pump and detection frequencies, such an effect is not visible for the spin label ^{ExIm}U, which can rotate around its N–O bond. Direct Tikhonov regularization leads to artifacts in the distance distribution in the first case, but such effects are not visible for the free rotatable spin label ^{ExIm}U (Fig. 4). Thus, a combination of both spin labels on the same nucleic acid molecule further simplifies the analysis of the data and the separation of orientation and distance information.

4.3 G-Band PELDOR (High Magnetic Field)

At high magnetic fields, PELDOR experiments of nitroxides appear at a first glance more challenging compared to standard frequencies, like X- or Q-band. This is due to a much broader EPR spectrum (about 250 G for nitroxides at 180 GHz, Fig. 3), and the fact that available excitation bandwidth of the pulses is much smaller (3 G at 180 GHz) due to limitations of available microwave power. Both effects decrease the sensitivity of the

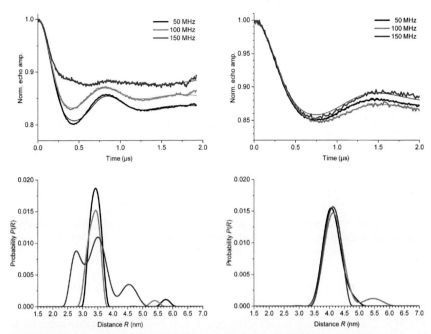

Figure 4 Orientation-selective PELDOR on a double-stranded DNA molecule measured at Q-band frequencies: upper row background corrected PELDOR time traces; lower row distance distribution estimated by Tikhonov regularization of PELDOR time trace. (A) ^{Im}U-spin-labeled DNA, eight residues apart. Different pump-detect frequency offsets at Q-band frequencies, due to the orientation selection is the distance analysis with standard Tikhonov regularization by using DeerAnalysis (Jeschke et al., 2006) total misleading. (B) ^{ExIm}U-spin-labeled but same DNA sequence and Q-band frequencies, all three distance distributions are equal, small differences are due to the direct interaction/overlap between the pump and detection pulses.

experiment because only very small fractions of the spins can be pumped and detected. On the other hand, only at high magnetic fields, all three molecular orientations can be distinguished by the g-tensor anisotropy, whereas at low fields the in-plane directions cannot be distinguished. Additionally, very high orientation selectivity can be achieved, overlap of pump and detection pulses is negligible and modulations by hyperfine interactions, which might lead to artifacts at low fields, are strongly suppressed at high fields.

Our available microwave power of 60 mW at 180 GHz, coupled to a cylindrical TE_{011} cavity with a $Q_{load} \sim 1000$, translates to an optimal $\pi/2$-pulse lengths of 30 ns at the detection frequency, the center of the cavity dip is tuned to that frequency. Due to the limited bandwidth of the resonator, the procedure to obtain 2D-PELDOR data sets differs from the variable detection frequency method that is used at X- and Q-band frequencies.

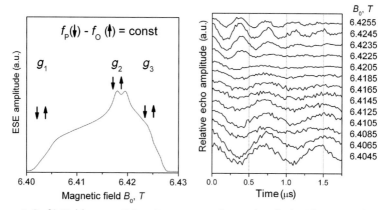

Figure 5 (Left) Field swept G-band spectrum of tyrosyl radical with pump–detection frequencies shown on the three main g-values. (Right) G-band PELDOR time traces recorded at different magnetic field values.

Here, the pump frequency f_P is set at a constant offset of 60 MHz from the detection frequency f_O, and PELDOR time traces are recorded for different external magnetic field values (Fig. 5). In this way, orientation-dependent dipolar modulation traces can be recorded throughout the whole EPR spectrum. This allows extraction of the mutual orientation of the two g-tensors of nitroxide spin labels that are rigidly attached to the nucleic acid molecule.

This procedure works nicely for flexible spin labels and in the case of collinear aligned spin labels. However, in a not collinear alignment of the two spins, offsets between pump and detection frequency in the full range of the spectral width need to be accessible. This is unfortunately not possible with a single-mode resonator. A possible solution to that problem is the use of a tunable double-mode resonator (Tkach, Sicoli, Höbartner, & Bennati, 2011) or by using a nonresonant sample holder (Cruickshank et al., 2009).

At our 180 GHz PELDOR setup, the time traces have only a few percent of modulation depth for nitroxide spin labels. One way to enlarge the modulation depth would be by using chirp pump pulses instead of fixed frequency rectangular pulses. A frequency chirp over 80 MHz of the pump pulse would already double the PELDOR modulation depth. The modulation depth and overall sensitivity of high-field PELDOR can be improved more significantly with a nonresonant sample holder and high transmitter output microwave power as demonstrated at the University of St. Andrews at W-band frequencies of 94 GHz (Cruickshank et al., 2009).

The first example demonstrating the power of orientation-selective PELDOR at high magnetic fields was the determination of the dimer structure of ribonucleotide reductase. Class I ribonucleotide reductase is an ideal enzyme for this type of studies, since radical states are involved in each of the three principal mechanistic steps of the enzymatic turnover, i.e., the generation of the tyrosyl radical in the subunit R2, the radical initiation process between R2 and R1, and the chemistry of nucleotide reduction in R1. One major advantage in performing distance measurements between endogenous protein radicals by PELDOR is that they are located at very precise distances and orientations in the protein frame as determined by their role in the protein function (Bennati et al., 2003). This leads to orientation-selective dipolar modulation traces at high magnetic fields where the g-tensor anisotropy of these radicals is resolved.

Precise determination of the mutual orientation between both tyrosyl radicals could be achieved by analyzing the 2D-PELDOR data set at G-band frequencies (Fig. 5) acquired as previously described (Denysenkov et al., 2006). The principal values of the g- and proton A-tensors have been independently measured by EPR and ENDOR experiments before (Bender et al., 1989; Gerfen et al., 1993; Hoganson, Sahlin, Sjöberg, & Babcock, 1996) and were kept fixed within the simulation, which included the distance R between the tyrosyl radicals, determined by X-band PELDOR (Bennati et al., 2003). Assuming C_2 symmetry of the dimer structure a single set of Euler angles (α,β,γ) describing the rotation matrix between the two tyrosyl axis systems had to be optimized to fit the field-dependent PELDOR data. The fit was performed by varying the Euler angles in steps of $0.1°$. The effective field strength of the pump pulse was set to $\omega_{1B} = 4.0 \times 10^7$ rad/s, the field strength at the detection frequency was $\omega_{1A} = 5.2 \times 10^7$ rad/s corresponding to a π-pulse of 60 ns length.

By this procedure in one case the unknown dimer structure was determined, because the location of the tyrosyl radical within the monomer unit was known from X-ray crystallography (Denysenkov et al., 2008). In another case, where the dimer structure was already known from X-ray crystallography, a fine-tuning of the tyrosyl molecule in its radical state in the protein environment was possible, and the results were in perfect agreement with ENDOR experiments performed on single crystals (Kolberg et al., 2005).

5. ANALYSIS OF MULTIFREQUENCY/MULTIFIELD PELDOR DATA

Methods to analyze PELDOR time traces are well developed in the case where no orientational effects occur. No orientational effect means that every orientation of the distance vector R of the spin pair with respect to the external magnetic field B_0 is detected with the same probability, leading to the well-known Pake-pattern of the dipolar coupling distribution (Borbat & Freed, 2007; Chiang et al., 2005; Martin et al., 1998; Pannier et al., 2000). Analyzing PELDOR time traces starts with the elimination of the background signal originating from the intermolecular dipole–dipole interaction between spin labels belonging to different proteins. The background corrected time domain signal is then converted into the distance distribution $P(R)$. This can be done by fitting the data with a model distance distribution function (e.g., one or more Gaussian distributions) or solving the integral equation employing the Tikhonov regularization of the ill-posed problem. The qualitative analysis of the time domain signal can already provide some information about the character of the distance distributions. If low-frequency oscillations are observable, then long distances can be expected between the spin labels. In such cases, observation time windows long enough to observe a full oscillation of the intramolecular interaction is necessary to quantitatively separate them from the intermolecular background signal.

In the case, where orientation selection exists, as explained above, the analysis procedure of the multifrequency/multifield PELDOR time traces has to include also the relative orientation between both spin labels (Marko et al., 2010; Schiemann et al., 2009). We have developed several approaches to analyze such data sets quantitatively. Simple geometrical modeling (Margraf et al., 2007; Schiemann et al., 2009) or comparison with MD simulations (Marko et al., 2009) can be used, if some prior knowledge about the molecule already exists. Analytical formula can be derived for specific cases and symmetries (Marko et al., 2010). A more general approach which does not require prior knowledge about the investigated molecule is to fit the PELDOR data with different pump/detection frequency offsets simultaneously to a calculated database containing each possible orientation and distance. This concept of a PELDOR signal database has proven to be very helpful for extracting the relative orientation between the spin labels from the experimental data set (Abé et al., 2012; Marko & Prisner, 2013). The signal database was generated for all possible relative orientations of the two radicals and stored. This is possible because all necessary parameters to fully describe the PELDOR signal of a nitroxide radical pair with a specific geometry are known. If the studied molecule is very rigid, then it is indeed possible to find unambiguously the corresponding signals from the database that fits best the experimental data set (Marko & Prisner, 2013). If the molecule can adopt several conformers, the program fits the experimental data by an ensemble of structures from the database until a quantitative agreement is found. Although in all cases reported solutions fitting nicely to the experimental traces can be found, it is more difficult to prove the uniqueness of these solutions. The uniqueness of the solution depends very much on the quality and completeness of the multifrequency/multifield PELDOR data set and on the degree of flexibility of the molecule under study. Experiments taken at both low fields and high fields are mandatory to be able to determine the conformational flexibility of the biomolecule. Of course, several different spin labels and different spin label positions give further independent information, which can be used to exclude wrong fits or prove the correctness of the fit. An extraction of the true distance distribution function $P(R)$ can be achieved by averaging of the PELDOR time signals measured for different detection frequencies at X-band as explained above. A second approach is the use of two spin labels that distinguish themselves with their rotational freedom around the nitroxide N–O axis, as the ^{Im}U and ^{ExIm}U spin label. In such cases, the orientation information can be chemically switched on and off, allowing more easy separation of these parameters.

An example of such an analysis was the investigation of the conformational dynamics of double-stranded DNA molecules. Different models for the twist-stretch-bend motion of double-stranded DNA molecules existed and were debated in the literature. Where single-molecule fluorescence measurements of long DNA molecules attached to a magnetic bead were interpreted by a twist-stretch motion where mainly the radius of the double helix was changed within this motion (Gore et al., 2006), small angle X-ray scattering data on short double-stranded DNA molecules, where gold particles were attached to the ends, were interpreted by a correlated motion with a change in the pitch height (Mathew-Fenn, Das, & Harbury, 2008). Modeling studies on the other side predicted a substantial bending of the double-stranded DNA molecules (Becker & Everaers, 2007, 2009) as shown schematically in Fig. 6.

We could show with multifrequency/multifield PELDOR experiments (performed at X-band and G-band frequencies) that only the model with variable radius is consistent with our experimental data set. For this analysis, a 15-base pair long double-stranded DNA molecule was single spin labeled in both complimentary strands with the rigid spin label Ç (at 10 different positions). The PELDOR data sets were simulated with three different models and could be fitted only with Model B. The best simulation for the model (Model B) with a flexible radius and one pair of spin labels is

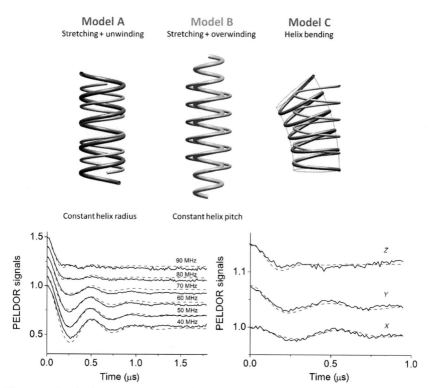

Figure 6 (Top) Three proposed conformation dynamics models of short double-stranded DNA helices. Model A describes double helix stretching with a constant radius and flexible pitch height. In Model B, the double helix shortens its radius and overwinds when stretched, keeping the pitch height constant. Model C allows the double helix to bend. (Bottom) Experimental data (solid) and best simulations (dashed) performed for the PELDOR signals of one single spin pair based on Model B showing good agreements for both X-band (left) and G-band measurements. (See the color plate.)

shown in Fig. 6. Analysis of the width of the distance distribution function $P(R)$ as a function of the number of base pairs between both spin labels additionally proved the correlation of the twist-stretch motion of Model B (Marko et al., 2011).

6. SUMMARY AND OUTLOOK

Acquiring a multifrequency/multifield data set is necessary for a comprehensive analysis of PELDOR time traces using rigid spin labels. Unfortunately, this makes the experiments more time consuming. However, this disadvantage is more than compensated by the much more precise distance restraints and the additional gained angular restraints. Such an effort is especially beneficial if the conformational flexibility of the biomolecule is of interest. By means of rigid spin labels, such information can be extracted from multidimensional data sets. In such cases, Tikhonov regularization cannot be used anymore, because not only the distance but additionally the orientation between both spin labels determines the PELDOR time traces.

We have shown that a procedure we developed for rigid labels at X-band frequencies, where detection frequency offsets ranging from 30 to 90 MHz (in steps of 10 MHz) are taken and summed up, leads to an almost perfect averaging out of the orientation information for any given geometry (Prisner et al., 2015). The measurement time to acquire such a multifrequency data set is not much longer compared to a standard 1D-PELDOR experiment with a flexible spin label. As explained above, the distance accuracy is much higher making such spin labels attractive even if only distance information is of interest. Additionally, orientation information is available and can be extracted from the data set by fitting procedures (Marko & Prisner, 2013). Fortunately, all of the parameters beside the distance R, the Euler angles (α, β, γ), and the polar angles between the dipolar vector R and the nitroxide molecular axis system (θ, φ) are well known or can be calibrated from independent measurements. Something like a κ factor or a Förster radius R_0 in FRET spectroscopy does not exist! The only small uncertainties in the spin Hamiltonian are the exact values of the small in-plane hyperfine coupling values A_{xx} and A_{yy}, and the residual inhomogeneous linewidth arising from unresolved hyperfine couplings to other nuclear spins. This gives some uncertainty in modeling the exited detection and pump spins and therefore some inaccuracies in the modeling of the PELDOR time traces—especially at low field (X- and Q-band). The uncertainty becomes much less important at higher magnetic field values, where the orientation selection is strongly enhanced.

Another somewhat more serious problem arises from the pulse excitation profiles in the frequency domain, which are influenced by the resonator bandwidth and leads to rise time artifacts as well as B_1 inhomogeneities over the sample volume. The problem is, again, most pronounced at lower magnetic fields, where more overlap of pump and detection pulse excitation profiles can occur. At higher magnetic fields other experimental problems occur, only a small number of spins are detected and, more importantly, pumped, leading to very small modulation depths. Compensating this by better signal-to-noise requires a very high amplitude and phase stability of the detected echo signal. Nevertheless, with the very small modulation depth (only few percent typically), the PELDOR time traces are very sensitive to experimental artifacts arising from the response of the detection channel to the excitation pulses.

Both problems described above may be overcome by hyperbolical pulses introduced recently as broadband pump pulses in PELDOR spectroscopy at X-band frequencies demonstrated improved performance (Spindler, Glaser, Skinner, & Prisner, 2013). The broader excitation profile of these pulses allows one to excite more spins, which could improve the modulation depth at high magnetic fields. Additionally, they also have a rectangular excitation profile in the frequency domain, which gives a well-defined selection of excited spins and can be used to avoid spectral overlap between pump and detection pulses. Recently, we carried out SIFTER dipolar experiment (Jeschke, Pannier, Godt, & Spiess, 2000; Schöps, Spindler, Marko, & Prisner, 2015) to show that such pulses have sufficient bandwidth to excite the full nitroxide spectrum at X-band frequencies without distortions. This offers a new and very promising perspective: to obtain the spectral dimension directly from the Fourier transformation of the echo signal. This would allow one to obtain all the orientation information without increased measurement time from the direct time domain Fourier transformation! Unfortunately, this dream is not reached yet; whereas a refocused and Hahn echo sequence can be created by such pulses using the Bohlen–Bodenhausen scheme (Bohlen, Rey, & Bodenhausen, 1989; Doll & Jeschke, 2014; Schöps et al., 2015), this does not hold true for the solid echo, which is used for the SIFTER experiment. The development of a phase coherent detection sequences with broadband pulses for SIFTER and DQC-EPR (Borbat, Mchaourab, & Freed, 2002) is in the focus of our current research efforts.

PELDOR experiments performed with rigid spin labels provide very detailed structural information on biomolecules. If these details are important and of particular interest, the extended effort in synthesis and data analysis is well spent. We believe that rigid spin labels will become more common for protein research as well. Because the spin label is rigidly linked to the much larger macromolecule, the overall tumbling rate is shifted into the slow-tumbling regime. This might allow to extend PELDOR experiments with such spin labels to room temperatures without the need of additional immobilization of the biomolecule (Babaylova et al., 2014; Meyer et al., 2015; Yang et al., 2012). Rigid spin label could be very attractive not only for distance measurements at physiological temperatures (avoiding freezing effects) but also to observe dynamics in the nanosecond to microsecond time range directly in the time domain PELDOR signal.

ACKNOWLEDGMENTS
Funding from the German Research Society (DFG) within the Collaborative Research Center CRC902 Molecular Principles of RNA-based Regulation is gratefully acknowledged as well as support from the Center of Biomolecular Magnetic Resonance (BMRZ) and the Center of Excellence Frankfurt Macromolecular Complexes (DFG).

REFERENCES
Abé, C., Klose, D., Dietrich, F., Ziegler, W. H., Polyhach, Y., Jeschke, G., et al. (2012). Orientation selective DEER measurements on vinculin tail at X-band frequencies reveal spin label orientations. *Journal of Magnetic Resonance, 216*, 53–61.

Altenbach, C., Marti, T., Khorana, H. G., & Hubbell, W. L. (1990). Transmembrane protein structure: Spin labeling of bacteriorhodopsin mutants. *Science, 248*(4959), 1088–1092.

Babaylova, E. S., Ivanov, A. V., Malygin, A. A., Vorobjeva, M. A., Venyaminova, A. G., Polienko, Y. F., et al. (2014). A versatile approach for site-directed spin labeling and structural EPR studies of RNAs. *Organic & Biomolecular Chemistry, 12*(19), 3129.

Barhate, N., Cekan, P., Massey, A. P., & Sigurdsson, S. T. (2007). A nucleoside that contains a rigid nitroxide spin label: A fluorophore in disguise. *Angewandte Chemie, International Edition, 46*(15), 2655–2658.

Becker, N. B., & Everaers, R. (2007). From rigid base pairs to semiflexible polymers: Coarse-graining DNA. *Physical Review E, 76*(2), 021923.

Becker, N. B., & Everaers, R. (2009). Comment on 'remeasuring the double helix'. *Science, 325*(5940), 538.

Becker, J. S., & Saxena, S. (2005). Double quantum coherence electron spin resonance on coupled Cu(II)–Cu(II) electron spins. *Chemical Physics Letters, 414*(1–3), 248–252.

Bender, C. J., Sahlin, M., Babcock, G. T., Barry, B. A., Chandrashekar, T. K., Salowe, S. P., et al. (1989). An ENDOR study of the tyrosyl free radical in ribonucleotide reductase from Escherichia coli. *Journal of the American Chemical Society, 111*(21), 8076–8083.

Bennati, M., Robblee, J. H., Mugnaini, V., Stubbe, J., Freed, J. H., & Borbat, P. (2005). EPR distance measurements support a model for long-range radical initiation in *E. Coli* ribonucleotide reductase. *Journal of the American Chemical Society, 127*(43), 15014–15015.

Bennati, M., Weber, A., Antonic, J., Perlstein, D. L., Robblee, J., & Stubbe, J. (2003). Pulsed ELDOR spectroscopy measures the distance between the two tyrosyl radicals in the R2 subunit of the *E. Coli* ribonucleotide reductase. *Journal of the American Chemical Society, 125*(49), 14988–14989.

Biglino, D., Schmidt, P. P., Reijerse, E. J., & Lubitz, W. (2006). PELDOR study on the tyrosyl radicals in the R2 protein of mouse ribonucleotide reductase. *Physical Chemistry Chemical Physics, 8*(1), 58–62.

Bode, B. E., Plackmeyer, J., Bolte, M., Prisner, T. F., & Schiemann, O. (2009). PELDOR on an exchange coupled nitroxide copper(II) spin pair. *Journal of Organometallic Chemistry, 694*(7–8), 1172–1179, Organo-Transition Metal Complexes Dedicated to Prof. Dr. Ch. Elschenbroich.

Bohlen, J.-M., Rey, M., & Bodenhausen, G. (1989). Refocusing with chirped pulses for broadband excitation without phase dispersion. *Journal of Magnetic Resonance (1969), 84*(1), 191–197.

Borbat, P. P., & Freed, J. H. (2007). Measuring distances by pulsed dipolar ESR spectroscopy: Spin-labeled histidine kinases. In A. C. Melvin I. Simon (Ed.), *Methods in enzymology: Vol. 423* (pp. 52–116). San Diego: Academic Press.

Borbat, P. P., Mchaourab, H. S., & Freed, J. H. (2002). Protein structure determination using long-distance constraints from double-quantum coherence ESR: Study of T4 lysozyme. *Journal of the American Chemical Society, 124*, 5304–5314.

Chiang, Y.-W., Borbat, P. P., & Freed, J. H. (2005). The determination of pair distance distributions by pulsed ESR using Tikhonov regularization. *Journal of Magnetic Resonance*, *172*(2), 279–295.

Cruickshank, P. A. S., Bolton, D. R., Robertson, D. A., Hunter, R. I., Wylde, R. J., & Smith, G. M. (2009). A kilowatt pulsed 94 GHz electron paramagnetic resonance spectrometer with high concentration sensitivity, high instantaneous bandwidth, and low dead time. *Review of Scientific Instruments*, *80*(10), 103102.

Denysenkov, V. P., Biglino, D., Lubitz, W., Prisner, T. F., & Bennati, M. (2008). Structure of the tyrosyl biradical in mouse R2 ribonucleotide reductase from high-field PELDOR. *Angewandte Chemie, International Edition*, *47*(7), 1224–1227.

Denysenkov, V. P., Prisner, T. F., Stubbe, J., & Bennati, M. (2006). High-field pulsed electron–electron double resonance spectroscopy to determine the orientation of the tyrosyl radicals in ribonucleotide reductase. *Proceedings of the National Academy of Sciences of the United States of America*, *103*(36), 13386–13390.

Doll, A., & Jeschke, G. (2014). Fourier-transform electron spin resonance with bandwidth-compensated chirp pulses. *Journal of Magnetic Resonance*, *246*, 18–26.

Edwards, T. E., Cekan, P., Reginsson, G. W., Shelke, S. A., Ferré-D'Amaré, A. R., Schiemann, O., et al. (2011). Crystal structure of a DNA containing the planar, phenoxazine-derived bi-functional spectroscopic probe Ç. *Nucleic Acids Research*, *39*(10), 4419–4426.

Edwards, T. E., & Sigurdsson, S. T. (2007). Site-specific incorporation of nitroxide spin-labels into 2′-positions of nucleic acids. *Nature Protocols*, *2*(8), 1954–1962.

Elsaesser, C., Brecht, M., & Bittl, R. (2005). Treatment of spin-coupled metal-centres in pulsed electron–electron double-resonance experiments. *Biochemical Society Transactions*, *33*, 15–19.

Endeward, B., Butterwick, J. A., MacKinnon, R., & Prisner, T. F. (2009). Pulsed electron–electron double-resonance determination of spin-label distances and orientations on the tetrameric potassium ion channel KcsA. *Journal of the American Chemical Society*, *131*(42), 15246–15250.

Fajer, M. I., Li, H., Yang, W., & Fajer, P. G. (2007). Mapping electron paramagnetic resonance spin label conformations by the simulated scaling method. *Journal of the American Chemical Society*, *129*(45), 13840–13846.

Gerfen, G. J., Bellew, B. F., Un, S., Bollinger, J. M., Jr., Stubbe, J., Griffin, R. G., et al. (1993). High-frequency (139.5 GHz) EPR spectroscopy of the tyrosyl radical in Escherichia coli ribonucleotide reductase. *Journal of the American Chemical Society*, *115*(14), 6420–6421.

Gophane, D. B., Endeward, B., Prisner, T. F., & Sigurdsson, S. T. (2014). Conformationally restricted isoindoline-derived spin labels in duplex DNA: Distances and rotational flexibility by pulsed electron–electron double resonance spectroscopy. *Chemistry: A European Journal*, *20*(48), 15913–15919.

Gophane, D. B., & Sigurdsson, S. T. (2013). Hydrogen-bonding controlled rigidity of an isoindoline-derived nitroxide spin label for nucleic acids. *Chemical Communications*, *49*(10), 999–1001.

Gore, J., Bryant, Z., Nöllmann, M., Le, M. U., Cozzarelli, N. R., & Bustamante, C. (2006). DNA overwinds when stretched. *Nature*, *442*(7104), 836–839.

Gromov, I., Shane, J., Forrer, J., Rakhmatoullin, R., Rozentzwaig, Y., & Schweiger, A. (2001). A Q-band pulse EPR/ENDOR spectrometer and the implementation of advanced one- and two-dimensional pulse EPR methodology. *Journal of Magnetic Resonance*, *149*(2), 196–203.

Höbartner, C., Sicoli, G., Wachowius, F., Gophane, D. B., & Sigurdsson, S. T. (2012). Synthesis and characterization of RNA containing a rigid and nonperturbing cytidine-derived spin label. *The Journal of Organic Chemistry*, *77*(17), 7749–7754.

Hoganson, C. W., Sahlin, M., Sjöberg, B.-M., & Babcock, G. T. (1996). Electron magnetic resonance of the tyrosyl radical in ribonucleotide reductase from Escherichia coli. *Journal of the American Chemical Society, 118*(19), 4672–4679.

Hubbell, W. L., Cafiso, D. S., & Altenbach, C. (2000). Identifying conformational changes with site-directed spin labeling. *Nature Structural Biology, 7*(9), 735–739.

Hubbell, W. L., López, C. J., Altenbach, C., & Yang, Z. (2013). Technological advances in site-directed spin labeling of proteins. *Current Opinion in Structural Biology, 23*(5), 725–733, Protein-carbohydrate interactions/Biophysical methods.

Jeschke, G. (2012). DEER distance measurements on proteins. *Annual Review of Physical Chemistry, 63*(1), 419–446.

Jeschke, G., Chechik, V., Ionita, P., Godt, A., Zimmermann, H., Banham, J., et al. (2006). DeerAnalysis2006—A comprehensive software package for analyzing pulsed ELDOR data. *Applied Magnetic Resonance, 30*, 473–498.

Jeschke, G., Koch, A., Jonas, U., & Godt, A. (2002). Direct conversion of EPR dipolar time evolution data to distance distributions. *Journal of Magnetic Resonance, 155*(1), 72–82.

Jeschke, G., Pannier, M., Godt, A., & Spiess, H. W. (2000). Dipolar spectroscopy and spin alignment in electron paramagnetic resonance. *Chemical Physics Letters, 331*(2–4), 243–252.

Karim, C. B., Kirby, T. L., Zhang, Z., Nesmelov, Y., & Thomas, D. D. (2004). Phospholamban structural dynamics in lipid bilayers probed by a spin label rigidly coupled to the peptide backbone. *Proceedings of the National Academy of Sciences of the United States of America, 101*(40), 14437–14442.

Kay, C. W. M., Elsässer, C., Bittl, R., Farrell, S. R., & Thorpe, C. (2006). Determination of the distance between the two neutral flavin radicals in augmenter of liver regeneration by pulsed ELDOR. *Journal of the American Chemical Society, 128*(1), 76–77.

Kolberg, M., Logan, D. T., Bleifuss, G., Potsch, S., Sjoberg, B. M., Graslund, A., et al. (2005). A new tyrosyl radical on Phe208 as ligand to the diiron center in Escherichia coli ribonucleotide reductase, mutant R2-Y122H: Combined X-ray diffraction and EPR/ENDOR studies. *Journal of Biological Chemistry, 280*(12), 11233–11246.

Larsen, R. G., Halkides, C. J., & Singel, D. J. (1993). A geometric representation of nuclear modulation effects: The effects of high electron spin multiplicity on the electron spin echo envelope modulation spectra of Mn^{2+} complexes of N-*ras* p21. *The Journal of Chemical Physics, 98*(9), 6704–6721.

Larsen, R. G., & Singel, D. J. (1993). Double electron–electron resonance spin-echo modulation: Spectroscopic measurement of electron spin pair separations in orientationally disordered solids. *Journal of Chemical Physics, 98*, 5134–5146.

Margraf, D., Bode, B. E., Marko, A., Schiemann, O., & Prisner, T. F. (2007). Conformational flexibility of nitroxide biradicals determined by X-band PELDOR experiments. *Molecular Physics, 105*(15–16), 2153–2160.

Marko, A., Denysenkov, V., Margraf, D., Cekan, P., Schiemann, O., Sigurdsson, S. T., et al. (2011). Conformational flexibility of DNA. *Journal of the American Chemical Society, 133*(34), 13375–13379.

Marko, A., Denysenkov, V., & Prisner, T. F. (2013). Out-of-phase PELDOR. *Molecular Physics, 111*(18–19), 2834–2844.

Marko, A., Margraf, D., Cekan, P., Sigurdsson, S. T., Schiemann, O., & Prisner, T. F. (2010). Analytical method to determine the orientation of rigid spin labels in DNA. *Physical Review E, 81*(2), 021911.

Marko, A., Margraf, D., Yu, H., Mu, Y., Stock, G., & Prisner, T. (2009). Molecular orientation studies by pulsed electron–electron double resonance experiments. *The Journal of Chemical Physics, 130*(6), 064102–064109.

Marko, A., & Prisner, T. F. (2013). An algorithm to analyze PELDOR data of rigid spin label pairs. *Physical Chemistry Chemical Physics, 15*(2), 619–627.

Martin, R. E., Pannier, M., Diederich, F., Gramlich, V., Hubrich, M., & Spiess, H. W. (1998). Determination of end-to-end distances in a series of TEMPO diradicals of up to 2.8 nm length with a new four-pulse double electron electron resonance experiment. *Angewandte Chemie, International Edition, 37*(20), 2833–2837.

Mathew-Fenn, R. S., Das, R., & Harbury, P. A. B. (2008). Remeasuring the double helix. *Science, 322*(5900), 446–449.

McNulty, J. C., & Millhauser, G. L. (2000). TOAC—The rigid nitroxide side chain. In L. Berliner, G. R. Eaton, & S. S. Eaton (Eds.), *Biological magnetic resonance: Distance measurements in biological systems by EPR: Vol. 19* (pp. 277–307). New York: Kluwer Academic/Plenum Publishers.

Meyer, V., Swanson, M. A., Clouston, L. J., Boratyński, P. J., Stein, R. A., Mchaourab, H. S., et al. (2015). Room-temperature distance measurements of immobilized spin-labeled protein by DEER/PELDOR. *Biophysical Journal, 108*(5), 1213–1219.

Milov, A. D., Maryasov, A. G., & Tsvetkov, Y. D. (1998). Pulsed electron double resonance (PELDOR) and its applications in free-radicals research. *Applied Magnetic Resonance, 15*, 107–143.

Milov, A. D., Ponomarev, A. B., & Tsvetkov, Y. (1984). Electron–electron double resonance in electron spin echo: Model biradical systems and the sensitized photolysis of decalin. *Chemical Physics Letters, 110*(1), 67–72.

Milov, A. D., Salikov, K. M., & Shirov, M. D. (1981). Application of the double resonance method to electron spin echo in a study of the spatial distribution of paramagnetic centers in solids. *Soviet Physics—Solid State, 23*, 565–569.

Milov, A. D., & Tsvetkov, Y. D. (1997). Double electron–electron resonance in electron spin echo: Conformations of spin-labeled poly-4-vinilpyridine in glassy solutions. *Applied Magnetic Resonance, 12*(4), 495–504.

Möbius, K., Savitsky, A., Wegener, C., Plato, M., Fuchs, M., Schnegg, A., et al. (2005). Combining high-field EPR with site-directed spin labeling reveals unique information on proteins in action. *Magnetic Resonance in Chemistry, 43*(S1), S4–S19.

Pannier, M., Veit, S., Godt, A., Jeschke, G., & Spiess, H. W. (2000). Dead-time free measurement of dipole–dipole interactions between electron spins. *Journal of Magnetic Resonance, 142*, 331–340.

Piton, N., Schiemann, O., Mu, Y., Stock, G., Prisner, T., & Engels, J. W. (2005). Synthesis of spin-labeled RNAs for long range distance measurements by PELDOR. *Nucleosides, Nucleotides & Nucleic Acids, 24*(5–7), 771–775.

Polyhach, Y., Bordignon, E., Tschaggelar, R., Gandra, S., Godt, A., & Jeschke, G. (2012). High sensitivity and versatility of the DEER experiment on nitroxide radical pairs at Q-band frequencies. *Physical Chemistry Chemical Physics, 14*(30), 10762.

Polyhach, Y., Godt, A., Bauer, C., & Jeschke, G. (2007). Spin pair geometry revealed by high-field DEER in the presence of conformational distributions. *Journal of Magnetic Resonance, 185*(1), 118–129.

Prisner, T. F., Marko, A., & Sigurdsson, S. T. (2015). Conformational dynamics of nucleic acid molecules studied by PELDOR spectroscopy with rigid spin labels. *Journal of Magnetic Resonance, 252*, 187–198.

Qin, P. Z., Haworth, I. S., Cai, Q., Kusnetzow, A. K., Grant, G. P. G., Price, E. A., et al. (2007). Measuring nanometer distances in nucleic acids using a sequence-independent nitroxide probe. *Nature Protocols, 2*(10), 2354–2365.

Qin, P. Z., Hideg, K., Feigon, J., & Hubbell, W. L. (2003). Monitoring RNA base structure and dynamics using site-directed spin labeling. *Biochemistry, 42*(22), 6772–6783.

Ramos, A., & Varani, G. (1998). A new method to detect long-range protein–RNA contacts: NMR detection of electron–proton relaxation induced by nitroxide spin-labeled RNA. *Journal of the American Chemical Society, 120*(42), 10992–10993.

Riplinger, C., Kao, J. P. Y., Rosen, G. M., Kathirvelu, V., Eaton, G. R., Eaton, S. S., et al. (2009). Interaction of radical pairs through-bond and through-space: Scope and limitations of the point–dipole approximation in electron paramagnetic resonance spectroscopy. *Journal of the American Chemical Society, 131*(29), 10092–10106.

Roessler, M. M., King, M. S., Robinson, A. J., Armstrong, F. A., Harmer, J., & Hirst, J. (2010). Direct assignment of EPR spectra to structurally defined iron-sulfur clusters in complex I by double electron–electron resonance. *Proceedings of the National Academy of Sciences of the United States of America, 107*, 1930–1935.

Sajid, M., Jeschke, G., Wiebcke, M., & Godt, A. (2009). Conformationally unambiguous spin labeling for distance measurements. *Chemistry: A European Journal, 15*(47), 12960–12962.

Sale, K., Song, L., Liu, Y.-S., Perozo, E., & Fajer, P. (2005). Explicit treatment of spin labels in modeling of distance constraints from dipolar EPR and DEER. *Journal of the American Chemical Society, 127*(26), 9334–9335.

Schiemann, O., Cekan, P., Margraf, D., Prisner, T. F., & Sigurdsson, S. T. (2009). Relative orientation of rigid nitroxides by PELDOR: Beyond distance measurements in nucleic acids. *Angewandte Chemie, International Edition, 48*(18), 3292–3295.

Schiemann, O., Piton, N., Mu, Y., Stock, G., Engels, J. W., & Prisner, T. F. (2004). A PELDOR-based nanometer distance ruler for oligonucleotides. *Journal of the American Chemical Society, 126*(18), 5722–5729.

Schiemann, O., Piton, N., Plackmeyer, J., Bode, B. E., Prisner, T. F., & Engels, J. W. (2007). Spin labeling of oligonucleotides with the nitroxide TPA and use of PELDOR, a pulse EPR method, to measure intramolecular distances. *Nature Protocols, 2*(4), 904–923.

Schöps, P., Spindler, P. E., Marko, A., & Prisner, T. F. (2015). Broadband spin echoes and broadband SIFTER in EPR. *Journal of Magnetic Resonance, 250*, 55–62.

Sezer, D., Freed, J. H., & Roux, B. (2009). Multifrequency electron spin resonance spectra of a spin-labeled protein calculated from molecular dynamics simulations. *Journal of the American Chemical Society, 131*(7), 2597–2605.

Shelke, S. A., & Sigurdsson, S. T. (2010). Noncovalent and site-directed spin labeling of nucleic acids. *Angewandte Chemie, International Edition, 49*(43), 7984–7986.

Shelke, S. A., & Sigurdsson, S. T. (2012). Site-directed spin labelling of nucleic acids. *European Journal of Organic Chemistry, 2012*(12), 2291–2301.

Sigurdsson, S. T., & Eckstein, F. (1996). Site specific labelling of sugar residues in oligoribonucleotides: Reactions of aliphatic isocyanates with $2'$ amino groups. *Nucleic Acids Research, 24*(16), 3129–3133.

Spaltenstein, A., Robinson, B. H., & Hopkins, P. B. (1989). Sequence-and structure-dependent DNA base dynamics: Synthesis, structure, and dynamics of site, and sequence specifically spin-labeled DNA. *Biochemistry, 28*(24), 9484–9495.

Spindler, P. E., Glaser, S. J., Skinner, T. E., & Prisner, T. F. (2013). Broadband inversion PELDOR spectroscopy with partially adiabatic shaped pulses. *Angewandte Chemie, International Edition, 52*(12), 3425–3429.

Strube, T., Schiemann, O., MacMillan, F., Prisner, T., & Engels, J. W. (2001). A new facile method for spin-labeling of oligonucleotides. *Nucleosides, Nucleotides and Nucleic Acids, 20*(4–7), 1271–1274.

Tkach, I., Sicoli, G., Höbartner, C., & Bennati, M. (2011). A dual-mode microwave resonator for double electron–electron spin resonance spectroscopy at W-band microwave frequencies. *Journal of Magnetic Resonance, 209*(2), 341–346.

Van Amsterdam, I., Ubbink, M., Canters, G. W., & Huber, M. (2003). Measurement of a Cu-Cu distance of 26 Å by a pulsed EPR method. *Angewandte Chemie, International Edition, 42*(1), 62–64.

Yang, Z., Liu, Y., Borbat, P., Zweier, J. L., Freed, J. H., & Hubbell, W. L. (2012). Pulsed ESR dipolar spectroscopy for distance measurements in immobilized spin labeled proteins in liquid solution. *Journal of the American Chemical Society, 134*(24), 9950–9952.

CHAPTER FIFTEEN

An Integrated Spin-Labeling/ Computational-Modeling Approach for Mapping Global Structures of Nucleic Acids

Narin S. Tangprasertchai*,[1], Xiaojun Zhang*,[1], Yuan Ding*, Kenneth Tham*,[2], Remo Rohs*,[†], Ian S. Haworth[‡], Peter Z. Qin*,[†],[3]

*Department of Chemistry, University of Southern California, Los Angeles, California, USA
[†]Molecular and Computational Biology Program, Department of Biological Sciences, University of Southern California, Los Angeles, California, USA
[‡]Department of Pharmacology and Pharmaceutical Sciences, School of Pharmacy, University of Southern California, Los Angeles, California, USA
[1]These authors contributed equally.
[3]Corresponding author: e-mail address: pzq@usc.edu

Contents

1. Introduction	428
2. SDSL of Nucleic Acids Using a Nucleotide-Independent Nitroxide Probe	431
2.1 Oligonucleotides with Site-Specific Phosphorothioate Modifications	434
2.2 Preparation of the R5 Precursor	435
2.3 Oligonucleotide Labeling	436
2.4 Purification of Labeled Oligonucleotides	437
2.5 Characterization of Labeled Oligonucleotides	438
3. Measuring Inter-R5 Distances Using Double Electron–Electron Resonance Spectroscopy	439
3.1 DEER Sample Preparation	439
3.2 DEER Data Acquisition	440
3.3 DEER Spectrum Analysis	443
4. Integration of DEER-Measured Distances with Computational Modeling	443
4.1 Model Generation	444
4.2 Computing Expected Inter-R5 Distances for the Model Pool	444
4.3 Model Selection and Characterization	446
4.4 Additional Considerations	448
5. Conclusions	449
Acknowledgment	450
References	450

[2] Current address: School of Pharmacy, University of California San Francisco, San Francisco, California, USA.

Abstract

The technique of site-directed spin labeling (SDSL) provides unique information on biomolecules by monitoring the behavior of a stable radical tag (i.e., spin label) using electron paramagnetic resonance (EPR) spectroscopy. In this chapter, we describe an approach in which SDSL is integrated with computational modeling to map conformations of nucleic acids. This approach builds upon a SDSL tool kit previously developed and validated, which includes three components: (i) a nucleotide-independent nitroxide probe, designated as R5, which can be efficiently attached at defined sites within arbitrary nucleic acid sequences; (ii) inter-R5 distances in the nanometer range, measured via pulsed EPR; and (iii) an efficient program, called NASNOX, that computes inter-R5 distances on given nucleic acid structures. Following a general framework of data mining, our approach uses multiple sets of measured inter-R5 distances to retrieve "correct" all-atom models from a large ensemble of models. The pool of models can be generated independently without relying on the inter-R5 distances, thus allowing a large degree of flexibility in integrating the SDSL-measured distances with a modeling approach best suited for the specific system under investigation. As such, the integrative experimental/computational approach described here represents a hybrid method for determining all-atom models based on experimentally-derived distance measurements.

1. INTRODUCTION

Rapid advances in a number of areas of biology, such as the discovery of ribozymes, riboswitches, noncoding RNAs, and noncanonical DNA structures, have clearly established that nucleic acids, including both DNA and RNA, are not just passive information carriers; instead, they play active and crucial roles as regulators and executors of biological functions. Similar to that of proteins, the ability of nucleic acids to fold into complex and compact three-dimensional structures is crucial for their functions. As such, information on tertiary structures of nucleic acids, as well as their complexes with proteins and small-molecule ligands, is essential for understanding many biological processes, and methodologies for obtaining such information are of great importance.

In this chapter, we describe an integrated approach in which long-range distances measured via site-directed spin labeling (SDSL) are combined with computational modeling to obtain structural models of nucleic acids. In SDSL, chemically stable radicals (i.e., spin labels) are covalently attached at specific sites of a macromolecule. The behavior of the spin labels is monitored using electron paramagnetic resonance (EPR) spectroscopy, from which local information on the macromolecule is obtained (Fig. 1; Hubbell & Altenbach, 1994). SDSL has been demonstrated as a powerful tool for investigating structure and dynamics of biomolecules, as evidenced by accompanying articles in this volume, as well as a number of excellent reviews (Ding et al., 2015; Fedorova & Tsvetkov, 2013; Hubbell & Altenbach, 1994; Hubbell, Cafiso, & Altenbach, 2000; Hubbell, López, Altenbach, & Yang, 2013; Krstic, Endeward, Margraf, Marko, & Prisner, 2012; Shelke & Sigurdsson, 2012; Sowa & Qin, 2008).

One of the main EPR observables used in SDSL studies is the distance between a pair of spin labels, which can be obtained by measuring magnetic dipolar coupling using either continuous-wave (cw) or, more recently, pulsed EPR techniques (Cafiso, 2012; Jeschke, 2012; Krstic et al., 2012; Polyhach, Bordignon, & Jeschke, 2011; Schiemann & Prisner, 2007). The measurable range of distances spans from 5 to up to 100 Å, providing direct structural constraints for monitoring conformational change and for mapping biomolecular structure. Compared to X-ray crystallography, SDSL distance measurements do not require crystalline samples, and therefore are suitable for studying systems that do not yield crystals or are susceptible to artifacts due to crystal packing. Compared to solution NMR measurements, SDSL is not limited by the molecular weight of the system, requires a smaller amount of sample, and provides longer-range distances. Last, but not least, in contrast to techniques, such as fluorescence resonance energy transfer, which generally use two chemically distinct fluorophores, with SDSL,

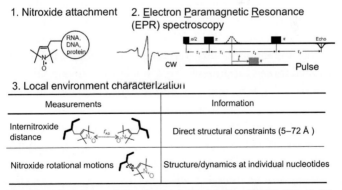

Figure 1 The general strategy of site-directed spin labeling (SDSL). Step 1: Attach a spin label (i.e., a nitroxide) to specific site(s) of a macromolecule. Step 2: Monitor behavior of the spin label using EPR spectroscopy. Step 3: Derive information about the target molecule. The table shows the two primary types of information that can be derived from EPR measurements on nucleic acids. *Adapted from Sowa and Qin (2008) with permission.*

distances can be measured using a pair of chemically identical spin labels (e.g., nitroxides), which simplifies the labeling procedure. In addition, in the majority of spin labels, the unpaired electron(s) is localized at a known position(s), thereby facilitating modeling that explicitly accounts for the label.

One should also take into account "limitations" when designing studies for SDSL distance measurements. Most spin labels are extrinsic probes tagged onto the target molecules, so it is important to assess perturbations to the native system due to the labels. The measured interspin distances (r_{spin}) are generally different from the actual distances at the parent molecule (r_{target}) and a number of methods are reported to correlate the two (e.g., Hagelueken, Ward, Naismith, & Schiemann, 2012; Hatmal et al., 2012; Polyhach et al., 2011; Qin et al., 2007; see also Section 4). Currently, most long-range SDSL distance measurements require the use of cryogenic temperatures, due to limitations imposed by spin relaxation, and one should assess potential impact on the particular system being studied (e.g., see Georgieva et al., 2012). It is also important to note that, in SDSL studies, the number of distances measured is relatively small. However, a number of studies have shown that with careful design of experiments these "sparse" long-range constraints can be highly informative (Duss, Michel, et al., 2014; Duss, Yulikov, Allain, & Jeschke, 2015; Duss, Yulikov, Jeschke, & Allain, 2014; Hirst, Alexander, McHaourab, & Meiler, 2011; Zhang et al., 2014, 2012).

In this chapter, we describe an integrated SDSL/computational-modeling approach that we have developed for mapping an RNA junction (Zhang et al., 2012) and examining sequence-dependent conformations of DNA duplexes (Chen et al., 2013; Zhang et al., 2014). This approach builds upon an SDSL tool kit previously developed and validated (Qin et al., 2007), which includes three components: (i) a nucleotide-independent nitroxide probe, designated as R5, that can be efficiently attached at defined sites within arbitrary nucleic acid sequences (Fig. 2); (ii) inter-R5 distances measured via EPR (Fig. 3); and (iii) a web-accessible program, called NASNOX, that efficiently computes inter-R5 distances on given nucleic acid structures (Fig. 4). The underlying philosophy of our approach is to use multiple measured inter-R5 distances to retrieve "correct" all-atom models from a large ensemble of models. This pool of models can be generated independently without relying on the measured inter-R5 distances, thus allowing a large degree of flexibility in integrating the SDSL-measured distances with a modeling approach that is best suited for the specific system

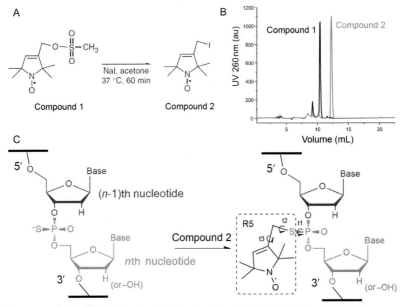

Figure 2 Nucleic acid labeling using the nucleotide-independent R5 nitroxide probe. (A) Synthesis of compound **2**, the reactive R5 precursor. (B) An example of reverse-phase HPLC characterization of compounds **1** (shown in black) and **2** (shown in red, light gray in the print version). Data shown were collected using a C18 column (Prosphere™ C18, Grace Davidson, Inc.) and a linear gradient generated with buffer A: 0.1 M triethylammonium acetate (TEAA, pH 7.0) and 5% (v/v) acetonitrile; and buffer B: 100% acetonitrile. (C) Site-specific attachment of compound **2** to a phosphorothioate-modified nucleic acid strand. As shown, R5 is attached to the S_p diastereomer of the nth nucleotide, and the three torsion angles about the single bonds connecting the pyrroline ring to the nucleic acid are indicated. Adapted from Qin et al. (2007) with permission.

under investigation. The use of experimentally-measured distance constraints in selecting an all-atom model from a large ensemble of pregenerated putative models represents a "hybrid" approach that combines experimental with computational methods in structure determination.

2. SDSL OF NUCLEIC ACIDS USING A NUCLEOTIDE-INDEPENDENT NITROXIDE PROBE

A variety of methods for attaching a spin label at a specific site of a target DNA or RNA have been reported and are summarized in a number of recent reviews (Fedorova & Tsvetkov, 2013; Shelke & Sigurdsson, 2012; Sowa & Qin, 2008), as well as in a number of chapters within this volume. Here, we focus on a family of nitroxides, called the R5-series, which can be

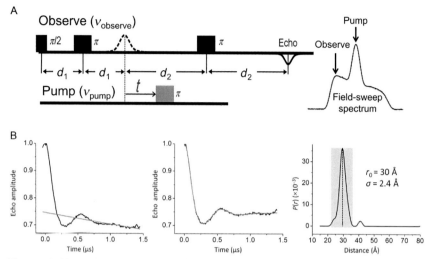

Figure 3 Measuring nanometer distances using DEER spectroscopy. (A) Pulse sequences for the four-pulse DEER. Three pulses are applied to the observe frequency and generate a refocused echo (shown as "Observe"). Another pulse is applied to a different frequency (pump frequency) to flip the pump spin. Shown on the right is a field-sweep spectrum acquired at the observe frequency with pump spin and observe spin indicated. (B) An example of DEER data acquired on a DNA duplex. Left: Measured normalized echo decay (black) overlaid with the simulated background decay (red, light gray in the print version); center: Measured background-corrected echo decay (black) overlaid with the simulated trace computed from the optimized distance distribution (red, light gray in the print version); right: The corresponding optimized distance distribution computed using DeerAnalysis2013. *Adapted from Qin et al. (2007), Sowa and Qin (2008), and Zhang et al. (2014) with permission.*

attached to a phosphorothioate group introduced at a defined location of the nucleic acid backbone during solid-phase chemical synthesis (Qin, Butcher, Feigon, & Hubbell, 2001; Qin et al., 2007). Among this family of probes, R5 (Fig. 2) has been primarily used for mapping nucleic acid structures and will be the focus of the method described herein. The other probes, R5a (Popova, Kálai, Hideg, & Qin, 2009) and R5c (Nguyen, Popova, Hideg, & Qin, 2015), are attached based on the same chemistry, but include substituent(s) that modulate motions of the nitroxide moiety with respect to the parent molecule. When used as single labels, R5a and R5c have been shown to be advantageous in probing local environments (Ding, Zhang, Tham, & Qin, 2014; Grant, Boyd, Herschlag, & Qin, 2009; Nguyen et al., 2015; Popova et al., 2012, 2009; Popova & Qin, 2010). It will be of interest to further explore whether they also offer enhancement in

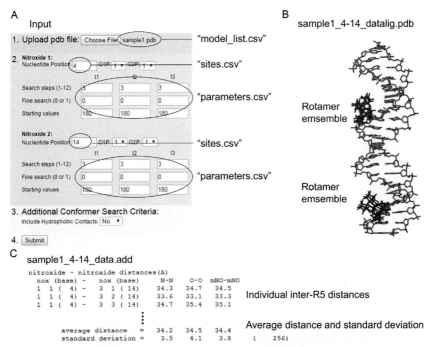

Figure 4 The NASNOX program for computing expected inter-R5 distances on given nucleic acid structures. (A) Examples of input parameters. The previously reported web interface of NASNOX_W (Qin et al., 2007) is shown to illustrate the organization of input information in the parameter files used to execute NASNOX in the batch mode. (B) An example of structure output in which a pair of R5's were modeled onto nucleotides 4 and 14 of the sample1.pdb input structure. The DNA is shown in red (light gray in the print version), and the allowed R5 rotamers at each site are shown in blue (black in the print version). (C) An example of text output showing the individual inter-R5 distances and the average distance and standard deviation for the entire ensemble. *Adapted from Qin et al. (2007) with permission.*

pairwise distance measurements, e.g., narrowing the distribution of the measured distances.

The R5 label has a number of advantages for mapping nucleic acid structures. Most importantly, the labeling site is not restricted by base identity, and the labeling method is highly effective and cost efficient (Qin et al., 2001, 2007). This enables scanning of the target molecule to obtain multiple distance measurements (Chen et al., 2013; Zhang et al., 2014, 2012), which is highly desirable in structural mapping. In addition, R5 has been shown to minimally perturb native nucleic acid duplexes (Cai et al., 2006, 2007), and an efficient program, called NASNOX (Price et al., 2007; Qin et al., 2007), has been developed to accurately correlate the measured inter-R5 distances to a given target structure.

In this section, we describe methods for attaching R5 to a specific site of a nucleic acid strand. The overall protocol described here largely follows what has been reported in Qin et al. (2007), with improvements noted where appropriate.

2.1 Oligonucleotides with Site-Specific Phosphorothioate Modifications

To achieve site-specific covalent attachment of a spin label(s), the phosphorodiester group(s) at the intended labeling site(s) is modified to a chemically more reactive phosphorothioate (Fig. 2). In this chapter (and throughout our work), we follow the convention that each phosphate belongs to the ribose on the $3'$-side of the P atom, such that, for an R5 labeled at the nth nucleotide, the phosphorothioate substitution is located between the $(n-1)$th and nth nucleotide (Fig. 2C). Furthermore, for RNA labeling, the $2'$-OH group at the $(n-1)$th position (i.e., $5'$ to the phosphorothioate modification, Fig. 2C) must be substituted with a nonnucleophilic functional group in order to prevent strand scission upon labeling (Qin et al., 2001). The simplest substitution is a $2'$-H (Fig. 2C), although others such as $2'$-O-methyl and $2'$-F, also suffice. It is worth noting that one, two, or even more phosphorothioates can be installed simultaneously within an oligonucleotide strand (Zhang et al., 2012, 2014), depending on the desired number of labels.

Oligonucleotides with site-specific phosphorothioate modifications are most easily obtained via chemical synthesis using the well-established phosphoramide scheme and are available from many commercial sources. In our work, we have been using custom-synthesized oligonucleotides from vendors such as Integrated DNA Technology (http://www.idtdna.com/). In general, a synthesis scale of 100 nmol usually yields greater than 50 nmol of products for oligonucleotides less than 30 nucleotides (nt) in length, which is sufficient for EPR studies. It is important to recognize that chemical synthesis results in two possible diastereomers (R_p and S_p) at each phosphorothioate-modified nucleotide. Once properly accounted for (see NASNOX modeling, Section 4.2.1), the presence of the diastereomers does not impact the use of the R5-series of probes (Qin et al., 2007).

When ordering a custom-synthesized oligonucleotide, one must adhere to the vendor's instructions on how to specify the desired location(s) of the internal phosphorothioate modification and check with the vendor regarding any additional in-house workup steps required. Typically, oligonucleotides that have undergone the basic postsynthesis deprotection and desalting procedures (e.g., "standard desalting" at Integrated DNA Technology) can be used for R5 labeling without further purification. However, we recommend first checking the quality of the synthesis products using high-performance liquid chromatography (HPLC) or gel electrophoresis. The phosphorothioate-modified oligonucleotides can be stored in powder form at $-20\ °C$ for 3 months or longer.

2.2 Preparation of the R5 Precursor

The R5 precursor, 3-iodomethyl-1-oxyl-2,2,5,5-tetramethylpyrroline (Hankovszky, Hideg, & Lex, 1980; compound **2**, Fig. 2A), is available commercially (e.g., Toronto Research Chemicals, cat. no. I709500). Alternatively, because compound **2** is light sensitive and may degrade upon long-term storage, one may desire to prepare compound **2** freshly before each labeling reaction. This can be done using a more stable compound, 1-oxyl-2,2,5,5-tetramethyl-3(methanesulfonyloxymethyl)pyrroline (compound **1**, Fig. 2A; Toronto Research Chemicals, cat. no. O872400), as a starting material, following the protocol described below.

Step 2.2.1 Prepare compound **2** from compound **1**

Thoroughly mix an equal amount of **1** and NaI in acetone. Incubate this reaction mixture at 37 °C for 60 min.

Notes

(a) The amount of **1** used can vary in a wide range, and is usually set based on the scale of the subsequent nucleic acid labeling reaction. For example, in our work a typical reaction starts with 2.48 mg of **1** (or 20 mM in a reaction volume of 500 μL), which generally yields sufficient product (i.e., compound **2**) for a 100 μL labeling reaction as described in Section 2.3.

(b) This reaction should contain equimolar amounts of NaI and **1**. An excess of NaI (e.g., >10%) will compete with compound **2** in the subsequent reaction, while an excess of **1** is simply a waste of the reagent.

(c) As the reaction progresses, a significant amount of white precipitate (by-product $NaOSO_2CH_3$) should become visible.

Step 2.2.2 Recover compound **2**

To separate **2** and by-products, spin down the reaction mixture in a benchtop centrifuge (typically at 14,000 rpm for 10 min at room temperature (20–25 °C)). Recover the supernatant, which contains the solubilized **2**.

Wash the precipitate with acetone, and repeat the spin-down procedure again, as described above. Combine all the supernatant fractions, then remove the solvent (e.g., using a benchtop speed vacuum) to obtain **2** as a red-brown oil.

Notes

(a) **2** is light sensitive, and from this point forward the tubes should be wrapped with aluminum foil to prevent excessive exposure to light.

(b) One may characterize conversion from **1** to **2** by methods such as reverse-phase HPLC (see example in Fig. 2B) or mass spectrometry. The conversion is usually quantitative.

(c) **2** can be stored at -20 or $-80\ °C$ for up to 6 months. However, this may result in reduced labeling efficiency.

2.3 Oligonucleotide Labeling

Step 2.3.1 Prepare a stock solution of compound **2**

Immediately before the labeling reaction, dissolve **2** (from Step 2.2.2 above or directly obtained from a vendor) in acetonitrile. Typically, a 1 M stock is prepared, with the total amount of compound **2** determined by the labeling scale according to the following Step 2.3.2.

Notes

(a) A solvent such as acetonitrile is used here and in the subsequent step to ensure solubility of **2** at the concentration required for the labeling reaction. We have found that formamide may also be used and may provide an additional benefit to denature the nucleic acids and enhance labeling.

Step 2.3.2 Assemble the labeling reaction

A typical reaction mixture contains 100 mM compound **2**, 100–150 μM oligonucleotide, 0.1 M 2-(N-morpholino)ethanesulfonic acid pH 5.8, and 20% (v/v) acetonitrile. Incubate the reaction mixture in the dark (wrap with aluminum foil) at room temperature for 12–24 h. Apply constant and gentle mixing during incubation.

Notes

(a) Oligonucleotides are generally dissolved in sterile ultrapure water or aqueous buffer (e.g., ME buffer: 10 mM MOPS, 1 mM EDTA, pH 6.5). As discussed above, crude oligonucleotides (deprotected and desalted) can be labeled directly.

(b) Studies have shown that **2** remains intact throughout the reaction period under conditions described here; however, the reaction rate is slow, on the order of $10^{-2}\ M^{-1}\ min^{-1}$ (data not shown). Conversely, prolonged incubation may lead to oligonucleotide degradation and/or off-target labeling. As such, we have found in most cases that using a concentration of **2** above 60 mM with an incubation time between 12 and 24 h gives the best results.

(c) Oligonucleotide concentration can be increased up to 1 mM, as needed. If a very large amount of labeled oligonucleotide is required, multiple reactions can be performed in tandem.

(d) One may reduce the reaction volume (e.g., 20 µL) in order to accommodate a smaller amount of **2**. There are no particular restrictions on reaction volume, as long as mixing of reagents is not impeded.

(e) No special accommodations are necessary to attach two labels, rather than one, to a single oligonucleotide strand containing two phosphorothioate modifications.

(f) The reaction mixture can be stored at -20 or -80 °C for up to 1 week. However, it is recommended that purification is performed as soon as possible after labeling.

2.4 Purification of Labeled Oligonucleotides

In choosing the appropriate purification scheme, multiple objectives should be considered, including: (i) removing excess unattached spin labels; (ii) separating labeled and unlabeled oligonucleotides; and (iii) in case crude oligonucleotides are used, separating full-length species from synthesis failures. In a majority of cases, we have found that HPLC or denaturing gel electrophoresis can achieve the desired degree of purification.

2.4.1 High-Performance Liquid Chromatography

We have successfully used anion-exchange HPLC to purify many oligonucleotides that are 10–55 nt long (Grant, Popova, & Qin, 2008; Qin et al., 2007). In particular, we have used a DNAPac PA-100 column from Dionex Inc. Upon successful anion-exchange HPLC purification, the samples usually need to be desalted, for example, using reverse-phase or size-exclusion chromatography. Following desalting, samples are generally lyophilized and stored at -20 or -80 °C.

Note that attachment of R5 results in the loss of one or more negative charges on the oligonucleotide strand and, in most cases, the labeled sample will elute earlier than the unlabeled one on anion-exchange HPLC (Qin et al., 2007). This is a very useful diagnostic feature for assessing the success of the labeling reaction. The PA-100 column also provides some degree of sample separation based on polarity. This enables separation of the R_p and S_p

diastereomers in some oligonucleotides, which are manifested as splitting of the sample peak in the HPLC trace (Grant et al., 2008; Popova & Qin, 2010).

A potential problem is that the HPLC trace shows broadly overlapping peaks, which indicates lack of separation between different species. This is particularly prone to occur with oligonucleotides that form stable secondary structures. In such cases, one should consider an alternative to HPLC, such as denaturing electrophoresis, discussed below.

2.4.2 Denaturing Polyacrylamide Gel Electrophoresis

Products of the labeling reaction can be purified with denaturing polyacrylamide gel electrophoresis (PAGE) using standard protocols for gel preparation, electrophoresis, and sample recovery. We have found that R5 is prone to degradation at elevated temperatures in the polyacrylamide gel matrix, possibly due to the involvement of nitroxides in radical-catalyzed polymerization reactions at an elevated temperature (Hawker, Bosman, & Harth, 2001). Thus, it is important to maintain a low temperature (e.g., 4 °C) during electrophoresis (Zhang et al., 2012).

2.5 Characterization of Labeled Oligonucleotides

The purified labeled oligonucleotides are generally resuspended in ME buffer or sterile ultrapure water, and their concentrations are determined by absorbance at 260 nm. Note that at 260 nm, absorption of the R5 label (extinction coefficient $\sim 1000\ M^{-1}\ cm^{-1}$; Qin et al., 2007) is negligible compared to that of nucleic acids (average extinction coefficient $\sim 10,000\ M^{-1}\ cm^{-1}$ per nucleotide). Therefore, UV–vis spectroscopy is not capable of characterizing the degree of labeling. Instead, one may use mass spectrometry, HPLC, or gel electrophoresis to characterize the labeling efficiency.

In addition, the labeled sample should be characterized using cw-EPR spectroscopy. The cw-EPR spectral lineshape, which reports rotational motions of the radical, is very informative in diagnosing whether the radical signal originates from a spin label attached to a macromolecule (Zhang, Cekan, Sigurdsson, & Qin, 2009). In addition, by comparing the observed signal to a standard (e.g., TEMPO) of known concentration, one can carry out a spin-counting procedure (Zhang et al., 2009) to quantify the amount of radical signal present within the sample, thereby determining the degree of labeling.

3. MEASURING INTER-R5 DISTANCES USING DOUBLE ELECTRON–ELECTRON RESONANCE SPECTROSCOPY

Interspin distances are measured by determining the strength of magnetic dipolar interaction by using cw-EPR (for interspin distances <20 Å) or pulsed-EPR (for distances from 20 to ~100 Å). In particular, developments of pulsed-EPR methodologies have been one of the major advances in SDSL and have been extensively reviewed (Cafiso, 2012; Jeschke, 2012; Krstic et al., 2012; Polyhach et al., 2011; Schiemann & Prisner, 2007), including in several chapters in this volume.

In our work, we have primarily used a four-pulse double electron–electron resonance (DEER) scheme (Pannier, Veit, Godt, Jeschke, & Spiess, 2000) to measure inter-R5 distances that are >20 Å. In this scheme (Fig. 3A), a three-pulse sequence is applied to a population of "observe spins," which generates a fixed refocused echo for detection. A fourth pulse, which is set to invert a different population of "pump spins," is applied at a time that is varying. Dipolar coupling between the observe and pump spins results in modulations of refocused echo amplitude, with the modulation frequency being a function of the interspin distance. Acquisition and analysis of DEER data on nucleic acids and protein–nucleic acid complexes follow the same methodologies used for protein and membrane studies, which are extensively described in other chapters in this volume. In the following section, we outline the key steps, and note issues that are specific to nucleic acid studies.

3.1 DEER Sample Preparation

Our DEER measurements were generally carried out on a Bruker ELEXSYS E580 X-band spectrometer equipped with a MS3 or a MD4 resonator, and a typical DEER sample is approximately 25 µL, with the concentration of the doubly labeled sample approximately 100 µM. Note that the signal-to-noise (S/N) ratio of the data depends on the number of spins within the active measuring volume, and it is possible to use smaller amounts of sample (e.g., lowering concentration to ~60 µM or reducing volume to ~20 µL), although a longer acquisition time may be required. On the other hand, sample concentration exceeding 250 µM should be avoided, as it could potentially lead to intermolecular spin–spin interactions that complicate data acquisition and analysis.

In the absence of sample degradation, the R5 label is usually stably attached to the target molecule during typical folding and preparation procedures, such as dialysis and overnight incubation. In particular, we have found that a majority of the R5 label (i.e., >95%) remains attached during typical heat denaturing (e.g., incubation at 95 °C for up to 5 min), used to ensure proper annealing or folding of nucleic acids.

Another important issue is the use of cryoprotective agents, such as glycerol, sucrose, or ficoll. In order to achieve the required spin phase memory time to measure distances >20 Å using a nitroxide label such as R5, DEER measurements are carried out at cryogenic temperatures (80 K or below), with the sample in a glassy state, in which the molecules are homogeneously distributed without aggregation (Zecevic, Eaton, Eaton, & Lindgren, 1998). As such, a sample is usually first assembled at physiological temperatures, then flash-frozen in the presence of cryoprotective agents. In our studies with nucleic acids, we typically use glycerol (in the range of 20–50% v/v) as a cryoprotectant, and flash-freeze the sample by directly immersing the capillary in which it is contained into liquid nitrogen. There are extensive investigations of the effects of various cryoprotectants and freezing procedures (e.g., Freed, Khan, Horanyi, & Cafiso, 2011; Galiano, Blackburn, Veloro, Bonora, & Fanucci, 2009; Georgieva et al., 2012; Kirilina, Grigoriev, & Dzuba, 2004; Sato et al., 2008), which readers are strongly advised to consult.

In assessing a specific procedure for preparing DEER samples, one particularly useful control is to measure single-labeled samples prepared by following identical procedures used for double-labeled samples. In a single-labeled sample, one expects to detect only intermolecular dipolar interaction, but not intramolecular interactions. Therefore, in a properly prepared sample, the resulting background-corrected DEER trace should show no (or very little) decay of the refocused echo. However, if one does observe a decay signal on the single-labeled sample, it could be due to excess crowding (i.e., sample concentration >250 μM) or local aggregation.

3.2 DEER Data Acquisition

DEER acquisition generally proceeds through three stages: (i) sample insertion and resonator and bridge tuning; (ii) field-sweep spectrum acquisition; and (iii) DEER setup and acquisition. As DEER acquisition procedures are described extensively within other chapters in this volume as well as in the

Bruker manuals (Weber, 2005, 2006), only a summary of each of these steps is presented here. Operation procedures may vary between different spectrometers, and readers are also strongly encouraged to consult instructions provided by the spectrometer vendor, which contain detailed descriptions of necessary protocols as well as tips and safety checks that are important for the handling of the spectrometer.

Step 3.2.1 Sample insertion and resonator and bridge tuning

Following vendor instructions, turn on the spectrometer, cool down the cryostat to the desired temperature (e.g., 80 or 50 K), and insert the flash-frozen sample. Care should be taken to minimize temperature fluctuation and air exposure to avoid water condensation within the resonator. Furthermore, to achieve optimal positioning of the sample within the active measuring volume, a common practice is to set the spectrometer to the cw-EPR "tune" mode, and follow the shift of the resonating frequency (i.e., the "dip"). Typically, when sample is optimally positioned, the dip is located farthest away from position observed for the empty resonator. Over-couple the resonator before switching the spectrometer to "operate" mode.

Step 3.2.2 Field-sweep spectrum acquisition

Following vendor instructions, over-couple the resonator, turn on the pulse traveling wave tube (TWT) amplifier, and switch the spectrometer to the "pulse" mode. Be sure to carry out the required safety tests. In particular, check the presence of the defense pulses before switching the TWT from "standby" to "operate," then set the TWT to "operate," slowly increasing the microwave (MW) power (decreasing MW attenuation) and checking cavity ringdown. Note: set the magnetic field away from the resonating field, so that the actual signal of the sample will not be mistaken for the ringdown.

Next, identify and optimize the refocused echo (Fig. 3A), which will be used as the "reporter" in DEER acquisition. Note that in a Bruker E580 system, the frequency of the detection channel ($\nu_{observe}$, as this channel is used for monitoring the observe spin) is not a user-controlled variable, and one should identify the proper sample signal by tuning magnetic field H (based on the relationship $h \cdot \nu_{observe} = g \cdot \beta \cdot H$; where h is Planck's constant, β is the Bohr magneton, and the electron g value is approximately 2 for a nitroxide sample). As such, the general procedure is to set up the proper three-pulse sequence for the observe spin (Fig. 3A), set the magnetic field position to the expected center field of the signal, and find the refocused echo. Note that phase cycling can be used to identify the correct refocused echo.

Once the refocused echo is identified, maximize its signal by adjusting acquisition parameters (e.g., field position, microwave power, signal phase). Then, obtain the field-sweep spectrum by scanning the magnetic field spanning the desired range. Note that phase cycling can be used to suppress artifacts and to eliminate unwanted echoes.

Step 3.2.3 DEER setup and acquisition

Step 3.2.3.1 Determine the pump frequency

The scheme used in our studies is such that the pump pulse is applied at the center manifold of the nitroxide field-sweep spectrum, while the observe pulse is set at its shoulder at the lower field (Fig. 3A). In a typical DEER measurement carried out on a Bruker E580 spectrometer, the sample is subjected to one magnetic field, and the pump and observe pulses are distinguished by a frequency offset.

Using the field-sweep spectrum obtained above, record the field positions corresponding to the maxima of the center (H_C) and low-field manifolds (H_L), as well as $\nu_{observe}$, the frequency at which the field-sweep spectrum is obtained. Then compute pump frequency (ν_{pump}) as:

$$\nu_{pump}(\text{GHz}) = \nu_{observe}(\text{GHz}) - 0.00283 \times [H_C(\text{Gauss}) - H_L(\text{Gauss})] \quad (1)$$

Step 3.2.3.2 Optimize the pump pulse

The pump pulse is applied via a separate channel (i.e., the ELDOR pulse channel), and needs to be optimized independently from the observe pulse. With the Bruker E580 system, we generally use the full power available for the ELDOR channel (i.e., ELDOR attenuation at 0 dB), then determine the proper ELDOR pulse length (π(ELDOR)) for a complete inversion of the spins. This can be achieved by performing a "nutation" measurement (Schweiger & Jeschke, 2001; Weber, 2006). An alternative way to optimize the pump pulse (Weber, 2006) is to first follow protocols described in Step 3.2.2 to identify the refocused echo. Then, set an ELDOR pulse at 100 ns after the first observe π pulse (see Fig. 3A) and at the same frequency as $\nu_{observe}$ (i.e., the frequency at which the refocused echo is observed). With the ELDOR attenuation set at 0 dB, increase the ELDOR pulse length, which should result in variations in refocused echo intensity. The ELDOR pulse length at which the refocused echo disappears for the first time is the optimized π(ELDOR).

Step 3.2.3.3 Acquire a DEER spectrum

Following steps described above, set the field position to H_L identified in Step 3.2.3.1. Set the ELDOR frequency to ν_{pump} calculated from Eq. (1)

in Step 3.2.3.1, ELDOR attenuation to 0 dB, and π(ELDOR) to the optimized value determined in Step 3.2.3.2. If necessary, readjust acquisition parameters (e.g., location and width of the integrator gate) for optimal detection of the refocused echo. Choose the appropriate phase cycling scheme (e.g., two-step), and set the number of scans as necessary. Acquire the DEER trace until desired S/N is reached, which generally takes 12–24 h depending on sample concentration, spin–spin distance, and other parameters.

3.3 DEER Spectrum Analysis

Two general approaches are used to derive the underlying interspin distance distribution, $P(r)$, from the measured DEER data. In the model-free approach, $P(r)$ is not assumed to adopt a particular functional form, and is retrieved by fitting using, for example, the Tikhonov regularization approach (Bowman, Maryasov, Kim, & DeRose, 2004; Chiang, Borbat, & Freed, 2005; Jeschke, Panek, Godt, Bender, & Paulsen, 2004). The other approach is based on specific model functions of $P(r)$; specifically, one can carry out the fitting with $P(r)$ assumed to be composed of one or more Gaussian functions (Sen, Logan, & Fajer, 2007).

One of the most commonly used programs for such spectral analysis is DeerAnalysis (http://www.epr.ethz.ch/software), developed by Jeschke et al. (2006). A recent version, DeerAnalysis2013, provides options for both model-free and model-based fitting. For details on using DeerAnalysis, readers are encouraged to consult the manuals associated with the program.

4. INTEGRATION OF DEER-MEASURED DISTANCES WITH COMPUTATIONAL MODELING

The DEER-measured distances serve as constraints in computational modeling in order to derive structural features of the target molecule. For this purpose, our approach is to use multiple sets of DEER-measured distances to data-mine a large ensemble of models for candidate structure(s) that are in best agreement with the experimental data (Zhang et al., 2012, 2014). The models are generated independently by any all-atom molecular modeling approach (e.g., Monte Carlo simulations). The viability of each model is then assessed by comparing expected inter-R5 distances in a given model to the corresponding DEER-measured values. Compared to other approaches in which the DEER-measured distances are directly incorporated into model building (e.g., Duss et al., 2015; Duss, Yulikov, et al., 2014), our method provides a large degree of flexibility in choosing the best-suited modeling approach for a specific system.

4.1 Model Generation

A model is defined by the Cartesian coordinates of all heavy atoms of the target molecule in the Protein Data Bank (PDB) file format (Berman et al., 2000). In our approach, any modeling method that outputs PDB files can be used to generate the ensemble of putative structural models. For example, in our work on an RNA three-way junction, the model pool was generated by rigid-body transformation (translation and rotation) of three A-form double helices, with concurrently imposed steric constraints (Zhang et al., 2012). Conversely, in our studies of sequence-dependent DNA duplex shapes, all-atom Monte Carlo simulations were employed to generate a large pool of 10,000 B-form DNA models for subsequent evaluation (Zhang et al., 2014).

4.2 Computing Expected Inter-R5 Distances for the Model Pool

4.2.1 The NASNOX Program

We have established and validated a program, called NASNOX, which computes expected inter-R5 distances in a given nucleic acid structure (Cai et al., 2006; Price et al., 2007; Qin et al., 2007). NASNOX models R5 at a pair of nucleic acid sites using experimentally-determined bond lengths and angles. With the nucleic acid coordinates (i.e., the PDB file) fixed, the program varies, in a stepwise fashion, the torsion angles (t1, t2, and t3, Fig. 2C) around the three single bonds connecting the pyrroline ring to the phosphorothioate, and identifies those allowed R5 conformers that have no steric clash between the nitroxide and the parent molecule. Once the ensemble of allowed conformers is identified, inter-R5 distances are calculated, and the corresponding mean and standard deviation are reported. The NASNOX program has been validated on DNA (Cai et al., 2006), RNA (Cai et al., 2007), and protein–DNA complexes (Chen et al., 2013; Zhang et al., 2014) with known high-resolution structures.

An Internet-accessible version of NASNOX, NASNOX_W (Qin et al., 2007), is available at http://pzqin.usc.edu/NASNOX/. The web server allows a user to upload a target structure (in the proper PDB format) and specify a set of search parameters (see example in Fig. 4), then carry out NASNOX calculations and generate two output files: (i) a PDB file, "data001lig.pdb," with the allowable R5 conformers modeled onto the original structure, and (ii) a text file, "data.add," which summarizes the input parameters, lists the allowable conformers, and reports the individual inter-R5 distances, and the average, for the ensemble. A unique feature of NAS-NOX is its speed—each R5 conformer distribution and corresponding inter-R5 distances can be computed in seconds.

4.2.2 Running NASNOX in a Batch Mode

To examine a large pool of models, NASNOX is run in a batch mode (Zhang et al., 2012, 2014), in which the program automatically computes inter-R5 distances based on user-specified labeling sites and structural models. We describe here procedures for batch-mode operation of NASNOX using a script written in Matlab (The MathWorks, Inc.), which will automatically output one user-specified distance for multiple models.

Step 4.2.2.1 Download and install relevant files

The script and relevant files are compressed as the "NASOX_batch2015.zip" file, which is available for download from http://pzqin.usc.edu/pzqhome/software. Unzip and place the files in a folder. The folder should contain:

(a) Matlab script, "BATCH_NASNOX.m." The script requires Matlab version 2010b or later.
(b) NASNOX program. An executable file compiled for the Windows operating system that has been successfully tested with Windows XP, 7, and 8.
(c) Nitroxide parameter files for NASNOX.
(d) A set of example models and a readme file.

Step 4.2.2.2 Prepare the input files

Prepare the following input files and place them in the same folder as the Matlab script and NASNOX files:

(a) Models of the target nucleic acid molecule:

These are a set of PDB files. Note that the NASNOX program requires a specific format, and the readers are strongly encouraged to consult the "sample.pdb" files provided. Particularly, the program identifies a specific nucleotide based on the order in which it is read; therefore, we suggest that the PDB files list the nucleotides in a consecutive fashion, with the first nucleotide set as "1." In addition, a

"TER" line is required at the end of the PDB file for it to be recognized by the NASNOX program. To check whether the input PDB files are formatted properly, one can use the web-based NASNOX_W (see Section 4.2.1) to execute the program on sample(s) of the input model pool.

(b) "model_list.csv":

This is a csv-format file that lists the names of the models described above (Fig. 4). Note that the model names in this file should not contain the ".pdb" suffix.

(c) "sites.csv":

This is a csv-format file listing positions of the two nucleotides at which the inter-R5 distance will be computed (Fig. 4). In the current version, only one pair of positions should be listed in this file. In addition, the program will model R5 onto both the R_p and S_p diastereomers at each selected nucleotide.

(d) "parameter.csv":

This is a csv-format file listing the search parameters for the torsion angles (Fig. 4).

Step 4.2.2.3 Batch execution

Run the "BATCH_NASNOX.m" program in Matlab. This will execute NASNOX based on information specified above.

Step 4.2.2.4 Retrieve output files

For each entry in "model_list.csv," corresponding "modelname_site1-site2_data.add" and "modelname_site1-site2_datalig.pdb" files are produced. These follow the same output format as those provided from the web-based NASNOX_W program (see Section 4.2.1), but with identifiers added for the respective model names and sites. In particular, for each model, the average inter-R5 distance is listed at the end of the ".add" file (see Fig. 4), which can be extracted and used for model characterization as described below. For more details, one should consult Qin et al., 2007 and the tutorial file provided for NASNOX_W.

4.3 Model Selection and Characterization

Here, the goal is to identify those models in which the expected inter-R5 distances match, collectively, to the set of DEER-measured distances. There are a variety of approaches to achieve this goal, and below we summarize criteria used in our previous work.

4.3.1 The RMSD Metric
One of the simplest approaches is to compute the root-mean-square-deviation (RMSD) between the measured and expected distances (Zhang et al., 2012):

$$\text{RMSD}_{\text{deer}}^{j} = \sqrt{\frac{1}{N}\sum_{i}\left(r^{\text{deer}} - r^{\text{model}-j}\right)_{i}^{2}} \qquad (2)$$

where N is the total number of distances measured, the summation index i designates a particular distance set in a model, $r^{\text{model}-j}$ is the NASNOX-computed expected distance for the model j, and r^{deer} is the DEER-measured distance. The lower the $\text{RMSD}_{\text{deer}}$ value, the better the model fits to the DEER measurement.

4.3.2 The Modified RMSD Metric
A DEER measurement not only renders the distance information but also provides the distance distribution profile (see Section 3.3). We have defined a modified RMSD metric (RMSD_{mod}) as (Zhang et al., 2012):

$$\text{RMSD}_{\text{mod}}^{j} = \sqrt{\frac{1}{N}\sum_{i}\left(\frac{r^{\text{deer}} - r^{\text{model}-j}}{\sigma}\right)_{i}^{2}} \qquad (3)$$

where the variables are defined as those in Eq. (2), with an additional parameter σ corresponding to the width of distance distribution obtained from the DEER measurement (see Section 3.3). The RMSD_{mod} metric gives higher weights to datasets with narrow distributions (which are considered to be better defined and of higher information content), and lower weights to those showing broad distributions. Comparing the ranking obtained between $\text{RMSD}_{\text{deer}}$ and RMSD_{mod} provides a means to assess impacts due to datasets with broad distance distributions (Zhang et al., 2012).

4.3.3 A Pseudo-Energy Scoring Function
To evaluate closely related models, such as DNA duplexes falling under the B-form category (Zhang et al., 2014), we further developed a scoring function defined as:

$$P_{t}^{j} = \prod_{i}\exp\left\{-\frac{\left(r^{\text{deer}} - r^{\text{model}-j}\right)^{2}}{2\sigma^{2}}\right\}_{i} \qquad (4)$$

P_t represents, under the assumption of an idealized normal distribution, the effective probability of a given set of r^{model} values matching the corresponding r^{deer} values, with a perfect match resulting in a maximum P_t score of 1. We have used the P_t score to assess B-DNAs with sequence-dependent shape (Zhang et al., 2014).

4.4 Additional Considerations

To enhance the chance of finding model(s) that satisfy the measured distances, one should ensure that the model pool has sufficient coverage of the conformational space allowed for the target molecule. In our work on an RNA junction, we generated the model pool using two different methods: stepwise search and Monte Carlo docking (Zhang et al., 2012). While the two pools were not identical, upon applying the DEER constraints, the resulting top models adopted the same configuration. While this is not definitive proof of a correct structure, it provides assurance, to a certain extent, that each of the model pools had sufficient coverage.

Another consideration is the diversity of the model pool. In our studies of DNA duplexes, we computed the heavy atom RMSD between structural models ($RMSD_{struct}$) to assess diversity within the pool of B-DNA models (Zhang et al., 2014). In addition, we devised an $RMSD_{struct}$ versus P_t plot encompassing all 10,000 entries within a given model pool (Fig. 5A). The plot shows a number of features that indicated the search was productive. First, the data points were spread along both axes of the plot, indicating sufficient diversity of the pool. Second, models that were more different (i.e., larger $RMSD_{struct}$) from the best-fit model (i.e., the one with the best P_t score) indeed deviated more from the measured distances (i.e., lower P_t score), indicating that the measured distances were capable of differentiating the models.

Furthermore, in the model search, one might wonder whether increasing the number of experimentally measured distances may improve the model. This may be addressed by analyzing the models retrieved using different combinations of the measured distances. For example, in our DNA work (Zhang et al., 2014), models with high P_t scores (e.g., those with the top 20 scores) are structurally very similar ($RMSD_{struct}$ <2.0 Å, Fig. 5B). Given the uncertainty present in both the DEER-measured and NASNOX-derived distances, we concluded that increasing the number of measured distances would not further improve the models. Furthermore, we have shown that searches in which a small number of distance constraints

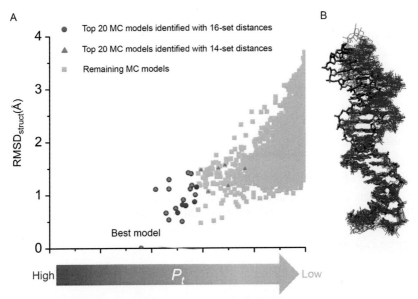

Figure 5 An example of analyzing a model pool using DEER-measured distances. Data shown are reproduced from reported work on the "p21-RE" duplex (Zhang et al., 2014). (A) The RMSD$_{struct}$ versus P_t score plot for the 10,000 models generated from MC simulations. Blue circles represent the top 20 ranked models, obtained using 16 sets of measured distances; red triangles represent the top 20 ranked models, obtained using 14 distances. Note that 14 models, including the best-fit model, are retrieved in both searches. (B) Overlay of the top 20 models of the unbound DNA (blue thin lines) and the bound DNA (red). The unbound DNA models are obtained using the integrated SDSL/MC approach, while the bound DNA is from a reported crystal structure, 3TS8.pdb (Emamzadah, Tropia, & Halazonetis, 2011). The analysis shows that the unbound DNA models converge and identifies the mode of DNA deformation upon protein binding (Zhang et al., 2014). *Adapted from Zhang et al. (2014) with permission.* (See the color plate.)

were omitted (i.e., using only 14 distances instead of 16) yield the same best fit model, and the majority of the top 20 ranked models show little structural deviation (i.e., RMSD$_{struct}$ <2.0 Å, see Fig. 5A). This indicates that the number of DEER-measured distances (in this case, 16 sets) was sufficient.

5. CONCLUSIONS

We present here an integrated method that combines distances measured using a nucleotide-independent nitroxide probe with computational modeling to derive all-atom models of nucleic acids. This approach allows a large degree of flexibility for interfacing with various molecular modeling methods, and its feasibility has been demonstrated in reported work. The approach presented here is a framework for the integrative modeling of nucleic acids with the goal to derive "hybrid models," by data mining computationally generated structural ensembles of a target molecule using experimental distance measurements from EPR spectroscopy. These hybrid models represent solution-state all-atom structures, which might provide important biological information to the structural biology community.

ACKNOWLEDGMENT

We gratefully acknowledge support from NSF (MCB-0546529 and CHE-1213673) and NIH (GM069557 and RR028992 to P.Z.Q.; GM106056 to R.R.).

REFERENCES

Berman, H. M., Westbrook, J., Feng, Z., Gilliland, G., Bhat, T. N., Weissig, H., et al. (2000). The protein data bank. *Nucleic Acids Research*, *28*(1), 235–242. http://dx.doi.org/10.1093/nar/28.1.235.

Bowman, M. K., Maryasov, A. G., Kim, N., & DeRose, V. J. (2004). Visualization of distance distribution from pulsed double electron–electron resonance data. *Applied Magnetic Resonance*, *26*(1–2), 23–39. http://dx.doi.org/10.1007/bf03166560.

Cafiso, D. S. (2012). Taking the pulse of protein interactions by EPR spectroscopy. *Biophysical Journal*, *103*(10), 2047–2048. http://dx.doi.org/10.1016/j.bpj.2012.10.005.

Cai, Q., Kusnetzow, A. K., Hideg, K., Price, E. A., Haworth, I. S., & Qin, P. Z. (2007). Nanometer distance measurements in RNA using site-directed spin labeling. *Biophysical Journal*, *93*(6), 2110–2117. http://dx.doi.org/10.1529/biophysj.107.109439.

Cai, Q., Kusnetzow, A. K., Hubbell, W. L., Haworth, I. S., Gacho, G. P., Van Eps, N., et al. (2006). Site-directed spin labeling measurements of nanometer distances in nucleic acids using a sequence-independent nitroxide probe. *Nucleic Acids Research*, *34*(17), 4722–4730. http://dx.doi.org/10.1093/nar/gkl546.

Chen, Y., Zhang, X., Dantas Machado, A. C., Ding, Y., Chen, Z., Qin, P. Z., et al. (2013). Structure of p53 binding to the BAX response element reveals DNA unwinding and compression to accommodate base-pair insertion. *Nucleic Acids Research*, *41*(17), 8368–8376. http://dx.doi.org/10.1093/nar/gkt584.

Chiang, Y. W., Borbat, P. P., & Freed, J. H. (2005). Maximum entropy: A complement to Tikhonov regularization for determination of pair distance distributions by pulsed ESR. *Journal of Magnetic Resonance*, *177*, 184–196.

Ding, Y., Nguyen, P., Tangprasertchai, N. S., Reyes, C. V., Zhang, X., & Qin, P. Z. (2015). Nucleic acid structure and dynamics: Perspectives from site-directed spin labeling. In B. C. Gilbert, V. Ghcechik, & D. M. Murphy (Eds.), *Electron paramagnetic resonance: Vol. 24.* (pp. 122–147). Cambridge, England (Thomas Graham House, Science Park, Cambridge Cb4 4WF): The Royal Society of Chemistry.

Ding, Y., Zhang, X., Tham, K. W., & Qin, P. Z. (2014). Experimental mapping of DNA duplex shape enabled by global lineshape analyses of a nucleotide-independent nitroxide probe. *Nucleic Acids Research*, *42*(18), e140. http://dx.doi.org/10.1093/nar/gku695.

Duss, O., Michel, E., Yulikov, M., Schubert, M., Jeschke, G., & Allain, F. H. T. (2014). Structural basis of the non-coding RNA RsmZ acting as a protein sponge. *Nature*, *509*, 588–592. http://dx.doi.org/10.1038/nature13271.

Duss, O., Yulikov, M., Allain, F. H. T., & Jeschke, G. (2015). Combining NMR and EPR to determine structures of large RNAs and protein–RNA complexes in solution. *Methods in Enzymology*, *558*, 279–331.

Duss, O., Yulikov, M., Jeschke, G., & Allain, F. H. T. (2014). EPR-aided approach for solution structure determination of large RNAs or protein–RNA complexes. *Nature Communications*, *5*(3669). http://dx.doi.org/10.1038/ncomms4669.

Emamzadah, S., Tropia, L., & Halazonetis, T. D. (2011). Crystal structure of a multidomain human p53 tetramer bound to the natural CDKN1A (p21) p53-response element. *Molecular Cancer Research*, *9*(11), 1493–1499. http://dx.doi.org/10.1158/1541-7786.mcr-11-0351.

Fedorova, O. S., & Tsvetkov, Y. D. (2013). Pulsed electron double resonance in structural studies of spin-labeled nucleic acids. *Acta Naturae*, *5*(1), 9–32.

Freed, D. M., Khan, A. K., Horanyi, P. S., & Cafiso, D. S. (2011). Molecular origin of electron paramagnetic resonance line shapes on beta-barrel membrane proteins: The local solvation environment modulates spin-label configuration. *Biochemistry*, *50*(41), 8792–8803. http://dx.doi.org/10.1021/bi200971x.

Galiano, L., Blackburn, M. E., Veloro, A. M., Bonora, M., & Fanucci, G. E. (2009). Solute effects on spin labels at an aqueous-exposed site in the flap region of HIV-1 protease. *The Journal of Physical Chemistry. B*, *113*(6), 1673–1680. http://dx.doi.org/10.1021/jp8057788.

Georgieva, E. R., Roy, A. S., Grigoryants, V. M., Borbat, P. P., Earle, K. A., Scholes, C. P., et al. (2012). Effect of freezing conditions on distances and their distributions derived from double electron electron resonance (DEER): A study of doubly-spin-labeled T4 lysozyme. *Journal of Magnetic Resonance*, *216*, 69–77. http://dx.doi.org/10.1016/j.jmr.2012.01.004.

Grant, G. P. G., Boyd, N., Herschlag, D., & Qin, P. Z. (2009). Motions of the substrate recognition duplex in a group I intron assessed by site-directed spin labeling. *Journal of the American Chemical Society*, *131*(9), 3136–3137. http://dx.doi.org/10.1021/ja808217s.

Grant, G. P. G., Popova, A., & Qin, P. Z. (2008). Diastereomer characterizations of nitroxide-labeled nucleic acids. *Biochemical and Biophysical Research Communications*, *371*(3), 451–455. http://dx.doi.org/10.1016/j.bbrc.2008.04.088.

Hagelueken, G., Ward, R., Naismith, J. H., & Schiemann, O. (2012). MtsslWizard: In silico spin-labeling and generation of distance distributions in PyMOL. *Applied Magnetic Resonance*, *42*(3), 377–391. http://dx.doi.org/10.1007/s00723-012-0314-0.

Hankovszky, H., Hideg, K., & Lex, L. (1980). Nitroxyls; VII1. Synthesis and reactions of highly reactive 1-Oxyl-2,2,5,5-tetramethyl-2,5-dihydropyrrole-3-ylmethyl sulfonates. *Synthesis*, 914–916.

Hatmal, M. M., Li, Y., Hegde, B. G., Hegde, P. B., Jao, C. C., Langen, R., et al. (2012). Computer modeling of nitroxide spin labels on proteins. *Biopolymers*, *97*(1), 35–44. http://dx.doi.org/10.1002/bip.21699.

Hawker, C. J., Bosman, A. W., & Harth, E. (2001). New polymer synthesis by nitroxide mediated living radical polymerizations. *Chemical Reviews*, *101*(12), 3661–3688. http://dx.doi.org/10.1021/cr990119u.

Hirst, S. J., Alexander, N., McHaourab, H. S., & Meiler, J. (2011). RosettaEPR: An integrated tool for protein structure determination from sparse EPR data. *Journal of Structural Biology*, *173*(3), 506–514. http://dx.doi.org/10.1016/j.jsb.2010.10.013.

Hubbell, W. L., & Altenbach, C. (1994). Investigation of structure and dynamics in membrane proteins using site-directed spin labeling. *Current Opinion in Structural Biology*, *4*, 566–573.

Hubbell, W. L., Cafiso, D. S., & Altenbach, C. (2000). Identifying conformational changes with site-directed spin labeling. *Nature Structural Biology*, *7*, 735–739. http://dx.doi.org/10.1038/78956.

Hubbell, W. L., López, C. J., Altenbach, C., & Yang, Z. (2013). Technological advances in site-directed spin labeling of proteins. *Current Opinion in Structural Biology, 23*(5), 725–733. http://dx.doi.org/10.1016/j.sbi.2013.06.008.

Jeschke, G. (2012). DEER distance measurements on proteins. *Annual Review of Physical Chemistry, 63*(1), 419–446. http://dx.doi.org/10.1146/annurev-physchem-032511-143716.

Jeschke, G., Chechik, V., Ionita, P., Godt, A., Zimmermann, H., Banham, J., et al. (2006). DeerAnalysis2006—A comprehensive software package for analyzing pulsed ELDOR data. *Applied Magnetic Resonance, 30*(3–4), 473–498. http://dx.doi.org/10.1007/BF03166213.

Jeschke, G., Panek, G., Godt, A., Bender, A., & Paulsen, H. (2004). Data analysis procedures for pulse ELDOR measurements of broad distance distributions. *Applied Magnetic Resonance, 26*(1–2), 223–244. http://dx.doi.org/10.1007/BF03166574.

Kirilina, E. P., Grigoriev, I. A., & Dzuba, S. A. (2004). Orientational motion of nitroxides in molecular glasses: Dependence on the chemical structure, on the molecular size of the probe, and on the type of the matrix. *The Journal of Chemical Physics, 121*(24), 12465–12471. http://dx.doi.org/10.1063/1.1822913.

Krstic, I., Endeward, B., Margraf, D., Marko, A., & Prisner, T. F. (2012). Structure and dynamics of nucleic acids. *Topics in Current Chemistry, 321*, 159–198. http://dx.doi.org/10.1007/128_2011_300.

Nguyen, P. H., Popova, A. M., Hideg, K., & Qin, P. Z. (2015). A nucleotide-independent cyclic nitroxide label for monitoring segmental motions in nucleic acids. *BMC Biophysics, 8*(6). http://dx.doi.org/10.1186/s13628-015-0019-5.

Pannier, M., Veit, S., Godt, A., Jeschke, G., & Spiess, H. W. (2000). Dead-time free measurement of dipole-dipole interactions between electron spins. *Journal of Magnetic Resonance, 142*, 331–340. http://dx.doi.org/10.1006/jmre.1999.1944.

Polyhach, Y., Bordignon, E., & Jeschke, G. (2011). Rotamer libraries of spin labelled cysteines for protein studies. *Physical Chemistry Chemical Physics, 13*(6), 2356–2366. http://dx.doi.org/10.1039/c0cp01865a.

Popova, A. M., Hatmal, M. M., Frushicheva, M. P., Price, E. A., Qin, P. Z., & Haworth, I. S. (2012). Nitroxide sensing of a DNA microenvironment: Mechanistic insights from EPR spectroscopy and molecular dynamics simulations. *The Journal of Physical Chemistry. B, 116*(22), 6387–6396. http://dx.doi.org/10.1021/jp303303v.

Popova, A. M., Kálai, T., Hideg, K., & Qin, P. Z. (2009). Site-specific DNA structural and dynamic features revealed by nucleotide-independent nitroxide probes. *Biochemistry, 48*(36), 8540–8550. http://dx.doi.org/10.1021/bi900860w.

Popova, A. M., & Qin, P. Z. (2010). A nucleotide-independent nitroxide probe reports on site-specific stereomeric environment in DNA. *Biophysical Journal, 99*(7), 2180–2189. http://dx.doi.org/10.1016/j.bpj.2010.08.005.

Price, E. A., Sutch, B. T., Cai, Q., Qin, P. Z., & Haworth, I. S. (2007). Computation of nitroxide-nitroxide distances for spin-labeled DNA duplexes. *Biopolymers, 87*, 40–50. http://dx.doi.org/10.1002/bip.20769.

Qin, P. Z., Butcher, S. E., Feigon, J., & Hubbell, W. L. (2001). Quantitative analysis of the GAAA tetraloop/receptor interaction in solution: A site-directed spin labeling study. *Biochemistry, 40*, 6929–6936. http://dx.doi.org/10.1021/bi010294g.

Qin, P. Z., Haworth, I. S., Cai, Q., Kusnetzow, A. K., Grant, G. P. G., Price, E. A., et al. (2007). Measuring nanometer distances in nucleic acids using a sequence-independent nitroxide probe. *Nature Protocols, 2*(10), 2354–2365. http://dx.doi.org/10.1038/nprot.2007.308.

Sato, H., Kathirvelu, V., Spagnol, G., Rajca, S., Rajca, A., Eaton, S. S., et al. (2008). Impact of electron–electron spin interaction on electron spin relaxation of nitroxide diradicals and tetraradical in glassy solvents between 10 and 300 K. *The Journal of Physical Chemistry. B, 112*(10), 2818–2828. http://dx.doi.org/10.1021/jp073600u.

Schiemann, O., & Prisner, T. F. (2007). Long-range distance determinations in biomacromolecules by EPR spectroscopy. *Quarterly Reviews of Biophysics, 40*(1), 1–53. http://dx.doi.org/10.1017/S003358350700460X.

Schweiger, A., & Jeschke, G. (2001). Nuclear modulation effect I: Basic experiments. *Principles of pulse electron paramagnetic resonance* (pp. 247–295). Oxford: Oxford University Press.

Sen, K. I., Logan, T. M., & Fajer, P. G. (2007). Protein dynamics and monomer-monomer interactions in AntR activation by electron paramagnetic resonance and double electron–electron resonance. *Biochemistry, 46*(41), 11639–11649. http://dx.doi.org/10.1021/bi700859p.

Shelke, S. A., & Sigurdsson, S. T. (2012). Site-directed spin labelling of nucleic acids. *European Journal of Organic Chemistry, 12*, 2291–2301. http://dx.doi.org/10.1002/ejoc.201101434.

Sowa, G. Z., & Qin, P. Z. (2008). Site-directed spin labeling studies on nucleic acid structure and dynamics. *Progress in Nucleic Acid Research and Molecular Biology, 82*, 147–197. http://dx.doi.org/10.1016/S0079-6603(08)00005-6.

Weber, R. T. (2005). *Bruker ELEXSYS E850 User's Manual (v 2.0)*. Billerica, MA, USA: Bruker BioSpin Corporation.

Weber, R. T. (2006). *Bruker pulsed ELDOR Option User's Manual (v 1.0)*. Billerica, MA, USA: Bruker BioSpin Corporation.

Zecevic, A. N. A., Eaton, G. R., Eaton, S. S., & Lindgren, M. (1998). Dephasing of electron spin echoes for nitroxyl radicals in glassy solvents by non-methyl and methyl protons. *Molecular Physics, 95*(6), 1255–1263. http://dx.doi.org/10.1080/00268979809483256.

Zhang, X., Cekan, P., Sigurdsson, S. T., & Qin, P. Z. (2009). Studying RNA using site-directed spin-labeling and continuous-wave electron paramagnetic resonance spectroscopy. *Methods in Enzymology, 469*, 303–328. http://dx.doi.org/10.1016/S0076-6879(09)69015-7.

Zhang, X., Dantas Machado, A. C., Ding, Y., Chen, Y., Lu, Y., Duan, Y., et al. (2014). Conformations of p53 response elements in solution deduced using site-directed spin labeling and Monte Carlo sampling. *Nucleic Acids Research, 42*(4), 2789–2797. http://dx.doi.org/10.1093/nar/gkt1219.

Zhang, X., Tung, C. S., Sowa, G. Z., Hatmal, M. M., Haworth, I. S., & Qin, P. Z. (2012). Global structure of a three-way junction in a phi29 packaging RNA dimer determined using site-directed spin labeling. *Journal of the American Chemical Society, 134*(5), 2644–2652. http://dx.doi.org/10.1021/ja2093647.

Section IV

EPR-NMR Methods and Applications

CHAPTER SIXTEEN

Overhauser Dynamic Nuclear Polarization Studies on Local Water Dynamics

Ilia Kaminker*, Ryan Barnes*, Songi Han*,†,1
*Department of Chemistry and Biochemistry, University of California Santa Barbara, Santa Barbara, California, USA
†Department of Chemical Engineering, University of California Santa Barbara, Santa Barbara, California, USA
[1]Corresponding author: e-mail address: songi@chem.ucsb.edu

Contents

1. Introduction	458
2. Theory	461
2.1 Relaxation Rates	463
2.2 Saturation Factor	465
2.3 Interpretation of Measured ODNP Parameters	467
3. Hardware	469
4. Data Acquisition	472
5. Data Analysis	475
5.1 Raw Data Analysis	475
5.2 Data Interpretation	476
6. Examples of ODNP	477
7. Summary	479
Acknowledgments	479
References	479

Abstract

Overhauser dynamic nuclear polarization (ODNP) is an emerging technique for quantifying translational water dynamics in the vicinity (<1 nm) of stable radicals that can be chemically attached to macromolecules of interest. This has led to many in-depth and enlightening studies of hydration water of biomolecules, revolving around the role of solvent dynamics in the structure and function of proteins, nucleic acids, and lipid bilayer membranes. Still to date, a complete and fully automated ODNP instrument is not commercialized. The purpose of this chapter is to share the technical know-how of the hardware, theory, measurement, and data analysis method needed to successfully utilize and disseminate the ODNP technique.

1. INTRODUCTION

Overhauser dynamic nuclear polarization (ODNP) has attracted renewed interest in recent years with the rediscovery of its potential as a tool to quantify local water dynamics on biological or soft material surfaces. The method is based on measurement of water dynamics within distances of 5–10 Å around stable-radical-based spin probes that are either tethered to a site and surface of interest or selectively imbibed within a local volume of interest (Armstrong & Han, 2007, 2009; Franck, Pavlova, Scott, & Han, 2013). While the principle of ODNP has been around for over 50 years (Bates & Drozdoski, 1977; Carver & Slichter, 1956; Hausser & Stehlik, 1968; Overhauser, 1953; Pedersen & Freed, 1974, 1975), new and exciting applications keep emerging (Gitti et al., 1988; Krummenacker, Denysenkov, Terekhov, Schreiber, & Prisner, 2012; Lingwood et al., 2012; McCarney, Armstrong, Lingwood, & Han, 2007; Neudert, Reh, Spiess, & Münnemann, 2015). The discovery of its usage at X-band (8–12 GHz) microwave (mw) frequencies as a ^{1}H relaxometry tool to reveal translational diffusion dynamics of hydration water in a region- and site-specific manner has excited a large community of researchers. This method is being applied to address the question of the role of water and solvation in biological structure, function, dynamics, and beyond (Armstrong et al., 2011; Cheng, Goor, & Han, 2012; Cheng, Olijve, Kauski, & Han 2014; Cheng, Varkey, Ambroso, Langen, & Han, 2013; Doll, Bordignon, Joseph, Tschaggelar, & Jeschke, 2012; Franck, Scott, & Han, 2013; Franck et al., 2014; Gitti et al., 1988; Kausik & Han, 2009). ^{1}H ODNP relaxometry is particularly powerful when it can naturally complement continuous-wave (CW) electron paramagnetic resonance (EPR) and other more advanced (pulsed) EPR measurements. A particularly rewarding combination is of ODNP with the pulse EPR technique of Electron Spin Echo Envelope Modulations (ESEEM) (Mims, 1972; Rowan, Hahn, & Mims, 1965) that can quantify the water density in the vicinity of the spin label of a frozen sample, and thus can powerfully complement the ODNP-derived information on local water diffusion dynamics that, in a first approximation, is uncoupled from the water density in the local environment (Franck et al., 2014).

Importantly, ODNP shares the broad application range, scope, and high sensitivity of EPR. Because the size and complexity of the macromolecular system of interest is not fundamentally limiting and only ~0.5 nmol of sample is required (3–5 μl of 50–100 μM), ODNP and EPR methods are typically applicable for the study of large and complex biomolecular systems (Altenbach, Kusnetzow, Ernst, Hofmann, & Hubbell, 2008; Cheng et al., 2013; Hussain, Franck, & Han, 2013; Kim et al., 2004). The main requirement is that a stable spin label, typically nitroxide or trityl radical, can be incorporated reliably and site specifically. This allows ODNP and EPR analysis to access a large range of interesting biological and synthetic systems that are often beyond the reach of light scattering, crystallography, or high-resolution NMR spectroscopy. Thus far, ODNP was successfully applied to studies of globular proteins (Franck et al., 2014), protein folding (Armstrong et al., 2011), intrinsically disordered proteins (Pavlova et al., 2009, 2015), lipid membrane systems (Franck, Scott, et al., 2013; Song, Han, & Han, 2015; Song, Kim, Kang, & Han, 2015), membrane proteins (Cheng & Han, 2013; Cheng et al., 2013; Doll et al., 2012; Hussain et al., 2013), peptides (Eschmann et al., 2015; Ortony, Hwang, Franck, Waite, & Han, 2013), polymer systems (Cheng, Wang, Kausik, Lee, & Han, 2012; Ortony et al., 2014; Song, Han, et al., 2015), and more recently, nucleic acids (Franck, Ding, Stone, Qin & Han, 2015). Nevertheless, these studies are barely beginning to scratch the surface of key questions that ODNP in combination with EPR methodology is capable of addressing, such as the functional role of solvent in protein binding, recognition or catalysis, or more fundamentally, the nature of protein–water coupling under ambient solution conditions.

Fortunately, for the experimentalist, an ODNP spectrometer operating at 0.35 T is in practice an add-on to an existing CW EPR spectrometer. An additional advantage is that a sample prepared for ODNP can also be used for CW EPR characterization and vice versa. As with most experimental techniques, the key for successful deployment of the technique and broad dissemination to a new user base relies on optimized and robust experimental and hardware designs. This chapter will describe the "how to" of ODNP experiments, focusing on techniques, analysis methods, and equipment employed in ODNP measurements.

The ODNP-derived water dynamics are most commonly analyzed with the unit-less electron–nuclear coupling factor, ξ, which is directly obtained from the NMR signal enhancement and ^{1}H T_1 relaxation data (Hausser & Stehlik, 1968). However, in order to translate ξ into a physical parameters that scale with known units, several assumptions are made and models applied to convert ξ into a correlation time τ_c and a local diffusion coefficient D_{local} of water within 5–10 Å of the spin label (see Section 2). The

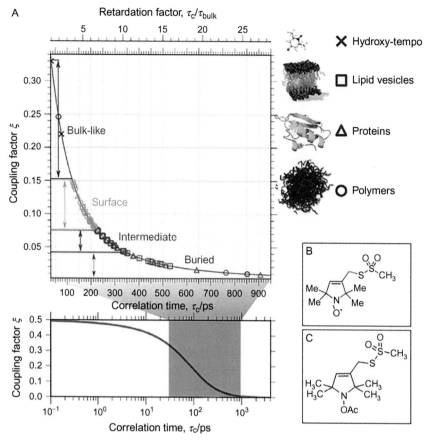

Figure 1 Data for the coupling factor ξ for a set of known biological spin probe (spin label) positions for a variety of systems appearing in the ODNP literature. The bottom panel shows the FFHS curve plotted over a larger range of logarithmically spaced correlation times. *Adopted from figure 12 of Franck, Pavlova, et al. (2013). (See the color plate.)*

relationship between ξ and τ_c is shown in Fig. 1, which illustrates the dynamic range and the sensitivity of the ξ parameter to the translational diffusion correlation time between a few tens of picoseconds to subnanosecond — the type of dynamics captured when ODNP is performed at 0.35 T. When compiling experimental ODNP data from the literature onto this $\xi(\tau_c)$ plot, distinct regimes for local water dynamics can be recognized. For example, ODNP provides a clear distinction between different regimes of water dynamics: (i) bulk-like water dynamics $(0.15 < \xi < 0.3)$ — as found not only in bulk solution systems, but also on the surface of DNA duplexes (Franck et al., 2015) and the inner surface of the chaperon GroEL/ES cavity (Franck et al., 2014) that displays unusually unhindered translational diffusion dynamics for water; (ii) surface dynamics $(0.07 < \xi < 0.15)$—as found on most hydrophilic biomolecular surfaces, such as the surface of lipid bilayers, peptides, proteins, and polymers; and finally (iii) buried dynamics $(\xi < 0.04)$—as found at sites within the core of a protein or lipid bilayer or strongly confined interfaces. The observation that the measured local water diffusivity across such a large range and classes of biological and soft material systems fall into distinctly identifiable regions implies that absolute characterization of solvent exposure or surface hydrophilicity/hydrophobicity is likely feasible, and also that events involving changes of surface–solvent interaction, such as protein folding, aggregation, and binding, can be probed by ODNP.

However, ODNP can do more than passively probe and report on events of macromolecular interaction, folding, and aggregation. More excitingly, ODNP at 0.35 T was found to be critically sensitive such that it has revealed a heterogeneous surface hydration dynamic topology on a globular protein (Armstrong et al., 2011), where the contrast in surface water diffusivity arises from different topology of the amino acid arrangement of entirely solvent-exposed sites, not simply from contrast due to differential solvent exposure or burial of different sites (Cheng et al., 2013). Thus, it has been proposed that this hydration dynamics landscape harbors information that has predictive values on the location of binding interfaces or enzymatically active sites (Song, Allison, & Han, 2014)—information that cannot be obtained even from a fully resolved 3D protein structure *a priori*.

2. THEORY

We briefly review the necessary theoretical basis for analyzing data obtained in an ODNP experiment for the purpose of extracting hydration dynamics. Particular attention will be given to the special case of using a nitroxide radical, as this is currently almost exclusively employed as the spin probe for ODNP-derived hydration dynamics studies. We do not intend to present a rigorous derivation of the ODNP theory, but rather to present a concise summary of the equations and relations needed for the basic understanding of the ODNP experiment and the associated data analysis. For more rigorous in-depth treatments, the reader is referred to the excellent reviews published throughout the years (Cheng & Han, 2013; Franck, Pavlova, et al., 2013; Griesinger et al., 2012; Hausser & Stehlik, 1968).

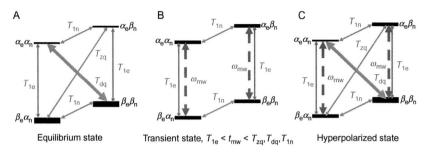

Figure 2 Energy level and populations diagram for the $S=1/2$, $I=1/2$ spin system. Thickness of the black bars represents the population of the corresponding level at each stage of the experiment. (A) At thermal equilibrium (no mw irradiation). (B) At transient state where irradiation time (t_{mw}) is short such that electron transitions are saturated but redistribution due to cross-relaxation is still negligible. (C) At steady state with mw irradiation on.

Figure 2 illustrates the change in the spin population of different levels in the course of a simplified ODNP experiment on an $S=1/2$, $I=1/2$ spin system. Figure 2A illustrates the thermal equilibrium state. The $\beta_e \alpha_n$ and $\beta_e \beta_n$ states have higher population as predicted by the Boltzmann distribution and depicted by the thicker black bars. All relaxation rates present in the system, namely electron spin lattice relaxation T_{1e}, nuclear spin lattice relaxation T_{1n}, and zero and double-quantum relaxation rates T_{zq} and T_{dq} are depicted on Fig. 2A as solid double headed arrows. Figure 2B illustrates a transient state that is achieved quickly after mw irradiation on resonance with the EPR transitions (denoted as dashed arrows) is applied, so that $T_{1e} < t_{mw} < T_{zq}, T_{dq}, T_{1n}$, where t_{mw} denotes the duration of the mw irradiation. Here, we assume the mw irradiation is strong enough to achieve full saturation of the $\beta_e \beta_n \rightarrow \alpha_e \beta_n$ and $\beta_e \alpha_n \rightarrow \alpha_e \alpha_n$ electron transitions, namely equalizing the populations of the connected energy levels. Note that the NMR signal is given by

$$S_{NMR} = P(\alpha_e \beta_n) - P(\alpha_e \alpha_n) + P(\beta_e \beta_n) - P(\beta_e \alpha_n). \qquad (1)$$

Thus, mw saturation in this regime does not alter the NMR signal from its equilibrium state (Fig. 2A). Figure 2C shows the redistribution of populations in the steady state achieved after prolonged mw irradiation, so that $t_{mw} > T_{1n}, T_{zq}, T_{dq}$. The hyperpolarized state is a result of simultaneous action of the mw irradiation and unequal zero-quantum and double-quantum relaxation rates. Note that if $T_{dq} < T_{zq}$, the resulting hyperpolarized NMR signal given by Eq. (1) is going to be inverted relative to the equilibrium signal. So, ODNP enhancement results from the difference in zero- and double-quantum relaxation rates. The source of these cross-relaxation rates in liquids lies in the high-frequency modulation of the dipolar interaction between water proton and the unpaired electron spin of the nitroxide radical-based spin probe. The extent of the inequality between zero- and double-quantum rates, which is the source of the ODNP enhancement, is thus a direct measurement of the spectral density function, which characterizes the local motion of water in the vicinity of the unpaired electron spin. The details are given in Section 2.1.

2.1 Relaxation Rates

The most common way to analyze ODNP data is by interpreting the unitless parameter ξ commonly referred to as the coupling factor that, as shown below, reports on the local (<1 nm) dynamics of the water molecules around the spin label.

$$\xi = \frac{\sigma}{\rho}. \qquad (2)$$

Here σ and ρ denote, respectively, the radical-induced self-relaxation $\rho = \frac{2}{T_{1n}} + \frac{1}{T_{zq}} + \frac{1}{T_{dq}}$ and the cross-relaxation $\sigma = \frac{1}{T_{dq}} - \frac{1}{T_{zq}}$ rates between the ^{1}H water nuclei and the electron spin. All above relaxation processes are driven by the time-dependent electron–nuclear dipolar interactions between the unpaired electron spin of the nitroxide and the water ^{1}H nuclei. The r^{-3} dependence of the dipolar interaction leads to rates governed by dipolar relaxation to scale as r^{-6}, where r denotes the electron–nuclear distance (Solomon, 1955). What this means is that the coupling factor reports on water dynamics in close vicinity (i.e., within <1 nm) of the spin probe (Armstrong & Han, 2007, 2009; Sezer, Prandolini, & Prisner, 2009). This localization effect is central to the high spatial sensitivity of the ODNP technique. It allows for the characterization of site-specific hydration dynamics by carefully positioning the spin labels on the biomacromolecules or by clever choice of tailored spin probes (by size, charge, hydrophobicity, etc.) with known partitioning properties.

To calculate the coupling factor, it is necessary to measure ρ, which includes the contribution to the relaxation rate of the nuclei due strictly to the presence of the unpaired electron. In practice, especially at low radical concentrations (<0.1 mM), other contributions to the water ^{1}H T_1 relaxation will typically dominate the experimental value of the T_1 relaxation as obtained in the inversion recovery experiment. To account for this, we measure the "background" relaxation rate $T_{1,0}$ of the sample of identical composition (i.e., same biomolecules, same buffer conditions, etc.) lacking the radical. Consequently, ^{1}H longitudinal relaxation time (T_1) in the presence of a radical is given by $\frac{1}{T_1} = \rho + \frac{1}{T_{1,0}}$, where one notes that the three different processes contributing to ρ will contribute together to the radical-induced nuclear T_1 relaxation as measured by an inversion recovery experiment. The self-relaxation rate is thus obtained from the difference of the relaxation rates obtained for the samples with and without the radical:

$$\rho = \frac{1}{T_1} - \frac{1}{T_{1,0}} \qquad (3)$$

The cross-relaxation rate σ is obtained from the ODNP experiment as derived in Franck, Pavlova, et al. (2013):

$$\sigma = \lim_{p \to \infty} \left(\frac{1 - E(p)}{T_1(p)} \left| \frac{\gamma_H}{\gamma_e} \right| \right) \frac{1}{s_{max}}. \qquad (4)$$

Here, $E(p)$ and $T_1(p)$ denote the enhancement and ^{1}H relaxation times, respectively, measured as function of applied mw power; γ_H and γ_e are proton and electron gyromagnetic ratios, respectively; s_{max} denotes the maximum achievable saturation of the unpaired electron EPR spectrum, as discussed in more detail below.

Note that Eq. (4) is different from the one that appears in the classical ODNP literature (Hausser & Stehlik, 1968) and in the earlier papers of the Han laboratory (Armstrong & Han, 2007, 2009). The early definition of the cross-relaxation rate assumes that T_1 is independent of temperature, or alternatively that mw-induced sample heating is negligible. We will be referring to this original definition as $\sigma_{uncorrected}$ in this text due to the fact that temperature correction is omitted in the classical definition of σ.

$$\sigma_{uncorrected} = \lim_{p \to \infty} \left(\frac{1 - E(p)}{T_1(p=0)} \left| \frac{\gamma_H}{\gamma_e} \right| \right) \frac{1}{s_{max}} \qquad (5)$$

A detailed description of the background and method of this heating correction necessary for the ODNP data analysis was introduced and provided in a review article by Franck, Pavlova, et al. (2013). It is important to keep this distinction in mind. Though the "temperature corrected" definition provides the "true" value of the coupling factor (Franck, Pavlova, et al., 2013), the "uncorrected" values are often useful for comparing the numbers with the available early ODNP literature, not only of Han and coworkers

(Armstrong et al., 2011; Armstrong & Han, 2007, 2009; Cheng et al., 2013, Cheng, Wang, et al., 2012; Kausik & Han, 2009) but also of Bennatti and coworkers (Bennati, Luchinat, Parigi, & Türke, 2010; Enkin et al., 2015; Enkin, Liu, Tkach, & Bennati, 2014; Lottmann et al., 2012; Türke, Tkach, Reese, Höfer, & Bennati, 2010), Münnemann and coworkers (Neudert et al., 2015), as well as Jeschke, Bordignon, and coworkers (Doll et al., 2012). Naturally, the $\sigma_{uncorrected}$ values depend on the amount of sample heating and thus on the experimental setup, such as the exact cavity and NMR probe design used, as well as sample conditions, including the amount of salt and nature of the sample. Fortunately, when the latter are kept reproducible the values can be compared between themselves and provide information on the relative differences in hydration dynamics. Note that the coupling factor values compiled in Fig. 1 are of the "uncorrected" type, due to the prevalence of these data in the literature.

2.2 Saturation Factor

From Eqs. (4) and (5), it is clear that accurate knowledge of the saturation factor s_{max} is essential for accurately determining the coupling factor and interpreting the ODNP results. This is not trivial, since in nitroxide radicals the unpaired electron spin is strongly coupled to the nuclear ^{14}N nitrogen spin ($I=1$), and consequently the EPR signal splits into three resonance lines by the hyperfine interaction. For an EPR spectrum consisting of a single line, the electron spin saturation as function of mw power is given by:

$$s = s_{max} \frac{P}{1+P}, \qquad (6)$$

where P is the applied mw power. Since in a typical ODNP experiment CW mw irradiation is applied on resonance with only one out of three EPR transitions of ^{14}N-nitroxide radicals, the extent of the cross talk between the three transitions will determine the value of s_{max}. Consequently, for ^{14}N nitroxide radical, the maximum saturation factor is given by:

$$s_{max} = \frac{1}{3} - \frac{2}{3}R, \qquad (7)$$

where $0 \geq R \geq -1$ characterizes the extent of cross talk between the three hyperfine lines under different sample conditions. Accordingly, the saturation factor can vary from $s_{max} = \frac{1}{3}$ (no cross talk) to $s_{max} = 1$ (complete cross talk) (Armstrong & Han, 2007; Bates & Drozdoski, 1977;

Franck, Pavlova, et al., 2013; Sezer, Gafurov, Prandolini, Denysenkov, & Prisner, 2009; Sezer, Prandolini, et al., 2009).

There are two different processes that can cause change in the ^{14}N quantum number and are responsible for the cross talk between different ^{14}N EPR lines: (i) Heisenberg exchange and (ii) ^{14}N nuclear relaxation (commonly abbreviated as T_{1N}). The Heisenberg exchange rate is proportional to the rate of collisions between the spin labels. Consequently, Heisenberg exchange plays a significant role only for spin probes in nonviscous samples at local concentrations above 0.5 mM (Bates & Drozdoski, 1977). In practice, such conditions are rarely encountered in biochemical applications of ODNP (Armstrong & Han, 2007). In turn, the effect of ^{14}N T_{1N} relaxation rate becomes significant ($T_{1N} \leq T_{1e}$) when spin labels are attached to macromolecules or immobilized on a surface, where the slowing of spin label tumbling motion has been shown to dramatically shorten the ^{14}N T_{1N} (Enkin et al., 2015; Robinson, Haas, & Mailer, 1994). This is the case in the majority of biochemical applications of ODNP, where the probing of water dynamics in the proximity of lipid membranes or the surfaces of proteins or nucleic acids is desired, for which tethered spin labels are employed.

A direct measurement of the saturation factor is complicated and requires either a CW EPR spectrometer capable of double resonance experiments or a pulsed electron–nuclear double resonance (ELDOR) equipment (Türke & Bennati, 2011). If the electron relaxation time T_{1e}, ^{14}N relaxation time T_{1N}, Heisenberg exchange rate, and strength of the mw irradiation are known it is possible to precisely calculate the saturation factor s_{max} (Armstrong & Han, 2007; Sezer, Gafurov, et al., 2009; Türke & Bennati, 2011). In practice, the determination of T_{1e} and T_{1n} for nitroxide radicals at room temperature and in aqueous solution requires sophisticated EPR instrumentation and is hardly a routine procedure (Hyde, Froncisz, & Mottley, 1984; Robinson et al., 1994). Fortunately, for many practical biochemical ODNP applications, the condition $T_{1N} \leq T_{1e}$ is fulfilled and consequently $s_{max} \approx 1$ (Armstrong & Han, 2007; Robinson et al., 1994).

An important point to make here, as should be obvious from the previous discussion, is that it is difficult to directly compare the hydration dynamics of the spin labeled macromolecule and a free spin probe under the same experimental conditions. As detailed in publications by Armstrong and Han (2007) and Franck, Pavlova, et al. (2013), characterization of the local water dynamics around a freely tumbling spin probe in solution requires a series of experiments with varying spin probe concentrations for careful determination of s_{max} using ODNP alone.

2.3 Interpretation of Measured ODNP Parameters

Here, we summarize the theory commonly applied for the interpretation of the relaxation rates and coupling factor values obtained in ODNP experiments. To translate the coupling factor into a physical picture of local water motion, a model is adopted that describes water diffusion around the spin label. Such model results in a spectral density function (either analytically or numerically) which in turn allows for derivation of the relaxation rates. In this chapter, we use the spectral density function derived based on the Force Free Hard Sphere (FFHS) model (Freed, 1978; Hwang & Freed, 1975), which is so far almost exclusively used for the analysis of the ODNP data in the context of obtaining water dynamics in the proximity of a spin label.

The self- and cross-relaxation of the ^{1}H water nuclei is driven by fluctuations of the electron–proton dipolar interactions. The time dependence of these interactions is commonly characterized by a single correlation time τ_c. In the case of water protons interacting with the spin label, this correlation time is dominated by the physical diffusion of water relative to that of the spin label or probe, as follows:

$$\tau_c = \frac{d^2}{D_{H_2O} + D_{SL}}, \quad (8)$$

where D_{SL} and D_{H_2O} are, respectively, the diffusivities of the spin label and water in the vicinity of the spin label, and d is the distance of minimum approach of water protons to the unpaired electron spin. The connection between relaxation rates and the correlation time is obtained through an analytical spectral density function $J(\omega; \tau_c)$ obtained from the FFHS model. It provides the transition probability of the given spin relaxation process (T_{1u}, T_{zq}, T_{dq}) at a given frequency. For dipolar interactions, σ and ρ relaxations take the form (Solomon, 1955):

$$\sigma(B_0; \tau_c) = \kappa C (6J(\omega_e - \omega_H; \tau_c) - J(\omega_e + \omega_H; \tau_c)), \quad (9)$$

$$\rho(B_0; \tau_c) = \kappa C (6J(\omega_e - \omega_H; \tau_c) - J(\omega_e + \omega_H; \tau_c) + 3J(\omega_H; \tau_c)), \quad (10)$$

where κ is a constant given by $\kappa = \frac{4}{10}\pi\gamma_H^2\gamma_e^2\hbar^2$ and C is the radical concentration.

The coupling factor is consequently given by:

$$\xi(B_0; \tau_c) = \frac{(6J(\omega_e - \omega_H; \tau_c) - J(\omega_e + \omega_H; \tau_c))}{(6J(\omega_e - \omega_H; \tau_c) - J(\omega_e + \omega_H; \tau_c) + 3J(\omega_H; \tau_c))}. \quad (11)$$

The dependence on B_0 arises from the linear dependence of ω_e and ω_H on B_0 following:

$$\omega_e = \gamma_e B_0 \quad \omega_H = \gamma_H B_0. \tag{12}$$

Note that the κ prefactor cancels out in the coupling factor parameter, as does the dependence on the radical concentration. Thus, if possible, a coupling factor-based analysis is preferred as it eliminates the systematic error that may be present when determining the radical concentration. This is especially important in biochemical preparations where spin labeling efficiency is not always precisely known. That said, it is possible to obtain information on hydration dynamics from σ alone, though from our experience, experimental errors resulting in such analysis are typically larger.

Up to the determination of σ and ξ, no approximations are made except for the well-justified assumption that the interaction of the water protons with the unpaired electron of the spin label is exclusively dipolar. However, to relate ξ to the actual local water diffusivity, one has to assume a model of motion that preferably results in a closed analytical form for the spectral density function $J(\omega; \tau_c)$. As mentioned, the most popular model that is almost exclusively used to date for the analysis of relaxivities, including ξ, in terms of hydration dynamics, is the FFHS model, with expression for $J(\omega; \tau_c)$ given in the following (Hwang & Freed, 1975):

$$J(\omega; \tau_c) = \Re e \left[\frac{1 + \left((i\omega\tau_c)^{1/2}/4\right)}{1 + (i\omega\tau_c)^{1/2} + \frac{4}{9}(i\omega\tau_c) + \frac{1}{9}(i\omega\tau_c)^{3/2}} \right]. \tag{13}$$

In practice, one looks for a correlation time value τ_c that, when using the FFHS model, results in the experimentally obtained coupling factor at a given magnetic field B_0 of typically 0.35 T. Several attempts were made throughout the years to go beyond the FFHS model (Ayant, Belorizky, Alizon, & Gallice, 1975; Ayant, Belorizky, Fries, & Rosset, 1977; Freed, 1978; Hwang & Freed, 1975) by both developing more complicated analytical expressions and more recently by obtaining spectral density function $J(\omega; \tau_c)$ numerically from molecular dynamics MD trajectories (Sezer, 2013, 2014; Sezer, Prandolini, et al., 2009).

3. HARDWARE

To date, no dedicated ODNP instrument is commercially available as a complete unit. Thus, essentially every laboratory has an ODNP spectrometer of a different design. Despite the differences, all are built along the same guidelines. At the heart of an ODNP spectrometer lies typically a commercial CW EPR instrument. An overall view of the ODNP spectrometer in the Han laboratory is shown in Fig. 3A. Modifications and additions to the CW EPR spectrometer required for ODNP experiments are listed below:

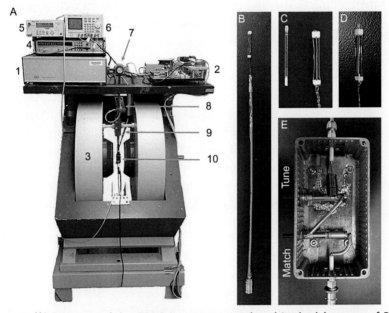

Figure 3 (A) Overview of the ODNP instrument employed in the laboratory of Prof. Songi Han, at the University of California Santa Barbara. 1, EPR (Bruker EMX) bridge; 2, ODNP bridge; 3, Electromagnet (Bruker EMX); 4, frequency counter (can be switched between EMX and ODNP bridges); 5, power meter (connected to ODNP bridge); 6, power supply for the ODNP bridge; 7, three port manual waveguide switch (used to switch the EPR resonator between the EPR and ODNP bridges); 8, grounding plate (used to ensure a common ground between the EPR transmission line and NMR RLC circuit); 9, NMR tuning box; 10, EPR cavity (Bruker 4123D resonator). (B) NMR probe. (C) NMR coil and sample. (D) ODNP sample mounted inside the coil. Note that EPR (EMX), NMR (Avance III) consoles, NMR preamplifier, electromagnet chiller, and power supply are not shown in the figure despite being part of the system.

1. An NMR console capable of low frequency (15 MHz) NMR experiments. Though dedicated low frequency NMR spectrometers are available from various vendors, utilizing a heteronuclear (X) channel of an obsolete NMR spectrometer is often a good practical solution.
2. An EPR cavity with high B_1 conversion is critical, so that relatively low mw power (3–10 W) can saturate the EPR transition of the nitroxide radical. The dielectric cavity 4123D from Bruker Biopsin is an excellent choice since it tends to minimize sample heating for a given strength of mw irradiation, followed by the standard Bruker EPR resonators, such as the super high-Q and the 4102ST cavities. The Bruker ENDOR cavity EN 801 suffers both from low mw B_1 conversion, as well as from low NMR filling factor, and thus is a poor choice for ODNP even after a resonant NMR circuit is added. However, this is by far not an exhaustive survey, and alternative cavity designs are conceivable (Doll et al., 2012; Hausser & Stehlik, 1968).
3. A low frequency NMR probe that is compatible with an existing EPR cavity. Several designs were described in the ODNP literature (Armstrong et al., 2008; Franck, Pavlova, et al., 2013; Hausser & Stehlik, 1968).
    a. The current design that has been in use in our laboratory for several years is shown in Fig. 3B and C. The coil wires are oriented along the long axis of the resonator to exert minimal distortion of the mw resonator mode, while concurrently designed to maximize the NMR filling factor. We found this design to perform well in both rectangular TE_{102} (Bruker 4102ST) and in dielectric cavities with a TE_{011}-like mode (ER 4119HS, ER 4122HSQE, and ER 4123D). The ODNP sample consisting of 3-4 µl of solution sealed inside the quartz capillary is mounted on the probe (Figure 3C,D). This NMR probe connects to the NMR tuning box (Fig. 3E) via an SMB connector. The piece of coaxial semirigid cable between the coil and the SMB (SubMiniature version B) RF connector allows for the insertion/removal of the probe without the need to unmount the tuning box. Despite the overall utility and efficiency of this probe, there are several drawbacks in this current design: (i) Sample exchange requires insertion and removal of the whole NMR probe. (ii) The probe position is set by eye relative to the EPR cavity, which reduces the reproducibility of the experiment especially when performed by different experimentalists. (iii) The ODNP sample is held in place by friction against the Teflon sample/coil holders (Fig. 3D), which results in their wear and tear and eventually requires rebuilding of the probes.

b. We have recently developed a more robust NMR probe where the tuning box and NMR coil are combined into a single unit that is mounted from the bottom on the EPR cavity (Fig. 4A–C, Ryan Barnes and Songi Han, unpublished data). In this new design, the sample is fixed in place by a separate sample holder that can be moved independently of the NMR probe, thus allowing for one to change

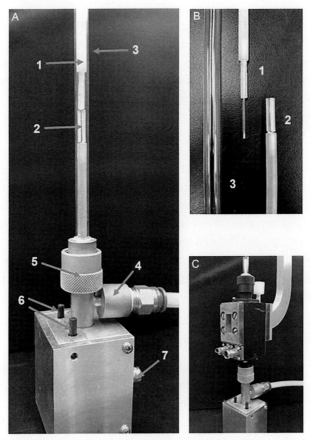

Figure 4 (A) Overview of the new NMR probe designed for ODNP. 1, sample holder and sample; 2, NMR coil; 3, quartz tube for aligning the sample holder and the NMR coil; 4, connector for air flow; 5, catch nut for fixing the assembly to the bottom of the microwave cavity; 6, adjustment rods for the match and tune capacitors of the RLC circuit; 7, BNC female jack for connecting to the NMR spectrometer. (B) 1, layout of the sample and sample holder; 2, the NMR probe windings in the saddle coil configuration; 3, the quartz tube used to align the sample into the NMR probe. (C) Picture of how the NMR probe assembly fits inside of the EPR cavity. Note, the cavity shown is Bruker 4102ST which is not the routine cavity used for ODNP.

the sample without removing the NMR probe (Fig. 4B). This design negates the need to carefully reposition the NMR probe each time the system is assembled, and consequently reduces operator error as well as wear and tear on the probe.

4. High power (3–10 W) CW mw source with a suitable computer interface for finely controlling the output mw power. The initial design used in our laboratory is described in detail in Armstrong et al. (2008). The current version differs only in minor details.
5. A mw power and frequency meter to carefully monitor the mw frequency and power during the course of the ODNP experiment. As the mw power is commonly varied by an attenuator and the absolute mw power is not set, one must measure the incident mw power in order to make accurate and reproducible measurements.

Putting together the hardware alone will only allow for a manually controlled ODNP experiment. For routine applications, software that allows for automation is extremely beneficial. In fact, for delicate biological samples, automation of the entire ODNP experiment is necessary to keep experimental time as short as possible to minimize sample damage. In practice, to fully automate the experiment, one must automatically: (i) loop through the NMR FID and inversion recovery acquisitions for the enhancement and T_1 measurements; (ii) set the mw power applied to the sample for each FID experiment in the enhancement series and for every inversion recovery experiment in the T_1 series; and (iii) log the mw power applied to the sample throughout the experiments.

4. DATA ACQUISITION

The specific goal of an ODNP experiment is to measure the ODNP enhancement and relaxation data for obtaining the coupling factor that in turn can be translated into physical parameters describing local hydration dynamics in close proximity (<1 nm) of an unpaired electron of a spin probe.

The ODNP sample contains 3–4 μl of solution sealed inside a 0.6 mm ID–0.84 mm OD quartz capillary (VitroCom). The common practice in the Han lab is to seal one side with the Critoseal™ tube sealant and the second side with melted capillary wax (Fig. 3C). However, a polymer-based capillary tubing has also been used (Ortony et al., 2013, 2014). It is very important to ensure good quality of the seal so that the sample does not evaporate due to the prolonged mw heating during an ODNP experiment.

In practice, ODNP experiments are preceded by a CW EPR measurement, with two goals: (i) to obtain the resonance frequency/field that is typically set to the maximum of the center peak of the nitroxide spectrum, i.e., the zero crossing of the center $m_I = 0$ line of the cw EPR spectrum and (ii) to quickly prove the sample's integrity and preparation quality. Here, since ODNP reports on the average local water dynamics in close proximity to the spin probes/spin labels, it is of utter importance that the spin is localized in the target environment. Similarly, it is necessary to ensure that all unattached spin label is carefully removed. It is important to keep in mind that when two components are present in the EPR spectrum, the ODNP measurement will report on an average value representing water dynamics within all spin label environments present in the sample. Consequently, samples exhibiting only a single component in the EPR spectrum will simplify interpretation of the results.

During the acquisition of an EPR spectrum, the EPR cavity is critically coupled and the mw frequency is set. Since the same transmission line is used to couple the EPR and ODNP bridges, the same mw frequency and cavity coupling can be used for both the EPR and ODNP experiments.

The full ODNP measurement consists of three separate series of experiments, two performed on the sample with a radical: (i) NMR enhancement as function of mw power (Fig. 5A) and (ii) series of inversion recovery experiments to obtain T_1 relaxation as function of applied mw power (Fig. 5B). A third experiment, inversion recovery to obtain $T_{1,0}$, is performed on a sample of identical composition, except without the paramagnetic radical. It is of great importance to ensure that the concentration of all the constituents is equal in both samples. Ideally in the "background" sample, the spin label is replaced by its reduced analog. For example, in case of spin labeled proteins, MTSL (Fig. 1B) is replaced by dMTSL (Fig. 1C). In practice, this can often be omitted if the dMTSL's perturbation on the physical properties and ^{1}H water relaxation of the system is negligible.

Typically, 15–20 experimental values are measured for the ODNP enhancement curve and 5–7 for the $T_1(p)$ series, both spanning the same range for the applied mw power. In addition, one $T_{1,0}$ experiment is performed on the sample without the radical. It is possible to acquire a full $T_{1,0}(p)$ series. The goal of such a series is to ensure sample integrity over the temperature range covered during the ODNP measurement. Both $T_1(p)$ and $T_{1,0}(p)$ are predicted to increase linearly with mw power with the same slope (Franck, Pavlova, et al., 2013; Hindman, Svirmickas, & Wood, 1973). Ensuring the latter provides for a good self-consistency check,

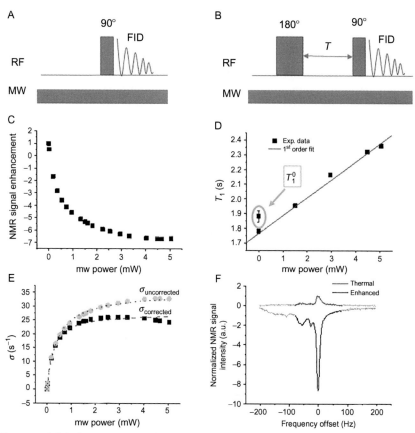

Figure 5 (A) Pulse sequence of the ODNP experiment (mw power is incremented). (B) Pulse sequence for the T_1 series experiment (note that both mw power and t delay are incremented for a pseudo 2D dataset). (C) Normalized ODNP enhancement series. (D) Processed data for the T_1 series and $T_{1,0}$ experiment. (E) $\sigma s(p)$ curves with (black squares) and without (green (light gray in the print version) circles) temperature corrections; fits to Eq. (14) are shown by red (gray in the print version) dashed line and blue (dark gray in the print version) dotted line, respectively. (F) Enhanced ($p=5$ mW) (black) and unenhanced (red (dark gray in the print version)) NMR spectra; part of the frequency range used for signal integration is highlighted in bold. The data shown are acquired from an MTSL labeled D41C mutant of Chemotaxis Y protein at 200 µM concentration.

which is especially recommended if phase transition, sample decomposition, or other changes to the sample are anticipated with temperature increase.

For all NMR experiments, we routinely employ a standard four-step CYCLOPS phase cycle. From our experience, this results in a SNR of ~40–50 for an unenhanced NMR signal. Note that the lineshape is determined by the homogeneity of the magnetic field inside a typical EPR magnet, so in practice the NMR lineshape is never an ideal Lorentzian. If the sample and coil are properly positioned inside the EPR resonator, the lineshape should be equal or very similar for the enhanced and unenhanced signals (Fig. 5F). Note that one should avoid a situation where only part of the NMR line is inverted during an ODNP experiment, as this will result in large errors when normalizing the spectra by the integral of the unenhanced signal.

The automated ODNP experiment typically takes ~2 to 3 h and will include a 5-min delay at the beginning of the experiment after the mw source is turned on for the first time to allow for the warm up and stabilization of the high power amplifiers, as well as a 10-s delay every time the mw power is changed to a different value to allow for the sample temperature to reach a steady state.

5. DATA ANALYSIS
5.1 Raw Data Analysis

This section describes the step-by-step procedure to analyze the experimental ODNP data. The procedure outlined below assumes that raw NMR enhancement spectra have been integrated, normalized to the unenhanced integral, and dimensioned with the corresponding mw power to yield $E(p)$. Also that the inversion recovery experiments were integrated and fit to extract the T_1 and that each T_1 value have been dimensioned with the corresponding mw power to yield $T_1(p)$.

An example of such processed raw ODNP dataset obtained for a spin label attached to a surface exposed cite of a globular protein is shown in Fig. 5 ([protein] ≈ [spin label] ≈ 200 μM). The ^{1}H signal enhancement plotted against applied mw power is shown in Fig. 5C, and T_1 and $T_{1,0}$ relaxation times as function of mw power shown in Fig. 5D. Note that the increase in T_1 with an increase in applied mw power is due to dielectric sample heating. To provide a concrete example, we provide in parenthesis the values obtained for the experimental data presented on Fig. 5 for each step of the analysis outlined below.

The first step is to fit the T_1 series using a first or second order polynomial to obtain the T_1 rate for each point in the enhancement series. For this particular dataset, the first order polynomial gave a satisfactory fit (red dashed (dark gray in the print version) line in Fig. 5D). The T_1 times are then interpolated for all mw power values used in the enhancement series.

1. a. The interpolated values are used to calculate the temperature corrected cross-relaxation rate as function of applied mw power according to $\sigma s(p) = \left(\frac{1-E(p)}{T_1(p)C}\left|\frac{\gamma_H}{\gamma_e}\right|\right)$; this is plotted in Fig. 5E as black squares.
   b. Alternatively, the uncorrected cross-relaxation rate $\sigma_{uncorrected}s(p) = \left(\frac{1-E(p)}{T_1(p=0)C}\left|\frac{\gamma_H}{\gamma_e}\right|\right)$ can be plotted as function of mw power as shown in Fig. 5E as green (gray in the print version) circles.
2. Since the mw power dependence of EPR saturation is known and given by Eq. (6), the $\sigma s(p)$ and $\sigma_{uncorrected}s(p)$ curves from step 2 can be fitted to:

$$\sigma s(p) = \frac{\sigma s_{max} p}{a+p} \tag{14}$$

to obtain σs_{max} and $\sigma_{uncorrected}s_{max}$, respectively ($a$ is a scaling factor and is set as a free parameter in the fit). Since in most cases of interest $s_{max} \approx 1$ is given, one obtains the values for σ and $\sigma_{uncorrected}$, respectively (here $\sigma = 27.5\,\text{s}^{-1}$; $\sigma_{uncorrected} = 35.0\,\text{s}^{-1}$).
3. The next step is to calculate the self-relaxation value ρ from $T_1(p=0)$ and $T_{1,0}(p=0)$ according to Eq. (3) (here $\rho = 149.5\,\text{s}^{-1}$).
4. Subsequently, the corrected ξ and uncorrected coupling factor $\xi_{uncorrected}$ is calculated according to Eq. (2) (here: $\xi = 0.18$; $\xi_{uncorrected} = 0.23$).

5.2 Data Interpretation

The coupling factor and the self- and cross-relaxation rates are parameters directly obtained from the ODNP experiment. What one would like to do next is to convert the coupling factor to a more intuitive and physical measurement of water dynamics, such as the correlation time or a local water diffusion coefficient.

In order to convert the coupling factor into a diffusion correlation time, one has to assume a model for the spectral density function. Here, we provide an example of ODNP data analysis using the FFHS model (Hwang & Freed, 1975). For a given coupling factor, a correlation time is obtained by solving Eq. (11) for a value of the magnetic field as used in the experiment, thus far typically $B_0 = 0.35$ T (here, for data shown in Fig. 5C–E: $\tau_c = 99\,\text{ps}$ from $\xi_{uncorrected} = 0.23$ for the uncorrected and $\tau_c = 71\,\text{ps}$ from $\xi = 0.18$ for the temperature corrected analysis).

For a better understanding of the water properties in a biochemical system, it is always illustrative to have a comparison with the properties of bulk water. For a nitroxide spin probe dissolved in bulk water, the coupling factor was shown to be $\xi^{bulk} = 0.27$, which corresponds to a correlation time $\tau_c^{bulk} = 54\,\text{ps}$, as derived from the FFHS model. Note that the typical value for the bulk water obtained without the temperature correction is $\xi^{bulk}_{uncorrected} = 0.33$ and consequently $\tau^{bulk}_{c,\,uncorrected} = 33.3\,\text{ps}$, as shown and discussed in Franck, Pavlova, et al. (2013).

One of the common ways to present ODNP data is by defining the retardation factor, defined as the ratio between the water diffusivities of the bulk water and that at the local or surface site of interest.

$$\rho_t = \frac{\tau_{c,\,local}}{\tau_{c,\,bulk}} \approx \frac{D_{H_2O} + D_{radical}}{D_{local}} \tag{15}$$

The assumption underlying Eq. (15) is that the distance of closest approach d between the ^{1}H of water and the electron spin of the probe in Eq. (8) remains the same regardless of the local environment of the spin, such as in bulk versus a macromolecular surface (here: $\rho_t = 0.55$; $\rho_{t,\,uncorrected} = 0.46$ for the spin labeled protein example shown in Fig. 5C–E). Substituting $D_{H_2O} = 2.3 \times 10^{-9}\,\text{m}^2\text{s}^{-1}$ for bulk water diffusion in the vicinity of the dissolved radical probe and $D_{radical} = 4 \times 10^{-9}\,\text{m}^2\text{s}^{-1}$ for diffusion of the free spin probe in water into Eq. (14), we arrive at local diffusivity for the water near the spin labeled site of the protein D_{local} (here: $D_{local} = 1.5 \times 10^{-9}\,\text{m}^2\text{s}^{-1}$ and $D^{uncorrected}_{local} = 1.24 \times 10^{-9}\,\text{m}^2\text{s}^{-1}$).

6. EXAMPLES OF ODNP

We conclude the chapter by showing how a succession of ODNP experiments performed on a series of carefully designed samples can provide new insights into the biomolecular systems of interest. We do this by reviewing two illustrative examples from the recent literature.

In a previous study, ODNP was employed to study water's role in the pH-dependent folding process of apomyoglobin (aMgb) (Armstrong et al., 2011). The study aimed to deduce whether or not the molten globule state, which for aMgb is the on-pathway folding intermediate, excludes solvent at the sites that will comprise the core of the folded protein. Five sites of aMgb were spin labeled, three at positions that are buried upon folding and two at positions that remain at the protein surface (Fig. 6A). In the molten

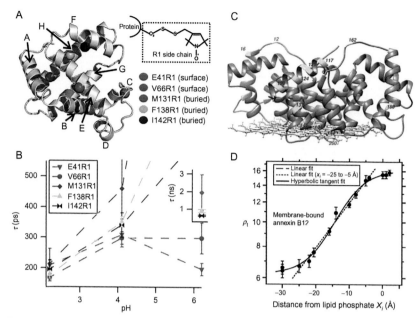

Figure 6 (A) The spin labeled sites of apomyoglobin are shown in the protein's native structure. (B) The measured translational correlation time as a function of pH, it is seen that in the native state the hydration dynamics show a large dispersion. (C) X-ray crystal structure of membrane-bound annexin B12 in the presence of Ca^{2+}. The residues subjected to site-directed spin labeling are marked by a red number and highlighted with yellow. (D) Distance dependence of retardation factor (ρ_t) at specific sites of annexin B12 bound on the surface of large unilamellar vesicles composed of POPC:POPS (1:2) in the presence of 1 mM Ca^{2+} at 25 °C. *(A and B) are reproduced with permission from Armstrong et al. (2011) and (C and D) are reproduced with permission and slight modifications from Cheng et al. (2013). (See the color plate.)*

globule state at pH 4.1, ODNP studies found that the core sites displayed intermediate hydration dynamic values, suggesting a "wet" interior of the molten globule and implying that solvent expulsion of the hydrophobic core is not necessarily the initial step in the folding process, at least for the sites investigated in aMgb (Fig. 6B). An additional observation was that the dispersion of site-specific protein hydration dynamics increases upon folding, even when taking only solvent-exposed surface sites into account. This implies that the 3D protein surface topology influences the surface hydration landscape, where the neighboring residues in the primary sequence as well as spatially proximal residues influence surface water diffusivity for a surface site. Differences in local water diffusivity reflect on the collective effect of water–protein adhesive and water–water cohesive interactions on the local surface water network that may be modulated by differences in surface topology or chemistry (Song et al., 2014).

In the next example, we demonstrate that when the globular protein, annexin, binds to a lipid membrane surface as induced by addition of Ca^{2+}, the surface hydration landscape of annexin "submits" itself to that of the lipid membrane surface (Cheng et al., 2013). This was revealed by the observation that the surface hydration dynamics at 13 different surface sites of annexin (Fig. 6C) display a strong correlation to the distance between the spin labeled site in a protein to the phosphate of the liposome's phosphatidylcholine headgroup (Fig. 6D). Thus, regardless of the specifics, any monotonic relationship of hydration dynamics with respect to the lipid surface implies that the lipid membrane surface, not the protein surface topology, dictates the protein surface hydration dynamics, so that the properties of the lipid membrane constitute and dictate the nature of local hydration in this case.

7. SUMMARY

The development in both theory and technique over the years has placed ODNP as a prominent technique in studying surface water dynamics. The techniques presented here are considered state-of-the-art for ^{1}H ODNP relaxometry at X-band EPR frequency for the study of hydration dynamics. We hope this chapter serves as a useful starting point for other laboratories interested in studying the role of water in different biological systems. We anticipate further refinement and development of the technique in years to come.

ACKNOWLEDGMENTS

We would like to thank the various members in our lab who have contributed to both improving and applying the ODNP technique, most notably John Franck, Chi-yuan Cheng, and Brandon Armstrong. In this and previous studies, we made use of the MRL Central Facilities, a member of the MRFN, supported by the NSF through the MRSEC (DMR 1121053). We also acknowledge support from the Dreyfus Teacher Scholar award to S.H., the Cluster of Excellence RESOLV (EXC 1069) funded by the Deutsche Forschungsgemeinschaft and a long-term postdoctoral fellowship by the Human Frontier Science Program awarded to I.K.

REFERENCES

Altenbach, C., Kusnetzow, A. K., Ernst, O. P., Hofmann, K. P., & Hubbell, W. L. (2008). High-resolution distance mapping in rhodopsin reveals the pattern of helix movement due to activation. *Proceedings of the National Academy of Sciences of the United States of America, 105*(21), 7439–7444.

Armstrong, B. D., Choi, J., López, C., Wesener, D. A., Hubbell, W., Cavagnero, S., et al. (2011). Site-specific hydration dynamics in the nonpolar core of a molten globule by dynamic nuclear polarization of water. *Journal of the American Chemical Society*, *133*(15), 5987–5995.

Armstrong, B. D., & Han, S. (2007). A new model for Overhauser enhanced nuclear magnetic resonance using nitroxide radicals. *The Journal of Chemical Physics*, *127*(10), 104508.

Armstrong, B. D., & Han, S. (2009). Overhauser dynamic nuclear polarization to study local water dynamics. *Journal of the American Chemical Society*, *131*(13), 4641–4647.

Armstrong, B. D., Lingwood, M. D., McCarney, E. R., Brown, E. R., Blümler, P., & Han, S. (2008). Portable X-band system for solution state dynamic nuclear polarization. *Journal of Magnetic Resonance*, *191*(2), 273–281.

Ayant, Y., Belorizky, E., Alizon, J., & Gallice, J. (1975). Calculation of spectral densities for relaxation resulting from random molecular translational modulation of magnetic dipolar coupling in liquids. *Journal de Physique*, *36*(10), 991–1004.

Ayant, Y., Belorizky, E., Fries, P., & Rosset, J. (1977). Effet des interactions dipolaires magnétiques intermoléculaires sur la relaxation nucléaire de molécules polyatomiques dans les liquides. *Journal de Physique*, *38*(3), 325–337.

Bates, R. D., & Drozdoski, W. S. (1977). Use of nitroxide spin labels in studies of solvent–solute interactions. *The Journal of Chemical Physics*, *67*(9), 4038.

Bennati, M., Luchinat, C., Parigi, G., & Türke, M.-T. (2010). Water 1H relaxation dispersion analysis on a nitroxide radical provides information on the maximal signal enhancement in Overhauser dynamic nuclear polarization experiments. *Physical Chemistry Chemical Physics*, *12*, 5902–5910.

Carver, T. R., & Slichter, C. P. (1956). Experimental verification of the Overhauser nuclear polarization effect. *Physical Review*, *102*(4), 975–980.

Cheng, C. Y., Goor, O. J. G. M., & Han, S. (2012). Quantitative analysis of molecular transport across liposomal bilayer by J-mediated 13C Overhauser dynamic nuclear polarization. *Analytical Chemistry*, *84*(21), 8936–8940. http://dx.doi.org/10.1021/ac301932h.

Cheng, C.-Y., & Han, S. (2013). Dynamic nuclear polarization methods in solids and solutions to explore membrane proteins and membrane systems. *Annual Review of Physical Chemistry*, *64*, 507–532.

Cheng, C., Olijve, L. L. C., Kausik, R., & Han, S. (2014). Cholesterol enhances surface water diffusion of phospholipid bilayers. *The Journal of Chemical Physics*, *141*, 22D513.

Cheng, C., Varkey, J., Ambroso, M. R., Langen, R., & Han, S. (2013). Hydration dynamics as an intrinsic ruler for refining protein structure at lipid membrane interfaces. *Proceedings of the National Academy of Sciences of the United States of America*, *110*(42), 1–6.

Cheng, C., Wang, J.-Y., Kausik, R., Lee, K. Y. C., & Han, S. (2012). Nature of interactions between PEO-PPO-PEO triblock copolymers and lipid membranes: (I) Effect of polymer hydrophobicity on its ability to protect liposomes from peroxidation. *Biomacromolecules*, *13*(9), 2616–2623.

Doll, A., Bordignon, E., Joseph, B., Tschaggelar, R., & Jeschke, G. (2012). Liquid state DNP for water accessibility measurements on spin-labeled membrane proteins at physiological temperatures. *Journal of Magnetic Resonance*, *222*, 34–43.

Enkin, N., Liu, G., Gimenez-Lopez, M. d. C., Porfyrakis, K., Tkach, I., & Bennati, M. (2015). A high saturation factor in Overhauser DNP with nitroxide derivatives: The role of ^{14}N nuclear spin relaxation. *Physical Chemistry Chemical Physics*, *17*(17), 11144–11149.

Enkin, N., Liu, G., Tkach, I., & Bennati, M. (2014). High DNP efficiency of TEMPONE radicals in liquid toluene at low concentrations. *Physical Chemistry Chemical Physics (PCCP)*, *16*(19), 8795–8800.

Eschmann, N. A., Do, T. D., LaPointe, N. E., Shea, J.-E., Feinstein, S. C., Bowers, M. T., et al. (2015). Tau aggregation propensity engrained in its solution state. (manuscript in preparation).

Franck, J. M., Ding, Y., Stone, K., Qin, P. Z., & Han, S. (2015). Anomalously rapid hydration water diffusion dynamics found near DNA surfaces. *Journal of the American Chemical Society* (accepted).

Franck, J. M., Pavlova, A., Scott, J. A., & Han, S. (2013). Quantitative cw Overhauser dynamic nuclear polarization for the analysis of local water dynamics. *Progress in Nuclear Magnetic Resonance Spectroscopy, 74*, 33–56.

Franck, J. M., Scott, J. A., & Han, S. (2013). Nonlinear scaling of surface water diffusion with bulk water viscosity of crowded solutions. *Journal of the American Chemical Society, 135*(11), 4175–4178.

Franck, J. M., Sokolovski, M., Kessler, N., Matalon, E., Gordon-Grossman, M., Han, S.-I., et al. (2014). Probing water density and dynamics in the chaperonin GroEL cavity. *Journal of the American Chemical Society, 136*(26), 9396–9403.

Freed, J. H. (1978). Dynamic effects of pair correlation functions on spin relaxation by translational diffusion in liquids. II. Finite jumps and independent T1 processes. *The Journal of Chemical Physics, 68*(9), 4034–4037.

Gitti, R., Wild, C., Tsiao, C., Zimmer, K., Glass, T. E., & Dorn, H. C. (1988). Solid–liquid intermolecular transfer of dynamic nuclear polarization. Enhanced flowing fluid 1H NMR signals via immobilized spin labels. *Journal of the American Chemical Society, 110*, 2294–2296.

Griesinger, C., Bennati, M., Vieth, H. M., Luchinat, C., Parigi, G., Höfer, P., et al. (2012). Dynamic nuclear polarization at high magnetic fields in liquids. *Progress in Nuclear Magnetic Resonance Spectroscopy, 64*, 4–28.

Hausser, K. H., & Stehlik, D. (1968). Dynamic nuclear polarization in liquids. *Advances in Magnetic Resonance, 3*, 79–139.

Hindman, J. C., Svirmickas, A., & Wood, M. (1973). Relaxation processes in water. *The Journal of Chemical Physics, 59*(3), 1517–1522.

Hussain, S., Franck, J. M., & Han, S. (2013). Transmembrane protein activation refined by site-specific hydration dynamics. *Angewandte Chemie (International Ed in English), 52*(7), 1953–1958.

Hwang, L.-P., & Freed, J. H. (1975). Dynamic effects of pair correlation functions of spin relaxation by translational diffusion in liquids. *Journal of Chemical Physics, 63*(9), 4017–4025.

Hyde, J. S., Froncisz, W., & Mottley, C. (1984). Pulsed ELDOR measurement of nitrogen T1 in spin labels. *Chemical Physics Letters, 110*(6), 621–625.

Kausik, R., & Han, S. (2009). Ultrasensitive detection of interfacial water diffusion on lipid vesicle surfaces at molecular length scales. *Journal of the American Chemical Society, 131*(51), 18254–18256.

Kim, J.-M., Altenbach, C., Kono, M., Oprian, D. D., Hubbell, W. L., & Khorana, H. G. (2004). Structural origins of constitutive activation in rhodopsin: Role of the K296/E113 salt bridge. *Proceedings of the National Academy of Sciences, 101*(34), 12508–12513.

Krummenacker, J. G., Denysenkov, V. P., Terekhov, M., Schreiber, L. M., & Prisner, T. F. (2012). DNP in MRI: An in-bore approach at 1.5 T. *Journal of Magnetic Resonance, 215*, 94–99.

Lingwood, M. D., Siaw, T. A., Sailasuta, N., Abulseoud, O. A., Chan, H. R., Ross, B. D., et al. (2012). Hyperpolarized water as an MR imaging contrast agent: Feasibility of in vivo imaging in a rat model. *Radiology, 265*(2), 418–425.

Lottmann, P., Marquardsen, T., Krahn, A., Tavernier, A., Höfer, P., Bennati, M., et al. (2012). Evaluation of a shuttle DNP spectrometer by calculating the coupling and global enhancement factors of l-tryptophan. *Applied Magnetic Resonance, 43*(1–2), 207–221.

McCarney, E. R., Armstrong, B. D., Lingwood, M. D., & Han, S. (2007). Hyperpolarized water as an authentic magnetic resonance imaging contrast agent. *Proceedings of the National Academy of Sciences of the United States of America, 104*(6), 1754–1759.

Mims, W. B. (1972). Envelope modulation in spin-echo experiments. *Physical Review B, 5*(7), 2409–2419.

Neudert, O., Reh, M., Spiess, H. W., & Münnemann, K. (2015). X-band DNP hyperpolarization of viscous liquids and polymer melts. *Macromolecular Rapid Communications, 36*(10), 885–889.

Ortony, J. H., Choi, S.-H., Spruell, J. M., Hunt, J. N., Lynd, N. A., Krogstad, D. V., et al. (2014). Fluidity and water in nanoscale domains define coacervate hydrogels. *Chemical Science, 5,* 58–67.

Ortony, J. H., Hwang, D. S., Franck, J. M., Waite, J. H., & Han, S. (2013). Asymmetric collapse in biomimetic complex coacervates revealed by local polymer and water dynamics. *Biomacromolecules, 14*(5), 1395–1402.

Overhauser, A. W. (1953). Polarization of nuclei in metals. *Physical Review, 92*(2), 411–415.

Pavlova, A., Cheng, C., Kinnebrew, M., Lew, J., Dahlquist, F. W., & Han, S. (2015). Early protein aggregation events revealed by mapping out surface water rearrangements. *Proceedings of the National Academy of Sciences,* (in review).

Pavlova, A., McCarney, E. R., Peterson, D. W., Dahlquist, F. W., Lew, J., & Han, S. (2009). Site-specific dynamic nuclear polarization of hydration water as a generally applicable approach to monitor protein aggregation. *Physical Chemistry Chemical Physics (PCCP), 11*(31), 6626–6637.

Pedersen, J. B., & Freed, J. H. (1974). Some theoretical aspects of chemically induced dynamic nuclear polarization. *The Journal of Chemical Physics, 61*(4), 1517.

Pedersen, J. B., & Freed, J. H. (1975). A hydrodynamic effect on chemically induced dynamic spin polarization. *Journal of Chemical Physics, 62*(June 1974), 1790–1795.

Robinson, B. H., Haas, D. A., & Mailer, C. (1994). Molecular dynamics in liquids: Spin–lattice relaxation of nitroxide spin labels. *Science (New York, NY), 263*(5146), 490–493.

Rowan, L. G., Hahn, E. L., & Mims, W. B. (1965). Electron-spin-echo envelope modulation. *Physical Review, 137*(1A), 61–71.

Sezer, D. (2013). Computation of DNP coupling factors of a nitroxide radical in toluene: Seamless combination of MD simulations and analytical calculations. *Physical Chemistry Chemical Physics (PCCP), 15*(2), 526–540.

Sezer, D. (2014). Rationalizing Overhauser DNP of nitroxide radicals in water through MD simulations. *Physical Chemistry Chemical Physics (PCCP), 16*(3), 1022–1032.

Sezer, D., Gafurov, M., Prandolini, M. J., Denysenkov, V. P., & Prisner, T. F. (2009). Dynamic nuclear polarization of water by a nitroxide radical: Rigorous treatment of the electron spin saturation and comparison with experiments at 9.2 Tesla. *Physical Chemistry Chemical Physics, 11*(31), 6638.

Sezer, D., Prandolini, M. J., & Prisner, T. F. (2009). Dynamic nuclear polarization coupling factors calculated from molecular dynamics simulations of a nitroxide radical in water. *Physical Chemistry Chemical Physics (PCCP), 11*(31), 6626–6637.

Solomon, I. (1955). Relaxation processes in a system of two spins. *Physical Review, 99*(2), 559–565.

Song, J., Allison, B., & Han, S. (2014). Local water diffusivity as a molecular probe of surface hydrophilicity. *MRS Bulletin, 39*(12), 1082–1088.

Song, J., Han, O. H., & Han, S. (2015). Nanometer-scale water- and proton-diffusion heterogeneities across water channels in polymer electrolyte membranes. *Angewandte Chemie, International Edition, 54,* 3615–3620.

Song, J., Kim, M., Kang, T. H., & Han, S. (2015). Ion specific effects: Decoupling ion-ion and ion-water interactions. *Physical Chemistry Chemical Physics*, *17*, 8306–8322.

Türke, M.-T., & Bennati, M. (2011). Saturation factor of nitroxide radicals in liquid DNP by pulsed ELDOR experiments. *Physical Chemistry Chemical Physics (PCCP)*, *13*(9), 3630–3633.

Türke, M.-T., Tkach, I., Reese, M., Höfer, P., & Bennati, M. (2010). Optimization of dynamic nuclear polarization experiments in aqueous solution at 15 MHz/9.7 GHz: A comparative study with DNP at 140 MHz/94 GHz. *Physical Chemistry Chemical Physics (PCCP)*, *12*(22), 5893–5901.

CHAPTER SEVENTEEN

Practical Aspects of Paramagnetic Relaxation Enhancement in Biological Macromolecules

G. Marius Clore[1]

Laboratory of Chemical Physics, National Institute of Diabetes and Digestive and Kidney Diseases, National Institutes of Health, Bethesda, Maryland, USA
[1]Corresponding author: e-mail address: mariusc@mail.nih.gov

Contents

1. Introduction	485
2. Paramagnetic Labels for PRE Measurements	487
3. Measurement of the PRE	487
4. Using the PRE in Structure Determination	489
5. Using the PRE to Detect Transient Sparsely Populated States	490
Acknowledgments	495
References	495

Abstract

In this brief review, we summarize various aspects of NMR paramagnetic relaxation enhancement (PRE). We discuss the types of spin labels used in NMR studies, describe the relevant theory used to accurately calculate PREs from coordinates, including how to take into account the fact that paramagnetic labels tend to be highly mobile and sample a wide range of conformational space, and outline methods to refine structures or ensembles of structures directly against PRE data using simulated annealing. Finally, we show how the PRE can be used to detect, characterize, and visualize sparsely populated states of proteins and their complexes that are invisible to all other biophysical techniques.

1. INTRODUCTION

The mainstay of classical macromolecular NMR structure determination resides in short (<6 Å) interproton distance restraints derived from nuclear Overhauser enhancement (NOE) measurements, supplemented by torsion angle restraints derived from three-bond homo- and heteronuclear scalar coupling constants (Clore & Gronenborn, 1989). While this approach has been very successful, especially for globular proteins where there are a large number of short interproton distances between residues far apart in the linear amino acid sequence (Clore & Gronenborn, 1989), long-range information can be extremely beneficial, especially in cases involving multidomain proteins, nucleic acids, or macromolecular complexes (Clore & Gronenborn, 1998a, 1998b; Clore & Venditti, 2013). There are two source of long-range information: the first is from residual dipolar couplings measured in weakly aligning media which provide bond vector orientations relative to an external alignment tensor (Bax & Grishaev, 2005); the second involves the application of paramagnetic relaxation enhancement (PRE) that can, in suitable cases, detect interactions between an unpaired electron of a paramagnetic site and protons up to ~35 Å away (Clore & Iwahara, 2009). In this brief review, I will summarize in nonmathematical terms various practical aspects of PRE measurements and show how the PRE can be used not only as an aid in structure determination but perhaps even more interestingly in the detection and characterization of transient sparsely populated states of proteins and their complexes. The latter has garnered considerable recent interest as such sparsely populated states, while invisible to conventional structural and biophysical methods (including crystallography, conventional NMR, cryoelectron microscopy, EPR, and single molecule spectroscopy), often play a key role in a variety of important biological processes, including molecular recognition, allostery, conformational selection and induced fit, and various assembly processes (Iwahara & Clore, 2006; Tang, Iwahara, & Clore, 2006; Tang, Louis, Aniana, Suh, & Clore, 2008; Tang, Schwieters, & Clore, 2007; Volkov, Worrall, Holtzmann, & Ubbink, 2006). For an in-depth review of the PRE, including a detailed description of the underlying physics and mathematical representations, I refer the reader to several extensive reviews that have recently appeared on the use of the PRE in both structure determination and the investigation of sparsely populated states (Anthis & Clore, 2015; Clore & Iwahara, 2009; Clore, Tang, & Iwahara, 2007).

The PRE arises from magnetic dipolar interactions between unpaired electrons in a paramagnetic center and a nucleus resulting in an increase in nuclear relaxation rates. The paramagnetic center can either be an intrinsic component of the system as in the case of metalloproteins or an extrinsic component that has to be added by chemical means (such as by attaching a paramagnetic label to a surface engineered cysteine via disulfide chemistry). Because the magnetic moment of an unpaired electron is very large, the PRE effect is measurable out to long paramagnetic center–proton distances.

2. PARAMAGNETIC LABELS FOR PRE MEASUREMENTS

In the case of PRE measurements, it is important to employ paramagnetic labels with an isotropic g tensor (Clore & Iwahara, 2009). This eliminates pseudo-contact shifts and minimizes Curie spin relaxation. There are generally two classes of paramagnetic labels: the first is based on nitroxide free radicals (Kosen, 1989) and the second on suitable metal ions, such as Mn^{2+}, Cu^{2+}, or Gd^{3+}, chelated to EDTA (Iwahara, Anderson, Murphy, & Clore, 2003). In both instances, site-directed spin labeling of proteins is generally achieved by conjugating the paramagnetic tag to a surface-exposed cysteine residue introduced by site-specific mutagenesis, making use of a methanethiosulfonate or pyridylthiol functional group to form a disulfide link between the cysteine and the paramagnetic tag (Altenbach, Marti, Khorana, & Hubbell, 1990). Alternatives to the use of disulfide chemistry include the addition of short metal-binding sequences to the protein of interest; for example, the N-terminal ATCUN motif (XXH) binds Cu^{2+} (Mal, Ikura, & Kay, 2002) and the N-terminal HHP sequences form a dimer between two peptides bound to Ni^{2+} (Jensen, Lauritzen, Dahl, Pedersen, & Led, 2004).

The linker connecting the surface cysteine with the paramagnetic label usually has several rotatable bonds. As a result, the paramagnetic center can sample a substantial amount of conformational space which has to be taken into account when interpreting PRE data quantitatively. There exist paramagnetic labels where the conformational space is more restricted: examples include the R1p side chain (Fawzi et al., 2011) which consists of a 4-pyridyl analog of the R1 side chain, and a caged lanthanide probe CLaNP which is attached to the protein via two cysteines (Keizers, Desreux, Overhand, & Ubbink, 2007; Keizers, Saragliadis, Hiruma, Overhand, & Ubbink, 2008).

3. MEASUREMENT OF THE PRE

The PRE is measured by taking the difference in nuclear relaxation rates between the paramagnetic sample and a diamagnetic control (e.g., Mn^{2+} vs. Ca^{2+} in the case of EDTA conjugates). While both longitudinal (Γ_1) and transverse (Γ_2) PRE rates can be measured, it is generally the case that measurements of Γ_2 provide the most reliable and accurate data (Clore & Iwahara, 2009; Iwahara, Schwieters, & Clore, 2004; Iwahara, Tang, & Clore, 2007). The reasons for this are several-fold. First, cross-relaxation and hydrogen exchange with water molecules reduce the accuracy of Γ_1 measurements (Iwahara & Clore, 2010). Second, in systems with an isotropic g tensor, the contribution of Curie spin relaxation to Γ_2 is negligible and Γ_2 is dominated by direct dipole–dipole interactions between the unpaired electrons of the paramagnetic tag and the nuclei of interest. As a result, Γ_2 does not exhibit cross-correlation with other relaxation mechanisms. Third, since there are no pseudo-contact shifts in isotropic systems, the exchange contributions to R_2 are identical for the paramagnetic and diamagnetic states, and are therefore canceled out when Γ_2 is measured by taking the difference in R_2 between the paramagnetic sample and the diamagnetic control. And fourth, Γ_2 is minimally impacted by fast internal motions, whereas Γ_1 is very sensitive to them.

Many PRE studies in the literature make use of a simplistic approach in which the ratio of peak intensities in a 2D correlation spectrum between paramagnetic and diamagnetic samples is used as a measure of the PRE. Determination of Γ_2 values from these ratios requires knowledge of the R_2 values in the diamagnetic state, the use of sufficiently long repetition delays between scans to ensure that magnetization recovery levels are identical for the paramagnetic and diamagnetic states, the assumption of Lorentzian lineshapes, and absolutely identical concentrations for the paramagnetic and diamagnetic samples (Iwahara et al., 2007). For a deuterated protein, the amide ^1H_N-T_1 relaxation time is very long and interscan delays in excess of 20 s may be required; the assumption of Lorentzian lineshapes precludes the use of all window functions other than an exponential and hence is not suitable for systems with any significant degree of spectral complexity and cross-peak overlap; and finally, it is virtually impossible to ensure equal concentrations for two separate samples. All these issues are readily overcome by measuring R_2 directly from a two-time point measurement, thereby enabling the direct determination of Γ_2 values and their associated errors without making use of any fitting procedures (Iwahara et al., 2004, 2007). Especially important is the fact that the two-time point approach does not require identical sample concentrations for the paramagnetic and diamagnetic states, nor does it necessitate the use of long interscan delays. As a result, quantitative measurement of Γ_2 using the two-time point approach is not only far more accurate but is also actually significantly faster than the single time point measurement (Iwahara et al., 2007).

Two-time point measurements of R_2 rates can be accomplished by a variety of 2D or 3D heteronuclear correlation-based experiments using either conventional heteronuclear single quantum correlation (HSQC) or transverse optimized (TROSY) correlation spectroscopy (Anthis, Doucleff, & Clore, 2011; Iwahara et al., 2007).

4. USING THE PRE IN STRUCTURE DETERMINATION

Early work using PREs for structure determination converted the PRE data into very approximate distance restraints with wide error ranges (Battiste & Wagner, 2000; Gaponenko et al., 2000; Kosen et al., 1986; Schmidt & Kuntz, 1984). While approximate distance restraints derived from NOE data are very effective since a restraint of say less than 6 Å between two protons that could potentially be 100 Å apart obviously places severe constraints on the conformational space consistent with the data, a PRE-derived restraint of say 20 ± 10 Å is obviously much less constraining. Effective use of PRE data in structure determination therefore requires direct refinement against the PRE data (Iwahara et al., 2004). To do this properly, several considerations have to be taken into account (Iwahara et al., 2004). First, as the paramagnetic label is generally not rigid and can sample a large region of conformational space, Γ_2 is an ensemble-averaged quantity related to the $\langle r^{-6} \rangle$ average of the electron–proton distance. Consequently, the paramagnetic label must be represented by an ensemble of states in any calculation. Practical experience indicates that a conformational ensemble with between three and six members is more than sufficient for this purpose. The second consideration is that the mobility of the paramagnetic label has to be taken into account which requires extension of the Solomon–Bloembergen (SB) theory of the PRE (Bloembergen & Morgan, 1961; Solomon, 1955) to a model-free formalism (referred to a SBMF) in which order parameters from the electron–proton vectors are calculated on the fly from the coordinates during the course of the structure calculations (Iwahara et al., 2004). Readily available simulated annealing structure determination protocols incorporating the SBMF framework have been implemented in Xplor-NIH (Schwieters, Kuszewski, & Clore, 2006).

An example of a structure of a protein–DNA complex involving the male sex-determining factor SRY determined from backbone amide and methyl proton PRE measurements using a DNA duplex labeled at three sites (individually) with dT-EDTA-Mn^{2+} is shown in Fig. 1 (Iwahara et al., 2004).

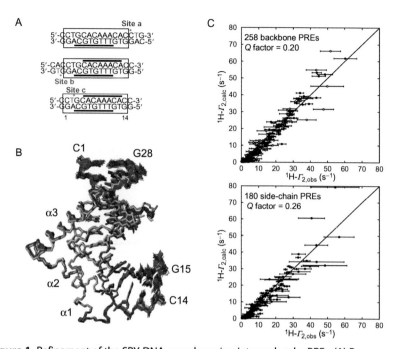

Figure 1 Refinement of the SRY-DNA complex using intermolecular PREs. (A) Paramagnetically labeled dT-EDTA-Mn^{2+} DNA duplexes used to measure intermolecular PREs on ^{15}N/^{13}C-labeled SRY. The positions of the paramagnetic tags are indicated by asterisks. (B) Superposition of 40 simulated annealing structures refined against 438 ^{1}H intermolecular PRE restraints (red, dark gray in the print version) superimposed on the regularized mean structure obtained from NOE, torsion angle, ^3J coupling, and residual dipolar coupling restraints (cyan, light gray in print version). (C) Correlation between observed and calculated PREs for backbone amide (top) and side chain (bottom) protons. The quality of agreement between experimental (Γ_2^{obs}) and calculated (Γ_2^{calc}) values is calculated by a Q-factor given by $\sqrt{\left\{\sum_i [\Gamma_2^{obs}(i) - \Gamma_2^{calc}(i)]^2\right\} / \sum_i \Gamma_2^{obs}(i)^2}$. Adapted from Iwahara et al. (2004).

5. USING THE PRE TO DETECT TRANSIENT SPARSELY POPULATED STATES

If two states exchange with one another, the PREs from the minor state can leave their footprint on the PREs measured on the observable major state, provided two conditions are fulfilled (Anthis & Clore, 2015; Clore & Iwahara, 2009; Iwahara & Clore, 2006): first, and most importantly, the paramagnetic center–proton distances in the minor state must be shorter than in the major state; and second, the overall interconversion rate (k_{ex})

between the two species must be fast enough to permit the transfer of PREs from the minor state to the observable major state. This is illustrated in Fig. 2. If exchange is slow on the PRE timescale (i.e., $\Gamma_2^{minor} - \Gamma_2^{major} \ll k_{ex}$), only the major species will contribute to the observed PRE; if, on the other hand, exchange is fast on the PRE timescale (i.e., $\Gamma_2^{minor} - \Gamma_2^{major} \gg k_{ex}$), then the observed PRE measured on the major species will be a populated-weighted average of the PREs in the major and minor states. If one considers a system in which a given paramagnetic center–proton distance varies from 30 Å in the major species to 8 Å in the minor one, this can translate into a substantial apparent PRE measured on the spectrum of the major species.

This effect was initially discovered by accident while carrying out PRE measurements on a tight ($K_D \sim 1$ nM) specific complex of the transcription factor HoxD9 with DNA in which the DNA was paramagnetically tagged at several sites with dT-EDTA-Mn^{2+} (Iwahara & Clore, 2006). At low salt (20 mM NaCl), all the PREs could be accounted for by the known crystal structure of the specific complex; but at higher salt concentrations (100–150 mM NaCl), PREs were observed at proton sites on the protein that were distant from the paramagnetic tags in the structure of the specific complex. These PREs were shown to arise from a combination of intramolecular sliding of HoxD9 along the DNA as well as from direct intermolecular translocation of HoxD9 between DNA molecules (without dissociating into free solution). Subsequently, PREs arising from minor states with occupancies as low as 0.5% have been used to probe transient encounter complexes in protein–protein association (Fawzi, Doucleff, Suh, & Clore, 2010; Tang et al., 2006, 2008; Volkov et al., 2006) and conformational sampling of domains in multidomain proteins (Anthis et al., 2011; Tang et al., 2007).

An example of the use of PREs to explore the conformational space sampled by sparsely populated states of a protein is illustrated with calmodulin (CaM)-4Ca^{2+} in Fig. 3 (Anthis et al., 2011). CaM-4Ca^{2+} has two domains connected by a linker; in the absence of a target peptide, the two domains reorient semi-independently of one another, but clamp down upon binding to a target peptide to form a compact globular structure in which the two domains envelope the peptide. No NOEs are observed between the two domains of CaM-4Ca^{2+}. When CaM-4Ca^{2+} is paramagnetically tagged at either the N- or C-terminal domains with a nitroxide label, the intradomain PRE profiles are consistent with the domain structures in the absence and presence of target peptide (Fig. 3A); however, whereas the interdomain PRE profiles obtained in the presence of the target peptide from myosin

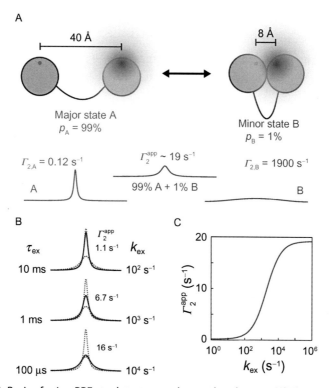

Figure 2 Basis of using PREs to detect sparsely populated states. (A) The unpaired electron of the paramagnetic label and the proton of interest are depicted as green (gray in the print version) and red (dark gray in the print version) dots on two separated domains of a protein. In the major species A (populated at 99%), the unpaired electron–proton distance is 40 Å but reduced to 8 Å in the minor species B (populated at 1%). For a nitroxide spin label, these distances correspond to Γ_2 PRE values of 0.12 and 1900 s^{-1}, respectively. If states A and B are in slow exchange on the PRE timescale, the observed transverse relaxation rates $R_{2,i}^{obs}$ for a given species will be the sum of its intrinsic transverse relaxation rate $R_{2,i}^{intrinsic}$ and $\Gamma_{2,i}$, resulting in a narrow linewidth for the resonance of species A (blue; light gray in the print version) but a line broadened beyond detection for species B (green; gray in the print version), which is already undetectable owing to its low population). In the fast-exchange limit, a single resonance will be observed at the position of species A, line broadened by the apparent Γ_2 (Γ_2^{app}) given by the population-weighted average of the Γ_2 values of species A and B (red; dark gray in the print version) line). (B) Impact of exchange (expressed as a lifetime τ_{ex} on the left and an exchange rate k_{ex} on the right) on the observed linewidth (black) of the resonance of species A; the resonance of species A in the absence of exchange and in the fast-exchange limit is depicted by the blue (light gray in the print version) and red (dark gray in the print version) dashed lines, respectively. (C) Dependence of Γ_2^{app} on k_{ex}. Adapted from Anthis and Clore (2015).

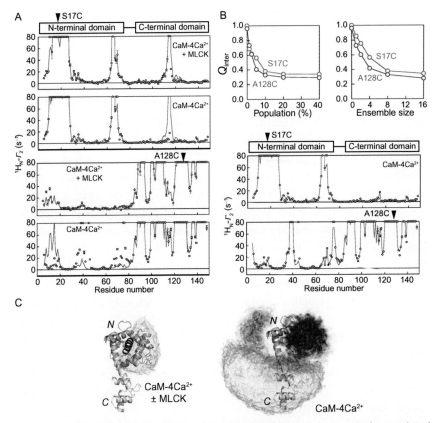

Figure 3 Exploring the conformational space sampled by transient sparsely populated closed states of calmodulin (CaM)-4Ca^{2+} from interdomain PRE data. (A) Observed (circles) and calculated PRE profiles (lines) observed for CaM-4Ca^{2+} in the presence (panels 1 and 3) and absence (panels 2 and 4) of the myosin light chain kinase (MLCK) target peptide. The paramagnetic nitroxide label is attached to S17C in panels 1 and 2 and to A128C in panels 3 and 4. Excellent agreement is obtained for the intradomain PREs in the presence and absence of the MLCK peptide. However, whereas the experimental interdomain PRE profiles obtained in the presence of the MLCK peptide agree well with those back-calculated from the crystal structure of the CaM-4Ca^{2+}-MLCK complex (red line), those in the absence of peptide do not agree with the back-calculated PREs derived from the crystal structures of either CaM-4Ca^{2+} (blue line) or CaM-4Ca^{2+}-MLCK (red line). (B) PRE-driven simulated annealing refinement of CaM-4Ca^{2+}. The top panels show the dependence of the PRE Q-factor on the population of closed states and on the ensemble size used to represent the sparsely populated species. The bottom two panels show the agreement between observed (circles) and calculated (red line) PRE data. (C) Visualization of the conformational space sampled by sparsely populated closed states of CaM-4Ca^{2+}. For reference, the dumbbell crystal structure of CaM-4Ca^{2+} (Babu et al., 1985) is shown as a ribbon diagram with the N- and C-terminal

(*Continued*)

light chain kinase (MLCK) are fully accounted for by the structure of CaM-$4Ca^{2+}$-MLCK complex (Ikura et al., 1992; Meador et al., 1992) (panels 1 and 3 of Fig. 3A), those obtained in the absence of target peptide are not consistent with either the crystal structure of CaM-$4Ca^{2+}$ (which predicts that no intermolecular PREs should be observed) (Babu et al., 1985) or the structure of the CaM-$4Ca^{2+}$-MLCK complex (panels 2 and 4 of Fig. 3A). The observed interdomain PREs for CaM-$4Ca^{2+}$ are indicative of the existence of transient sparsely populated closed conformations in which the N- and C-terminal domains of CaM come into close proximity to one another. Using PRE-driven simulated annealing, it was shown that the population of such closed states is about 10% (Fig. 3B), and the majority (about half) of these states sample a region of conformational space that closely overlaps that found in various crystal structures CaM-$4Ca^{2+}$-peptide complexes (Fig. 3C). Thus, even in the absence of target peptide there exists a small population of closed or partially closed states that are similar but not identical to the closed states observed in complexes with target peptides. The population of the partially closed states correlates with linker length between the two domains as well as with the affinity of CaM-$4Ca^{2+}$ for target peptide (Anthis & Clore, 2013). These results highlight the complementarity and interplay of conformational selection and induced fit in ligand binding, and suggest that the existence of a small population of partially closed states in the absence of ligand may facilitate the transition to the full closed ligand-bound state (Clore, 2014).

Figure 3—Cont'd domains shown in dark and light green, respectively; a ribbon of the C-terminal domain in the CaM-$4Ca^{2+}$-MLCK structure (Meador, Means, & Quiocho, 1992) is depicted in blue, and on the left panel the gray atomic probability map depicts the location of the C-terminal domain (relative to a fixed N-terminal domain) in a variety of CaM-peptide complex (Anthis et al., 2011). On the right panel, the conformational space sampled by the C-terminal domain (relative to a fixed N-terminal domain) in the minor (~10%) closed species of CaM-$4Ca^{2+}$ is depicted by an atomic probability map plotted at contour levels ranging from 0.1 (blue) to 0.5 (red); the gray probability map depicts the region of conformational space consistent with interdomain Γ_2 values less than $2\ s^{-1}$ and is representative of the major species ensemble. The predominant region of conformational space (red probability map) sampled by the transient closed states of CaM-$4Ca^{2+}$ overlaps with that sampled in the CaM-peptide complexes. *Adapted from Anthis et al. (2011).* (See the color plate.)

ACKNOWLEDGMENTS

This work was supported by the Intramural Program of the National Institute of Diabetes and Digestive and Kidney Diseases at the National Institutes of Health, and by the AIDS Targeted Antiviral Program of the Office of the Director of the National Institutes of Health.

REFERENCES

Altenbach, C., Marti, T., Khorana, H. G., & Hubbell, W. L. (1990). Transmembrane protein structure: Spin labeling of bacteriorhodopsin mutants. *Science, 248*(4959), 1088–1092.

Anthis, N. J., & Clore, G. M. (2013). The length of the calmodulin linker determines the extent of transient interdomain association and target affinity. *Journal of the American Chemical Society, 135*(26), 9648–9651.

Anthis, N. J., & Clore, G. M. (2015). Visualizing transient dark states by NMR spectroscopy. *Quarterly Reviews of Biophysics, 48*(1), 35–116.

Anthis, N. J., Doucleff, M., & Clore, G. M. (2011). Transient, sparsely populated compact states of apo and calcium-loaded calmodulin probed by paramagnetic relaxation enhancement: Interplay of conformational selection and induced fit. *Journal of the American Chemical Society, 133*(46), 18966–18974.

Babu, Y. S., Sack, J. S., Greenhough, T. J., Bugg, C. E., Means, A. R., & Cook, W. J. (1985). Three-dimensional structure of calmodulin. *Nature, 315*(6014), 37–40.

Battiste, J. L., & Wagner, G. (2000). Utilization of site-directed spin labeling and high-resolution heteronuclear nuclear magnetic resonance for global fold determination of large proteins with limited nuclear overhauser effect data. *Biochemistry, 39*(18), 5355–5365.

Bax, A., & Grishaev, A. (2005). Weak alignment NMR: A hawk-eyed view of biomolecular structure. *Current Opinion in Structural Biology, 15*(5), 563–570.

Bloembergen, N., & Morgan, L. O. (1961). Proton relaxation times in paramagnetic solutions effects of electron spin relaxation. *Journal of Chemical Physics, 34*(3), 842–850.

Clore, G. M. (2014). Interplay between conformational selection and induced fit in multidomain protein-ligand binding probed by paramagnetic relaxation enhancement. *Biophysical Chemistry, 186*, 3–12.

Clore, G. M., & Gronenborn, A. M. (1989). Determination of three-dimensional structures of proteins and nucleic acids in solution by nuclear magnetic resonance spectroscopy. *Critical Reviews in Biochemistry and Molecular Biology, 24*(5), 479–564.

Clore, G. M., & Gronenborn, A. M. (1998a). New methods of structure refinement for macromolecular structure determination by NMR. *Proceedings of the National Academy of Sciences of the United States of America, 95*(11), 5891–5898.

Clore, G. M., & Gronenborn, A. M. (1998b). Determining the structures of large proteins and protein complexes by NMR. *Trends in Biotechnology, 16*(1), 22–34.

Clore, G. M., & Iwahara, J. (2009). Theory, practice, and applications of paramagnetic relaxation enhancement for the characterization of transient low-population states of biological macromolecules and their complexes. *Chemical Reviews, 109*(9), 4108–4139.

Clore, G. M., Tang, C., & Iwahara, J. (2007). Elucidating transient macromolecular interactions using paramagnetic relaxation enhancement. *Current Opinion in Structural Biology, 17*(5), 603–616.

Clore, G. M., & Venditti, V. (2013). Structure, dynamics and biophysics of the cytoplasmic protein-protein complexes of the bacterial phosphoenolpyruvate: Sugar phosphotransferase system. *Trends in Biochemical Sciences, 38*(10), 515–530.

Fawzi, N. L., Doucleff, M., Suh, J. Y., & Clore, G. M. (2010). Mechanistic details of a protein-protein association pathway revealed by paramagnetic relaxation enhancement titration measurements. *Proceedings of the National Academy of Sciences of the United States of America, 107*(4), 1379–1384.

Fawzi, N. L., Fleissner, M. R., Anthis, N. J., Kálai, T., Hideg, K., Hubbell, W. L., et al. (2011). A rigid disulfide-linked nitroxide side chain simplifies the quantitative analysis of PRE data. *Journal of Biomolecular NMR, 51*(1–2), 105–114.

Gaponenko, V., Howarth, J. W., Columbus, L., Gasmi-Seabrook, G., Yuan, J., Hubbell, W. L., et al. (2000). Protein global fold determination using site-directed spin and isotope labeling. *Protein Science, 9*(2), 302–309.

Ikura, M., Clore, G. M., Gronenborn, A. M., Zhu, G., Klee, C. B., & Bax, A. (1992). Solution structure of a calmodulin-target peptide complex by multidimensional NMR. *Science, 256*(5057), 632–638.

Iwahara, J., Anderson, D. E., Murphy, E. C., & Clore, G. M. (2003). EDTA-derivatized deoxythymidine as a tool for rapid determination of protein binding polarity to DNA by intermolecular paramagnetic relaxation enhancement. *Journal of the American Chemical Society, 125*(22), 6634–6635.

Iwahara, J., & Clore, G. M. (2006). Detecting transient intermediates in macromolecular binding by paramagnetic NMR. *Nature, 440*(7088), 1227–1230.

Iwahara, J., & Clore, G. M. (2010). Structure-independent analysis of the breadth of the positional distribution of disordered groups in macromolecules from order parameters for long, variable-length vectors using NMR paramagnetic relaxation enhancement. *Journal of the American Chemical Society, 132*(38), 13346–13356.

Iwahara, J., Schwieters, C. D., & Clore, G. M. (2004). Ensemble approach for NMR structure refinement against (1)H paramagnetic relaxation enhancement data arising from a flexible paramagnetic group attached to a macromolecule. *Journal of the American Chemical Society, 126*(18), 5879–5896.

Iwahara, J., Tang, C., & Clore, G. M. (2007). Practical aspects of ^{1}H transverse paramagnetic relaxation enhancement measurements on macromolecules. *Journal of Magnetic Resonance, 184*(2), 185–195.

Jensen, M. R., Lauritzen, C., Dahl, S. W., Pedersen, J., & Led, J. J. (2004). Binding ability of a HHP-tagged protein towards Ni^{2+} studied by paramagnetic NMR relaxation: The possibility of obtaining long-range structure information. *Journal of Biomolecular NMR, 29*(3), 175–185.

Keizers, P. H. J., Desreux, J. F., Overhand, M., & Ubbink, M. (2007). Increased paramagnetic effect of a lanthanide protein probe by two-point attachment. *Journal of the American Chemical Society, 129*(30), 9292–9293.

Keizers, P. H. J., Saragliadis, A., Hiruma, Y., Overhand, M., & Ubbink, M. (2008). Design, synthesis, and evaluation of a lanthanide chelating protein probe: CLaNP-5 yields predictable paramagnetic effects independent of environment. *Journal of the American Chemical Society, 130*(44), 14802–14812.

Kosen, P. A. (1989). Spin labeling of proteins. *Methods in Enzymology, 177*, 86–121.

Kosen, P. A., Scheek, R. M., Naderi, H., Basus, V. J., Manogaran, S., Schmidt, P. G., et al. (1986). Two-dimensional ^{1}H NMR of three spin-labeled derivatives of bovine pancreatic trypsin inhibitor. *Biochemistry, 25*(9), 2356–2364.

Mal, T. K., Ikura, M., & Kay, L. E. (2002). The ATCUN domain as a probe of intermolecular interactions: Application to calmodulin-peptide complexes. *Journal of the American Chemical Society, 124*(47), 14002–14003.

Meador, W. E., Means, A. R., & Quiocho, F. A. (1992). Target enzyme recognition by calmodulin: 2.4 Å structure of a calmodulin-peptide complex. *Science, 257*(5074), 1251–1255.

Schmidt, P. G., & Kuntz, I. D. (1984). Distance measurements in spin-labeled lysozyme. *Biochemistry, 23*(18), 4261–4266.

Schwieters, C. D., Kuszewski, J. J., & Clore, G. M. (2006). Using Xplor-NIH for NMR molecular structure determination. *Progress in Nuclear Magnetic Resonance Spectroscopy, 48*(1), 47–62.

Solomon, I. (1955). Relaxation processes in a system of 2 spins. *Physical Review, 99*(2), 559–565.

Tang, C., Iwahara, J., & Clore, G. M. (2006). Visualization of transient encounter complexes in protein-protein association. *Nature, 444*(7117), 383–386.

Tang, C., Louis, J. M., Aniana, A., Suh, J. Y., & Clore, G. M. (2008). Visualizing transient events in amino-terminal autoprocessing of HIV-1 protease. *Nature, 455*(7213), 693–696.

Tang, C., Schwieters, C. D., & Clore, G. M. (2007). Open-to-closed transition in apo maltose-binding protein observed by paramagnetic NMR. *Nature, 449*(7165), 1078–1082.

Volkov, A. N., Worrall, J. A., Holtzmann, E., & Ubbink, M. (2006). Solution structure and dynamics of the complex between cytochrome c and cytochrome c peroxidase determined by paramagnetic NMR. *Proceedings of the National Academy of Sciences of the United States of America, 103*(50), 18945–18950.

SECTION V

In Vivo EPR Oxymetry and Imaging

CHAPTER EIGHTEEN

In Vivo pO$_2$ Imaging of Tumors: Oxymetry with Very Low-Frequency Electron Paramagnetic Resonance

Boris Epel, Howard J. Halpern[1]

Center for Electron Paramagnetic Resonance Imaging *In Vivo* Physiology, Department of Radiation and Cellular Oncology, University of Chicago, Chicago, Illinois, USA
[1]Corresponding author: e-mail address: h-halpern@uchicago.edu

Contents

1. Introduction: The Importance of Imaging Molecular Oxygen in Cancer Therapy — 502
   1.1 Imaging Physiology in Living Animals — 502
   1.2 The Nature of Molecular Oxygen — 503
   1.3 Measurement of Molecular Oxygen — 503
   1.4 Principles of Electron Paramagnetic Resonance — 504
   1.5 Oxymetry with EPR — 507
   1.6 Spin Probe Sensitivity to Molecular Oxygen (O$_2$ or pO$_2$) and Other Environmental Factors — 509
   1.7 Radiofrequency Magnetization Excitation Is Necessary for *In Vivo* EPR Imaging of Large Animals and Humans — 510
   1.8 Dissolved Spin Probe Oxymetry — 510
   1.9 Soluble Spin Probes — 512
   1.10 Trityls — 512
   1.11 Particulates — 513
   1.12 Techniques for EPR Imaging pO$_2$ — 514
   1.13 Oxygen Imaging as Longitudinal Relaxation (R_1) Parametric Imaging — 515
   1.14 EPR Imaging in Cancer Biology — 516
   1.15 Validation of the Oxymetry in Animals — 516
   1.16 Transient Hypoxia can be Imaged with EPR pO$_2$ Images — 519
   1.17 Effectiveness of Localizing Radiation to Regions of the Tumor with High Hypoxic Fraction Using EPR pO$_2$ Images Needs to be Tested — 520
2. Summary — 521
Acknowledgments — 522
References — 522

Abstract

For over a century, it has been known that tumor hypoxia, regions of a tumor with low levels of oxygenation, are important contributors to tumor resistance to radiation therapy and failure of radiation treatment of cancer. Recently, using novel pulse electron paramagnetic resonance (EPR) oxygen imaging, near absolute images of the partial pressure of oxygen (pO$_2$) in tumors of living animals have been obtained. We discuss here the means by which EPR signals can be obtained in living tissues and tumors. We review development of EPR methods to image the pO$_2$ in tumors and the potential for the pO$_2$ image acquisition in human subjects.

1. INTRODUCTION: THE IMPORTANCE OF IMAGING MOLECULAR OXYGEN IN CANCER THERAPY

A near universal characteristic of solid human tumors is hypoxia. Hypoxia is defined as clinically significant regions of a malignant tumor with low levels of molecular oxygen, or low values of pO$_2$. Resistance of human solid tumors to radiation induced by such regions of hypoxia has been recognized for over a century (Thomlinson & Gray, 1955). Similar resistance has more recently been noted for chemotherapy (Kennedy, Teicher, Rockwell, & Sartorelli, 1980; Teicher, 1994). Human trials investigating hyperbaric oxygen and hypoxic sensitizers to overcome tumor hypoxia have shown promise but limited success (Henk, Kunkler, & Smith, 1977; Henk & Smith, 1977). However, the extensive variability of tumor oxygenation, both in overall extent and in the location of hypoxic regions, may confound such attempts. The results demonstrate the importance of assessment not only of overall tumor oxygenation but of locating, through imaging, significant regions of hypoxia in the tumor. We will briefly review this data, and then discuss various alternative methods of measuring tissue and imaging tumor oxygenation.

1.1 Imaging Physiology in Living Animals

Short of transillumination, the first images of deep structures in living animals began with Wilhelm Roentgen's discovery of the X-ray in 1895 (Rontgen, 1896). Since then, imaging modalities have evolved sensitivity to varied soft-tissue states of animals and humans (Cormack, 1963; Lauterbur, 1973). In parallel, spectroscopies have evolved, revealing the physical aspects of solid and liquid states of matter (Condon & Shortley, 1935). These states are the environment in which biologic processes evolve.

However, living samples are heterogeneous. The full definition of physiologic states, thus, requires spectroscopic imaging. Various forms of spectroscopic imaging have evolved in the past three decades. This makes available from the spectroscopic process information that is localized within a subvolume of a living sample (Lauterbur, Levin, & Marr, 1984; Maltempo, 1986).

1.2 The Nature of Molecular Oxygen

The oxygen molecule, O_2, is a diradical with two unpaired electrons in the triplet state that define its interaction with other molecules bearing unpaired electron and nuclear spins and with electromagnetic radiation. It is a rapidly tumbling diatomic molecule with its two unpaired spins rapidly relaxing each other. Thus, the oxygen molecule in solution at room temperature has a nearly unmeasurably fast electron relaxation rate. When interacting with another molecule bearing a single highly stable unpaired electron spin—a spin probe—oxygen increases the relaxation rates of the probe, mainly via Heisenberg exchange (Eastman, Kooser, Pas, & Freed, 1969; Molin, Salikhov, & Zamaraev, 1980). The solution interaction rates between the spin probe and O_2 are described by the Smoluchowski diffusion equation, which predicts a linear relation between pO_2 and spin probe relaxation rate, validated for multiple free radicals (Swartz & Glockner, 1991).

1.3 Measurement of Molecular Oxygen

A number of techniques for measurement of molecular oxygen have been developed. The chemical and physical properties of oxygen enable a variety of methods, each with their own applicability and advantages. The "gold standard" of oxymetry in cells and live animal tissues is the platinum electrode (Whalen, Riley, & Nair, 1967). A recent enhancement of the standard platinum electrode is the Eppendorf electrode. This is inserted into tissues of living animals and human subjects with a highly regular advance and retreat pattern for measurement consistency. It has a 200–300 μm diameter tip that is inserted into the tissue, which measures oxygen along a series of tracks (Harrison, 2003). The OxyLite™ probe (Oxford Optronix, Oxford, UK) utilizes fluorescence quenching of fluorophore by oxygen (Griffiths & Robinson, 1999). OxyLite™ can be used for repetitive measurements in the same location or along tracks, although the 200-μm glass fiber will wander from a straight-line sampling. The electrode and Oxylite provide highly local samples and do not provide an overall inventory of the pO_2 distributions in tissues.

There are a number of noninvasive qualitative oxymetries available. These include near-infrared spectroscopy, typically of blood saturation, fluoromisonidozole retention PET (Krause, Beck, Souvatzoglou, & Piert, 2006), and blood oxygenation level-dependent magnetic resonance imaging (MRI) (Ogawa & Lee, 1990; Ogawa, Lee, Kay, & Tank, 1990). These methods are subject to confounding variation that frustrates quantitative measurement, and, particularly if the animal is stressed or otherwise changes its state, prevents reliable repeated measurement. Phosphorescence quenching (Rumsey, Vanderkooi, & Wilson, 1988), ^{19}F (Busse, Pratt, & Thomas, 1988; Hunjan et al., 2001), proton–electron double resonance imaging (PEDRI) (Efimova et al., 2011; Ogawa & Lee, 1990) and Electron Paramagnetic Resonance (EPR) are more quantitative pO_2 imaging modalities. Despite their quantitative imaging capabilities, phosphorescence quenching, ^{19}F MRI (Ogawa et al., 1990), and PEDRI are still subject to confounding variation, most significantly from the effect of the concentration of the sensing molecule (the phosphorescent compound, the ^{19}F for MRI, or the self-relaxation of the electron-bearing spin probe and the relaxation time of the water proton in PEDRI) on the relaxation rate dependence on O_2. Solutions to this problem have included the use of EPR with particulates or application of longitudinal or spin–lattice relaxation (SLR) rate EPR imaging of soluble spin probes that provide near absolute measurements of local pO_2, with order-of-magnitude reduction in confounding variation.

1.4 Principles of Electron Paramagnetic Resonance

The EPR technique detects molecular species with one or more unpaired electrons: paramagnetic complexes, radicals, lattice defects, etc. Principles of magnetic resonance are covered in a number of books (Abragam & Bleaney, 1970; Atherton, 1993; Schweiger & Jeschke, 2001; Slichter, 1996). Here, we discuss essential relevant aspects of the technique for understanding EPR imaging and oxymetry.

An unpaired electron possesses an intrinsic magnetic moment not associated with its orbital angular momentum. The spin moment is $m = -\frac{1}{2\hbar}g\mu_B$ where g is the electron g-factor equal to 2.0023 for a free electron, μ_B is the Bohr electron magneton, and $\hbar$ is Planck's constant divided by 2π.

A crucial difference between the technique of EPR and that of nuclear magnetic resonance and their respective imaging techniques is the difference in the magnitude of the magnetic moments. This three order-of-magnitude difference completely transforms the respective methods. Relativistic quantum mechanics predicts the spin magnetic moment μ_B to be $\mu_B = \hbar e/2\pi mc$ (Dirac, 1958), where e is the magnitude of the charge of the electron or proton, c is the speed of light, and m is the mass of either the electron or, in the case of a hydrogen nucleus, the mass of the proton. The magnetic moment of the electron is 1836 times larger than that of the proton. Due to the anomalous magnetic moment of the proton, which is 2.79 times larger than the pure relativistic prediction because of the finite size or nonpoint charge distribution of the proton, the actual magnetic moment of the electron is 658 times or still nearly three orders of magnitude larger than that of the proton.

In the presence of a static magnetic field (B_0), the energy (E) of a magnetic moment depends on its orientation relative to B_0. For a spin 1/2 particle ($S=1/2$), quantum mechanics counterintuitively requires orientations to be parallel (−) or antiparallel (+) to the imposed magnetic field B_0.

$$E = \pm \frac{1}{2} g \mu_B B_0 \tag{1}$$

The difference between these energy levels (ΔE) can be described in terms of angular frequency (ω_0) via Planck's energy–frequency relationship.

$$\Delta E = \hbar \omega_0 \tag{2}$$

Combining these two equations gives the relation between frequency and applied field as:

$$\omega_0 = \gamma_e B_0 \tag{3}$$

where $\gamma_e = g\mu_B/\hbar$ is the electron gyromagnetic ratio. This is equal to 1.76×10^{11} s^{-1} T^{-1} for the free electron. The three order-of-magnitude smaller proton gyromagnetic ratio γ_p is 2.68×10^8 s^{-1} T^{-1}. ω_0 is known as the Larmor frequency, the Zeeman frequency, or the basic resonance frequency.

The major instrumental consequences of these results are

(1) The magnetic field at which an electron magnetic absorbs significantly more energy from the oscillating electromagnetic field—resonance—is approximately 658 times smaller ($=\gamma_p/\gamma_e$) than that for a proton at a given excitation frequency, ω_0. The "approximately" is due to the inevitable modification of the local environment of the resonating electron by the magnetic fields of nearby nuclei or, rarely, unpaired

electrons. *This means that at high-field MRI frequencies, EPR imagers use much lower magnetic fields.*

(2) The rates at which excited electron spins relax are proportional to γ^2 (Abragam, 1961). Electrons relax $\sim (658(=\gamma_e/\gamma_p))^2$ or nearly six orders of magnitude faster than water protons. Electrons relax in times of nanoseconds to microseconds. Water protons and common nuclei relax in tens of milliseconds to many tens of seconds. Excited ^{13}C-enriched pyruvate, for example in human subjects, relaxes in ~ 1 min.

The numbers of unpaired spins in common samples measured with magnetic resonance, either electron spins or nuclear spins, are very large, from a billion billion to a trillion trillion spins, which are thermodynamic numbers. These are subject to thermodynamic energy distributions, the Boltzmann distribution, with smaller numbers of spins in the higher energy orientation than in the lower energy orientation. The magnitude of the equilibrium magnetization is proportional to the imposed magnetic field, B_0. The overall magnetic moment summed over all the magnetic moments of the spins is referred to as the magnetization of the spin system. The magnetization in a sample can be manipulated by addition of a second, oscillating magnetic field (B_1). To be effective for $S = 1/2$ states, B_1 is oriented orthogonal to B_0 and oscillates at frequencies close to the Larmor frequency. The time-varying magnetic field B_1 can add energy to the magnetization system. It accomplishes this by inverting the orientations of individual electronic spins and changing the direction of the overall sum of the unpaired spins, the magnetization. The phenomenological Bloch equation describes the time evolution of magnetization orientation in the presence of both B_0 and B_1 magnetic fields (Schweiger & Jeschke, 2001). In a coordinate system rotating at the spectrometer operating frequency where B_1 is not changing, the return of transverse (M_T) and longitudinal (M_Z) components of magnetization to equilibrium in the absence of B_1 is described by Schweiger and Jeschke (2001):

$$M_T(t) \propto \exp(i\Omega t) \exp\left(-\frac{t}{T_2}\right) \quad (4)$$

$$M_Z(t) \propto \left(1 - M \cdot \exp\left(-\frac{t}{T_1}\right)\right) \quad (5)$$

Here, Ω is the difference between the Larmor frequency of the electron and the EPR spectrometer's operating frequency. The longitudinal relaxation time, T_1, and the transverse relaxation time, T_2, correspond to the relaxation of M_T and M_Z, respectively. T_1 and T_2 are the inverses of the respective longitudinal relaxation rate, R_1, and transverse relaxation rate, R_2. The conventional EPR time-domain signal, $s(t)$, is proportional to M_T. M describes an initial state of longitudinal magnetization. The longitudinal magnetization can be encoded into the EPR signal by using special pulse sequences. The transverse magnetization relaxation rate, R_2, is commonly referred to as the phase memory relaxation rate. This is the rate at which spins that have been aligned by a very short duration pulse of oscillating magnetic field, which prepares them in a particular magnetization state, lose their coherent alignment in the time after the pulse.

The first EPR experiments were continuous-wave (CW) EPR experiments. Here, the B_1 excitation is applied continuously during an experiment or image. In a CW spectrometer, the EPR signal is produced by applying B_1 at a fixed frequency ω_0 and sweeping B_0 through the resonance condition of Eq. (3). (Atherton, 1993). In principle, the field B_0 could be fixed and the frequency swept through the resonance condition of Eq. (3), but the former approach simplifies the electronics and allows a simpler amplification of signal. An alternate strategy applied often to simpler, more robust systems, involves subjecting the unpaired electron spin-bearing sample, prepared in a magnetic field, to a high-power pulse of radiation, of duration shorter than T_2. This pulse aligns the magnetization in the magnetic field. Detecting the rate of decay of the magnetization signal is the basis of this latter experiment. This is referred to as a time-domain experiment. It is important to note that the CW magnetic field (B) -domain spectrum is related to the time-domain signal through a Fourier transformation.

$$S(B) = \int_{-\infty}^{\infty} s(t) \exp(-i\gamma_e Bt) dt \qquad (6)$$

Note that the factor γ_e converts magnetic field units into inverse time, or rate.

1.5 Oxymetry with EPR

Injectable, water-soluble spin label or spin probe EPR oxymetry is a minimally invasive method that can report absolute pO_2 deep in tissues (Ahmad, Khan, Vikram, Bratasz, & Kuppusamy, 2010; Khan, Williams, Hou, Li, & Swartz, 2007; Tatum et al., 2006). In the 1980s, EPR detection of oxygen using the broadening of the width of the EPR spectrum from a nitroxide spin probe was first reported by Backer, Budker, Eremenko, and Molin

(1977) and Popp and Hyde (1981) and later extensively investigated by Swartz and coworkers (Khan et al., 2007; Swartz & Clarkson, 1998; Swartz & Glockner, 1991), by using various classes of spin probes. In the nineties, a few groups pioneered multidimensional CW imaging on rodents *in vivo* and *ex vivo*, enabling repeated measurements of oxygen concentrations in living tissues (Halpern et al., 1994; Kuppusamy et al., 1994; Zweier, Thompsongorman, & Kuppusamy, 1991).

The underlying mechanism of interaction between spin probe and oxygen is predominantly Heisenberg spin exchange (Dirac, 1958). In this mechanism, the electrons from the rapidly relaxing oxygen environment, during the encounter with a spin probe, are not distinguishable from the electrons of the spin probe. The spin probe electrons thereby share their environment with that of oxygen during which time they relax more rapidly. The increased rate loss of both phase and energy of the spin probe electron is proportional to the rate of encounter with oxygen. This is directly proportional to the oxygen concentration and the oxygen partial pressure.

EPR techniques like CW or time-domain free induction decay pulse imaging measure total EPR decay rates or linewidths (LWs) instead of the transverse or longitudinal relaxation rates. The total EPR decay rate/LW of free radicals is due to two major components: (1) homogeneous broadening due to transverse relaxation and (2) inhomogeneous broadening due to the interaction of the electron with neighboring paramagnetic nuclei (Abragam & Bleaney, 1970; Slichter, 1996). In the case of inhomogeneous broadening, the electron experiences the magnetic fields of multiple neighboring nuclei, which variably shifts the resonance frequency, creating hyperfine (HF) spectral structure. A more precise method for obtaining spin packet LW, that is used in CW oxymetry, explicitly fits the EPR line to spectral models that include HF structure, and accounts for the effect of the magnetic field modulation (Mailer, Robinson, Williams, & Halpern, 2003; Robinson, Mailer, & Reese, 1999a, 1999b).

In addition to partial pressure of molecular oxygen, details of EPR lineshape provide an insight into the thiol–disulfide balance or redox state, the related local bioreduction capability, the acidity (pH), the temperature, the presence of specific oxygen-centered free radicals, and many others (Swartz et al., 2004).

Characterizing the spin relaxation by relaxation rates is the inverse of characterizing it by relaxation times. Relaxation rates are related to relaxation times, T: $R=1/T$. Relaxation rates can be multiplied by $(\gamma_e)^{-1}$, the inverse of the gyromagnetic ratio, to allow expression of the rates directly in magnetic field units (μT). Thus, line broadening is directly related to relaxation rate increase. Relaxation rates naturally demonstrate the relationship between increasing numbers of relaxation mechanisms and increasing relaxation rates, or linewidths. This process also provides a conceptual link between parameters determined by CW and pulse methods. For Lorentzian lineshapes, the half width at half maximum is equal to $1/(\gamma_e T_2) = (1/\gamma_e)R_2$. For EPR lines with multiple broadening mechanisms, $(1/\gamma_e)R_2$ describes the homogeneous broadening of the EPR line (spin packet LW). In early CW oxymetry, the spin packet LW allowed the determination of pO_2.

Transverse relaxation-based oxymetry is more susceptible to variation of other physical parameters beside oxygen partial pressure. Parameters such as temperature, viscosity, and salinity are tightly controlled in the body of a living animal (Wilson et al., 1991). This allows a quantitative enumeration of their effects on relaxation, allowing correction and calibration of oxygen measurements. However, spin probe concentration may vary and affect the accuracy of LW/transverse relaxation rate-based oxymetry.

1.6 Spin Probe Sensitivity to Molecular Oxygen (O_2 or pO_2) and Other Environmental Factors

Heisenberg spin exchange between a spin probe and oxygen acts on the spin probe's longitudinal relaxation in a manner nearly identical to action on transverse relaxation (Eaton & Eaton, 2000; Molin et al., 1980). Importantly, other relaxation processes affect transverse and longitudinal relaxation differently. Electron spin exchange between two trityl spin probe molecules increases R_2 (Molin et al., 1980). The effect of the spin exchange between spin probes on R_2 is dependent on the number of collisions per unit time, and, therefore, on the concentration of spin probe. R_2 cannot distinguish between the dephasing effect of an interaction of a spin probe with oxygen and a dephasing effect of an interaction of a spin probe with another spin probe. The exchange of energies between two spin probes, however, does not alter the total energy of the interacting spin pair, and therefore does not affect the spin probe longitudinal magnetization component. Thus, R_1 is much less susceptible to self-broadening. R_1 is nearly an absolute measure of oxygen concentration, or pO_2. Figure 2 shows the effect of O_2 concentration, or pO_2, on the relaxation rates, R_1 and R_2, to have identical slopes, but there is nearly an order-of-magnitude reduction in the dependence of R_1 on spin probe concentration relative to R_2. R_1 is a near absolute measure of pO_2 with precision exceeding 1 torr. This increase in probe specificity is a distinguishing feature of EPR oxymetry, in comparison with other noninvasive methods (Tatum et al., 2006).

The ability of an EPR image to quantify molecular oxygen dissolved in the life supporting solvents of a living animal and eventually a human is among the most important of its abilities. The absence of oxygen in the heart, brain, and limbs affected by peripheral vascular disease or diabetes, ischemic bowel, and portions of cancers of human patients rendered resistant to therapy by the absence of oxygen, is responsible for the death of greater than half of our species. The evaluation of treatments and pharmaceutical agents to ameliorate the hypoxic state promises to prolong the useful and involved lives of all of us (Longo et al., 2011).

1.7 Radiofrequency Magnetization Excitation Is Necessary for *In Vivo* EPR Imaging of Large Animals and Humans

EPR imaging needs to be done under unusual conditions for eventual human application. Conventional EPR spectroscopy is usually carried out at frequencies of gigahertz to hundreds of gigahertz. Lower frequencies are required to penetrate deeply into the body of a human composed two thirds by weigh of salt water. The nonresonant loss of signal amplitude caused by conductive loss at low frequency, or dipolar loss at higher frequency, requires magnetization excitation frequencies of the order of hundreds of MHz for large human-size animals. This is the excitation frequency of a high-field, whole-body MRI. For small animals, frequencies of up to approximately 1 GHz can be used (Bottomley & Andrew, 1978; Halpern et al., 1989). As we mentioned above, unlike MRI, which requires multi-Tesla superconducting magnets, at 250 MHz, the B_0 for EPR images is 9 milliTesla. These stationary magnetic fields can be generated by simple copper air-core magnets. Small animal experiments to rapidly and efficiently evaluate treatment and pharmaceutical effectiveness, may use higher frequencies near 1 GHz (L-band), that will give higher signals and will not suffer paralyzing loss that would make large animal experiments difficult. Because relaxation rates of even the most slowly relaxing aqueous spin probes are five to six orders of magnitude faster than those of a water hydrogen nucleus, fixed stepped gradients, tomographic image reconstruction, and modest power requirements make the technique relatively inexpensive.

1.8 Dissolved Spin Probe Oxymetry

In general, quantification of aspects of the tissue microenvironment requires an environmental reporter and a readout technique. The reporter can be endogenous, such as water protons or sodium ions in MRI, or exogenous, such as implanted particulates or injected soluble spin probes. Endogenous reporters typically have much higher concentrations than exogenous ones and therefore, are easier to detect, but their localization cannot be controlled. Consequently, they are nonspecific and sometimes provide an overwhelming background signal that masks a signal of interest. On the contrary, exogenous spin probes can be specifically targeted to the areas of interest.

Endogenous paramagnetic species found in mammalian bodies include hemoglobin, metalloenzymes, but extremely low concentrations of unbound diffusible species except molecular oxygen. Metal centers and oxygen at animal body temperature have very short relaxation times, broad lines and thus are not easily and directly measurable at low EPR frequencies. Moreover, unbound diffusible paramagnetic metal species can interact with and transform covalent carbon–carbon and carbon–hydrogen molecular bonds as they do as enzymatic reaction centers. Therefore, living systems have developed arrays of binding proteins to maintain endogenous diffusible concentrations below 1 nmol because of this threat to covalent bonds of the molecules of living systems. At present, exogenous spin probes are really the only practical reporters, and appropriate spin probes are the key to successful imaging. The LW and relaxation times of the probe, and their sensitivity to oxygen, largely define sensitivity of methodology and pO_2 accuracy. Probes with narrower LWs and longer relaxation times allow higher resolution imaging. High fractional sensitivity of LW to oxygen ensures better imaging accuracy. Finally, the spin probes should be minimally toxic and metabolically stable.

Summarizing the above, EPR oxymetric imaging requires a spin probe to sample the fluid environment and report the oxygen partial pressure through increase in its spin packet LW or, equivalently, its relaxation rates. Measurement of line broadening of nitroxide spin probes has, until recently, been the principle means of dissolved spin probe oxymetry. Hypoxic nitroxide relaxation rates, $1-2$ μs^{-1}, are still too rapid for imaging at several hundred MHz excitation frequencies. Triarylmethyl (TAM) radicals, specifically, partially deuterated methyl-tris[8-carboxy-2,2,6,6-tetrakis[2-hydroxyethyl]benzo[1,2-d:4,5-d′]bis[1,3]dithiol-4-yl]-trisodium salt, OX063, have an order-of-magnitude smaller hypoxic R_1 value of $1/6$ μs^{-1} ($T_1 = 6$ μs). These relaxation rates enable pulse measurement and imaging at 250 MHz (Mailer, Sundramoorthy, Pelizzari, & Halpern, 2006). R_1, measured with inversion recovery pulse sequences, increases signal-to-noise by nearly a factor of two relative to electron spin-echo imaging, which measures R_2, and has reduced the confounding sensitivity of spin probe to the self-relaxation or broadening described above, by nearly an order of magnitude. This makes the spin probe measurement accurate to within 1–2 torr, an absolute measurement or image for animal application (*vide infra*) (Epel, Bowman, Mailer, & Halpern, 2014).

At present, two large classes of spin probes are used for *in vivo* oxymetry: soluble free radicals and insoluble paramagnetic particles (particulates) (Fuchs et al., 2003; Kameneva, Watach, & Borovetz, 2003). They are introduced into animals in different ways and require different imaging methods.

1.9 Soluble Spin Probes

Physiologic EPR imaging has been enabled by the synthesis of free radical reporter molecules. These molecules distribute in specific physiologic compartments. Desirable characteristics for these probes include water solubility, kinetic and metabolic stability, a single narrow line resonance, LWs or relaxation rates directly related to pO_2, persistence of signal from tumors longer than imaging times, and low toxicity (Tatum et al., 2006). Historically, the first probes applied for oxymetry were nitroxides (Fig. 1A; Ardenkjaer-Larsen et al., 1998; Berliner & Fujii, 1985; Biller et al., 2011; Pandian, Parinandi, Ilangovan, Zweier, & Kuppusamy, 2003).

1.10 Trityls

The recent success of spin probe oxymetry was enabled by TAM radicals, which are also referred to as trityls (Fig. 1B). Trityl compounds were developed by Nycomed innovation (later acquired by GE Healthcare; Little Chalfont, Buckinghamshire, United Kingdom). Trityls have extremely narrow, single EPR lines (Ardenkjaer-Larsen et al., 1998). Those which are used for *in vivo* imaging are OX063 (16 µT peak-to-peak, p–p) and its partially deuterated form Ox63d$_{24}$ or OX071 (8 µT p–p) (Fig. 1B). The trityl

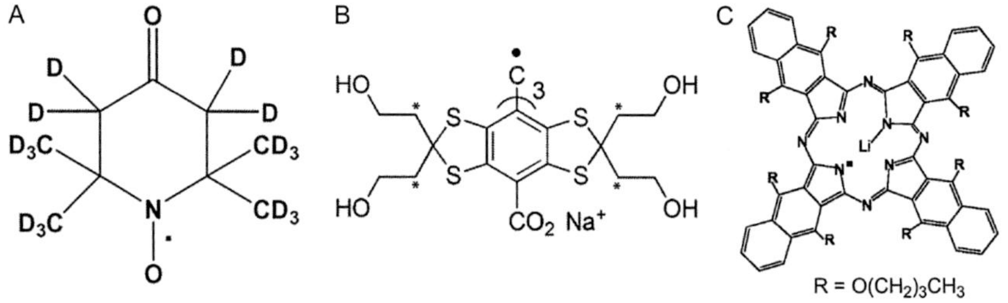

Figure 1 Chemical structures of typical spin probes: (A) water-soluble fully deuterated six-member ring nitroxide; (B) water-soluble OX071 trityl, the deuterated methylene groups are marked with *; (C) particulate LiNc-BuO spin probe.

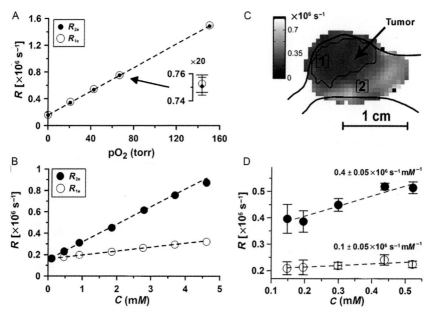

Figure 2 (A and B) Relaxation rates of OX063 dissolved in saline at 37 °C. The range of the measured spin probe concentrations is considerably larger than the one observed *in vivo*. (C) Relaxation rate image. (D) Relaxation rates in position marked as [1] in (C). Rates are measured as a function of spin probe concentration while spin probe is infused at different rates, from low to high. The slopes of concentration dependence are given in the plot.

R_1 and R_2 are linearly dependent on pO_2 (Ardenkjaer-Larsen et al., 1998; Fig. 2). These molecules are triacid, charge 3− anionic and distribute in the extracellular fluid compartment (Tatum et al., 2006; Williams et al., 2002). In the blood stream of a mouse, the clearance halftime of these probes is 9–10 min, while in tumors they remain and provide strong signals for 40–50 min (Matsumoto et al., 2004). The dose at which 50% of animals die, the LD_{50}, of OX063 is large (8 mmol/kg), which allows high-dose injections (Krishna et al., 2002). Typically, 0.5 ml of 80–100 mM solutions at neutral pH are continuously injected intravenously into 20–25 g animals to give tumor average concentrations of several hundred μM.

1.11 Particulates

A number of different solid, crystalline particulates have been used for EPR oxymetry: activated charcoal, and lithium phthalocyanine (Liu et al., 1993) and its derivatives with higher oxygen sensitivity, such as octa-*n*-butoxy-naphthalocyanine (Pandian et al., 2003; Fig. 1C). These insoluble particulate spin probes can be inserted surgically as several tens or hundreds of micron large polycrystals; injected in a form of slurry of finely ground powder; fed to an animal (He et al., 1999); or implanted with tumor cells during inoculation in mice (Ilangovan, Bratasz, & Kuppusamy, 2005; Ilangovan et al., 2004). One of their major advantages is that the oxygen sensing is physically decoupled from the local environment. O_2 must diffuse into a pore or channel where it interacts with an individual phthalocyanine molecule, which is part of a helical stack. The interaction locally breaks the symmetry of the exchange-narrowed, electron cloud of the stack of phthalocyanine molecules lining the channel, increasing its relaxation rate or linewidth.

Limited mobility of particulates enables repeated measurements of oxygen concentration for an extended time. The higher concentration of unpaired electrons or spin density in comparison with a soluble probe produces higher signal and local sensitivity, although spectroscopy in the absence of imaging gives reduced knowledge of the location of the signal. Spin probe migration and degradation leads to loss of EPR signal intensity. There remain potential biocompatibility concerns with prolonged exposure of particulates to tissue, unless the spin probe is excised. Biocompatible coating of the crystalline probes may overcome some of these problems and enhance their clinical applicability (Meenakshisundaram, Pandian, Eteshola, Lee, & Kuppusamy, 2010).

1.12 Techniques for EPR Imaging pO_2

For imaging, the location in space of a paramagnetic species is encoded using magnetic fields that vary linearly in space, or magnetic field gradients, denoted by a vector along the gradient direction, G. Gradients are designed to alter only local amplitude and not direction of the constant magnetic field. The additional magnetic field experienced by a species at position x in the sample is then $\Delta B = G \cdot x$. Here, x is a three-dimensional spatial coordinate in vector form. Since, from Eq. (3), $\omega_0 = \gamma_e B_0$, $k = \Delta \omega_0 = \gamma_e \Delta B$, this is referred to as frequency encoding. The space of frequencies is referred to as k-space (Epel & Halpern, 2012; Epel, Sundramoorthy, Mailer, & Halpern, 2008; Mailer et al., 2006). Each measurement represents a trajectory in k space. The trajectory for a static gradient is a radial line. In the process of obtaining a CW spectroscopic image, B_0 is swept. Fourier transform of the resulting spectra lies along the radial line. This is referred to as a projection, whose angles are defined by the directional part of G. Alternately, a very short pulse of radiofrequency is applied to the sample. This broadband pulse in the presence of a gradient excites all of the spins in the sample. The time trace of the pulse experiment can be mapped to k-space directly, as $k = \gamma Gt$. It is, thus, a very efficient imaging acquisition strategy. This strategy is referred to as radial imaging and requires filtered backprojection tomographic image reconstruction to recover the images. For oxygen imaging, pulse techniques are used with different detection times after a magnetization inversion to determine the SLR rate.

A second method for encoding spatial position origin of signal is referred to as phase encoding. Because the relaxation of electrons is very rapid, a variation of the pulse phase encoding technology used in solid-state MRI is used. This involves measuring a signal at a single time point, t_p, after a pulse that rotates the magnetization into a plane perpendicular to the magnetic field. Very rapid gradient vector angle and magnitude stepping to cover a 3D Cartesian grid is necessary. Coverage of k-space is produced by selection of gradient $k_{ijk} = \gamma G_{ijk} t_p$, one for each k-space point. The technique is referred to as single point imaging (Subramanian et al., 2002). It produces images on a Cartesian grid with a much larger number of acquisitions than tomographic imaging (Epel & Halpern, 2015). As a result, it is less efficient than filtered back projection, but freer from artifact imposed by radial k-space coverage.

1.13 Oxygen Imaging as Longitudinal Relaxation (R_1) Parametric Imaging

Oxygen images are derived from three-dimensional images of the average longitudinal relaxation rate of the trityl spin probe in each voxel or subvolume that is resolved in the image. This relaxation rate is related to the oxygen concentration or partial pressure through a relevant calibration.

$$pO_2 = aR_i + b \tag{7}$$

where $i = 1$ for longitudinal relaxation and $i = 2$ for transverse relaxation. As shown in Fig. 2, the constant, b is far less sensitive to the concentration of spin probe for R_1 than for R_2. One of the main advantages of pulse methods, where multiple spatial images are obtained, each at a different delay time relative to a magnetization orientation pulse, is their ability to determine the relaxation rates directly. The 3D spatial images are reconstructed separately. Then in each image, the voxel in the same location is selected. Pulse sequences that have a relaxation sensitive component (for example, the delay) are used for imaging. This forms a time dependence of the signal for each voxel. Finally, this time dependence is fitted to exponential decay or recovery, and relaxation rate is extracted. The information obtained to reconstruct each image is obtained from different delay times between the inversion pulse for R_1 images, and from the echo time for R_2 images. These times are so thoroughly interleaved (separated by a microsecond or less) that the reconstructed voxel locations are remarkably stable and reproducible. One of the advantages of the use of highly water-soluble trityl spin probes is that their distribution volume is extracellular. Use of R_1-sensitive pulse sequences virtually eliminates confounding variation from spin probe concentration-dependent effects.

1.14 EPR Imaging in Cancer Biology

Several laboratories report various spin probe EPR oxygen measurement techniques, including spectral–spatial CW imaging and localized spectroscopies and pulse-based imaging (Ellis et al., 2001; Khan et al., 2005, 2007; Salikhov et al., 2005; Williams et al., 2010; Zweier, Chzhan, Wang, & Kuppusamy, 1996). Here, we describe work in our laboratory focusing on images of tumors in animals mice, although we have demonstrated both pulse and CW EPR images from rats and rabbits (Epel et al., 2010; Epel, Sundramoorthy, Barth, Mailer, & Halpern, 2011). We will also describe our ideas for the progression of the technique of SLR-based imaging of oxygen tumors to human subjects.

1.15 Validation of the Oxymetry in Animals

Validation of the biologic relevance of oxygen images in animals has required three sets of experiments. We first compared the pO_2 from image voxels (volume elements) with point fluorescence quenching measurements using an Oxylite fiberoptic probe (Elas, 2006). We used a stereotactic needle insertion device to locate the end of the 200 μm diameter optical fiber in an FSa mouse fibrosarcoma tumor and registered it with the EPR oxygen image of the tumor. The Oxylite fiber tip was inserted just after tumor pO_2 imaging, inside the imaging resonator/sample holder without disturbing the tumor location. The tumor born by the mouse leg in our resonator with the stereotactically registered fiberoptic probe is shown in Fig. 3A. Figure 3B shows examples of 0.7 mm thick pO_2 image slices of a mouse leg bearing a tumor. Two orthogonal planes are shown, with a thick

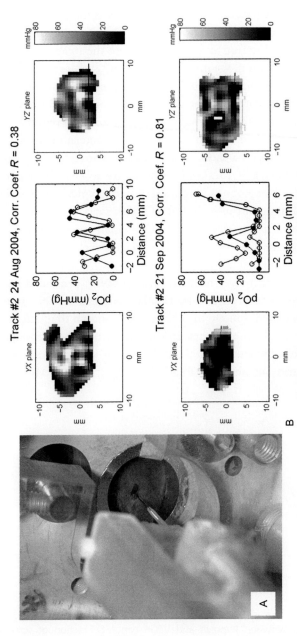

Figure 3 (A) Leg-born mouse tumor in an EPR pO$_2$ imager with an Oxylite fiberoptic pO$_2$ probe being launched into a tumor from a stereotactic frame. (B) Two examples of the correlation between EPR and fiberoptic pO$_2$ data. *Left and right panels*: sagittal and coronal slices showing tumor pO$_2$. The assumed probe track is shown as a black line in the images. *Middle panel*: plot of the oxygen values from the Oxylite™ track (filled circles) and pO$_2$ image (open circles). The remarkable correlation between these values was observed. (See the color plate.)

line indicating the path of the fiberoptic Oxylite probe, assuming the fiber to pass from the truly registered entrance straight along the axis of the resonator. The pO_2 values measured in the Oxylite tracks as the probe was withdrawn from the tumor and the corresponding EPR pO_2 values from voxels attributed to the track location are shown between the images.

For the second validation, we measured voxel pO_2 values within regions of tumors that were subsequently sampled by a number 12 breast biopsy needle and compared them with the average levels of hypoxia-induced proteins produced by the tumor cells. pO_2 images of the tumor were registered with quantitative enzyme-linked immunosorbent assay determinations of the concentration of the hypoxia-induced protein vascular endothelial growth factor from stereotactically localized biopsies from FSa fibrosarcomas. Figure 4 presents a scatter plot of 17 biopsies, showing typical biological scatter and high statistical significance. About 2/3 of the variation in the protein concentration is associated with the mean pO_2, which arises from the image of the volume sampled. Approximately, 100 image voxels were localized to each biopsy volume (Elas et al., 2011).

Locally administered tumor necrosis factor (TNF), produced by viral vectors carrying genes to promote the cellular synthesis of TNF (referred to as TNFerade) injected directly into tumors, has shown dramatic sensitization of tumors to radiation therapy in early human trials. Although this method of treatment of tumors failed a phase 3 trial, there were a number of remarkable cures in phase 2 trials (Mundt et al., 2004). This is

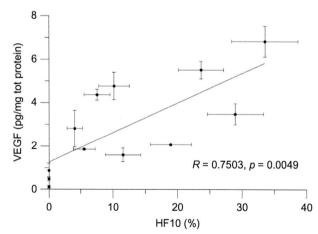

Figure 4 Correlation between hypoxia and VEGF content. Fraction of volumes in the biopsy with pO_2 less than 10 torr, HF10 from the EPR oxygen image versus VEGF concentration in picograms per microgram of total protein.

contradictory to a history of tumor biology showing that hypoxia in tumors creates resistance to radiation therapy (Thomlinson & Gray, 1955). This prompts inquiry as to the contradictory obliteration of tumor vasculature, which, in principle, would increase tumor hypoxia and increase tumor sensitivity to radiation. EPR pO_2 images showed (Haney et al., 2009) that tumor pO_2 increased after TNF administration. These findings were consistent with the difficulty of chaotic, dysfunctional tumor vasculature to convey the 51 kD TNF protein away from the injection site. Increasing vascular obliteration of *intact* microvessels, on the other hand, can rid the tumor of TNF. Thus, TNFerade pruned the tumor of chaotic vessels as predicted by Jain et al. (Carmeliet & Jain, 2011). Consistently, tumor sensitization with the antiangiogenic drug, Sunitinib, which also interferes with tumor angiogenesis, has been shown to be associated with a similarly paradoxical increase in EPR image-based oxygenation from this anti-blood vessel agent (Matsumoto et al., 2011). EPR pO_2 images provide *in situ* information defining the molecular biologic response to the microenvironment pO_2. EPR imaging is a powerful tool in defining graded molecular biologic response to microenvironment.

The pO_2 images that have been obtained with longitudinal relaxation rate pulsed technology have spatial resolution of approximately 1 mm. A major question to be answered is the biological relevance of this resolution. This third validation asks whether or not these images can show, based on the fractions of tumor voxels with pO_2, less than a particular threshold, and also, that they can predict outcomes of radiation treatment of tumors of a given size to a dose sufficient to cure 50% of the tumors. Figure 5 shows the results of treating two different tumor types, a syngeneic mouse mammary tumor, MCa4, and a syngeneic fibrosarcoma, FSa, grown in the legs of C3H mice. The data demonstrate that a threshold of 10% of voxels with pO_2 less than 10 torr for FSa tumors, and 15% for MCa4 tumors, separates the tumor cure probability. Figure 5 dramatically shows that, for both tumor types, the probability for tumor control was significantly better, a factor of two or more better, for the hypoxic fraction less than the threshold than for the hypoxic fraction larger than the threshold. Therefore, the biomedical relevance of images with the 1 mm spatial resolution appears strong (Elas et al., 2013).

1.16 Transient Hypoxia can be Imaged with EPR pO_2 Images

If tumor hypoxic regions are chronic and unchanging with time, then direction of local therapy such as radiation to such regions would be crucially

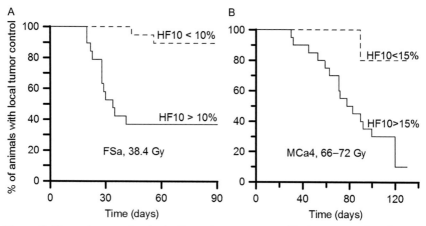

Figure 5 The outcome of a single X-ray dose treatment. Kaplan–Meier plots for two groups of animals separated according to their tumor hypoxic fraction (HF10), a percent of voxels in the pO_2 image with pO_2 below 10 torr. (A) FSa tumors treated with 33.8 Gy. Wilcoxon test shows that HF10 > 10% threshold is a significant predictor of tumor failure ($p = 0.0138$). (B) MCa4 tumors treated with a single dose in the range of 66–72 Gy. Wilcoxon test shows that HF10 > 15% threshold is a significant predictor of tumor failure ($p = 0.0193$).

important. However, if tumor pO_2 distributions fluctuate wildly, such images become irrelevant. There is data showing fluctuations in tumor blood flow. This can influence oxygen concentrations (Brown, 1979). This may be linked to therapeutic resistance. Dynamic, rapidly acquired images of pO_2 are necessary to establish this. Yasui et al. (2010) showed that pulsed EPR can monitor fluctuations in oxygen concentrations in mouse models. Oxygen images acquired every 3 min for a total of 30 min revealed large fluctuations in pO_2 in some tumor regions. Redler et al. demonstrated that peripheral regions of tumors with intermediate levels of hypoxia, and which presumably have more *intact* vasculature, have larger spontaneous pO_2 variations then more central regions of FSa fibrosarcomas (Redler, Epel, & Halpern, 2014). This EPR imaging technique, registered with tumor locating MRI, may offer a powerful clinical tool to noninvasively detect variable oxygenation in tumors.

1.17 Effectiveness of Localizing Radiation to Regions of the Tumor with High Hypoxic Fraction Using EPR pO_2 Images Needs to be Tested

The success noted above in identification of regions within a tumor whose voxels are hypoxic, and showing that large fractions of these voxels induce

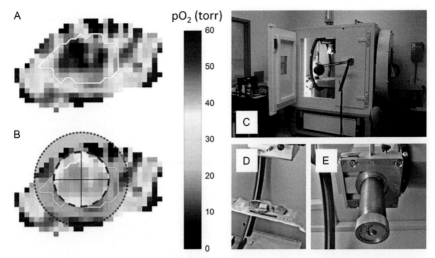

Figure 6 EPR oxygen image of the leg tumor. (A) Oxygen map with tumor contour transferred from the registered MRI image. (B) Boost (red line) and antiboost (shaded area) as determined by the boost planning software. (C) XRad225Cx small animal cone beam CT imager and irradiator. (D) Animal bed. (E) Collimator. (See the color plate.)

resistance to a 50% tumor control dose, leads to the question: does treating only these regions of tumors with extra "boost" dose increase the tumor cure? This needs to be demonstrated in animal models prior to human applications. Figure 6A and B show an hypoxic region defined in a mouse MCa4 breast tumor, and a spherical volume for radiation boost treatment. This boost would be added to a 50% control dose given to the entire tumor. Figure 6C–E show an XRAD225Cx system to precisely deliver the boost dose to the indicated region. A trial comparing the tumor control using such a boost with treatment to a well-oxygenated shell of similar volume (an "antiboost") is underway (Epel, Redler, Pelizzari, Tormyshev, & Halpern, 2015).

2. SUMMARY

We have brought the technology for the EPR imaging of tumor pO_2 to a point where quantitative images of oxygen in the tissues and tumors of living animals, with 1–2 torr pO_2 resolution and 1 mm spatial resolution, can be obtained in 10 min or faster. We have shown the biologic relevance of the oxygen images and demonstrated that they predict cure from radiation therapy. Rapid acquisition of EPR pO_2 images will validate the relevance of transient hypoxia. We are in the process of validation of the potential use of EPR oxygen images in guiding radiation to resistant portions of animal tumors. A path to human applications may be possible.

ACKNOWLEDGMENTS
This work was supported by NIH grants P41 EB002034 and R01 CA098575.

REFERENCES
Abragam, A. (1961). *Principles of nuclear magnetism*. Oxford: Oxford University.
Abragam, A., & Bleaney, B. (1970). *EPR of transition ions*. Oxford: Clarendon.
Ahmad, R., Khan, M., Vikram, D., Bratasz, A., & Kuppusamy, P. (2010). EPR oximetry: Method and application. In D. Das (Ed.), *Methods in redox signaling* (pp. 19–23). New York: Mary Ann Liebert Inc.
Ardenkjaer-Larsen, J. H., Laursen, I., Leunbach, I., Ehnholm, G., Wistrand, L. G., Petersson, J. S., et al. (1998). EPR and DNP properties of certain novel single electron contrast agents intended for oximetric imaging. *Journal of Magnetic Resonance, 133*(1), 1–12. http://dx.doi.org/10.1006/jmre.1998.1438.
Atherton, N. M. (1993). *Principles of electron spin resonance*. Chichester: Ellis Horwood Ltd.
Backer, J. M., Budker, V. G., Eremenko, S. I., & Molin, Y. N. (1977). Detection of the kinetics of biochemical reactions with oxygen using exchange broadening in the ESR spectra of nitroxide radicals. *Biochimica et Biophysica Acta, 460*(1), 152–156. http://dx.doi.org/10.1016/0005-2728(77)90161-X.
Berliner, L. J., & Fujii, H. (1985). Magnetic-resonance imaging of biological specimens by electron-paramagnetic resonance of nitroxide spin labels. *Science, 227*(4686), 517–519.
Biller, J. R., Meyer, V., Elajaili, H., Rosen, G. M., Kao, J. P. Y., Eaton, S. S., et al. (2011). Relaxation times and line widths of isotopically-substituted nitroxides in aqueous solution at X-band. *Journal of Magnetic Resonance, 212*(2), 370–377. http://dx.doi.org/10.1016/j.jmr.2011.07.018.
Bottomley, P. A., & Andrew, E. R. (1978). RF magnetic field penetration, phase shift and power dissipation in biological tissue: Implications for NMR imaging. *Physics in Medicine and Biology, 23*(4), 630–643.
Brown, J. M. (1979). Evidence for acutely hypoxic cells in mouse-tumors, and a possible mechanism of re-oxygenation. *British Journal of Radiology, 52*(620), 650–656.
Busse, L. J., Pratt, R. G., & Thomas, S. R. (1988). Deconvolution of chemical-shift spectra in two-dimensional or 3-dimensional F-19 MR imaging. *Journal of Computer Assisted Tomography, 12*(5), 824–835. http://dx.doi.org/10.1097/00004728-198809010-00020.
Carmeliet, P., & Jain, R. K. (2011). Molecular mechanisms and clinical applications of angiogenesis. *Nature, 473*(7347), 298–307. http://dx.doi.org/10.1038/nature10144. nature10144 [pii].
Condon, E. U., & Shortley, G. H. (1935). *The theory of atomic spectra*. Cambridge, UK: Cambridge University Press.
Cormack, A. M. (1963). Representation of a function by its line integrals with some radiological applications. *Journal of Applied Physics, 34*(9), 2722–2727. http://dx.doi.org/10.1063/1.1729798.
Dirac, P. A. M. (1958). *Principles of quantum mechanics* (4th (paperback) ed.). Oxford, UK: Oxford University Press.
Eastman, P. E., Kooser, R. G., Pas, M. R., & Freed, J. H. (1969). Studies of Heisenberg spin exchange in ESR spectra I. Linewidth and saturation effects. *Journal of Chemical Physics, 54*, 2690.

Eaton, S. S., & Eaton, G. R. (2000). Relaxation times of organic radicals and transition metal ions. *Biological magnetic resonance*: Vol. 19 (pp. 29–154). New York: Kluwer Academic/Plenum Publishers.

Efimova, O. V., Caia, G. L., Sun, Z. Q., Petryakov, S., Kesselring, E., Samouilov, A., et al. (2011). Standard-based method for proton-electron double resonance imaging of oxygen. *Journal of Magnetic Resonance*, 212(1), 197–203. http://dx.doi.org/10.1016/j.jmr.2011.06.030.

Elas, M. (2006). Electron paramagnetic resonance oxygen images correlate spatially and quantitatively with oxylite oxygen measurements. *Clinical Cancer Research*, 12(14), 4209–4217. http://dx.doi.org/10.1158/1078-0432.ccr-05-0446.

Elas, M., Hleihel, D., Barth, E. D., Haney, C. R., Ahn, K. H., Pelizzari, C. A., et al. (2011). Where it's at really matters: In situ in vivo vascular endothelial growth factor spatially correlates with electron paramagnetic resonance pO_2 images in tumors of living mice. *Molecular Imaging and Biology*, 13(6), 1107–1113. http://dx.doi.org/10.1007/s11307-010-0436-4.

Elas, M., Magwood, J. M., Butler, B., Li, C., Wardak, R., Barth, E. D., et al. (2013). EPR oxygen images predict tumor control by a 50% tumor control radiation dose. *Cancer Research*, 73(17), 5328–5335. http://dx.doi.org/10.1158/0008-5472.CAN-13-0069. 0008-5472.CAN-13-0069 [pii].

Ellis, S. J., Velayutham, M., Velan, S. S., Petersen, E. F., Zweier, J. L., Kuppusamy, P., et al. (2001). EPR oxygen mapping (EPROM) of engineered cartilage grown in a hollow-fiber bioreactor. *Magnetic Resonance in Medicine*, 46(4), 819–826.

Epel, B., Bowman, M. K., Mailer, C., & Halpern, H. J. (2014). Absolute oxygen R_1 imaging in vivo with pulse electron paramagnetic resonance. *Magnetic Resonance in Medicine*, 72, 362–368. http://dx.doi.org/10.1002/mrm.24926.

Epel, B., & Halpern, H. J. (2012). Electron paramagnetic resonance oxygen imaging in vivo. In B. C. Gilbert, V. Chechik, & D. M. Murphy (Eds.), *Electron paramagnetic resonance*: Vol. 23 (pp. 180–208). Cambridge, UK: RSC Publishing.

Epel, B., & Halpern, H. J. (2015). Comparison of pulse sequences for R-based electron paramagnetic resonance oxygen imaging. *Journal of Magnetic Resonance*, 254, 56–61. http://dx.doi.org/10.1016/j.jmr.2015.02.012. S1090-7807(15)00046-4 [pii].

Epel, B., Haney, C. R., Hleihel, D., Wardrip, C., Barth, E. D., & Halpern, H. J. (2010). Electron paramagnetic resonance oxygen imaging of a rabbit tumor using localized spin probe delivery. *Medical Physics*, 37(6), 2553–2559.

Epel, B., Redler, G., Pelizzari, C., Tormyshev, V. M., & Halpern, H. J. (2015). Approaching oxygen-guided intensity-modulated radiation therapy. *Advances in Experimental Medicine and Biology*. in press.

Epel, B., Sundramoorthy, S. V., Barth, E. D., Mailer, C., & Halpern, H. J. (2011). Comparison of 250 MHz electron spin echo and continuous wave oxygen EPR imaging methods for in vivo applications. *Medical Physics*, 38(4), 2045–2052. http://dx.doi.org/10.1118/1.3555297.

Epel, B., Sundramoorthy, S. V., Mailer, C., & Halpern, H. J. (2008). A versatile high speed 250-MHz pulse imager for biomedical applications. *Concepts in Magnetic Resonance Part B: Magnetic Resonance Engineering*, 33B(3), 163–176. http://dx.doi.org/10.1002/cmr.b.20119.

Fuchs, M., Groth, N., & Herrling, N. (2003). Applications of in vivo EPR spectroscopy and imaging to skin. In L. J. Berliner (Ed.), *Biological magnetic resonance*: Vol. 18. *In vivo EPR (ESR): Theory & applications* (pp. 483–515). New York: Kluwer Academic.

Griffiths, J. R., & Robinson, S. P. (1999). The OxyLite: A fibre-optic oxygen sensor. *British Journal of Radiology*, 72(859), 627–630.

Halpern, H. J., Spencer, D. P., Vanpolen, J., Bowman, M. K., Nelson, A. C., Dowey, E. M., et al. (1989). Imaging radio-frequency electron-spin-resonance spectrometer with high-resolution and sensitivity for in vivo measurements. *Review of Scientific Instruments*, *60*(6), 1040–1050.

Halpern, H. J., Yu, C., Peric, M., Barth, E., Grdina, D. J., & Teicher, B. A. (1994). Oxymetry deep in tissues with low-frequency electron-paramagnetic-resonance. *Proceedings of the National Academy of Sciences of the United States of America*, *91*(26), 13047–13051.

Haney, C. R., Parasca, A. D., Fan, X., Bell, R. M., Zamora, M. A., Karczmar, G. S., et al. (2009). Characterization of response to radiation mediated gene therapy by means of multimodality imaging. *Magnetic Resonance in Medicine*, *62*(2), 348–356.

Harrison, D. K. (2003). Physiological oxygen measurements using oxygen electrodes. In D. F. Wilson, S. M. Evans, J. Biaglow, & A. Pastuszko (Eds.), *Advances in experimental medicine and biology: Vol. 510. Oxygen transport to tissue volume XXIII: Oxygen measurements in the 21st century: Basic techniques and clinical relevance. Adv Exp Med Biol.* (pp. 163–167). New York: Plenum Press.

He, G. L., Shankar, R. A., Chzhan, M., Samouilov, A., Kuppusamy, P., & Zweier, J. L. (1999). Noninvasive measurement of anatomic structure and intraluminal oxygenation in the gastrointestinal tract of living mice with spatial and spectral EPR imaging. *Proceedings of the National Academy of Sciences of the United States of America*, *96*(8), 4586–4591.

Henk, J. M., Kunkler, P. B., & Smith, C. W. (1977). Radiotherapy and hyperbaric oxygen in head and neck cancer. Final report of first controlled clinical trial. *Lancet*, *2*(8029), 101–103. S0140-6736(77)90116-7 [pii].

Henk, J. M., & Smith, C. W. (1977). Radiotherapy and hyperbaric oxygen in head and neck cancer. Interim report of second clinical trial. *Lancet*, *2*(8029), 104–105. S0140-6736(77)90117-9 [pii].

Hunjan, S., Zhao, D., Constantinescu, A., Hahn, E. W., Antich, P. P., & Mason, R. P. (2001). Tumor oximetry: Demonstration of an enhanced dynamic mapping procedure using fluorine-19 echo planar magnetic resonance imaging in the Dunning prostate R3327-AT1 rat tumor. *International Journal of Radiation Oncology, Biology, Physics*, *49*(4), 1097–1108.

Ilangovan, G., Bratasz, A., & Kuppusamy, P. (2005). Non-invasive measurement of tumor oxygenation using embedded microparticulate EPR spin probe. In P. Okunieff, J. Williams, & Y. Chen (Eds.), *Advances in experimental medicine and biology: Vol. 566. Oxygen transport to tissue XXV* (pp. 67–73).

Ilangovan, G., Bratasz, A., Li, H., Schmalbrock, P., Zweier, J. L., & Kuppusamy, P. (2004). In vivo measurement and imaging of tumor oxygenation using coembedded paramagnetic particulates. *Magnetic Resonance in Medicine*, *52*(3), 650–657. http://dx.doi.org/10.1002/mrm.20188.

Kameneva, M. W., Watach, M. J., & Borovetz, H. S. (2003). Rheologic dissimilarities in female and male blood: Potential link to development of cardiovascular diseases. In J. W. Dunn & H. M. Swartz (Eds.), *Vol. 530. Oxygen transport to tissue XXIV* (pp. 689–696). New York: Kluwer Academic/Plenum Publishers.

Kennedy, K. A., Teicher, B. A., Rockwell, S., & Sartorelli, A. C. (1980). The hypoxic tumor cell: A target for selective cancer chemotherapy. *Biochemical Pharmacology*, *29*(1), 1–8. 0006-2952(80)90235-X [pii].

Khan, N., Hou, H., Hein, P., Comi, R., Buckey, J., Grinberg, O., et al. (2005). *Black magic and EPR oximetry: From lab to initial clinical trials*. New York: Plenum Publishers.

Khan, N., Williams, B. B., Hou, H., Li, H., & Swartz, H. M. (2007). Repetitive tissue pO_2 measurements by electron paramagnetic resonance oximetry: Current status and future potential for experimental and clinical studies. *Antioxidants & Redox Signaling*, *9*(8), 1169–1182. http://dx.doi.org/10.1089/ars.2007.1635.

Krause, B. J., Beck, R., Souvatzoglou, M., & Piert, M. (2006). PET and PET/CT studies of tumor tissue oxygenation. *Quarterly Journal of Nuclear Medicine and Molecular Imaging*, *50*(1), 28–43.

Krishna, M. C., English, S., Yamada, K., Yoo, J., Murugesan, R., Devasahayam, N., et al. (2002). Overhauser enhanced magnetic resonance imaging for tumor oximetry: Coregistration of tumor anatomy and tissue oxygen concentration. *Proceedings of the National Academy of Sciences of the United States of America, 99*(4), 2216–2221. http://dx.doi.org/10.1073/pnas.042671399.

Kuppusamy, P., Chzhan, M., Vij, K., Shteynbuk, M., Lefer, D. J., Giannella, E., et al. (1994). 3-Dimensional spectral spatial EPR imaging of free-radicals in the heart—A technique for imaging tissue metabolism and oxygenation. *Proceedings of the National Academy of Sciences of the United States of America, 91*(8), 3388–3392.

Lauterbur, P. C. (1973). Image formation by induced local interactions—Examples employing nuclear magnetic-resonance. *Nature, 242*(5394), 190–191. http://dx.doi.org/10.1038/242190a0.

Lauterbur, P. C., Levin, D. N., & Marr, R. B. (1984). Theory and simulation of NMR spectroscopic imaging and field plotting by projection reconstruction involving an intrinsic frequency dimension. *Journal of Magnetic Resonance, 59*, 536–541.

Liu, K. J., Gast, P., Moussavi, M., Norby, S. W., Vahidi, N., Walczak, T., et al. (1993). Lithium phthalocyanine—A probe for electron-paramagnetic-resonance oximetry in viable biological-systems. *Proceedings of the National Academy of Sciences of the United States of America, 90*(12), 5438–5442.

Longo, D. L., Fauci, A. S., Kasper, D. L., Hauser, S. L., Jameson, J. L., & Loscalzo, J. (Eds.). (2011). *Harrison's principles of internal medicine.* (18 ed. Vol. 1). New York, NY: McGraw Hill.

Mailer, C., Robinson, B. H., Williams, B. B., & Halpern, H. J. (2003). Spectral fitting: The extraction of crucial information from a spectrum and a spectral image. *Magnetic Resonance in Medicine, 49*(6), 1175–1180. http://dx.doi.org/10.1002/Mrm.10474.

Mailer, C., Sundramoorthy, S. V., Pelizzari, C. A., & Halpern, H. J. (2006). Spin echo spectroscopic electron paramagnetic resonance imaging. *Magnetic Resonance in Medicine, 55*(4), 904–912.

Maltempo, M. M. (1986). Differentiation of spectral and spatial components in EPR imaging using 2-D image reconstruction algorithms. *Journal of Magnetic Resonance, 69*, 156–161.

Matsumoto, S., Batra, S., Saito, K., Yasui, H., Choudhuri, R., Gadisetti, C., et al. (2011). Antiangiogenic agent sunitinib transiently increases tumor oxygenation and suppresses cycling hypoxia. *Cancer Research, 71*(20), 6350–6359. http://dx.doi.org/10.1158/0008-5472.CAN-11-2025. 0008-5472.CAN-11-2025 [pii].

Matsumoto, K., English, S., Yoo, J., Yamada, K., Devasahayam, N., Cook, J. A., et al. (2004). Pharmacokinetics of a triarylmethyl-type paramagnetic spin probe used in EPR oximetry. *Magnetic Resonance in Medicine, 52*(4), 885–892. http://dx.doi.org/10.1002/Mrm.20222.

Meenakshisundaram, G., Pandian, R. P., Freshola, E., Lee, S. C., & Kuppusamy, P. (2010). A paramagnetic implant containing lithium naphthalocyanine microcrystals for high-resolution biological oximetry. *Journal of Magnetic Resonance, 203*(1), 185–189. http://dx.doi.org/10.1016/j.jmr.2009.11.016.

Molin, Y. N., Salikhov, K. M., & Zamaraev, K. I. (1980). *Spin exchange: Principles and applications in chemistry and biology.* Berlin: Springer-Verlag.

Mundt, A. J., Vijayakumar, S., Nemunaitis, J., Sandler, A., Schwartz, H., Hanna, N., et al. (2004). A phase I trial of TNFerade biologic in patients with soft tissue sarcoma in the extremities. *Clinical Cancer Research, 10*(17), 5747–5753.

Ogawa, S., & Lee, T. M. (1990). Magnetic-resonance-imaging of blood-vessels at high fields—In vivo and in vitro measurements and image simulation. *Magnetic Resonance in Medicine, 16*(1), 9–18. http://dx.doi.org/10.1002/mrm.1910160103.

Ogawa, S., Lee, T. M., Kay, A. R., & Tank, D. W. (1990). Brain magnetic-resonance-imaging with contrast dependent on blood oxygenation. *Proceedings of the National*

Academy of Sciences of the United States of America, 87(24), 9868–9872. http://dx.doi.org/10.1073/pnas.87.24.9868.

Pandian, R. P., Parinandi, N. L., Ilangovan, G., Zweier, J. L., & Kuppusamy, P. (2003). Novel particulate spin probe for targeted determination of oxygen in cells and tissues. Free Radical Biology and Medicine, 35(9), 1138–1148. http://dx.doi.org/10.1016/S0891-5849(03)00496-9.

Popp, C. A., & Hyde, J. S. (1981). Effects of oxygen on electron-paramagnetic-resonance of nitroxide spin-label probes of model membranes. Journal of Magnetic Resonance, 43(2), 249–258. http://dx.doi.org/10.1016/0022-2364(81)90036-6.

Redler, G., Epel, B., & Halpern, H. J. (2014). EPR image based oxygen movies for transient hypoxia. Advances in Experimental Medicine and Biology, 812, 127–133. http://dx.doi.org/10.1007/978-1-4939-0620-8_17.

Robinson, B. H., Mailer, C., & Reese, A. W. (1999a). Linewidth analysis of spin labels in liquids—I. Theory and data analysis. Journal of Magnetic Resonance, 138(2), 199–209.

Robinson, B. H., Mailer, C., & Reese, A. W. (1999b). Linewidth analysis of spin labels in liquids—II. Experimental. Journal of Magnetic Resonance, 138(2), 210–219.

Rontgen, W. C. (1896). On a new kind of rays. Science, 3(59), 227–231. http://dx.doi.org/10.1126/science.3.59.227. 3/59/227 [pii].

Rumsey, W. L., Vanderkooi, J. M., & Wilson, D. F. (1988). Imaging of phosphorescence—A novel method for measuring oxygen distribution in perfused tissue. Science, 241(4873), 1649–1651. http://dx.doi.org/10.1126/science.3420417.

Salikhov, I., Walczak, T., Lesniewski, P., Khan, N., Iwasaki, A., Comi, R., et al. (2005). EPR spectrometer for clinical applications. Magnetic Resonance in Medicine, 54(5), 1317–1320.

Schweiger, A., & Jeschke, G. (2001). Principles of pulse electron paramagnetic resonance. Oxford, UK: Oxford University Press.

Slichter, C. P. (1996). Principles of magnetic resonance. New York: Springer-Verlag.

Subramanian, S., Devasahayam, N., Murugesan, R., Yamada, K., Cook, J., Taube, A., et al. (2002). Single-point (constant-time) imaging in radiofrequency Fourier transform electron paramagnetic resonance. Magnetic Resonance in Medicine, 48(2), 370–379. http://dx.doi.org/10.1002/Mrm.10199.

Swartz, H. M., & Clarkson, R. B. (1998). The measurement of oxygen in vivo using EPR techniques. Physics in Medicine and Biology, 43(7), 1957–1975. http://dx.doi.org/10.1088/0031-9155/43/7/017.

Swartz, H. M., & Glockner, J. F. (1991). Measurements of oxygen by EPRI and EPRS. Boca Raton, FL: CRC Press, Inc.

Swartz, H. M., Khan, N., Buckey, J., Comi, R., Gould, L., Grinberg, O., et al. (2004). Clinical applications of EPR: Overview and perspectives. NMR in Biomedicine, 17(5), 335–351.

Tatum, J. L., Kelloff, G. J., Gillies, R. J., Arbeit, J. M., Brown, J. M., Chao, K. S. C., et al. (2006). Hypoxia: Importance in tumor biology, noninvasive measurement by imaging, and value of its measurement in the management of cancer therapy. International Journal of Radiation Biology, 82(10), 699–757. http://dx.doi.org/10.1080/09553000601002324.

Teicher, B. A. (1994). Hypoxia and drug resistance. Cancer Metastasis Reviews, 13(2), 139–168.

Thomlinson, R. H., & Gray, L. H. (1955). The histological structure of some human lung cancers and the possible implications for radiotherapy. British Journal of Radiology, 9, 539–563.

Whalen, W. J., Riley, J., & Nair, P. (1967). A microelectrode for measuring intracellular pO_2. Journal of Applied Physiology, 23, 798–801.

Williams, B. B., al Hallaq, H., Chandramouli, G. V., Barth, E. D., Rivers, J. N., Lewis, M., et al. (2002). Imaging spin probe distribution in the tumor of a living mouse with 250 MHz EPR: Correlation with BOLD MRI. *Magnetic Resonance in Medicine*, 47(4), 634–638.

Williams, B. B., Khan, N., Zaki, B., Hartford, A., Ernstoff, M. S., & Swartz, H. M. (2010). Clinical electron paramagnetic resonance (EPR) oximetry using India ink. *Advances in Experimental Medicine and Biology*, 662, 149–156. http://dx.doi.org/10.1007/978-1-4419-1241-1_21.

Wilson, J. D., Braunwald, E., Isselbacher, K. J., Petersdorf, R. G., Martin, J. B., Fauci, A. S., et al. (1991). *Harrison's principles of internal medicine* (12th ed.). New York: McGraw-Hill.

Yasui, H., Matsumoto, S., Devasahayam, N., Munasinghe, J. P., Choudhuri, R., Saito, K., et al. (2010). Low-field magnetic resonance imaging to visualize chronic and cycling hypoxia in tumor-bearing mice. *Cancer Research*, 70(16), 6427–6436. http://dx.doi.org/10.1158/0008-5472.can-10-1350.

Zweier, J. L., Chzhan, M., Wang, P. H., & Kuppusamy, P. (1996). Spatial and spectral-spatial EPR imaging of free radicals and oxygen in the heart. *Research on Chemical Intermediates*, 22(6), 615–624.

Zweier, J. L., Thompsongorman, S., & Kuppusamy, P. (1991). Measurement of oxygen concentrations in the intact beating heart using electron-paramagnetic resonance spectroscopy—A technique for measuring oxygen concentrations in situ. *Journal of Bioenergetics and Biomembranes*, 23(6), 855–871.

CHAPTER NINETEEN

Direct and Repeated Measurement of Heart and Brain Oxygenation Using In Vivo EPR Oximetry

Nadeem Khan, Huagang Hou, Harold M. Swartz, Periannan Kuppusamy[1]

Department of Radiology, EPR Center for the Study of Viable Systems, Geisel School of Medicine at Dartmouth, Norris Cotton Cancer Center, Dartmouth-Hitchcock Medical Center, Lebanon, New Hampshire, USA
[1]Corresponding author: e-mail address: kuppu@dartmouth.edu

Contents

1. Introduction	530
2. Principles of EPR Oximetry	532
3. Paramagnetic Oxygen-Sensitive Probes for pO_2 Measurements by EPR	535
3.1 India Ink	535
3.2 Particulate Oximetry Probes	537
3.3 Implantable Resonators for Oximetry Beyond 10 mm from Surface	538
4. Application of EPR Oximetry in Ischemic Pathologies	540
4.1 Ischemic Stroke	540
4.2 Procedures for the Implantation of Particulate Probes in the Brain of Rats for Oximetry Following MCAO	541
4.3 Procedures for Oximetry in the Heart	544
5. Summary	546
Acknowledgments	546
References	547

Abstract

Low level of oxygen (hypoxia) is a critical factor that defines the pathological consequence of several pathophysiologies, particularly ischemia, that usually occur following the blockage of a blood vessel in vital organs, such as brain and heart, or abnormalities in the microvasculature, such as peripheral vascular disease. Therefore, methods that can directly and repeatedly quantify oxygen levels in the brain and heart will significantly improve our understanding of ischemic pathologies. Importantly, such oximetry capability will facilitate the development of strategies to counteract low levels of oxygen and thereby improve outcome following stroke or myocardial infarction.

In vivo electron paramagnetic resonance (EPR) oximetry has the capability to monitor tissue oxygen levels in real time. The method has largely been tested and used in experimental animals, although some clinical measurements have been performed. In this chapter, a brief overview of the methodology to repeatedly quantify oxygen levels in the brain and heart of experimental animal models, ranging from mice to swine, is presented. EPR oximetry requires a one-time placement of an oxygen-sensitive probe in the tissue of interest, while the rest of the procedure for reliable, accurate, and repeated measurements of pO_2 (partial pressure of oxygen) is noninvasive and can be repeated as often as desired. A multisite oximetry approach can be used to monitor pO_2 at many sites simultaneously. Building on significant advances in the application of EPR oximetry in experimental animal models, spectrometers have been developed for use in human subjects. Initial feasibility of pO_2 measurement in solid tumors of patients has been successfully demonstrated.

1. INTRODUCTION

Although the discovery of oxygen was made in the eighteenth century, measurements of oxygen concentration (oximetry) *in vivo* have only been achieved in the twentieth century. Tissue pO_2 (partial pressure of oxygen) is a key parameter in various physiological and pathological processes of biological systems (Brahimi-Horn & Pouyssegur, 2007; Kulkarni, Kuppusamy, & Parinandi, 2007; Taylor & Pouyssegur, 2007). Therefore, it is desirable to develop techniques for oximetry, i.e., measurement of tissue pO_2, with sufficient sensitivity, accuracy, and ease. Such techniques will enhance our understanding of pathological processes, such as stroke, which results from the occlusion of a cerebral artery, and is a leading cause of long-term disability and mortality in the United States (Mozaffarian et al., 2015; Rink & Khanna, 2011; Shi & Liu, 2007). Likewise, an obstruction in the blood flow to specific regions of the heart leads to a decline in tissue pO_2 from physiological levels (>20–25 mmHg), which plays a decisive role in the extent of vital tissue loss (infarction) and subsequent pathological events, e.g., congestive heart failure (Frank et al., 2012; Mozaffarian et al., 2015; Turer & Hill, 2010). These ischemic pathologies change tissue pO_2 in a temporal manner during disease progression. Treatment outcome primarily depends on the extent of recovery of tissue pO_2 up to a level that can restrict tissue damage. Consequently, temporal monitoring of tissue pO_2 in the brain and heart of experimental animal models is of great interest in order to develop appropriate strategies for improving treatment outcome.

Direct measurement of tissue pO_2 was first made by polarographic needle electrodes (Eppendorf) in the 1950s and has been extensively used to measure pO_2, particularly in solid tumors, in animal models and a few studies in cancer patients (Doll, Milosevic, Pintilie, Hill, & Fyles, 2003; Kallinowski, Zander, Hoeckel, & Vaupel, 1990; Lally et al., 2006; O'Hara et al., 2004; Sauer, Weber, Peschke, & Eble, 2000; Vaupel, Hockel, & Mayer, 2007). However, all the techniques based on electrodes require a direct insertion of the electrode in the tissue of interest for making pO_2 measurement (Griffiths & Robinson, 1999; O'Hara et al., 2005; Tran et al., 2012; Vikram, Bratasz, Ahmad, & Kuppusamy, 2007). Such procedure does not fulfill the need for repeated measurement of pO_2 during the progression of ischemic diseases and therapy. There are also additional concerns about the potential for significant disadvantage, as the invasiveness of these techniques may cause artifacts at the time of measurements. Other widely available noninvasive techniques to assess oxygenation provide data on parameters that, while related to oxygen levels, do not measure tissue pO_2 directly. These include techniques such as blood oxygen level-dependent magnetic resonance imaging (MRI), nuclear magnetic resonance proton spectroscopy, duplex Doppler ultrasound, positron emission tomography (PET) based on metabolism or hypoxia localizing drugs, and near-infrared measurements of hemoglobin (Baete, Vandecasteele, & De Deene, 2011; Blockley, Griffeth, Simon, & Buxton, 2013; Christen, Bolar, & Zaharchuk, 2013; Dunn, Zaim-Wadghiri, & Kida, 1997; Gaertner, Souvatzoglou, Brix, & Beer, 2012; Keddie & Rohman, 2012; Krishna et al., 2002; Krishna, Subramanian, Kuppusamy, & Mitchell, 2001; Krohn, Link, & Mason, 2008; Liu et al., 2011; Morgalla et al., 2012; Vikram, Zweier, & Kuppusamy, 2007). These techniques can provide valuable information on the saturation of hemoglobin within the vascular system or redox-related metabolites, but do not provide direct quantification of pO_2 in a tissue of interest.

We have implemented EPR oximetry for direct and repeated quantification of tissue pO_2 in intact animals, and isolated functioning organs. This has been made possible by the development of low-frequency EPR spectrometers (1200–250 MHz) (Ahmad & Kuppusamy, 2010; Halpern et al., 1994; Salikhov et al., 2005). The EPR spectrometers operating at 1200 MHz (L-band) have recently been adapted at our Center for measurement of tissue pO_2 in human subjects (Khan, Williams, & Swartz, 2006; Swartz, Hou, et al., 2014; Swartz, Williams, et al., 2014; Williams et al., 2010). The keys to these developments have been improved technology, highly sensitive oximetry probes, and capability to measure pO_2 at multiple sites simultaneously at any depth from the surface. A multi-institutional clinical study is underway to quantify tumor pO_2 in patients undergoing radio—and/or chemotherapy, with the overall goal to optimize outcomes by scheduling treatment at times of optimal tumor oxygenation. Extensive reviews of EPR oximetry with comprehensive information on the principles, and technical aspects are available in several reports (Ahmad & Kuppusamy, 2010; Dunn & Swartz, 2003; Gallez, Baudelet, & Jordan, 2004; Grinberg, Smirnov, & Swartz, 2001; Ilangovan, Zweier, & Kuppusamy, 2004; Salikhov et al., 2005; Swartz & Clarkson, 1998; Swartz, Dunn, Grinberg, O'Hara, & Walczak, 1997). The low-frequency EPR technique has also been applied successfully to investigate anticancer therapies, hyperoxic strategies, burn injury, pathologies associated with kidneys, liver, lungs, and skeletal muscle in various animal models (Fisher, Khan, Salisbury, & Kuppusamy, 2013; Hou, Dong, et al., 2011; Hou et al., 2012, 2015; Hou, Khan, et al., 2014; Ibragimova et al., 2008; James et al., 1996; James, Madhani, Roebuck, Jackson, & Swartz, 2002; Khan et al., 2008, 2010; Khan, Mupparaju, Hou, Williams, & Swartz, 2012; Madhani et al., 2002; Mupparaju, Hou, Lariviere, Swartz, & Khan, 2011; Rivera et al., 2014).

2. PRINCIPLES OF EPR OXIMETRY

EPR oximetry refers to the measurement of tissue pO_2 by EPR spectroscopy (Ahmad & Kuppusamy, 2010). The basis of EPR oximetry is the interaction between molecular oxygen (O_2), which is paramagnetic, and paramagnetic probes that are introduced into the tissue of interest. The physical interaction of O_2 with the probe leads to spin exchange, which results in broadening of the spectral features (line-width) primarily through shortening of the relaxation times of the paramagnetic probe. When the broadening (i.e., change in line-width) of the EPR spectrum of the probe is calibrated against a range of gases of known pO_2, the resulting calibration plot can be used to quantify pO_2 in the tissue surrounding the probe (Fig. 1B).

Two types of probes are used for oximetry: (i) soluble probes (such as nitroxides and trityls) that report the concentration of dissolved oxygen and (ii) particulate probes (such as lithium phthalocyanine crystals or their analogues) that measure pO_2 (Fig. 1A). The soluble probes are used for imaging, but have limited oxygen sensitivity and lack the potential to repeat the measurements over days, as they are metabolized *in vivo* and also excreted from the system. Nevertheless, they have been used successfully for oxygen imaging in preclinical models (Epel, Redler, & Halpern, 2014; Krishna et al., 2012;

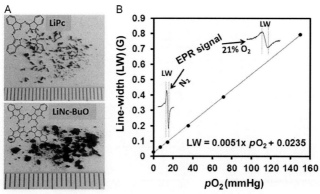

Figure 1 Particulate probes and calibration. (A) Lithium phthalocyanine (LiPc) and lithium octa-n-butoxy-naphthalocyanine (LiNc-BuO) crystals; (B) typical increase of line-width (LW) of the EPR spectrum of LiPc crystals with increase in pO_2 under different perfused gas mixtures (0% O_2 to 21% O_2 balance N_2). R (BuO): [O(CH$_2$)$_3$CH$_3$].

Subramanian, Matsumoto, Mitchell, & Krishna, 2004; Vikram et al., 2007). We have primarily focused on the development of particulate probes in which the unpaired electrons are distributed in stable crystalline solids to provide a suitable configuration for measuring pO_2 (Ilangovan et al., 2002, 2004; Liu et al., 1993; Meenakshisundaram, Eteshola, Pandian, Bratasz, Selvendiran, et al., 2009; Meenakshisundaram, Pandian, Eteshola, Lee, & Kuppusamy, 2010; Pandian, Parinandi, Ilangovan, Zweier, & Kuppusamy, 2003). The region that is measured directly is that immediately surrounding the oximetry probe. If the probe is a macroscopic particle, then it reflects the pO_2 in the tissue that is in contact with the surface of the particle. If the probe is a slurry of small particles (such as India ink), then it reports the average pO_2 of the sum of the surfaces that are in contact with the individual components of the slurry.

The initial placement of the oximetry probe requires an insertion via a 25–23 gauge needle in a tissue of interest that is directly accessible from the surface. For inaccessible tissue, such as brain and heart, a surgical procedure is required for the placement of the oximetry probe using needle or catheters. All subsequent procedures for pO_2 measurement are entirely noninvasive and can be repeated as often as desired. In order to measure tissue pO_2, a detector loop (also referred to as a surface coil resonator) is placed on the surface of the skin above the tissue injected with the oximetry probe (Fig. 2). By using an electromagnetic field (excitation frequency near 1200 MHz for L-band EPR spectrometers), the scanning of magnetic field [~400 Gauss (G)] produces a single-line EPR signal as shown in Fig. 1B.

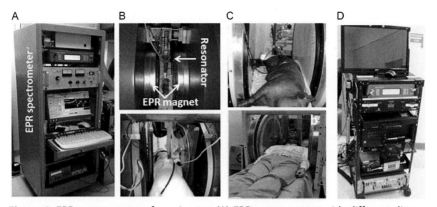

Figure 2 EPR spectrometers for oximetry. (A) EPR spectrometer with different dimensions of magnets to enable tissue pO_2 measurement in (B) rodents and rabbits, and (C) swine and human subjects. (D) A portable version of the EPR spectrometer designed for dosimetry, which can be adapted for oximetry at bedside in patients.

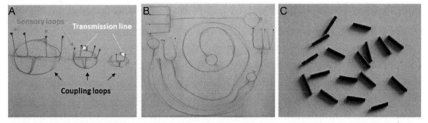

Figure 3 Encapsulated probes for oximetry. Different designs of implantable resonators with 1–4 sensory tips for tissue pO_2 measurement in (A) the brain and (B) deep-sited tissue (>10 mm depth from surface) in animal models. (C) OxyChips prepared by encapsulating LiNc-BuO particulate probe in biocompatible polymer (5 mm length, 0.6 mm diameter) for clinical applications.

Over the years, EPR spectrometers have been refined to facilitate pO_2 measurements, in preclinical animal models to human subjects, including EPR spectrometers that can be potentially wheeled to the bedside for oximetry in patients (Fig. 2). EPR oximetry using particulates has been successfully used for pO_2 measurements up to 10 mm from the surface (restricted by the penetration depth of the microwave energy in tissues). For depths greater than 10 mm, such as in large animals, tissue pO_2 measurements can be made using a novel approach of implantable resonators (Fig. 3A and B), wherein the deep-sited oximetry probe is connected to the coupling loop through a wire.

Under appropriate conditions, EPR oximetry using particulates can measure pO_2 levels in tissues over a wide range from the extremes of very low (<0.1%) to very high (up to 100%) oxygen levels. In order to provide additional spatial resolution of pO_2 within a tissue, procedures for simultaneous spatially resolved measurements from multiple implants, as well as implantable resonators with multiple sensing tips that are spatially distributed within the tissue, have been developed (Grinberg et al., 2005; Hou et al., 2012, 2015; Hou, Khan, et al., 2014; Hou, Li, et al., 2011, 2014; Li et al., 2010; Swartz, Hou, et al., 2014; Williams, Hou, Grinberg, Demidenko, & Swartz, 2007). The multisite measurement technique uses magnetic field gradients to spectrally resolve their respective EPR signals. The conventional multisite approach provides an accurate measurement of pO_2 from the oximetry probes implanted in a tissue of interest at a distance of approximately twice the size of each implant (Smirnov, Norby, Clarkson, Walczak, & Swartz, 1993). In the case of closely placed implants (~1 mm), accurate measurements of pO_2 can be acquired by using high-spatial resolution multisite oximetry capability (Grinberg et al., 2005; Williams et al., 2007). This approach utilizes two consecutively acquired spectra with different magnetic field gradients and an analytic relationship between these spectra is then used to determine the intrinsic line-widths of multiple closely spaced implants, or edges of single implants that are distributed spatially along one dimension. In circumstances where weak EPR signals are observed, likely due to greater depths of the implants, the strategy of overmodulation using modulation amplitudes considerably larger than the intrinsic line-width of the probes can be used to increase signal-to-noise ratio (SNR) of spectra (Williams et al., 2007). The increase in SNR of the EPR spectra by overmodulation also facilitates oximetry without unacceptable decrease in the precision of the pO_2 measurements.

3. PARAMAGNETIC OXYGEN-SENSITIVE PROBES FOR pO_2 MEASUREMENTS BY EPR

3.1 India Ink

India ink is a suspension of carbon black particles in a medium (such as ethylene glycol) and is commonly used in pens for writing, drawing, or epidermal tattooing. India ink also has a long history of clinical use as an anatomic marker for surgery and radiotherapy. Fortuitously, carbon black particles in some of the India ink formulations contain stable radical species at sufficient concentrations with EPR signals that are sensitive to the presence of oxygen

(Goda et al., 1995; Jiang et al., 1996; Jordan, Baudelet, & Gallez, 1998; Khan et al., 2005; Nakashima, Goda, Jiang, Shima, & Swartz, 1995; O'Hara, Goda, Demidenko, & Swartz, 1998; Swartz, Hou, et al., 2014; Swartz, Liu, Goda, & Walczak, 1994; Swartz, Williams, et al., 2014; Williams et al., 2010) (Fig. 4). The preparation and biocompatibility testing of oxygen-sensitive inks for EPR oximetry have been described previously (Charlier, Beghein, & Gallez, 2004; Gallez et al., 1998; Gallez & Mader, 2000; Jordan et al., 1998). Given the fact that some of the India inks have already been used as a tissue marker in patients for several decades, they can also be used for oximetry in patients provided they have oxygen-sensitive carbon particles. In preclinical studies, we have extensively used oxygen-sensitive India ink as a probe to measure the effects of split-dose radiation on the oxygenation of RIF-1 and MTG-B tumors in mice (Jiang et al., 1996; Nakashima et al., 1995; O'Hara et al., 1998). The suitability of the clinical EPR system for pO_2 measurements in superficial tumors (<10 mm depth) using India ink has been demonstrated for locations ranging from the feet, to the anterior and posterior surfaces of the torso, and scalp (Khan, Williams, Hou, Li, & Swartz, 2007; Khan et al., 2006; Swartz, Hou, et al., 2014; Swartz, Williams, et al., 2014; Williams et al., 2010). Despite the advantages for clinical applications, India ink has some limitations, including a tendency to diffuse and moderate EPR sensitivity.

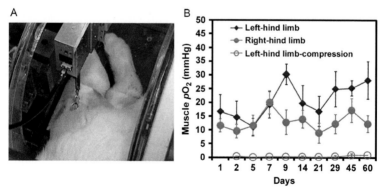

Figure 4 Tissue pO_2 measurement in rabbit skeletal muscle using India ink. (A) Hind limb of the rabbit injected with India ink and surface resonator placement for pO_2 measurement. (B) Temporal changes in tissue pO_2 of the skeletal muscle measured weekly for 8 consecutive weeks. The hind limb was constricted by using a rubber band to investigate tissue pO_2 when the blood flow was restricted in the hind limb. Data represent mean ± SEM, n (number of rabbits) = 4.

3.2 Particulate Oximetry Probes

Over the years, we have developed several particulate and crystalline probes with a range of oxygen sensitivity and applicability to a broader range of oxygen levels. Charcoal, a naturally occurring carbonaceous material, has been extensively characterized for EPR oximetry in animal models (Hou, Li et al., 2014; Meenakshisundaram et al., 2010; Swartz et al., 1994). Charcoal has been approved for clinical use (as a tissue marker) in Europe. Despite their availability for clinical use, the carbon-based sensors have some limitations, such as a modest sensitivity to oxygen and tendency to diffuse in the tissue, especially if used as a suspension. On the other hand, synthetic paramagnetic crystalline materials based on lithium phthalocyanine and derivatives (Fig. 1A) have excellent sensitivity to oxygen (5–10 mG change in line-width per mmHg pO_2), which make them ideally suited for preclinical oximetry (Ilangovan et al., 2002, 2004; Liu et al., 1993; Pandian et al., 2003). These probes are synthesized in our laboratory under strictly controlled conditions to maintain the quality and sensitivity to oxygen, and are available upon request for use at other EPR laboratories.

The line-width of these probes varies linearly with pO_2 and is independent of local metabolic processes, temperature, and pH. Additionally, these probes equilibrate with pO_2 in less than 10 s, and the response of the line-width to changes in pO_2 is stable for several weeks to months and longer. High density of unpaired spins combined with a small intrinsic line-width allows measurements of pO_2 in a tissue of interest using only 40–60 μg of the particulate probe. The length of these implants is usually 0.8–1.0 mm with a diameter of less than 0.5 mm. The implant provides an average pO_2 in the region of approximately 1.0–2.4 mm^2 and therefore samples a region that includes many capillary segments.

There are several advantages in using synthetic particulate probes, especially lithium derivatives: (i) high resolutions in the range of 0.1 mmHg (torr) can be obtained; (ii) suitability for repeated measurements *in vivo* without reintroduction of the probe into the tissue; (iii) noninvasive measurement-one-time introduction of the probe using a small gauge needles (23–25 gauge) is required, however, subsequent measurements are performed under noninvasive conditions; (iv) accuracy-measured values are highly reproducible and correlate closely with measurements by other methods; (v) localized measurements-the measurement is made from a single voxel/region containing the particulate, thus the spatial resolution is the size of the particulate deposit; (vi) insolubility in aqueous solvents; (vii) no effect of biological oxidoreductants, pH, and temperature; (viii) nontoxic – the probes are very inert in biological systems; (ix) temporal response is very good, usually less than 1 s; (x) the response is to pO_2, rather than concentration of oxygen. Oxygen concentration may be quite heterogeneous in cellular/tissue environments because of the solubility of oxygen in lipophilic materials, and hence, calibration can be difficult.

For a bench-to-bedside translation of oximetry using particulates, we have developed polydimethylsiloxane (PDMS)-encapsulated probes (called "OxyChips," Fig. 3C) (Meenakshisundaram, Eteshola, Pandian, Bratasz, Lee, et al., 2009; Meenakshisundaram, Eteshola, Pandian, Bratasz, Selvendiran, et al., 2009; Meenakshisundaram et al., 2010; Pandian et al., 2010). PDMS is a biocompatible and oxygen-permeable silicone polymer, and has been used in a wide range of medical devices and health care applications. Further, PDMS has been approved for use in human subjects and is one of the reference materials provided by the National Heart Lung and Blood Institute for standardized biocompatibility testing (Kim, Moon, Kim, Jeong, & Lee, 2011; Mata, Fleischman, & Roy, 2005). Biocompatible encapsulation effectively shields the probes from interaction with the biological milieu that may result in biochemical degradation and breakdown, as well as limiting the probability of local and/or systemic toxicity effects from interactions with the tissues. These probes can be left in the tissue or removed when no longer needed using biopsy procedures. These OxyChips can be prepared in different lengths and diameters. As with particulates, 2–4 implants of OxyChips can be used to quantify pO_2 at several sites in a tissue of interest. We have obtained an investigational device exemption from the United States Food and Drug Administration (FDA) to test the safety and feasibility of oximetry using OxyChips in patients with head and neck cancer.

3.3 Implantable Resonators for Oximetry Beyond 10 mm from Surface

India ink, particulates, and OxyChips can be used only for pO_2 measurements in tissues at depths of less than 10 mm from the surface, owing to nonresonant absorption of microwave energy by conventional surface coil resonators at L-band frequencies (~1200 MHz). The penetration of microwave energy can be increased up to approximately 7 cm by using lower frequencies, but this decreases the SNR of the EPR signal, thus compromising the accuracy of measurements (Halpern et al., 1994).

To resolve this problem, we have developed a new class of probes called implantable resonators (Fig. 3A and B) for tissue pO_2 measurements at any depth from the surface, and at multiple sites, simultaneously (Hou et al., 2012, 2015; Hou, Li, et al., 2011, 2014; Khan et al., 2015; Li et al., 2010; Rivera et al., 2014). The principle of the implantable resonator is to connect the particulate probe by a wire to a coupling loop that is placed subcutaneously, and then inductively coupled to a surface resonator to acquire EPR spectra. Because the microwaves are transmitted via the wire, there is minimal loss of microwave energy with depth, and therefore the nonresonant loss of microwave in tissues is overcome. The implantable resonators are assembled with very thin enameled copper or alloy wires (thickness: 0.1–0.15 mm) as transmission lines and contain two sets of loops: (i) sensory loop(s) of 0.2–0.3 mm diameter at one end and (ii) a coupling loop of about 10 mm diameter at the other end. The sensory loop is loaded with the particulate probe (40–80 μg) and then coated with a gas-permeable polymer such as Teflon AF2400 or PDMS. Alternatively, the sensory loop(s) are loaded with particulate probes already embedded in the polymer. The entire resonator is then coated with a biocompatible polymer for *in vivo* applications. The length of the transmission line can be anywhere from a few mm to more than 20 cm, depending on the measurement depth that is needed. Implantable resonators with 2–4 sensory loops are used for oximetry at multiple sites, simultaneously, by using magnetic field gradients (Hou et al., 2012, 2015; Hou, Khan, et al., 2014; Hou, Li, et al., 2011, 2014; Khan et al., 2015; Li et al., 2010). The implantable resonators are compatible with MRI and will not change the dose distribution of radiation. The implantable resonators can be placed during initial surgery or biopsy prior to treatment and removed via a simple incision to grasp the coupling loop or in subsequent surgery for other purposes.

Results obtained in animal models indicate that the implantable resonators are safe and enable pO_2 measurements in the brain, skeletal muscle, and heart with high SNR that is approximately 6–10 times greater than for traditional oximetry using the direct implants of particulate probes. Based on our studies in phantoms and rodents, we found the implantable resonators to have several advantages: (i) the implantable resonators are very well tolerated by the animals, retaining normal function; (ii) they enable repeated measurement of pO_2 at depths greater than 10 mm; (iii) they display robust insensitivity to motion allowing pO_2 measurement in sites subject to motion, as the coupling loop can be placed away from the organ such as heart; (iv) they have biocompatibility that should be clinically acceptable, because it is achieved by encapsulating the implantable resonator in an oxygen-permeable, clinically utilized, biocompatible polymer; and (v) tissue pO_2 measurements can be made at 2–4 (or more) sites simultaneously, with sufficient spatial resolution in tissues. We are currently seeking Investigational Device Exemption from FDA for the use of implantable resonators in patients with deep-sited tumors, to investigate temporal changes in pO_2 following chemo- and radiotherapy.

The oximetry probes (particulates and implantable resonators) are typically implanted 24–48 h prior to pO_2 measurement in order to minimize the effects of trauma from the implantation. Once the probe is introduced, repeated measurements of tissue pO_2 can be made at the same location, a particularly useful capability for monitoring tissue oxygenation during the development of disease or in response to a therapy. Implantable resonators with multiple sensor loops can be used for multisite oximetry using appropriate magnetic field gradients.

4. APPLICATION OF EPR OXIMETRY IN ISCHEMIC PATHOLOGIES

4.1 Ischemic Stroke

Ischemic stroke usually occurs due to an obstruction in the blood flow to a region of the brain. Subsequently, the primary pathological event following stroke is a temporal decline in the oxygen levels and nutrients to specific regions of the brain. The outcome of stroke depends on the size of the severely ischemic and irreversibly damaged central infarct core and the potential to salvage the cells in the penumbra, which is hypoperfused, and therefore at risk of infarction, but still viable. Such viable penumbral tissue can be rescued by quick interventions that can increase oxygen levels and/or slowing metabolism in the ischemic area and minimize oxidative injury on reperfusion. One of the rate-limiting steps in developing strategies to counteract low levels of oxygen following stroke has been the lack of a technique that can quantify tissue pO_2 in well-defined regions of the brain in a minimally invasive manner, and measurements that can be repeated as often as desired.

In this context, EPR oximetry is an extremely valuable technique in monitoring temporal decline in pO_2 following stroke and to test strategies to restore oxygen levels in the ischemic regions of the brain. We have systematically investigated temporal changes in the pO_2 following ischemic stroke using the middle cerebral arterial occlusion (MCAO) model in rats and the embolic clot model in rabbits (Hou, Grinberg, Grinberg,

Demidenko, & Swartz, 2005; Hou, Li, et al., 2011, 2014; Khan et al., 2015). The guidelines for making oximetry in the brain of rats are described below. These procedures can also be adapted for use in large animal models. However, for studies involving rabbits or larger species, implantable resonators are the probe of choice, due to depth limitations with particulates and OxyChips.

4.2 Procedures for the Implantation of Particulate Probes in the Brain of Rats for Oximetry Following MCAO

The following describes the procedures commonly used in our laboratory. We use 2–2.5% isoflurane with 30% oxygen through a nose cone to anesthetize the animals and then the head is immobilized on a stereotaxic apparatus (ASI Instruments, MI) (Fig. 5A). The surgical area (head) should be

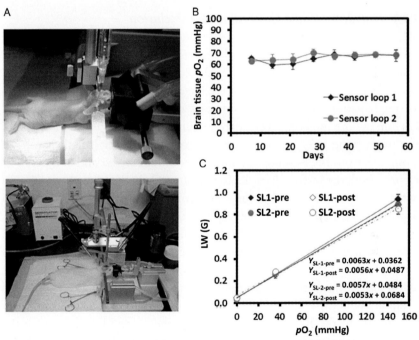

Figure 5 Tissue pO_2 measurement in mouse brain using implantable resonator with LiPc particulate probe. (A) An anesthetized mouse with head positioned in the stereotactic frame for the implantation of oximetry probe. A 25–23 gauge needle is used to create burr holes in the skull for lowering the implantable resonators in the brain. (B) Temporal monitoring of normal brain pO_2 in each hemisphere using implantable resonators for 8 consecutive weeks. (C) Calibration plot of implantable resonators pre- and postimplantation in the mouse brain. Data represent mean±SEM, $n=8$. (See the color plate.)

aseptically prepared as per Institutional Animal Care and Use Committee (IACUC) guidelines. The skin is then incised at the midline and burr holes drilled using 23-gauge needles. The coordinates of burr holes for the implantation of oximetry probes (particulates, OxyChips, or implantable resonators) should be predetermined based on the expected location of ischemic regions (infarct core and penumbra) following stroke. Additionally, the number of burr holes depends on the number of sites desired for monitoring oxygen levels in the brain. The probes are then inserted at the desired sites through the burr holes. The animals should be allowed to recover for at least 24–48 h prior to pO_2 measurements, which can be repeated as often as desired (Fig. 5B). The response of oximetry probes to change in pO_2 is linear and does not change after implantation in the brain for atleast 8 weeks (Fig. 5C).

4.2.1 Transient Middle Cerebral Artery Occlusion

MCAO is the most commonly used model to investigate ischemic stroke and has been described in several reports (Bederson, Pitts, Tsuji, et al., 1986; Hou et al., 2007; Hou, Li, et al., 2014; Liu & McCullough, 2011). For the stroke experiment in rats, we have used the following coordinates for implantation of probes in the infarct core: 3.5 mm left from midline, 2.0 mm behind the bregma at a depth 4.5 mm from the surface of the skull. For monitoring pO_2 in the penumbra, we typically implant the probe at 2 mm left from the midline, 2.0 mm behind the bregma and at a depth of 1.5 mm from the surface. In order to monitor pO_2 in the contralateral region of the brain, another probe can be implanted 1.0 mm behind bregma and at a depth of 1.5 mm in the right hemisphere of the brain. To induce MCAO, the left common carotid artery (CCA), internal carotid artery (ICA), and external carotid artery (ECA) are exposed through a midline incision of the neck in anesthetized animals. A silicone rubber-coated monofilament nylon suture (tip diameter 0.26 mm) is then inserted into the ECA through a stump, while the CCA is kept open and intact. The suture is slowly advanced into the ICA 18–20 mm beyond the carotid bifurcation for the occlusion of the MCA, to induce ischemia. After the initiation of ischemia, the neck incision can be closed and the rats can be kept anesthetized while transient ischemia is induced for 90–120 min. For oximetry, the anesthetized animal is gently moved to the EPR spectrometer for real-time monitoring of pO_2 during transient ischemia (Fig. 6A). To induce reperfusion, the incision is reopened and the intraluminal suture is carefully withdrawn. The CCA and ICA should be inspected to ensure the return of pulsation

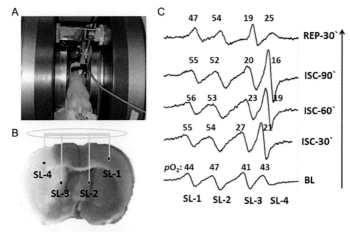

Figure 6 Tissue pO_2 measurement in rat brain using implantable resonator with LiPc. (A) A rat positioned in the EPR magnet for the acquisition of pO_2 data. The surface resonator is positioned on the head for pO_2 measurements in the rat. (B) TTC-stained section of rat brain showing the location of the sensor loops in the ischemic core (SL-4), penumbra (SL-3) and contralateral striatum (SL-2), and cortex (SL-1). (C) Typical EPR spectra acquired from an implantable resonator with four sensor loop (SL) placed in the brain of a rat for pO_2 measurements at two sites in each hemisphere. Tissue pO_2 (mmHg) acquired from each sensor loop (SL) using multisite EPR oximetry is shown. (See the color plate.)

indicative of reperfusion. The incision is then closed with a silk suture and the animals are allowed to recover from anesthesia. Depending on the experimental protocol, pO_2 measurements can be repeated following reperfusion to study the dynamics of tissue pO_2 in the infarct core, penumbra, and contralateral regions of the brain (Fig. 6C).

4.2.2 EPR Acquisition Parameters

The parameters of the EPR spectrometer including magnetic field gradients should be optimized for cerebral pO_2 measurements in these experiments. Typical spectrometer settings that we have used for oximetry are incident microwave power of 10 mW or less, scan range of 1–2.5 G (depends on number of probes), and modulation amplitude of less than one-third of the EPR line-width. The typical magnetic field center and modulation frequency of L-band EPR spectrometers are 410 G and 27 kHz respectively. Three to five scans each of 10 s scan time can be averaged to acquire a temporal profile of pO_2 at every minute during the experiments. In order to obtain well-resolved spectra from three to four sites simultaneously (multisite oximetry), a field gradient (1–3 G cm^{-1}) should be applied. However, precaution must be exercised to avoid any undesirable broadening of the line-width due to high-incident microwave power, high modulation amplitude, or high magnetic field gradients. The direction of the gradient should be chosen to maximize the separation of the spectra from multiple implants. If the size of the implanted probe is sufficiently small compared with the distance between the implants, then the magnitude of magnetic field gradient should be minimal to resolve individual EPR spectra without undesirable increase of the line-width. It has been estimated that for implanted probes of rectangular shapes and 0.2 mm length, optimally resolved spectra with minimal (<1%) distortion of Lorentzian line shapes are obtained if the distance between the implants exceeds 1.8 mm (Smirnov et al., 1993).

The animals should be maintained under physiological conditions to maintain blood pressure, heart rate, and blood gas during the experiments. Additionally, the body temperature of the animals should be maintained at 37 ± 0.5 °C, by using a heated water pad and/or warmth during surgery and EPR measurements. It is important to ascertain that the probes are located in the regions of interest, i.e., infarct core and penumbra, to facilitate appropriate analysis of the data acquired. The coordinates of the probes can be verified either by using 2,3,5-triphenyltetrazolium chloride (TTC) staining of the brain posteuthanasia, or noninvasively by MRI (Bederson, Pitts, Germano, et al., 1986; Duong, 2013; Milidonis, Marshall, Macleod, & Sena, 2015; Wey, Desai, & Duong, 2013; Yang, Shuaib, & Li, 1998). The measurement of tissue pO_2 in the nonischemic contralateral brain can be used as internal control and can also provide valuable information on potential adaptive response to ischemic stroke.

4.3 Procedures for Oximetry in the Heart

Given the anatomical location of the heart, thoracotomy must be performed in anesthetized animals for the implantation of particulate probes into the left ventricular myocardial wall. A 25-gauge needle plunger is suggested for loading the particulate probes for injection into the heart. The implantation of the probes also can be done during other surgeries, for example, during the implantation of a pneumatic occluder to induce ischemia–reperfusion. The animals should be allowed to recover for at least 2–3 days prior to any experiments to measure pO_2. The particulate oximetry probes (particulates, OxyChips) will provide sufficient SNR for experiments in small rodents. The implantable resonators should be used in larger animal models, e.g., rabbits or swine, for oximetry in the heart. For the placement of implantable resonators, approximately 0.5–1.0 mm long channels should be created in the myocardial tissue by gently inserting the bevel of a sterile 23-gauge needle at 45° angle to the surface of the heart. The sensor loop can be inserted through the channel in the myocardium of the animals. In this procedure, extreme care is advised to avoid perforation of the myocardial wall. The transmission line should be gently tied to the surface of the heart by using a suture, in order to prevent the sensory tip from becoming dislodged due to cardiac motion. The coupling loop should be secured by sutures in the subcutaneous tissue, and preferably away from the heart, to minimize noise in data collection induced by beating heart. Additionally, implantable resonators with multiple sensory loops can be used for myocardial pO_2 measurement at several sites simultaneously.

For pO_2 measurements, the anesthetized animal is placed between the magnet poles of the EPR spectrometer, and the surface resonator of the EPR spectrometer is positioned on the skin above the myocardial tissue implanted with the oximetry probe or coupling loop of the implantable resonator (Fig. 7A). The body temperature of the animals is maintained at 37 °C using a warm water pad and warm air blower. The spectrometer

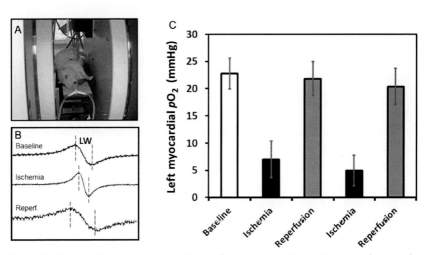

Figure 7 Tissue pO_2 measurement in rat heart using LiNc-BuO particulate probe. (A) A rat positioned between the poles of an EPR magnet for pO_2 measurement in the heart. (B) EPR spectra acquired from the particulate probe implanted in the left myocardial wall during baseline (prior to ischemia), ischemia and reperfusion. (C) Typical change in the left myocardial pO_2 at the baseline, ischemia induced by transient occlusion of coronary artery, and then released for reperfusion. The baseline pO_2 was acquired for 10 min, 5 min of ischemia, and 10 min of reperfusion. Data represent mean ± SD.

parameters are optimized and the EPR spectra can be averaged to obtain better SNR for quantification of myocardial pO_2 (Fig. 7B). Several surgical procedures to induce ischemia in the myocardium by ligating the coronary artery transiently or permanently have been reported (Frank et al., 2012; Khan, Kwiatkowski, Rivera, & Kuppusamy, 2010; Patten & Hall-Porter, 2009; Rivera et al., 2014; Rodriguez, Trayanova, & Noble, 2006; Verdouw, van den Doel, de Zeeuw, & Duncker, 1998; Zornoff, Paiva, Minicucci, & Spadaro, 2009). We have implemented a transient-ischemia model by placing a fine ligature around the coronary artery, which can be tightened remotely to temporarily occlude the blood flow. The ischemia–reperfusion can be repeated as desired to investigate temporal changes in myocardial pO_2 in animal models (Fig. 7C).

5. SUMMARY

In vivo EPR oximetry using oxygen-sensitive and biocompatible probes provides a direct and repeated quantification of tissue pO_2 in experimental animal models. Tissue pO_2 can be measured in a few seconds and can be repeated for several hours, days, weeks, or longer. Temporal information about cerebral and myocardial pO_2 has improved our understanding of ischemic insult and has facilitated the development of potential strategies for improving treatment outcome. The implantable resonators have extended oximetry capability, from rodents to large animal models, such as rabbits and swine, which are considered clinically pertinent for translational research. The application of EPR oximetry in the brain can also be used to investigate the consequence of other pathologies, such as primary or metastatic brain tumors, traumatic brain injury, and cold injury, on the oxygen levels in animal models. In addition to the application of EPR for oximetry using oxygen-sensitive probes, a variety of other probes are also available to investigate redox status, pH, thiols, and free radicals *in vitro* and *in vivo*. The procedures described herein should be used as a guideline for oximetry in the brain and heart of animal models. Similar procedures can be adapted to investigate various pathologies in other organs, such as liver, kidney, lung, and skeletal muscle. The readers are encouraged to contact the authors for specific needs and potential collaborative projects using the EPR facility at our Center.

ACKNOWLEDGMENTS

The work was supported by National Institutes of Health Grants R01 EB004031 and R21 NS082585.

REFERENCES

Ahmad, R., & Kuppusamy, P. (2010). Theory, instrumentation, and applications of electron paramagnetic resonance oximetry. *Chemical Reviews*, *110*(5), 3212–3236.

Baete, S. H., Vandecasteele, J., & De Deene, Y. (2011). ^{19}F MRI oximetry: Simulation of perfluorocarbon distribution impact. *Physics in Medicine and Biology*, *56*(8), 2535–2557.

Bederson, J. B., Pitts, L. H., Germano, S. M., Nishimura, M. C., Davis, R. L., & Bartkowski, H. M. (1986). Evaluation of 2,3,5-triphenyltetrazolium chloride as a stain for detection and quantification of experimental cerebral infarction in rats. *Stroke*, *17*(6), 1304–1308.

Bederson, J. B., Pitts, L. H., Tsuji, M., Nishimura, M. C., Davis, R. L., & Bartkowski, H. (1986). Rat middle cerebral artery occlusion: Evaluation of the model and development of a neurologic examination. *Stroke*, *17*(3), 472–476.

Blockley, N. P., Griffeth, V. E., Simon, A. B., & Buxton, R. B. (2013). A review of calibrated blood oxygenation level-dependent (BOLD) methods for the measurement of task-induced changes in brain oxygen metabolism. *NMR in Biomedicine*, *26*(8), 987–1003.

Brahimi-Horn, M. C., & Pouyssegur, J. (2007). Oxygen, a source of life and stress. *FEBS Letters*, *581*(19), 3582–3591.

Charlier, N., Beghein, N., & Gallez, B. (2004). Development and evaluation of biocompatible inks for the local measurement of oxygen using in vivo EPR. *NMR in Biomedicine*, *17*(5), 303–310.

Christen, T., Bolar, D. S., & Zaharchuk, G. (2013). Imaging brain oxygenation with MRI using blood oxygenation approaches: Methods, validation, and clinical applications. *American Journal of Neuroradiology*, *34*(6), 1113–1123.

Doll, C. M., Milosevic, M., Pintilie, M., Hill, R. P., & Fyles, A. W. (2003). Estimating hypoxic status in human tumors: A simulation using Eppendorf oxygen probe data in cervical cancer patients. *International Journal of Radiation Oncology Biology Physics*, *55*(5), 1239–1246.

Dunn, J. F., & Swartz, H. M. (2003). In vivo electron paramagnetic resonance oximetry with particulate materials. *Methods*, *30*(2), 159–166.

Dunn, J. F., Zaim-Wadghiri, Y., & Kida, I. (1997). BOLD MRI vs NIR spectroscopy: Will the best technique come forward? In *Paper presented at the ISOTT XXV, Milwaukee, WI*.

Duong, T. Q. (2013). Magnetic resonance imaging of perfusion-diffusion mismatch in rodent and non-human primate stroke models. *Neurological Research*, *35*(5), 465–469.

Epel, B., Redler, G., & Halpern, H. J. (2014). How in vivo EPR measures and images oxygen (Research Support, N.I.H., Extramural Review) *Advances in Experimental Medicine and Biology*, *812*, 113–119.

Fisher, E. M., Khan, M., Salisbury, R., & Kuppusamy, P. (2013). Noninvasive monitoring of small intestinal oxygen in a rat model of chronic mesenteric ischemia. *Cell Biochemistry and Biophysics*, *67*(2), 451–459.

Frank, A., Bonney, M., Bonney, S., Weitzel, L., Koeppen, M., & Eckle, T. (2012). Myocardial ischemia reperfusion injury: From basic science to clinical bedside. *Seminars in Cardiothoracic and Vascular Anesthesia*, *16*(3), 123–132.

Gaertner, F. C., Souvatzoglou, M., Brix, G., & Beer, A. J. (2012). Imaging of hypoxia using PET and MRI. *Current Pharmaceutical Biotechnology*, *13*(4), 552–570.

Gallez, B., Baudelet, C., & Jordan, B. F. (2004). Assessment of tumor oxygenation by electron paramagnetic resonance: Principles and applications. *NMR in Biomedicine*, *17*(5), 240–262.

Gallez, B., Debuyst, R., Dejehet, F., Liu, K. J., Walczak, T., Goda, F., et al. (1998). Small particles of fusinite and carbohydrate chars coated with aqueous soluble polymers: Preparation and applications for in vivo EPR oximetry. *Magnetic Resonance in Medicine*, *40*(1), 152–159.

Gallez, B., & Mader, K. (2000). Accurate and sensitive measurements of pO$_2$ in vivo using low frequency EPR spectroscopy: How to confer biocompatibility to the oxygen sensors. *Free Radical Biology & Medicine, 29*(11), 1078–1084.

Goda, F., Liu, K. J., Walczak, T., O'Hara, J. A., Jiang, J., & Swartz, H. M. (1995). In vivo oximetry using EPR and India ink. *Magnetic Resonance in Medicine, 33*(2), 237–245.

Griffiths, J. R., & Robinson, S. P. (1999). The OxyLite: A fibre-optic oxygen sensor. *British Journal of Radiology, 72*(859), 627–630.

Grinberg, V. O., Smirnov, A. I., Grinberg, O. Y., Grinberg, S. A., O'Hara, J. A., & Swartz, H. M. (2005). Practical experimental conditions and limitations for high-spatial-resolution multisite EPR oximetry. *Applied Magnetic Resonance, 28*, 69–78.

Grinberg, O. Y., Smirnov, A. I., & Swartz, H. M. (2001). High spatial resolution multi-site EPR oximetry: The use of a convolution-based fitting method. *Journal of Magnetic Resonance, 152*, 247–258.

Halpern, H. J., Yu, C., Peric, M., Barth, E., Grdina, D. J., & Teicher, B. A. (1994). Oxymetry deep in tissues with low-frequency electron paramagnetic resonance. *Proceedings of the National Academy of Sciences of the United States of America, 91*(26), 13047–13051.

Hou, H., Dong, R., Lariviere, J. P., Mupparaju, S. P., Swartz, H. M., & Khan, N. (2011). Synergistic combination of hyperoxygenation and radiotherapy by repeated assessments of tumor pO$_2$ with EPR oximetry. *Journal of Radiation Research, 52*(5), 568–574.

Hou, H., Dong, R., Li, H., Williams, B., Lariviere, J. P., Hekmatyar, S. K., et al. (2012). Dynamic changes in oxygenation of intracranial tumor and contralateral brain during tumor growth and carbogen breathing: A multisite EPR oximetry with implantable resonators. *Journal of Magnetic Resonance, 214*(1), 22–28.

Hou, H., Grinberg, O. Y., Grinberg, S. A., Demidenko, E., & Swartz, H. M. (2005). Cerebral tissue oxygenation in reversible focal ischemia in rats: Multi-site EPR oximetry measurements. *Physiological Measurement, 26*(1), 131–141.

Hou, H., Grinberg, O., Williams, B., Grinberg, S., Yu, H., Alvarenga, D. L., et al. (2007). The effect of oxygen therapy on brain damage and cerebral pO$_2$ in transient focal cerebral ischemia in the rat. *Physiological Measurement, 28*(8), 963–976.

Hou, H., Khan, N., Lariviere, J., Hodge, S., Chen, E. Y., Jarvis, L. A., et al. (2014). Skeletal muscle and glioma oxygenation by carbogen inhalation in rats: A longitudinal study by EPR oximetry using single-probe implantable oxygen sensors. *Advances in Experimental Medicine and Biology, 812*, 97–103.

Hou, H., Krishnamurthy Nemani, V., Du, G., Montano, R., Song, R., Gimi, B., et al. (2015). Monitoring oxygen levels in orthotopic human glioma xenograft following carbogen inhalation and chemotherapy by implantable resonator-based oximetry. *International Journal of Cancer, 136*(7), 1688–1696.

Hou, H., Li, H., Dong, R., Khan, N., & Swartz, H. (2014). Real-time monitoring of ischemic and contralateral brain pO$_2$ during stroke by variable length multisite resonators. *Magnetic Resonance Imaging, 32*(5), 563–569.

Hou, H., Li, H., Dong, R., Mupparaju, S., Khan, N., & Swartz, H. (2011). Cerebral oxygenation of the cortex and striatum following normobaric hyperoxia and mild hypoxia in rats by EPR oximetry using multi-probe implantable resonators. *Advances in Experimental Medicine and Biology, 701*, 61–67.

Ibragimova, M. I., Petukhov, V. Y., Zheglov, E. P., Khan, N., Hou, H., Swartz, H. M., et al. (2008). Quinoid radio-toxin (QRT) induced metabolic changes in mice: An ex vivo and in vivo EPR investigation. *Nitric Oxide, 18*(3), 216–222.

Ilangovan, G., Manivannan, A., Li, H., Yanagi, H., Zweier, J. L., & Kuppusamy, P. (2002). A naphthalocyanine-based EPR probe for localized measurements of tissue oxygenation. *Free Radical Biology & Medicine, 32*(2), 139–147.

Ilangovan, G., Zweier, J. L., & Kuppusamy, P. (2004). Mechanism of oxygen-induced EPR line broadening in lithium phthalocyanine microcrystals. *Journal of Magnetic Resonance*, *170*(1), 42–48.

James, P. E., Bacic, G., Grinberg, O. Y., Goda, F., Dunn, J. F., Jackson, S. K., et al. (1996). Endotoxin-induced changes in intrarenal pO_2, measured by *in vivo* electron paramagnetic resonance oximetry and magnetic resonance imaging. *Free Radical Biology & Medicine*, *21*(1), 25–34.

James, P. E., Madhani, M., Roebuck, W., Jackson, S. K., & Swartz, H. M. (2002). Endotoxin-induced liver hypoxia: Defective oxygen delivery versus oxygen consumption. *Nitric Oxide*, *6*(1), 18–28.

Jiang, J., Nakashima, T., Liu, K. J., Goda, F., Shima, T., & Swartz, H. M. (1996). Measurement of PO_2 in liver using EPR oximetry. *Journal of Applied Physiology*, *80*(2), 552–558.

Jordan, B. F., Baudelet, C., & Gallez, B. (1998). Carbon-centered radicals as oxygen sensors for in vivo electron paramagnetic resonance: Screening for an optimal probe among commercially available charcoals. *Magma*, *7*(2), 121–129.

Kallinowski, F., Zander, R., Hoeckel, M., & Vaupel, P. (1990). Tumor tissue oxygenation as evaluated by computerized-pO_2-histography. *International Journal of Radiation Oncology Biology Physics*, *19*(4), 953–961.

Keddie, S., & Rohman, L. (2012). Reviewing the reliability, effectiveness and applications of Licox in traumatic brain injury. *Nursing in Critical Care*, *17*(4), 204–212.

Khan, N., Hou, H., Eskey, C. J., Moodie, K., Gohain, S., Du, G., et al. (2015). Deep-tissue oxygen monitoring in the brain of rabbits for stroke research. *Stroke*, *46*(3), e62–e66.

Khan, N., Hou, H., Hein, P., Comi, R. J., Buckey, J. C., Grinberg, O., et al. (2005). Black magic and EPR oximetry: From lab to initial clinical trials. *Advances in Experimental Medicine and Biology*, *566*, 119–125.

Khan, M., Kwiatkowski, P., Rivera, B. K., & Kuppusamy, P. (2010). Oxygen and oxygenation in stem-cell therapy for myocardial infarction. *Life Sciences*, *87*(9–10), 269–274.

Khan, N., Mupparaju, S., Hekmatyar, S. K., Hou, H., Lariviere, J. P., Demidenko, E., et al. (2010). Effect of hyperoxygenation on tissue pO_2 and its effect on radiotherapeutic efficacy of orthotopic F98 gliomas. *International Journal of Radiation Oncology Biology Physics*, *78*(4), 1193–1200.

Khan, N., Mupparaju, S., Hou, H., Williams, B. B., & Swartz, H. (2012). Repeated assessment of orthotopic glioma pO_2 by multi-site EPR oximetry: A technique with the potential to guide therapeutic optimization by repeated measurements of oxygen. *Journal of Neuroscience Methods*, *204*(1), 111–117.

Khan, N., Mupparaju, S. P., Mintzopoulos, D., Kesarwani, M., Righi, V., Rahme, L. G., et al. (2008). Burn trauma in skeletal muscle results in oxidative stress as assessed by *in vivo* electron paramagnetic resonance. *Molecular Medicine Reports*, *1*(6), 813–819.

Khan, N., Williams, B. B., Hou, H., Li, H., & Swartz, H. M. (2007). Repetitive tissue pO_2 measurements by electron paramagnetic resonance oximetry: Current status and future potential for experimental and clinical studies. *Antioxidants & Redox Signaling*, *9*(8), 1169–1182.

Khan, N. W., Williams, B. B., & Swartz, H. M. (2006). Clinical applications of in vivo EPR: Rationale and initial results. *Applied Magnetic Resonance*, *30*, 185–199.

Kim, S. H., Moon, J., Kim, J. H., Jeong, S., & Lee, S. (2011). Flexible, stretchable and implantable PDMS encapsulated cable for implantable medical device. *Biomedical Engineering Letters*, *1*(3), 199–203.

Krishna, M. C., English, S., Yamada, K., Yoo, J., Murugesan, R., Devasahayam, N., et al. (2002). Overhauser enhanced magnetic resonance imaging for tumor oximetry: Coregistration of tumor anatomy and tissue oxygen concentration. *PNAS*, *99*(4), 2216–2221.

Krishna, M. C., Matsumoto, S., Yasui, H., Saito, K., Devasahayam, N., Subramanian, S., et al. (2012). Electron paramagnetic resonance imaging of tumor pO$_2$. *Radiation Research*, *177*(4), 376–386.

Krishna, M. C., Subramanian, S., Kuppusamy, P., & Mitchell, J. B. (2001). Magnetic resonance imaging for *in vivo* assessment of tissue oxygen concentration. *Seminars in Radiation Oncology*, *11*(1), 58–69.

Krohn, K. A., Link, J. M., & Mason, R. P. (2008). Molecular imaging of hypoxia. *Journal of Nuclear Medicine*, *49*(Suppl. 2), 129S–148S.

Kulkarni, A. C., Kuppusamy, P., & Parinandi, N. (2007). Oxygen, the lead actor in the pathophysiologic drama: Enactment of the trinity of normoxia, hypoxia, and hyperoxia in disease and therapy. *Antioxidants and Redox Signaling*, *9*(10), 1717–1730.

Lally, B. E., Rockwell, S., Fischer, D. B., Collingridge, D. R., Piepmeier, J. M., & Knisely, J. P. (2006). The interactions of polarographic measurements of oxygen tension and histological grade in human glioma. *Cancer Journal*, *12*(6), 461–466.

Li, H., Hou, H., Sucheta, A., Williams, B. B., Lariviere, J. P., Khan, M. N., et al. (2010). Implantable resonators: A technique for repeated measurement of oxygen at multiple deep sites with *in vivo* EPR. *Advances in Experimental Medicine and Biology*, *662*, 265–272.

Liu, K. J., Gast, P., Moussavi, M., Norby, S. W., Vahidi, N., Walczak, T., et al. (1993). Lithium phthalocyanine: A probe for electron paramagnetic resonance oximetry in viable biological systems. *Proceedings of the National Academy of Sciences of the United States of America*, *90*(12), 5438–5442.

Liu, F., & McCullough, L. D. (2011). Middle cerebral artery occlusion model in rodents: Methods and potential pitfalls. *Journal of Biomedicine & Biotechnology*, *2011*, 464701.

Liu, S., Shah, S. J., Wilmes, L. J., Feiner, J., Kodibagkar, V. D., Wendland, M. F., et al. (2011). Quantitative tissue oxygen measurement in multiple organs using ^{19}F MRI in a rat model. *Magnetic Resonance in Medicine*, *66*(6), 1722–1730.

Madhani, M., Barchowsky, A., Klei, L., Ross, C. R., Jackson, S. K., Swartz, H. M., et al. (2002). Antibacterial peptide PR-39 affects local nitric oxide and preserves tissue oxygenation in the liver during septic shock. *Biochimica et Biophysica Acta*, *1588*(3), 232–240.

Mata, A., Fleischman, A. J., & Roy, S. (2005). Characterization of polydimethylsiloxane (PDMS) properties for biomedical micro/nanosystems. *Biomedical Microdevices*, *7*(4), 281–293.

Meenakshisundaram, G., Eteshola, E., Pandian, R. P., Bratasz, A., Lee, S. C., & Kuppusamy, P. (2009). Fabrication and physical evaluation of a polymer encapsulated paramagnetic probe for biomedical oximetry. *Biomedical Microdevices*, *11*(4), 773–782.

Meenakshisundaram, G., Eteshola, E., Pandian, R. P., Bratasz, A., Selvendiran, K., Lee, S. C., et al. (2009). Oxygen sensitivity and biocompatibility of an implantable paramagnetic probe for repeated measurements of tissue oxygenation. *Biomedical Microdevices*, *11*(4), 817–826.

Meenakshisundaram, G., Pandian, R. P., Eteshola, E., Lee, S. C., & Kuppusamy, P. (2010). A paramagnetic implant containing lithium naphthalocyanine microcrystals for high-resolution biological oximetry. *Journal of Magnetic Resonance*, *203*(1), 185–189.

Milidonis, X., Marshall, I., Macleod, M. R., & Sena, E. S. (2015). Magnetic resonance imaging in experimental stroke and comparison with histology: Systematic review and meta-analysis. *Stroke*, *46*(3), 843–851.

Morgalla, M. H., Haas, R., Grozinger, G., Thiel, C., Thiel, K., Schuhmann, M. U., et al. (2012). Experimental comparison of the measurement accuracy of the Licox® and Raumedic® Neurovent—PTO brain tissue oxygen monitors. *Acta Neurochirurgica. Supplement*, *114*, 169–172.

Mozaffarian, D., Benjamin, E. J., Go, A. S., Arnett, D. K., Blaha, M. J., Cushman, M., et al. (2015). Heart disease and stroke statistics-2015 update: A report from the American Heart Association. *Circulation, 131*, e29–e322.

Mupparaju, S., Hou, H., Lariviere, J. P., Swartz, H. M., & Khan, N. (2011). Tumor pO_2 as a surrogate marker to identify therapeutic window during metronomic chemotherapy of 9 L gliomas. *Advances in Experimental Medicine and Biology, 701*, 107–113.

Nakashima, T., Goda, F., Jiang, J., Shima, T., & Swartz, H. M. (1995). Use of EPR oximetry with India ink to measure the pO_2 in the liver *in vivo* in mice. *Magnetic Resonance in Medicine, 34*(6), 888–892.

O'Hara, J. A., Goda, F., Demidenko, E., & Swartz, H. M. (1998). Effect on regrowth delay in a murine tumor of scheduling split-dose irradiation based on direct pO_2 measurements by electron paramagnetic resonance oximetry. *Radiation Research, 150*(5), 549–556.

O'Hara, J. A., Hou, H., Demidenko, E., Springett, R. J., Khan, N., & Swartz, H. M. (2005). Simultaneous measurement of rat brain cortex PtO2 using EPR oximetry and a fluorescence fiber-optic sensor during normoxia and hyperoxia. *Physiological Measurement, 26*(3), 203–213.

O'Hara, J. A., Khan, N., Hou, H., Wilmo, C. M., Demidenko, E., Dunn, J. F., et al. (2004). Comparison of EPR oximetry and Eppendorf polarographic electrode assessments of rat brain PtO_2. *Physiological Measurement, 25*(6), 1413–1423.

Pandian, R. P., Meenakshisundaram, G., Bratasz, A., Eteshola, E., Lee, S. C., & Kuppusamy, P. (2010). An implantable Teflon chip holding lithium naphthalocyanine microcrystals for secure, safe, and repeated measurements of pO_2 in tissues. *Biomedical Microdevices, 12*(3), 381–387.

Pandian, R. P., Parinandi, N. L., Ilangovan, G., Zweier, J. L., & Kuppusamy, P. (2003). Novel particulate spin probe for targeted determination of oxygen in cells and tissues. *Free Radical Biology & Medicine, 35*(9), 1138–1148.

Patten, R. D., & Hall-Porter, M. R. (2009). Small animal models of heart failure: Development of novel therapies, past and present. *Circulation. Heart Failure, 2*(2), 138–144.

Rink, C., & Khanna, S. (2011). Significance of brain tissue oxygenation and the arachidonic acid cascade in stroke. *Antioxidants and Redox Signaling, 14*(10), 1889–1903.

Rivera, B. K., Naidu, S. K., Subramanian, K., Joseph, M., Hou, H., Khan, N., et al. (2014). Real-time, in vivo determination of dynamic changes in lung and heart tissue oxygenation using EPR oximetry. *Advances in Experimental Medicine and Biology, 812*, 81–86.

Rodriguez, B., Trayanova, N., & Noble, D. (2006). Modeling cardiac ischemia. *Annals of the New York Academy of Sciences, 1080*, 395–414.

Salikhov, I., Walczak, T., Lesniewski, P., Khan, N., Iwasaki, A., Comi, R., et al. (2005). EPR spectrometer for clinical applications. *Magnetic Resonance in Medicine, 54*(5), 1317–1320.

Sauer, G., Weber, K. J., Peschke, P., & Eble, M. J. (2000). Measurement of hypoxia using the comet assay correlates with preirradiation microelectrode pO_2 histography in R3327-at rodent tumors. *Radiation Research, 154*, 439–446.

Shi, H., & Liu, K. J. (2007). Cerebral tissue oxygenation and oxidative brain injury during ischemia and reperfusion. *Frontiers in Bioscience, 12*, 1318–1328.

Smirnov, A. I., Norby, S. W., Clarkson, R. B., Walczak, T., & Swartz, H. M. (1993). Simultaneous multi-site EPR spectroscopy *in vivo*. *Magnetic Resonance in Medicine, 30*, 213–220.

Subramanian, S., Matsumoto, K., Mitchell, J. B., & Krishna, M. C. (2004). Radio frequency continuous-wave and time-domain EPR imaging and overhauser-enhanced magnetic resonance imaging of small animals: Instrumental developments and comparison of relative merits for functional imaging. *NMR in Biomedicine, 17*(5), 263–294.

Swartz, H. M., & Clarkson, R. B. (1998). The measurement of oxygen in vivo using EPR techniques. *Physics in Medicine and Biology, 43*(7), 1957–1975.

Swartz, H. M., Dunn, J., Grinberg, O., O'Hara, J., & Walczak, T. (1997). What does EPR oximetry with solid particles measure and how does this relate to other measures of pO_2? *Advances in Experimental Medicine and Biology, 428*, 663–670.

Swartz, H. M., Hou, H., Khan, N., Jarvis, L. A., Chen, E. Y., Williams, B. B., et al. (2014). Advances in probes and methods for clinical EPR oximetry. *Advances in Experimental Medicine and Biology, 812*, 73–79.

Swartz, H. M., Liu, K. J., Goda, F., & Walczak, T. (1994). India ink: A potential clinically applicable EPR oximetry probe. *Magnetic Resonance in Medicine, 31*(2), 229–232.

Swartz, H. M., Williams, B. B., Zaki, B. I., Hartford, A. C., Jarvis, L. A., Chen, E. Y., et al. (2014). Clinical EPR: Unique opportunities and some challenges. *Academic Radiology, 21*(2), 197–206.

Taylor, C. T., & Pouyssegur, J. (2007). Oxygen, hypoxia, and stress. *Annals of the New York Academy of Sciences, 1113*, 87–94.

Tran, L. B., Bol, A., Labar, D., Jordan, B., Magat, J., Mignion, L., et al. (2012). Hypoxia imaging with the nitroimidazole ^{18}F-FAZA PET tracer: A comparison with OxyLite, EPR oximetry and ^{19}F-MRI relaxometry. *Radiotherapy and Oncology, 105*(1), 29–35.

Turer, A. T., & Hill, J. A. (2010). Pathogenesis of myocardial ischemia-reperfusion injury and rationale for therapy. *The American Journal of Cardiology, 106*(3), 360–368.

Vaupel, P., Hockel, M., & Mayer, A. (2007). Detection and characterization of tumor hypoxia using pO_2 histography. *Antioxidants and Redox Signaling, 9*(8), 1221–1235.

Verdouw, P. D., van den Doel, M. A., de Zeeuw, S., & Duncker, D. J. (1998). Animal models in the study of myocardial ischaemia and ischaemic syndromes. *Cardiovascular Research, 39*(1), 121–135.

Vikram, D. S., Bratasz, A., Ahmad, R., & Kuppusamy, P. (2007). A comparative evaluation of EPR and OxyLite oximetry using a random sampling of pO_2 in a murine tumor. *Radiation Research, 168*(3), 308–315.

Vikram, D. S., Zweier, J. L., & Kuppusamy, P. (2007). Methods for noninvasive imaging of tissue hypoxia. *Antioxidants and Redox Signaling, 9*(10), 1745–1756.

Wey, H. Y., Desai, V. R., & Duong, T. Q. (2013). A review of current imaging methods used in stroke research. *Neurological Research, 35*(10), 1092–1102.

Williams, B. B., Hou, H., Grinberg, O. Y., Demidenko, E., & Swartz, H. M. (2007). High spatial resolution multisite EPR oximetry of transient focal cerebral ischemia in the rat. *Antioxidants and Redox Signaling, 9*(10), 1691–1698.

Williams, B. B., Khan, N., Zaki, B., Hartford, A., Ernstoff, M. S., & Swartz, H. M. (2010). Clinical electron paramagnetic resonance (EPR) oximetry using India ink. *Advances in Experimental Medicine and Biology, 662*, 149–156.

Yang, Y., Shuaib, A., & Li, Q. (1998). Quantification of infarct size on focal cerebral ischemia model of rats using a simple and economical method. *Journal of Neuroscience Methods, 84*(1–2), 9–16.

Zornoff, L. A., Paiva, S. A., Minicucci, M. F., & Spadaro, J. (2009). Experimental myocardium infarction in rats: Analysis of the model. *Arquivos Brasileiros de Cardiologia, 93*(4), 434–440, 426–432.

CHAPTER TWENTY

Free Radical Imaging Using *In Vivo* Dynamic Nuclear Polarization-MRI

Hideo Utsumi[1], Fuminori Hyodo
Innovation Center for Medical Redox Navigation, Kyushu University, Fukuoka, Japan
[1]Corresponding author: e-mail address: hideo.utsumi.278@m.kyushu-u.ac.jp

Contents

1. Introduction	554
2. Principle of DNP	556
3. Apparatus	558
3.1 Custom-Built DNP-MRI from Philips	558
3.2 Home-Made DNP-MRI System	560
3.3 Commercially Available DNP-MRI	560
3.4 Spectroscopic DNP-MRI	561
4. Spin-Probe Technique for DNP-MRI	562
4.1 Synthesis of Nitroxyl Probe	562
4.2 Reactivity of Nitroxyl Radical Toward ROS	564
4.3 Tissue Distribution of Nitroxyl Radical Injected Intravenously	564
4.4 Application of *In Vivo* DNP-MRI to Animal Models	564
5. Application of DNP-MRI to the Intrinsic Molecular Imaging	566
6. Conclusion	568
Acknowledgments	568
References	568

Abstract

Redox reactions that generate free radical intermediates are essential to metabolic processes, and their intermediates can produce reactive oxygen species, which may promote diseases related to oxidative stress. The development of an *in vivo* electron spin resonance (ESR) spectrometer and its imaging enables us noninvasive and direct measurement of *in vivo* free radical reactions in living organisms. The dynamic nuclear polarization magnetic resonance imaging (DNP-MRI), also called PEDRI or OMRI, is also a new imaging method for observing free radical species *in vivo*. The spatiotemporal resolution of free radical imaging with DNP-MRI is comparable with that in MRI, and each of the radical species can be distinguished in the spectroscopic images by changing the frequency or magnetic field of ESR irradiation.

    Several kinds of stable nitroxyl radicals were used as spin probes to detect *in vivo* redox reactions. The signal decay of nitroxyl probes, which is determined with *in vivo* DNP-MRI, reflects the redox status under oxidative stress, and the signal decay is suppressed by prior administration of antioxidants. In addition, DNP-MRI can also visualize various intermediate free radicals from the intrinsic redox molecules.

This noninvasive method, in vivo DNP-MRI, could become a useful tool for investigating the mechanism of oxidative injuries in animal disease models and the in vivo effects of antioxidant drugs.

1. INTRODUCTION

The *in vivo* redox reaction plays a central role in living organisms. As shown in Fig. 1, mitochondria maintain redox homeostasis via endogenous molecules, such as flavin mononucleotide (FMN), flavin adenine dinucleotide (FAD), and coenzyme Q_{10} (CoQ_{10}), which are converted to the free radical intermediates FMNH and FADH, and $CoQ_{10}H$, respectively, by the obligate one-electron transfer reactions (Lin & Beal, 2006).

These free radical intermediates produce the superoxide anion radical, $O_2^{-\cdot}$ (Murphy, 2009), which is responsible for various oxidative diseases (Lin & Beal, 2006; Schapira, 2006). In the well-known reactive oxygen species (ROS), hydroxyl radical ($\cdot OH$), has extremely high reactivity and commonly reacts with the surrounding biomolecules at a diffusion-limited rate. Such reactions could lead to the severe redox-related diseases, including lifestyle-related diseases, cancer, and neurodegenerative disorders.

The developments of *in vivo* electron spin resonance (ESR) spectroscopy and imaging have demonstrated the involvement of *in vivo* redox processes

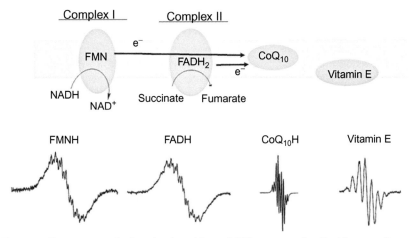

Figure 1 Electron transfer in mitochondria and ESR spectra of radical intermediates.

in oxidative stress-mediated pathologies (Elas et al., 2006; He et al., 2007; Utsumi et al., 2002; Yasukawa, Yamada, Ichikawa, & Utsumi, 2010). The largest difficulty of *in vivo* ESR is low sensitivity for detection of ROS, and/or redox status, generated in oxidative diseases. To resolve these problems, the nitroxyl probe technique (Miura, Utsumi, & Hamada, 1992; Takeshita, Utsumi, & Hamada, 1991; Utsumi, Muto, Masuda, & Hamada, 1990; Utsumi & Yamada, 2003; Yasukawa et al., 2010) is utilized in our laboratory to monitor directly free radical reactions in living animals. This technique provides us noninvasive and real-time information on free radical reactions. Nitroxyl radicals are highly redox sensitive, and those exogenously administered to animals are gradually converted to the reduced form, hydroxylamine, under oxidative conditions. However, the spatiotemporal resolution of ESR imaging is much less than that of magnetic resonance imaging (MRI), because of the short transverse relaxation time, T_2, of free radicals.

In vivo DNP-MRI, also called proton–electron double-resonance imaging (PEDRI) (Lurie, Bussell, Bell, & Mallard, 1988) or Overhauser-enhanced MRI (OMRI) (Krishna et al., 2002), is a new imaging method for observing free radical species *in vivo*. Major advantages of *in vivo* DNP-MRI are its high spatiotemporal resolution of free radical imaging, which rivals that of conventional MRI. About the development of *in vivo* DNP-MRI, Philips gave some comments in the book, "50 Years History of Philips": In 1996, the Nycomed pharmaceutical company approached Philips about research cooperation on DNP-MRI. Realizing the potential, Philips took up the challenge and Philips Research Laboratories (PBL)-Hamburg provided a first prototype scanner in 1997 and a more advanced one in 1998. Unfortunately, the enhancement factors were somewhat disappointing for clinical approaches, so the technique has not caught on yet. PBL-Hamburg has supplied similar systems (for animal experiments) to the National Institutes of Health (NIH) (USA), Remagen (Germany), and Kyushu University (Japan).

In vivo DNP MRI from Philips is operated at low magnetic field (~20 mT), resulting in less sensitivity. Most studies to date have been performed via *in vivo* DNP-MRI, where obtaining sufficient image intensity requires the use of synthetic stable radicals having simple and narrow resonance lines as a spin-probe technique, and these studies have provided unique information regarding pharmacokinetics (PK) and redox processes of cancer and stroke in living animals (Kosem et al., 2012; Krishna et al., 2002; Li et al., 2002; Lurie et al., 1988; Matsumoto et al., 2009; Shiba et al., 2011; Utsumi et al., 2006; Yamato et al., 2011; Yamato, Shiba, Yamada, Watanabe, & Utsumi, 2009).

As, shown in Fig. 1, the ESR spectra of most endogenous free radical intermediates exhibit multiple hyperfine splitting and inhomogeneous broadening. The imaging of these free radicals has not previously been attempted via *in vivo* ESR imaging. The *in vivo* DNP-MRI offered the possibility of imaging free radical intermediates and the potential to add metabolic/biochemical information to anatomic images.

In this chapter, we describe the principle, apparatus, spin-probe technique, and application of *in vivo* DNP-MRI.

2. PRINCIPLE OF DNP

There are four DNP mechanisms, the Overhauser effect, the solid effect, the cross effect, and thermal mixing. The Overhauser effect has been used extensively in studies of liquids. The enhancement factor, ε, of DNP is defined as $\varepsilon = \langle I_z \rangle / I_0$, where $\langle I_z \rangle$ denotes the expectation value of the DNP, and I_0 is its thermal equilibrium value.

Theoretical principles of Overhauser DNP are well documented (Benial, Ichikawa, Murugesan, Yamada, & Utsumi, 2006; Grucker et al., 1995; Lurie et al., 1988; Zotev et al., 2010). Here, the Overhauser effect is, at first approximation, considered to be based as the coupling of an electron spin, S, of free radical having a singlet resonance line with proton spin, I, in water. Figure 2 gives the four energy-level diagram with the transition probabilities, w_n, among the levels.

Taking into account all the cross-transition probabilities, the basic equation of DNP of the ^{1}H nuclei ($I=1/2$) of water molecules with couplings to an unpaired electron spin $S=1/2$ of a dissolved free radical is given by

$$\langle I_z \rangle = I_0 - \rho f (\langle S_z \rangle - S_0) \tag{1}$$

Here, $\langle S_z \rangle$ denotes the electron spin polarization during ESR irradiation, S_0 is its thermal equilibrium value, ρ is the coupling factor, and f is the leakage factor. The two factors, ρ and f are defined by the following expressions:

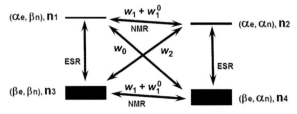

Figure 2 The four energy-level diagram with the transition probabilities (w_n) in DNP.

$$\rho = (w_2 - w_0)/(w_0 + 2w_1 + w_2) \qquad (2)$$

$$f = (w_0 + 2w_1 + w_2)/(w_0 + 2w_1 + w_2 + w_1^0) = 1 - T_1/T_{10} \qquad (3)$$

The electron–nuclear interactions to generate the relaxation processes of S and I are composed of the dipolar interaction and scalar terms. The coupling factor ρ ranges from -1 for pure scalar coupling ($w_0 > w_2$) to 0.5 for pure dipole coupling ($w_2 > w_0$). The leakage factor takes values between 0 and 1 and depends on the ratio of the longitudinal relaxation times for nuclear spins in the presence (T_1) and in the absence of free radicals (T_{10}). ESR irradiation of the free radicals causes the saturation of the ESR resonance. In Eq. (1), ($\langle S_z \rangle - S_0$) is defined with the saturation factor, s, as follows:

$$s = (S_0 - \langle S_z \rangle)/S_0 \qquad (4)$$

The saturation factor changes from 0 to 1 for complete saturation of all ESR transitions, depending on the ESR irradiation power, relaxation properties of electron spins, the ESR spectrum profiles, and other unknown factors, at a given magnetic field strength.

Hence, the enhancement factor, ε, of DNP due to the Overhauser effect is expressed as follows:

$$\varepsilon = \langle I_z \rangle / I_0 = 1 - \rho f s \frac{|\gamma_e|}{\gamma_N} \qquad (5)$$

Here, γ_e and γ_N are, respectively, the electron (28.0 GHz T^{-1}) and nuclear gyromagnetic ratios (42.6 MHz T^{-1} for the proton), resulting in 660× larger enhancement as the calculated maximum.

In *in vivo* DNP-MRI, the ε value for DNP is not as large as calculated, above. Thus, for the convenience of the reader, $E = (I_0 - \langle I_z \rangle)/I_0$ is used, instead of ε in this brief outline of the principles of the Overhauser effect relevant to *in vivo* DNP-MRI. E is defined as follows:

$$E = 1 - \varepsilon = (I_0 - \langle I_z \rangle)/I_0 = \rho f s \frac{|\gamma_e|}{\gamma_N} \qquad (6)$$

The leakage factor f in Eq. (6) depends upon the concentration of the free radical, as given by

$$f = 1 - \frac{T_1}{T_{10}} = \frac{kCT_{10}}{1 + kCT_{10}} \qquad (7)$$

Here, the concentration of the free radical is given by C and k denotes the relaxivity constant. As the concentration of the free radical is increased, the leakage factor is increased and then approaches to unity, because with increasing C, $kC \gg 1/T_{10}$.

The factor more critical to the sensitivity of DNP-MRI is the degree of saturation of the electron spin, which depends on the electron spin relaxation rates of the free radical. For on-resonance irradiation by a rotating magnetic field of amplitude B_1, the saturation factor, s, is given by

$$s = \frac{\gamma_e^2 B_1^2 T_{1e} T_{2e}}{1 + \gamma_e^2 B_1^2 T_{1e} T_{2e}} \quad (8)$$

Here, T_{1e} and T_{2e} are the electron spin–lattice and spin–spin relaxation times, respectively. Thus, the value of s is expected to be dependent on both the resonance peak height of free radical at ESR irradiation and the ESR irradiation power B_1, but not on free radical concentration. If the free radicals have hyperfine splitting, s decreases as the number of resonance lines increase, due to hyperfine splitting, as $(2I+1)$. In *in vivo* DNP-MRI, the synthetic stable free radicals having simple and narrow resonance lines are utilized to obtain sufficient DNP enhancement.

3. APPARATUS

3.1 Custom-Built DNP-MRI from Philips

Nycomed (Sweden) has succeeded in the development of oxygen-concentration sensitive probes, which are now widely used for enhancement of ^{13}C signals in *in vitro* DNP-MRI instruments from General Electric (GE) Corporation. Nycomed asked for research cooperation for development of oxygen imaging with DNP-MRI to PBL-Hamburg, but a prototype DNP-MRI scanner developed by PBL-Hamburg has not proven efficacious for clinical diagnostic imaging, because of low enhancement factors of DNP and the spatiotemporal resolution due to low external magnetic field of MRI. Thus, we obtained a similar system for an animal experiment from PBL-Hamburg.

As shown in Fig. 3, the DNP-MRI instrument from Philips consists of the human whole-body magnet (magnet bore: 79 cm diameter, 125 cm length) and low magnetic field scanner operating in a field-cycled mode to avoid excess RF power deposition during the ESR cycle (Lurie et al., 1989).

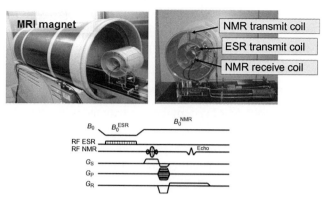

Figure 3 *In vivo* DNP-MRI from Philips instrument and pulse sequence.

The magnetic field for MRI, B_0^{NMR} was set at 14.5 mT, and the NMR resonator assembly, consisting of a transmit saddle coil (25 cm diameter, 23 cm length) and a receive solenoidal coil (5 cm diameter, 6 cm length), was tuned to 620 kHz with a band width of 1.5 kHz. The ESR irradiation frequency used was 220 MHz. A saddle coil (13.5 cm diameter, 23.5 cm length) was used as the ESR resonator. The efficiency parameter of the ESR coil used was measured to be 5.2 µT W$^{-1/2}$. In order to obtain the high enhancement, a home-made surface coil was also utilized (Matsumoto et al., 2007).

The DNP spectra of free radical solutions were recorded by sweeping the magnetic field (B_0^{ESR}) from 5 to 10 mT, in steps of 0.01–0.1 mT. As shown in Fig. 3, imaging experiments were performed by using standard spin warp gradient echo sequence for MRI, except that each phase-encoding step was preceded by an ESR saturation pulse to elicit DNP. The pulse sequence started with the ramping of the B_0 field to B_0^{ESR}, followed by switching on the ESR irradiation. Then, the B_0 was ramped up to 14.5 mT before the NMR pulse and the associated field gradients were turned on. At the beginning or end of the cycle, a conventional (native) NMR signal intensity (with ESR OFF) was measured for computing the enhancement factors. A Hewlett-Packard PC (operating system, LINUX 5.2) was used for data acquisition. The images were reconstructed from the echoes by using standard software, and were stored in DICOM format (digital imaging and communications in medicine). MATLAB codes were used for the computation of DNP parameters and curve fitting. Typical scan conditions were as follows, repetition time (T_R)/echo time (T_E): 2000 ms/25 ms; ESR irradiation time (T_{ESR}): 50 to ~800 ms, in steps of 50 or 100 ms; ESR power,

5 to ~60 W; phase-encoding steps, 64; and slice thickness, 20 mm. The image field of view (48 mm) was represented by a 64 × 64 matrix, with a pixel size of 0.63 mm × 0.63 mm.

3.2 Home-Made DNP-MRI System

The home-made DNP-MRI system was constructed using the external magnet of a commercial ESR spectrometer/imaging system (JES-ES20, JEOL Ltd., Akishima, Japan, Fig. 4), and two-axis field gradient coils for CW-ESR imaging. The detail was described by Naganuma, (Naganuma, 2009; Naganuma, Nakao, Ichikawa, & Utsumi, 2015).

The external magnetic field B_0 for ESR irradiation and MRI is fixed at 20 mT. NMR coil assembly consists of the NMR transmit saddle coil and a solenoidal receiver coil at 850 kHz with a bandwidth of 1 kHz. One-turned surface coil was used for ESR irradiation, which was set to 500–580 MHz, and DNP-MRI was performed using a standard spin echo for MRI, where each phase-encoding step was preceded by an ESR irradiation pulse to elicit the DNP enhancement. Typical scan conditions for the DNP-MRI experiment were: 12 W of ESR irradiation; a 90° flip angle; T_{ESR} × repetition time (T_R) × echo time (T_E) = 500 × 1000 × 40 ms; number of averages = 1; slice thickness 30 mm; and 64 phase-encoding steps. The image field of view (32 × 32 mm) was represented by a 64 × 64 matrix. The DNP-MRI data are analyzed using the Image J software package.

3.3 Commercially Available DNP-MRI

A custom-made DNP-MRI is also commercially available from Japan Redox Co. Ltd. (Fukuoka, Japan), but some improvement is still needed under the custom requests. Permanent ferrite magnets (16–18 mT) are used, and the ESR irradiation and MRI for protons are done at 450 MHz and 700 kHz, respectively.

Figure 4 *In vivo* ESR imaging system from JEOL instrument.

3.4 Spectroscopic DNP-MRI

Under our request, the pulse sequence of DNP-MRI from Philips was modified to make possible the imaging of plural free radicals, as shown in Fig. 5 (Utsumi et al., 2006).

The pulse sequence starts with the ramping of the B_0 field to B_0^{ESR} (normally 6.09 mT for ^{14}N labeled nitroxyl radical or 6.57 mT for ^{15}N labeled nitroxyl radical), followed by switching on the ESR irradiation. Then, the B_0 is ramped up to 14.5 mT before the NMR pulse (617 KHz) and the associated field gradients were turned on. At the beginning or end of the cycle, a conventional (native) NMR signal intensity (with ESR OFF) was measured for computing the enhancement factors.

Home-made spectroscopic DNP-MRI experiments were performed by changing the microwave frequency for ESR irradiation between 500 and

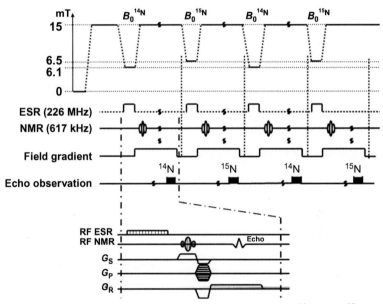

Figure 5 Sequence of DNP-MRI for simultaneous imaging of ^{14}N- and ^{15}N-labeled nitroxyl radicals. The pulse sequence started with ramping of the B_0 field to either 6.103 or 6.563 mT, corresponding to the ESR irradiation for ^{14}N- and ^{15}N-labeled nitroxyl radicals and followed by the ESR irradiation for a period of about 600 ms. Then B_0 was ramped to 14.552 mT before the radiofrequency (RF) pulse and the associated field gradients were turned on. Also, at the beginning or end of the cycle, a conventional MRI (without ESR irradiation) was collected for computing the enhancement factors. Typical scan conditions in MRI are T_R/T_E: 1200 ms/25 ms; 64 phase-encoding steps. The image field of view was represented by a 64 × 64 matrix.

580 MHz, because the external magnetic field is fixed at 20 mT. To keep the conditions of ESR irradiation constant during spectroscopic imaging experiments, the coupling of the ESR coil and its output power were carefully adjusted at every ESR frequency. Images for the individual free radicals can be performed by changing the electromagnetic wave frequency of ESR resonance, in our home-made DNP-MRI (Hyodo, Ito, Yasukawa, Kobayashi, & Utsumi, 2014).

4. SPIN-PROBE TECHNIQUE FOR DNP-MRI

In *in vivo* DNP-MRI, the synthetic stable free radicals, tetrathiatriarylmethyl and nitroxyl radicals, are utilized to obtain sufficient DNP enhancement, because of their simple and narrow resonance lines. Since the 1980s, reactions of nitroxyl radicals with redox enzymes (Swartz, Sentjurc, & Kocherginsky, 1995), with ascorbic acid (Couet et al., 1985; Swartz, Sentjurc, & Morse, 1986), and with metal ions (Bar-On et al., 1999; Kocherginsky, Swartz, & Sentjurc, 1995) have been studied. As shown in Fig. 6, nitroxyl radicals are highly redox sensitive, and are converted to the corresponding nonparamagnetic species by interacting with redox enzymes, ROS, and antioxidants, depending upon the redox status in living body. The structures and properties of nitroxyl probes are demonstrated in another review (Yasukawa et al., 2010).

4.1 Synthesis of Nitroxyl Probe

From the last half of 1950s to the end of 1960s, a great number of nitroxyl radicals, such as 2,2,6,6-tetramethyl-piperidine-1-oxyl (TEMPO) derivatives and 2,2,5,5-tetramethyl-pyrrolidine-1-oxyl (PROXYL) derivatives, were prepared in solution by oxidation of the corresponding aromatic,

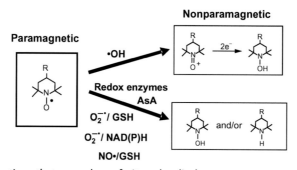

Figure 6 Reactions that cause loss of nitroxyl radicals.

aromatic–aliphatic, aliphatic, and alicyclic amines (Rozantse & Sholle, 1971; Rozantsev, 1970).

Some of these TEMPO and PROXYL compounds are now commercially available. However, adequate modification should increase the adaptability of nitroxyl radicals as probes. Figure 7 demonstrates our strategy of the modification for nitroxyl radicals: (1) Substitution at the 3-position of PROXYL to increase the site specificity, (2) modification of 2,2,6,6-position of TEMPO to increase the reaction sensitivity, and (3) labeling with ^{15}N of TEMPO for dual imaging of nitroxyl probes.

Membrane permeable methoxycarbonyl (MC)-PROXYL, which can penetrate through the blood–brain barrier (BBB), is prepared by esterifying 3-carboxy-PROXYL with methyl alcohol (Sano, Matsumoto, & Utsumi, 1997). A nitroxyl radical having high retentivity in cells is also prepared by the conversion of 3-position of PROXYL to the acetoxymethoxycarbonyl (AMC) group (Sano, Naruse, Matsumoto, Oi, & Utsumi, 2000). This new nitroxyl probe with high retentivity in brain has been applied to brain imaging (Utsumi et al., 2002). Recently, Sakai et al. (2010) developed a new method to substitute the 2,6-position of TEMPO under mild reaction condition. This method was also useful to exchange ^{14}N of TEMPO to form ^{15}N-labeled TEMPO. Kinoshita et al. (2010) found that the 2,2,6,6-ethyl substitution of TEMPO led to resistance toward reaction with ascorbic acid. Yamasaki et al. (2011) clarified the relation of the reactivity toward ascorbic acid with the redox potential, by synthesizing various derivatives.

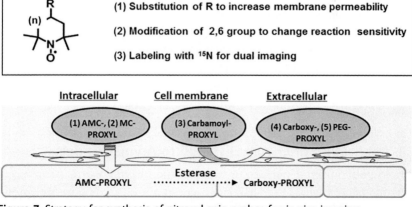

Figure 7 Strategy for synthesis of nitroxyl spin probes for *in vivo* imaging.

4.2 Reactivity of Nitroxyl Radical Toward ROS

Willson et al. reported that the second-order rate constant for the reaction of ·OH, which was produced by pulse radiolysis, with norpseudopelletierine-N-oxyl was $7.0 \pm 1.0 \times 10^9\ M^{-1}\ s^{-1}$ (Willson, Greensto, Adams, Wageman, & Dorfman, 1971). They used a competition method with carbonate ions. Asmus, Nigam, & Willson (1976) also reported a similar rate constant for nitroxyl radicals, $4.0 \times 10^9\ M^{-1}\ s^{-1}$ for oxo-TEMPO and $3.4 \times 10^9\ M^{-1}\ s^{-1}$ for hydroxy-TEMPO, by using a pulse-radiolysis conductivity method. Yasukawa et al. (2010) confirmed the relation between *in vitro* signal decays of nitroxyl radical and ·OH generation. The amount of ·OH, which was generated using the Fenton reaction, was estimated from the signal intensity of ·OH adduct of 5,5-dimethyl-1-pyrroline-N-oxide (DMPO). The plots of the amount of DMPO-OH generation against the signal decay rate gave a good correlation. The presence of DMPO competitively suppressed the *in vitro* signal decay of carbamoyl-PROXYL and the plot of signal decay rate against DMPO concentration showed a sigmoidal curve. The facts indicated that carbamoyl-PROXYL reacts with ·OH derived from Fenton reaction, and the reaction rate constant between the nitroxyl radical and ·OH is 2–3 × higher than that between DMPO and ·OH.

4.3 Tissue Distribution of Nitroxyl Radical Injected Intravenously

Figure 8 demonstrates 2D-images, obtained with a home-made ESR-CT system, of the mouse head after intravenous injection of the nitroxyl radicals, carboxy-, MC-, and AMC-PROXYL (Utsumi et al., 2002). The *in vivo* 2D-image of AMC-PROXYL gave distinguished contrast in the encephalon region, while carboxy-PROXYL distributed only in the extraencephalon region, and MC-PROXYL in both regions. The image of MC-PROXYL with *in vivo* DNP-MRI demonstrates clear contrast in brain after intravenous injection. The image clearly demonstrates that both MC-PROXYL probes penetrate through the BBB to brain, and are retained in the encephalon region after intravenous injection.

4.4 Application of *In Vivo* DNP-MRI to Animal Models

Whole-body imaging was also obtained as the PK and pharmacodynamics imaging using *in vivo* DNP-MRI (Kosem et al., 2012). Figure 9 demonstrates the whole-body imaging after intravenous injection of carboxy-PROXYL.

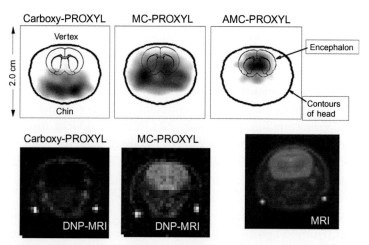

Figure 8 ESRI and DNP-MRI of nitroxyl probes having various membrane permeabilities in the mouse head.

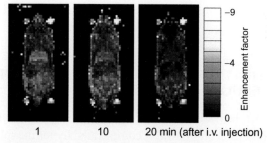

Figure 9 Whole-body images of carboxy-PROXYL by DNP-MRI. An anesthetized mouse was secured on the holder (stomach-side down) with four fiducial markers (0.25, 0.5, 0.75, and 1 mM carboxy-PROXYL in a 4 mm inner diameter × 8.5 mm height marker tube). Carboxy-PROXYL (1 mmol kg^{-1} body weight) was intravenously injected and images were visualized using 15 mT DNP-MRI with a small coil (including: NMR receiver coil, 30 mm diameter; ESR transmit coil, 137 mm diameter; and NMR transmit coil, 252 mm diameter) and the following parameters: time of repetition (T_R), 1100 ms; time of echo delay (T_E), 25; ESR irradiation time (T_{ESR}), 500 ms; scanning time, 72 s; NMR frequency, 617 kHz; ESR irradiation frequency, 220.6 MHz, FOV, 85 × 85 mm and pixel size, 1.33 × 1.33 mm. The first OMRI image was performed after 1 min of injection and continued until the image intensity was invisible.

Whole-body image from the anterior to posterior ends was obtained with resolution of 1.33 × 1.33 mm per pixel, and 50-mm-thick slice of coronal entire. The image intensity gradually decreased until the signal disappeared at approximately 25 min postinjection. The coregistration technique of DNP-MRI with MRI images provided the precise location of inner organs and showed the feasibility to observe these organs simultaneously. The different level of apparent intensity was observed in four main organs including heart and lung of the thoracic region, liver in the upper abdominal region, and kidney in the lower abdominal region, whereas no intensity of the brain region was observed.

To estimate the concentration of carboxy-PROXYL in living animals, a standard concentration curve was constructed with an external standard of four actual concentrations of carboxy-PROXYL in fiducial markers. The local concentrations of ROIs were directly evaluated using the standard curve from the markers. The parent form of carboxy-PROXYL as paramagnetic compound, mainly distributing in the serum, provided the apparent intensity prior to sudden reduction to diamagnetic hydroxylamine inside the cells.

In vivo imaging of redox status in brain was also carried out using the DNP-MRI/nitroxyl probe technique, demonstrating the metabolic imaging of redox status in brain injuries (Shiba et al., 2011; Yamato et al., 2009, 2011). The changes of redox status were noninvasively demonstrated in mice with dextran sodium sulfate-induced colitis, which might be good animal model for DNP-MRI (Yasukawa, Miyakawa, Yao, Tsuneyoshi, & Utsumi, 2009).

5. APPLICATION OF DNP-MRI TO THE INTRINSIC MOLECULAR IMAGING

Typical first derivative ESR spectra of free radical intermediates from the intrinsic redox molecules exhibit complicated hyperfine splitting lines, as reported previously (Swartz, Bolton, & Borg, 1972). To demonstrate the capabilities of *in vivo* DNP-MRI to simultaneously image free radical intermediates generated through redox reactions, phantoms filled with FMNH, FADH, $CoQ_{10}H$, vitamins E and K radicals, or carbamoyl-PROXYL were simultaneously imaged using a home-made *in vivo* DNP-MRI system (Hyodo et al., 2014). The ESR irradiation for DNP-MRI was performed at 527.5 MHz, which is the central peak of carbamoyl-PROXYL. As shown in Fig. 10, all of the free radical intermediates gave distinct images of DNP-MRI with high signal intensity, while the MRI image without ESR irradiation was of poorer quality.

DNP-MRI images were derived from the intensity-enhanced images of the solvent protons (water protons for FMNH, FADH, and carbamoyl-PROXYL; and nonaqueous hydrocarbon protons for $CoQ_{10}H$, and vitamins E and K_1 radicals). The enhancements due to FADH and FMNH were

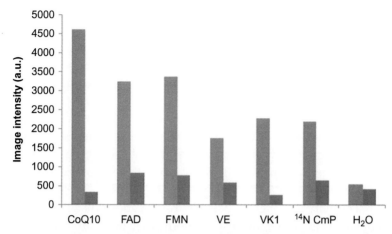

Figure 10 DNP-MRI intensities of free radical intermediates from FMN, FAD, CoQ$_{10}$, vitamin E, vitamin K$_1$, and the synthetic stable radical ^{14}N-carbamoyl-PROXYL obtained by using DNP-MRI. The frequency of ESR irradiation for DNP-MRI (527.5 MHz) is the resonance frequency of the central peak of carbamoyl-PROXYL. MRI spectra with and without ESR irradiation were obtained with a spin-echo sequence at 850 kHz using a home-made *in vivo* DNP-MRI system kept at room temperature.

similar to that induced by carbamoyl-PROXYL, even though the ESR absorption spectra of both FMNH and FADH exhibit 13-fold broader line width. The enhancement was much higher for the vitamin K$_1$ radical and the vitamin E radical than for carbamoyl-PROXYL. These high enhancements could not be explained with the principle of Overhauser effect as demonstrated in the previous section. Further physico–chemical studies are needed to confirm the mechanism for giving the high enhancement. Other DNP mechanisms besides the Overhauser effect might also be considered.

It is critically important for molecular imaging to distinguish a variety of free radical species. Spectral and spatial ESR imaging is one of the methods for imaging the plural radical species (Biller et al., 2014; Matsumoto & Utsumi, 2000; Pawlak, Ito, Fujii, & Hirata, 2011), but the spatiotemporal resolution of ESR imaging was very poor. DNP-MRI gave clear images for plural free radical species (Hyodo et al., 2014; Utsumi et al., 2006). By changing the magnetic field or the frequency of ESR irradiation in DNP-MRI, distinct images of plural free radicals were visualized in a single experiment. This fact demonstrates that DNP-MRI can characterize individual free radicals simultaneously. Such an imaging functionality can be applied to monitor free radical intermediates and assess metabolic profiles *in vivo*.

6. CONCLUSION

In vivo DNP-MRI can simultaneously visualize not only synthetic free radicals, but also free radical intermediates of endogenous redox molecules. In vivo DNP-MRI system performed as a "spectroscopic molecular imaging scanner," achieving separable visualization of multiple free radical intermediates. This is the largest advantage of DNP-MRI as an important technology for imaging free radical intermediates in biological redox reactions.

A specific application of DNP-MRI would be to visualize the multiple redox reactions involved in human metabolic processes. The development of high-sensitivity in vivo DNP-MRI instrumentation will allow preclinical research to obtain redox/metabolic information in addition to anatomic images.

ACKNOWLEDGMENTS

This work was jointly supported by the "Formulation of Advanced Collaborative Medical Innovation Center" as Special Coordination Funds for Promoting Science and Technology, Ministry of Education, Culture, Sports, Science and Technology, Japan; and by Development of Advanced Measurement and Analysis Systems from Japan Science and Technology Agency. Additional funds were provided by a Grant-in-Aid for Scientific Research and the Core-to-Core Program for Scientific Research from the Japan Society for the Promotion of Science.

REFERENCES

Asmus, K. D., Nigam, S., & Willson, R. L. (1976). Kinetics of nitroxyl radical reactions a pulse-radiolysis conductivity study. *International Journal of Radiation Biology and Related Studies in Physics, Chemistry, and Medicine, 29*(3), 211–219.
Bar-On, P., Mohsen, M., Zhang, R. L., Feigin, E., Chevion, M., & Samuni, A. (1999). Kinetics of nitroxide reaction with iron(II). *Journal of the American Chemical Society, 121*(35), 8070–8073.
Benial, A. M., Ichikawa, K., Murugesan, R., Yamada, K., & Utsumi, H. (2006). Dynamic nuclear polarization properties of nitroxyl radicals used in Overhauser-enhanced MRI for simultaneous molecular imaging. *Journal of Magnetic Resonance, 182*(2), 273–282. http://dx.doi.org/10.1016/j.jmr.2006.06.025.
Biller, J. R., Tseitlin, M., Quine, R. W., Rinard, G. A., Weismiller, H. A., Elajaili, H., et al. (2014). Imaging of nitroxides at 250 MHz using rapid-scan electron paramagnetic resonance. *Journal of Magnetic Resonance, 242*, 162–168. http://dx.doi.org/10.1016/j.jmr.2014.02.015.
Couet, W. R., Brasch, R. C., Sosnovsky, G., Lukszo, J., Prakash, I., Gnewuch, C. T., et al. (1985). Influence of chemical-structure of nitroxyl spin labels on their reduction by ascorbic-acid. *Tetrahedron, 41*(7), 1165–1172.
Elas, M., Ahn, K. H., Parasca, A., Barth, E. D., Lee, D., Haney, C., et al. (2006). Electron paramagnetic resonance oxygen images correlate spatially and quantitatively with oxylite oxygen measurements. *Clinical Cancer Research, 12*(14 Pt. 1), 4209–4217.

Grucker, D., Guiberteau, T., Eclancher, B., Chambron, J., Chiarelli, R., Rassat, A., et al. (1995). Dynamic nuclear polarization with nitroxides dissolved in biological fluids. *Journal of Magnetic Resonance, Series B, 106*, 101–109.

He, G., Dumitrescu, C., Petryakov, S., Deng, Y., Kesselring, E., & Zweier, J. L. (2007). Transverse oriented electric field re-entrant resonator (TERR) with automatic tuning and coupling control for EPR spectroscopy and imaging of the beating heart. *Journal of Magnetic Resonance, 187*(1), 57–65.

Hyodo, F., Ito, S., Yasukawa, K., Kobayashi, R., & Utsumi, H. (2014). Simultaneous and spectroscopic redox molecular imaging of multiple free radical intermediates using dynamic nuclear polarization-magnetic resonance imaging [Research Support, Non-U.S. Gov't]. *Analytical Chemistry, 86*(15), 7234–7238. http://dx.doi.org/10.1021/ac502150x.

Kinoshita, Y., Yamada, K., Yamasaki, T., Mito, F., Yamato, M., Kosem, N., et al. (2010). In vivo evaluation of novel nitroxyl radicals with reduction stability [Evaluation Studies Research Support, Non-U.S. Gov't]. *Free Radical Biology & Medicine, 49*, 1703–1709. http://dx.doi.org/10.1016/j.freeradbiomed.2010.08.027.

Kocherginsky, N., Swartz, H. M., & Sentjurc, M. (1995). Chemical reactivity of nitroxides. In N. Kocherginsky & H. M. Swartz (Eds.), *Nitroxide spin labels: Reactions in biology and chemistry* (pp. 27–66). New York, NY: CRC Press.

Kosem, N., Naganuma, T., Ichikawa, K., Phumala Morales, N., Yasukawa, K., Hyodo, F., et al. (2012). Whole-body kinetic image of a redox probe in mice using Overhauser-enhanced MRI. *Free Radical Biology & Medicine, 53*(2), 328–336.

Krishna, M. C., English, S., Yamada, K., Yoo, J., Murugesan, R., Devasahayam, N., et al. (2002). Overhauser enhanced magnetic resonance imaging for tumor oximetry: Coregistration of tumor anatomy and tissue oxygen concentration. *Proceedings of the National Academy of Sciences of the United States of America, 99*(4), 2216–2221.

Li, H., Deng, Y., He, G., Kuppusamy, P., Lurie, D. J., & Zweier, J. L. (2002). Proton electron double resonance imaging of the in vivo distribution and clearance of a triaryl methyl radical in mice. *Magnetic Resonance in Medicine, 48*(3), 530–534. http://dx.doi.org/10.1002/mrm.10222.

Lin, M. T., & Beal, M. F. (2006). Mitochondrial dysfunction and oxidative stress in neurodegenerative diseases. *Nature, 443*(7113), 787–795.

Lurie, D. J., Bussell, D. M., Bell, L. H., & Mallard, J. R. (1988). Proton-electron double magnetic resonance imaging of free radical solutions. *Journal of Magnetic Resonance, 76*, 366–370.

Lurie, D. J., Hutchison, J. M. S., Bell, L. H., Nicholson, I., Bussell, D. M., & Mallard, J. R. (1989). Field-cycled proton electron double-resonance imaging of free-radicals in large aqueous samples. *Journal of Magnetic Resonance, 84*(2), 431–437.

Matsumoto, K., & Utsumi, H. (2000). Development of separable electron spin resonance-computed tomography imaging for multiple radical species: An application to .OH and .NO. *Biophysical Journal, 79*(6), 3341–3349.

Matsumoto, S., Yamada, K., Hirata, H., Yasukawa, K., Hyodo, F., Ichikawa, K., et al. (2007). Advantageous application of a surface coil to EPR irradiation in Overhauser-enhanced MRI. *Magnetic Resonance in Medicine, 57*, 806–811.

Matsumoto, S., Yasui, H., Batra, S., Kinoshita, Y., Bernardo, M., Munasinghe, J. P., et al. (2009). Simultaneous imaging of tumor oxygenation and microvascular permeability using Overhauser enhanced MRI. *Proceedings of the National Academy of Sciences of the United States of America, 106*(42), 17898–17903. http://dx.doi.org/10.1073/pnas.0908447106.

Miura, Y., Utsumi, H., & Hamada, A. (1992). Effects of inspired oxygen concentration on in vivo redox reaction of nitroxide radicals in whole mice. *Biochemical and Biophysical Research Communications, 182*(3), 1108–1114.

Murphy, M. P. (2009). How mitochondria produce reactive oxygen species. *The Biochemical Journal, 417*(1), 1–13.

Naganuma, T. (2009). *Development of multi bio-functional imaging system using magnetic resonance*. Fukuoka, Japan: Kyushu University. Ph.D. Thesis of Doctor Degree.

Naganuma, T., Nakao, M., Ichikawa, K., & Utsumi, H. (2015). Development of a new redox molecular imaging method. *Yakugaku Zasshi: Journal of the Pharmaceutical Society of Japan, 135*(5), 733–738.

Pawlak, A., Ito, R., Fujii, H., & Hirata, H. (2011). Simultaneous molecular imaging based on electron paramagnetic resonance of 14 N- and 15 N-labelled nitroxyl radicals. *Chemical Communications (Cambridge, England), 47*(11), 3245–3247. http://dx.doi.org/10.1039/c0cc03581e.

Rozantse, E. G., & Sholle, V. D. (1971). Synthesis and reactions of stable nitroxyl radicals I. Synthesis. *Synthesis,* (4), 190–202.

Rozantsev, E. G. (1970). Synthesis of some stable radicals and the most important intermediates. In H. Ulrich (Ed.), *Free nitroxyl radicals* (pp. 203–246). New York, NY: Plenum Press.

Sakai, K., Yamada, K., Yamasaki, T., Kinoshita, Y., Mito, F., & Utsumi, H. (2010). Effective 2,6-substitution of piperidine nitroxyl radical by carbonyl compound. *Tetrahedron, 88*, 2311–2315.

Sano, H., Matsumoto, K., & Utsumi, H. (1997). Synthesis and imaging of blood-brain-barrier permeable nitroxyl-probes for free radical reactions in brain of living mice. *Biochemistry and Molecular Biology International, 42*(3), 641–647.

Sano, H., Naruse, M., Matsumoto, K., Oi, T., & Utsumi, H. (2000). A new nitroxyl-probe with high retention in the brain and its application for brain image. *Free Radical Biology & Medicine, 28*, 959–969.

Schapira, A. H. (2006). Mitochondrial disease. *Lancet, 368*(9529), 70–82.

Shiba, T., Yamato, M., Kudo, W., Watanabe, T., Utsumi, H., & Yamada, K. (2011). In vivo imaging of mitochondrial function in methamphetamine-treated rats. *NeuroImage, 57*(3), 866–872. http://dx.doi.org/10.1016/j.neuroimage.2011.05.041.

Swartz, H. M., Bolton, J. R., & Borg, D. C. (1972). *Biological application of electron spin resonance*. New York, NY: Wiley-Interscience.

Swartz, H. M., Sentjurc, M., & Kocherginsky, N. (1995). Metabolism of nitroxides and their products in cells. In N. Kocherginsky & H. M. Swartz (Eds.), *Nitroxide spin labels: Reactions in biology and chemistry* (pp. 113–147). New York, NY: CRC Press.

Swartz, H. M., Sentjurc, M., & Morse, P. D. (1986). Cellular-metabolism of water-soluble nitroxides—Effect on rate of reduction of cell nitroxide ratio, oxygen concentrations and permeability of nitroxides. *Biochimica et Biophysica Acta, 888*(1), 82–90.

Takeshita, K., Utsumi, H., & Hamada, A. (1991). ESR measurement of radical clearance in lung of whole mouse. *Biochemical and Biophysical Research Communications, 177*(2), 874–880.

Utsumi, H., Muto, E., Masuda, S., & Hamada, A. (1990). In vivo ESR measurement of free radicals in whole mice. *Biochemical and Biophysical Research Communications, 172*(3), 1342–1348.

Utsumi, H., Sano, H., Naruse, M., Matsumoto, K., Ichikawa, K., & Oi, T. (2002). Nitroxyl probes for brain research and their application to brain imaging. *Methods in Enzymology, 352*, 494–506.

Utsumi, H., & Yamada, K. (2003). In vivo electron spin resonance-computed tomography/nitroxyl probe technique for non-invasive analysis of oxidative injuries. *Archives of Biochemistry and Biophysics, 416*(1), 1–8.

Utsumi, H., Yamada, K., Ichikawa, K., Sakai, K., Kinoshita, Y., Matsumoto, S., et al. (2006). Simultaneous molecular imaging of redox reactions monitored by Overhauser-enhanced

MRI with 14 N- and 15 N-labeled nitroxyl radicals. *Proceedings of the National Academy of Sciences of the United States of America, 103*(5), 1463–1468.

Willson, R. L., Greensto, C. L., Adams, G. E., Wageman, R., & Dorfman, L. M. (1971). Standardization of hydroxyl radical rate data from radiation chemistry. *International Journal for Radiation Physics and Chemistry, 3*(3), 211–220.

Yamasaki, T., Mito, F., Ito, Y., Pandian, S., Kinoshita, Y., Nakano, K., et al. (2011). Structure-reactivity relationship of piperidine nitroxide: Electrochemical, ESR and computational studies [Research Support, Non-U.S. Gov't]. *The Journal of Organic Chemistry, 76*(2), 435–440. http://dx.doi.org/10.1021/jo101961m.

Yamato, M., Shiba, T., Naganuma, T., Ichikawa, K., Utsumi, H., & Yamada, K. (2011). Overhauser-enhanced magnetic resonance imaging characterization of mitochondria functional changes in the 6-hydroxydopamine rat model. *Neurochemistry International, 59*(6), 804–811. http://dx.doi.org/10.1016/j.neuint.2011.08.010.

Yamato, M., Shiba, T., Yamada, K., Watanabe, T., & Utsumi, H. (2009). Noninvasive assessment of the brain redox status after transient middle cerebral artery occlusion using Overhauser-enhanced magnetic resonance imaging. *Journal of Cerebral Blood Flow and Metabolism, 29*(10), 1655–1664.

Yasukawa, K., Miyakawa, R., Yao, T., Tsuneyoshi, M., & Utsumi, H. (2009). Non-invasive monitoring of redox status in mice with dextran sodium sulphate-induced colitis. *Free Radical Research, 43*(5), 505–513.

Yasukawa, K., Yamada, K., Ichikawa, K., & Utsumi, H. (2010). In vivo ESR/spin probe technique. In D. Das (Ed.), *Methods in redox signaling* (pp. 90–99). Mary Ann Liebert, Inc.

Zotev, V. S., Owens, T., Matlashov, A. N., Savukov, I. M., Gomez, J. J., & Espy, M. A. (2010). Microtesla MRI with dynamic nuclear polarization. *Journal of Magnetic Resonance, 207*, 78–88.

AUTHOR INDEX

Note: Page numbers followed by "*f*" indicate figures and "*t*" indicate tables.

A

Abé, C., 331, 335–336, 406, 416–417
Abragam, A., 248–249, 504, 506, 508
Abu-Baker, S., 80, 293–294
Abulseoud, O.A., 458
Adams, G.E., 564
Adin, I., 193–194
Agafonov, R.V., 102
Agniswamy, J., 154–156
Ahmad, R., 507–508, 530–532
Ahn, K.H., 518, 554–555
Aihara, T., 105
Aina, O.H., 220
Airola, M.V., 322–323
Ajaj, Y., 324
Akahane, H., 105
Akasaka, K., 30–32, 50
Akyuz, N., 351–353
al Hallaq, H., 512–513
Alaouie, A.M., 228
Albericio, F., 294
Alexander, N.S., 292, 322–323, 367–370, 373, 430
Alexanian, V., 221
Alguel, Y., 351
Alizon, J., 468
Allain, F.H.T., 430, 443–444
Al-Lazikani, B., 351
Allison, B., 461, 477–479
Allouch, A., 49
Almén, M.S., 290
Al-Mestarihi, A., 292, 367–368
Altenbach, C., 4–23, 30–53, 60–64, 68–75, 80–81, 82*f*, 83–86, 84*f*, 90, 92–94, 102, 113, 156–158, 221, 227–228, 236–239, 242–244, 248, 262–263, 270–272, 274–276, 292, 316–323, 328–329, 338–339, 354–360, 379, 390–391, 394–398, 404–406, 428–429, 458–459, 487
Altman, R.B., 351–353
Alvarenga, D.L., 542–543

Ambroso, M.R., 260–285, 262*f*, 458–459, 461, 464–465, 478*f*, 479
Anderson, D.E., 487
Anderson, D.J., 225–226
Anderson, J.R., 4–5
Anderson, L.L., 322–323
Ando, N., 32, 49
Andrew, E.R., 510
Andrews, A.J., 141–142
Andrews, S.S., 192–193
Angert, I., 110–111
Aniana, A., 154–156, 485–486, 491
Ann, H.S., 261
Anthis, N.J., 15–17, 63–64, 485–494, 492–493*f*
Antich, P.P., 504
Antonic, J., 405–406, 414–415
Antonny, B., 260–261
Antson, A., 248, 320–322, 335–336
Arachchige, R.J., 15
Arata, T., 105
Arbeit, J.M., 507–509, 512–513
Ardenkjaer-Larsen, J.H., 512–513
Armstrong, B.D., 458–459, 461, 463–466, 470, 472, 477–479, 478*f*
Armstrong, F.A., 405
Arnett, D.K., 530
Arnold, K., 225–226
Arrey, T.N., 297
Aruksakunwong, O., 155–156, 160–161
Ashford, P., 350
Asmus, K.D., 564
Athanasoula, E.A., 32
Atherton, N.M., 504, 507
Aveyard, R., 234–236
Ayant, Y., 468
Aznar, C.P., 354–355

B

Baase, W.A., 32, 49
Babaylova, E.S., 420
Babcock, G.T., 415

Baber, J.L., 135–136, 149, 164, 292
Babu, Y.S., 491–494, 493f
Bacic, G., 531–532
Backer, J.M., 507–508
Baete, S.H., 530–531
Baeumlisberger, D., 297
Bah, A., 30, 49–52
Bahar, I., 291
Bakan, A., 291
Baker, C.M., 248, 322
Baker, M., 290–291
Baldo, G., 234–236
Baldwin, A.J., 30
Baldwin, E.P., 49
Bales, B.L., 192–193
Balog, M.R., 102–103, 106–107, 225–226
Banachewicz, W., 30
Banham, J.E., 50, 141, 149, 170–173, 248, 322, 366–367, 404–405, 413f, 443
Baoukina, S., 192–193
Barabasi, A.L., 351
Baranova, T.Y., 225–226
Baranski, T.J., 322–323
Barbosa, M.P., 208–210
Barbosa, S.R., 225–226
Barbuto, S., 394–395
Barchowsky, A., 531–532
Barfield, M., 357–358
Barhate, N., 407
Barnes, J.P., 292, 319
Barnes, R., 458–477, 479
Bar-On, P., 562
Barr, D.P., 105, 116
Barratt, M.D., 192–193
Barry, B.A., 415
Barstow, B., 32, 49
Barth, E.D., 507–508, 512–513, 516, 518–519, 531–532, 538, 554–555
Barthe, P., 32
Bartkowski, H.M., 542–544
Bartucci, R., 262–263, 293
Basus, V.J., 489
Bates, R.D., 458, 465–466
Batra, S., 518–519, 555
Batt, C.A., 30–31
Battiste, J.L., 489
Baudelet, C., 531–532, 535–536
Bauer, C., 405, 410–411

Bax, A., 485–486, 491–494
Bayburt, T.H., 377–379
Beal, M.F., 554
Beck, R., 504
Becker, J.S., 15–18, 405
Becker, N.B., 417
Beckert, D., 301
Beck-Sickinger, A.G., 225–226, 294
Beckstein, O., 380
Bederson, J.B., 542–544
Beer, A.J., 530–531
Beer, L., 136, 137f
Beghein, N., 535–536
Behnke, C.A., 379
Behrmann, E., 260–261, 263, 266, 272, 274–275
Beier, C., 319
Beis, K., 351
Belaya, M., 208–210
Belchenko, O.I., 195–198
Belford, R.L., 199–200, 230–231, 239–240, 244–246
Bell, L.H., 555, 558
Bell, R.M., 395–397, 518–519
Bellew, B.F., 415
Belorizky, E., 468
Bender, A., 126–128, 142–143, 158, 168–170, 443
Bender, C.J., 415
Benial, A.M., 556
Benjamin, E.J., 530
Bennati, M., 225–226, 405–406, 414–415, 461, 464–466
Bennett, V.J., 225–226
Benson, D.R., 317
Berente, Z., 225–226
Berleur, F., 192–193
Berliner, L.J., 5–6, 5f, 10, 22, 37–38, 45, 62, 65–68, 126–128, 139–140, 146, 194–195, 222–223, 225–226, 239, 316–319, 356, 406, 512
Berlow, R.B., 45
Berman, H.M., 444
Bernardo, M., 555
Berndsen, C.E., 141–142
Bertini, I., 12, 15
Beschiaschvili, G., 210–211
Besserer, G.M., 350

Beth, A.H., 61–64, 80, 319–320, 365–367
Bettio, A., 225–226
Betzler, M., 391–392
Bezanilla, F., 290–291
Bhabha, G., 350
Bhargava, K., 230
Bhat, T.N., 444
Bhatia, V.K., 260–261, 266
Bhatnagar, J., 322–323
Bhattacharjee, S., 354–355
Biaglow, J., 503–504
Biglino, D., 405–406, 415
Biller, J.R., 512, 567
Bilwes, A.M., 322–323
Binder, B.P., 102–120
Bittl, R., 225–226, 405
Blaber, M., 49
Blackburn, M.E., 155–170, 163f, 166f, 440
Blaha, M.J., 530
Blakely, S.E., 84–85, 102
Blanchard, S.C., 351–353
Bleaney, B., 504, 508
Bleifuss, G., 415
Blinco, J.P., 4–6, 12–14
Blockley, N.P., 530–531
Bloembergen, N., 489
Blümler, P., 470, 472
Bobko, A.A., 204
Bode, B.E., 249–250, 405, 407, 410–411, 416–417
Bodenhausen, G., 420
Bodo, G., 126–128
Boehr, D.D., 60–61
Boelens, R., 129
Bohlen, J.-M., 420
Bol, A., 530–531
Bolar, D.S., 530–531
Bolen, D.W., 63
Bollinger, J.M., 415
Bolte, M., 405
Bolton, D.R., 142–143, 414
Bolton, J.R., 354–355, 566
Bonangelino, C.J., 141–142
Bonifačić, M., 301
Bonifacino, J.S., 141–142
Bonnet, P.A., 192–193
Bonney, M., 530, 545–546
Bonney, S., 530, 545–546

Bonora, M., 155, 157–160, 162, 163f, 165, 440
Bonvin, A.M.J., 129
Boohaker, R.J., 220
Boone, C.D., 155, 158–160, 165–168
Boratynski, P.J., 14, 42, 365, 420
Borbas, J., 317
Borbat, P.P., 14–17, 32–34, 72, 142–143, 158, 160, 162, 164–165, 168–170, 221, 249–250, 319, 322–323, 353–354, 361, 363, 365–367, 370, 379, 404–406, 415–416, 420, 430, 440, 443
Bordignon, E., 15–17, 60–61, 78, 316, 321–324, 326, 328–330, 335–336, 340, 353–354, 366–368, 412, 429–430, 439, 458–459, 464–465, 470
Borg, D.C., 566
Bornemann, S., 297
Borovetz, H.S., 512
Bortolus, M., 248, 292, 363, 367–368, 378–379
Bosco, D.A., 19
Bosman, A.W., 438
Bottle, S.E., 4–6, 12–14
Bottomley, P.A., 510
Boucrot, E., 260–261
Boudker, O., 351–354, 370
Boussac, A., 293
Boussard, G., 357–358
Bouteiller, J.-C., 49
Bouvignies, G., 30, 49–52
Bowers, M.T., 458–459
Bowman, A., 128–130, 135–138, 141–142, 164, 322
Bowman, M.K., 443, 510–512
Boxer, S.G., 192–193, 212
Boyd, N., 431–433
Bozelli, J.C., 225–226
Bracher, S., 337f, 339–340
Bracho-Sanchez, E., 154–156
Bracken, C., 86–88
Brahimi-Horn, M.C., 530
Braide, O., 242
Brandenburg, D., 248, 320–322, 335–336
Brandon, S., 361, 365–367
Brasch, R.C., 562
Bratasz, A., 507–508, 513–514, 530–533, 538

Braunwald, E., 509
Brecht, M., 405
Bretscher, L.E., 274
Breukink, E., 233–234
Brewer, C.F., 224–225
Bridges, M.D., 4–23, 13f, 16f, 23f, 63–64, 68–69, 72, 84–85, 90, 92, 102–104, 109–110, 112–113, 115, 118–119, 141–142, 224, 318–319, 393–394
Britt, R.D., 354–355
Britto, M.D., 155, 158–160, 165–168
Brix, G., 530–531
Brockman, H., 208–210
Brooks, E.K., 6–7, 15–18, 37, 39, 43–44, 43f, 51f, 63–64, 78–79, 91–92, 102–104, 109–110, 112–113, 115, 118–119, 141–142, 224, 393–394
Brophy, P.J., 192–193
Brown, E.R., 470, 472
Brown, J.M., 507–509, 512–513, 519–520
Brown, L., 366–367
Brown, L.J., 78
Brown, L.S., 321
Brückner, A., 292
Brudvig, G.W., 17–18
Brunger, A.T., 351–353
Bruno, G.V., 105
Brustad, E.M., 15–17
Bryant, R.G., 12, 193–194
Bryant, Z., 417
Bublitz, G.U., 212
Buchaklian, A.H., 274
Buckey, J., 508, 516
Buckey, J.C., 535–536
Budil, D.E., 22, 45, 65–66, 109f, 113, 142–143, 194–195, 250, 319, 329–330, 378
Budker, V.G., 507–508
Buffy, J.J., 112–113
Bugg, C.E., 491–494, 493f
Büldt, G., 329–330
Bund, T., 320
Bunton, C.A., 208–210
Burdette, D., 390–399
Busse, L.J., 504
Bussell, D.M., 555, 558
Bustamante, C., 417
Butcher, S.E., 142–143, 158, 431–434

Butler, B., 519
Butler, P.J.G., 260–261, 263, 266, 270–271
Butterwick, J.A., 142–143, 322–323, 406
Buxton, R.B., 530–531
Bykov, I.P., 39

C

Cabral, J.M., 395–396
Cadene, M., 397–398
Cadieux, N., 380
Cafiso, D.S., 60–61, 68, 86, 91–92, 156–158, 160, 162–163, 192–194, 228, 230, 236–239, 243, 248, 262–263, 270–275, 292, 351, 353–354, 356–357, 380, 390–391, 404–405, 428–430, 439–440
Cai, K., 84–85
Cai, Q., 407, 430–434, 431–433f, 437–438, 444–446
Caia, G.L., 504
Calderon, E., 204
Calle, C., 300
Calvo, R., 354–355
Camacho-Pérez, M., 260
Camdere, G., 260–261
Camenisch, T.G., 4–5, 358–360
Camera, E., 104, 118–119
Camilli, P.D., 260–261
Campbell, K.A., 354–355
Campelo, F., 262f
Canaan, S., 320
Candelaria, M.B., 138–139
Canters, G.W., 405
Caporini, M.A., 112
Carbonaro, M., 291–292
Cardon, T.B., 112
Carlsson, U., 135–136, 147–148
Carmeliet, P., 518–519
Carmieli, R., 293
Caro, J.A., 32
Carr, P.A., 86–88
Carrascosa, J.L., 192–193
Cartailler, J.-P., 270–271
Cartaud, J., 227
Carter, J.D., 155–156, 158–160, 163–168
Carver, T.R., 458
Cascio, D., 4, 6–7, 15–17, 32–34, 47–48, 62–64, 66–68, 90, 92, 102–104, 109–110,

112–113, 115, 118–119, 141–142, 224, 270–271, 277–279, 339, 350, 357–358, 393–394
Casey, J.R., 358–360
Casey, T.M., 154–182, 292
Cavagnero, S., 458–459, 461, 464–465, 477–479, 478f
Cavalli, A., 30
Cavanagh, J., 86–88
Cekan, P., 405, 407, 416–418, 438
Ceola, S., 248, 322
Chadwick, T.G., 212
Chait, B.T., 397–398
Chakrabarty, T., 225
Chakrapani, S., 390
Chalikian, T.V., 41
Chan, H.R., 458
Chandra, S.S., 260–261
Chandramouli, G.V., 512–513
Chandrashekar, T.K., 415
Chandrudu, S., 294
Chantratita, W., 155–156, 160–161
Chao, K.S.C., 507–509, 512–513
Charlier, N., 535–536
Chatterjee, S., 354–355
Chattopadhyay, A., 291
Chavez, E.M., 141–142
Che, K.P., 238–239
Chechik, V., 50, 141, 149, 170–173, 333–335, 366–367, 404–405, 413f, 443, 514–515
Chemla, D.S., 126–128
Chen, C.T., 41–42
Chen, E.Y., 531–532, 535–536, 539
Chen, J., 243, 261, 263–264, 266, 270–272, 274–275, 277–279, 281–283, 353–354, 391–392, 397–398
Chen, L., 353 354, 380 381
Chen, M.L., 220, 243
Chen, P.S., 227
Chen, P.Y., 294–298
Chen, R., 292
Chen, Y., 430–431, 432f, 433–434, 443–445, 447–449, 449f, 513–514
Chen, Z., 430–431, 433–434, 444
Cheng, C., 458–459, 461, 464–465, 478f, 479
Cheng, C.-Y., 458–459, 461

Cheng, G., 354–355
Cheung, J.C., 290
Cheung, M.S., 63
Chevion, M., 562
Chiang, Y.-W., 168–170, 366–367, 404–405, 415–416, 443
Chien, J.C.W., 7
Chik, W.W.C., 138–139
Chizhov, I., 329–330
Cho, H.S., 112
Choe, H.W., 379
Choi, J., 458–459, 461, 464–465, 477–479, 478f
Choi, S.-H., 458–459, 472
Chothia, C., 290–291
Choudhuri, R., 518–520
Choung, S.Y., 233–234
Christen, T., 530–531
Chu, S., 351–353
Chua, K.C., 92
Chui, A.J., 92
Chzhan, M., 507–508, 513–514, 516
Cieslak, J.A., 293, 390, 397–398
Clarkson, R.B., 230–231, 239–240, 244–246, 507–508, 531–532, 535, 543–544
Claxton, D.P., 336, 350–381
Cleghorn, W.M., 322–323
Clore, G.M., 135–136, 149, 164, 291–292, 485–494, 490f, 492–493f
Closs, G.L., 79–80
Clouston, L.J., 14, 42, 365, 420
Coalson, R., 15–18
Cobb, N.J., 243
Cofiell, R., 260
Coggshall, K.A., 380
Cohen, S.L., 395–396
Collier, P.J., 11, 68–69, 82f, 83–85, 390–391
Collingridge, D.R., 530–531
Collins, M.D., 32, 42
Colnago, L.A., 556
Columbus, L., 4, 7, 19, 34–36, 62, 64–68, 85–89, 105–107, 160, 243–244, 270–271, 317–318, 322–323, 356–358, 390–391, 394–398, 489
Coman, R.M., 154–156
Combariza, J.E., 357–358

Comi, R.J., 508, 516, 531–532, 535–536
Condon, E.U., 502–503
Congreve, M., 290
Conn, P.M., 290, 317
Constantinescu, A., 504
Cook, J.A., 512–513, 515
Cook, W.J., 491–494, 493f
Cooke, J.A., 78
Cooke, R., 102, 103f, 105–106, 107f, 225
Cooper, A., 36, 318–319
Cordero-Morales, J.F., 322–323, 380
Cormack, A.M., 502–503
Cornish, V.W., 317
Correa, A.M., 290–291
Correia, B.E., 30, 49–52
Correia, J.J., 194–195, 236–237, 316
Cortes, D.M., 84–85, 322–323, 351, 353–354, 380, 390–392, 395–398
Costa-Filho, A.J., 319
Cotter, R.J., 290
Couet, W.R., 562
Cowieson, N.P., 291
Cozzarelli, N.R., 417
Craig, A., 292
Crane, B.R., 322–323
Crisma, M., 249–250
Cross, T.A., 351
Crowell, K.J., 192–193
Cruickshank, P.A.S., 142–143, 414
Cuello, L.G., 84–85, 322–323, 351, 353–354, 380, 390–392, 397–398
Cuendet, M.A., 351–353
Cuervo, A., 192–193
Cuff, A., 350
Cui, C., 11, 82f, 83–85
Cullis, P.R., 204
Cunningham, T.F., 15, 64, 68
Curi, R., 208–210
Cushman, M., 530
Cusick, M.E., 351

D

Dabney-Smith, C., 293–294
Dahl, S.W., 487
Dahlquist, F.W., 49, 458–459
Dalmas, O., 353–354, 380
Dalton, L.A., 274
D'Amore, P.W., 155, 158–160, 165–167

Dancel, M.C., 155, 157, 159–160, 163–168
Dans, P.D., 192–193
Dantas Machado, A.C., 430–431, 432f, 433–434, 443–445, 447–449, 449f
Das, B.B., 290
Das, D., 507–508, 562, 564
Das, R., 417
Das, S., 350
Daumke, O., 260–261, 266
Davis, J.H., 142–143, 158
Davis, M.W., 260
Davis, R.L., 542–544
Dawson, N.L., 350
Day, I.J., 248, 322
De Angelis, A.A., 112
De Camilli, P., 260–261
de Carvalho, F.D., 336, 360, 372–373, 377–379
de Castries, A., 225–226
De Deene, Y., 530–531
De Fabritiis, G., 154–156
de Kruijff, B., 233–234
de Sousa, P.L., 556
de Souza, R.E., 556
de Vera, I.M.S., 155–170
de Zeeuw, S., 545–546
De Zorzi, R., 353–354, 380–381
De Zotti, M., 248–250
Deber, C.M., 290
Debuyst, R., 535–536
DeGrado, W.F., 292
Dejehet, F., 535–536
Deligiannakis, Y., 293, 298–299
Demidenko, E., 530–532, 535–536, 540–541
Deng, Y., 554–555
Denysenkov, V.P., 404–420, 458, 465–466
DeRose, V.J., 443
Desai, A., 15
Desai, V.R., 544
DeSensi, S.C., 319, 361
Desreux, J.F., 487
Deterding, L.J., 354–355
Detrich, H.W., 194–195, 236–237, 316
Deupi, X., 225–226
Devasahayam, N., 512–513, 515, 519–520, 530–533, 555
Devaux, P.F., 227

Dicus, M.M., 354–355
Diederich, F., 408, 415–416
Diederichsen, U., 225–226
Dietrich, F., 406, 416–417
Ding, F., 154–163, 165–167
Ding, Y., 428–450, 432*f*, 449*f*, 458–461
Dintzis, H.M., 126–128
Dirac, P.A.M., 504–505, 508
Do, T.D., 458–459
Do Cao, M.A., 92
Dockter, C., 320
Dodt, A., 326
Doehner, J., 192–193
Doherty, G.J., 260–261
Doll, A., 420, 458–459, 464–465, 470
Doll, C.M., 530–531
Dombrowsky, O., 224–225, 321–322
Dominick, J.L., 112
Dong, J., 293, 360, 363, 377–379
Dong, R., 531–532, 535, 539–543
Dorfman, L.M., 564
Dorn, H.C., 458
Doster, W., 44
Doucleff, M., 488–489, 491–494, 493*f*
Doudna, J.A., 194
Doupey, T.G., 92
Dowey, E.M., 510
Doyle, D.A., 395–396
Drescher, M., 249, 317
Drew, D., 290–291
Drin, G., 260–261
Drozdoski, W.S., 458, 465–466
Drummond, C.J., 208–210
Du, G., 531–532, 535, 539–541
Duan, Y., 430–431, 432*f*, 433–434, 443–445, 447–449, 449*f*
Duerner, G., 249–250
Dumitrescu, C., 554–555
Dunagan, M.M., 115, 142–143, 292
Duncker, D.J., 545–546
Dunkel, S., 336, 338, 340
Dunn, B.M., 155–160, 163–168
Dunn, J.F., 530–532
Dunn, J.W., 512
Duong, T.Q., 544
Durr, K.L., 353–354, 380–381
Durr, U.H.N., 112
Duss, O., 430, 443–444

Dyson, H.J., 45–47, 60–61
Dzikovski, B.G., 221, 319–320
Dzuba, S.A., 147–148, 262–263, 293, 440

E

Earle, K.A., 109*f*, 113, 156–157, 160, 162, 164–165, 319, 430, 440
Eastman, P.E., 503
Eaton, G.R., 4–6, 10, 12–17, 37–38, 105, 116, 135–136, 138–139, 147–148, 225–226, 239, 405–406, 440, 509
Eaton, S.R., 37–38
Eaton, S.S., 4–6, 10, 12–17, 105, 116, 135–136, 138–139, 147–148, 225–226, 239, 405–406, 440, 509, 512
Eble, M.J., 530–531
Eckle, T., 530, 545–546
Eckstein, F., 407
Edwards, D.T., 292
Edwards, T.E., 407
Efimova, O.V., 504
Ehnholm, G., 512–513
Ehrenshaft, M., 354–355
Eisenmesser, E.Z., 19
Eisenstein, M., 293
Ekiert, D.C., 350
El Mahdi, O., 294
El Mkami, H., 126–150, 137*f*, 164
Elajaili, H., 512, 567
Elas, M., 516–519, 554–555
Eliezer, D., 45–47
Ellena, J.F., 380
Ellis, S.J., 516
El-Mkami, H., 130, 135–138, 141–142, 322
Elsaesser, C., 405
Elsasser, C., 225–226, 405
Els-Heindl, S., 294
El'yanov, B.S., 11
Emamzadah, S., 449*f*
Enderle, T., 126–128
Endeward, B., 142–143, 322–323, 404–420, 428–430, 439
Engelhard, M., 78, 225–226, 321–323, 328–330
Engels, J.W., 407
Engelsberg, M., 556
English, S., 512–513, 530–531, 555
Enkin, N., 464–466

Epel, B., 511–512, 514–516, 519–521, 532–533
Eremenko, S.I., 195–198, 507–508
Eriksson, A.E., 49
Erilov, D.A., 249–250, 262–263
Ermolieff, J., 155–156, 160–161
Ernst, O.P., 72–73, 321–323, 379, 458–459
Ernstoff, M.S., 516, 531–532, 535–536
Eschmann, N.A., 458–459
Eskey, C.J., 539–541
Esmann, M., 293
Espinoza-Fonseca, L.M., 113–115
Esquiaqui, J.M., 156–157, 292
Eteshola, E., 514, 532–533, 537–538
Etzkorn, M., 261–262
Evans, D.F., 208–210
Evans, G., 290
Evans, P.A., 294–298
Evans, P.R., 260–261, 263, 270–271
Evans, S.M., 503–504
Everaers, R., 417
Evergren, E., 260–261
Everitt, L., 160–161

F

Faham, S., 350
Faingold, O., 293
Fajer, M.I., 404–405
Fajer, P.G., 63–64, 113, 319, 366–367, 404–405, 443
Falke, J., 238–239
Fan, C., 350
Fan, X., 518–519
Fang, C.J., 72, 78–79, 79f, 244–246, 361
Fanucci, G.E., 92, 154–182, 163f, 166f, 236–237, 242, 292, 380, 390–391, 440
Farahbakhsh, Z.T., 238, 320–321, 338–339
Farmer, B.T., 129
Farrell, S.R., 405
Farrens, D.L., 72–73, 321–323, 340
Farsad, K., 260–261
Fasshauer, D., 270–271
Fatome, M., 192–193
Fauci, A.S., 509–510
Fava, A., 104, 118–119
Fawzi, N.L., 15–17, 63–64, 487, 491
Fedorova, O.S., 428–429, 431–433
Feher, G., 354–355

Feigin, E., 562
Feigon, J., 407, 431–434
Feiner, J., 530–531
Feinstein, S.C., 458–459
Feix, J.B., 9–10, 94, 194–195, 222–223, 230, 236–237, 239, 242, 316, 320
Feng, W., 291
Feng, Z., 444
Fernandez, M.A., 154–156
Fernandez, M.S., 192–193, 204, 207–210
Ferré-D'Amaré, A.R., 194, 407
Fersht, A.R., 30
Fields, A., 32, 49
Fine, R.A., 41–42
Finn, M.G., 317
Fiori, W.R., 79–80
Fischer, D.B., 530–531
Fisher, A.J., 110–111
Fisher, E.M., 531–532
Fleischer, S., 274
Fleischman, A.J., 538
Fleissner, M.R., 4, 6–7, 15–19, 21–22, 32–34, 39, 63–64, 66–68, 73–75, 78–79, 90–92, 102–104, 109–110, 112–113, 115, 118–119, 141–142, 156–157, 224, 270–271, 318–319, 340, 393–394, 487
Flitsch, S.L., 113, 292, 316, 320
Flores Jimenez, R.H., 92
Floyd, S.R., 260–261
Focia, P.J., 293, 390
Fogassy, E., 225–226
Folea, I.M., 353–354, 380–381
Folli, F., 260
Foltz, G.N., 92
Font, J., 30–31
Forbes, M.D.E., 79–80
Formaggio, F., 225–226, 248–250
Forman-Kay, J.D., 194
Fornes, J.A., 208–210
Forrer, J., 412
Forrest, L.R., 336, 350
Fourme, R., 30–31
Fox, B.A., 379
Franck, J.M., 458–461, 460f, 464–466, 470, 472–474, 477
Frank, A., 530, 545–546
Fraser, R.R., 357–358
Frauenfelder, H., 47, 318–319

Frazier, A.A., 238–239, 262–263, 272–273
Freed, D.M., 68, 92, 351, 353–354, 380, 440
Freed, J.H., 6–7, 14–17, 22, 32–34, 45,
　65–68, 72, 105, 113, 142–143, 158,
　168–170, 221, 249–250, 292, 319,
　322–323, 329–330, 353–354, 361, 363,
　365–367, 370, 379, 404–406, 415–416,
　420, 443, 458, 467–468, 476, 503
Freedberg, D.I., 154–156
Fried, S.D., 192–193
Fries, P., 468
Fritzsching, K.J., 291–292
Fromherz, P., 192–194, 207–210
Froncisz, W., 4–7, 9–12, 68–69, 71–72,
　80–81, 82f, 83, 236–237, 239, 320, 326,
　338–339, 358–360, 466
Frushicheva, M.P., 431–433
Fu, G., 155–156
Fuchs, M., 201–202, 409, 512
Fuglestad, B., 32
Fujii, H., 512, 567
Fumagalli, L., 192–193
Fyles, A.W., 530–531

G

Gacho, G.P., 433–434, 444
Gadisetti, C., 518–519
Gaertner, F.C., 530–531
Gaffney, B.J., 192–193
Gafurov, M., 465–466
Gagnon, D.G., 353–354, 380
Galiano, L., 155, 157–170, 163f, 440
Gallez, B., 531–532, 535–536
Gallice, J., 468
Gallop, J.L., 260–261, 263–264, 266,
　270–272, 274–275, 277–279, 283
Gandra, S., 326, 412
Ganesh, T., 354–355
Gao, M., 15
Gaponenko, V., 489
Garber, S.M., 112
Garman, E.F., 291
Gary-Bobo, C.M., 192–193
Gasmi-Seabrook, G., 489
Gast, P., 513–514, 532–533, 537
Gekko, K., 30–31
Gelb, M.H., 238–239, 320

Georgieva, E.R., 160, 162, 164–165,
　351–354, 370, 430, 440
Gerfen, G.J., 415
Germano, S.M., 544
Gerwert, K., 326, 328–329
Gesell, J.J., 294–298
Getz, E.B., 225
Ghcechik, V., 428–429
Ghimire, H., 80, 142–143, 225–226, 250,
　292, 378
Giacomini, K.M., 351
Giannella, E., 507–508
Gilbert, B.C., 428–429, 514–515
Gilchrist, M.L., 354–355
Gill, S., 192–193
Gillies, R.J., 507–509, 512–513
Gilliland, C.T., 154–156
Gilliland, G., 444
Gimenez-Lopez, M.d.C., 464–466
Gimi, B., 531–532, 535, 539
Ginter, C., 353–354, 370
Girard, E., 30–31
Gitti, R., 458
Gladden, J.A., 320
Glaser, R., 193–194
Glaser, S.J., 420
Glass, T.E., 458
Glinchuk, M.D., 39
Glockner, J.F., 503, 507–508
Gnewuch, C.T., 562
Go, A.S., 530
Goda, F., 531–532, 535–537
Godt, A., 32–34, 50, 113–115, 126–128,
　132–133, 141–143, 148–149, 158,
　168–173, 322, 332–335, 361, 366–367,
　404–406, 408, 410–412, 413f, 415–416,
　420, 439, 443
Goh, K.I., 351
Gohain, S., 539–541
Golczak, M., 379
Goldfarb, D., 292–293
Goldman, S.A., 105
Gomila, G., 192–193
Góngora-Benítez, M., 294
Gonzales, E.G., 155, 158–160, 165–167
Gonzalez-Bonet, G., 322–323
Gonzalezmartinez, M.T., 204
Goodenow, M.M., 155–156

Goor, O.J.G.M., 458
Gophane, D.B., 407–408
Gordeliy, V.I., 329–330
Gordon-Grossman, M., 458–461
Gore, J., 417
Gorka, J., 297
Goto, Y., 30–31
Gouaux, E., 350–351, 380
Gould, L., 508
Gramlich, V., 408, 415–416
Grant, G.H., 248, 322
Grant, G.P.G., 407, 430–434, 431–433f, 437–438, 444–446
Graslund, A., 415
Gray, L.H., 502, 518–519
Grdina, D.J., 507–508, 531–532, 538
Green, B., 292
Green, P.R., 395–397
Greenfield, N.J., 291–292
Greenhalgh, D.A., 80, 83–85, 84f, 238–239, 262–263, 272, 274, 292, 319–320, 360
Greenhough, T.J., 491–494, 493f
Greensto, C.L., 564
Grieser, F., 208–210
Griesinger, C., 461
Griffeth, V.E., 530–531
Griffin, L., 351
Griffin, R.G., 415
Griffith, O.H., 65–66, 222–223
Griffiths, J.R., 503–504, 530–531
Grigor'ev, I.A., 194–202, 440
Grigoryants, V.M., 94, 160, 162, 164–165, 430, 440
Grigoryev, I.A., 147–148
Grinberg, O.Y., 508, 516, 531–532, 535–536, 540–543
Grinberg, S., 542–543
Grinberg, S.A., 535, 540–541
Grinberg, V.O., 535
Grishaev, A., 485–486
Grobner, G., 192–193
Groenen, E.J.J., 248, 322
Gromov, I., 412
Gronenborn, A.M., 155–156, 485–486, 491–494
Gross, A., 11, 156–157, 227–228, 243–244, 263, 272, 274–275, 292–293, 316, 322–323, 390–399

Gross, M., 290–291
Grote, M., 322–323, 335–336, 340
Groth, N., 512
Grozinger, G., 530–531
Grucker, D., 556
Gruner, S.M., 32, 42, 49
Grunwald, J., 5f, 62, 126–128, 222–223, 316–317, 356
Gryczynski, Z., 116–118
Guiberteau, T., 556
Gulbis, J.M., 395–396
Gulla, A.F., 194–195
Gunasekara, L., 192–193
Guo, L., 297
Guo, Z., 7, 19, 21–22, 47–48, 63–64, 66–68, 73–75, 90–92, 94, 270–271, 277–279, 318–319, 340
Guskov, A., 350
Gustchina, A., 154–155, 154f
Guzzi, R., 262–263, 293

H

Ha, T., 126–128
Haas, D.A., 4–7, 466
Haas, R., 530–531
Hadden, J.M., 380
Haehnel, W., 225–226
Hafner, J.H., 192–193
Hagelueken, G., 324, 430
Hahn, E.L., 458
Hahn, E.W., 504
Haigler, H.T., 11–12, 69–70, 82f, 83–85, 270–271, 273, 317–318
Halazonetis, T.D., 449f
Haldar, S., 291
Halkides, C.J., 404
Hall-Porter, M.R., 545–546
Halpern, H.J., 507–508, 510–512, 514–516, 519–521, 531–533, 538
Hamada, A., 554–555, 563
Hamann, S.D., 41
Hamelberg, D., 155
Hamm, H.E., 93–94
Hammarstrom, P., 135–136, 147–148
Hammond, C.M., 135–136, 141–142, 322
Han, O.H., 458–459
Han, S.-I., 292, 458–479, 460f, 478f
Hänelt, I., 322–323, 353–354

Author Index

Haney, C.R., 516, 518–519, 554–555
Hankovszky, H.O., 5f, 62, 222–223, 316–317, 356, 435
Hanna, N., 518–519
Hannongbua, S., 160–161
Hansen, D.F., 30, 49–52
Hanson, P., 225–226
Harbury, P.A.B., 417
Harlos, K., 205
Harmer, J.R., 14, 32–34, 142–143, 249–250, 405
Harris, J.R., 291
Harrison, D.K., 503–504
Harrison, R.W., 155–156
Hartford, A.C., 516, 531–532, 535–536
Harth, E., 438
Hartmann, H., 47
Hartsuck, J.A., 155–156, 160–161
Hassner, A., 221
Hata, K., 30
Hatmal, M.M., 430–434, 438, 443–445, 447–448
Hauser, S.L., 510
Hausser, K.H., 458–461, 464, 470
Havelka, J.J., 262–263, 272–273
Hawker, C.J., 438
Haworth, I.S., 260–285, 353–354, 407, 428–450, 431–433f
Haydon, D.A., 234–236
He, G.L., 513–514, 554–555
Hecht, J.L., 192–194
Heerklotz, H., 41
Hegde, B.G., 260–261, 262f, 263–264, 266, 270–275, 277–279, 281–283, 353–354, 430
Hegde, P.B., 260–261, 263–264, 266, 270–275, 277–279, 283, 430
Hein, P., 516, 535 536
Heinz, D.W., 49
Hekmatyar, S.K., 531–532, 535, 539
Hemker, R., 5–6, 11–12, 68–69, 71–72, 80–81, 320, 338–339, 358–360
Hemminga, M.A., 139–140, 146, 316
Henderson, I.R., 136, 137f
Hengstenberg, C.S., 324
Henk, J.M., 502
Henry, E.R., 49
Henzler-Wildman, K.A., 19, 30, 60

Herberhold, H., 30–31
Hernández-Guzmán, J., 293
Herrling, N., 512
Herrmann, C., 324
Herschlag, D., 431–433
Hideg, K., 4–7, 5f, 11–17, 13f, 19–22, 23f, 32–35, 47–48, 60–94, 82f, 102–107, 109–110, 112–113, 115, 118–119, 141–142, 156–157, 160, 194–197, 222–228, 243–244, 248, 270–271, 276–279, 316–319, 321–323, 328–329, 355–358, 390, 393–398, 407, 431–435, 444, 487
Hilger, D., 19, 30, 249–250, 336, 337f, 338–340, 378
Hill, J.A., 530
Hill, R.P., 530–531
Hiller, S., 261–262
Hills, R., 78
Hilser, V.J., 19
Hinderliter, A., 262–263, 272–273
Hindman, J.C., 473–474
Hinnah, S., 391–392
Hirata, H., 567
Hirayama, B.A., 350
Hirsh, D.J., 17–18
Hirst, J., 405
Hirst, S.J., 292, 368–370, 430
Hiruma, Y., 487
Hleihel, D., 516, 518
Höbartner, C., 407, 414
Hockel, M., 530–531
Hodge, S., 531–532, 535, 539
Hoeckel, M., 530–531
Höfer, P., 461, 464–465
Hoff, A.J., 71–72
Hoffman, B.M., 298–299
Hofmann, K.P., 72 73, 321 323, 379, 458–459
Hofrichter, J., 49
Hoganson, C.W., 415
Hogg, N., 49
Holden, H.M., 110–111
Holmes, K.C., 110–111
Holt, A., 249
Holterhues, J., 78, 321, 328–329
Holton, J.M., 141–142
Holtzmann, E., 485–486, 491

Hon, B., 49
Hon, W.-C., 260–261
Hong, H., 12
Hong, L., 155–156, 160–161
Hong, M., 291–292
Honig, B., 192–194
Hope, M.J., 204
Hopkins, A.L., 351
Hopkins, P.B., 407
Horanyi, P.S., 64, 68, 92, 351, 358, 380, 440
Hori, T., 379
Hornak, V., 154–156
Horne, W.S., 15, 64, 68
Horváth, L.I., 262–263
Horwitz, A.F., 222–223, 227
Horwitz, J., 21–22, 32, 37–39, 38f, 42–44, 46f, 49–52, 51f
Hou, H., 507–508, 516, 530–546
Houk, K.N., 14–18, 16f, 64, 72
Howard, E.C., 102, 105–106
Howard, K.P., 112
Howarth, J.W., 489
Hruby, V.J., 357–358
Hu, D.H., 192–193
Hu, N.J., 351
Hu, R.B., 192–193
Huang, C.H., 227
Huang, S., 292
Huang, X., 155–160, 163–168
Hubbell, C.M., 60–61, 270–271, 379
Hubbell, W.L., 4–23, 13f, 16f, 23f, 30–53, 33f, 38f, 43f, 46f, 48f, 51f, 60–94, 79f, 82f, 84f, 89f, 102, 105–107, 113, 156–158, 160, 192–197, 221, 227–228, 230, 236–239, 242–246, 248, 262–263, 270–279, 292, 316–323, 326, 338–340, 354–361, 365, 379, 390–391, 394–398, 404–407, 420, 428–429, 431–434, 444, 458–459, 461, 464–465, 477–479, 478f, 487, 489
Huber, M., 135–136, 147–148, 320–321, 405
Huber, T., 291
Hubrich, M., 408, 415–416
Huisjen, M., 4
Hulse, R.E., 396–397
Hume, D.A., 291
Hummer, G., 32

Hunjan, S., 504
Hunt, J.N., 458–459, 472
Hunter, R.I., 142–143, 414
Hussain, S., 458–459
Hustedt, E.J., 61–64, 80, 225–226, 292, 319, 361, 365–367
Huster, D., 225–226
Hutchison, C.A., 160–161
Hutchison, J.M.S., 558
Hwa, J., 84–85
Hwang, D.S., 458–459, 472
Hwang, L.-P., 467–468, 476
Hyde, J.S., 4–11, 61–64, 71–72, 82f, 83, 230, 236–237, 239, 320, 326, 358–360, 466, 507–508
Hyodo, F., 555, 561–562, 564, 566–567

I

Ibragimova, M.I., 531–532
Ichikawa, K., 554–556, 561–564, 566–567
Ikura, M., 487, 491–494
Ilangovan, G., 512–514, 531–533, 537
Iliceto, A., 104, 118–119
Ilnicki, J., 5–7, 12, 68–69, 239, 320
Inbaraj, J.J., 80, 112, 225–226, 292
Inman, J.K., 225
Inoue, J., 233–234
Inoue, K., 233–234
Ionita, P., 50, 141, 149, 170–173, 333–335, 366–367, 404–405, 413f, 443
Isaacs, N.W., 290–291
Isaacson, R.A., 354–355
Isas, J.M., 11–12, 69–70, 82f, 83–85, 260–261, 263, 266, 270–271, 273, 317–318
Ishima, R., 154–156
Ishitsuka, H., 233–234
Islam, S.M., 104, 115, 350–351, 353–354, 367–373, 380
Isselbacher, K.J., 509
Itaya, K., 227
Ito, A.S., 208–210
Ito, J., 290
Ito, R., 567
Ito, S., 561–562, 566–567
Ito, Y., 563
Itoh, T., 41, 42t
Ivanov, A.V., 420

Iwahara, J., 485–491, 490f
Iwarsson, E., 260
Iwasaki, A., 516, 531–532

J

Jackson, S.K., 531–532
Jacob, J., 154–156
Jahn, R., 270–271
Jahn, W., 110–111
Jain, R.K., 518–519
James, L.C., 19
James, P.E., 531–532
James, Z.M., 102, 105–106, 112
Jameson, J.L., 510
Janovick, J.A., 290
Janzen, E.G., 230–231
Jao, C.C., 260–261, 263–264, 266, 270–272, 274–275, 277–279, 281–283, 353–354, 430
Jardetzky, O., 336
Jaroniec, C.P., 64, 68
Jarvis, L.A., 531–532, 535–536, 539
Jastrzebska, B., 379
Javitch, J.A., 336, 351–353, 360, 372–373, 376–379
Jayasinghe, S., 270–271
Jeko, J., 4, 7, 34–35, 62, 64–66, 68, 85, 88, 102–103, 105–107, 160, 225–226, 270–271, 317–318, 356–358
Jennewein, M., 350
Jensen, M.B., 260–261, 266
Jensen, M.R., 487
Jeong, S., 538
Jeschke, G., 14–17, 32–34, 37, 50, 51f, 72, 113–116, 126–128, 132–133, 139–143, 146, 148–149, 158, 162–164, 168–173, 182, 249–250, 263, 277, 298–301, 320, 322 324, 326, 332 336, 338, 340, 361, 363–368, 404–406, 408, 410–412, 413f, 415–417, 420, 429–430, 439, 442–444, 458–459, 464–465, 470, 504, 506
Jiang, E., 141–142
Jiang, J., 354–355, 535–537
Jiang, Y., 391–392, 397–398
Jiao, D., 357–358
JiJi, R.D., 291–292
Jimenez, Y.L., 155–156
Jimenez-Oses, G., 14–18, 16f, 72

Jin, Y., 351, 380
Jogini, V., 322–323, 353–354, 380
Johnson, M.A., 249–250
Jonas, U., 171–172, 404–405
Jones, A., 291
Jonsson, B.H., 135–136
Jordan, B.F., 530–532, 535–536
Jordi, W., 192–193
Joseph, B., 458–459, 464–465, 470
Joseph, J., 354–355
Joseph, M., 531–532, 539, 545–546
Jost, M., 225–226
Jost, P.C., 65–66
Jun, S., 15–18
Jung, H., 249–250, 337, 337f, 339–340, 366–367, 378
Jung, K., 337

K

Kaback, H.R., 395
Kadner, R.J., 380
Kálai, T.K., 4, 6–7, 15–17, 34–35, 47–48, 62–68, 85, 88, 90, 92, 102–107, 109–110, 112–113, 115, 118–119, 141–142, 160, 224–226, 270–271, 317–318, 356–358, 393–394, 431–433, 487
Kallinowski, F., 530–531
Kalyanaraman, B., 49
Kameneva, M.W., 512
Kaminker, I., 458–477, 479
Kang, H.J., 290–291
Kang, T.H., 458–459
Kao, J.P.Y., 405, 512
Kaptein, R., 129
Karczmar, G.S., 518–519
Karim, C.B., 102, 105–106, 116, 224–225, 321–322, 406
Karplus, M., 47, 60
Kasper, D.L., 510
Kast, D., 113–115
Kathirvelu, V., 138–139, 405, 440
Katoh, E., 155–156
Kaufman, A., 395–396
Kaufman, P.D., 141–142
Kausik, R., 458–459, 464–465
Kauzmann, W., 41, 42t
Kawasaki, K., 4–6
Kawate, T., 351, 380

Kay, A.R., 504
Kay, C.W.M., 405
Kay, L.E., 30, 49, 60–61, 194, 351–353, 487
Kazmier, K., 32–34, 350–381, 390–391
Kear, J.L., 155–160, 163–168
Kear-Scott, J.L., 155, 158–160, 165–168
Keddie, S., 530–531
Keizers, P.H.J., 487
Kelloff, G.J., 507–509, 512–513
Kelly, B.L., 394–395
Kendrew, J.C., 126–128
Kennedy, K.A., 502
Kent, H.M., 260–261, 263, 270–271
Kent, S.B., 155
Kern, D., 19, 30, 60
Kerns, S.J., 60
Kerr, I.D., 292
Kesarwani, M., 531–532
Kesselring, E., 504, 554–555
Kessels, M.M., 260–261
Kessler, N., 458–461
Keszler, A., 49
Kevan, L., 7–8, 10, 300, 302–303
Khairy, K., 113, 319
Khaled, A.R., 220
Khan, A.K., 68, 440
Khan, M.N., 507–508, 531–532, 535, 539, 545–546
Khan, N.W., 507–508, 516, 530–546
Khanna, S., 530
Khelashvili, G., 351–353, 376
Kheterpal, I., 242–243
Khorana, H.G., 60–62, 69–70, 72–73, 80–81, 82f, 83–85, 84f, 102, 113, 238–239, 262–263, 270–272, 274, 292, 316, 319–323, 328–329, 354–355, 360, 379, 404–405, 458–459, 487
Khramtsov, V.V., 15–17, 62–64, 192–193, 195–198, 208–210
Kida, I., 530–531
Kiihne, S., 12
Killian, J.A., 249
Kim, C.S., 78
Kim, C.U., 42
Kim, J.H., 538
Kim, J.-M., 458–459
Kim, M., 92, 157–158, 160, 162–163, 322–323, 380, 458–459

Kim, N., 443
Kim, P.S., 192–193
Kim, R.B., 351
Kim, S.H., 538
Kim, T.H., 19, 30
Kim, Y., 294–298
Kim, Y.J., 379
King, G.J., 291
King, M.S., 405
Kinnebrew, M., 458–459
Kinoshita, Y., 555, 561, 563, 567
Kirby, T.L., 105–106, 116, 406
Kirilina, E.P., 440
Kirilyuk, I.A., 194–197, 199, 204, 208–210
Kisselev, O.G., 322–323
Kitahara, R., 30–31, 50
Kitamura, Y., 41, 42t
Klare, J.P., 60–61, 78, 316–340, 355–357
Klebba, P.E., 94
Klee, C.B., 491–494
Klei, L., 531–532
Klein, J.C., 102, 105
Klein, M.B., 160–161
Klein, M.P., 227
Kleinschmidt, J.H., 208–210
Klein-Seetharaman, J., 84–85
Klimenko, V.P., 39
Kloda, A., 380, 390
Klose, D., 321, 328–331, 335–336, 406, 416–417
Klug, C.S., 94, 194–195, 236–237, 242, 274, 316
Knisely, J.P., 530–531
Kobayashi, R., 561–562, 566–567
Kobayashi, T., 233–234
Kobe, B., 291
Koch, A., 171–172, 404–405
Koch, D., 260–261
Kocherginsky, N., 562
Koda, S., 227
Kodibagkar, V.D., 530–531
Koeppen, M., 530, 545–546
Koharudin, L.M.I., 129
Koide, S., 30
Kolb, H., 317
Kolberg, M., 415

Kolodziej, A.F., 322–323
Kono, M., 458–459
Konstantinova, T.E., 39
Kooser, R.G., 503
Korsmeyer, S.J., 394–395
Korzhnev, D.M., 19, 30
Kosem, N., 555, 563–564
Kosen, P.A., 487, 489
Koshland, D.E., 322–323
Koshy, C., 350
Kotake, Y., 230–231
Kozlov, M.M., 260–261, 262f
Krahn, A., 464–465
Kramer, R., 350
Krasnaberski, A., 78, 321
Krause, B.J., 504
Krauss, N., 379
Krebs, M.P., 321, 328–329
Kreitman, M.J., 32, 43–44, 49–52, 51f
Kretz, C., 260
Krishna, M.C., 512–513, 530–533, 555
Krishnamurthy, H., 350, 380
Krishnamurthy Nemani, V., 531–532, 535, 539
Krogstad, D.V., 458–459, 472
Krohn, K.A., 530–531
Kroncke, B.M., 64, 88–89, 292, 358
Kroon, G., 350
Krstic, I., 428–430, 439
Krummenacker, J.G., 458
Krusong, K., 380
Kubota, T., 290–291
Kuchinka, E., 234–236
Kudo, W., 555, 566
Kuech, T.R., 192–193
Kuipers, O.P., 233–234
Kulik, L.V., 147–148
Kulkarni, A.C., 530
Kull, F.J., 110–111
Kumasaka, T., 379
Kummerlen, R., 391–392
Kunkler, P.B., 502
Kuntz, I.D., 47, 489
Kuo, A., 395–396
Kuppusamy, P., 507–508, 512–514, 516, 530–546, 555
Kuriyan, J., 47
Kurochkina, N., 290–291

Kusnetzow, A.K., 19, 21–22, 63, 72–75, 90–92, 321–323, 379, 407, 430–434, 431–433f, 437–438, 444–446, 458–459
Kusnetzow, A.N., 318–319, 340
Kustanovich, I., 154–156
Kusumi, A., 4–6, 239
Kuszewski, J.J., 489
Kuwata, K., 30–31
Kweon, D.H., 78
Kwiatkowski, P., 545–546

L

Labahn, J., 329–330
Labar, D., 530–531
Labeikovsky, W., 19
Lacroix, J.J., 290–291
Ladizhansky, V., 291
Ladokhin, A.S., 291
Laeremans, T., 351
Laggner, P., 192–193
Lai, C.-L., 260–261, 263, 266, 272
Lakowicz, J.R., 116–118
Lally, B.E., 530–531
Lam, K.S., 220
Lambert, J.B., 357–358
Lamy-Freund, M.T., 192–193
Landreh, M., 290
Lange, O., 30, 49–52
Langen, J., 260–261, 263, 266, 270–271, 273
Langen, R., 11–12, 69–70, 82f, 83–85, 156–157, 227–228, 242–243, 260–285, 262f, 292, 316–318, 339, 353–354, 357–358, 390–391, 430, 458–459, 461, 464–465, 478f, 479
Lanyi, J.K., 321
Lao, K.Q., 192–193
LaPointe, N.E., 458–459
Lardinois, O., 354–355
Lariviere, J.P., 531–532, 535, 539
Larrabee, A.L., 227
Larsen, R.G., 404, 409–410
Laryukhin, M., 112
Lasic, D.D., 227
Lauricella, R., 49
Lauritzen, C., 487
Laursen, I., 512–513
Lauterbur, P.C., 502–503
Lawson, A., 351

Layten, M., 154–156, 158–161, 165–167
Le, M.U., 417
le Coutre, J., 395
Le Trong, I., 379
Led, J.J., 487
Lee, A., 391–392, 397–398
Lee, C., 290–291
Lee, D., 554–555
Lee, J.Y., 92, 162–163, 380
Lee, K.Y.C., 458–459, 464–465
Lee, M.W., 220
Lee, R.T., 80
Lee, S., 22, 41, 45, 65–66, 113, 141–142, 329–330, 538
Lee, S.C., 514, 532–533, 537–538
Lee, T.M., 504
Lees, A., 225
Lefer, D.J., 507–508
Lei, M., 60
Lemke, E.A., 317
Lenz, V., 248, 320–322, 335–336
Leonenko, Z., 192–193
Lerch, M.T., 4–23, 30–53, 38f, 43f, 46f, 51f, 61–62, 68–69, 84–85, 90, 92
Lesniewski, P., 516, 531–532
Leunbach, I., 512–513
Levadny, V., 208–210
Levin, D.N., 502–503
LeVine, M.V., 376
Levitt, M., 290–291, 350
Lew, J., 458–459
Lewis, A., 380
Lewis, M., 512–513
Lewis, T.E., 350
Lex, L., 435
Lezon, T.R., 291
Li, C., 519
Li, H., 30–31, 404–405, 507–508, 513–514, 516, 531–533, 535–537, 539–543, 555
Li, M., 154–156
Li, Q., 393–394, 397–398, 544
Li, Y., 430
Liang, Z., 156–157, 292, 319
Liang, Z.C., 113
Lichte, B., 260
Lietzow, M.A., 11, 60–64, 68, 73–75, 80, 83–85, 94, 156–157, 194–197, 227–228, 263, 270–272, 274–275, 292, 316–318, 339, 355–358, 360, 390–391

Likhtenshtein, G.I., 193–194
Lin, A.Y., 102
Lin, L., 351
Lin, M.T., 554
Lin, Q., 317
Lin, X., 155–156, 160–161
Lin, Y., 12, 69–70
Lindgren, M., 135–136, 138–139, 147–148, 440
Lindner, S.E., 141–142
Lindstrom, F., 192–193
Lingwood, M.D., 458, 470, 472
Link, J.M., 530–531
Lipiszko, K., 336, 338, 340
Liska, N., 260–261
Liu, F., 542–543
Liu, G., 464–466
Liu, J., 94
Liu, K.J., 513–514, 530, 532–533, 535–537
Liu, L., 290–308
Liu, S., 530–531
Liu, Y., 14–17, 365, 420
Liu, Y.-S., 78, 390–391, 396–397, 404–405
Liu, Z., 292
Liu, Z.L., 156–157
Livshits, V.A., 319–320
Lo, R.H., 88–89
Lockhart, D.J., 192–193
Lodowski, D.T., 379
Loeb, D.D., 160–161
Logan, D.T., 415
Logan, T.M., 443
Lohse, S.E., 192–193
Long, J.R., 242
Longo, D.L., 510
López, C., 458–459, 461, 464–465, 477–479, 478f
López, C.J., 14–19, 16f, 21–22, 32, 33f, 34–37, 38f, 39, 42–44, 47, 49–52, 51f, 60–94, 89f, 156–157, 292, 318–319, 340, 356–357, 406, 428–429
Lorigan, G.A., 112, 142–143, 225–226, 250, 290–308, 378
Loscalzo, J., 510
Lottmann, P., 464–465
Louis, J.M., 135–136, 149, 154–156, 164, 292, 485–486, 491
Lovett, B.W., 142–143
Lovett, J.E., 142–143

Lu, J.X., 112
Lu, P., 350
Lu, Y., 430–431, 432f, 433–434, 443–445, 447–449, 449f
Lubitz, W., 405–406, 415
Luchinat, C., 12, 15, 461, 464–465
Ludwig Brand, M.L.J., 49
Luecke, H., 270–271
Luger, K., 141–142
Lukac, S., 208–210
Lukasik, S.M., 353–354
Lukszo, J., 562
Lundmark, R., 260–261, 266
Lundström, P., 30
Luo, C., 141–142
Luptak, A., 194
Lurie, D.J., 555–556, 558
Lybrand, T.P., 319
Lyles, D., 107–108
Lyman, E., 260–261, 263, 266, 272
Lynd, N.A., 458–459, 472
Lynn, D.G., 293

M

Ma, D., 350
Ma, Z., 292
Macdonald, P.M., 192–193
MacKinnon, R., 142–143, 322–323, 391–392, 395–398, 406
Macleod, M.R., 544
MacMillan, F., 407
Mäde, V., 294
Mader, K., 535–536
Madhani, M., 531–532
Maeno, A., 30
Magat, J., 530–531
Magnani, F., 351
Magwood, J.M., 519
Mailer, C., 4–7, 12, 69–70, 320, 466, 508, 511–512, 514–516
Mainali, L., 358–360
Mal, T.K., 487
Mallard, J.R., 555, 558
Maltempo, M.M., 502–503
Malygin, A.A., 420
Manchester, M., 160–161
Mandal, K., 155
Manglik, A., 19, 30
Maniero, A.L., 248

Manivannan, A., 532–533, 537
Manivannan, K., 234–236
Manogaran, S., 489
Marassi, F.M., 294–298
Marchal, S., 30–31
Marchetto, R., 225–226
Margittai, M., 242–243, 270–271, 274–275
Margraf, D., 115–116, 249–250, 405, 410–411, 416–418, 428–430, 439
Marien-Cortez, D., 322–323
Marín, N., 225–226
Markley, J.L., 32
Marko, A., 115–116, 404–420, 428–430, 439
Marmorstein, R., 141–142
Marquardsen, T., 464–465
Marr, R.B., 502–503
Marra, J., 192–193
Marsh, D., 66–68, 192–193, 195, 205, 208–210, 225–226, 233–234, 262–263, 293, 319–320
Marshall, F., 290
Marshall, G.R., 322–323
Marshall, I., 544
Martell, S., 78, 321, 328–329
Martens, C., 353–354, 380
Martens, S., 260–261, 266
Martensson, L.G., 135–136, 147–148
Marti, T., 62, 69–70, 81, 82f, 102, 316, 354–355, 404–405, 487
Martin, F.J., 227
Martin, J.B., 509
Martin, J.L., 291
Martin, P., 155–156
Martin, R.E., 408, 415–416
Martinac, B., 380, 390
Martinez, T.J., 249–250
Martinez-Cajas, J.L., 154–155, 160–161
Maryasov, A.G., 249–250, 409–410, 443
Maschke, W., 205
Mason, R.P., 504, 530–531
Massey, A.P., 407
Masuda, S., 554–555
Masureel, M., 19, 30, 353–354, 380
Mata, A., 538
Matalon, E., 293, 458–461
Mathew-Fenn, R.S., 417
Mathias, R.T., 234–236

Matsumoto, K., 512–513, 532–533, 554–555, 563–564, 567
Matsumoto, S., 518–520, 532–533, 555, 561, 567
Matsuoka, T., 227
Matthews, B.W., 32, 49
Mayer, A., 530–531
Mayer, K.M., 192–193
Mayer, L.D., 204
Mayo, D.J., 80, 225–226, 290–308
Mazaleyrat, J.P., 225–226
McCaffrey, J.E., 102–120
McCarney, E.R., 458–459, 470, 472
McCarrick, R.M., 80, 115, 142–143, 225–226, 250, 290–308, 378
McConnel, H.M., 222–223
McConnell, H.M., 105f
McCoy, J.J., 21–22, 34–39, 38f, 42, 45, 46f, 47–50, 48f, 318–319, 340
McCullough, L.D., 542–543
McDevitt, C.A., 292
McGoff, M.S., 64, 68
McHaourab, H.S., 5–6, 11–12, 14, 32–34, 42, 60–64, 68–69, 71–75, 78–81, 79f, 94, 156–158, 194–197, 227–228, 230, 243–246, 270–271, 292, 316–320, 338–339, 350–381, 390–391, 420, 430
McIntosh, T., 239
McIntyre, J.O., 274
McLaughlin, A., 192–194
McLaughlin, S., 192–194, 234–236
McLuskey, K., 290–291
McMahon, H.T., 260–261, 262f, 263–264, 266, 270–272, 274–275, 277–279, 283
McMahon, R.M., 292
McNulty, J.C., 225–226, 406
Meade, T.J., 292
Meador, W.E., 491–494, 493f
Means, A.R., 491–494, 493f
Meenakshisundaram, G., 514, 532–533, 537–538
Meeth, K., 141–142
Mehlhorn, R.J., 192–193
Mehta, A.K., 293
Meiler, J., 292, 367–370, 373, 430
Meinecke, M., 260–261
Melby, E.S., 192–193
Mello, R.N., 102, 105

Melnyk, O., 294
Metzler, W.J., 129
Meyer, V., 14, 42, 365, 420, 512
Micallef, A.S., 4–6, 12–14
Mich, R.J., 192–193
Michaels, D.M., 227
Michel, E., 430
Mielke, T., 260–261, 263, 266, 272, 274–275
Mignion, L., 530–531
Miick, S.M., 79–80
Milidonis, X., 544
Miller, C., 391–392
Millero, F.J., 41–42
Millet, O., 19
Millhauser, G.L., 79–80, 225–226, 406
Mills, F.D., 242
Mills, I.G., 260–261
Milosevic, M., 530–531
Milov, A.D., 126–128, 132–133, 249–250, 404, 408–410
Milshteyn, E., 156–157, 292
Mims, W.B., 458
Minicucci, M.F., 545–546
Mintzopoulos, D., 531–532
Mirza, O., 380
Mishra, S., 350–381
Mitchell, J.B., 530–533
Mito, F., 563
Mitrikas, G., 300
Mittal, R., 260–261
Mittermaier, A.K., 49, 60–61, 351–353
Miura, Y., 554–555
Miyakawa, R., 566
Mizuno, N., 260–261, 266
Möbius, K., 201–202, 320–321, 409
Moen, R.J., 102, 105
Moenke, G., 260–261, 263, 266, 272, 274–275
Mohsen, M., 562
Mokdad, A., 353–354
Molin, Y.N., 243, 503, 507–509
Mollaaghababa, R., 321, 328–329
Monaco, V., 225–226
Monien, B., 225–226
Montano, R., 531–532, 535, 539
Monticelli, L., 192–193
Mooberry, E.S., 32

Moodie, K., 539–541
Moon, J., 538
Moore, T.A., 510
Moradi, M., 396–397
Moraes, I., 290
Morais-Cabral, J.H., 395–396
Morash, J., 394–395
Morén, B., 260–261, 263, 266, 272, 274–275
Morgalla, M.H., 530–531
Morgan, L.O., 489
Morizumi, T., 379
Morse, P.D., 199–200, 562
Moscicki, J.K., 113, 156–157, 319
Motlagh, H.N., 19
Motoshima, H., 379
Mottley, C., 466
Mouradov, D., 291
Moussavi, M., 513–514, 532–533, 537
Mozaffarian, D., 530
Mu, Y., 115–116, 405, 407, 416–417
Mueller, L., 129
Mugnaini, V., 406
Mulder, F.A., 49
Muller, D., 391–392
Müller, M., 319
Munasinghe, J.P., 519–520, 555
Mundt, A.J., 518–519
Münnemann, K., 458, 464–465
Mupparaju, S.P., 531–532, 535, 539–541
Murphy, D.M., 428–429, 514–515
Murphy, E.C., 487
Murphy, M.P., 554
Murray, D., 12, 69–70, 157–158, 160
Murugesan, R., 512–513, 515, 530–531, 555–556
Muto, E., 554–555

N
Naber, N., 102, 103f, 105–106, 107f
Naderi, H., 489
Naganuma, T., 555, 560, 564, 566
Naidu, S.K., 531–532, 539, 545–546
Nair, P., 503–504
Naismith, J.H., 324, 430
Nakagawa, K., 138–139
Nakaie, C.R., 225–226
Nakamura, M., 105

Nakano, K., 563
Nakashima, T., 535–537
Naruse, M., 554–555, 563–564
Nascimento, O.R., 192–193
Needleman, R., 321
Negrashov, I.V., 102
Nelson, A.C., 510
Nelson, W.D., 84–85
Nemunaitis, J., 518–519
Nesmelov, Y.E., 84–85, 105–106, 116, 292, 406
Neudecker, P., 30
Neudert, O., 458, 464–465
Neuman, R.C., 41, 42t
Newstadt, J.P., 225–226
Newstead, S., 290, 351
Neyton, J., 391–392
Nguyen, P.H., 428–429, 431–433
Nicholson, I., 556, 558
Nicot, A.-S., 260
Nielsen, R.D., 4–5, 12, 69–70, 238–239, 320
Nietschke, M., 366–367
Nigam, S.K., 351, 564
Nilsson, L., 192
Ninham, B.W., 208–210
Nishimura, M.C., 542–544
Nissen, P., 376
Nji, E., 351
Noble, D., 545–546
Nöllmann, M., 417
Nomura, M., 227
Norberg, J., 192
Norberto, D.R., 32
Norby, S.W., 513–514, 532–533, 535, 537, 543–544
Norel, R., 192
Norman, D.G., 126–150, 164
Novoselsky, A., 193–194
Nucara, A., 291–292
Nucci, N.V., 32
Nussinov, R., 60

O
Oblatt-Montal, M., 294–298
Oesterhelt, D., 326, 328–329
Oga, S., 19, 33f, 34–37, 38f, 47, 88, 89f
Oganesyan, V.S., 319

Ogawa, S., 222–223, 504
Ogletree, D.F., 126–128
Oh, K.J., 11, 32–34, 68–69, 72–75, 78–79, 79f, 82f, 83–85, 243–246, 248, 270–271, 276, 321–322, 339, 357–358, 361, 390–391, 394–395
O'Hara, J.A., 530–532, 535–536
Ohmae, E., 30–31
Oi, T., 554–555, 563–564
Okamura, M.Y., 354–355
Okunieff, P., 513–514
Okur, A., 154–156
Oldham, W.M., 93–94
Olenick, L.L., 192–193
Olijve, L.L.C., 458
Olkhova, E., 337f, 339–340
O'Mara, M.L., 292
Opella, S.J., 112, 290, 294–298
Oprian, D.D., 458–459
Orban-Glaß, I., 321, 328–330
Orozco, M., 192–193
Ortony, J.H., 458–459, 472
Ottemann, K.M., 322–323
Overhand, M., 487
Overhauser, A.W., 458
Overington, J.P., 351
Owen-Hughes, T., 128–130, 135–138, 141, 164
Owenius, R., 4–5, 15–17
Ozarowski, A., 212

P

Packer, L., 192–193
Padan, E., 249–250, 378
Padmanabhan, A., 112
Pai, E.F., 379
Pai, N.P., 160–161
Paiva, S.A., 545–546
Pake, G.E., 77
Palczewski, K., 379
Palmer, A.G., 86–88
Palmer, T., 136, 137f
Pan, A.C., 353–354, 380
Pan, L., 291
Pandian, R.P., 512–514, 532–533, 537–538
Pandian, S., 563
Panek, G., 158, 168–170, 443

Pannier, M., 32–34, 113–115, 132–133, 148, 158, 322, 332–333, 361, 408, 415–416, 420, 439
Pant-Pai, N., 160–161
Papasotiriou, D.G., 297
Papo, N., 293
Parak, F.G., 47, 318–319
Parasca, A.D., 518–519, 554–555
Parasuk, V., 155–156, 160–161
Parce, J., 107–108
Pardon, E., 351
Parigi, G., 12, 15, 461, 464–465
Parinandi, N.L., 512–514, 530, 532–533, 537
Park, E.S., 192–193
Park, J.H., 320, 379
Park, S.H., 290
Park, S.Y., 322–323
Park, Y.J., 141–142
Parrish, R.G., 126–128
Pas, M.R., 503
Pasenkiewicz-Gierula, M., 4, 320
Passeggi, M.C., 354–355
Pastuszko, A., 503–504
Patel, D.R., 270–271
Patten, R.D., 545–546
Paulsen, H., 126–128, 142–143, 158, 168–170, 320, 443
Pavlova, A., 458–459, 460f, 461, 464–466, 470, 473–474, 477
Pawlak, A., 567
Pedersen, J., 487
Pedersen, J.B., 458
Peggion, C., 225–226, 248–250
Pelizzari, C.A., 511–512, 514–515, 518, 520–521
Peloquin, J.M., 354–355
Penmatsa, A., 351
Percival, P.W., 4, 6, 8
Perez, C., 350
Perez, J.M., 220
Peric, M., 192–193, 507–508, 531–532, 538
Perlstein, D.L., 405–406, 414–415
Perozo, E., 78, 84–85, 155, 322–323, 351, 380, 390–398, 404–405
Perutz, M.F., 126–128
Peschke, P., 530–531
Peter, B.J., 260–261, 263, 266, 272

Peters, F.B., 15–17
Petersdorf, R.G., 509
Petersen, E.F., 516
Peterson, D.W., 458–459
Petersson, J.S., 512–513
Petryakov, S., 504, 554–555
Petukhov, V.Y., 531–532
Peulen, T.O., 324
Pfeiffer, M., 319–321, 326
Pfuetzner, R.A., 395–396
Phillips, D.C., 126–128
Phillips, G.N., 42, 47, 88
Phumala Morales, N., 555, 564
Piccolo, G., 260
Pick, A., 260–261
Piepmeier, J.M., 530–531
Piert, M., 504
Pink, D.A., 208–210
Pink, M., 138–139
Pintilie, M., 530–531
Piotrowiak, P., 79–80
Pirman, N.L., 156–157, 292
Piscitelli, C.L., 350
Pistolesi, S., 222–223
Piton, N., 407
Pitter, K., 394–395
Pitts, L.H., 542–544
Plackmeyer, J., 249–250, 405, 407
Plato, M., 320–321, 409
Pogni, R., 222–223
Poletti, E.F., 225–226
Polienko, J.F., 204
Polienko, Y.F., 420
Pollard, A.M., 322–323
Poluektov, O., 212
Polyhach, Y., 72, 158, 162–164, 170, 182, 263, 277, 322–324, 326, 335–336, 338, 340, 361, 363–368, 405–406, 410–412, 416–417, 429–430, 439
Ponomarev, A.B., 249–250, 404, 408
Popova, A.M., 431–433, 437–438
Popp, C.A., 4, 320, 507–508
Porfyrakis, K., 464–466
Potapov, A., 293
Potsch, S., 415
Pouyssegur, J., 530
Prakash, I., 562
Prandolini, M.J., 463, 465–466, 468

Pratt, R.G., 504
Pratt, V., 81
Prehoda, K.E., 32
Price, E.A., 407, 430–434, 431–433f, 437–438, 444–446
Prisner, T.F., 115–116, 142–143, 249–250, 322–323, 404–420, 428–430, 439, 458, 463, 465–466, 468
Prochniewicz, E., 102, 108
Procopio, J., 208–210
Proteasa, G., 155–156
Ptak, M., 192–193
Putterman, M.R., 15
Pyka, J., 5–7, 12, 68–69, 239, 320

Q

Qin, P.Z., 407, 428–450, 429f, 431–433f, 458–461
Qin, Z.H., 238–239
Qualmann, B., 260–261
Quick, M., 336, 351–353, 360, 367–368, 370–373, 376–379
Quillin, M.L., 32
Quine, R.W., 567
Quiocho, F.A., 491–494, 493f

R

Raap, J., 248–250, 293
Raba, M., 336, 337f, 338–340
Rabenstein, M.D., 32–34, 72–75, 113–115, 248, 321–322, 361
Radzwill, N., 248, 320–322, 335–336
Raghuraman, H., 155
Ragulya, A.V., 39
Rahme, L.G., 531–532
Raibaut, L., 294
Raine, A., 294–298
Rajca, A., 138–139, 440
Rajca, S., 138–139, 440
Rakhmatoullin, R., 412
Ramamoorthy, A., 112
Ramil, C.P., 317
Ramos, A., 407
Rangel, D.P., 319
Raschle, T., 261–262
Rask-Andersen, M., 290
Rasmussen, S.G., 351
Rasper, J., 41

Rassat, A., 225–226
Ratke, J.J., 4–5
Rato, M., 201–202
Rayment, I., 110–111
Razzaghi, S., 15–17
Recouvreur, M., 227
Redler, G., 519–521, 532–533
Reese, A.W., 508
Reese, M., 464–465
Reginsson, G.W., 142–143, 407
Reh, M., 458, 464–465
Reijerse, E.J., 405
Religa, T.L., 30
Reuben, J., 317–318
Rey, M., 420
Rey, P., 225–226
Reyes, C.V., 428–429
Reznikov, V.A., 192–202, 208–210
Ribo, M., 30–31
Richardson, D., 290–291
Rickert, C., 321, 328–330
Riehm, J.P., 224–225
Riesle, J., 328–329
Rigaud, J.L., 192–193
Righi, V., 531–532
Rijkers, D.T.S., 249
Riley, J., 503–504
Rinard, G.A., 567
Ring, A., 293–294
Ringstad, N., 260–261
Rink, C., 530
Rink, T., 326, 328–329
Riplinger, C., 405
Riske, K.A., 192–193
Rivera, B.K., 531–532, 539, 545–546
Rivera-Rivera, I., 194–195, 204–206, 208–210, 234–236
Rivers, J.N., 512–513
Rizzo, R.C., 154–156
Roach, C.A., 291–292
Robb, F.J.L., 556
Robbins, A.H., 154–156
Robblee, J.H., 405–406, 414–415
Robertson, D.A., 414
Robinson, A.J., 405
Robinson, B.A., 6–7
Robinson, B.H., 4–7, 12, 69–70, 238–239, 320, 407, 466, 508

Robinson, C.V., 290
Robinson, S.P., 503–504, 530–531
Robustelli, P., 30
Rocca, J.R., 155–156, 158–160, 164–168
Roche, J., 32
Rockwell, S., 502, 530–531
Rodriguez, B., 545–546
Roebuck, W., 531–532
Roessler, M.M., 405
Rohman, L., 530–531
Rohmer, M., 297
Rohs, R., 428–450
Roller, C.R., 262–263, 272–273
Roman, V., 192–193
Romero, J.G., 390–392
Rontgen, W.C., 502–503
Roopnarine, O., 108
Rose, G.D., 63
Rose, K., 260–261
Rosen, G.M., 405, 512
Rosenmund, C., 260
Ross, B.D., 458
Ross, C.R., 531–532
Rosset, J., 468
Rost, B.R., 260
Roszak, A.W., 290–291
Rottenberg, H., 192–193
Roubaud, V., 49
Roumestand, C., 32
Rouviere, C., 78
Roux, B., 104, 115, 319, 322–323, 350–351, 353–354, 367–373, 380, 390, 406
Roversi, P., 19
Rowan, L.G., 458
Roy, A., 373–374
Roy, A.S., 160, 162, 164–165, 430, 440
Roy, S., 538
Roy, S.K., 138–139
Rozantse, E.G., 562–563
Rozantsev, E.G., 562–563
Rozentzwaig, Y., 412
Ruan, K.C., 30–31
Rudnick, G., 336
Ruf, A., 351
Rumsey, W.L., 504
Ruta, V., 391–392, 397–398
Rutherford, A.W., 293, 298–299

Rutherford, N.G., 380
Rutters-Meijneke, T., 249
Ruuge, A., 194–202
Ruysschaert, J.M., 353–354, 380
Ryan, C.J., 261

S

Sachleben, J.R., 396–397
Sack, J.S., 491–494, 493f
Sadiq, S.K., 154–156
Saen-oon, S., 155–156, 160–161
Sahl, H.G., 233–234
Sahlin, M., 415
Sahu, I.D., 80, 115, 142–143, 225–226, 290–308
Sailasuta, N., 458
Saito, K., 518–520, 532–533
Sajid, M., 406
Sakai, K., 555, 561, 563, 567
Sale, K.L., 63–64, 78, 319, 404–405
Salikhov, I., 516, 531–532
Salikhov, K.M., 126–128, 132–133, 243, 503, 509
Salikov, K.M., 404, 408
Salisbury, R., 531–532
Salnikov, E.S., 249–250
Salom, D., 379
Salowe, S.P., 415
Salwinski, L., 84–85
Samanta, D., 322–323
Samaranayake, C.P., 41, 42t
Samoilova, R.I., 249–250
Samouilov, A., 504, 513–514
Samuni, A., 562
Sanchez-Weatherby, J., 290
Sandler, A., 518–519
Sankaram, M.B., 192–193
Sano, H., 554–555, 563–564
Sanson, A., 192–193
Sar, C., 63–64
Saragliadis, A., 487
Sarciaux, M., 225–226
Sareth, S., 30–31
Sarkar, S., 138–139
Sartorelli, A.C., 502
Sarver, J., 18–19
Sasaki, D.Y., 261
Sastry, S.K., 41, 42t

Sato, H., 4–6, 12–14, 242, 440
Sauer, G., 530–531
Saunders, J.K., 357–358
Savitsky, A., 201–202, 320–321, 409
Saxena, S., 15–19, 22, 45, 64–66, 68, 72, 113, 329–330, 405
Sayer, J.M., 154–156
Scarpelli, F., 249
Schapira, A.H., 554
Schastnev, P.V., 195–198
Scheek, R.M., 489
Scheidt, H.A., 225–226
Schiemann, O., 249–250, 322, 324, 405, 407, 410–411, 416–418, 429–430, 439
Schindler, J., 225–226
Schiöth, H.B., 290
Schlessman, J.L., 32
Schlick, S., 109f, 113
Schmalbrock, P., 513–514
Schmid, E.M., 261
Schmid, R., 337
Schmidt, M.J., 317
Schmidt, O., 391–392
Schmidt, P.G., 489
Schmidt, P.P., 405
Schmidt-Rohr, K., 291–292
Schnegg, A., 201–202, 409
Schneider, E., 322–323, 331, 335–336, 340
Schneiderhahn, C., 224–225, 321–322
Scholes, C.P., 94, 160, 162, 164–165, 430, 440
Schöps, P., 420
Schreiber, L.M., 458
Schreier, S., 225–226
Schrempf, H., 391–392
Schroder, R.R., 110–111
Schubert, M., 430
Schuhmann, M.U., 530–531
Schulten, K., 322–323, 380, 390
Schultz, P.G., 317
Schwartz, H., 518–519
Schwartz, R.N., 7–8, 10, 300, 302–303
Schwarz, G., 210–211
Schweiger, A., 126–128, 298–301, 412, 442, 504, 506
Schwieters, C.D., 485–489, 490f, 491
Scott, J.A., 458–459, 460f, 461, 464–466, 470, 473–474, 477

Seelig, J., 234–236
Seidel, C.A.M., 324
Selth, L., 141–142
Selvendiran, K., 532–533, 538
Selvin, P.R., 126–128, 225
Semenov, S.V., 204
Sen, K.I., 443
Sena, E.S., 544
Senapati, S., 155
Sengupta, I., 15, 64, 68
Sentjurc, M., 562
Serbulea, L., 64
Serrano-Vega, M.J., 351
Sethaphong, L., 361
Sezer, D., 319, 406, 463, 465–466, 468
Shafer, A.M., 225–226
Shafer, D.E., 225
Shah, C., 260–261, 263, 266, 272, 274–275
Shah, S.J., 530–531
Shai, Y., 293
Shames, A., 193–194
Shane, J., 412
Shang, W.F., 135–136, 141–142, 322
Shankar, R.A., 513–514
Shannon, M.D., 15
Shao, W., 160–161
Sharma, M., 351
Sharma, S., 350–351, 353–354, 367–368, 370–373, 380
Sharp, K.A., 63–64
Sharples, D., 380
Sharpless, K.B., 317
Shea, J.-E., 458–459
Sheinerman, F.B., 192
Shelke, S.A., 406–407, 428–429, 431–433
Shen, C.H., 154–156
Sherman, M.B., 261
Sherwin, M.S., 292
Shi, H., 297, 530
Shi, L., 336, 351–353, 360, 372–373, 376–379
Shiba, T., 555, 566
Shibata, Y., 351
Shih, W.M., 116–118
Shima, T., 535–537
Shimamura, T., 380
Shin, Y.-K., 12, 32–34, 72–75, 78, 113–115, 192–194, 248, 321–323, 361
Shirov, M.D., 126–128, 132–133, 404, 408

Sholle, V.D., 562–563
Shortley, G.H., 502–503
Shrivastava, I.H., 291
Shteynbuk, M., 507–508
Shuaib, A., 544
Shukla, H.D., 290
Siaw, T.A., 458
Sicoli, G., 225–226, 407, 414
Sidney Fleischer, B.F., 192–194
Sidney, T.J.M., 234–236
Sigurdsson, S.T., 142–143, 404–420, 428–429, 431–433, 438
Sikora, A., 353–354
Sillitoe, I., 350
Silva, K.I., 18–19
Simerska, P., 294
Simmerling, C., 154–168
Simon, A.B., 234–236, 530–531
Simorellis, A.K., 30
Simpson, J.V., 291–292
Singel, D.J., 404, 409–410
Singh, S.K., 351, 380
Sippach, M., 331, 335–336
Sjöberg, B.-M., 415
Skinner, T.E., 420
Slichter, C.P., 458, 504, 508
Sligar, S.G., 377–379
Slipenyuk, A.M., 39
Slotboom, D.J., 322–323, 350, 353–354
Smirnov, A.I., 194–202, 204–210, 212, 228, 230–231, 233–236, 239–240, 244–246, 531–532, 535, 543–544
Smirnova, T.I., 199–200, 212, 230–231, 244–246
Smith, A.N., 155, 157–160, 163–168
Smith, C.A., 110–111
Smith, C.W., 502
Smith, G.M., 142–143, 414
Smith, H.J., 115, 142–143, 292
Smith, R., 110–111
Sochet, A.A., 155–156
Soderman, O., 208–210
Söhl-Kielczynski, B., 260
Sokolovski, M., 458–461
Solomon, I., 463, 467, 489
Solomon, T.L., 88–89
Sompornpisut, P., 78, 155–156, 160–161, 380, 390–391, 393–394, 396–398

Song, J., 458–459, 461, 477–479
Song, L., 78, 366–367, 404–405
Song, R., 531–532, 535, 539
Sönnichsen, F.D., 243
Sonoda, Y., 351
Soong, R., 112
Sosnovsky, G., 562
Sotomayor, M., 322–323, 380, 390
Souvatzoglou, M., 504, 530–531
Sowa, G.Z., 428–434, 429f, 432f, 438, 443–445, 447–448
Sozudogru, E., 130, 135–136, 141
Spadaro, J., 545–546
Spagnol, G., 440
Spaltenstein, A., 407
Specht, A., 350
Spence, M.M., 112
Spencer, D.P., 510
Spicer, L.D., 129
Spiess, H.W., 32–34, 113–115, 132–133, 148, 158, 322, 332–333, 361, 408, 415–416, 420, 439, 458, 464–465
Spindler, P.E., 420
Sportelli, L., 262–263
Springett, R.J., 530–531
Spruell, J.M., 458–459, 472
Spudich, J.A., 108, 116–118
Squier, T.C., 105
Sroka, T.C., 220
Stachowiak, J.C., 261
Stagg, L., 63
Stargell, L.A., 141–142
Steed, P.R., 32–34, 353–354, 356–357, 366, 390–391
Steffen, M.A., 192–193
Stehlik, D., 458–461, 464, 470
Stein, R.A., 14, 42, 353–354, 361, 365–370, 378–381, 420
Steinhoff, H.J., 60–61, 224–226, 248–250, 316–340, 353–356, 366–367, 378
Stenkamp, R.E., 379
Stewart, P.D., 290
Stirling, A., 135–136, 141–142, 322
Stock, G., 115–116, 405, 407, 416–417
Stoll, S., 300
Stoller, S., 225–226
Stolzenberg, S., 351–353
Stone, K., 458–461

Strangeway, R.A., 4–5
Strop, P., 351–353
Strube, T., 407
Stubbe, J., 405–406, 414–415
Su, W., 94
Su, W.Y., 242
Subbaraman, N., 225–226
Subczynski, W.K., 4–6, 239, 358–360
Subramanian, K., 531–532, 539, 545–546
Subramanian, S., 515, 530–533
Sucheta, A., 535, 539
Sudhoff, K.B., 141–142
Sugano, T., 6–7
Sugata, K., 105
Suh, J.Y., 485–486, 491
Sukomon, N., 322–323
Summerer, D., 317
Sun, A., 354–355
Sun, L., 293
Sun, Z.Q., 504
Sundramoorthy, S.V., 511–512, 514–516
Surek, J.T., 102, 193–194
Surendhran, K., 363, 379
Surewicz, W.K., 243
Sutch, B.T., 433–434, 444
Sutoh, K., 110–111
Suydam, I.T., 192–193
Suzuki, S., 380
Svejstrup, J.Q., 141–142
Svensson, B., 102–120
Svensson, M., 135–136
Svirmickas, A., 473–474
Swanson, M.A., 14, 42, 365, 420
Swanstrom, R., 160–161
Swartz, H.M., 503, 507–508, 512, 516, 530–546, 562, 566
Syryamina, V.N., 248

T

Tajkhorshid, E., 353–354, 366–367, 378–379, 396–397
Takei, K., 260–261
Takemoto, K., 233–234
Takeshita, K., 554–555, 563
Tang, C., 291, 485–489, 491
Tang, J., 155–156, 160–161
Tang, Y., 141–142
Tangprasertchai, N.S., 428–429

Tank, D.W., 504
Tarabek, P., 301
Tate, C.G., 351
Tatum, J.L., 507–509, 512–513
Taube, A., 515
Tavernier, A., 464–465
Tawfik, D.S., 19
Taylor, C.T., 530
Tebbe, S., 337
Teicher, B.A., 502, 507–508, 531–532, 538
Teng, C.-L., 12, 193–194
ter Beek, J., 350
Terekhov, M., 458
Terry, D.S., 351–353
Terry, G.E., 4–5
Thai, V., 60
Tham, K.W., 428–450
Thevis, W., 248, 320–322, 335–336
Thiel, C., 530–531
Thiel, K., 530–531
Thoden, J., 110–111
Thomas, A., 260
Thomas, D.D., 61–64, 84–85, 102–120, 103f, 105f, 107f, 193–194, 406
Thomas, L., 225–226
Thomas, M., 107–108
Thomas, S.R., 504
Thomaston, J., 292
Thomlinson, R.H., 502, 518–519
Thompson, A.R., 102–120, 103f, 107f
Thompson, M., 292
Thompsongorman, S., 507–508
Thomson, A.J., 292
Thorgeirsson, T.E., 321–323
Thorpe, C., 405
Tiburu, E.K., 112
Timmel, C.R., 14, 32–34, 249–250
Tipikin, D.S., 113, 156–157
Titus, M.A., 102
Tkach, I., 414, 464–466
Tkachev, Y.V., 102
Toffoletti, A., 225–226
Toledo Warshaviak, D., 15–17, 62–64
Tollinger, M., 194
Tominaga, M., 225–226
Tonelli, M., 155
Toniolo, C., 225–226, 248–250
Torbeev, V.Y., 155

Torchia, D.A., 154–156
Torchilin, V.P., 220
Torgersen, K.D., 102, 105–106
Toribara, T.Y., 227
Tormyshev, V.M., 520–521
Torrent, J., 30–31
Törring, J.T., 320–321
Tosch, V., 260
Toth, I., 294
Toussaint, A., 260
Traaseth, N.J., 112–113
Trabanino, R.J., 32–34, 72–75, 243, 248, 276, 321–322
Traikia, M., 227
Tran, L.B., 530–531
Trayanova, N., 545–546
Triest, S., 351
Troiano, J.M., 192–193
Tropia, L., 449f
Troxel, K.R., 115, 142–143, 292–294, 304
Tschaggelar, R., 326, 412, 458–459, 464–465, 470
Tseitlin, M., 567
Tsiao, C., 458
Tsubota, T., 141–142
Tsuji, M., 542–543
Tsuneyoshi, M., 566
Tsvetkov, Y.D., 147–148, 249–250, 404, 408–410, 428–429, 431–433
Tuccio, B., 49
Tulla-Puche, J., 294
Tung, C.S., 430–431, 433–434, 438, 443–445, 447–448
Turer, A.T., 530
Turk, E., 395
Türke, M.-T., 464–466
Turner, A.L., 242
Tuttle, L.M., 350
Tyagi, S., 317

U

Ubbink, M., 405, 485–487, 491
Ueki, S., 105
Ui, M., 227
Ulloa-Aguirre, A., 290
Ulrich, H., 562–563
Un, S., 415
Urayama, P., 42

Urban, L., 320
Utsumi, H., 554–556, 561–564, 566–567

V

Vahidi, N., 513–514, 532–533, 537
Vaisbuch, I., 193–194
Vaitiekunas, P., 290
Valente, A.P., 294–298
Vallis, Y., 260–261, 266
Vallurupalli, P., 30, 49–52
Vamvouka, M., 390, 397–398
Van Amsterdam, I., 405
van den Doel, M.A., 545–546
Van Eps, N., 93–94, 322–323, 390, 397–398, 433–434, 444
van Kraaij, C., 233–234
van Tol, J., 212
Van Vleck, J.H., 248–249
van Wonderen, J.H., 292
Vandecasteele, J., 530–531
Vanderkooi, J.M., 504
Vanpolen, J., 510
Varani, G., 407
Varkey, J., 260–261, 266, 458–459, 461, 464–465, 478f, 479
Vasquez, V., 322–323, 380, 390
Vaupel, P., 530–531
Veglia, G., 112–113
Veit, S., 32–34, 113–115, 132–133, 148, 158, 322, 332–333, 361, 408, 415–416, 439
Velan, S.S., 516
Velayutham, M., 516
Veloro, A.M., 155–168, 163f, 166f, 440
Venditti, V., 485–486
Venkataramu, S.D., 320
Venters, R.A., 129
Venyaminova, A.G., 420
Verdouw, P.D., 545–546
Verhalen, B., 353–354, 366–367, 378–379
Veselov, A.V., 94
Vickrey, J.F., 155–156
Victor, K.G., 228, 238–239
Vidal, M., 351
Vieira, R.F.F., 225–226
Vieth, H.M., 461
Vij, K., 507–508
Vijayakumar, S., 518–519

Vikram, D.S., 507–508, 530–533
Vishnivetskiy, S.A., 322–323
Vishnubhotla, P., 220
Voet, D., 63
Voet, J.G., 63
Vogel, K.P., 249–250, 378
Voinov, M.A., 194–202, 204–206, 208–210, 212, 234–236
Volkov, A.N., 320, 485–486, 491
Volodarsky, L.B., 195–197
von Heijne, G., 290
Vondrasek, J., 154–155
Vöpel, T., 324
Vorobjeva, M.A., 420
Voskoboynikova, N., 321, 328–330
Voss, J.C., 225–226
Vrljic, M., 351–353

W

Wachowius, F., 407
Wageman, R., 564
Wagner, G., 261–262, 489
Wainberg, M.A., 154–155, 160–161
Waite, J.H., 458–459, 472
Wakselman, M., 225–226
Walczak, T., 513–514, 516, 531–533, 535–537, 543–544
Walensky, L.D., 394–395
Wallace, B.A., 291–292
Wallgren-Pettersson, C., 260
Walsh, P., 30
Walton, P., 293–294
Walz, T., 353–354, 380–381
Wand, A.J., 32
Wanderling, S., 393–394, 397–398
Wang, J., 350
Wang, J.-Y., 458–459, 464–465
Wang, K.H., 351
Wang, P.H., 516
Wang, S., 291
Wang, Y.X., 154–156
Ward, R., 128–130, 135–138, 137f, 141–142, 164, 322, 324, 430
Wardak, R., 519
Wardrip, C., 516
Warncke, K., 293
Warner, H., 227
Warr, G.G., 208–210

Warschawski, D.E., 227
Warshaviak, D.T., 64
Watach, M.J., 512
Watanabe, A., 350
Watanabe, S., 260
Watanabe, T., 555, 566
Watts, A., 205
Watts, K.J., 322–323
Wawrzak, Z., 155–156
Weber, A., 405–406, 414–415
Weber, I.T., 154–156
Weber, K.J., 530–531
Weber, R.T., 105, 116, 440–442
Wegener, A.A., 329–330
Wegener, C., 201–202, 249–250, 320–321, 366–367, 378, 409
Weidlich, D., 331, 335–336
Weil, J.A., 354–355
Weiner, L.M., 192–193, 195–198, 208–210
Weinstein, H., 336, 351–353, 360, 367–368, 370–373, 376–379
Weismiller, H.A., 567
Weiss, S., 126–128
Weissig, H., 444
Weitzel, L., 530, 545–546
Wen, P.C., 353–354, 366–367, 378–379, 396–397
Wendland, M.F., 530–531
Weninger, K.R., 351–353
Wesener, D.A., 458–459, 461, 464–465, 477–479, 478f
Westbrook, J., 444
Westler, W.M., 155
Westphal, C.H., 260–261
Wetzel, R., 242–243
Wey, H.Y., 544
Weyand, S., 380
Whalen, W.J., 503–504
Whiddon, C.R., 208–210
White, S.H., 290–291
Whitelegge, J.P., 395
Whitmore, L., 291–292
Widomska, J., 239
Wiebcke, M., 406
Wiedemann, I., 233–234
Wiener, M.C., 92, 351, 380
Wieprecht, T., 234–236
Wilcox, M., 107–108

Wild, C., 458
Wilkens, S., 353–354
Williams, B.B., 507–508, 512–513, 516, 531–532, 535–536, 539, 542–543
Williams, D.H., 294–298
Williams, J., 513–514
Williams, M.J., 4–5
Williamson, P.T.F., 192–193
Willson, R.L., 564
Wilmes, L.J., 530–531
Wilmo, C.M., 530–531
Wilson, C., 102, 103f, 105–106, 107f
Wilson, D.F., 503–504
Wilson, J.D., 509
Wimley, W.C., 290–291
Winiski, A., 192–194
Winters, M.A., 155–156
Winther, A.M., 376
Wisner, M., 238–239
Wisniewska, A., 239
Wistrand, L.G., 512–513
Wittayanarakul, K., 155–156, 160–161
Wittung-Stafshede, P., 63
Wlodawer, A., 154–155, 154f
Wohlkonig, A., 351
Woldman, Y.Y., 204
Wolf-Watz, M., 19
Wood, M., 473–474
Worrall, J.A., 485–486, 491
Wrabl, J.O., 19
Wright, E.M., 395
Wright, K., 225–226
Wright, P.E., 45–47, 50, 60–61
Wunnicke, D., 322–323, 353–354
Wyckoff, H., 126–128
Wylde, R.J., 414

X

Xiao, M., 225
Xiao, W., 321
Xu, Q., 157–158, 160, 162–163, 380

Y

Yajima, S., 380
Yamada, H., 30–31, 50
Yamada, K., 512–513, 515, 530–531, 554–556, 561–564, 566–567
Yamaguchi, T., 227

Yamasaki, T., 563
Yamashita, A., 351, 380
Yamato, M., 555, 563, 566
Yamazaki, T., 155–156
Yan, C., 350
Yanagi, H., 532–533, 537
Yang, G., 360, 363, 377–379
Yang, K., 72–73, 321–323
Yang, W., 404–405
Yang, X., 297
Yang, Y., 192–193, 291–292, 544
Yang, Z., 4–23, 16f, 30–53, 43f, 51f, 61–62, 68–69, 72, 84–85, 90, 92–94, 156–157, 292, 356–357, 365, 406, 420, 428–429
Yao, J., 45–47
Yao, T., 566
Yasui, H., 518–520, 532–533, 555
Yasukawa, K., 555, 561–562, 564, 566–567
Yee, S.W., 351
Yi, C., 113–115
Yi, M., 351
Yildirim, M.A., 351
Yildiz, O., 350
Yin, J.J., 4–6, 9–11, 239, 320
Yin, P., 350
Yokoyama, S., 30–31
Yonkunas, M., 15–18
Yoo, J., 512–513, 530–531, 555
Young, R.D., 318–319
Yu, C., 507–508, 531–532, 538
Yu, H., 115–116, 405, 416–417, 542–543
Yu, X., 290–291
Yuan, J., 297, 489
Yulikov, M., 15–17, 430, 443–444

Z

Zaharchuk, G., 530 531
Zaim-Wadghiri, Y., 530–531
Zaki, B.I., 516, 531–532, 535–536
Zamaraev, K.I., 243, 503, 509
Zamoon, J., 112–113
Zamora, M.A., 518–519
Zander, R., 530–531
Zarrine-Afsar, A., 30
Zavoisky, E., 126–128

Zecevic, A., 138–139
Zecevic, A.N.A., 440
Zehnpfennig, B., 376
Zeng, H., 297
Zerella, R., 294–298
Zhan, H., 11, 82f, 83–85
Zhang, M., 291
Zhang, R., 115, 290–308
Zhang, R.F., 142–143
Zhang, R.L., 562
Zhang, S.Q., 63
Zhang, X., 78, 428–450, 432f, 449f
Zhang, X.C., 155–156, 160–161
Zhang, X.J., 49
Zhang, Z., 102, 105–106, 116, 156–157, 406
Zhang, Z.W., 113
Zhao, D., 504
Zhao, J., 297
Zhao, Y., 322–323, 351–353
Zheglov, E.P., 531–532
Zhou, A., 80, 290–308
Zhou, H.X., 351
Zhou, K.H., 194
Zhou, Y., 395–396
Zhou, Z., 351–353, 361
Zhu, G., 491–494
Zhu, Q., 358–360
Zhu, Y., 290–291
Ziegler, C., 350
Ziegler, W.H., 406, 416–417
Zielonka, J., 354–355
Zielonka, M., 354–355
Zilm, K.W., 194
Zimmer, K., 458
Zimmermann, H., 50, 126–128, 141–143, 149, 170–173, 293, 333–335, 366–367, 404 405, 413f, 443
Zipp, A., 41, 42t
Zmasek, C.M., 350
Zoltner, M., 136, 137f
Zornoff, L.A., 545–546
Zou, P., 363, 378–379
Zukerman-Schpector, J., 225–226
Zweier, J.L., 14–17, 365, 420, 507–508, 512–514, 516, 530–533, 537, 554–555

SUBJECT INDEX

Note: Page numbers followed by "*f*" indicate figures, "*t*" indicate tables, and "*s*" indicate schemes.

A

Antimicrobial peptides, 220
Apomyoglobin, 477–479, 478*f*
ATP-binding cassette (ABC) transporters, 331
ATP hydrolysis, 335

B

Backbone dynamics, 85–89, 87*f*, 89*f*
Backbone fluctuations, 63
Bacterial toxins, 220
BAR. See BIN/Amphiphysin/Rvs (BAR)
BEER. See Bifunctional electron–electron resonance (BEER)
Bicelles, 103–104, 112
Bifunctional electron–electron resonance (BEER), 103–104
Bifunctional spin label (BSL)
  bifunctional electron–electron resonance (BEER), 103–104, 119–120
  general labeling procedure, 104–105
  of muscle proteins
    DEER, distance measurements with, 113–115, 115*f*
    labeling specificity and protein function, 116–119, 117–118*f*
    magnetically aligned systems, 111–113, 112*f*, 114*f*
    mechanically aligned systems, 108–111, 109–111*f*
    methods, 104–105
    orientation selection, problem of, 115–116
    rotational dynamics, 105–108, 105–107*f*
    saturation transfer EPR (STEPR), 103–104
BIN/Amphiphysin/Rvs (BAR), 260–261
Binding isotherms, 233–236
Boltzmann constant, 36
Boltzmann distribution, 4–5
Brain, oximetry procedures, 530, 533–534, 541–544
BSL. See Bifunctional spin label (BSL)

C

Calmodulin(CaM)-4Ca^{2+}, 491–494, 493*f*
Cancer therapy, imaging molecular oxygen in
  in cancer biology, 516
  electron paramagnetic resonance oximetry, 507–509
  pO_2, 514–515
  principles of, 503–504
  electron paramagnetic resonance, principles of, 503–504
  hypoxic fraction, 520–521, 521*f*
  in living animals, 502–503
  longitudinal relaxation, 515–516
  molecular oxygen
    measurement of, 503–504
    nature of, 503
  oxymetry, validation of, 516–519, 517–518*f*
  particulates, 513–514
  radiofrequency magnetization excitation, 510
  soluble spin probes, 512
  spin probe oximetry, 510–512
  spin probe sensitivity, 509–510
  transient hypoxia, 516–520, 517*f*, 520*f*
  trityls, 512–513, 512–513*f*
Circular dichroism (CD), 291–292
Coenzyme Q_{10} (CoQ_{10}), 554
Common carotid artery (CCA), 542–543
Complement continuous-wave (CW) electron paramagnetic resonance (EPR), 458–459
Computational modeling, nucleic acids
  additional considerations, 448–449, 449*f*
  inter-R5 distances
    NASNOX in batch mode, 445–446
    NASNOX program, 444–445

603

Computational modeling, nucleic acids (*Continued*)
  model generation, 444
  model selection and characterization
    modified RMSD metric, 447
    root-mean-squaredeviation (RMSD), 447
Computational refinement
  membrane-curving proteins, 261–263, 262*f*
  protein–lipid assembly, reconstruction of, 283–284, 284*s*
  simulated annealing, 280–283, 281*t*, 282*f*
  starting structure, construction of, 277–280, 278*f*, 279*s*
Continuous-wave (CW) and pulsed EPR spectroscopy, HIV-1PR, 156–160
Continuous wave-electron paramagnetic resonance (CW-EPR), 4–5
  applications of, 93–94
  backbone dynamics, 85–89, 87*f*, 89*f*
  conformational exchange and conformational changes, 90–92, 91*f*, 93*f*
  distances measurements, 275
  interspin distance measurements
    bond or space spin exchange interactions, 79–80
    relaxation dipolar broadening, 78–79, 79*f*
    static dipolar interaction, 73–78, 74*f*, 76*f*
  peptides, aggregation and agglomeration, 242–249, 244–245*f*, 247*f*
  practical considerations
    buffer selection, 41, 42*t*
    water compressibility, 41–42
  R1 nitroxide motion
    EPR timescale, 63–64
    internal motion, local protein dynamics, 66–68, 67*f*
    protein rotational diffusion, 63–64
    tertiary interactions, 64–66, 65*f*
  R1 solvent accessibility with power saturation, 68–72
  secondary structure and features, 80–85, 82*f*, 84*f*
  theoretical power saturation curves, 70–71, 70*f*

variable-pressure
  pressure generation and regulation, 39–41, 40*f*
  SDSL-EPR, high-pressure cells, 37–39, 38*f*
Continuous wave (CW) technique, 131
Continuous wave (CW) X-band EPR spectra, 227–228
Cryogenic temperatures, ligand binding, 164–165
Crystalline probes, oximetry probes, 537
Crystallographic or computational mechanistic models, 368
Cysteine (CYS), 161–162
Cysteine-mutated nitroxide spin label, 293–294
Cysteines, 355–356

D

Dead-time-free four-pulse DEER, 408, 408*f*
DEERconstruct
  Gaussian reconstruction, 173–174
  populations, suppressions of, 174–175
  procedure, 170–175
  processing data and generating distance profiles, 171
  suppression combinations, 175
  time domain representation, 171–173
Deuteration
  deuterated proteins, production of, 144
  electron spin echo envelope modulation (ESEEM), 139–140
  ESE decay curves, 137–138, 140*f*
  histone chaperone Vps75, 136, 137*f*
  histone core octamer, 136, 136*f*
  measurement of, 144–145
  neutron diffraction, 143
  NMR, 143
  nuclear–electron spin interactions, 137–138
  POTRA domains, 137*f*
  on relaxation and distance measurements, 136–143
  relaxation time, 137–141
  sensitivity and distance measurements, 141–143
  spin–lattice relaxation mechanism, 137–138

Subject Index

605

W-band PELDOR, 139–142
(S-4-(4-(dimethylamino)-2-ethyl-5,5-dimethyl-1-oxyl-2,5-dihydro-1H-imidazol-2-yl) benzyl methanesulfonothioate (IKMTSL), 196–197, 196f, 198f
5,5-Dimethyl-1-pyrroline-N-oxide (DMPO), 564
1,2-Dipalmitoleoyl-sn-glycero-3-phosphoethanolamine (DPoPE), 240–242
Distance measurements, 136–143
  deuteration on, 136–143
  PELDOR experiment, 135–136
  and sensitivity, 141–143
Distance profile definition, 170–171
Dithiothreitol (DTT), 225
D25N mutation, 165
Double electron–electron resonance (DEER), 113–115, 115f, 322, 331–336
  data analysis, 333–336, 334f
  distances measurements, 276
  and double-quantum coherence, 14
  experiment setup, 146–147
  HIV-1PR, 158
  interspin distance measurements, 332–333
  membrane protein
    data analysis and interpretation, 363–367
    distance distribution, 361–363, 362f, 364f
    low-temperature acquisition of, 365
    structure and thermodynamics, 370–372, 371f
    structure, distance distributions, 367–370, 369f
  nucleic acids
    data acquisition, 440–443
    sample preparation, 439–440
    spectrum analysis, 443
Double-quantum coherence (DQC), 14, 322

E

Electron–nuclear double resonance (ELDOR/ENDOR), 131, 466
Electron–nuclear interactions, 557

Electron paramagnetic resonance (EPR), 4, 102, 126–128, 292, 390
  accessibility
    basic considerations for, 272–273
    protocol, 273–274
  distances measurements
    application of, 275
    basic considerations for, 274–275
    protocol, 275–277
  membrane-curving proteins, 261–263, 262f
  membrane protein
    distance measurements and structure elucidation, 368, 369f
    site-directed spin labeling, 354–356, 355f
    spin labels, 357–361, 359f
    tool kit, 356–357, 357f
  mobility measurements
    basic considerations for, 270–271
    protocol, mobility parameters, 271–272
  pH-sensitive thiolspecific nitroxide labels
    chemical structures, 196–197, 196f
    electrically neutral reference interface, 207–210, 208f
    Gouy–Chapman theory, 210–211
    IMTSL- and IKMTSL-labeled thiols, 199–200, 200f
    ionizable nitroxides, methanethiosulfonate derivatives of, 200–202, 201t
    lipid bilayers, titration experiments with, 204–205
    spin-labeled lipid vesicles, 204
    surface lipid bilayers electrostatics, 202–211, 203f
    two-site slow-exchange model, titration experiments, 205–207, 205f
  protein structure, 351–353
Electron spin echo (ESE), 126–128
Electron spin echo envelope modulation (ESEEM)
  data analysis and interpretation, 301
  experimental setup, 301
  membrane peptides into lipid bilayer
    experimental setup and data analysis, 301
    peptide purification and validation, 295

Electron spin echo envelope modulation (ESEEM) (*Continued*)
  peptide secondary structure validation, 297, 298f
  principles, 298–301, 299f
  solid-phase peptide synthesis (SPPS), 294, 296t
  spin label, attachment of, 297
  synthetic peptide into lipid bilayers, 297
  membrane protein secondary structure, 290–293
  secondary structure determination approach
    α-helical secondary structure, 302–304, 302–303f, 305–306f
    α-Helices *vs.* β-sheet, 304–307, 307f
Electron spin–lattice relaxation time (T1e), 4
  electron–electron double resonance (ELDOR) technique, 7
  free induction decay (FID), 8
  Heisenberg exchange, 6–7
  nitrogen nuclear relaxation, 6–7
  nitroxide, correlation time of, 7
  with SR, 6
Electrostatic potential, 193–196
External carotid artery (ECA), 542–543

F

Flavin adenine dinucleotide (FAD), 554
Flavin mononucleotide (FMN), 554
Fluorescence resonance energy transfer (FRET), protein structure, 351–353
Force Free Hard Sphere (FFHS) model, 467
Förster resonance energy transfer (FRET), 126–128, 404
Free radical imaging, DNP-MRI
  apparatus
    commercially available, 560
    home-made system, 560, 560f
    from Philips, 558–560, 559f
    spectroscopic DNP-MRI, 561–562, 561f
  intrinsic molecular imaging, application, 566–567, 567f
  principle of, 556–558, 556f
  spin-probe technique
    application of, 564–566, 565f
    nitroxyl probe, synthesis of, 562–563, 563f
    nitroxyl radical injected intravenously, 564, 565f
    nitroxyl radicals, loss of, 562f
    reactivity of nitroxyl radical, 564
Full width at half maximum (FWHM), 173

G

Gaussian reconstruction, 173–174
Gaussian-shaped functions, 165–167
G-band PELDOR (high magnetic field), 412–415, 413–414f
Gouy–Chapman theory, 210–211

H

Heart, oximetry procedures, 544–546
Heisenberg exchange mechanism, 320
Hemoglobin, 126–128
Heteronuclear single quantum correlation (HSQC), 488–489
1-Hexadecanoyl-2-(9Z-octadecenoyl)-*sn*-glycero-3-phosphocholine (POPC), 229, 229f
Histone core octamer, 136, 136f
Human immunodeficiency virus-1 protease (HIV-1PR), 154–155, 154–155f
  conformational states
    minor components, 167–168
    statistical analysis, 168–170, 169s
    Tikhonov regularization distance profiles, Gaussian reconstruction of, 165–167, 166f
  molecular dynamics (MD) simulations, 155–156
  numerous crystal structures of, 155–156
Hydration dynamics, 461, 463
Hydration water, 458
Hydrostatic pressure, 30–31
Hydroxyl radical (·OH), 554
Hypoxic fraction, 520–521, 521f

I

Imaging molecular oxygen, in cancer therapy
  in cancer biology, 516
  electron paramagnetic resonance oximetry, 507–509
  pO_2, 514–515
  principles of, 503–504

electron paramagnetic resonance,
principles of, 503–504
hypoxic fraction, 520–521, 521f
in living animals, 502–503
longitudinal relaxation, 515–516
measurement of molecular oxygen,
503–504
nature of molecular oxygen, 503
oxymetry, validation of, 516–519,
517–518f
particulates, 513–514
radiofrequency magnetization
excitation, 510
soluble spin probes, 512
spin probe oxymetry, 510–512
spin probe sensitivity, 509–510
transient hypoxia, 516–520, 517f, 520f
trityls, 512–513, 512–513f
Imidazolidine, 194–195
Imidazoline, 194–195
Implantable resonators, 534f, 539–540
India ink, 535–536, 536f
Institutional Animal Care and Use
Committee (IACUC) guidelines,
541–542
Internal carotid artery (ICA), 542–543
Interspin distance measurements
bond or space spin exchange interactions,
79–80
relaxation dipolar broadening, 78–79, 79f
static dipolar interaction, 73–78,
74f, 76f
In vivo dynamic nuclear polarization
(DNP)-MRI
apparatus
commercially available, 560
home-made system, 560, 560f
from Philips, 558 560, 559f
spectroscopic DNP-MRI, 561–562,
561f
intrinsic molecular imaging, application,
566–567, 567f
principle of, 556–558, 556f
spin-probe technique
application of, 564–566, 565f
nitroxyl probe, synthesis of, 562–563,
563f
nitroxyl radical injected intravenously,
564, 565f

nitroxyl radicals, loss of, 562f
reactivity of nitroxyl radical, 564
In vivo electron paramagnetic resonance
(EPR) oximetry
application
EPR acquisition parameters,
543–544
in heart, 544–546, 545f
ischemic stroke, 540–541
particulate probes, implantation of,
541–544, 541f
paramagnetic oxygen-sensitive probes
implantable resonators, 538–540
India ink, 535–536, 536f
particulate oximetry probes, 537–538
principles, 532–535, 533–534f
Ionizable nitroxides, methanethiosulfonate
derivatives of, 200–202, 201t

K
K55C, reporter site, 157–158, 157f

L
Labeled oligonucleotides
characterization of, 438
purification of
high-performance liquid
chromatography (HPLC), 437–438
polyacrylamide gel electrophoresis
(PAGE), 438
Labeling methods. *See also* Spin labeling
prokaryotic KcsA potassium channel
procedure, 395–397
solutions requirement, 397
prokaryotic KvAP potassium channel
procedure, 397–398
solutions requirement, 398–399
Langmuir adsorption isotherm, 234
Local water dynamics, 458

M
Macroscopic-order microscopicdisorder
(MOMD), 65–66
Matrix-assisted laser desorption/ionization
time-of-flight (MALDI-TOF), 295
Membrane-associated peptides
accessibility measurements, 237–239
experimental considerations, 239–242,
241f

Membrane-curving proteins, 260–261
Membrane mimetic systems
  liposomes preparation, 226–227
  MTSL, 224
  natural amino acid side chains, covalent modification of, 221–225, 222–223f
  nitroxides, maleimido- and iodo-derivatives of, 224–225
  unnatural amino acid via peptide synthesis, 225–226
Membrane protein
  cellular signaling and signal transduction pathways, 351
  crystal structures, 351
  DEER spectroscopy
    data analysis and interpretation, 363–367
    distance distribution, 361–363, 362f, 364f
    structure and thermodynamics, 370–372, 371f
    structure, distance distributions, 367–370, 369f
  EPR primer
    site-directed spin labeling, 354–356, 355f
    spin labels, 357–361, 359f
    tool kit, 356–357, 357f
  mechanistic descriptions of, 351
  sample preparation
    detergent, labeling and handling of, 374–377, 375f
    EPR data, interpretation of, 373–374
    labeling sites, selection of, 372–373
    lipid environments, 377–379, 377f
Membrane proteins and peptides
  ESEEM secondary structure determination approach
    α-helical secondary structure, 302–304, 302–303f, 305–306f
    α-Helices vs. β-sheet, 304–307, 307f
  ESEEM spectroscopy, 293–294
  lipid bilayer
    ESEEM experimental setup and data analysis, 301
    ESEEM principles, 298–301, 299f
    peptide purification and validation, 295
    peptide secondary structure validation, 297, 298f
    solid-phase peptide synthesis (SPPS), 294, 296t
    spin label, attachment of, 297
    synthetic peptide into lipid bilayers, 297
  secondary structure, 290–293
Membrane scaffold protein (MSP), 377–378
(1-oxy-2,2,5,5-tetramethylpyrroline-3-methyl)-methane-thiosulfonate (MTSL), 103f, 106–107, 110f, 130, 196–197, 196f, 222–223, 223f, 262–263, 263f, 316–317, 316f, 355f, 356, 404–405
Middle cerebral arterial occlusion (MCAO), 540–544
  EPR acquisition parameters, 543–544
  transient middle cerebral artery occlusion, 542–543, 543f
MLCK. See Myosin light chain kinase (MLCK)
Molecular oxygen. See also Imaging molecular oxygen, in cancer therapy
  measurement of, 503–504
  nature of, 503
  spin probe sensitivity, 509–510
Muscle proteins, BSL of
  DEER, distance measurements with, 113–115, 115f
  labeling specificity and protein function, 116–119, 117–118f
  magnetically aligned systems, 111–113, 112f, 114f
  mechanically aligned systems, 108–111, 109–111f
  methods, 104–105
  orientation selection, problem of, 115–116
  rotational dynamics, 105–108, 105–107f
Mutagenesis, 324
Myoglobin, 126–128
Myosin, 103–104, 108
Myosin light chain kinase (MLCK), 491–494, 493f

N

NASNOX program
  batch mode, 445–446
  inter-R5 distances, 433f
National Biomedical Research Center, 157–158
National Institutes of Health (NIH), 555
Native-like system, 160
Natural and synthetic peptides, 220
Ni(II)ethylenediamine diacetate (NiEDDA), 69–70, 236–237, 240, 242, 272, 319–320, 360
Nitroxide side chains, 61–62, 61f
Nitroxide spin labels, 409
NOE. *See* Nuclear overhauser enhancement (NOE)
Nuclear magnetic resonance (NMR), 112, 126–128
  protein structure, 351–353
Nuclear overhauser enhancement (NOE), 485–486
Nucleic acids
  with computational modeling
    additional considerations, 448–449, 449f
    inter-R5 distances, 444–446
    model generation, 444
    model selection and characterization, 446–448
  double electron–electron resonance (DEER) spectroscopy
    data acquisition, 440–443
    sample preparation, 439–440
    spectrum analysis, 443
  multifrequency/multifield PELDOR
    analysis of, 415–418, 418f
    G-band PELDOR (high magnetic field), 412–415, 413–414f
    Q-band PELDOR (medium magnetic field), 412
    X-band PELDOR (low magnetic field), 410–412, 410f
  orientation-selective PELDOR, 408f, 410f
  of site-directed spin labeling (SDSL), 431–438
    labeled oligonucleotides, 437–438
    oligonucleotide labeling, 436–437
    oligonucleotides with site-specific phosphorothioate modifications, 434–435
    R5 precursor, 435–436
    spin labels, 407–408, 407f
Nucleotide-binding domains (NBDs), 331
Nucleotide-independent nitroxide probe, 431–438

O

Oligonucleotides
  labeling, 436–437
  with site-specific phosphorothioate modifications, 434–435
Orientation
  muscle proteins, BSL of
    magnetically aligned systems, 111–113, 112f, 114f
    mechanically aligned systems, 108–111, 109–111f
  selection, problem of, 115–116
Osmolytes, 162–164
Overhauser dynamic nuclear polarization (ODNP)
  biological spin probe, 459–461, 460f
  data acquisition, 472–475, 474f
  data analysis
    data interpretation, 476–477
    examples of, 477–479, 478f
    raw data analysis, 475–476
  hardware, 469–472, 469f, 471f
  theory, 461–468, 462f
    interpretation, 467–468
    relaxation rates, 463–465
    saturation factor, 465–466
Overhauser-enhanced MRI (OMRI), **555**
Oxidative disease, 554–555
Oximetry
  deep-sited oximetry probe, 533–534, 534f
  spin probe oximetry, 510–512
  types of probes, 532–533
  validation of, 516–519, 517–518f
Oxygen concentration, 530. *See also* In vivo electron paramagnetic resonance (EPR)

P

1-Palmitoyl-2-oleoyl-sn-glycero-
3-phosphocholine (POPC),
240–242
Paramagnetic relaxation enhancement
(PRE)
  detect transient sparsely populated states,
  490–494, 492–493f
  measurement, 487–489
  paramagnetic labels, 487
  structure determination, 489, 490f
PDB. See Protein Data Bank (PDB)
pEPR spectroscopy, 131–132
Peptide–membrane interactions,
  spin-labeling EPR
  aggregation and agglomeration, peptides
    CW EPR method, 242–249,
    244–245f, 247f
    pulsed electron–electron double
    resonance (PELDOR), 249–250
  binding isotherms, 233–236
  membrane-associated peptides
    accessibility measurements, 237–239
    experimental considerations, 239–242,
    241f
  membrane mimetic systems
    liposomes preparation, 226–227
    MTSL, 224
    natural amino acid side chains, covalent
    modification of, 221–225, 222–223f
    nitroxides, maleimido- and iodo-
    derivatives of, 224–225
    unnatural amino acid via peptide
    synthesis, 225–226
  partition coefficients
    experimental considerations, 227–229,
    229f
    signal amplitude method, 230
    spectral simulations, 230–233, 232f
Peptides
  critical micelle concentration (CMC),
  228–229
  natural amino acid side chains, covalent
  modification of, 221–225, 222–223f
  purification and validation, 295
  secondary structure validation, 297, 298f
  signal amplitude method, 230
  by spectral simulations, 230–233

spin labeling of
  by MTSL, 224
  nitroxides, maleimido- and iodo-
  derivatives of, 224–225
Phospholamban (PLB), 103–104
Photosynthetic reaction centers, 354–355
pH-sensitive thiolspecific nitroxide labels
  chemical structures, 196–197, 196f
  electrically neutral reference interface,
  207–210, 208f
  Gouy–Chapman theory, 210–211
  IMTSL- and IKMTSL-labeled thiols,
  199–200, 200f
  ionizable nitroxides,
  methanethiosulfonate derivatives of,
  200–202, 201t
  lipid bilayers, titration experiments with,
  204–205
  spin-labeled lipid vesicles, 204
  surface lipid bilayers electrostatics,
  202–211, 203f
  two-site slow-exchange model, titration
  experiments, 205–207, 205f
Polydimethylsiloxane (PDMS)-
  encapsulated probes, 538
Potassium channels, spin labeling
  general considerations
    flask vs. fermentor growth, 392–393
    labeling reactions and reagents,
    393–394, 393f
    mass spectrometry, 395
    quality control, 394–395
    quantification of, 394–395
    reconstitution, 395
    target DNA, 391
    toxicity to host organism, 391–392
  labeling methods
    prokaryotic KcsA potassium channel,
    395–397
    prokaryotic KvAP potassium channel,
    397–399
PRE. See Paramagnetic relaxation
  enhancement (PRE)
Pressure of oxygen (pO$_2$), 502.
  See also Imaging molecular oxygen,
  in cancer therapy
Pressure-resolved DEER (PR DEER)
  equipment, 43–44, 43f

high-pressure conformational ensemble, 50–52, 51f
practical considerations, 44
Prokaryotic KcsA potassium channel
procedure, 395–397
solutions requirement, 397
Prokaryotic KvAP potassium channel
procedure, 397–398
solutions requirement, 398–399
Protein conformational exchange, T_1 exchange spectroscopy, 19–22, 20f
Protein Data Bank (PDB), 350
Protein purification and spin labeling, 325
Protein rotational diffusion, 63
Proteins
nuclear magnetic resonance (NMR) spectroscopy, 60–61
role in, 60
Pulsed dipolar spectroscopy (PDS), 14
Pulsed electron–electron double resonance (PELDOR), 128f, 129f, 126–128, 128–129, 132–135, 249–250.
See also Double electron-electron resonance (DEER)
data analysis, 148–149
echo and the modulation depth, 133–135
EPR spectrum, 133–135, 134f
four-pulse PELDOR signal, 132–133, 132f
Hahn echo, 132–133
multifrequency/multifield
analysis of, 415–418, 418f
G-band PELDOR (high magnetic field), 412–415, 413–414f
Q-band PELDOR (medium magnetic field), 412
X-band PELDOR (low magnetic field), 410–412, 410f
orientation-selective, 408f, 410f
and relaxation, 135–136
three-pulse PELDOR, 132–133, 132f
Tikhonov regularization, 405
time trace, 405
Pulsed microwave technology, 126–128
Pulse electron double resonance (PELDOR), 322
Pulse EPR experiment
shot repetition time (SRT), 145–146

T_m measurements, 146
traveling wave tube (TWT) amplifier, 145
PutP, spin labeling site scan reveals
data analysis, 339
EPR experiments, 338–339
homology model, 336, 337f
sample preparation, 337–338
secondary structure determination, 339–340
2,2,5,5-tetramethyl-pyrrolidine-1-oxyl (PROXYL), 562–563

Q

Q-band PELDOR (medium magnetic field), 412

R

Reactive oxygen species (ROS), 554
Reciprocal correlation time, 35
Rigid spin label, 406, 407f, 418–419
R1 nitroxide motion
EPR timescale, 63–64
internal motion, local protein dynamics, 66–68, 67f
protein rotational diffusion, 63–64
tertiary interactions, 64–66, 65f
Root-mean-square-deviation (RMSD), 447
R5 precursor, preparation of, 435–436

S

SAMD. See Simulated annealing molecular dynamics (SAMD)
Saturation recovery (SR), 4, 5f
CW observe power, 10
instrumentation and practical considerations, 9–11
long pulse SR
intermediate exchange, 22, 23f
interspin distances with relaxation enhancement, 14–19, 16f
protein secondary structure, 11–14, 13f
Saturation transfer EPR (STEPR), 103–104
SB theory. See Solomon–Bloembergen (SB) theory
SDSL. See Site-directed spin labeling (SDSL)
Side chain internal motions, 63
Signal-to-noise ratio (SNR), 168–170, 535

Simulated annealing molecular dynamics
 (SAMD), 277, 278f, 279s, 281t
Site-directed spin labeling (SDSL), 4, 102
 cysteine, 62
 general strategy of, 429f
 membrane-curving proteins, 261–263,
  262–263f
 nitroxide side chain, 63
 of nucleic acids, 431–438
 protein–lipid complex preparation
  basic considerations for, 267–268
  optimizing, 268–270
 protein purification
  basic considerations for, 266
  spin-labeling protocol, 267
 quantitative aspects of, 62–63
 R1 side chain, 62
 sample preparation methodology
  basic considerations for, 264, 265f
  cysteine mutant constructs,
   264–266
Site-directed spin labeling electron
 paramagnetic resonance
 (SDSL EPR), 60–61
 applications and perspectives
  helical proteins, compressibility and
   structural fluctuations in, 45–47, 46f
 click chemistry, 317
 distance measurements, 321–322
 HisQMP, 331–336
 instrumentation, 326
 membrane proteins, applications to,
  322–340
  mutagenesis, 324
  protein purification, 325
  spin labeling, 325
 proteins, pressure effects on, 32–37, 33f
 PutP, 336–340
 site-directed mutagenesis, 316–317
 spin label accessibility, 319–320
 spin label mobility, 317–319, 318f
 SRII/HtrII, static and time-resolved
  detection, 326–330, 327f
 time-resolved cw EPR measurements,
  321
Site-directed spin labeling-electron
 paramagnetic resonance
 (SDSL-EPR) spectroscopy
EPR studies
 inactivating D25N mutation, 165
 ligand binding, cryogenic
  temperatures, 164–165
 osmolytes, conformational sampling,
  162–164, 163f
 spin label, incorporation, 160–162
HIV-1PR, 156–160, 157f, 159f
measurable range, profiles extending
 outside of, 182
minor components, 179–181
peptide–membrane interactions
 aggregation and agglomeration,
  peptides, 242–250
 binding isotherms, 233–236
 membrane-associated peptides, 236–242
 membrane mimetic systems, 221–227
 partition coefficients, 227–233
signal-to-noise ratio (SNR), 181
SNR. See Signal-to-noise ratio (SNR)
Solid-phase peptide synthesis (SPPS), 294,
 296t
Solid-phase synthesis, 407
Solomon–Bloembergen (SB) theory, 489
Solvent accessibility, 68–72
S-(1-oxyl-2,2,3,5,5-
 pentamethylimidazolidin-4-yl)
 methyl methanesulfonothioate
 (IMTSL), 196–197, 196f
Spin label, attachment of, 297
Spin label distance heterogeneity, 160–161
Spin-labeled proteins
 accessibility, 319–320
 distance measurements, 321–322
 environment, 320–321
 mobility, 317–319, 318f
 time-resolved cw EPR measurements,
  321
Spin-labeled T4 lysozyme crystal structures,
 366
Spin labeling
 potassium channels
  flask vs. fermentor growth, 392–393
  labeling reactions and reagents,
   393–394, 393f
  mass spectrometry, 395
  prokaryotic KcsA potassium channel,
   395–397

prokaryotic KvAP potassium channel, 397–399
quality control, 394–395
quantification of, 394–395
reconstitution, 395
target DNA, 391
toxicity to host organism, 391–392
Spin-label solvent, 358–360, 359f
Spin population, 462, 462f
SPPS. See Solid-phase peptide synthesis (SPPS)
SSL. See Succinimidyl-2,2,5,5-tetramethyl-3-pyrroline-1-oxyl-3-carboxylate (SSL)
Static and time-resolved EPR measurements, 328–330
Static dipolar interaction, 73–78, 74f, 76f
S-(2,2,5,5-tetra ethyl-2,5-dihydro-1H-pyrrol-3-yl)methyl methanesulfonothioate (MTSL), 297
Stokes–Einstein behavior, 63
Stripline pulse-forming unit (SPFU), 146–147
Structural biology, protein, 350
Substrate-binding protein (SBP), 331
Succinimidyl-2,2,5,5-tetramethyl-3-pyrroline-1-oxyl-3-carboxylate (SSL), 221, 222f
Sulfhydryl-reactive nitroxide labels, 222–223
Synthetic particulate probes, oximetry probes, 537–538

T

Target DNA, potassium channels spin labeling, 391
Taylor's expansion, 31–32
3,4-bis-(methanethiosulfonylmethyl)-2,2,5,5-tetramethyl-2,5-dihydro-1H-pyrrol-1-yloxy spin label (HO-1944 or BSL), 102–104
2,2,6,6-Tetramethyl-N-oxyl-4-amino-4-carboxylic acid (TOAC), 222f, 225–226
2,2,6,6-Tetramethyl-piperidine-1-oxyl (TEMPO), 562–563

2,2,6,6-Tetramethyl-piperidine-1-oxyl-4-amino-4-carboxyl (TOAC), 105–107
2,2,5,5-Tetramethylpyrroline-1-yloxy-3-acetylene (TPA), 407
Thermodynamics, proteins under pressure, 31–32
Three-pulse ESEEM, 300, 303f, 305f, 307f
Tikhonov regularization (TKR), 168–170
Tissue pO_2 (partial pressure of oxygen)
direct measurement of, 530–531
direct quantification of, 530–532
ischemic pathologies change, 530
TKR. See Tikhonov regularization (TKR)
T4 lysozyme (T4L), 162
T_m measurements, 146
data fitting, 147–148
Transient absorption measurements, 328
Transient hypoxia, 516–520, 517f, 520f
Transient middle cerebral artery occlusion, 542–543, 543f
Transmembrane domains (TMDs), 331
Transverse optimized (TROSY) correlation spectroscopy, 488–489
Triarylmethyl (TAM), 14
Tris(2-carboxyethyl)phosphine (TCEP), 225
Trityls, 512–513, 512–513f

V

Variable-pressure continuous wave EPR, 47–49, 48f

W

Water dynamics, 458–459
Water-soluble protein T4 lysozyme, 358
W-band ESEEM-derived oscillation, 147–148

X

X-band ESEEM-derived oscillation, 147–148
X-band PELDOR (low magnetic field), 410–412, 410f
X-ray crystallography, protein structure and dynamics, 351–353, 352f
X-ray diffraction, 126–128

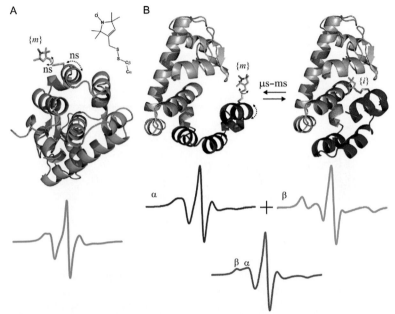

Michael T. Lerch et al., Figure 1 Protein dynamical modes and their manifestation in SDSL-EPR. (A) In a well-ordered protein with a single dominant conformational state (green ribbon), the CW spectral lineshape for R1 at noninteracting surface sites reflects weakly anisotropic motion of the nitroxide (green spectrum), which varies from site-to-site due to variations in local backbone motion (López et al., 2012). The structure of R1 is shown in an inset, and the side chain is shown in stick representation attached to the protein model. The nitroxide motion is directly influenced by backbone fluctuations, here illustrated by a helix rocking mode (dashed curved arrow) and internal R1 motions (solid curved arrow) on the nanosecond timescale. (B) Equilibrium exchange between conformations α and β on the microsecond and longer timescale. Provided that R1 is placed such that it has distinct motions and spectra in α and β, the spectrum of the equilibrium mixture is the weighted sum (purple spectrum) of the spectra reflecting each state. In this example, the nitroxide exhibits a weakly anisotropic motion in α (red), and is immobilized due to interaction with the local protein environment in β (blue). The contributions from α and β in the low-field resonance line of the composite spectrum are indicated.

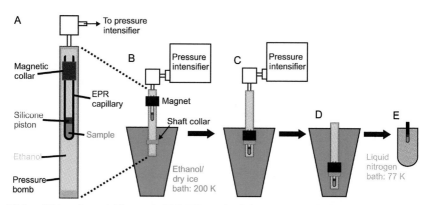

Michael T. Lerch et al., Figure 4 PR DEER methodology. (A) Detail of sample in the pressure bomb. A borosilicate capillary is modified by the addition of a magnetic collar near the top. A silicone piston (red) separates the sample (green) from the ethanol pressurization fluid (light blue) that fills the pressure bomb. (B) The bomb is connected to the pressure intensifier with the lower portion immersed in dry ice/ethanol (dark blue) at 200 K. The sample is held at the top of the bomb using the magnet and the temperature is controlled during pressurization. (C) Rapid cooling to 200 K is triggered while under pressure, when the sample is moved quickly to the bottom of the bomb. Addition of a shaft collar on the pressure bomb helps ensure proper positioning of the cell during cooling. (D) The bomb is depressurized, disconnected from the pressure intensifier, and submerged in the dry ice/ethanol bath. (E) The sample capillary is transferred to liquid nitrogen (purple) for cooling to 77 K in preparation for DEER data acquisition at 80 K. The magnetic collar is removed prior to transfer to the resonator. *Figure is adapted from Lerch et al. (2014).*

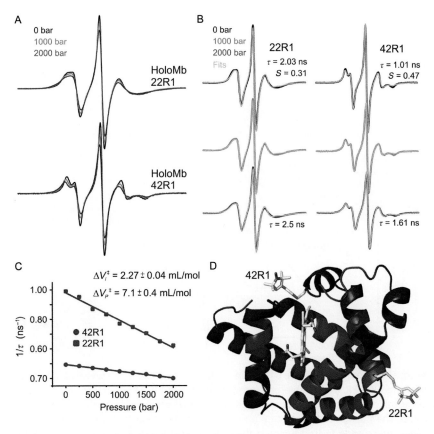

Michael T. Lerch et al., Figure 5 Protein backbone fluctuations revealed in variable-pressure CW EPR. (A) Variable-pressure CW spectra of holoMb 22R1 (top) and 42R1 (bottom) at pH 6, normalized to the same number of spins. (B) Pressure-dependence of the nitroxide τ was determined from fits to the spectra, and (C) a plot of $1/\tau$ versus pressure was fit to extract the indicated $\Delta V_i^{\ddagger}$ and $\Delta V_p^{\ddagger}$. (D) Models of the R1 side chain are shown at residue 22 and 42 of holoMb (PDB: 2MBW) in stick representation. The backbone is color coded according to crystallographic thermal factor from lowest (blue) to highest (red). In (A) and (B), spectra are color coded as indicated. HoloMb 22R1 and 42R1 were prepared and variable-pressure CW data were collected according to methods in Lerch et al. (2013). HoloMb 42R1 0 and 2000 bar spectra were previously reported in Lerch et al. (2013).

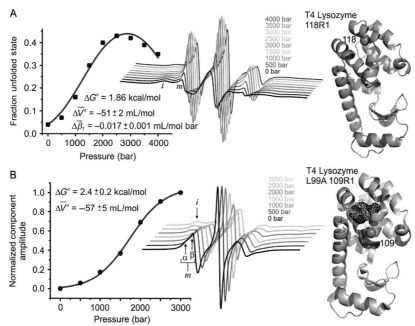

Michael T. Lerch et al., Figure 6 Protein conformational exchange revealed in variable-pressure CW EPR. (A) Variable-pressure CW spectra of T4L 118R1 in 2 M urea at pH 6.8. Contributions from *i* and *m* spectral components in the low-field resonance line are indicated. The fraction unfolded state (*m*) determined from spectral simulation is plotted versus pressure. A fit (red trace) to this plot using the two-state model of (1) yielded the indicated values of $\Delta G°$, $\Delta \bar{V}°$, and $\Delta \bar{\beta}_T$. (B) Variable-pressure CW spectra of T4L L99A 109R1 at pH 6.8. The normalized first component amplitude from SVD of the spectra is plotted versus pressure, and a fit (red trace) to this plot yielded the indicated values of $\Delta G°$ and $\Delta \bar{V}°$. EPR spectra are normalized to the same number of spins for each pressure series. Spectra are color coded as indicated. Structures are shown of wild-type T4L (PDB: 3LZM) and T4L L99A (PDB: 3DMV) with the spin-labeled sites indicated by orange spheres. EPR data in (B) were reported in McCoy (2011). *Panel (A) is adapted from McCoy and Hubbell (2011).*

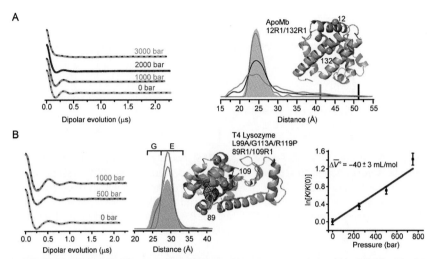

Michael T. Lerch et al., Figure 7 Pressure-populated structural changes reported by PR DEER. (A) (Left) Background-corrected dipolar evolution functions with fits to the data overlaid (dotted black); (right) area-normalized PR DEER distance distributions for apoMb 12R1/132R1 at pH 6. A crystal structure of holoMb (PDB: 2MBW) is shown in an inset. The spin-labeled sites are indicated with orange spheres. (B) (Left) Background-corrected dipolar evolution functions with fits to the data overlaid (dotted black); (center) area-normalized PR DEER distance distributions for T4 lysozyme L99A/G113A/R119P 89R1/109R1 at pH 6.8. The inset shows structures of wild-type T4L (gray, PDB: 3DMV) and the excited state of T4L L99A (green, PDB: 2LCB) overlaid to illustrate the difference in helix F position in the ground and excited states. The spin-labeled sites on both protein models are indicated with orange spheres at residue 109 and 89; (right) a plot of $\ln(K/K(0))$ versus pressure was fit (red trace) using a two-state model of (1) to determine $\Delta \bar{V}^{\circ}$ for the $G \leftrightarrow E$ exchange. In (A) and (B), the distance distributions and dipolar evolutions are color coded as indicated. Dipolar evolutions are vertically offset for clarity. The green and black bars in (A) indicate the upper limit of reliable shape and distance of the distribution (Jeschke, 2012). Absence of these bars in (B) indicates that the limit of reliability is beyond the maximum distance shown. *Panels (A) and (B) are adapted from Lerch et al. (2014, 2015).*

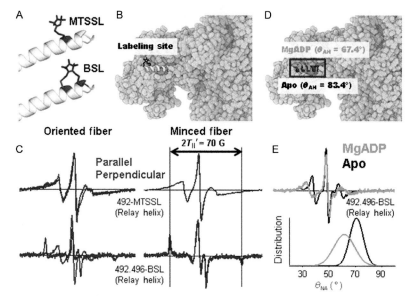

Andrew R. Thompson et al., Figure 7 (A) MTSSL and BSL attached to the myosin relay helix. (B) Myosin bound to actin within an oriented muscle fiber. (C) Conventional EPR spectra of actomyosin complexes labeled with MTSSL and BSL within oriented muscle fibers demonstrate orientation sensitivity. Minced (randomly oriented) fiber samples show nanosecond dynamics of MTSSL but not BSL. (D and E) Addition of ADP reveals a previously undetected rotation of the relay helix within actin-bound myosin.

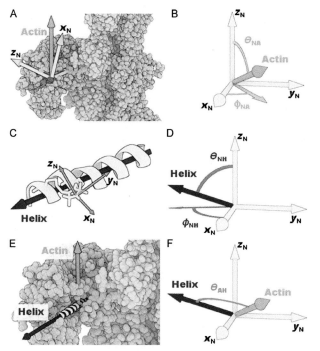

Andrew R. Thompson et al., Figure 8 (A–D) Visualization of coordinate transformations used in the analysis of oriented muscle fiber EPR. Projection of the actin and helix vectors on the nitroxide frame yield the relative angle θ_{AH}, the axial tilt of a myosin helix relative to actin (E and F).

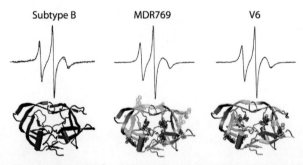

Thomas M. Casey and Gail E. Fanucci, Figure 3 Selected CW EPR spectra for site K55R1 [R1 = Cys labeled with S-(1-oxyl-2,2,5,5-tetramethyl-2,5-dihydro-1H-pyrrol-3-yl)methyl-methanesulfonothioate (MTSL)] on three different HIV-1PR constructs in the absence and presence of inhibitor (scan width, 100 Gauss). The overlain spectra are nearly identical. Ribbon diagrams show the location of the amino acid substitutions relative to subtype B.

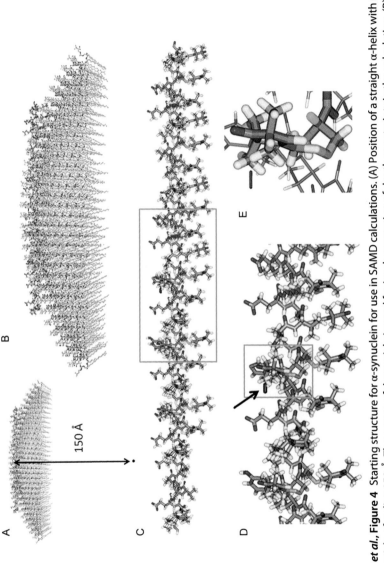

Mark R. Ambroso et al., Figure 4 Starting structure for α-synuclein for use in SAMD calculations. (A) Position of a straight α-helix with respect to the surface of a lipid vesicle of radius 150 Å. The center of the vesicle is used for implementation of depth constraints in the calculation. (B) A closer view of the α-helix and lipid surface. (C) The 81-amino acid α-helix including 26 labels. (D) A close-up view of the central part of the α-helix (blue box in C), showing several spin labels in more detail. The purple bar (indicated by an arrow) indicates the "surface" of the lipid vesicle. (E) A detailed view of a single label (blue box in D). This and all other labels are added in a "m,m" conformation for χ_1 and χ_2. This conformation is favored for an MTSL-labeled amino acid in an α-helix due to the interaction between Sδ (lighter yellow atom in the label) and Hα (green atom).

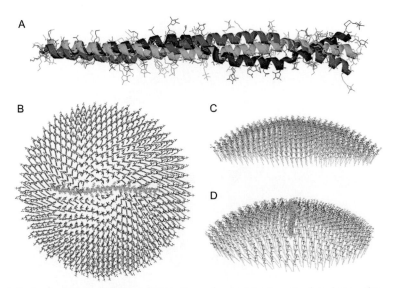

Mark R. Ambroso et al., Figure 5 Structures formed during simulated annealing and reconstruction of the protein–lipid surface. (A) Deviation of structures at low temperature (green) and higher temperature (red and blue). (B–D) Reconstructed protein–lipid surface from three perspectives, with the helix shown in green.

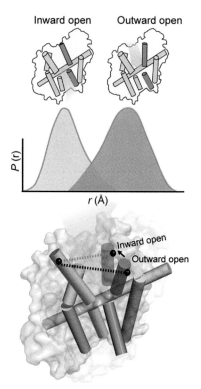

Derek P. Claxton et al., Figure 2 EPR methods report on the ensemble of conformations in solution. In a DEER experiment, each molecule in solution reports a characteristic interprobe distance consistent with its conformation. The distance distribution reports these distances as a function of their frequency within the ensemble. Thus, discrete conformations undergoing equilibrium fluctuations in solution (inward facing and outward facing) at ambient temperatures are represented as distinct distance populations in the DEER distance distribution (yellow and green) in the solid state. Therefore, individual conformations can be described using distance parameters generated from ensemble-based measurements.

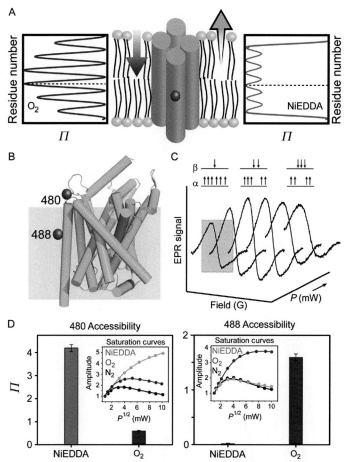

Derek P. Claxton et al., Figure 5 Spin-label solvent accessibility and the correlation with local environment. (A) The differential solubility of fast-relaxing PRAs (NiEDDA and O_2) allows the determination of spin label environment. Nitroxide scanning of an α-helix that is asymmetrically solvated between aqueous and hydrocarbon milieu will report the gradient of oxygen accessibility toward the center of the bilayer in accordance with helix periodicity. The dotted line highlights the site of expected maximum in O_2 accessibility. The NiEDDA accessibility profile, which probes water exposure, is 180° out-of-phase with the O_2 profile. (B) Two spin-labeled sites are shown on a model of LeuT (PDB 2A65), which are used to probe the membrane–water interface. An approximate position for this interface is outlined by an orange box. (C) Power saturation experiments showing the reduction in signal intensity as a function of microwave power. (D) The high NiEDDA accessibility at site 480 in LeuT relative to O_2 and N_2 determined from power saturation curves (inset) suggests a water-exposed position of the spin label. In contrast, the high O_2 accessibility at site 488 indicates that the spin label samples the lipid bilayer.

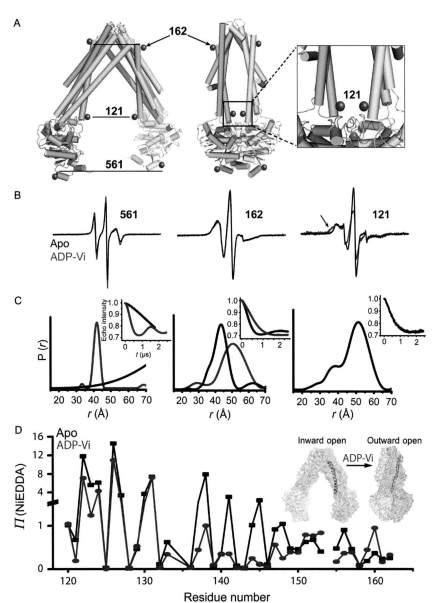

Derek P. Claxton et al., Figure 7 Correlation of global structural rearrangements with local helix packing in MsbA. (A) Model of the MsbA homodimer in the open, Apo (PDB 3B5W) and the closed, AMP-PNP-bound (PDB 3B60) states showing symmetry-related sites for spin label incorporation. Individual monomers are identified by the color scheme. (B) EPR spectra of spin labels at these positions and the corresponding distance distributions (C) in the Apo and ADP-Vi-bound states (trapped posthydrolysis). Labels at 561 and 162 show opposite distance changes between states, consistent with rigid body movement of helices in an alternating access mechanism. Although separated by ∼50 Å in the Apo state (C), spin labels at site 121 are within 20 Å in the ADP-Vi-bound state as indicated by broadening of the EPR lineshape (arrow in B). (D) Formation of a closed conformation on the intracellular side according to distance analysis is consistent with changes in the NiEDDA accessibility profile of transmembrane helix 3 induced by ADP-Vi.

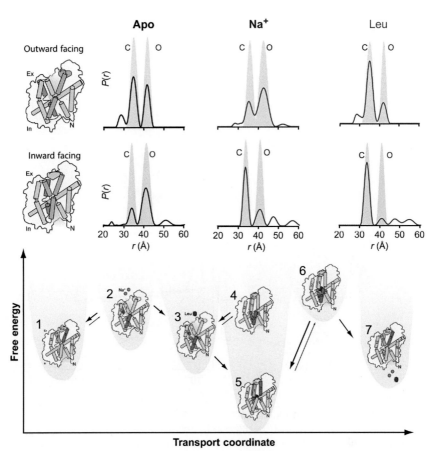

Derek P. Claxton et al., Figure 9 EPR reveals equilibria that can be used to describe energy landscapes and mechanism. LeuT is a Na$^+$-coupled amino acid transporter. We conducted DEER experiments that monitored the conformational transitions on the extracellular and intracellular sides, shown here for helix 6/intracellular loop 3 (orange). We observed conformational equilibria between inward-facing, outward-facing, and occluded conformations associated with apo (ligand-free, black), Na$^+$-bound (blue), and Na$^+$/Leu-bound (red) conditions. These were used along with structural characterization of intermediate states (numerically identified) and a biochemical description of transport to produce a novel description of alternating access in LeuT.

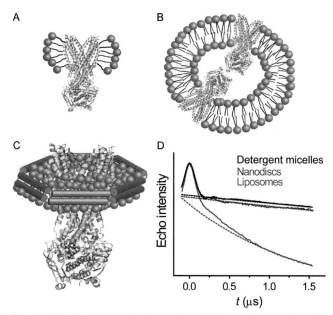

Derek P. Claxton et al., Figure 11 Detergent and lipid environments and the consequence on the DEER signal background. Solvation of membrane protein in (A) detergent micelles, (B) liposomes, and (C) nanodiscs. Liposome reconstitution often introduces more than one protein copy per liposome. As a result, the higher effective spin concentration increases the contribution of the background decay in the DEER signal relative to proteomicelles and nanodiscs (D).

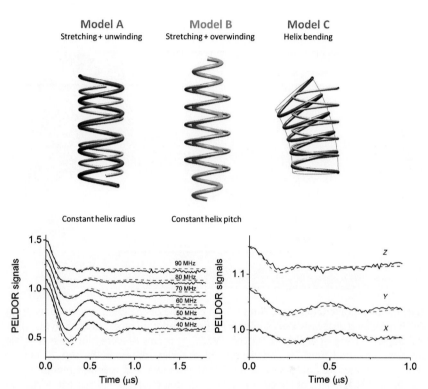

B. Endeward *et al.*, Figure 6 (Top) Three proposed conformation dynamics models of short double-stranded DNA helices. Model A describes double helix stretching with a constant radius and flexible pitch height. In Model B, the double helix shortens its radius and overwinds when stretched, keeping the pitch height constant. Model C allows the double helix to bend. (Bottom) Experimental data (solid) and best simulations (dashed) performed for the PELDOR signals of one single spin pair based on Model B showing good agreements for both X-band (left) and G-band measurements.

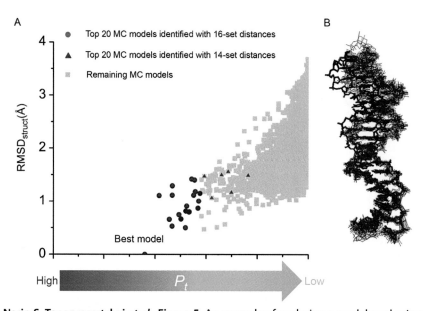

Narin S. Tangprasertchai et al., Figure 5 An example of analyzing a model pool using DEER-measured distances. Data shown are reproduced from reported work on the "p21-RE" duplex (Zhang et al., 2014). (A) The RMSD$_{struct}$ versus P_t score plot for the 10,000 models generated from MC simulations. Blue circles represent the top 20 ranked models, obtained using 16 sets of measured distances; red triangles represent the top 20 ranked models, obtained using 14 distances. Note that 14 models, including the best-fit model, are retrieved in both searches. (B) Overlay of the top 20 models of the unbound DNA (blue thin lines) and the bound DNA (red). The unbound DNA models are obtained using the integrated SDSL/MC approach, while the bound DNA is from a reported crystal structure, 3TS8.pdb (Emamzadah, Tropia, & Halazonetis, 2011). The analysis shows that the unbound DNA models converge and identifies the mode of DNA deformation upon protein binding (Zhang et al., 2014). *Adapted from Zhang et al. (2014) with permission.*

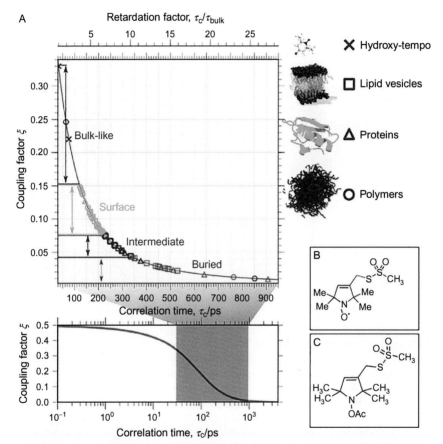

Ilia Kaminker et al., Figure 1 Data for the coupling factor ξ for a set of known biological spin probe (spin label) positions for a variety of systems appearing in the ODNP literature. The bottom panel shows the FFHS curve plotted over a larger range of logarithmically spaced correlation times. *Adopted from figure 12 of Franck, Pavlova, et al. (2013).*

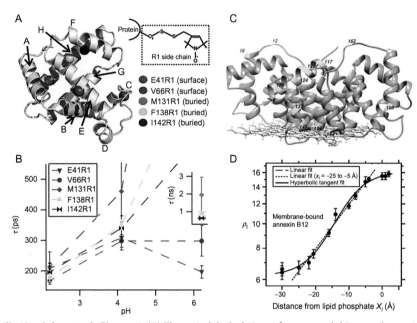

Ilia Kaminker et al., Figure 6 (A) The spin labeled sites of apomyoglobin are shown in the protein's native structure. (B) The measured translational correlation time as a function of pH, it is seen that in the native state the hydration dynamics show a large dispersion. (C) X-ray crystal structure of membrane-bound annexin B12 in the presence of Ca^{2+}. The residues subjected to site-directed spin labeling are marked by a red number and highlighted with yellow. (D) Distance dependence of retardation factor (ρ_t) at specific sites of annexin B12 bound on the surface of large unilamellar vesicles composed of POPC:POPS (1:2) in the presence of 1 mM Ca^{2+} at 25 °C. *(A and B) are reproduced with permission from Armstrong et al. (2011) and (C and D) are reproduced with permission and slight modifications from Cheng et al. (2013).*

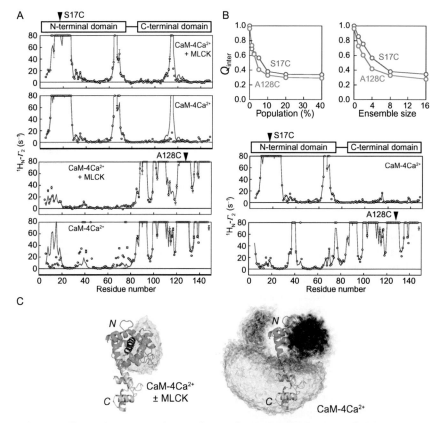

G. Marius Clore, Figure 3 Exploring the conformational space sampled by transient sparsely populated closed states of calmodulin (CaM)-4Ca^{2+} from interdomain PRE data. (A) Observed (circles) and calculated PRE profiles (lines) observed for CaM-4Ca^{2+} in the presence (panels 1 and 3) and absence (panels 2 and 4) of the myosin light chain kinase (MLCK) target peptide. The paramagnetic nitroxide label is attached to S17C in panels 1 and 2 and to A128C in panels 3 and 4. Excellent agreement is obtained for the intradomain PREs in the presence and absence of the MLCK peptide. However, whereas the experimental interdomain PRE profiles obtained in the presence of the MLCK peptide agree well with those back-calculated from the crystal structure of the CaM-4Ca^{2+}-MLCK complex (red line), those in the absence of peptide do not agree with the back-calculated PREs derived from the crystal structures of either CaM-4Ca^{2+} (blue line) or CaM-4Ca^{2+}-MLCK (red line). (B) PRE-driven simulated annealing refinement of CaM-4Ca^{2+}. The top panels show the dependence of the PRE Q-factor on the population of closed states and on the ensemble size used to represent the sparsely populated species. The bottom two panels show the agreement between observed (circles) and calculated (red line) PRE data. (C) Visualization of the conformational space sampled by sparsely populated closed states of CaM-4Ca^{2+}. For reference, the dumbbell crystal structure of CaM-4Ca^{2+} (Babu et al., 1985) is shown as a ribbon diagram
(*Continued*)

G. Marius Clore, Figure 3—Cont'd with the N- and C-terminal domains shown in dark and light green, respectively; a ribbon of the C-terminal domain in the CaM-4Ca^{2+}-MLCK structure (Meador, Means, & Quiocho, 1992) is depicted in blue, and on the left panel the gray atomic probability map depicts the location of the C-terminal domain (relative to a fixed N-terminal domain) in a variety of CaM-peptide complex (Anthis et al., 2011). On the right panel, the conformational space sampled by the C-terminal domain (relative to a fixed N-terminal domain) in the minor (~10%) closed species of CaM-4Ca^{2+} is depicted by an atomic probability map plotted at contour levels ranging from 0.1 (blue) to 0.5 (red); the gray probability map depicts the region of conformational space consistent with interdomain Γ_2 values less than 2 s^{-1} and is representative of the major species ensemble. The predominant region of conformational space (red probability map) sampled by the transient closed states of CaM-4Ca^{2+} overlaps with that sampled in the CaM-peptide complexes. *Adapted from Anthis et al. (2011).*

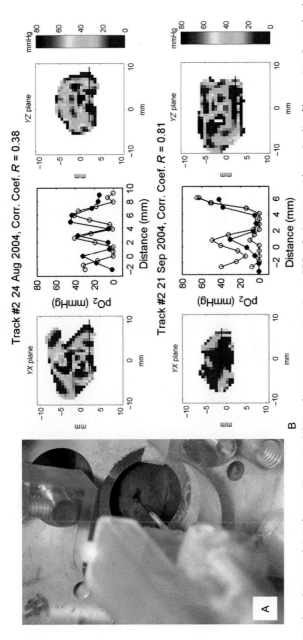

Boris Epel and Howard J. Halpern, Figure 3 (A) Leg-born mouse tumor in an EPR pO_2 imager with an Oxylite fiberoptic pO_2 probe being launched into a tumor from a stereotactic frame. (B) Two examples of the correlation between EPR and fiberoptic pO_2 data. *Left and right panels*: sagittal and coronal slices showing tumor pO_2. The assumed probe track is shown as a black line in the images. *Middle panel*: plot of the oxygen values from the Oxylite™ track (filled circles) and pO_2 image (open circles). The remarkable correlation between these values was observed.

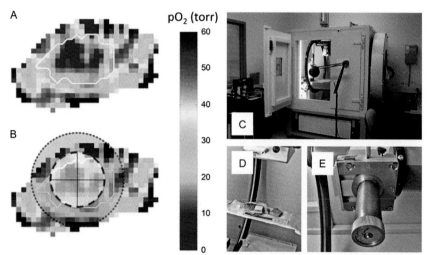

Boris Epel and Howard J. Halpern, Figure 6 EPR oxygen image of the leg tumor. (A) Oxygen map with tumor contour transferred from the registered MRI image. (B) Boost (red line) and antiboost (shaded area) as determined by the boost planning software. (C) XRad225Cx small animal cone beam CT imager and irradiator. (D) Animal bed. (E) Collimator.

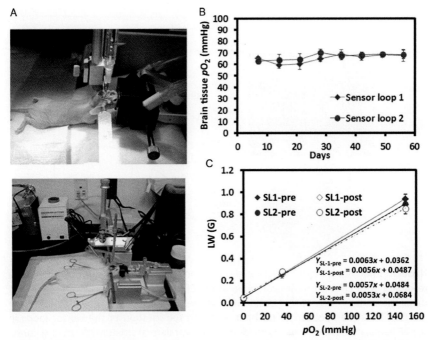

Nadeem Khan et al., **Figure 5** Tissue pO_2 measurement in mouse brain using implantable resonator with LiPc particulate probe. (A) An anesthetized mouse with head positioned in the stereotactic frame for the implantation of oximetry probe. A 25–23 gauge needle is used to create burr holes in the skull for lowering the implantable resonators in the brain. (B) Temporal monitoring of normal brain pO_2 in each hemisphere using implantable resonators for 8 consecutive weeks. (C) Calibration plot of implantable resonators pre- and postimplantation in the mouse brain. Data represent mean $\pm$ SEM, $n = 8$.

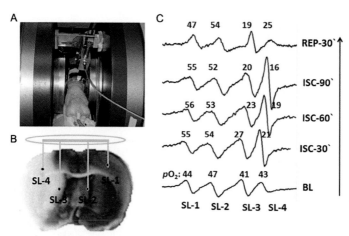

Nadeem Khan et al., Figure 6 Tissue pO_2 measurement in rat brain using implantable resonator with LiPc. (A) A rat positioned in the EPR magnet for the acquisition of pO_2 data. The surface resonator is positioned on the head for pO_2 measurements in the rat. (B) TTC-stained section of rat brain showing the location of the sensor loops in the ischemic core (SL-4), penumbra (SL-3) and contralateral striatum (SL-2), and cortex (SL-1). (C) Typical EPR spectra acquired from an implantable resonator with four sensor loop (SL) placed in the brain of a rat for pO_2 measurements at two sites in each hemisphere. Tissue pO_2 (mmHg) acquired from each sensor loop (SL) using multisite EPR oximetry is shown.

Edwards Brothers Malloy
Thorofare, NJ USA
October 8, 2015